John Pell (1611–1685) and his correspondence with Sir Charles Cavendish

John Pell (1611–1685) and his correspondence with Sir Charles Cavendish

The mental world of an early modern mathematician

Noel Malcolm
All Souls College, Oxford

and

Jacqueline Stedall
The Queen's College, Oxford

OXFORD
UNIVERSITY PRESS

OXFORD
UNIVERSITY PRESS

Great Clarendon Street, Oxford OX2 6DP

Oxford University Press is a department of the University of Oxford.
It furthers the University's objective of excellence in research, scholarship,
and education by publishing worldwide in

Oxford New York

Auckland Bangkok Buenos Aires Cape Town Chennai
Dar es Salaam Delhi Hong Kong Istanbul Karachi Kolkata
Kuala Lumpur Madrid Melbourne Mexico City Mumbai Nairobi
São Paulo Shanghai Taipei Tokyo Toronto

Oxford is a registered trade mark of Oxford University Press
in the UK and in certain other countries

Published in the United States
by Oxford University Press Inc., New York

© Oxford University Press 2005

The moral rights of the authors have been asserted
Database right Oxford University Press (maker)

First published 2005

All rights reserved. No part of this publication may be reproduced,
stored in a retrieval system, or transmitted, in any form or by any means,
without the prior permission in writing of Oxford University Press,
or as expressly permitted by law, or under terms agreed with the appropriate
reprographics rights organization. Enquiries concerning reproduction
outside the scope of the above should be sent to the Rights Department,
Oxford University Press, at the address above

You must not circulate this book in any other binding or cover
and you must impose this same condition on any acquirer

A catalogue record for this title is available from the British Library

Library of Congress Cataloging in Publication Data

(Data available)

ISBN 0 19 856484 8 (Hbk)

10 9 8 7 6 5 4 3 2 1

Typeset by Julie M. Harris
Printed in Great Britain
on acid-free paper by
Antony Rowe, Chippenham

Contents

General introduction	1
A note on dates and transcriptions	8
Acknowledgements	9
Abbreviations	10
Part I The life of John Pell *by Noel Malcolm*	11
1 Sussex, Cambridge, 1611–1629	13
2 Sussex, London, Sussex, 1629–1638	25
3 London, 1638–1643	61
4 Amsterdam, Breda, 1643–1652	102
5 London, Zurich, 1652–1658	139
6 London, Essex, London, 1658–1665	177
7 Cheshire, 1665–1669	198
8 London, 1669–1685	213
Part II The mathematics of John Pell *by Jacqueline Stedall*	245
Part III The Pell–Cavendish correspondence	329
Textual introduction	331
The correspondence	335
List of manuscripts	587
Bibliography	601
Index	639

General introduction

Writing in 1939, the historian of mathematics Herbert Turnbull described John Pell as 'a mysterious figure' and commented: 'he may well prove to be an unsuspected genius, for his manuscripts appear to survive, unexamined, in the British Museum.'[1] Turnbull's expectations were pitched a little too high: although Pell was a man of unusual intellectual abilities, and one of the leading mathematicians of his day, his work does not merit any use of the term 'genius'. The term 'mysterious', on the other hand, was well deserved; and no less mysterious is the fact that the great majority of Pell's voluminous manuscripts have remained almost entirely unexamined for more than sixty years after Turnbull published those comments.

Among the many talented individuals who belonged to the Royal Society in its early years, there were several who left large personal archives of papers and correspondence to posterity – Boyle, Oldenburg, Petty, Evelyn, and Aubrey being the most prominent. All of these have been the subject of serious research, and materials from their archives have appeared in major multi-volume publications: the correspondences of Oldenburg and Boyle, the 'Petty papers', Evelyn's diary, and Aubrey's 'Brief Lives' (as well as many other items by him). John Pell belonged to the same intellectual and social world as these people: he was a valued friend of Boyle, Oldenburg, and Aubrey, and it was the introductions furnished by Pell in the mid-1640s that had helped launch Petty's career. Pell left a personal archive which runs to approximately fifty bound volumes of manuscripts; unlike his friends and colleagues, however, he has remained almost completely ignored. Indeed, out of the whole galaxy of talent that came together in the early Royal Society, he is the *only* person, among those that left large bodies of material, to have suffered such near-total neglect. Even the cataloguers of the British Library have baulked at their task, giving only very summary descriptions of the contents of his manuscripts.

Only two groups of documents from among Pell's papers (representing rather less than six out of those fifty volumes) have received some measure of serious attention. One is his correspondence from Zurich in the mid-1650s, especially with John Thurloe and Samuel Morland: a significant part of this was printed in 1839 by the historian Robert Vaughan. However, even though he also printed some personal letters (including a few items from the Pell–Cavendish correspondence), Vaughan's primary focus was not on Pell; his aim was, rather – in the words of his title – to 'illustrate' the story of 'The Protectorate of Oliver Cromwell, and

---

[1] H. W. Turnbull, ed., *James Gregory Tercentenary Memorial Volume* (London, 1939), p. 21.

the State of Europe'.[2]

The other is Pell's correspondence with Sir Charles Cavendish. In addition to the eleven letters (eight by Cavendish, three by Pell) published in Vaughan's book, fifteen more (all of them by Cavendish) were printed by the historian of science James Orchard Halliwell in 1841.[3] Twenty-six letters (out of a total of 115 or 116) were thus made available in print. These were occasionally noticed by scholars working on related topics: for example, the Danish Hobbes scholar Frithiof Brandt, writing in 1921, made use of information about Hobbes in these published letters (while mistakenly supposing that no other letters from Pell to Cavendish had survived).[4] In 1952 Helen Hervey published an article about this correspondence, in which she printed the full texts of eight more letters, as well as extracts from several others; but her stated aim was not to characterize the correspondence as a whole, nor to investigate the range of matters discussed in it, but merely to extract from it information about Hobbes and Descartes.[5] Between 1967 and 1986 some of the letters were printed in successive volumes of the Mersenne correspondence; however, as the rationale for printing them there was simply to supply illustrative material relating to Mersenne and his circle, the letters were in some cases represented only by short extracts.[6] In 1986 Jan van Maanen published a very thorough and scholarly study of one key episode in Pell's career as a mathematician, his dispute with the Danish circle-squarer Longomontanus; in it, he not only printed substantial extracts from several of Pell's letters, but also included an extremely valuable listing of the Pell–Cavendish correspondence.[7] Once again, however, the focus of interest in this study was necessarily a narrow one, and the new material contained in the extracts was confined to discussions of the dispute with Longomontanus. Ten years later a few more extracts were published by the historian of mathematics G. J. Toomer; this time the topic under discussion was Pell's translation of an Arabic manuscript of the *Conics* of Apollonius. However, Toomer did pause to comment in a footnote: 'A complete annotated edition of this correspondence

---

[2] R. Vaughan, *The Protectorate of Oliver Cromwell, and the State of Europe during the Early Part of the Reign of Louis XIV, Illustrated in a Series of Letters*, 2 vols. (London, 1839). Vaughan included in an 'Appendix' some letters between Pell and Cavendish, Pell and his wife, and Pell and Hartlib, as well as a few other items: ii, pp. 345–478.

[3] J. O. Halliwell, ed., *A Collection of Letters Illustrative of the Progress of Science in England from the Reign of Queen Elizabeth to that of Charles the Second* (London, 1841), pp. 72–88.

[4] F. Brandt, *Thomas Hobbes' Mechanical Conception of Nature*, tr. V. Maxwell and A. I. Fausbøll (Copenhagen, 1928), p. 179: 'Pell's letters have unfortunately been lost.' This work was first published in Danish in 1921.

[5] H. Hervey, 'Hobbes and Descartes in the Light of some Unpublished Letters of the Correspondence between Sir Charles Cavendish and Dr John Pell', *Osiris*, 10 (1952), pp. 67–90.

[6] M. Mersenne, *La Correspondance*, ed. C. de Waard *et al.*, 17 vols. (Paris, 1933–88) [hereafter: *MC*]; the letters appear in vols. x–xiv, xvi.

[7] J. A. van Maanen, 'The Refutation of Longomontanus' Quadrature by John Pell', *Annals of Science*, 43 (1986), pp. 315–52 (listing: pp. 347–52) [reprinted in J. A. van Maanen, *Facets of Seventeenth Century Mathematics in the Netherlands* (Utrecht, 1987), pp. 109–46].

would be most useful.'[8] He seems to have been the first researcher to express this desire in print; but he was surely not the first to feel it.

In the third part of this book, accordingly, readers will find an edition of the Pell–Cavendish correspondence that is both complete (incorporating some material not considered or listed by van Maanen) and fully annotated. The comments just made about the particular purposes for which previous scholars have approached this correspondence are not meant to deprecate in any way the mining of it for information about any particular theme; it is indeed a rich mine, full of ores of different kinds, and it is to be hoped that it will be exploited by historians interested in such matters as philosophy, mathematics, astronomy and telescopy, Arabic scholarship, higher education in the Netherlands, the Mersenne circle, the Royalist exiles, the Republic of Letters, and the nature of social and intellectual patronage in the early modern period. But, above all, the intended benefit of publishing this correspondence as a whole is that it can be read as a whole; this is one of the comparatively rare cases in which both sides of a lengthy correspondence have survived, and it therefore becomes possible for the reader to get a sense, in the round, of the interests, contacts, and sources of information of these two individuals, of the character of the relationship between them, and of the nature of the mental world (or worlds) they inhabited.

It was our original intention to publish an edition of these letters prefaced by two brief introductory essays, one setting the correspondence in the context of Pell's life, the other relating it to his mathematical work more generally. But, given the unexplored nature of Pell's papers, we could not produce even short essays (at least, with any confidence in their accuracy and value) without undertaking the very lengthy business of exploring the whole range of Pell's manuscript materials. And so – predictably, perhaps – the work that was needed to produce two short essays turned out to be the same as the work involved in preparing two much longer ones. The results are the biography of Pell and the essay on his mathematics which form the first and second parts of this book. In each case, the reader is offered an overview of a subject which has never been presented in any detail before. We hope that the intrinsic interest of Pell's life and work will justify the degree of attention given to them here. Each of these essays is a self-contained piece of work, and, whilst the authors coordinated their efforts and benefited from each other's work-in-progress, each essay can be read independently of the other. In one sense, the edition of the correspondence may also be consulted independently: for the convenience of the reader, information needed for the explication of the text is given directly in the notes to the correspondence, so that the reader is not constantly being asked to

---

[8] Toomer, G. J., *Eastern Wisedome and Learning: The Study of Arabic in Seventeenth-Century England* (Oxford, 1996), pp. 183–5 (here p. 183, n. 169).

re-read or search through pages in the earlier sections of the book.[9] Nevertheless, it is hoped that the study of any one of the three components of this volume will be complemented and enriched by the other two; in this way, readers will have the opportunity, for the first time, to consider both the breadth and the depth of Pell's intellectual life.

Anyone who has read widely in seventeenth-century intellectual history will already have at least a shadowy sense of the range of Pell's activities and interests. It will be known, for example, that he was a member of Samuel Hartlib's inner circle; that he was therefore one of the people who both welcomed Comenius to England and propagated his ideas; that he was for a while Mersenne's most important English correspondent; that he had contacts with many of the leading mathematicians in Europe; that he was a friend of Oldenburg and an active member of the early Royal Society; and that he was well acquainted with Boyle, Hooke, Wallis, Aubrey, and Collins. As modern scholars have recognized, understanding both the history of science and intellectual history more generally in this period does not only involve studying key publications, but also depends, to a large extent, on mapping the networks of association and communication through which knowledge was generated and circulated. Pell's presence in many of these networks is well attested. Indeed, it is difficult to think of any other person (apart from his somewhat self-effacing friend Theodore Haak) who was so closely connected with all three of the greatest 'intelligencers' of the age – Samuel Hartlib, Marin Mersenne, and Henry Oldenburg. And yet his involvement in these networks has always seemed somewhat shadowy and marginal; many standard works note his presence, add a brief comment to identify him, and then move on. It may be that a vicious circle is at work here: the lack of attention paid to Pell makes him appear insignificant, which makes a full-length study of him seem unnecessary; and in the absence of any full-length study, researchers who stumble on any particular aspect of Pell's activities have no way of relating it to a larger picture of his life or work, and may therefore be inclined to treat it – and him – as marginal, thus reinforcing the assumptions behind the lack of attention paid to him. The main purpose of the present book is not to make exaggerated claims about Pell's importance, but merely to constitute the sort of detailed study by which, perhaps, that unfairly vicious circle might at last be broken.

A subsidiary purpose should also be mentioned here. One of the consequences of the lack of attention paid to John Pell is that the small existing literature about him – especially the brief accounts of his life in standard works of ref-

---

[9]Inevitably, there is much cross-referencing in the notes to the correspondence; but these cross-references are only to other notes. They are included on the assumption that an edition of correspondence will not always be read from start to finish, but may be opened and consulted at any page: therefore the editors cannot presume familiarity with information they have presented in the notes to previous letters, but must supply references to it. The only exceptions to this rule are that a handful of well-known names are assumed not to require any identification (Descartes, Hobbes, Mersenne, Gassendi), and that the place-names or addresses from which letters are written are annotated only on their first occurrence.

erence – contains an extraordinary quantity of factual errors and unwarranted statements. Often these errors have been reinforced by the Unoriginal Sin of all reference works, the copying of information from previous reference works; in some cases they have multiplied and proliferated almost as if they had a life of their own. A few examples may illustrate the nature of the problem. In the *Dictionary of National Biography* we are informed by Agnes Clerke that Pell's father was the 'incumbent' of the parish of Southwick (he was not); that Pell was the younger son (he was the elder); that Bathsua Makin was his sister (she became his sister-in-law); that he proceeded MA at Cambridge in 1630 (he did graduate BA at Cambridge in 1629, but there is no evidence of an MA); and that he succeeded Hortensius in the Chair of Mathematics at the Amsterdam Athenaeum in 1643 (the appointment was in April 1644).[10] The entry by P. J. Wallis in the *Dictionary of Scientific Biography* tells us that his father was vicar of Southwick, that he proceeded MA in 1630, that he became Professor at the Amsterdam Athenaeum in December 1643, that he read his inaugural oration at Breda in 1647 (he did so in 1646), and that the tricolumnar algebraic method he invented was probably the 'joint responsibility' of Pell and his pupil Johann Heinrich Rahn (to whom, in fact, he taught it).[11] The entry by Cornelis de Waard in the *Nieuw nederlandsch biografisch woordenboek* states that he was born in Southwick in Essex (it is in Sussex), that he proceeded MA in 1630, that he moved to London in 1639 (he did so in 1638), and that he was appointed to Amsterdam in 1643.[12]

De Waard's entry also lists five early publications by Pell, giving places and dates of publication; in fact, not one of these works was published. His is not the only account of Pell's life to suffer from this type of error. It seems that a list of manuscript works given by the eighteenth-century antiquary Thomas Birch in his own brief account of Pell's life was the source of all such confusions; Birch printed the titles of these manuscripts in the same format as the titles of printed works, and no fewer than nine of them would eventually make their way, as ghost publications, into the secondary literature.[13] On the other hand, one listing of Pell's published works includes a book which was indeed printed, but which was written by his arch-enemy Longomontanus.[14] As for the most famous work

---

[10] A. Clerke, 'John Pell', in *The Dictionary of National Biography*, 2nd edn., 22 vols. (London, 1909) [hereafter: *DNB*], xv, pp. 706–8.

[11] P. J. Wallis, 'John Pell', in *The Dictionary of Scientific Biography*, 18 vols. (New York, 1970–90) [hereafter: *DSB*], x, pp. 495–6.

[12] C. de Waard, 'John Pell', in *Nieuw nederlandsch biografisch woordenboek*, 10 vols. (Leiden, 1911–37) [hereafter: *NNBW*], iii, cols. 961–6.

[13] For a listing and discussion of these ghost publications, and of their origins in Birch, see N. Malcolm, 'The Publications of John Pell, F.R.S. (1611–1685): Some New Light, and some Old Confusions', *Notes and Records of the Royal Society of London*, 54 (2000), pp. 275–92, esp. pp. 276–7.

[14] Bom, G. D., *Het hooger onderwijs te Amsterdam van 1632 tot onze dagen: bibliographische bijdragen* (Amsterdam, 1882), p. 90, item 1232 (referring to Longomontanus, *Rotundi in plano* (Amsterdam, 1644)).

that Pell did both write and publish, his *Idea of Mathematics*, the dating of it offered by modern writers varies widely, but is almost invariably false. According to Wallis it was written before 1630; de Waard thinks it was written in 1639; Clerke claims that it was written in 'about 1639' but not published until 1650; and other writers have dated it to 1634. (It was in fact composed in the summer of 1638 and printed in October of that year.)[15] And Pell's other best-known work, the *Controversiae pars prima*, which appeared for the first time in Latin in 1647, has been described as a translation of a work previously published in English in 1646.[16]

Meanwhile other errors have proliferated, about both Pell's personal life and his career. The mistaken identification of Bathsua Makin as his sister (already mentioned) has found its way into many reference works, sometimes embroidered with further details (such as 'Southwick' as her fictitious place of birth). Her sister, whom Pell married, was named either Ithamar or Ithamara; but in several reference works she is 'Ithamaria' or 'Ithumaria', and in one book she is both 'Tehamaria' and 'Ilhamarie'.[17] (Even Pell's own name, which is surely less problematic, has been misreported: one modern writer refers to him as 'Theodore Pell'.)[18] Ithamar's father, the schoolmaster Henry Reynolds, has been routinely misidentified with a different Henry Reynolds, a courtier and littérateur. Pell's younger brother has been declared the elder not only in the *DNB* but also in the *American National Biography*; Pell's son John has been granted a non-existent knighthood, and Pell himself has been awarded, on the most nugatory of evidence, a second wife. According to some writers, Pell's father-in-law worked with him at a school in Chichester in 1630 (there is no evidence for this).[19] Ferdinand Sassen has Pell in a teaching post in Oxford in 1631 (no such appointment is recorded).[20] Robert Kargon describes Pell as a member of a 'circle of natural philosophers under the patronage of William, Earl of Newcastle' in the 1630s (Pell was never under Newcastle's patronage, and his acquaintance with the Earl's brother dates from 1640).[21] W. C. Abbott, in his monumental work on Cromwell, has Pell occupying a lectureship in Cambridge in 1653 (again, there is no evidence). And whereas Pell did finally obtain from Cromwell an important appointment, as official envoy to the Swiss cantons – a post for which he departed in the company of his friend John Dury, who was pursuing a different mission,

---

[15] See below, Part I, ch. 3, at nn. 28–34.

[16] See below, Part I, ch. 4, n. 88.

[17] See below, Part I, ch. 2, n. 38.

[18] W. Hine, 'Athanasius Kircher and Magnetism', in J. Fletcher, ed., *Athanasius Kircher und seine Beziehungen zum gelehrten Europa seiner Zeit* (Wiesbaden, 1988), pp. 79–97; here p. 97, n. 65.

[19] See below, Part I, ch. 2, n. 39.

[20] F. L. R. Sassen, 'Levensberichten van de hoogleraren der Illustre School te Breda', *Jaarboek van de geschied- en oudheidkundige kring van stad en land van Breda "De Oranjeboom"'*, 19 (1966), pp. 123–57; here p. 136.

[21] R. H. Kargon, *Atomism in England from Hariot to Newton* (Oxford, 1966), p. 41.

largely private and at best semi-official – it is dispiriting to read, in the most authoritative recent study of Cromwellian foreign policy, that 'John Dury was chosen for the 1654 Swiss embassy', and that 'The Oxford [sic] mathematician John Pell ... accompanied him.'[22]

The purpose of citing these specimen errors here is not to castigate previous authors; all writers are fallible, and the writers of entries in reference works, in particular, are seldom given the time or opportunity to engage in primary research.[23] Nor is it to imply that the present book must be entirely free of errors; some, no doubt, will be discovered, both by alert readers and by future researchers. (The authors will be grateful for corrections of both sorts.) Rather, it is to suggest the reason why the essays in the first part of this book are so closely referenced to their archival sources. The evidence that would enable one to give a reasonably full and accurate account of Pell's life and thought does exist, but it lies buried within the mountainous masses of his manuscripts. We have tried not only to put forward such an account, but also to present, so far as possible, its evidential basis.

Finally, a word of explanation about the division of labour in this volume. This project was originally undertaken by Noel Malcolm; but he soon realized that he could not present an adequate edition of the Pell–Cavendish correspondence, nor an adequate account of Pell himself, without enlisting the help of a specialist in the history of mathematics. He therefore regards Jacqueline Stedall's contribution as no less essential than his own, even though there is some disproportion between them in quantitative terms. The biography of Pell is by Noel Malcolm; the essay on Pell's mathematics is by Jacqueline Stedall; the transcription of the correspondence is by Noel Malcolm; the majority of the notes to the correspondence are also by him, but all notes on technical mathematical matters, and some of the notes relating to mathematical authors and publications, are by Jacqueline Stedall. Noel Malcolm is also responsible for this Introduction, the Textual Introduction to the correspondence, and the compilation of the list of manuscripts and bibliography.

---

[22] See below, Part I, ch. 5, n. 54.

[23] Human error is one thing; recklessly ignorant denigration, however, is another. The historian of mathematics Eric Bell wrote: 'Pell mathematically was a nonentity, and humanly an egregious fraud. It is long past time that his name be dropped from the textbooks ... By almost stupid diligence he earned his B.A. degree (1623) [sic] and his M.A. (1630) [sic]; he was fairly competent in Greek and Latin. It is not clear how these modest attainments got him a professorship at the University [sic] of Amsterdam, where he lectured on, of all things, Diophantine analysis, about which he knew practically nothing ... the alleged mathematician ... just kept his mouth shut and left his fortune to the ignorance and stupidity of his patrons' (*The Last Problem*, first published New York, 1961; 2nd edn. ed. U. Dudley (n.p., 1990), pp. 244–5). It is dismaying to find this farrago in a work specially reissued by the Mathematical Association of America.

A note on dates and transcriptions

In the seventeenth century England followed the Julian calendar ('Old Style'); this was ten days behind the Gregorian calendar ('New Style'), which was used on most of the Continent. The English traditionally dated the year from 25 March, although dating from 1 January was also quite common. In the edition of the correspondence presented here, most of the material was written by two Englishmen on the Continent, which means that potential ambiguities are constantly present. For this reason, all dates are given in double format. Where a writer has used only one format, the other – which is the product of an editorial inference – is placed in square brackets (thus: '10 [/20] January'); but if the writer has used the double format himself, or has eliminated the ambiguity by stating that he is using Old Style, no square brackets are used. In the earlier parts of the book, however, the double format is used only in cases of possible ambiguity (for example, in the dating of letters, or travel, between England and the Continent). Where the narrative describes only events in England, Old Style is used; where it describes only events on the Continent which are ambiguity-free, the dates are in New Style; but, as an exception to this last rule, events in the Protestant cantons of Switzerland are given in Old Style, since those territories retained the Julian calendar. (Pell's own letters from Zurich were also dated in Old Style.) In all cases, the year is dated from 1 January.

The conventions used in transcribing manuscript material are explained in the Textual Introduction to the edition of the correspondence. The same conventions have been used for manuscript material quoted elsewhere in this book.

Acknowledgements

The authors are grateful to the Trustees of the British Library for permission to print the texts of the letters by Pell and Cavendish, and to all the other libraries and archives that have allowed them to study and quote from materials in their collections. They owe a special debt of thanks to Mr Eddie Smith for his help and advice at Westminster School, and to Mr Adam Perkins at the Cambridge University Library. They are also very grateful to the staff of the Bodleian Library, where much of the research for this book was carried out. And they are especially grateful to two outstanding scholars, Prof. Mordechai Feingold and Prof. Gerald Toomer, who read the entire typescript and were exceptionally generous with both their time and their learning, contributing many valuable suggestions and corrections.

Noel Malcolm wishes to thank Peter Anstey, Amy Erickson, Lynn Hulse, Dorothea McEwen, Larry Principe, Timothy Raylor, Charles Webster, and Lucy Worsley for their valuable help on specific points.

Jacqueline Stedall is indebted to Jan van Maanen for sharing his interest in Pell over several years; to him and to Danny Beckers for help with translating Dutch; to Angus Bowie for help with Greek; and to Peter Neumann for his typically meticulous and helpful reading of Part II.

Abbreviations

The following abbreviations are used in notes and references:

Add.: Additional

A&T: R. Descartes, *Oeuvres*, ed. C. Adam and P. Tannery, rev. edn., 11 vols. (Paris, 1974)

BL: British Library

Bodl.: Bodleian Library, Oxford

CUL: Cambridge University Library

DNB: *The Dictionary of National Biography*, 2nd edn., 22 vols. (London, 1909)

DSB: *The Dictionary of Scientific Biography*, 18 vols. (New York, 1970–90)

ERO: Essex Record Office

ESRO: East Sussex Record Office

Harl.: Harleian

HP: Hartlib Papers, University of Sheffield (CD-ROM edition: Ann Arbor, Mich., 1995; 2nd edn. 2002)

MC: M. Mersenne, *La Correspondance*, ed. C. de Waard *et al.*, 17 vols. (Paris, 1933–88)

MCA: Mercers' Company Archives

MS: manuscript

NNBW: *Nieuw nederlandsch biografisch woordenboek*, 10 vols. (Leiden, 1911–37)

n.d.: no date of publication

n.p.: no place of publication

PRO: Public Record Office (The National Archives), London

SAZ: Staatsarchiv, Zurich

SUBH: Staats- und Universitätsbibliothek, Hamburg

WSRO: West Sussex Record Office

Part I

The life of John Pell

by Noel Malcolm

1
Sussex, Cambridge, 1611–1629

John Pell was born in the village of Southwick, in West Sussex (between Brighton and Shoreham-on-Sea), on 1 March 1611; the Bishop's Transcript of the records of the parish states that 'Jhon Pell. the son of Jhon' was baptized there on 3 March.[1] In his manuscript account of Pell's life, John Aubrey wrote at first that Pell's father was 'Rector' of Southwick; but Pell himself, when offered the manuscript to correct, deleted the word. Aubrey later added a phrase describing the father as 'a Divine but a kind of Non-Conf[ormist]'; whether Pell saw or sanctioned this comment is not known.[2] Anthony Wood, influenced no doubt by Aubrey's account, called Pell's father 'minister' of Southwick, and many subsequent writers have claimed that he was the vicar of that parish; but these formal titles go beyond anything claimed by Pell and Aubrey, and are incorrect.[3] The nineteenth-century Sussex antiquary M. A. Lower observed that 'Southwick possesses no record of his [sc. Pell's] family in tombstone, parish register, or tradition', and more than fifty years ago the local historian Ernest Salmon pointed out that although John Pell Sr may have officiated in some capacity in the parish, 'he was certainly not Rector'.[4] His name does not feature in any of the published lists of clergy in the Chichester diocese, and there is indeed no reason to believe that he was ever ordained.[5]

---

[1] West Sussex Record Office, Chichester [hereafter: WSRO], Microfilm 608 (Bishop's Transcripts, Southwick, 1606–1812), fourth page. The date of birth is given in the life of Pell (corrected by Pell himself) in Bodleian Library, Oxford [hereafter: 'Bodl.'], MS Aubrey 6, fo. 52^r (printed in J. Aubrey, *'Brief Lives', chiefly of Contemporaries, set down by John Aubrey, between the years 1669 & 1696*, ed. A. Clark, 2 vols. (Oxford, 1898), ii, p. 121), and is confirmed in a note by Pell in BL, MS Lansdowne 754, fo. 481^v.

[2] Bodl., MS Aubrey 6, fo. 52^r. Clark's edition attributes the word 'Divine' here to Pell himself (Aubrey, *'Brief Lives'*, ii, p. 121); but the whole phrase (a later interlineation) is in the same hand, which, though less rough than the main body of the notes, does appear to be Aubrey's.

[3] A. Wood, *Fasti oxonienses; or, Annals of the University of Oxford*, 2 vols., appended to vols. ii and iv of A. Wood, *Athenae oxonienses*, ed. P. Bliss, 4 vols. (London, 1813–20); i, col. 462. Agnes Clerke calls him the 'incumbent' of Southwick (*DNB*); P. J. Wallis describes him as the 'vicar' (*DSB*).

[4] M. A. Lower, *The Worthies of Sussex* (Lewes, 1865), p. 177 (Lower was evidently unaware of the Bishop's Transcript of the register); E. Salmon, 'Southwick', *Sussex Archaeological Collections*, 63 (1922), pp. 87–111; here, p. 107 (n.). Pell's baptism was in fact recorded by Cornelius Tinley or Tinsley, who had been Rector of Southwick since 1608.

[5] See G. L. Hennessy, *Chichester Diocese Clergy Lists* (London, 1900), and J. Le Neve, *Fasti ecclesiae anglicanae 1541–1857*, ii, 'Chichester Diocese', compiled by J. M. Horn (London, 1971).

It is safe to assume, nevertheless, that John Pell Sr was both an educated man and – as we shall see – a pious one. When his son John matriculated at Cambridge, the father's status was recorded as 'clerk'; John Pell Sr is possibly to be identified with the John Pell who matriculated at Magdalene College, Cambridge, in c.1596, and received his BA from Jesus College in 1599.[6] In the last few years of his life he was working as a schoolmaster. By early 1612 he had moved from Southwick to Eastbourne, in East Sussex; there, on 28 February, the parish register recorded the baptism of his second son, 'Pell, Thomas, son of Mr Pell, Schoolmaster'.[7]

Little is known of the mother of these two boys. *Née* Mary Holland, she was laconically described in Pell's corrections to Aubrey's brief life of him as 'of the Hollands of Kent'; but her own immediate family background seems to have been local to Sussex, as she was presumably related to the prominent alderman of Chichester, William Holland, who founded the school at Steyning (near Southwick) which the young John Pell would attend.[8] She died in February 1615.[9] By mid-September of that year, John Pell Sr was married again, this time to Joan, *née* Gravett, the sister of a Richard Gravett who lived in Steyning. This information is derived from John Pell Sr's will, which was drawn up at that time; and only seven months after that, on 14 April 1616, the burial of 'John Pell, Schoolmaster' was recorded at Eastbourne.[10]

---

[6] J. Venn and J. A. Venn, eds., *Alumni cantabrigienses*, part 1, 2 vols. (Cambridge, 1922), ii, s.v. 'Pell'. The family background of John Pell Sr is not known, beyond Aubrey's statement that he was 'of the Pells of Lincolnshire' (*'Brief Lives'*, ii, p. 121). It has been claimed that he was one of six sons of the John Pell (1527–1607) who was Lord Mayor of King's Lynn, but no evidence to support this has been presented (see R. Bolton, *The History of the Several Towns, Manors, and Patents of the County of Westchester, from its First Settlement to the Present Time*, 2nd edn., 2 vols. (New York, 1881), ii, pp. 40–2). Certainly there is no trace among Pell's voluminous papers of references to any of the children of the other five brothers.

[7] East Sussex Record Office, Lewes [hereafter: ESRO], 'Eastbourne baptisms, 1558–1898' (typescript transcript), entry for that date. Clerke's entry in the *DNB* mistakenly makes John Pell the younger of the two sons. Likewise, Rebecca Tannenbaum's entry on Thomas Pell in the *American National Biography*, 24 vols. (New York, 1999), xvii, pp. 263–4, incorrectly gives Thomas's date of birth as 1608.

[8] Bodl., MS Aubrey 6, fo. 52^r (Aubrey, *'Brief Lives'*, ii, p. 121). On William Holland and the school, see below, at n. 15. One modern writer has identified Pell's mother with the Mary Holland who was baptized at Steyning on 12 December 1596 (R. Pell, 'Thomas Pell: First Lord of the Manor of Pelham, Westchester Co., New York', *Pelliana*, n.s., 1, no. 1 (1962), p. 4). This would mean that she was aged only thirteen years and six months at the time of Pell's conception. Marriage at such an early age was, though legally possible, extremely rare (except for dynastic and/or financial purposes, which cannot have applied in this case), so it seems more reasonable to suppose that the 1596 baptism was of a cousin or niece. (I am very grateful to Dr Amy Erickson for advice on this matter.)

[9] On 18/28 June 1657 Pell wrote to Samuel Morland: 'My mother died above 42 yeares agoe; & my father, 14 months after her' (BL, MS Lansdowne 748, fo. 51^r). For the date of the father's burial, see below, at n. 10.

[10] ESRO, 'Eastbourne burials, 1558–1843' (typescript transcript), entry for that date. For the will, see below, n. 11.

PART I: THE LIFE OF JOHN PELL 15

The will he had drawn up provides some evidence not only of John Pell Sr's socio-economic standing, but also of his religious feelings and his attitude to the Church. It begins with a standard formula favoured by the 'godly': 'First I bequeath my Soule to Almighty God my maker verilie trusting to bee saved by the death and passion of Christ Jesus...' John Pell Sr's close relationship with the Vicar of Eastbourne, Richard Vernon, is clearly attested: 'Item I give vnto Mr Vernon Vicar of this parrishe Tenn shillings out of Thirty shillings he oweth mee for schoolinge his children'; Vernon's name also appears as a witness to the will. To ensure that his widow would deliver the legacies he bequeathed to his two sons on their reaching the age of twenty-one, John Pell Sr ordained that she 'shall enter into bonds of double the value of my childrens porcions to the Lord Byshopp of this diocesse'; whatever Aubrey may have meant by calling Pell's father 'a kind of Non-Conf[ormist]', he was obviously not a radical anti-episcopalian.[11]

As for the legacies, their main components were cash sums of £70 for John Pell and £40 for Thomas. The elder son was also to receive 'three silver spoones' and various other domestic items, including 'my best featherbed and bolster'. Other bequests consisted of ten shillings to John Pell Sr's brother James, five shillings to his sister Elizabeth, and 'vnto my twoo maydservants eyther of them Twelve pence'; everything else was to pass to his wife.[12] (Aubrey records that Pell's father 'left him an excellent library', but this is not specified in the will.)[13] These bequests, and the requirement to place £220 in bond, indicate that the Pell family was quite comfortably off. The other witness to the will, apart from Vernon, was Pelham Burton, a member of a prominent local family: he was the younger brother of Sir Edward Burton, whose son, Dr William Burton, would become a chaplain to Charles I.[14]

The young John Pell was sent off at an early age to the newly founded grammar school at Steyning – a small town (scarcely larger than a village, but with 'borough' status) near Southwick, five miles north of Shoreham-on-Sea. More than one family connection was probably at work here. The school had been founded in 1614 by William Holland, the presumed relative of Pell's mother,

---

[11] ESRO, Lewes Archdeaconry wills, W/A 15, fos. 151^v–152 (on microfilm); the Bishop of Chichester at this time was Lancelot Andrewes. I am very grateful to Roger Davey, the County Archivist, for his help in locating this will.

[12] Ibid.

[13] Aubrey, *'Brief Lives'*, ii, p. 121.

[14] ESRO, Lewes Archdeaconry wills, W/A 15, fo. 152^v. On the Burton family see W. Budgen, *Old Eastbourne: Its Church, its Clergy, its People* (London, n.d. [c.1913]), pp. 218–19. Pelham Burton's youngest sister, Susan, married James Pell (John Pell Sr's brother) in 1624; pride in this family connection is indicated by the fact that Pell's brother Thomas would later give the name 'Pelham' to the manor he acquired in Westchester County, New York (see R. Pell. *Thomas Pell*, p. 5). (It is thus not necessary to suppose, as Bolton has done (*History of the County of Westchester*, ii, p. 28), that 'the Pells were a younger branch of the illustrious house of Pelham'.)

so Pell may have been entered in it as founder's kin.[15] And Pell's step-mother's brother, Richard Gravett, who lived in Steyning, was one of the two 'overseers' appointed by John Pell Sr to supervise the execution of his will.[16] So, during his time at the school, Pell would at least have had a quasi-uncle to act *in loco parentis*.[17]

Of his education at Steyning grammar school, almost nothing is known apart from what can be deduced from the statutes laid down by its founder. The school was to have a maximum of six boarders and no more than fifty pupils altogether; each pupil was to pay one shilling on entry, and eight pence per annum. As was increasingly common during this period, there was a strong emphasis on acquiring fluency in Latin: 'the Scholars of the four chief forms shall in all their speeches within this School use the Latin tongue, and none other, except the Schoolemaster shall license or appoint them to speak English.'[18] Clearly the education Pell received, under the first master of Steyning grammar school, John Jeffreys, was a good one – supplemented, no doubt, by the studious child's explorations in his father's library. As Aubrey admiringly put it: 'At 13 yeares and a quarter old he went as good a scholar to Cambridge, to Trinity Colledge, as most Masters of Arts in the University (he understood Latin, Greek, and Hebrew)...'[19]

Pell matriculated as a 'sizar' of Trinity College, Cambridge, in the Easter (i.e. summer) term of 1624.[20] Sizars were defined in the College statutes as 'poor scholars'; they functioned as valets to the Fellows, and to other students, and

---

[15] On William Holland and the foundation of the school see the *Victoria County History of Sussex* (London, 1905–), ii, p. 424, and A. M. Butler, *Steyning, Sussex: The History of Steyning and its Church from 700–1913* (Croydon, n.d.), pp. 93–6.

[16] ESRO, Lewes Archdeaconry wills, W/A 15, fo. 152^r: 'Mr Richard Gravett my brother in law of Steaninge'. The other overseer was 'Mr Henry Panton of Lewes the elder'; both were described as 'my lovinge frends'.

[17] How close Pell's relations with Richard Gravett were, and what sort of a family atmosphere he might have enjoyed in Steyning, are difficult to judge. There are no references to Gravetts in his later correspondence. Gravett's wife, Deborah, had died in childbirth in mid-September 1615; the daughter who was born on that occasion (also Deborah) apparently survived (WSRO, Microfilm 874 (Bishop's Transcripts, Steyning, 1591–1812), entries for 14 Sept. 1615 (baptism) and 15 Sept. (burial)).

[18] *Victoria County History of Sussex*, ii, p. 424. One recent study of educational practice in this period comments: 'By the late sixteenth century, however, the greater part of language training was expected to be acquired at grammar school and not, as had previously been the case, at university. Indeed, instruction in grammar at the university was now officially prohibited, fluency in Latin having effectively become "the basic entrance requirement for Oxford and Cambridge"' (M. Feingold, 'The Humanities', in N. Tyacke, ed., *Seventeenth-Century Oxford* (*The History of the University of Oxford*, iv) (Oxford, 1997), pp. 211–357; here p. 243).

[19] Aubrey, *'Brief Lives'*, ii, p. 122. Anthony Wood borrowed this phrase, but, perhaps out of his greater respect for the University, downgraded it from 'most' Masters of Arts to 'some' (*Fasti*, i, col. 462).

[20] Venn and Venn, *Alumni*.

their duties could include waiting at table in the Hall.[21] Entrance ages at Cambridge were lower than they are now – matriculation at sixteen, or even fifteen, was quite common – but a thirteen-year-old was a precocious entrant even in those days. It is not clear why Pell's guardians should have chosen this college, rather than Magdalene or Jesus (assuming that those have been correctly identified as the colleges attended by his father), or, for that matter, Magdalen College, Oxford, where his uncle James matriculated in 1620.[22] Trinity College was a good choice for someone with mathematical interests (its statutes provided for two mathematical lecturers); but at this stage it had nothing like the reputation as a special centre of mathematical or scientific studies which it acquired one or two generations later, through the activities of Fellows such as John Ray, Isaac Barrow, and Isaac Newton. Nor is it easy to tell what sort of stimulus Pell may have received at Trinity towards his future mathematical and philosophical pursuits. There does seem to have been a particular concentration of admirers of Bacon at Trinity in the 1620s: the volume of memorial poems, *Manes verulamiani*, published after Bacon's death in 1626, includes contributions by no fewer than seven Trinity men.[23] One of these, the poet Thomas Randolph, also matriculated as an undergraduate in 1624; another, James Duport, who was later to become a famous tutor and Regius Professor of Greek, had entered the college two years earlier and became a Fellow in 1627. Neither of these could be described, however, as an enthusiast for scientific innovation: Randolph later used one of his poems to celebrate the centrality of Aristotle to a Cambridge education, and Duport would write a famous 'Apology for Aristotle, against the new philosophers'.[24]

One of the other Trinity elegists was, however, a skilled mathematician: Herbert Thorndike (Fellow, 1620–45), whom a competent judge was later to describe

---

[21] For a detailed account of the duties and status of a sizar, given in connection with Newton's subsizarship at Trinity in 1661, but equally applicable to conditions there 37 years earlier, see R. Westfall, *Never at Rest: A Biography of Isaac Newton* (Cambridge, 1980), pp. 71–7. However, I am grateful to Prof. Feingold for pointing out that one reason for matriculating as a sizar was simply to avoid paying higher admission fees – and that if, like Hooke and Newton, Pell entered servitor to a family friend, he might have been assigned few, if any, menial duties.

[22] On the father's colleges see above, at n. 6; for James Pell's matriculation see J. Foster, *Alumni oxonienses: The Members of the University of Oxford, 1500–1714*, 4 vols. (Oxford, 1891–2), iii, p. 1139.

[23] See W. G. C. Grundy, ed., *Manes verulamiani* (London, 1950), esp. the biographical notes on pp. 56–9. Of course the inclusion of a poem in this volume is not unequivocal evidence of an interest in Bacon's philosophy: some of the writers would have been selected primarily for their skill in writing Latin poetry. And the Trinity connection here may also reflect not only the fact that Bacon was a Trinity man, but also the role played by George Herbert (a Fellow of the college) in the preparation of the volume.

[24] I. Mullick, *The Poetry of Thomas Randolph* (Bombay, 1974), pp. 17–18; M. Feingold, 'Isaac Barrow: Divine, Scholar, Mathematician', in M. Feingold, ed., *Before Newton: The Life and Times of Isaac Barrow* (Cambridge, 1990), pp. 1–104; here, pp. 11–13 (though Feingold gives a more balanced account of Duport than the traditional depiction of him as an extreme reactionary).

as 'one of the best scholars and mathematicians of this age'.[25] In the early 1650s Thorndike would be in touch with Pell about a project (never realized) to publish the mathematical papers of Walter Warner; this contact may possibly have been based on an early acquaintance, or perhaps even lessons from Thorndike, during Pell's Cambridge years, though direct evidence for this is lacking.[26] Nor do we know what contacts the young Pell may have had with the other Fellow of Trinity known to have had a special interest in mathematics and astronomy at this time, Thomas Whalley (Fellow, 1591–1637), who would later bequeath to the college his valuable 1566 edition of Copernicus.[27]

Recent research has shown that the mathematical and experimental sciences were widely cultivated in Oxford and Cambridge during this period; the traditional view, which was that students with special interest in those fields had to go elsewhere (above all, to London) for guidance or tuition, has largely been abandoned.[28] This whole debate is of particular relevance to Pell's intellectual biography, since Pell's own case has been used in support of the traditional interpretation: the fact that in 1628 he corresponded about logarithms with Henry Briggs has been offered as proof that on such matters 'he could not get help in Cambridge'.[29] Such a conclusion is too categorical. The evidence that has come down to us of Pell's intellectual activities during his Cambridge years is so fragmentary that no sure argument can be built on it. And, though fragmentary and random, it may at the same time be systematically skewed in favour of his contacts outside the University: the type of evidence most likely to be retained by Pell was correspondence, which by its very nature is conducted with people at a distance. What sorts of discussion or tuition Pell may have had in Cambridge, we do not know; that he 'could not get help' there is surely an exaggeration; that he did also seek and receive various kinds of intellectual stimulus from London and elsewhere is, nevertheless, quite clear.

The earliest dated mathematical or scientific text to have survived among Pell's papers does indeed emanate from outside the University: it is a set of tables of latitude calculated by 'Arthur Pollard, vicar of Eastdeane and Frithston in

---

[25] Seth Ward, recorded in Aubrey, *'Brief Lives'*, ii, p. 257. Thorndike is best remembered for his writings on theology and ecclesiology, in defence of Anglican doctrine and church government, against Presbyterianism, published between 1641 and 1670. In the early part of his career, however, his theological position was more Calvinist: see T. A. Lacey, *Herbert Thorndike, 1598–1672* (London, 1929), p. 20.

[26] See below, ch. 5, at nn. 18, 123.

[27] On Whalley see M. Feingold, *The Mathematicians' Apprenticeship: Science, Universities and Society in England, 1560–1640* (Cambridge, 1980), p. 79.

[28] Much of the path-breaking work here has been done by Mordechai Feingold: see his *Mathematicians' Apprenticeship*, and his 'The Mathematical Sciences and New Philosophies', in N. Tyacke, ed., *Seventeenth-Century Oxford (The History of the University of Oxford*, iv) (Oxford, 1997), pp. 359–448.

[29] M. H. Curtis, *Oxford and Cambridge in Transition, 1558–1642* (Oxford, 1959), p. 242. Curtis's support for the traditional view is, however, heavily qualified.

Sussex', marked 'Ex dono Authoris 1627'.[30] Pollard was a family friend (he had been a witness to a codicil to Pell's father's will) and near neighbour (Eastdean and Friston are two villages just to the west of Eastbourne), with whom Pell was evidently able to share some of his intellectual interests during the vacations.[31] Possibly it was Pollard who had first introduced him to the study of astronomy. (If Pollard had a teacher–pupil relationship with Pell, however, the roles were eventually reversed. A later letter from Pollard, dated 21 December 1635, begins: 'Mr Pell, pray at your leasure shew me how I shall find out ye true logathrime [sic] of a mixt number...'.)[32] On the other hand, it is quite likely that Arthur Pollard's own mathematical interests had been stimulated during his studies at Merton College, Oxford, in the late 1580s, under the Wardenship of Sir Henry Savile.[33] And the expert on logarithms to whom Pell turned for guidance in 1628, Henry Briggs, had himself been appointed by Savile to the latter's newly founded Chair of Astronomy at Oxford in 1619. Whether Briggs and Pell actually met is not known, but the one surviving letter (from Briggs in Oxford, dated 25 October 1628) is a testimony to the Professor's good-natured encouragement of learning: he concluded with the remark, 'If yet there remaine any scruple, if you please to write the thirde time I shall be desirous to give you further satisfaction', and signed himself 'Your very lovinge frende'.[34]

That Pell did have some contacts, during his Cambridge years, with mathematical practitioners in London is indicated by two pieces of evidence. The first is a letter written by him (the surviving draft is dated 23 February 1629, from Trinity) to John Tapp, who was both the leading publisher of practical mathematical works in London and a well-known teacher of the art of navigation.[35] The nature and tone of the letter (recommending a treatise for publication) suggest that the 17-year-old student was already well known to Tapp: in it, Pell made no attempt to introduce himself, and implicitly assumed that Tapp would have some respect for his judgement. Presumably he had made some visits to Tapp's shop at St Magnus Church Corner, near London Bridge, which was the

---

[30] British Library [hereafter: BL], MS Add. 4397, fos. 48–65. One brief text in Pell's hand refers to an earlier year: 'The signe Leo as it was varied by the motions of the planets in ye Yeare 1623' (BL, MS Add. 4365, fo. 27^r). However, as Dr Stedall has pointed out to me, the handwriting (which is too mature for 1623) and the subject matter (which is complicated) suggest that it was a later work, constructed retrospectively for 1623.

[31] For the codicil see ESRO, Lewes Archdeaconry wills, W/A 15, fo. 152^v.

[32] BL, MS Add. 4414, fo. 106 (with draft reply on fo. 108).

[33] Pollard matriculated in 1585, proceeding BA in 1588 and MA in 1590 (Foster, *Alumni*). On Savile's mathematical and scientific interests in the 1580s and 1590s see Feingold, *Mathematicians' Apprenticeship*, pp. 124–30.

[34] BL, MS Add. 4398, fo. 137 (printed in Halliwell, ed., *Collection of Letters*, pp. 55–7, but with the reference wrongly given as MS 4395).

[35] BL, MS Add. 4387, fo. 40^r (draft). On Tapp see E. G. R. Taylor, *The Mathematical Practitioners of Tudor and Stuart England* (Cambridge, 1954), pp. 55–7, 193. Tapp's most popular work was his *The Path-Way to Knowledge; Containing the Whole Art of Arithmeticke* (London, 1613), a primer written in simple English in the form of a dialogue between teacher and scholar.

best place in London for mathematical publications, especially ones on matters such as navigation and surveying, and almanacs of astronomical data. And if Pell made occasional visits there – perhaps while travelling between Sussex and Cambridge – he may also have visited Gresham College, where special attention was paid to the practical applications of the new advances in mathematics. The second suggestive piece of evidence is a note by John Aubrey: 'Dr Pell told me, that one Jeremiah Grinken (a Mathematical-Instrument maker) frequented Mr Gunters Lectures at Gresham-College: He used an Instrument called a Mathematical Jewell, by which he did speedily performe all Operations in Arithmetick, without writing any figures, by little Sectors of Brasse (or some semi-circles) that did turne every one of them upon a Center.'[36] If Pell had personally encountered Grinken at one of Edmund Gunter's lectures, as those words seem to imply, it must have been during Pell's early years at Cambridge: Gunter died (in Gresham College) in December 1626.

Finally, one other contact with a mathematician outside Cambridge can also be mentioned, although the earliest dated evidence of it comes from after Pell's departure from Trinity: on 7 June 1631 Pell wrote 'To his much respected freind Mr Edmund Wingate', suggesting an improvement to the way he presented logarithms in his popular work, *Arithmetique made Easie* (London, 1630).[37] Once again the tone of the letter suggests that he was already quite well acquainted with his correspondent (who, having spent part of the 1620s in Paris, now lived either in London or on his family estate in Bedfordshire). Possibly Pell had made use of personal connections here: Edward Wingate, the younger brother of the mathematician, was a near-contemporary of Pell's at Trinity, having entered the

---

[36] Bodl., MS Aubrey 10, fo. 37^r (printed, in an inaccurate and modernized transcription, in J. E. Stephens, ed., *Aubrey on Education: A Hitherto Unpublished Manuscript by the Author of Brief Lives* (London, 1972), p. 105). Jeremiah was presumably a son or brother of the instrument-maker Robert Grinken (d. 1626): see G. L'E. Turner, *Elizabethan Instrument Makers: The Origins of the London Trade in Precision Instrument Making* (Oxford, 2000), pp. 281–3, and the note on the Grinkin family in A. Mundy [Munday], *Chruso-thriambos: The Triumphs of Gold*, ed. J. H. P. Pafford (London, 1962), pp. 52–4. Aubrey followed this note with a reference to a book by William Pratt: 'The Doctor [*sc.* Pell] hath the booke ... he told me he thought his name Prat. It is in 4to' (Bodl., MS Aubrey 10, fo. 37^r), and A. J. Turner has written, accordingly, that the instrument was the one devised by William Pratt, who obtained a patent for it on 27 March 1616 (*Calendar of State Papers, Domestic* [hereafter: *CSPD*] *1611–1618*, p. 357) and described it in his *The Arithmeticall Jewell* (London, 1617) ('Mathematical Instruments and the Education of Gentlemen', *Annals of Science*, 30 (1973), pp. 51–88, esp. pp. 83–4; on Pratt see also D. J. Bryden, 'Evidence from Advertising for Mathematical Instrument Making in London, 1556–1714', *Annals of Science* 49 (1992), pp. 301–36, esp. p. 308). However, the instrument described in Pratt's book was simply a 'table' on which numbers were to be written in certain positions to facilitate addition, subtraction, multiplication, and division; it was not a calculating machine. Either Pell's memory was faulty, or (more probably) Aubrey had conflated two quite separate remarks by him.

[37] BL, MS Add. 4280, fo. 233.

college in 1623.[38]

These few fragments of evidence have already indicated some of the areas of Pell's special interest during this period: astronomical calculations, mathematical instruments, and the use (and calculation) of logarithms. Scattered among his papers are various writings on these topics from his Cambridge years: minor treatises or sets of tables, drafted, partly written or, in some cases, completed and fair-copied. The earliest items are meticulously penned fair copies of two little texts, 'A New Almanacke and Prognostication For the yeare of ... Redemption 1628' and 'A Prognostication for the yeare of our Lord God 1628'; the former has details of phases of the moon and positions of the planets, while the latter has a text explaining chronological technicalities such as the golden number and the epact.[39] Each bears an elaborate mock titlepage, with a decorative border modestly incorporating the initials 'I. P.'; the design in the former case includes the wording 'horizon londinensis', and in the latter 'horizon willindonensis'. Willingdon, near Eastbourne, was the home of Pell's uncle James, so perhaps this was a New Year's present for him; whether the other treatise was intended as a personal gift or a work for publication is not clear, though it does not seem to have been published.[40] Another short work, completed not long after those two items, also survives in a fair copy: 'The Description and Vse of the Quadrant', with a prefatory epistle dated 19 May 1628.[41] And from the same year there is a brief draft of 'Tabulae Directoriae ad praxin Mathematicam nonnihil conferentes' ('Directory tables, contributing something to the practice of mathematics'), as well as a partial fair copy (ending in the middle of chapter 3) of 'The best Way & manner of calculating of Tables for all kinde of Dialls & in all Countryes whatsoeuer By Mr Gunters Canon of Artificiall Sines & Tangents'.[42] Also possibly from this period, though undated, is a draft of the beginning of a treatise on an improved version of 'Blagrave's Jewel', a type of compact astrolabe (that is, a

---

[38] Venn and Venn, *Alumni*, identify the Edward Wingate who matriculated in 1623 as the son of Edward, of Harlington, Beds., who was the younger brother of Edmund. (The brothers' father was Roger Wingate: see F. A. Blaydes, ed., *The Visitations of Bedfordshire, annis Domini 1566, 1582, and 1634*, Harleian Society Publications, xix (London, 1884), p. 151.) But it is chronologically almost impossible for a younger brother of Edmund (b. 1596) to have had a son of undergraduate age by 1623; hence my assumption that the undergraduate was the younger brother himself.

[39] BL, MS Add. 4403, fos. 120–34, 135–45.

[40] BL, MS Add. 4428, fo. 197^r is a draft by Pell of a letter 'To his very loving uncle Mr [James Pell Minister *deleted*] at his house in [Willingdon *deleted*]', dated 'Jan. 4' (the year is not given, but internal evidence suggests 1638).

[41] BL, MS Add. 4401, fos. 3–14. Eva Taylor has written that Pell's treatise on the quadrant 'gained him the notice of Henry Briggs and Walter Warner' (*Mathematical Practioners*, p. 215), but there is no evidence for this claim, or for the idea that the treatise was circulated to anyone.

[42] BL, MSS Add. 4397, fos. 1–4; Add. 4387, fos. 7–11.

disc with a universal astronomical projection engraved on it).[43]

These texts by a sixteen- or seventeen-year-old author already display many of the most constant features of Pell's literary activities throughout his life: his industry, his devotion to the demanding but mind-numbing work of compiling mathematical tables, his eagerness to promote practical applications of mathematics for the benefit of humankind, and – in apparent contradiction of that last tendency – his repeated failure to publish his work (or, in some cases, even to complete it). Admittedly, it is not clear whether any of the treatises just listed was seriously intended for publication; their elaborate titlepages and prefatory epistles may have been no more than elements of a game-like imitation of published work, elaborated for private satisfaction. But there is at least one piece of evidence of an attempt to publish: the letter (mentioned above) to John Tapp in London. And what this evidence also shows is that Pell was already grappling, at this early age, with the moral or psychological difficulties over public authorial status which would plague him for the rest of his life.

'These are to let you understand,' Pell wrote on 23 February 1629, 'yt a freind of mine hath a booke of dialling to set forth, as it were a new Edition of Fale [the standard work: Thomas Fale, *Horologiographia: The Art of Dialling* (London, 1593)], and is at length perswaded to let it be printed, though not in his name...'. The *nom de plume* chosen by this 'freind' was 'Redivivus Falus' ('Fale renewed'). And among Pell's papers there is a fair copy, in his hand, of the first part of this treatise, 'Horologiographia. The Art of Dialling: Teaching an Easye and Perfect way to make all kinds of Dialls vpon any plat whatsoeuer ... by Redivivus Falvs', with a mock titlepage giving the publication details as 'London, 1629, Printed by B. A. and T. Fawcet, for Iohn Tap'.[44] One modern historian, Eva Taylor, has identified the pseudonymous author of this unpublished work with George Atwell, a surveyor and teacher of mathematics who lived in Cambridge; but her evidence for the attribution is insubstantial, and appears to be based mainly on a confusion about Pell's own letter to Tapp.[45] Stronger evidence for the identification of Redivivus Falus with Pell can be found in Pell's notes for another projected treatise, also dated 1629. This was an account of his improved version of the

---

[43] BL, MS Add. 4409, fo. 385^r (cf. also notes on fo. 383^r). On John Blagrave see Taylor, *Mathematical Practitioners*, p. 181; for his 'Jewel' see his own account, *The Mathematical Jewel, Shewing the Making and most Excellent Use of a Singular Instrument so called* (London, 1584), and the details in Turner, *Elizabethan Instrument Makers*, pp. 187–90.

[44] BL, MS Add. 4387, fos. 40^r (letter), 1–6 (treatise).

[45] Taylor, *Mathematical Practioners*, p. 360: 'It is not clear whether this work was ever printed, for in 1658 the author [sc. Atwell] complained of his inability to find a publisher for his works, mentioning one on dialling ... it appears to have been refused publication in 1628/9 by John Tapp, to whom Atwell wrote from Trinity College, Cambridge. He had then proposed to publish it under the pseudonym of *Redivivus Falus*.' Although Taylor gives no reference, the last claim is presumably a reference to Pell's letter; Atwell was not a member of Trinity. The letter itself is a rough draft in Pell's hand, bearing clear indications that he was composing it as he wrote. It is true that in the dedicatory epistle to *The Faithfull Surveyour* (Cambridge, 1658), Atwell did refer to some of his own unpublished writings, including 'a fourth piece on *Dialling*' (sig. *2^v); but that minimal description can scarcely suffice to justify the identification.

PART I: THE LIFE OF JOHN PELL   23

logarithmic scale which Edmund Wingate had developed for use in arithmetical calculations (a prototype of the slide rule, though his scale did not in fact slide). Wingate's method, first announced by him in Paris in 1624, was set out in his *The Use of the Rule of Proportion* (London, 1626) and *The Construction and Use of the Line of Proportion* (London, 1628).[46] As Pell explained in the preface to his treatise: 'Having met with a discourse of ye finding out of Logarithmes ... written by Mr Wingate under the name of the Line of Proportion: I have after the likenesse of it, invented another Scale, wch as it is like in the construction, so it differeth not much in the name, for I call it the Proportioned Line.'[47] Two mock titlepages survive. The first gives the title as 'The Construction and Vse of the Proportioned Line. By helpe whereof the hardest Questions of Arithmetique and Geometry, as well in broken, as whole numbers, are resolued by Addition and Subtraction'; after the word 'by', a blank is left for the author's name. In the second, the title begins 'Linea Proportionata. The Construction... [*etc.*]'; the publication details are given as 'London ... 1629'; and the author's name is 'R. Fale, Gent'.[48] Self-concealment would become a *Leitmotif* of Pell's later career – taking the form either of anonymous publication or, sadly, of no publication at all. But he would never cease to obey the motto which he had chosen to place on the first of those confected titlepages: 'Nulla dies sine Linea' ('No day without a line' – in other words, 'Let no day pass without practising your art').[49]

John Pell proceeded BA in 1629. It has sometimes been claimed that he also proceeded to the degree of MA in the following year. Thomas Birch stated this; but the University records are silent on the subject, and Anthony Wood, who was scrupulous in such matters of detail, made no such claim, having checked the point with John Aubrey a few years after Pell's death.[50] An MA would have been a necessary next step had Pell wished to pursue an academic career in the University, and a useful one if he had aimed at preferment in the Church. Both of those lines of advancement had clearly been ruled out by the young Pell, who had decided to follow in his father's footsteps as a schoolmaster – though we can only speculate about his reasons for doing so.

---

[46] See F. Cajori, *A History of the Logarithmic Slide Rule and Allied Instruments* (London, 1909), pp. 8–9. Wingate did, however, describe a slide rule in a subsequent publication: *Of Natural and Artificial Arithmetic* (London, 1630).

[47] BL, MS Add. 4431, fo. 71^r. There is also a draft of the preface in BL, MS Add. 4426, fo. 96^r (with a list of its four chapters on fo. 96^v).

[48] BL, MS Add. 4431, fos. 70^r, 73^r. The first few pages of the text are given on fos. 71–2.

[49] Ibid., fo. 70^r. For the origin of the phrase, see the account of the Greek artist Apelles in Pliny, *Historia naturalis*, 35.10.36, sect. 8.

[50] T. Birch, *The History of the Royal Society of London*, 4 vols. (London, 1756–7), iv, p. 444; A. Wood, *The Life and Times of Anthony Wood, Antiquary, of Oxford, 1632–1695, Described by Himself*, ed. A. Clark, 5 vols. (Oxford, 1891–1900), iii, p. 295, and Wood, *Fasti*, i, col. 461. W. W. Rouse Ball and J. A. Venn note only the BA (*Admissions to Trinity College, Cambridge*, 5 vols. (London, 1911–16), ii, p. 306). Venn and Venn, presumably influenced by Birch, note: 'Perhaps MA 1630' (*Alumni*). They also record that Pell incorporated at Oxford (as graduates of Cambridge were entitled to do) in 1631.

Fifty years later, when Pell was looking through Aubrey's biographical sketch of him, he made a small but revealing correction: where Aubrey had written 'He was never fellow or scholar at Trin. Coll.', Pell changed the wording to 'He never stood at any election of Fellows or scholars of the House at Trinity College.'[51] Pell's point, evidently, was that it had been his own choice not to aim for a fellowship. Perhaps his interest in the practical applications of mathematics was so all-consuming that he did not want to have to relegate it to second place in an academic position officially devoted to teaching courses in the humanities. (Teaching positions in the mathematical sciences, while not unknown, were still a great rarity at Cambridge; Herbert Thorndike, for example, was officially a lecturer in Hebrew.) Perhaps his reluctance to take holy orders – which, as we shall see, was deep-rooted – arose from a disaffection from the established Church which was strong enough to prejudice him also against the quasi-ecclesiastical world of a Cambridge college, where the overwhelming majority of Fellows were ordained. Or perhaps, while reluctant to become a priest, he was also keen to get married – something that Church of England priests could do, but Fellows of Cambridge colleges could not. Whatever the reason, Pell returned to his native Sussex, where in the spring or early summer of 1629 he obtained his first employment, as a schoolmaster.

---

[51] Bodl., MS Aubrey 6, fo. 52^r (Aubrey, *'Brief Lives'*, ii, p. 122). On the procedure at Trinity, where election to a scholarship was a necessary first step towards election as a Fellow, see Westfall, *Never at Rest*, pp. 100–3.

2
Sussex, London, Sussex, 1629–1638

Pell's first employment was at Collyer's School, a grammar school at Horsham founded by Richard Collyer in 1533. Collyer had appointed the Mercers' Company as the governing body of the school; and it is among that Company's records that the slender evidence of Pell's involvement is preserved. Teaching at the school was conducted by a Master (i.e. headmaster) and an Usher (his deputy): the former was paid £20 per annum and the latter, £10.[1] Since 1617 the Master had been Richard Nye: a member of a local family, he had been a popular choice at the outset, but it seems that his relations with his staff, and with the local community, had eventually deteriorated. An Usher of two years' standing resigned in the summer of 1628; in May 1629 Nye was locked in a dispute with 'the minister and parishioners of Horsham' over the choice of a new Usher (rejecting their preferred candidate on the grounds that he was 'a drunkard, and of ill carriage'); Nye was also being asked to explain to the Mercers' Company why it was that 'the number of his schollers, which ought to be 60, is now reduced to 11'; and in December 1629 Richard Nye was removed from his post.[2] It was during this difficult period that Pell's short stint as an Usher took place. The only record of it is an entry in the Mercers' Company accounts for the year from midsummer 1628 to midsummer 1629: 'Paid to John Whitell, vsher there for halfe a yeares sallary of the ffounders allowance iii.^{li} vi.^{s} viij.^{d}, and to him & John Pell for the last halfe yeares sallary due at midsomer 1629 of the ffounders allowance & companies augmentacion together v.^{li} [Total:] viij.^{li} vj.^{s} viij.^{d}'[3] At some point during the first six months of 1629, therefore, Pell had come to Horsham either to teach together with Whitell, or (more probably) to take over from him.[4]

---

[1] See A. N. Willson, *A History of Collyer's School* (London, 1965).

[2] Mercers' Company Archives, Mercers' Hall, London [hereafter: 'MCA'], Renterwarden's Accounts, 1618–1629, fo. 211^v; Acts of Court, vol. 4 (1595–1629), fo. 384^v; Renterwarden's Accounts, 1629–1639, entries for midsummer 1630. I am very grateful to Ursula Carlyle, Curator and Deputy Archivist of the Company, for her assistance during my visit to Mercers' Hall.

[3] MCA, Renterwarden's Accounts, 1618–1629, fo. 233^v. As this entry shows, the income from the founder's endowment was supplemented by the Company to bring the salary up to £10 per annum.

[4] A. N. Willson concluded that 'Whithell [*sic*] acted during Autumn and Winter of 1628–29 and Pell during the Spring and Summer of 1629': MCA, Willson notes, file 1, 'Guide to Sources of the History of Collyer's School, Horsham, Part One' (typescript), p. 10. However, Willson's statement in his published history that Pell received £2 10s (*History of Collyer's School*, p. 54) seems unwarranted: we do not know what proportion of the six months was served by Pell, except that, given his presence in Cambridge in late February (see above, ch. 1, at n. 44), it must have been four months or less.

This brief experience, in what may have been awkward circumstances, did not put Pell off schoolteaching as a profession; and although his employment was not renewed after June 1629 there is some indication, as we shall see, that his relations with Richard Nye remained good. One small fragment of evidence suggests that he was looking around for similar employment in July and August: a draft of a letter, dated 1 August 1629, addressed to a 'Louinge Cousin', insists that 'I haue made no compact with Mr Ro:', and refers to Pell's uncle James, who was himself a schoolmaster.[5] So perhaps he now spent some time working as an assistant to his uncle, at a school in or near Eastbourne. Another piece of evidence, however, suggests a different possibility. In a letter to Samuel Hartlib written on 12 October 1642, Pell referred to Hartlib as 'you that have so well knowen me above these 13 yeeres'.[6] Since Pell was usually extremely accurate in matters of chronology, especially where his own life was concerned, we can take this as strong evidence that his acquaintance with Hartlib began some weeks or months before October 1629. It thus seems likely that, within a few months of the termination of his employment in Horsham, Pell was put in touch with Samuel Hartlib, who was then living in London. Most probably it was Richard Nye who had arranged the introduction, through the good offices of his brother, the Puritan minister Philip Nye (later famous as the leader of the Independents in the Westminster Assembly) who was a supporter of Hartlib's projects from an early stage.[7] Whether Pell worked for Hartlib in some capacity in the latter part of 1629 is not known; the earliest evidence of their collaboration comes from the summer of the following year. What is clear is that his introduction to Hartlib, whenever precisely it occurred, was to have a huge and far-reaching influence not only on Pell's career, but also on his entire intellectual development.

Samuel Hartlib was born in 1600 in the Prussian port of Elbing (now Elbląg, in Poland); his father was a prominent Prussian merchant who also owned a dyeing factory there, and his mother was English. His early life is obscure. He studied first at Brieg in Silesia and then (probably) at Königsberg in eastern Prussia; it is known that he was in Cambridge in 1625–6, and it is possible that he was studying there as early as 1621. Having spent the winter of 1627–8

---

[5] BL, MS Add. 4431, fo. 342^v. James Pell was described as a 'schoolmaster' at the time of his marriage to Susan Burton in 1624 (Budgen, *Old Eastbourne*, p. 210). 'Mr Ro:' has not been identified. Nor has the 'Cousin' to whom Pell wrote, the term being hugely extensible in 17th-century usage.

[6] BL, MS Add. 4280, fo. 206^v. Hartlib is not named, but internal evidence renders the identification quite certain.

[7] G. H. Turnbull, *Hartlib, Dury and Comenius: Gleanings from Hartlib's Papers* (London, 1947), pp. 133, 139. Richard Nye was born at Slinfold (near Horsham) on 11 July 1591; Philip was born at Slinfold on 19 Jan. 1594: International Genealogical Index (copy in Guildhall Library, London), Sussex, microfiche A2122. Philip Nye had also held a curacy at Steyning since 1627 (*DNB*, 'Philip Nye'), so may have had some other contacts with Pell, the Gravetts, or the Hollands.

in Elbing, he returned to England by September 1628.[8] He would spend the rest of his life there, becoming known as one of the most tireless 'intelligencers' (promoters and circulators of knowledge) of his age – 'a person,' in Milton's words, 'sent hither by some good providence from a farre country to be the occasion and the incitement of great good to this Iland'.[9]

In the fragmentary evidence that survives of Hartlib's activities in the 1620s, four of the dominant themes of his later life are already apparent. One is his interest in utopianism – that is, in plans either for the establishment of separate, model communities, or for the gradual transformation (morally and intellectually) of the whole of human society. In Elbing he had come into contact with members of a network of idealistic North German intellectuals (centred on the port of Rostock) who were labouring to put into practice the utopian ideas of the philosopher Johann Valentin Andreae. Their ideal society was known as 'Antilia', and a detailed description of it – a quasi-monastic community devoted to godliness and 'the scrutiny of nature' – survives among Hartlib's papers. On his return to England in 1628, Hartlib was asked to investigate the possibility of acquiring land, for the establishment of Antilia, in the English colony of Virginia.[10]

The second theme in Hartlib's life was his involvement in the cause of the Palatinate, the Calvinist principality to which the only daughter of James I, Princess Elizabeth, had travelled to become the bride of the Prince Elector Frederick V. The Palatinate (consisting of two main territories, one on the Rhine, the other in western Bohemia) had been overrun by the Catholic Habsburg army in 1620–2, and many of its inhabitants had fled into exile. The cause of the Palatinate quickly became a popular one in England, not only because of the romantic story of Elizabeth, the 'Winter Queen', but also because this issue was regarded as a test of James I's willingness to pursue a robustly pro-Protestant foreign policy (a test he conspicuously failed). Hartlib had many contacts with the Palatinate – some of them through his brother Georg, who had studied there, at Heidelberg University. In 1625 Hartlib was corresponding with a former Professor at Heidelberg, now at Oxford, Mathias Pasor, who was an old friend of his brother, and the person who carried one of these letters from Cambridge to

---

[8] G. H. Turnbull's *Samuel Hartlib: A Sketch of his Life and his Relations to J. A. Comenius* (Oxford, 1920) is still valuable, but on the early chronology it is superseded by his *Hartlib, Dury, Comenius*: see esp. pp. 11–16 of the latter work. For the family background, and the evidence in favour of Königsberg, see M. Rozbicki, *Samuel Hartlib z dziejów polsko-angielskich związków kulturalnych w XVII wieku* (Warsaw, 1980), pp. 10–11.

[9] 'Of Education', in J. Milton, *Complete Prose Works*, ed. D. W. Wolfe et al., 8 vols. (New Haven, Conn., 1953–86), ii, p. 363.

[10] See D. R. Dickson, *The Tessera of Antilia: Utopian Brotherhoods & Secret Societies in the Early Seventeenth Century* (Leiden, 1998), esp. pp. 114–38 (Virginian plan: p. 124); on Hartlib's utopianism more generally, see Rozbicki, *Samuel Hartlib*, pp. 21–31. The description of Antilia, entitled 'Leges Societatis Christianae', is in the Hartlib Papers, University of Sheffield (CD-ROM edition: Ann Arbor, Mich., 1995; 2nd edn. 2002) [hereafter: HP], 55/20/1–5; the text has been published by G. H. Turnbull, 'Johann Valentin Andreae's *Societas Christiana*', *Zeitschrift für deutsche Philologie*, 73 (1954), pp. 407–14, and a translation is given in Dickson, *Tessera of Antilia*, pp. 251–6.

Oxford may have been Theodore Haak, a Palatinate-born scholar and translator who would later work as an official agent for the Palatine exiles in England.[11]

The third theme, Protestant irenicism, was intimately connected with the second: concern for the Palatinate was a powerful stimulus to irenicist initiatives within the Protestant camp, on the understanding that the victories of Catholic armies in central Europe could be reversed only if Calvinists, Lutherans, and Anglicans could resolve or minimize their differences and learn to work together. The most tireless campaigner for the union – or, at the very least, mutual acceptance – of the Protestant Churches was an old friend of Hartlib's: John Dury, the son of a Scottish Calvinist, who studied at Leiden and served (from 1624 to 1630) as minister to a small congregation of English and Scottish Presbyterians among the merchant community in Elbing. Hartlib had made his acquaintance by 1627, and would remain a loyal friend and an active supporter of his irenicist schemes.[12]

The fourth theme, already apparent in Hartlib's papers from the 1620s, is his special interest in the theory and practice of education. As early as 1625 there is a reference in one of his letters to the educationalist Dr Joseph Webbe; two letters from Webbe to Hartlib survive from 1628, and from the following year there is a long letter to Hartlib by another English educationalist, William Brookes or Brooke, commenting on Webbe's schemes.[13] (These two theorists would become the principal English influences on Hartlib's educational thinking: their ideas are discussed below.) Hartlib was not, and does not seem ever to have planned to become, a professional teacher; but he was passionately interested in the potential of new pedagogical methods for spreading enlightenment, both intellectual and spiritual. Under the influence both of Baconianism and (from the early 1630s onwards) of the writings of Comenius, Hartlib's educational projects began to merge with his utopianism, as he devoted his energies to furthering the 'great instauration' of natural and spiritual knowledge that would pave the way for the

---

[11] See Turnbull, *Hartlib, Dury, Comenius*, pp. 13–14; O. P. Grell, *Dutch Calvinists in Early Stuart London: The Dutch Church in Austin Friars, 1603–1642* (Leiden, 1989), pp. 183–4; G. J. Toomer, *Eastern Wisedome and Learning: The Study of Arabic in Seventeenth-Century England* (Oxford, 1997), pp. 98–100 (Pasor); P. R. Barnett, *Theodore Haak (1605–1690): The First German Translator of Paradise Lost* (The Hague, 1962), pp. 13–23.

[12] On Dury see J. M. Batten, *John Dury: Advocate of Christian Reunion* (Chicago, 1944) (p. 17 and n. on his first acquaintance with Hartlib); Turnbull, *Hartlib, Dury, Comenius*, pp. 127–341.

[13] See Turnbull, *Hartlib, Dury, Comenius*, pp. 13–14 (letter of 21 Dec. 1625, commenting on Webbe); BL, MS Sloane 1466, fos. 301–2, 377–8 (Webbe to Hartlib, 1 Sept. and 13 Dec. 1628), 396–7, 402 (Brookes to Hartlib, dated only '1 May'; assigned to 1629 by Turnbull, *Samuel Hartlib*, p. 7 (n.)).

Second Coming of Christ.[14]

The earliest surviving evidence of Hartlib's activities in England, after his return from Elbing in 1628, is a letter from Joseph Webbe which includes some encouraging remarks about Hartlib's plans to create a college.[15] Those plans were eventually realized in the summer of 1630, when Hartlib set up a school in Chichester and employed John Pell to teach in it. Modern scholars have speculated about the special purpose of this school. One suggestion is that it was 'in part an attempt to put into practice Webbe's theories'; this seems quite plausible, although, as it happens, there is no concrete evidence to support it.[16] Webbe's special interest was in the teaching of languages: he believed that the rules of grammar were best learned by practice, not by memorizing declensions and conjugations. His most novel idea was that the basic unit of speech was not the word but the clause, and he developed a so-called 'clausulary' method in which Latin texts would be printed in a special typographical arrangement that broke them up into numbered clauses. (The equivalent clauses in the English translation could then be numbered in the same way.) In 1626 he was granted a patent, permitting him to print texts in his special format (two were issued in the following year) and to license teachers to use his method.[17] It is possible that John Pell was one of the people so licensed – though no 'clausulary' texts survive among his papers.

Another interpretation of the Chichester venture is that it was an attempt to carry out the Puritan project of creating alternatives to the existing Oxford and Cambridge colleges.[18] This too has a certain plausibility. Hartlib's choice of Chichester is probably to be explained by the involvement in his schemes of a Puritan minister there, William Speed, who appears to have been co-responsible

---

[14] On Hartlib's educational concerns see C. Webster, ed., *Samuel Hartlib and the Advancement of Learning* (Cambridge, 1970); A. Cagnolati, *Il circolo di Hartlib: riforme educative e diffusione del sapere (Inghilterra, 1630–1660)* (Bologna, 2001); and the comments by Ernest Sirluck in Milton, *Complete Prose Works*, ii, pp. 184–212. On Baconianism and the whole range of Hartlib's projects, see the indispensable work by Charles Webster, *The Great Instauration: Science, Medicine and Reform, 1626–1660* (London, 1975).

[15] BL, MS Sloane 1466, fo. 301^r.

[16] V. Salmon, 'Problems of Language Teaching: A Discussion among Hartlib's Friends', *Modern Language Review*, 59 (1964), pp. 13–24; here p. 23.

[17] See V. Salmon, 'Joseph Webbe: Some Seventeenth-Century Views on Language-Teaching and the Nature of Meaning', *Bibliothèque d'humanisme et renaissance: travaux et documents*, 23 (1961), pp. 324–40, and her 'An Ambitious Printing Project of the Early Seventeenth Century', *The Library*, 3rd ser., 16 (1961), pp. 190–6. One teacher so licensed was James Shirley, the future playwright: see S. A. Burner, *James Shirley: A Study of Literary Coteries and Patronage in Seventeenth-Century England* (New York, 1988), p. 46. On Webbe's career more generally see W. Munk *et al.*, eds., *The Roll of the Royal College of Physicians of London* (London, 1861–), i, p. 159, and R. C. Evans, *Jonson and the Contexts of his Time* (London, 1994), pp. 132–47. Some confusion has been caused by Foster Watson, 'Dr Joseph Webbe and Language Teaching (1622)', *Modern Language Notes*, 26 (1911), pp. 40–6, who attributes most of Webbe's didactic works to a George Webbe; Charles Webster refers mistakenly to 'William Webbe' (*Great Instauration*, p. 105).

[18] Webster, *Great Instauration*, p. 37.

for the administration of the school. (Possibly the original connection between Hartlib and Speed was through Joseph Mede, an influential millenarian theologian and Fellow of Christ's College, Cambridge, whom Hartlib had apparently got to know during his Cambridge years.)[19] The idea that the school was intended as an alternative to the universities is strengthened by Hartlib's use of the phrase 'Illustre Collegium' to describe it: this term (or its equivalent, 'Ecole Illustre') was used in the Netherlands and northern Germany for an institution of higher education which taught most of the same subjects as a university but was of lesser status, lacking the formal *jus promovendi*, the right to award degrees.[20] But, on the other hand, there is clear evidence that Pell's pupils at Chichester were being taught at a level well below that of a university education: soon after the closure of the school, William Speed wrote to Hartlib that 'I wish your Schollers from Sussex may appear according to promise & expectation, I feare Calloway will require more time ... before he be called vnto the Vniversity'.[21] In the end, the best account we have of the purpose of the school is the one Hartlib gave, many years later, in a petition to Parliament: he stated that he had tried to serve the public good 'by erecting a little academie for the education of the gentrie of this nation, to advance pietie, learning, moralitie, and other exercises of industrie, not usual then in common schools'.[22]

For the young John Pell, whose experience of schooling (both as pupil and as teacher) up to the summer of 1630 may have been confined to the traditional methods of the grammar schools, his time in Chichester was probably quite liberating. He was able to devise a mathematics course of his own: two problems in a little booklet of problems and methods compiled by Pell are annotated 'Dictated to George Schwarts at his going away Chichester 1630' and 'Given to William More at his going away from Chichester A° 1630'.[23] He also became a participant in Hartlib's on-going discussions of pedagogical methods with the educational theorists he knew – among them, John Dury, whom at this stage Hartlib regarded no less as an educationalist than as a promoter of Protestant unity. Dury sailed from Elbing to England in June 1630, and seems to have visited Chichester at some time during the next two months; he was personally acquainted with both John Pell and William Speed by the autumn of that year.[24]

---

[19] On Speed see Turnbull, *Hartlib, Dury, Comenius*, pp. 16–18. One of Speed's letters to Hartlib of 1631 refers to 'Mr Meades' as a mutual friend: HP 46/6/20A. On Mede's millenarianism and its influence see Webster, *Great Instauration*, pp. 9–11.

[20] See Hartlib's letter to an unnamed correspondent, Oct. 1632, cited in n. 34 below. For the term 'Ecole Illustre' or 'Illustre School', see P. Dibon, *La Philosophie néerlandaise au siècle d'or*, i (Paris, 1954), pp. 197, 220.

[21] HP 46/6/8–9, Speed to Hartlib, 9 Jan. 1631.

[22] BL, MS Add. 6269, fos. 30^v–31^r.

[23] BL, MS Add. 4430, fos. 22^r, 30^r (the booklet is fos. 21–32). The name 'Schwarts' (Schwarz) suggests that Hartlib had brought in some pupils from the German – including, probably, the Palatine – community in London.

[24] See for example HP 7/16/1B, Hartlib to Dury, 20 Sept. 1630 ('Mr Speed & Mr Pell remember their love to you'), and the letter quoted in the following note.

In one of his letters to Dury from Chichester, dated 31 August 1630, Hartlib wrote: 'As for Mr Pell hee remembers his special loue to you and sendes you a rude draught of his Method.'[25] G. H. Turnbull, the first modern scholar to investigate the Hartlib papers, supposed that this 'Method' of Pell's was 'probably his *Idea of Mathematics*'; but there is no basis for such an identification.[26] In the context of Hartlib's other letters to Dury from this period, it is clear that Pell was participating in a wide-ranging discussion of educational techniques and systems; these can have had little connection with his later *Idea of Mathematics*, which is not so much an educational project as a proposal for the codification of existing mathematical knowledge and for the promotion of mathematical studies by the state. One example of a 'method' devised by Pell during this period survives in very rudimentary form: entitled 'Methodus docendi omnes linguas universalis' ('A universal method for teaching all languages'), it consists of a circle with the Latin verb 'amare' at the centre, and radiating lines creating segments which are labelled 'who?', 'how many?', 'why?', and so on.[27]

The nature of Pell's exchange of ideas with Dury during his months at Chichester is suggested by another letter to Dury from Hartlib, sent on 13 September:

> When you goe about to answer Mr Pelles Questions which j sent you last time let mee entreate you to resolue also these following. As first what course to be taken with a Scollar after hee understandes the precepts of Logick & the common vse of it in Genesis & Analysis. Whether he were best to bee taught the fullest & best ordered Systemes which as yet wee can haue ... Or rather to follow my Lord Verulams directions in his so much comended Aphorismes as the onliest way for deliverie of Knowledge ... Or Thirdly whether the maine Principles or Seeds of all Artes and Sciences only should bee proposed, with a compleat After-Art of drawing Consequences[28]

Pell's 'Questions' may perhaps have been more narrowly focussed, but presumably touched also on fundamental questions of pedagogical method. Some undated notes by Pell survive on a manuscript by Dury, dealing with the best way of explaining the basic concepts of geometry; that manuscript (itself lost) may possibly have been sent in reply to some of Pell's 'Questions'.[29] It probably dates from late 1630 or early 1631, as a similar set of notes by Pell on how to

---

[25] HP 7/11/1A.

[26] Turnbull, *Hartlib, Dury, Comenius*, p. 37. For the dating of the *Idea of Mathematics* (1638), see below, ch. 3, at nn. 31–4.

[27] BL, MS Add. 4428, fo. 35^r (dated 'March 8 1630').

[28] HP 7/12/2A–B. The full text of this important letter is printed in Webster, *Hartlib and the Advancement of Learning*, pp. 75–8 (here pp. 76–7); Webster's comment, however (p. 202, n. 3), that Pell's 'Questions' related to his *Idea of Mathematics* is incorrect.

[29] BL, MS Add. 4429, fos. 29^r (notes on MS by 'Mr Durry'), 45^r ('Haec Durraeus ex Manuscripto'). On fo. 29^r Pell comments: 'And thus I acknowledge it good to set all ye Definitions together as it were in a Lexicon Philosophicum'.

classify concepts in the science of 'measuring' (line, plane, surface, etc.), written possibly in response to Dury, is dated 3 February 1631.[30] Dury had also written a more wide-ranging treatise 'De cura paedagogica' ('On the responsibilities of a teacher'); but as he was becoming more and more immersed in his reunionist project (trying to organize support in England for a kind of personal embassy to the Protestant powers of Europe, on which he eventually embarked in July 1631), he never found time to prepare a copy and send it to Hartlib, despite the latter's entreaties.[31]

By November 1630, in any case, Hartlib's Chichester venture was on the point of collapse. The financial backing which he believed to have been promised by a group of sympathizers – he described them as those 'whom I thought to be fellow-feeling members of one Publicke-Bodie' – never materialized. Heavily in debt, he closed down the school and, on Speed's advice, returned to London before the end of that month.[32] With him went a few of his pupils, together with John Pell. By February they were established in Hackney (possibly under the patronage of Robert Greville, Lord Brooke, who had a house there); on 21 February 1631 William Speed wrote to Hartlib in Hackney that 'I joy in the present state you shall have comfort by. & Mr Pell must be tender & studious in sweetening your behaviour & all your actions in the hearts of those you take into both Schoole & House, yea & vnto all men'.[33] Eventually Hartlib salvaged his finances; but his plans for an 'academy' had to be suspended. As he wrote to an unnamed German correspondent in October 1632: 'Otherwise, I have already informed you of the reasons why we cannot go ahead with our "illustrious college" – or rather, why we wish to postpone it to a more convenient opportunity and time.'[34] At some point in 1632, according to the later recollection of Comenius, one of Hartlib's patrons made a new proposal: 'A certain person offered to Samuel Hartlib his country house, with provision for the cost of his board, so that he could live there with twenty well-born youths, supervising their

---

[30] Ibid., fo. 38^r, 'Methodus metricae nostrae. Feb. 3. 1630' (cf. notes on fo. 38^v, dated 4 Feb. 1630/1).

[31] Turnbull, *Hartlib, Dury, Comenius*, pp. 36, 143. Hartlib would continue to chivvy Dury for his pedagogical treatises, finally obtaining one from him eight years later: see P. G. Westin, 'Brev från John Durie åren 1636–1638', *Kyrkhistorisk årsskrift*, 33 (1933), pp. 193–349, esp. pp. 261, 324–5, 348.

[32] Turnbull, *Hartlib, Dury, Comenius*, pp. 17–18.

[33] HP 46/6/15A. Hartlib was certainly staying in Lord Brooke's house by the end of 1631: see Turnbull, *Hartlib, Dury, Comenius*, p. 20.

[34] J. Kvačala, ed., *Die pädagogische Reform des Comenius in Deutschland bis zum Ausgange des XVII Jahrhunderts*, 2 vols., Monumenta Germaniae paedagogica, xxvi, xxxii (Berlin, 1903–4), i, p. 26 ('Sonsten habe ich dir schon zu vohr angezeiget, aus was ursachen wir unser Illustre Collegium nicht fortstellen können oder vielmehr biss auf eine bequemere gelegenheit und zeit aufschieben wollen').

PART I: THE LIFE OF JOHN PELL   33

studies.'[35] This would-be benefactor may perhaps have been Lord Brooke; or, more probably, it was the munificent John Williams, Bishop of Lincoln, the Baconian, anti-Laudian, and later chief patron of Comenius in England, who entertained numerous scholars at his country residence of Buckden, in Hertfordshire.[36] But it seems that Hartlib, once bitten, was twice shy; he remained in London, pursuing his various educational, religious, and philosophical interests mainly through the medium of correspondence.

How long Pell stayed with him is unclear. His tuition of the remaining scholars from Sussex probably came to an end some time in 1631; during that year he must have made at least one visit to Oxford, where he incorporated as a BA; but thereafter his places of employment and residence are unknown, until he reappears – probably working as a schoolmaster again – at Billingshurst in Sussex in 1634.[37] There is some reason, however, to think that he stayed on in London until at least the summer of 1632. On 3 July of that year he married the daughter of a London schoolmaster; so it is possible that he had made her acquaintance while working as a junior master under her father. The name of Pell's bride was Ithamar or Ithamara: she had been baptized 'Ithamar' – a man's name, oddly enough, in Hebrew, being that of one of the sons of Aaron – but Pell himself used the feminized version, recording for Aubrey's benefit that he had married 'Ithamara Reginalds [>second] daughter to Mr Henry Reginalds of London'.[38] The surname also appears as 'Reginolles'; both this and 'Reginalds' were variant forms of 'Reynolds' (usually Latinized as 'Reginaldus').

Little is known about Henry Reynolds, and several of the claims made about him by modern scholars seem unfounded. It has been alleged, for example, that he 'ran' Hartlib's school in Chichester with John Pell; but there is simply no

---

[35] J. A. Comenius, *Veškeré spisy*, vols. i, iv, vi, ix, x, xv, xvii, xviii (Brno, 1911–38); iv, p. 431 (*Didaktika česká*, ch. 28): ' léta 1632 N. N. Samuelovi Hartlibovi, že jemu zámek svůj N. i s opatřením z důchodů vyživeni, aby tam se s dvadcíti urozenými mládenečky bydleje, studií hleděl, pustil'. I translate 'zámek' (the Czech equivalent of 'château' or 'Schloss') as 'country house', since 'castle' in English has excessively military overtones.

[36] Turnbull, *Hartlib, Dury, Comenius*, p. 20 (suggesting Brooke); R. F. Young, *Comenius in England* (London, 1932), pp. 34(n.), 86 (suggesting Williams).

[37] See above, ch. 1, n. 50 (Oxford); below, n. 62 (Billingshurst).

[38] Bodl., MS Aubrey 6, fo. 52^r (Aubrey, *'Brief Lives'*, ii, p. 122); for the baptismal record see below, n. 43. A recent article on the seventh-century Bishop Ithamar of Rochester comments, with forgivable inaccuracy: 'Now Ithamar is a very unusual name, and no one else of this name is found in Christian history' (R. Sharpe, 'The Naming of Bishop Ithamar', *English Historical Review*, 117 (2002), pp. 889–94; here p. 891). Several modern writers on Ithamar Pell (including Vivian Salmon and Frances Teague) call her 'Ithamaria'; this is incorrect. Stranger misunderstandings have also arisen: Robert Bolton referred to her once as 'Tehamaria', and once as 'Ilhamarie' (*History of the County of Westchester*, ii, pp. 50, 56).

evidence for this in Hartlib's correspondence from that period.[39] It has also been claimed that he was the Henry Reynolds who was a poet, a friend of Drayton and Chapman, and the author of a literary treatise, *Mythomystes*. This poet has been identified (on quite convincing grounds) with the courtier Henry Reynolds who, as secretary to the Lord Chamberlain in the period 1608–11, was responsible for organizing Ben Jonson's *Masque of Queens* in 1609, running up huge personal debts in the process; the claim is therefore made that the courtier and Ithamar's father were one and the same person.[40] This identification is not impossible, but the balance of probabilities is set heavily against it. From the memoirs of Sir Simonds d'Ewes it is known that Ithamar's father was keeping a small school in the parish of St Mary Axe, London, in 1614; d'Ewes described him as a 'mere pretender' to learning, but added that he had 'a pleasing way of teaching, contrary to all others of that kind; for the rod and ferular stood in his school rather as ensigns of his power than as instruments of his anger'.[41] Vivian Salmon deduced from these comments that 'he had little experience or competence as a schoolmaster, and such lack of skill is precisely what we would expect from a man previously employed in administration'.[42] However, the record of Ithamar's christening clearly shows that he was working as a schoolmaster as early as 1601: the entry, for 11 January 1601, refers to 'Ithamar daughter of Henry Reignolds of Bowe scholem^r'.[43] The only certain evidence of his career consists of this reference to schoolteaching in 1601; d'Ewes's account of the tuition he received from him in 1614–16; a description of him as 'gymnasiarcha' ('schoolmaster', or 'headmaster' – the term implies that he 'ruled' the school, not merely that he taught there) in a publication by his elder daughter, Bathsua, in 1616; and another description of him as 'gymnasiarcha' in a manuscript work by him, dated 1625.[44] His few known writings – one printed broadside, and two very

---

[39] The claim is made by Mary Hobbs in her 'Drayton's "Most Dearely-Loved Friend Henery Reynolds Esq"', *Review of English Studies*, n.s., 24 (1973), pp. 414–28, here p. 419, and is endorsed by Frances Teague (*Bathsua Makin, Woman of Learning* (Lewisburg, PA, 1998), p. 51). It appears to arise from a misreading of a grammatically ambiguous statement by Vivian Salmon, made while suggesting that Hartlib had got to know Reynolds via Pell: 'Hartlib knew Reynolds because he and Dr. John Pell had been colleagues together at a school in Chichester' ('The Family of Ithamaria (Reginolles) Reynolds Pell', *Pelliana*, n.s., 1, no. 3 (1965), pp. 1–24, 105–6, here p. 18).

[40] See Salmon, 'Family of Ithamaria', and Hobbs, 'Drayton's "Friend"'.

[41] Sir Simonds d'Ewes, *The Autobiography and Correspondence of Sir Simonds d'Ewes, Bart., during the Reigns of James I and Charles I*, ed. J. O. Halliwell, 2 vols. (London, 1845), i, pp. 63–4.

[42] Salmon, 'Family of Ithamaria', p. 16.

[43] London Metropolitan Archives, microfilm X24/66 (St Dunstan's, Stepney, baptisms 1568–1656), entry for 11 Jan. 1600/01. Frances Teague gives the date of the baptism (from the International Genealogical Index), but the record itself appears not to have been consulted by any of the writers on Ithamar or Henry Reynolds.

[44] The last two references are to B. Reynolds (later Makin or Makins), *Musa virginea graeco-latino-gallica, Bathsuae R. (filiae Henrici Reginaldi gymnasiarchae et philogloti apud londinenses) anno aetatis suae decimo sexto edita* (London, 1616); BL, MS Lansdowne 684, fo. 5^r.

minor treatises in manuscript – are intellectually shallow, thus chiming with d'Ewes's comments and discouraging any identification with the learned author of *Mythomystes* and the respected friend of Chapman and Drayton; and the disparity in social status between this minor schoolmaster and a courtier who could run up personal debts of many hundreds of pounds seems overwhelming.[45]

The only piece of evidence that might serve to identify Ithamar's father as a courtier consists of a phrase he wrote in the dedicatory epistle of one of his little manuscript treatises: the epistle, dated 7 February '1603' (i.e. 1604), was addressed to James I and signed 'Regiae Tuae Majestati: Ex intimis deditissimus Henricus Reginaldus'.[46] This has been taken to mean 'Henry Reynolds, the most devoted to your Royal Majesty out of all your intimate friends'. The construction is very odd, however ('intimorum deditissimus' would be the natural way of putting it), and seems much more likely that the intended meaning of 'Ex intimis' was 'from my innermost feelings', 'from the bottom of my heart'.[47] It is very hard to believe that someone who had been a schoolmaster in Bow or Stepney only three years earlier could by now have become an 'intimate friend' of the King; and the earliest record of any activities at court by the Henry Reynolds who worked for the Lord Chamberlain dates from October 1606, when that person was paid for providing coaches and post-horses for a visitor to the court – a role that does not seem to imply much royal intimacy either.

Ithamar was the second of three daughters. The youngest was called Mespira (another unusual name, probably also derived from that of a character in the Hebrew Bible: Mispereth, who was one of those who returned to Jerusalem

---

[45] For the treatises see below, n. 46. The printed broadside, entitled *Magnae Britanniae chronographia* and dated 1625, survives in multiple copies in Pell's papers: BL, MSS Add. 4409, fos. 310, 311, 392; Add. 4416, fo. 186; Add. 4426, fo. 234. (Pell had been given, or had inherited, a stock of copies, and used them as rough paper.) A revised version of the text survives in manuscript: BL, MS Lansdowne 684, fos. 3–4. The text contains poems and chronograms celebrating members of the royal family from 1602 to 1625. Jean Brink has argued that this work proves that its author cannot be identified with the courtier Henry Reynolds, on the grounds that he 'regarded himself as a schoolmaster from 1602 to 1625' ('Bathsua Reginald Makin: "Most Learned Matron"', *Huntington Library Quarterly*, 54 (1991), pp. 313–26, here p. 315); but no statement to that effect is contained in the text or suggested by it.

[46] BL, MS Add. 4384, fo. 70^v. This MS – a volume of Pell's papers – contains two copies of this treatise, the first on fos. 67–77, the second, differing from it in some respects, on fos. 78–89. The other treatise by Reynolds, entitled 'Nuncius volucris' or 'Macrolexis', expands on one of the ideas set out in the first treatise. There are two different versions of 'Nuncius volucris': BL, MSS Lansdowne 684, fos. 5–15 (where the prefatory materials are dated 1 Jan. 1625: fo. 9^r), and Add. 4403, fos. 154–64: the latter is annotated, in Pell's hand, 'Ex dono soceri' ('gift from my father-in-law': fo. 164^v).

[47] Vivian Salmon writes: 'Whether "intimus" means friend or courtier does not matter; the author was undoubtedly at Court in some capacity' ('Family of Ithamaria', p. 16). She cites the wording incorrectly as 'Rex intimus deditissimus', and refers, also incorrectly, to the treatise in BL, MS Lansdowne 684, fos. 5–15. Of the two copies of the treatise in BL, MS Add. 4384, only the first has the phrase; in the second, the dedicatory epistle bears the same date but is signed simply 'Hen. Reginaldus' (fo. 82^v).

with Zerubbabel).[48] She married first a William Morrell, and then, after his death (some time before March 1636) a Francis Rogers, who lived in London; otherwise, however, her life is sunk in obscurity.[49] Much more is known about the eldest of the three sisters, Bathsua, who became famous (under her married name, Bathsua Makin or Makins) as one of the most learned women in England: she was appointed 'Tutress' to Princess Elizabeth (the daughter of Charles I), corresponded in Greek with the learned Anna Maria van Schurman of Utrecht in the 1640s, probably gave tuition to members of the Hastings family (the Earls of Huntingdon) between the late 1640s and the 1660s, and taught at a 'school for Gentlewomen' to the north of London in the 1670s.[50] Until recently, it was widely assumed that she was John Pell's sister (an assumption arising from a misunderstanding of seventeenth-century usage, in which 'sister' was also used for sister-in-law); the error recurs in many modern reference works, even though

---

[48] See Nehemiah 7: 7. I am very grateful to Prof. Toomer for this suggestion, and for the following comment: 'This character must indeed be male (he is called Mizpar in the parallel passage, Ezra 2: 2), but his name is feminine in form, which may have been what attracted the attention of Reynolds. The consonantal structure is exactly right for Mespira, and the difference in vocalization means little in the context of 17th-century Hebrew.'

[49] Details of the first marriage are supplied by Ithamar's deposition of 14 Nov. 1640, in the Public Record Office, London [hereafter: PRO] C 24/654 (Town depositions, Chancery), in which Ithamar also certified that she and Mespira were 'naturall sisters both by father & mother'. A letter from Mespira Rogers to Ithamar, dated 7 March 1636 and beginning 'Loving Sister, My loue and my husbands remembered vnto you and yor husband', is in BL, MS Add. 4416, fo. 26^v. The existence of Mespira has been ignored by all writers on the Reynolds family; Frances Teague merely speculates that there might have been 'a third unrecorded Reginald sister' (*Bathsua Makin*, p. 83).

[50] The fullest modern study is Teague, *Bathsua Makin*. Teague's dating of the tuition of the Princess should, however, be corrected in the light of the comments in A. M. van Schurman, *Whether a Christian Woman should be Educated and Other Writings from her Intellectual Circle*, ed. and tr. J. L. Irwin (Chicago, 1998), pp. 68–9 (n.); Teague's assumption that Bathsua wrote the treatise *An Essay to revive the Antient Education of Gentlewomen* (London, 1673) must be questioned (see below, ch. 8, n. 91); and her statement that the last sign of Bathsua's activities is a letter dated 1675 (p. 103) can now be supplemented by the evidence of Bathsua's letter of *c.*1680 to Boyle (R. Boyle, *The Correspondence*, ed. M. Hunter, A. Clericuzio and L. M. Principe, 6 vols. (London, 2001), v, pp. 282–3), in which she describes herself as in her 81st year. Van Schurman's two letters to Bathsua are in her *Opuscula hebraea, graeca, latina, gallica prosaica & metrica* (Leiden, 1648), pp. 165–6; Bathsua's three letters to the Countess of Huntingdon (1664, 1667 and 1668) are in the Huntington Library, San Marino, California, MSS 8799, 8801, and 8800, and her Latin elegy for the Countess's son (d. 1649) is in the same library, Hastings Collection, Literature, box 1, folder 1.

it was corrected by Vivian Salmon as long ago as 1965.[51] Bathsua – the name is a Latinized version of the Hebrew 'Bathshua', a variant form of 'Bathsheva' or 'Bathsheba': see I Chronicles 3: 5 – was probably born in 1600: the title page of the little book of poems by her published in 1616 described her as 'in her sixteenth year' (i.e. fifteen years old), but since we know that her younger sister Ithamar was born in January 1601, it seems likely that Bathsua's precocity was being ever so slightly exaggerated.[52] That same title page also described her father as 'philoglottus', a lover of languages; the knowledge of Latin, Greek, Hebrew, French, and Spanish displayed by the young authoress of that book was doubtless intended, among other things, as an advertisement of her father's skills as a teacher. Similarly, another early production by her must also reflect her father's interests: it is a small engraving, containing a brief exposition of a system of shorthand, described as 'the invention of Radiography, which is a speedy and short writing, wth great facility to be practized in any languag', and a roundel of $c.\frac{1}{4}$ inch radius, in which she had written the entire text of the Lord's Prayer.[53] Henry Reynolds had a special interest in the development of shorthand ('brachygraphy' or stenography): his pupils were taught, as Simonds d'Ewes recalled, 'to take notes in writing at sermons' – the main purpose for which shorthand was developed in England during this period – and while attending the school d'Ewes even 'invented a strange handwriting consisting of an

---

[51] Salmon, 'Family of Ithamaria', pp. 18–19. The error appears to stem from a reference in J. Evelyn, *Numismata: A Discourse of Medals, Ancient and Modern* (London, 1697), to 'Mrs. *Makins*, the Learned Sister of the Learned Dr. Pell' (p. 265), and has been reinforced by similar references to her in Pell's MSS. Standard works identifying her as Pell's sibling include M. Reynolds, *The Learned Lady in England, 1650–1760* (New York, 1920), pp. 276–7; M. R. Mahl and H. Koon, eds., *The Female Spectator: English Women Writers before 1800* (Bloomington, Indiana, 1977), p. 115 (also printing, as a specimen of her work, a text on imprisonment for debt which was not written by her and not connected with her in any way: pp. 118–24); J. R. Brink, 'Bathsua Makin: Educator and Linguist (1608?–1675?)', in J. R. Brink, ed., *Female Scholars: A Tradition of Learned Women before 1800* (Montreal, 1980), pp. 86–100, here p. 87; P.L. Barbour, 'Introduction' to 'B. Makin' (attrib.), *An Essay to Revive the Antient Education of Gentlewomen* (Los Angeles, 1980), p. iii; P. Schlueter and J. Schlueter, *An Encyclopedia of British Women Writers* (New York, 1988), p. 307; J. Todd, ed., *Dictionary of British Women Writers* (London, 1989), p. 433 (giving her date of birth as '1612?'); and M. Parry, ed., *Chambers Biographical Dictionary of Women* (London, 1996), p. 431 (giving her name as 'Bathsua Pell Makin', and confidently asserting that she was born, like Pell, in Southwick).

[52] '*Anno aetatis suae decimo sexto edita*': see the full title, above, n. 44. Possibly the title had merely borrowed this information from the heading of the opening poem in the book, which describes it as 'anno aetatis suae 16° ... compositum' (sig. A1^r). D'Ewes described Bathsua as Reynolds's 'eldest' daughter: *Autobiography*, i, p. 63.

[53] London University Library, Carlton Shorthand Collection, item 1615. This item is undated, but its dedication to Anne of Denmark (who died in 1619) supplies a *terminus ad quem*. See also E. H. Butler, *The Story of British Shorthand* (London, 1951), pp. 212–14, and R. C. Alston, *Treatises on Short-hand* (Leeds, 1966), plate 4.

alphabet of strange letters' of his own.[54]

In the early years of his marriage, Pell was made a participant in his father-in-law's interests and enthusiasms. A note in Samuel Hartlib's 'Ephemerides' (diary-notebook) for late 1634 comments that Pell and 'Mr Reinolds' might together 'perfect' the latter's new method of stenography.[55] In the spring of the following year Hartlib recorded, a little cryptically, that 'Mr Reynolds hase revealed his Macro-Lexia to Mr Pell and his other feate'.[56] This was a reference to one of his manuscript treatises: when Pell marked his copy 'gift of my father-in-law', he dated it 24 March 1635. The treatise, under the general title 'Nuncius volucris' ('The Flying Messenger'), concerns – in the words of its English-language mock titlepage – 'Two Mathematicall Invention[s] for briefe speedie and secret Intelligence without messenger or Letter': the first is called 'Macrolexis', the second, 'Scenolexis'. (The methods involved signalling over large distances by means such as fireworks.)[57] Pell also acquired copies of the much earlier little treatise, dedicated to James I, 'Architectiones seu Inventiones sex' ('Six Constructions or Inventions'), which described, rather vaingloriously, six similar 'feats' of secret communication, including the transmission of messages by fireworks at night. (The treatise described only the feats, not how they were accomplished: the secrets were revealed on a separate page, copies of which Pell also acquired. The methods turn out to be disappointingly prosaic: voices were to be transmitted down pipes, letters conveyed into a chamber through a 'long case of wood', and so on.)[58] The Latin mock titlepage of 'Nuncius volucris', given to Pell on 24 March 1635, described Reynolds as 'Gy. [sc. Gymnasiarcha] apud Londinenses': if this description referred to the present, not merely to the past, it would imply that he had continued to work as a schoolmaster to the very end of his life, as he died less than two weeks later, and was buried on 4 April.[59] His intellectual influence on the young Pell was probably not very great; the main benefit Pell might have gained, during visits to London, would have come from consulting his well-stocked library, which Bathsua inherited.[60] Nevertheless, Reynolds's interests in stenography and cryptography may have stimulated in some ways Pell's own thinking about such matters as 'real characters' and universal language

---

[54] D'Ewes, *Autobiography*, i, p. 95. A letter survives from d'Ewes to his mother, written at Reynolds's school on 16 Dec. 1615, in which he proudly declares: 'I haue heare sent you a Sermon, of my owne collecting' (BL, MS Harl. 379, fo. 1^r).

[55] HP 29/2/45A.

[56] HP 29/3/19B.

[57] BL, MS Add. 4403, fos. 154^r (titlepage), 164^v (Pell's note).

[58] For the copies of the treatise, see above, n. 46. The copies of the page containing the 'key' are in BL, MSS Add. 4384, fo. 96^v, and Add. 4431, fo. 42^r.

[59] Teague, *Bathsua Makin*, p. 52, citing the burial record for that date of St Botolph's Church, London, where Reynolds is described as 'Schoolm^r in the Minories Streete bur. at Bedlam'. Teague assumes, surely unnecessarily, that Reynolds had been incarcerated as a madman at the Bethlehem ('Bedlam') hospital.

[60] A letter or memorandum by Pell, dated 16 Sept. 1637 and copied by Hartlib, referred to 'my Sister Makins Library, for shee had all her Fathers bookes': HP 22/15/1A.

PART I: THE LIFE OF JOHN PELL   39

schemes. And Hartlib, apparently, continued to discuss Reynolds's ideas with other correspondents: as late as 1646 his friend Abraham von Franckenberg was writing from Danzig to request from him a copy of Reynolds's 'Macrolexis'.[61]

Of the practical details of Pell's life in the period 1632–8, little is known. In the spring of 1634 he and Ithamar were living in Billingshurst: their twins, Richard and Mary, were christened there in April. By 1635 they had moved to Eastbourne, where their daughter Judith was christened in January.[62] The Visitation records for Eastbourne in 1635 state that 'John Pell BA' produced his license as a schoolmaster; he was apparently teaching in the small school next to the church, where his father's friend Richard Vernon was still vicar.[63] Pell was thus back on home ground, close to the Burtons (the leading gentry family of Eastbourne) and to his uncle James, vicar of the neighbouring parish of Willingdon. (One family connection was lost to him at this time, however: it was in 1635 that his younger brother Thomas sailed to America, settling first in Boston, then in Connnecticut, where he served as a military surgeon.)[64] By 1638 Pell also had a link with another gentry family, the Colepepers of Folkington or Fowington (which lies just to the north of Willingdon): he accompanied the ailing Sir Thomas Colepeper on a journey towards London in April of that year.[65] Otherwise, however, almost nothing is known about his conditions of life or his social world. We can presume that he worked during most or all of this period as a schoolmaster, and there is one chronologically vague note by John Aubrey which suggests that, for a time at least, he was in a position to make some pedagogical experiments of his own devising: 'Dr Pell told me (who taught a private schoole a yeare or two) that he taught his Scholars Greeke before their Latin: and that they learned their Latin the easier for it.'[66]

Of Pell's mental life, on the other hand, much more can be discovered, thanks to the survival of many of his papers (including some of his correspondence with Samuel Hartlib) from this period. With Hartlib's encouragement, much of Pell's energy was concentrated on questions of pedagogical method. Hartlib's

---

[61] BL, MS Sloane 648, fo. 90^r: 'Desidero Henr. Reginaldi Μαχρολεξίαν' (mistranscribed as 'Μαχφλεξίαν' in A. von Franckenberg, *Briefwechsel*, ed. J. Telle (Stuttgart, 1995), p. 199).

[62] International Genealogical Index (copy in Guildhall Library, London), microfiche A2124. Mary and Judith survived into adulthood, but Richard did not; details of his burial are not available, but it seems likely that he died in infancy.

[63] Budgen, *Old Eastbourne*, pp. 28–30 (Vernon), 210 (Visitation). Among Pell's papers there is the cover-sheet of a letter addressed 'To his very loving Friend. Mr Pell at Eastbourne' (BL, MS Add. 4416, fo. 182^v); this is datable to the summer of 1635 by the jottings Pell has added on the trisection of an angle (cf. MS Add. 4415, fos. 158–61).

[64] Tannenbaum, 'Thomas Pell', p. 263.

[65] BL, MS Add. 4424, fos. 359–60. Sir Thomas was a close relative of Sir Cheney Colepeper or Culpeper (their fathers were half-brothers), who later became a friend and supporter of Hartlib: see F. W. T. Attree and J. H. L. Booker, 'The Sussex Colepepers', *Sussex Archaeological Collections*, 47 (1904), pp. 47–81, and 48 (1905), pp. 65–98, here vol. 47, pp. 72–3; and M. J. Braddick and M. Greengrass, eds., 'The Letters of Sir Cheney Culpeper (1641–1657)', *Camden Miscellany*, xxiii (Camden Society, ser. 5, vol. 7) (London, 1996), pp. 105–402.

[66] Bodl., MS Aubrey 10, fo. 26^r.

connections with Germany were particularly important here, as it was in the German lands that a new approach to teaching – above all, language-teaching – developed between the 1610s and the 1630s. In 1612 Wolfgang Ratke (Ratichius) had proposed a method of teaching Latin which, he said, was rapid, effective, and pleasant for the child. Specifically, it depended on learning a short passage in the vernacular first, and then listening to its Latin version over and over again. Generally, the method emphasized learning from use and example (not the rote-learning of grammatical rules); guaranteeing comprehension (by the use of the vernacular too); making the memorizing of examples come naturally, through frequent repetition; and extending the pupil's knowledge step-by-step, reinforcing what was known and adding new material in small gradations. These ideas were not radically original: they crystallized objections shared by many pedagogues to traditional methods of grammar-teaching. The 'clausulary' method proposed by Joseph Webbe was in some ways a similar attempt to make the learning of Latin follow a more natural course, with less dependence on the grammar text-book. But it was in Germany that such ideas were most influentially developed into a self-consciously new system of pedagogy, especially by teachers who had taken an interest in Ratke's work, such as Christoph Helwig (Helvicus), Professor of Oriental Languages at Giessen.[67] In the early summer of 1631 Pell made notes on Helwig's 'didactica'; we may assume that it was either Hartlib or Dury – both of whom had many contacts with educationalists in northern Germany – who had brought this writer to his attention.[68]

By far the most influential writer in this tradition was the Moravian churchman Jan Amos Comenius (Komenský), who lived and taught in a small community of his co-religionists (the Unity of Brethren) in exile, in the Polish town of Leszno.[69] He was influenced by Ratke and by Bacon; and from a minor German theorist, Elias Bodinus, he took not only the idea of composing a complete didactic in the vernacular, but also the principle that the ordering of its material

---

[67] On Ratke and Helwig see J. W. Adamson, *Pioneers of Modern Education, 1600–1700* (Cambridge, 1905), pp. 32–43; J. E. Sadler, *J. A. Comenius and the Concept of Universal Education* (London, 1966), pp. 104–10; G. Michel, *Die Welt als Schule: Ratke, Comenius und die didaktische Bewegung* (Berlin, 1978); and U. Kordes, *Wolfgang Ratke (Ratichius, 1571–1635): Gesellschaft, Religiosität und Gelehrsamkeit im frühen 17. Jahrhundert* (Heidelberg, 1999).

[68] BL, MS Add. 4428, fo. 147 (dated 14 May 1631). The work Pell consulted was probably C. Helvicus, *Libri didactici, grammaticae universalis ... in usum scholarum editi* (Giessen, 1619).

[69] The standard modern biography is M. Blekastad, *Comenius: Versuch eines Umrisses von Leben, Werk und Schicksal des Jan Amos Komenský* (Oslo, 1969); also valuable is D. Murphy, *Comenius: A Critical Reassessment of his Life and Work* (Dublin, 1995).

should reflect the ordering of reality in the universe.[70] As a student at Herborn, Comenius had already come under the influence of the great encyclopaedist Johann Heinrich Alsted; as his pedagogical theories developed, they came to involve not just an encyclopaedistic plan for the ordering of universal knowledge, but a project of 'pansophy' – the imparting to humankind of universal wisdom, both natural and spiritual.[71] Such ideas resonated powerfully with someone such as Hartlib, who had already been deeply committed to the utopianism of Andreae and the 'Antilians'.

The work that made Comenius famous throughout Europe was his *Janua linguarum reserata* ('The door of languages unlocked'), published in 1631. This was a book designed to teach children Latin while at the same time – and by means of – teaching them about the world around them. Each of its ninety-nine sections contained a simple discussion of a feature of the world: first astronomy, then geology, botany, zoology, human beings, physiology, psychology, mechanical things, gardens, agriculture, cookery, navigation, trade, and so on. The method was equally applicable to any language (Comenius himself produced a Czech version in 1633), or to bilingual editions in Latin and the vernacular.[72] Before the end of 1631 (or by 25 March 1632 at the latest) a trilingual edition, in Latin, English and French, had been issued in England by a little-known educationalist, John Anchoran; it appeared with commendatory verses expressing the approval of Westminster School.[73] It was probably soon thereafter – in 1632 or 1633 – that Hartlib was stimulated to make contact with Comenius, perhaps through the good offices of his brother Georg Hartlib, who had got to know the Czech thinker when they were both students in Heidelberg. The details here are uncertain, but it is known that Samuel Hartlib was communicating with Comenius by the

---

[70] There is a huge literature on Comenius's educational theories; see, above all, K. Schaller, *Die Pädagogik des Johann Amos Comenius und die Anfänge des pädagogischen Realismus im 17. Jahrhundert*, 2nd edn. (Heidelberg, 1967). On his debt to Ratke see A. Israel, 'Das Verhältnis der Didactica magna des Comenius zu der Didaktik Ratkes', *Monatshefte der Comenius-Gesellschaft*, 1 (1892–3), pp. 173–95, 242–74; on his Baconianism see Adamson, *Pioneers*, pp. 46–79; on the impact of Bodinus see Sadler, *J. A. Comenius*, p. 109, and Blekastad, *Comenius*, p. 139.

[71] On Alsted see H. Hotson, *Johann Heinrich Alsted, 1588–1638: Between Renaissance, Reformation, and Universal Reform* (Oxford, 2000), and on his tuition of Comenius see Blekastad, *Comenius*, pp. 32–3. On the relationship between Comenius's pansophy and pedagogy see E. Sturm, 'Pansophie und Pädagogik bei Jan Amos Komenský', in K. Gossmann and C. T. Schielke, eds., *Jan Amos Comenius, 1592–1992: theologische und pädagogische Deutungen* (Gütersloh, 1992), pp. 101–25. Jan Patočka argues that although the didactic theories were published first, the basic principles of pansophy were contemporary with them in Comenius's thinking: 'Komenského názory a pansofické literární plány od spisů útěšných ke "Všeobecné poradě"', in D. Gerhardt et al., eds., *Orbis scriptus: Dmitrij Tschižewskij zum 70. Geburtstag* (Munich, 1966), pp. 594–620.

[72] On the early printing history, and the Czech version, see Blekastad, *Comenius*, pp. 171–4.

[73] [J. A. Comenius,] *Porta linguarum trilinguis reserata et aperta*, ed. J. Anchoran (London, 1631). The titlepage implied that Anchoran was the author; the preface appeared over two names, 'Ioh. Anchor.' and 'I. A. Comenius'.

summer of 1633.[74] And before long, Hartlib's enthusiasm for Comenius's work – not only for the encyclopaedistic nature of his language-teaching, but also for his larger plans of educational reform, and for his whole 'pansophical' project – was shared by John Pell: a list of promised writings by Pell, noted expectantly by Hartlib in the summer of 1634, included 'Letters recommendatory for Comenius et his Works'.[75]

Pell's own pedagogical work had hitherto taken a rather different course, though it had also been moving in the direction of a kind of encyclopaedism, albeit one more closely tied to Holy Scripture. By Whitsun (25 May) 1634 he had completed a little book, designed to help children (and others) to learn to read. It consisted, quite simply, of lists of all the words in the King James version of the Bible, arranged by the number of syllables they contained: first all the monosyllables, then all the bisyllabic words, and so on. The first surviving reference to this book comes in a letter to Hartlib of 23 February 1635, in which Pell remarked that 'If Mr Bellamy had not been from home last Whitsuntide I had gone with him for it, to ye Licencer, & it may be, I may have ye like hindrance [>this] yeare'; it was entered in the registers of the Stationers' Company by the bookseller John Bellamy as 'a booke called *The English Schoole &c.* by J. P.' on 23 March 1635, licensed the following day, and printed not long thereafter.[76] The full title was *The English Schoole, Teaching to reade English distinctly, and to write it both faire and true. By a perfect Table of all the words in the Bible of the last translation. Disposed in a naturall order; So that all (whether Strangers or others, though but of meane capacity) may with very little helpe, soone learne to reade* English *perfectly, by this booke onely,* and the author's name was given – in accordance with Pell's habitual scruples – only as 'I. P.'[77] In his preface, Pell set out his reasons for compiling the book: 'if you learne but this booke onely, it will more sufficiently inable you to read the Bible, then the farre greater bookes

---

[74] In August 1633 Hartlib was communicating with Comenius via Jan Jonston (the Polish-born son of a Scottish merchant) in Leiden: see HP 44/1/2 (printed in W. J. Hitchens, A. Matuszewski, and J. Young, eds., *The Letters of Jan Jonston to Samuel Hartlib* (Warsaw, 2000), p. 67). The first surviving letter from Comenius to Hartlib is dated 1633: Kvačala, *Die pädagogische Reform*, i, pp. 37–9. Otakar Odložilík has written that the contact began in 1633 ('Z pansofických studií J. A. Komenského', *Časopis Matice Moravské*, 52 (1928), pp. 125–98, here p. 140); Young has argued for 1632 (*Comenius in England*, p. 35(n.)); Blekastad has written in one study that the contact was made in c.1633 (*Menneskenes sak: den tsjekkiske tenkeren Comenius i kamp om en universal reform av samfunnslivet* (Oslo, 1977), p. 97), and in another that the correspondence between them in that year was merely a 'renewal' of previous contacts (*Comenius*, pp. 213–14).

[75] HP 29/2/30B (Ephemerides, 1634, part 3, before 11 July).

[76] BL, MS Add. 4425, fo. 68^v (Pell to Hartlib, 23 Feb. 1635); E. Arber, ed., *A Transcript of the Registers of the Company of Stationers of London, 1554–1640* AD, 5 vols. (London, 1875–94), iv, p. 309; the imprimatur at the end of the book is signed 'Guil. Bray. Dat. Lamethi. Mar. 24. 1634' ('William Bray, given at Lambeth Palace, 24 March 1634/5': Bray was chaplain to Archbishop Laud); Pell had a printed copy by 10 June (see below, at n. 79).

[77] 'Printed by T. C. for Iohn Bellamy' (London, 1635). This book is not attributed to Pell in any modern reference work, and his responsibility for it was apparently unknown to all previous writers on Pell.

of others ... Heereby therefore may much time labour and cost bee spared, and many both in the city of *London* and elsewhere, as well servants as others, may bee brought to reade, that would otherwise never have learned.' And he took pains to explain the pedagogical principles – incrementalism, and the rejection of rote-learning – that underlay his work: 'As for the easie learning of it, the order is such as I think, an easier cannot possibly be invented, for first you learne to spell & reade words of one syllable, then of 2 syllables, &c. & then the other are so set, that one word helpeth another very much, & yet not so as to make the learner reade by rote.'[78]

The book was printed by early June 1635; writing to Hartlib on the 10th of that month, Pell referred to 'my English Schoole, in the next edition whereof I hope the printer will have more care to satisfy me, as I shall have more to satisfy my selfe'.[79] Unfortunately there would be no more editions of this book; instead, the existing edition was suppressed, so successfully that the only known copy today is the one in Westminster School, which evidently came from Pell's own library.[80] The reason for the suppression of the work was that the Stationers' Company operated a monopoly on the publishing of school primers, in the form of a joint-stock arrangement between members of the Company, known as 'the English stock'. Evidently the Company authorities had not realized that this book was a reading primer when they allowed it to be entered in their registers. One primer already published as part of the English stock, Edmund Coote's *The English Schoole-Maister*, was quite similar in conception to Pell's work, also beginning with a list of monosyllabic words. Less than eighteen months after the publication of Pell's book, the Court of the Stationers' Company met to consider the matter, and recorded its decision as follows:

> The booke intituled the English Schoole entred to m^r Bellamy about two yeares since is by order of this Co^{rt}. and by m^r Bellamies Consent to be crossed out of the Hall booke for that it is prejudiciall to the Stocks book called The English Schoolemaster. And what bookes he hath vnsold is to be brought into the hall. And moreover to pay xx^s. for a fine for the same and the said Booke to belong to the English stock.[81]

At the time of the book's publication, however, Pell had no inkling that such problems would arise, and was hard at work on other projects that would treat the Bible not just as a word-stock, but as a quasi-encyclopaedic stock of

---

[78] Pell, *The English Schoole*, sigs. A2^v–A3^r.

[79] BL, MS Add. 4425, fo. 10^v.

[80] Busby Library, Westminster School, pressmark K. B. 6. I am very grateful to Mr Peter Holmes and Mr Eddie Smith for their help in enabling me to consult this and other works in the Busby Library. This book does not bear an ownership inscription, but occasional pen-and-ink corrections in it appear to be in Pell's hand.

[81] W. A. Jackson, ed., *Records of the Court of the Stationers' Company, 1602 to 1640* (London, 1957), p. 301 (entry for 13 Nov. 1637). On the English stock see C. Blagden, *The Stationers' Company: A History, 1403–1959* (London, 1960), pp. 92–106.

concepts, the proper organization of which would give readers better access to the spiritual truths it contained. On a visit to London in the early summer of 1635 he was shown by Hartlib a copy of a Hebrew lexicon-encyclopaedia, *Moses omniscius*, by the German scholar Zacharias Rosenbach: this was a dictionary of all the words in the Old Testament, arranged in six sections (universals; natural things; human things; divine things; pronouns and particles; proper names) and divided within those sections by subject matter (for example, human life; the body; generation; nutrition; health; clothing; and so on).[82] Rosenbach had also published a specimen of an equivalent work (somewhat differently organized) dealing with the New Testament, *Methodus omniscientiae Christi*.[83] Drafting a letter to Hartlib on 10 June 1635, Pell wrote:

> I [could have wished also *deleted* >desired] ... to have read over Rosenbachius y^t I [may *deleted* >might] throughly [>have] understood his project, [>to] which I thinke mine [goes beyond *deleted* >somewhat likes], For if he can draw [an Omniscientia Mosaica *deleted* >a Moses omniscius] out of ye words of ye Old Testament & Christus Omniscius out of ye New. I suppose I may draw an Omniscientia Biblica out of [y^m both more *deleted* >ye words of y^m both more] perfect then either of his: But in both his project and mine, the greatest difficulty is to have all. for when this is done, seeing y^t words have a body a soule & a spirit, I first make use of their bare cadavers [>which are about 11000] hence my English Schoole ... If ye words as they have a soule or common signification, Rosenbach thinkes that there may be made an Encyclopaedic Lexicon, but I feare it is but a conceit & imperfect ... But of their psychic & spiritual sense there may be made in our Language a profitable [interpretation *deleted*] explication (as I signify in ye porch of my schoole) yea surely for ye most part of men, more profitable then his omniscientia.[84]

Four months earlier, Pell had referred in another letter to Hartlib not only to an 'Index' he had compiled of words in the New Testament (probably part of his

---

[82] Z. Rosenbach, *Moses omniscius, sive omniscientia mosaica, sectionibus VI* (Frankfurt am Main, 1633). Rosenbach (1595–1638) was Professor of Medicine and Oriental Languages at the Hohe Schule, Herborn; he was a friend and colleague there of Alsted, and of the scholar Georg Pasor (father of Hartlib's friend Mathias Pasor). See the entry on Rosenbach by F. Otto in *Allgemeine deutsche Biographie*, 56 vols. (Leipzig, 1875–1912), xxix, pp. 199–200; G. Menk, *Die Hohe Schule Herborn in ihrer Frühzeit (1584–1660)* (Wiesbaden, 1981), pp. 270–1; and P. Dibon, 'Le Fonds néerlandais de la bibliothèque académique de Herborn', in his *Regards sur la Hollande du siècle d'or* (Naples, 1990), pp. 267–89, esp. p. 276.

[83] Z. Rosenbach, *Methodus omniscientiae Christi, cum specimine omniscientiae Gentilis, & indice corollarium physicorum Novi Testamenti* (Herborn, 1634). In his preface Rosenbach commented that he had been encouraged to publish this book by 'the praise of many good people, especially of the English nation' for his *Moses omniscius* (p. 3: 'multorum bonorum, inprimis Anglicae nationis, applausu').

[84] BL, MS Add. 4425, fo. 10^v.

'Omniscientia Biblica' project), but also to 'my Grammar for ye N.T.' And in that letter he also discussed a more general encyclopaedic project, designed to facilitate the 'interrogation' of nature by experimenters and explorers; although this was inspired by Bacon, not the Bible, he must have been aware of the similarity (and, to some extent, overlap) between this and his biblical encyclopaedia.

> To say the truth it is the worke of one y^t knowes much & the more we know, the more will our Inquiryes grow, for as Verulam saith [>Novum cap. II.X.] experiments will bring forth axioms & those Axioms will require new experiments, I have beene labouring at this part also, it consists in a perfect catalogue of all entia disposed in the most naturall order, so that the labour is double 1 to get all 2 to dispose them the best way. Verulams Termini Inquisitionis sive Synopsis omnium Naturalium in Universo (Nov. Org. II. Aph. XXI) is of lesse compasse yⁿ my desyderatum.[85]

Pell was also well aware that such a 'perfect catalogue of all entia' was a goal towards which others in Hartlib's circle of contacts and correspondents were also striving. As he added in his letter: 'Doctor Web it may be could furnish us with [>such] an Index ready drawen, I thinke I have heard you speake of such a thing ... And it may be that Comenius his Pansophy will save us all a labour.'[86]

But of all Hartlib's friends, the one whose work was closest in spirit to Pell's at this time was the educationalist William Brookes, who was active as a schoolmaster in London throughout the 1630s. (He died some time before the end of 1640.)[87] He too was attempting to anchor his pedagogy firmly in the text of the Bible. As Hartlib noted in late 1634: 'Mr Brooke is now contriving an English Clausulary of the Scripture which will in some sort bee in stead of a commentary by reason of the vses of the clauses. The fabrick of it is to containe not only phrases, formes, common-expressions but give also all the Biblicall wayes of Amplifications, Insinuations or workings vpon the Affections, hoc est a biblical Logick Rhetorick et Oratory.'[88] More generally, Brookes was a strong advocate of the incremental approach to learning; he developed a method of 'precognitions', by which he meant that the knowledge imparted to the child at each step should contain the means by which the child could recognize, or work out for

---

[85] BL, MS Add. 4425, fo. 68^v (Pell to Hartlib, 23 Feb. 1635).

[86] Ibid. One anonymous and undated MS (not in Pell's hand) that survives among Hartlib's papers may be related to these projects: 'An Imperfect Enumeration of Naturall Things', HP 22/6/1–6.

[87] On Brookes see Webster, *Great Instauration*, p. 106; Hezekiah Woodward referred to him as recently deceased in the preface (written as a letter to Hartlib) to *A Light to Grammar and All Other Arts and Sciences* (London, 1641), sig. a6^v. Both Webster and Turnbull call him John Brooke (probably on the basis of a document incorrectly headed 'Johannis Brukii': BL, MS Sloane 649, fo. 285^r), but he himself signed one of his letters to Hartlib 'Gulielm. Brookes' (BL, MS Sloane 1466, fo. 401^r).

[88] HP 29/2/58A (Ephemerides, 1634, part 6).

itself, the things contained in the next step.[89] How exactly his method operated is not clear, and the account of it which he eventually provided for Hartlib, Dury and Pell in 1637 seems to have caused nothing but disappointment; but there must have been something original about it, since, as Hartlib noted in 1635, 'All Brook's Scollers can teach others how themselves were taught which Scollers of other didactiks cannot doe commonly.'[90]

Pell had direct personal contacts with Brookes (he refers in a letter of 1635 to having visited him in London), and Hartlib's memoranda from the period 1634–5 show how closely intertwined their pedagogical projects were.[91] In a note on infant education, Hartlib referred to 'an encyclopedia of individual things, to be perfected by Brookes, Pell, Evenius and Docemius'; other jottings call it an 'Encyclopaedia Brukiana-Pelliana' or 'Encyclopaedia Singularium Comeniana-Brukiana-Lamplovi-Pelleana', or 'Encyclopaedia singularium Bruko-Hornio-Pelliana'.[92] In a memorandum on ways of teaching 'Children et people to reade vnderstandingly', Hartlib noted: 'to Observe and teach per praecognitions Pels or Brook's Method of Construing'.[93] And another note, referring to the schoolteacher Thomas Hodges, recorded: 'Brook per Hodges hase caused to

---

[89] Cf. the explanation of the term 'precognition' by his friend Hezekiah Woodward: 'It is an *anticipation* of the understanding, that is, a stealing upon it, and catching of it, unfolding unto it, that [which] the childe *knowes not*, by that *medium* or *meanes* he *knew before*' (*A Light to Grammar*, p. 20). Brookes may have taken the term from the writings of Alsted, where it had a more general philosophical meaning, signifying the principles on which any discipline depended (see Hotson, *Johann Heinrich Alsted*, pp. 31–2, 79–80).

[90] HP 29/3/12A (Ephemerides, 1635, part 2). For the reactions of Dury and Pell in August 1637 see HP 22/3/1, where Dury complains that 'In a discourse of Praecognitions ... every thing should bee so ordered that what goeth before should bee a Praecognition to that which followeth and also more cleare by itself then anything else, but here in these generalities and superficial kinds of divisions the definition of every member canot be vnderstood till you come to the particulars', and Pell comments that 'I doe not remember that ever j saw any Mans writings (of matters that j vnderstood) so cloudy and tedious as his are.'

[91] BL, MS Add. 4425, fo. 68^v (Pell to Hartlib, 23 Feb. 1635): 'Hodges method of reading I have seene a slight draught of at Mr Brookes.'

[92] HP 29/2/57A, 61A, 63B (Ephemerides, 1634, part 6); BL, MS Sloane 653, fo. 111^r (Hartlib, list of 'desiderata', printed in Kvačala, *Die pädagogische Reform*, i, p. 44). On Sigismund Evenius (d. 1639), Rector of the Gymnasium in Halle, who had published a *Methodi linguarum artiumque compendiosioris scholasticae demonstrata veritas* (Wittenberg, 1621) and was an advocate of direct learning from pictures of objects, see Schaller, *Die Pädagogik des Comenius*, pp. 332–4, and Michel, *Die Welt als Schule*, pp. 84–5, 95–100. On Johann Docemius (d. 1638), a teacher in Hamburg who translated Comenius's *Janua linguarum* into German, see Blekastad, *Comenius*, pp. 200–2. William Lamplugh (d. 1636 or 1637), a Hebrew scholar, engaged in research connecting Greek, Latin, and English with Hebrew (see the comments in T. Hayne, *Linguarum cognatio, seu de linguis in genere, & de variarum linguarum harmonia dissertatio* (London, 1639), p. 46), and left what Hartlib called a 'Hebrew MS. of his Harmonical and Etymological Lexicon' (HP 30/4/77B (Ephemerides, 1641)). On Thomas Horne, a schoolteacher who corresponded with Hartlib in 1630 and later became headmaster of Tonbridge School and Eton College, see Turnbull, *Hartlib, Dury, Comenius*, pp. 37–8, and S. Rivington, *The History of Tonbridge School* (London, 1925), p. 137; he worked in the 1630s as deputy to Brookes ('his second': Woodward, *A Light to Grammar*, sig. a6^v).

[93] HP 29/3/9B (Ephemerides, 1635, part 2).

contrive Ianuam Syllabarum ... which they should learne first and afterwards learne to spel by sounds and make perfect. 1. promising to spel Syllabically. 2. then for Conformation Pels books. 3. then Textual Reading of some selected annexed Psalms and the Bibel.'[94]

As some of those notes indicate, the educational ideas of Comenius were already becoming interwoven with those of the Hartlib circle. It has long been known that Samuel Hartlib was the main channel through which Comenianism flowed into English intellectual life; but Pell's correspondence sheds additional light on the transmission of Comenius's writings to Hartlib and his friends. Within two years of the publication of his *Janua linguarum*, Comenius issued two other books aimed at teaching children at a more elementary level. The first was an even simpler introduction, for younger children, to the Latin language, entitled *Januae linguarum reseratae vestibulum* ('The porch to the unlocked door of languages'), published in Leszno in 1633; this was being studied by Brookes – who found it unsatisfactory – before the end of 1634.[95] Doubtless Pell read it too: his reference in his letter of 10 June 1635 to 'ye porch of my schoole' suggests that he prepared a similar preparatory work for his *English Schoole*.[96] The second work by Comenius, which also appeared at Leszno in 1633, was concerned with infant education, up to the age of six; written originally in Czech, it was published in a German translation, *Informatorium der Mutterschule*.[97] Pell's correspondence with Hartlib shows that by February 1635 Hartlib had commissioned a rough version of an English translation, which was given to Pell to be put into its final form. As Pell wrote on the 23rd of that month:

> I am sorry y^t I learned Dutch no better whilst I was with you, that so I might ha[*page torn*] labour of translating. [*page torn*] hope I shall understand ye Translators meaning so well y^t I shall [*page torn*] Materna speake English, how good I know not, it had neede be plaine [*page torn*] Mothers, [*page torn*] I doubt, some things in it will scarce passe the Licencing And therefore twere good you considered it well before you goe about to print it.[98]

By June of that year Pell was able to tell Hartlib that 'The Mothers-Schoole you shall have as soone as I have copied it out'; but the work – like so many

---

[94] HP 29/3/42A (Ephemerides, 1635, part 4). For Hodges see above, n. 91, and Turnbull, *Hartlib, Dury, Comenius*, p. 77. The reference to 'Pels books' indicates that Pell was planning at least one more book to supplement his *English Schoole*.

[95] HP 29/2/58A (Ephemerides, 1634, part 6): 'Brukius likes not so well didacticam Linguae Latinae as it is described in Vestibulo.'

[96] See above, at n. 84.

[97] See J. A. Comenius, *School of Infancy*, ed. W. S. Monroe (n.p., n.d. [1895?]), 'Introduction', pp. ix–xi, and Blekastad, *Comenius*, p. 212. For the original text see J. A. Comenius, *Mutterschule*, ed. A. Richter (Leipzig, 1891).

[98] BL, MS Add. 4425, fo. 68^v. ('Dutch' here means 'High Dutch', i.e. German.) Possibly Pell thought that the licensers might object to the simplified theology which Comenius recommended to be taught to children, including the principle that if they were good, God would take them to heaven (see Comenius, *School of Infancy*, p. 18).

other publication projects in which Pell would be involved – never came to fruition.[99] Nor, for that matter, did the guide to infant education which Pell was apparently planning to write at the same time: a list of desiderata for 'Didactica Eruditionis Infantilis et Puerilis' drawn up by Hartlib in early 1635 begins with 'Schola Materna Comeni' and includes a work described as 'Status praeparatorius Pelleanus'.[100]

Pell's letter to Hartlib of 23 February 1635 also refers to other works by Comenius which were known about, and expected, by Hartlib. After discussing his own project of 'a perfect catalogue of all entia', Pell commented (as quoted above) that 'it may be Comenius his Pansophy will save us all a labour.' He continued: 'His 4th Libellus Bohem[*page torn*] will doe much towards it, which with the rest when they come to your hands I doubt not but I shall see.'[101] The identity of this fourth 'little Czech book' is hard to establish. It may have been part of the 'Didaktika Česká' which Comenius had written in 1628–32 (but which remained unpublished in that language until the nineteenth century), or it may have been a version of some of the lectures about which Hartlib had been informed, by another correspondent in Leszno, in the latter part of 1634: 'Mr Comenius gives public lectures about his pansophy, in the Czech language, every Wednesday and Saturday.'[102] Later in his letter Pell referred to yet another promised work by Comenius, adding a comment which shows that his admiration for the Czech thinker was not unqualified: 'As for Comenius his linguarum Harmonia, let him doe it at leisure. I have a conceit yt it will be ye most uselesse booke that ever he wrote.'[103] This was apparently a reference to an early version of at least one part of Comenius's *Linguarum methodus novissima*, a text that would be completed in the mid-1640s (Hartlib would receive a manuscript version compiled in 1645–7) and published in 1648.[104] In chapter 3 of that work,

---

[99] BL, MS Add. 4425, fo. 10^v (Pell to Hartlib, 10 June 1635). Joseph Müller's listing of Comenius's works includes a reference to an English translation published in 1641 ('Zur Bücherkunde des Comenius: chronologisches Verzeichnis der gedruckten und ungedruckten Werke des Johann Amos Comenius', *Monatshefte der Comenius-Gesellschaft*, 1 (1892–3), pp. 19–53; here p. 25); but I have not found any trace of such a publication. Writing from London to his friends in Leszno in Oct. 1641, Comenius observed that a version of his *Informatorium* 'had been prepared here before my arrival, but it has not yet been printed' (Young, *Comenius in England*, p. 66); possibly this comment was the origin of Müller's claim.

[100] HP 29/3/3B (Ephemerides, 1635, part 1); cf. a reference in a similar (undated) list to 'Eruditio Praeparatoria Pelleana' (HP 22/8/1A). This may perhaps be the work also described in another undated list as 'Liber Elementarius Pell' (HP 22/10/1A).

[101] BL, MS Add. 4425, fo. 68^v.

[102] HP 29/2/44B (Ephemerides, 1634, part 5): 'Ex literis Iohannis Decani Lesna scriptis. Dnus Comenius publice suam Pansophiam Bohemica Lingua praelegit diebus Mercurii et Saturni.' For the 'Didaktika Česká' see Comenius, *Veškeré spisy*, iv, pp. 12–454, and the comments in J. Brambora, *Znižní dílo Jana Amose Komenského: studie bibliografická* (Prague, 1954), p. 48.

[103] BL, MS Add. 4425, fo. 68^r.

[104] For the published text see Comenius, *Veškeré spisy*, vi, pp. 183–530. Hartlib's MS (which differs from it) is HP 35/5/1–180; for the dating see Turnbull, *Hartlib, Dury, Comenius*, pp. 403–6.

Comenius argued that there were only four 'cardinal languages' in Europe, to which all others could be reduced – Greek, Latin, Germanic, and Slavonic – and that all four could be shown to be derived from Hebrew.[105] Pell's critical comments show that he must have received an account of this part of Comenius's argument:

> For our Europe hath besides those 4 [>wch he speaks of Greeke, Latin, Slavonick German] other distinct mother tongues (12 Irish Welsh, Pyrenian, [Arabique *deleted* >Alpuxarran] Finnique, East Frisian, Veggian. Epirotique, Hungarian Jazygian. Tartarian,) which will require as large a proofe (& yet then [would *deleted* >will] nothing be prooved but y^t All ye European languages have much affinity with ye Hebrew. & I suppose I can counterpoise it with [>wide] differences in those languages. I would have him reserve that worke to ye very last.[106]

Despite the independence of spirit shown by those remarks, it remains possible that Pell was nevertheless influenced in some ways by Comenius's work on these topics. For if the account he had received covered the whole subject-matter of chapter 3 of the published work, it would have stimulated him to think about such matters as the possibility of a 'natural language', and the use of mathematical combinatorics to calculate the number of possible words and possible languages that could be formed – topics which, as we shall see, he was one of the first people to consider in England.[107]

By the end of 1636 Hartlib had received from Comenius two separate versions of an introductory work on his pansophical project, setting out its underlying principles and arguing for its necessity, its possibility and its ease of performance. The shorter text (which survives among Hartlib's papers) was known as the

---

[105] Comenius, *Veškeré spisy*, vi, pp. 237–9.

[106] BL, MS Add. 4425, fo. 68^r. The source of Pell's list (actually of 11 languages, plus Comenius's four) must have been Edward Brerewood's popular work, *Enquiries touching the Diversity of Languages and Religions through the Cheife Parts of the World* (London, 1614), which listed these 15 languages (pp. 21–2). There is a copy of this work among Pell's books in the Busby Library, Westminster School, pressmark L B 7, annotated by Pell: 'June 14. 1633. J. P. Donum Soceri mei' ['a gift from my father-in-law']. 'Pyrenian' ('Cantabrian' in Brerewood) means Basque; 'Alpuxarran' means Arabic spoken in the Alpujarra region of Spain; 'East Frisian' is in fact Germanic; 'Veggian' is 'Veglian', i.e. Vegliot (a Latinate language then spoken in Dalmatia: the mis-spelling derived from Brerewood, who referred to 'the Isle of *Veggia*'); 'Epirotique' means Albanian; 'Jazygian' refers to a small population in the Hungarian region of Jazygia (Jászság), descended from Alans and Sarmatians and originally speaking Alan, an Indo-European language. On the sources of Brerewood's information see G. Bonfante, 'Una descrizione linguistica d'Europa del 1614', *Paideia*, 10 (1955), pp. 22–7.

[107] Comenius, *Veškeré spisy*, vi, pp. 236 (giving the number of possible words as 1 561 327 494 111, and calculating the number of actually existing words in the world's languages as only 200 000 000), 240 (suggesting that something similar to a natural language could be created, 'namely, one which by the quality of the sound and by the structure of the parts may express in some way the nature of things' ('nempe quae ipsa soni qualitate partiumque structura rerum naturas quodam modo exprimat')).

'Praecognita', the longer as the 'Praeludia'. Both these works were passed by Hartlib to a young German scholar, Joachim Hübner, who had come to study at Oxford.[108] (Hübner, whose father was an official in Brandenburg, had previously studied at Leiden University; he was roughly the same age as Pell, and over the next few years would become a key member of Hartlib's inner circle of friends and collaborators.)[109] In early 1637 Hartlib also sent Hübner a copy of a treatise by Pell, which does not survive, entitled 'De numero et ordine disciplinarum' ('Of the number and arrangement of disciplines'); the young German was 'extremely pleased' by it, and commented that 'It seems that Comenius is working on the same principles.'[110] During that year, however, Pell's own trust in Comenius's judgement was further weakened by reading an astronomical text by him (now lost) in which the Czech writer confidently dismissed the theory that the earth moved round the sun.[111] By September, when Pell in turn was consulted by Hartlib about one or other of Comenius's manuscript introductions to pansophy, Pell's response was notably cautious: 'you desire to haue freely my judgment on the Comenian and Mr. Hubner's papers ... I love not to bee forward in censuring of other Mens labours especially their Ideas ...' He said he was reluctant 'to guesse of his Pansophy', but ended by giving Hartlib, albeit in very unspecific terms, the encouragement he no doubt craved: 'vnles he be overwhelmed and dazled with brightnes I make noe doubt but that he will reflect it to vs in such a measure as shall beyond all proportion exceede that luster which first drewe mens eyes towards him.'[112]

---

[108] See the discussion in J. A. Comenius, *Dva spisy vševědné: Two Pansophical Works*, ed. G. H. Turnbull (Prague, 1951), pp. 19–22.

[109] On Hübner see *MC* vi, p. 291; Blekastad, *Comenius*, pp. 249–51. In later life he became historiographer and chief librarian to the Elector of Brandenburg: for a detailed account of his later career see K. Tautz, *Die Bibliothekare der churfürstlichen Bibliothek zu Cölln an der Spree: ein Beitrag zur Geschichte der Preussischen Staatsbibliothek im siebzehnten Jahrhundert* (Leipzig, 1925), pp. 4–16.

[110] BL, MS Sloane 417, fos. 105^v (Hübner to Hartlib, early 1637), 'Mr Pells Discours De Numero et Ordine Disciplinarum gefahllet mir uber alle massen wohl'; 106^r (Hübner to Hartlib, 3 Feb. 1637), 'Es scheinet Comenius gehet auff die selbige fundamenten.' The author of these letters is incorrectly identified as Christoph Schloer in Barnett, *Theodore Haak*, p. 30. The work by Pell may possibly have been related to an early (undated) text by him on method, logic and 'the teaching of disciplines', which includes the note: 'NB. I would teach Logicke not by it selfe but by some discipline (as Geometry) he should see all done practicè ['practically']' (BL, MS Add. 4388. fos. 1–2).

[111] BL, MS Sloane 417, fos. 164^v (Hübner to Hartlib, 5 July 1637), 'Comenij Paradoxa Astronomica, wan sie der Hr. Von Mr Pelle wieder bekommen'; 166^r (Hübner to Hartlib, 13 July 1637), 'Finde etliche gute sachen darin, bin aber wegen Unterschiedlicher Paradoxorum mit Mr Pellen einig. In refutatione Terrae motus braucht er Vnd Sorellus fast einerley argument.'

[112] HP 36/1/3A–B (Pell to Hartlib (extract), 16 Sept. 1637). 'Idea' here is used almost as a technical term, meaning an advance summary of the principles and aims of a projected work (see below, ch. 3, at nn. 24–6). Possibly the 'papers' by Hübner mentioned here represented an early draft of his 'Idea politicae', of which a scribal copy, addressed to Hartlib and dated 1 July 1638, is HP 26/24/1–8 (for Hübner's authorship of this text see J. T. Young, *Faith, Medical Alchemy and Natural Philosophy: Johann Moriaen, Reformed Intelligencer, and the Hartlib Circle* (Aldershot, 1998), p. 65, n. 19).

Before the end of that year, Hübner succeeded in getting the longer of the two Comenius texts printed in Oxford: it was published under the title *Conatuum comenianorum praeludia ex bibliotheca S.H.*, and copies were sent to Comenius himself (who, having had no inkling that his manuscript would be printed, received them with mixed feelings).[113] By the summer of 1638 even Hübner's enthusiasm for this work seems to have waned: writing to Johann Heinrich Bisterfeld (Alsted's son-in-law), he commented that he had studied the *Conatuum praeludia* and had come to the conclusion that it was impossible for Comenius to complete what he promised in it; he also observed that 'in political theory he will bring forth less than nothing, and in mathematical matters Mr Pell also has poor expectations of him'.[114] Nevertheless, Hartlib's enthusiasm for Comenius remained as strong as ever, and in 1639 he arranged the publication of a revised version of the *Conatuum praeludia*, approved this time by the author, under the title *Pansophiae prodromus*. A note among Hartlib's papers, entitled 'The New Comenian Booke given away', lists recipients of this volume; among them is John Pell.[115] Another memorandum, entitled 'Comenia', is a set of accounts that seems to relate to the printing of either this edition or its predecessor: in the list of expenses, there are entries for 'Bills' (£24), 'Printing' (£22 10s 7d), and 'Pell' (£23).[116] Such a large sum must imply that Pell had been closely involved in the production of the work (or works), though what exactly his function was remains uncertain.

Most of the topics, interests, and activities outlined above related more or less directly to education. But it will already be evident that the general aims of both Hartlib and Comenius were far, far broader than that: nothing less, indeed, than the reformation and perfection of all human knowledge. Hartlib's enthusiasm was tireless, and his curiosity omnivorous. He was equally interested in the development of agricultural techniques, chemical processes, or mechanical devices, all of which were relevant to the improvement of human life on earth; and he was constantly circulating news of new inventions, and stimulating the inventors to improve them. One typical document among Pell's papers is a list of inventions (in a scribal hand), which begins, 'My first Mechanicall Inuention is for the liftinge vp and transportinge from place to place the greatest weights...', and is annotated in Pell's hand, 'Received of Mr Hartlieb. Aprill 27 A° 1636'.[117]

---

[113] See Blekastad, *Comenius*, pp. 251–6.

[114] BL, MS Sloane 417, fo. 202^r (Hübner to Bisterfeld, 1638), 'dass er in politicis weniger dan nichts praestiren werde, Vnd in Mathematischen Sachen hat Mr Pelle ... auch schlechte hoffnung Von ihme'.

[115] See Turnbull, *Hartlib, Dury, Comenius*, p. 343. Turnbull identifies the book as the *Conatuum praeludia*; but the term 'New' surely implies that it is the second of the two editions. The identification offered here is confirmed by Richard Serjeantson, who notes that the copy 'given away' to Edward Herbert survives among Herbert's books: 'Herbert of Cherbury before Deism: The Early Reception of the *De veritate*', *The Seventeenth Century*, 16 (2001), pp. 217–38, here p. 234, n. 52.

[116] HP 23/2/20A (accounts, May 1637–Feb. 1639).

[117] BL, MS Add. 4429, fos. 389–91 (annotation: fo. 391^v).

Hartlib was always interested in knowledge that had a practical application; and, in many cases, what interested him about the application was that it in turn was a way of generating new knowledge. This is true, for example, of the work that was being carried out in the 1630s on the magnetic compass and navigation (a subject in which the English, with their close conjunction of mathematicians and mariners at Gresham College and Deptford, were at the forefront of European research): any breakthrough in the science of navigation would facilitate the gathering of knowledge from all corners of the earth, in a more or less literal fulfilment of Daniel's prophecy, 'Many shall run to and fro ... and knowledge shall be increased.'[118] The leading scientist in this field was the Professor of Astronomy at Gresham College, Henry Gellibrand; in his *Discourse Mathematicall on the Variation of the Magneticall Needle*, published at the beginning of 1635, he used logarithmic calculations to analyse the secular variation of magnetic north. Writing to Hartlib on 23 February of that year, Pell asked for a copy of Gellibrand's book; he explained that he too had been working on 'The Art of Navigation', which he hoped to establish more firmly as an 'art of Mathematicall practise'. (This research was in turn related to his Baconian scheme, mentioned above, of 'a perfect catalogue of all entia': the latter was intended to assist explorers in investigating and categorizing phenomena. As he explained to Hartlib, 'perceiving a great cause of our ignorance of ye World, to be the ignorance of our Saylors to whom it belongs to observe abroad & report at home, I perceived y^t they wanted a twofold instruction, [first in their owne profession *deleted*] 1 in Navigation & 2 [in all other things *deleted*] in Inquisition.')[119] Hartlib was quick to respond to Pell's request: on 28 February Pell received a copy of Gellibrand's work, and immediately began to write a 'discourse' of his own, 'making ye pith of his [booke *deleted* >discourse] ye foundation of mine'.[120] A copy of Pell's discourse was then transmitted, via Hartlib, to Gellibrand, and on 13 May Pell was able to write to Hartlib: 'I am glad that Mr Gellibrand likes my discourse of ye magneticall needle.'[121] Characteristically, however, Pell must have requested that his name be omitted or concealed; for it was only in the following month, after Hartlib had told him that Gellibrand wanted to learn his identity, that Pell conceded: 'You may if you thinke fit satisfy his desire to know who was ye Author.'[122] Five years later, Pell would send a copy of his work to the French scholar Marin Mersenne, who in turn would send it to Descartes; sadly, no copy of it survives, and the only related material by Pell is a rough draft of some

---

[118] Daniel 12: 4. For comments on the importance of this text, see Webster, *Great Instauration*, pp. 9–12. On the Hartlib circle's special interest in the magnetic compass see S. Pumfrey, '"These 2 hundred years not the like published as Gellibrand has done de Magnete"': the Hartlib Circle and Magnetic Philosophy', in M. Greengrass, M. Leslie, and T. Raylor, eds., *Samuel Hartlib and Universal Reformation: Studies in Intellectual Communication* (Cambridge, 1994), pp. 247–67.
[119] BL, MS Add. 4425, fo. 68^v.
[120] Ibid., fo. 10^v (Pell to Hartlib, 3 June 1635).
[121] Ibid., fo. 10^r (Pell to Hartlib, 13 May 1635).
[122] Ibid., fo. 10^v.

further considerations, written in July 1636.[123] (This last text was written not long after Pell had visited Gellibrand at his home in Mayfield, Sussex.)[124]

Also of special interest to Hartlib was any invention that would facilitate the transmission of knowledge.[125] His hope that Pell might 'perfect' Henry Reynolds's 'stenography' has already been mentioned; possibly Reynolds's so-called inventions were of more interest to Hartlib than they were to Pell, as the former made special mention, in one of his lists of recent inventions, of 'Reynolds's method of speaking at a distance'.[126] Another item on that list was 'Pell's new typography'; this apparently involved writing or drawing directly onto a specially prepared surface of tin or pewter, which could then be used as a plate for printing. As early as the summer of 1634 Hartlib noted that 'Mr Pel hase perfected the Printery of Writing et hase largely described it which I expect from him'.[127] He recorded more details later that year, and in early 1635 he reproduced what were evidently further promises made by Pell: 'Great benefits in respect of Mathematical diagrams etc. 2. in Musical bookes. 3. stenographical etc. Pell. 4. Item for Arabick prints.'[128] What practical use was ever made of this invention – if any – is not known.

Pell was not, on the whole, an experimentalist or an inventor of mechanical devices. But he did share Hartlib's interest in the idea of facilitating human communication, and some of his more theoretical work was also directed at this goal. Indirectly, as suggested above, the efforts of his father-in-law may have had some influence here. Reynolds's 'inventions' were probably inspired by a famous text written by the fifteenth-century German cryptographer Johannes Trithemius, who had boasted that he had devised ways of transmitting messages over great distances 'without words, without writings, and even without signs', and of secretly communicating thoughts to a person in the company of others.[129]

---

[123] *MC* ix, pp. 59, 262; the other text, dated 18 July 1636 and beginning, 'Out of Mr Gellibrands discourse, we may gather these 3 particular experiments concerning ye Magneticall needle...', is in BL, MS 4408, fo. 384^r.

[124] BL, MS 4408, fo. 384^r: 'with Mr Gellibrand in a conference at Mayfield this Mid sommer'; cf. Hartlib's note, dated 16 Sept. 1637, of an account by Pell of a manuscript he saw 'in Mr Gelebrands study' (HP 22/15/1A), and Pell's reference to his meeting with Gellibrand (*MC* ix, p. 60).

[125] See the valuable comments on this topic in M. Greengrass, 'Samuel Hartlib and Scribal Communication', *Acta comeniana*, 12 (1997), pp. 47–62, esp. pp. 56–7.

[126] HP 8/60 (undated): 'Ars Loquendj per Interualla Reinolds.'

[127] HP 29/2/30B (Ephemerides, 1634, part 3).

[128] HP 29/2/37B (Ephemerides, 1634, part 4); 29/3/19A (Ephemerides, 1635, part 2). I am grateful to Prof. Toomer for pointing out that a similar invention for facilitating the printing of oriental languages was later announced by Christian Ravius in his *Panegyrica prima orientalibus linguis dedicata* (Utrecht, 1643); it thus seems possible that Ravius had borrowed the idea from Pell. For an account of Pell's invention see HP 71/8A.

[129] The text was a letter of 1499 from Trithemius to Arnold Bostius, published in J. Trithemius, *Polygraphiae libri VI* (Basel, 1518), and reprinted by subsequent writers: see N. L. Brann, *Trithemius and Magical Theology: A Chapter in the Controversy over Occult Studies in Early Modern Europe* (New York, 1999), pp. 85–6.

Trithemius's main work on cryptography, his *Steganographia*, was printed for the first time in 1606; it generated both interest and controversy, not least because of the way in which the author had dressed up his methods (complex systems of substitution and transposition) in a spurious garb of incantations and communications with spirits and demons.[130] Pell's interest in this type of encryption – which required an understanding of combinations and permutations to decode it – is attested to by an early booklet of notes in his hand, entitled 'Logarithmes. Extraction of rootes. Steganographia Trithemij'; and in 1634 Hartlib included, in a list of works under preparation (and/or works promised or suggested), two items by Pell: 'Stenograp. Latina' and 'Clavis Trithemij'.[131] Two years earlier Pell had also studied, at Hartlib's request, a mathematical text on combinatorics, called the 'Tables of Variation', by Joseph Webbe.[132]

In the latter part of the 1630s Pell began to consider new ways of applying these methods to human language. A note jotted down in February 1638 referred to 'y^t seeming mad conceit of mine of Turris Babel, which should comprehend all ye possible words y^t all ye sonnes of men could speake or devise'; this idea (which, as we have seen, may have been stimulated by Comenius's suggestions) does not seem to have been put to practical use.[133] However, Pell was also proposing at this time a kind of philosophical language for place-names, in which the components of the name would come together in a unique combination that expressed the geographical location of the place – somewhat in the manner of a modern grid-reference.[134] Here two important lines of thought converged: the idea of using combinations and permutations (which, if correctly constructed, would map the entire field of reality to which the reference system applied), and the idea of a philosophical language (that is, one in which the natures of the words or signs would systematically express the natures of the things for which they stood). Pell appears to have been one of the very first people in England to think along these lines. In France there had been some discussion of these matters between Mersenne and Descartes in 1629, and Mersenne had published

---

[130] J. Trithemius, *Steganographia, hoc est, ars per occultam scripturam animi sui voluntatem absentibus aperiendi certa* (Frankfurt am Main, 1606). On the 17th-century reception of this work see Brann, *Trithemius*, pp. 186–237. The work had circulated in manuscript before its publication: John Dee had a copy of it. For a good summary of the cryptographic methods employed in it, see W. Shumaker, 'Johannes Trithemius and Cryptography', in his *Renaissance curiosa* (Binghampton, NY, 1982), pp. 91–131.

[131] BL, MS Add. 4430, fos. 21–32; MS Sloane 653, fo. 135^r. (The first of Hartlib's references is apparently to stenography (shorthand), not steganography (encrypted writing), though it is also possible that the latter was intended.)

[132] Pell's notes on this are in BL, MS Add. 4394, fos. 14–21 (dated 16 Jan. 1632: fo. 20^v). His draft of a covering letter begins: 'Sr You desire my annotations upon Dr Webs tables of variation of which you have a copy' (fo. 21^r).

[133] BL, MS Add. 4415, fo. 27^r, dated '9853' (i.e. 19 Feb. 1638 in Pell's personal chronological system, which counted the days of his life).

[134] BL, MS Add. 4423, fo. 376^v. Pell notes that he mentioned this scheme 'to Mr Frost at ye Councell chamber Jan. 1638' (this might mean 1638/9, but Pell usually dated the year from 1 Jan.).

some thoughts on the construction of a philosophical language (also basing it on combinatorics) in his *Harmonie universelle* of 1636; but there is no sign that Pell or any member of Hartlib's circle had any knowledge of Mersenne or his work until 1639.[135] Possibly Comenius's hints at the idea of a 'natural' language had had some influence on Pell (though we do not know whether they were contained in the Comenian text he had read.)[136]

But, however he had got there, it is evident that by late 1638 Pell had arrived at a very clear understanding of the task that would preoccupy other theorists of 'philosophical' or 'universal' languages for many years to come. On 29 October of that year he wrote: 'A *philosophical language* is one which names things as they are – that is, it names simple notions with simple sounds, different notions with different sounds, composite notions with composite sounds...' The word for 'fire', he noted, would thus be composed of the terms for 'hot thing' and 'shining'. But, he continued, 'no one, so far as I know, has dared to attempt this task, namely, to register in a catalogue all the simple sounds of human language, and all the simple concepts of the mind, and to express the complications of the latter by means of the combinations of the former. May God – the Light, Love, the Way, the Truth, and the Life – permit me to begin, complete, and publish that work, for the glory of his name and the unspeakable benefit of the human race, Amen!'[137]

Just a few months earlier, Pell had drafted a longer text on the organization of human knowledge which contained a rather similar passage. The idea of a universal language, as such, was not considered here; what Pell was concerned with was the reduction of knowledge into its elementary units – a process of

---

[135] On Mersenne's work and his earlier correspondence with Descartes see J. Knowlson, *Universal Language Schemes in England and France, 1600–1800* (Toronto, 1975), pp. 65–72. For the first mention of Mersenne in Hartlib's papers see HP 30/4/5A (Ephemerides, 1639, part 1).

[136] The distinction must be noted between a 'natural' language (as discussed by Comenius), in which the sounds 'naturally' express qualities of the things, and a 'philosophical' language (as suggested by Pell), in which the relation between elementary sounds and things is arbitrary, but the organization of the sounds systematically reflects the organization or classification of the natures of the things. Comenius's comment that 'the structure of the parts may express in some way the nature of things' (above, n. 107) could be taken to point, nevertheless, in the direction of a philosophical language. In a later work (*Via lucis*, which he was composing in England in the winter of 1641–2) Comenius did discuss the possibility of a philosophical language: see B. DeMott, 'Comenius and the Real Character in England', *Publications of the Modern Language Association of America*, 70 (1955), pp. 1068–81. The best modern discussion of these issues is R. Lewis, 'John Wilkins's *Essay* (1668) and the Context of Seventeenth-Century Artificial Languages in England' (Oxford University D.Phil. dissertation, 2003); this includes a valuable discussion of Pell (pp. 17–29).

[137] BL, MS Add. 4409, fo. 254^r, dated '10105': '*Lingua Philosophica* est quae ita nominat res ut sunt viz simplices notiones, simplicibus sonis, diversis diversis, compositas compositis ...'; 'ignis ... calidum lucens'; 'nemo quod sciam hoc ausus fuit aggredi, scilicet omnes linguae humanae simplices sonos, mentisque simplices conceptus in Catalogum referre, atque horum complicationes, illorum compositionibus exprimere. Deus, Lux, Amor, Via, Veritas, Vita det mihi rem istam incipere perficere & publicare in nominis sui gloriam, generisque humani commodum ineffabile, Amen!'

'analysis' – and the formation of new combinations of those units by a process of 'synthesis'. This text too deserves to be quoted, not only because of its similar tone of religious invocation, but also because it shows so clearly how these ideas were connected with Pell's earlier encyclopaedic projects of forming 'a perfect catalogue of all entia'.

> Perfect knowledge requires nothing but a perfect Synthesis of perfectly-enumerated simples.
> a Perfect enumeration of simples requires nothing but a perfect analysis of *naturall* compounds & (which we will still be too lazy to doe) a compleate setting y^m downe ...
> Father of lights, God only wise, which enablest men to will & to doe, make me able to gather, set downe & publish for ye good of mankind, a complete enumeration of simple notions with the art of combining & synthesizing of y^m for ye enlightening of those y^t sit in darknesse & to ye glory of thy name! Amen.[138]

Here, then, was the central project on which all Pell's interests at this time converged. It was the ultimate key to the reformation of pedagogy; it aimed at a kind of universal knowledge, similar to Comenian 'pansophy'; it offered the possibility of a new way of embodying and transmitting that knowledge, a philosophical language. And, last but absolutely not least, the methodological model on which it was based – that of a single, all-powerful system of reasoning in which individual units could be combined and manipulated in possibly infinite variations, thereby generating new knowledge – was that of Pell's first and most abiding interest: mathematics.[139]

Pell's work on mathematics had continued unabated throughout the 1630s. This was a good decade in which to develop a knowledge of pure mathematics, especially algebra: 1631 had seen the publication of two important treatises, the algebra textbook by William Oughtred, generally known as his *Clavis*, and the posthumous work by Thomas Harriot (edited by his friend Walter Warner), *Artis analyticae praxis*.[140] Pell made a careful study of these, and of a wide range of other writers, among them Bartholomaeus Pitiscus, Gemma Frisius, Albert Girard, Pierre Hérigone, and the late sixteenth-century mathematician, sometimes

---

[138] BL, MS Add. 4409, fo. 251^r. In the same set of notes there is one item dated 'July 2. 9986' (i.e. 1638), and another which says 'I have lived just 10007 dayes' (i.e. 23 July 1638).

[139] This whole pattern of thought makes Pell a suitable candidate for inclusion in the tradition of Lullian-influenced encyclopaedism discussed by Paolo Rossi (see his *Logic and the Art of Memory: The Quest for a Universal Language*, tr. S. Clucas (London, 2000), pp. 130–44 – esp. pp. 143–4, on the theories of Bisterfeld which, published in the 1650s, closely match Pell's). But Pell himself showed no interest in Lull, and expressed a low opinion of the most explicitly Lullian writer in this tradition, Alsted (see below, at n. 142).

[140] W. Oughtred, *Arithmeticae in numeris et speciebus institutio: quae tum logisticae, tum analyticae, atque adeo totius mathematicae, quasi clavis est* (London, 1631); T. Harriot, *Artis analyticae praxis, ad aequationes algebraicas nova, expedita, & generali methodo, resolvendas: tractatus*, ed. W. Warner (London, 1631).

described as the father of modern algebra, François Viète or Vieta.[141] Clearly, much of his time was spent mastering and analysing particular techniques and problems – many of them at the frontiers of current mathematical research. He did occasionally draft little treatises, or critical comments, on the general nature of mathematical reasoning: some of these may have been prompted by his pedagogical concerns (such as the work he sketched in May 1635, entitled 'Logistica or the Art of reckoning'), and some were probably responses to requests by Hartlib (such as his strongly critical comments on Alsted's text 'Fundamenta disciplinarum mathematicarum').[142] But his interests were mainly concentrated in two areas: the improvement of methods of problem-solving, especially in algebra; and the compiling of logarithmic and trigonometrical tables as aids to problem-solving more generally.

Examples of the first of these would include Pell's undated notes 'on the Vietan-Harriotic way of solving affected Equations', which probably date from the latter part of the 1630s.[143] Pell was increasingly conscious of the ways in which recent progress in algebra had opened up the possibility of further advances and improvements. In an important (but, unfortunately, undated) draft of a letter, possibly to Hartlib, in response to 'quaeres' about the algebraic work of Oughtred and Harriot, he observed that 'none of those things were known to ye Ancients', and added: 'this whole new doctrine of Aequations is yet but very imperfectly taught though you put all our new maisters together. And I have some reason to think that Mr Oughtred will not adde much perfection to these things.' His conclusion was straightforwardly negative: 'And therefore whereas you inquire what bookes doe *most easily* & *best* teach *all* sorts of aequations and the extraction of *all sorts* of Algebraicall rootes. I answer *None* at all.'[144] But the positive implication of this can hardly have escaped him, or his correspondent – namely, that if none of the 'new maisters' was going to perform this task, Pell should do it himself. Here were the origins, perhaps, of the planned work on 'analytics' to which his friend and patron Sir Charles Cavendish would so frequently and imploringly refer.

Of the second category of Pell's special interests, the compilation of tables, there is copious evidence. This activity was in fact a constant feature of his intellectual life: among his manuscripts there are many hundreds, probably thou-

---

[141] See, for example, his comments on Oughtred and Harriot (BL, MS Add. 4409, fos. 248–9), notes on Pitiscus (BL, MS Add. 4400, fos. 16–39), notes on Hérigone and Gemma Frisius (BL, MS Add. 4431, fos. 517–20, 524–8, 530^r, 535–9), notes on Girard (BL, MS Add. 4431, fo. 157^r), and notes on Viète (BL, MSS Add. 4417, fos. 451–8; Add. 4423, fos. 86–145, 152–3; Add. 4431, fo. 157^r).

[142] The 'Logistica' (dated 8 May 1635) is in BL, MS Add. 4420, fos. 47–8; cf. notes entitled 'Logistica generalis' (dated 14 Sept. 1636), followed by 'Mechanica generalis' (24 Oct. 1636), in BL, MS Add. 4415, fos. 9–15. The Alsted text is given (in a scribal hand) in BL, MS Add. 4429, fos. 31–35^r, followed by Pell's hostile comments (fos. 35^v–36); cf. also his criticisms of Alsted's *Cosmographia* and *Uranometria* in BL, MS Add. 4407, fos. 197–8.

[143] BL, MSS Add. 4409, fo. 369^r.

[144] BL, MS Add. 4409, fo. 248.

sands, of pages of mathematical tables of all kinds and from all periods of his life. In the evidence surviving from these early years, some particular projects stand out. His early interest in logarithms has already been noticed: this was the principal topic on which he corresponded with Henry Briggs in 1628. Pell planned to continue the logarithmic tables printed in Briggs's book, but, in the words of a later account given by Pell to John Collins, 'having, Anno 1628, received a letter from Mr Briggs, wherein he told me that his *Arithmetica Logarithmica* was reprinted with 100 000 Logarithmes, by one Adrien Vlacq a Hollander, I turned my thoughts another way.'[145] He certainly did not lose interest in the development of logarithms, however; as we have seen, he was corresponding three years later with Edmund Wingate, suggesting improvements to the presentation of logarithms in the latter's *Arithmetique*.[146] Notes dated October 1634 show that he was then planning to produce a book of his own (which, he calculated, with characteristic attention to detail, would come to 451 pages): its full title, given in the mock titlepage he drew up for it, was 'The Manual of Logarithmicall Trigonometry, Breifly containing the compend of both Mr Brigs his great bookes viz his Logarithmicall Arithmeticke, & his Brittish Trigonometry Affording the Logarithmes of every number under 100 thousand and of ye chords, Sines, Tangents and Secants for every Degree and centesim, of degrees ... to a Radius of 10 000 000 000 particles. As also the making & use of the aforesaid Logarithmes, expressed in CCL Aphorismes.'[147] Another such work, of which only an undated initial fragment survives, was to be called 'Trigonometria logistica Αὐτάρκης id est, Omnium triangulorum planorum rectilineorum solutio non per instrumenta sed per supputationem' ('Self-sufficient calculatory trigonometry, that is, the solution of all plane rectilinear triangles, not by instruments but by calculation').[148] And in February 1635 Pell informed Hartlib that he was planning to produce a comprehensive book of tables, entitled 'Comes Mathematicus ['the mathematical companion'] or the Mathematicians pocket booke Containing a briefe collection of all such tables as are requisite for ye exact & easy solution of any Mathematicall question in ordinary practise, With ye uses of [*page torn*] said Tables in Arithmetic, Geometry, Staticks; Optics Geodesy. Geography. Astronomy. Navigation. Architecture. Fortification.'[149]

One project to which Pell devoted special attention in the period 1637–8 represented, in effect, a combination of his two special interests, tabulation and the development of algebra: it was a table of equations, entitled 'Progymnasma Analytico-logisticum' ('preparatory exercise for algebraic calculation'). One complete copy, dated 16 August 1637, bears the note: 'I coppied out this faire August 17. and sent it to Mr Hartlib. August 19.' The accompanying explanatory text

---

[145] BL, MS Add. 4414, fo. 10^r; the letter from Briggs is in MS Add. 4398, fo. 137.
[146] BL, MS Add. 4280, fo. 233^r.
[147] BL, MS Add. 4431, fos. 30^r (calculation of pages), 33^v (titlepage).
[148] BL, MS Add. 4409, fo. 356^v.
[149] BL, MS Add. 4425, fo. 68^r; cf. also notes for 'Comes mathematicus' in MS Add. 4408, fos. 30–1, 34.

begins, 'This foregoing table containeth 1892 aequations, expressed after ye way y^t Mr Oughtred useth in his Clavis Mathematicae'; it praises those 'moderne Christian writers' (such as Viète, Harriot, and Oughtred) who chose to 'make common this high way of invention for all mankind', unlike 'the [>heathen] Ancients' who had 'reserved it to themselves'.[150] The table itself has eight columns, for the two quantities ('a' and 'b'), their sum, their difference, their product, their quotient, the sum of their squares, and the difference of their squares; in his accompanying text, Pell explains that the 'greate Probleme' which his tabulation makes it easy to solve is, 'Any two of these 8 quantities being given, to finde the other 6.'[151]

Eight months later, after further efforts to perfect his 'Progymnasma', Pell penned a self-examination, in question and answer form, about his reasons for taking such trouble over this project. It is a remarkable document (indeed, it is hard to think of a comparable statement by any other seventeenth-century mathematician), giving an unusually intimate and vivid portrayal of Pell's intellectual character at this time.

> Why did I take all this paines to make & write downe [>all] these aequations? why did I not rather sleepe etc.
>
> A. [sc. 'Answer'] 1 It is not good to be idle, our soule loves to be in perpetuall motion & agitation...
>
> But why did you take in hand such difficulties? You might have beene busy in things y^t required not so much intention of spirit? Make songs, reade some merry bookes, or write some slighter matters.
>
> A2 It is a singular pleasure to overcome & vanquish difficulties, to subdue things y^t were before too hard for us, or seeme impossible to other men...
>
> But there were other difficulties which you knew not, Forraine languages, etc. Many other things to learne... Many which you knew already & might have taught to others.
>
> A3 We love not to meddle with things y^t be too hard for us, We wrestle with our Matches, not with children, nor steeples.
>
> A4 We labour most profitably, when we doe y^t which for ye present our mind is aptest for...

---

[150] BL, MS Add. 4398, fo. 170. Pell had clearly changed his mind about the 'ancients' since writing the letter cited above at n. 144. For similar suspicions that the ancient Greek mathematicians had possessed a knowledge of algebra which they concealed, see Descartes, *Regulae ad directionem ingenii*, IV, in his *Oeuvres*, ed. C. Adam and P. Tannery, rev. edn, 11 vols. (Paris, 1974) [hereafter: A&T], x, pp. 376–7.

[151] BL, MS Add. 4398, fo. 170^r.

A5 Variety delights, when we are weary of one thing, we may exercise another power of our soule upon another thing & refresh our selves. I came from other matters nothing like this.

But what made you pitch upon y^{se} difficulties rather y^n some other which you might as soone hope to finde out.

A6 Our thoughts have a kind of casualty in y^m, they sometimes light upon one thing, sometimes on another... My mind fell upon y^s by occasion of Mr Oughtreds booke.

But what made you labour in these first things, seeing you might have beene upon higher

A7 It is a great folly to adventure upon greater things before we be sufficiently versed in smaller...

But why doe you write it downe?

A8 The memory of man is a thing so slippery...

A9 It is commendable to be ready to communicate any good thing to one another...

But why doe you propose your selfe so large a scope?

A10 The most refined soules aime at perfection, & therefore desire as much as y^{ei} are capable, to be omniscient omnipotent, most prudent, most holy, most perfect as y^{eir} father in heaven is perfect

But why doe you seeke ye prime truths & put y^m together?

A11 It is ye surest way to have all ye knowledge y^t can be to marry together all y^t we know, by which meanes the truthe, doe wondrously increase & multiply by ye facultas synthetica which is naturally implanted in us.[152]

Captured in this document is the strange interaction of Pell's humility (his willingness to work on 'smaller' things when he 'might have beene upon higher') and his ambition (his hopes 'to have all ye knowledge y^t can be'). And, not for the last time, anyone who studies Pell's biography will be struck by the contrast between his sense of a duty 'to communicate any good thing to one another' and his actual failure to publish the work he was here discussing. There was, it seems, a tension between promoting the good of mankind and promoting the good of John Pell – the latter being a task for which he was unfitted by various scruples, including perfectionism as well as humility. Nevertheless, before the end of 1638 he would at last begin to acquire – thanks to the efforts of his friend Samuel Hartlib to publicize his work – an international reputation as a mathematical thinker.

---

[152] BL, MS Add. 4415, fo. 24^r (dated 27 April 1637).

3
London, 1638–1643

For someone exploring the frontiers of research across a range of disciplines, it must have been peculiarly frustrating to be confined to a small country town. Describing his work (of 1635) on Gellibrand's theory of magnetic declination in a later letter to Mersenne, Pell explained that once he had written his response to Gellibrand's book, 'I was indeed forced to stop there ... because of the lack of further experiments, consideration of which was impossible for me, leading a rustic existence among people who had not the slightest interest in such things, and insofar as I was completely lacking in the means to make any further investigation of them.'[1] His letters to Hartlib from Sussex show how much he depended on his 'intelligencer' friend for information, books, and copies of letters to Hartlib from like-minded people on the Continent.[2] They also show that, as we have seen, Pell did visit London when school holidays (Whitsuntide, for example) made it possible to do so. But the impetus for his move to London – an uncertain and long-drawn-out process, which seems to have started at the beginning of 1638 – came very much from Hartlib, not from Pell himself. Hartlib's first task was to overcome Pell's scruples about self-promotion. The line of argument he used was summarized by Pell in a letter to a benefactor in the summer of 1638: 'I was drawn up higher by ye importunityes of some frends who told me I buried my selfe in y^t retirednesse & hid my candle under a bushel which if a little more oile were infused & it set upon a Candle sticke would give a farre cleerer light...'[3] Modesty was not the only problem; there were financial practicalities to consider, if Pell, who had a wife and three children, were to give up his paid employment in Sussex. When Theodore Haak wrote in a later memoir that Hartlib, thinking that Pell's talents 'might be farre more usefully employed and improved for the publick advancement of learning ... never left soliciting and engaging frends heer to perswade Mr Pell, instead of keeping scool ... to come up to London', the persuasion he had in mind must have involved pledges of financial support.[4] Despite all Hartlib's solicitations, few such promises seem to have been made, and those that were made were evidently not kept for long.

---

[1] *MC* ix, p. 59 (Pell to Mersenne, [14/] 24 Jan. 1640): 'sistere nimirum ibi cogebar ... ulteriorum experimentorum defectu. De quibus cogitare tum non licebat, inter ... homines talium minimè curiosos, rusticanti, adeoque ab omnibus ulterioris inquisitionis subsidijs destituto'.

[2] See for example BL, MS Add. 4425, fo. 68^v, a set of notes preparatory to the drafting of his letter to Hartlib of 23 Feb. 1635, which begins: 'I received Febr. 21 at night Saturday 3 printed bookes 9 letters 5 from Norimberg 2 from Poleland, 2 from you'.

[3] BL, MS Add. 4415, fo. 14^r (Pell to Thomas Goad, 7 Aug. 1638). On Goad see below, n. 19.

[4] Aubrey, *'Brief Lives'*, ii, p. 129.

The first sign of Pell's resolve to give up teaching is a draft of a letter to his 'very loving uncle Mr [James Pell Minister *deleted*] at his house in [Willingdon *deleted*]', written from London on 4 January 1638. It begins: 'S^r, I doubt not but y^t you have already heard by my wife, of my journey hither, & what helpes I had from Mr Croply & Mr Hartlib. My hopes are here great [>enough] to hinder me from thinking of returning to my former imploiment.' He referred also to an 'annuity' promised to him by Dr Burton, and commented, 'seeing I can not conveniently remove my family from y^{nce} till Easter I hope they will not thinke of a lesse summe y^n y^s quarters stipend.'[5] There is also an undated letter from Ithamar (who was evidently still in Sussex) to her husband, which probably relates to this visit to London: 'I was sorrowed to heare of your cold Jorney and am very glad you came saife to your Jorney end it beeing such slipery wether ... I rejoyce that [>you are] soe kindely entertained and that Mr Hartlib furnished you with monie and for the hopes of a better charge ... [*signed:*] Your louing wife to comand IP'.[6] The 'Mr Croply' referred to by Pell was a friend or employee of Sir Thomas Colepeper, who lived at Folkington or Fowington, near Willingdon.[7] In late April Pell wrote a letter to his 'much respected friend Mr Thomas Croply At Fowington House', describing a journey he had recently made to London with Sir Thomas: from this it appears that Pell did at least remove himself from Sussex at Eastertide, though not his family. 'I left him on Munday, Aprill 9. ... I would have accompanyed [him] longer if I had knowen y^t my businesse heere in London had moved so slowly as now I finde it, yet they make me beleeve, y^t it is slow & sure.'[8]

One of the attractions of life in the capital was the prospect of close acquaintance with other members of Hartlib's inner circle. During the first half of 1638 Pell got to know the two young Germans who were among the most active supporters of Hartlib's projects. Theodore Haak, whom Hartlib may have known since 1625, had studied at Oxford during the period 1628–31; since then he had been ordained a deacon in the Church of England, had been appointed an official agent in England for the 'collections' on behalf of the exiled Palatine ministers, and had spent some time travelling on the Continent. Pell probably made his acquaintance in early January 1638; Haak was abroad again by the 19th of that

---

[5] BL, MS Add. 4428, fo. 197^r. The reason for the deletion of the name and place of the addressee is not apparent. The date is given only as 'Jan. 4.', but the letter can be confidently assigned to 1638: the sequence of events that follows indicates that it must date from either 1638 or 1639, but a letter of 3 Jan. 1639 from Pell to Hartlib (MS Add. 4419, fo. 136) shows that Pell was not in London at that time. On Dr Burton see above, ch. 1, n. 14.

[6] BL, MS Add. 4431, fo. 301^r.

[7] On Sir Thomas see above, ch. 2, n. 65. Little information is available about Croply, who was mentioned in a letter from Dury to Hartlib of 5 Sept. 1642 as a possible conduit for letters to Dury in Amsterdam (HP 2/9/15). Possibly he was the 'Thomas Croplie' who denounced radical elements in the Army in 1648: see the broadside *The Resolutions of the Army, against the King, Kingdome and City. July 15 1648* (n.d. [London,] 1648), and the pamphlet *Look to it London, Threatned to be fired by Wilde-fire-zeal, Schismatical-faction, & Militant-mammon. Discovered July 15 1648 in a Discourse with one Croply and Hide* (n.d. [London,] 1648).

[8] BL, MS Add. 4424, fos. 359–60 (Pell to Croply, 26 Apr. 1638).

month, matriculating at Leiden University in April and returning to England only in the autumn.[9] A generous and sweet-natured man, Haak would remain a loyal friend for the rest of Pell's life. The other young Hartlibian, Joachim Hübner, moved from Oxford to London some time in the first half of 1638, and seems to have joined Pell as a lodger in Hartlib's house. Writing to Bisterfeld in the summer (probably late July or early August) of that year, he explained that 'I have been on quite familiar terms with this man Pell for a long time, especially because we are of the same age, and, what is more, dining companions at Mr Hartlib's table.'[10] As we have seen, Hübner had already formed a high opinion of Pell from reading the manuscript texts which Hartlib had circulated. He admired his mathematical skills and respected his judgement; he also shared his long-term aims. A note made by Hartlib at the beginning of the following year compared the complementary concerns of Pell, Hübner, and Comenius (in that order) as follows: 'These 3. are very fit to be imploied about the Reformation of Learning. The one vrges mainly a perfect Enumeration of all things. The other is all for that which hase an evident vse in vita humana. The third is all for Methodizing...'[11] Yet, despite the emphasis given in that comment to Hübner's utilitarian concerns, there is evidence from this period that his interest was most strongly engaged by contemporary philosophy: in 1638 he wrote to Comenius praising Descartes, and in 1640 he would write a long letter – in effect, a short treatise – addressed to Edward Herbert, discussing the theories of his *De veritate*.[12]

Pell certainly enjoyed friendship, encouragement, and intellectual stimulus in Hartlib's entourage; what he lacked was paid employment. Only two small scraps of evidence survive from the first half of 1638 to indicate what sort of position Hartlib was trying to obtain for him. One has already been mentioned: Pell's note about the philosophical language for place-names which he explained to 'Mr Frost at ye Councell chamber Jan. 1638'.[13] This suggests that Hartlib was trying to find some sort of salaried governmental post for Pell, perhaps in connection with matters of navigation, shipping, or communication. The 'Mr Frost' mentioned here was probably not himself in a position to grant such a

---

[9] Barnett, *Theodore Haak*, pp. 19–28. Barnett is not aware of Haak's presence in England in early Jan., which I deduce from two facts: that Haak himself wrote that he was introduced to Pell by Hartlib in 1638 (Aubrey, *'Brief Lives'*, ii, p. 129), and that Hübner, writing to Bisterfeld in the summer of 1638, referred to 'Mr Pell (that is the name of the mathematician, who is very well-known to Mr Haak)' ('Mr Pelle (dass ist dess ... Mathematici Dn.^{o} Haakio optimè noti nahme)': BL, MS Sloane 417, fo. 202^r).

[10] BL, MS Sloane 417, fo. 201^v: 'Mit diesem Mann bin ich eine geraume Zeit hero gantz familiariter, sondermahl wir beide eines alters Vnd über dass noch dess Hr Hartliebs tischgesellen sein'. Pell's draft letter to Goad of 7 Aug. 1638 (MS Add. 4415, fo. 14) shows that he was then staying at Hartlib's house.

[11] HP 30/4/5B (Ephemerides, 1639, part 1).

[12] BL, MS Sloane 639, fo. 155^v, Hübner to Comenius, 14/24 Dec. 1638; MS Sloane 639, fos. 89–97^r, Hübner to Herbert, 4 Id. Apr. 1640.

[13] BL, MS Add. 4423, fo. 376^v. As discussed above (ch. 2, n. 134), this refers probably (but not certainly) to 1638, rather than 1639.

post (he was presumably Gualter Frost, who would later become Secretary to the Council of State, but is not known to have held any office at this time); what is significant is that Pell had been brought to 'ye Councell chamber' to present his scheme.[14] The other piece of evidence is a brief memorandum by Pell about a mathematical problem which was put to him by Sir Thomas Aylesbury 'to solve extempore' on 29 January 1638.[15] Sir Thomas would have been a fine catch as a patron, for several reasons. He was an important official and a rich man, being both Master of Requests and Master of the Mint, and having previously been Surveyor of the Navy; what is more, he was an enthusiast for mathematics, one of Thomas Harriot's executors, and the chief patron of the scientist and mathematician Walter Warner.[16] Hartlib may have hoped either that Aylesbury would find a salaried post for Pell, or that he would take him on, like Warner, as a personal retainer. But although there is some evidence of further contacts between him and Pell during the following year, it seems that neither kind of employment was ever offered by him.[17]

Hartlib's efforts on Pell's behalf were, nevertheless, unflagging. This is a reflection not only of Hartlib's generosity of spirit and enthusiasm for the promotion of knowledge, but also of the fact (generally neglected or underplayed by modern studies of the Hartlib group) that Pell was clearly the most talented of all the members of Hartlib's inner circle. He was the intellectual star of whom the greatest things were expected. Writing on 10 [/20] August 1638 to Johann Adolf Tassius, the mathematician and colleague of Joachim Jungius in Hamburg, Hartlib commented on a recent attempt by a French scientist to solve the problem of determining longitude, and then added these remarks about Pell: 'If he enjoyed complete tranquillity (which I hope will come about soon), he would doubtless compose many excellent things, not only on this topic, but also in other areas of mathematical studies.' In the same letter he deferred to Pell's judgement, not only on the question of longitude, but also on recent claims about the invention of a perpetual motion machine, and about Descartes's 'Géométrie',

---

[14] Frost was employed as a secret courier between English opposition leaders and the Scottish covenanters in 1639–40; he later became a Parliamentary comissary and intelligence agent, before being appointed Secretary to the Committee of Both Kingdoms (1644) and to the Council of State (1649) (see G. E. Aylmer, *The State's Servants: The Civil Service of the English Republic, 1649–1660* (London, 1973), pp. 19, 254–5). In January 1652 Frost asked the Dutch Ambassador's secretary if he had known 'his old friend Pell' in Breda: L. Huygens, *The English Journal, 1651–1652*, ed. A. G. H. Bachrach and R. G. Collmer (Leiden, 1982), p. 212 ('sijnen ouden vrind Pellius').

[15] BL, MS Add. 4419, fo. 139^r (the problem: 'A line being given to divide it by extreame & meane proportion'). The year here is not in doubt: Pell first wrote '1637', then deleted the '7' and substituted '8'.

[16] See G. E. Aylmer, *The King's Servants: The Civil Service of Charles I, 1625–1642* (London, 1961), pp. 77–8, 163; Aubrey, *'Brief Lives'*, ii, pp. 291–2.

[17] HP 30/4/2B (Hartlib, Ephemerides, 1639, part 1, following a note dated 14 Jan.): note of a comment by Aylesbury on Warner's preface to his edition of Harriot, reported to Hartlib by Pell.

which Pell had 'explored very assiduously, deriving great satisfaction from it.'[18]

By this stage, however, Hartlib's hopes that Pell would 'soon' find a suitable position were no longer shared by Pell himself. A long letter drafted by Pell just three days earlier (probably to Thomas Goad, though the name has been deleted and is not easily legible) contains a graphic account of his situation and the disappointments he had encountered.

> 'Tis true, I have beene taken notice of by divers, not of ye meanest since I came to y^s city, who, having seene some [extracts of my letters, & some Mathematicall exercitations of mine *deleted* >thing of mine] desired to speake with me & to appropriate me to y^{m}selves promising to entertaine me as a Gentleman, as an Artist &c. But as soone as y^{ey} heare y^t I have a wife & 3 young children, that heate is presently allayed. And this hath befallen me, so often y^t my welwillers now beleeve y^t if ever any *one man* make me meerely *his alone*, it must be some Noble man, whose munificence is not so straightened as y^t it permits [not] any of his dependants to be married men.
>
> But not knowing when or where to finde any such & wearied with long expectation, & fearing y^t holding my candle at y^s great fire I should melt it all away before I should get it lighted, I determined to see it rest as it is & retire to my former obscurity of a Country ScholeM^r, counting y^t better y^n to make my selfe one of y^{ese} City-teachers of Mathematickes, who by y^m y^t looke most narrowly into their doings are found to be [>but] Mathematickasters & a more artificiall sort of cheaters, making a confederacy with ye instrument makers to fleece our Nobility & Gentry who for their love to these Arts & liberality to these pretenders, deserved better usage at their hands. I shall never be able to bring my selfe to teach so superficially, or cheate so artificially as they doe.[19]

---

[18] Staats- und Universitätsbibliothek, Hamburg [hereafter: SUBH], MS Sup. ep. 100. fo. 62^r: 'Wen er könte gantzlichen zur ruhe kommen (welches ich hoffe in kurzen geschehen soll) wurde Er ausser zweiffel viel herliche sachen so wohl in diesem als auch in anderen stucken des Studii Mathematicj elaboriren'; 'gar fleissig durchsuchet vnd grosse satisfaction darin bekomen'. (I am very grateful to the SUBH for supplying a copy of this letter, and to Dr Dorothea McEwen for her help in deciphering it.) In this letter Hartlib referred to work on perpetual motion by Edward Herbert and Caspar Kalthoff. The latter (a Dutch mechanic employed by Edward Somerset, the future Marquis of Worcester) set up a perpetually turning wheel in the Tower of London. A brief account of this is given in E. Somerset, *A Century of the Names and Scantlings of such Inventions, as at Present I can Call to Mind to have Tried and Perfected* (London, 1663), pp. 36–8; a more detailed description was written by Pell, after he was shown the device by Kalthoff on 16 Apr. 1639: BL, MS Add. 4423, fo. 17^r.

[19] BL, MS Add. 4415, fo. 14 (Pell to Goad, 7 Aug. 1638). This Thomas Goad may have been either the Regius Professor of Law at Cambridge or his cousin, the theologian (and former domestic chaplain to Archbishop Abbot) who lived in Hadleigh, Suffolk. If it was the latter, Pell's ill fortune continued: he died the day after this letter was written.

Some time in the next few weeks Pell returned to his family in Sussex; he came back to London in the last week of September, and resumed his search for a patron. But his prospects seemed no better by 6 December 1638, when he drafted a despairing letter to his wife:

> I have beene now heere 10 weekes and yet see no end of my expectations, and therefore begin to determine to make an end, and come home ye weeke before Christmas, with an intent to returne no more, but [>to] betake my selfe to my [former *deleted*] old employment, In ye meane time as I shall strive to let some of these promisers know, that my patience is worne out, so doe you let Mr Reinalds know y^t I am inclined to listen after y^t of Rye, & tell me what answere he gives...[20]

What it was that turned Pell aside from this resolution – some new undertaking by one of his 'promisers', perhaps – is not known.

From the *cri de coeur* of his letter to Goad of 7 August, the difficulty of the task Pell had set himself is fully apparent: he was looking not for someone who would employ him as a mathematics tutor, but for a patron who would support him 'as an Artist' – in other words, pay him to get on with his own mathematical projects. In order to attract such patronage, it was necessary to set out some sort of prospectus, explaining what those projects might be. Two such prospectuses survive from the summer of 1638. One is a rough draft, which seems not to have been taken any further; the other is a finished text, which, printed and circulated by Hartlib in the autumn of that year, did eventually attract international attention.

The rough draft appears in a set of notes dated 2 July 1638. It presents a scheme for twelve 'tomes', of which the first six would cover pure mathematics: 1) quantity; 2) figure; 3) number; 4) figured quantity; 5) numbered quantity; 6) figured number. The other six tomes would 'apply' the methods of the first, both analytically and synthetically: thus 'Tome 7 should contain A naturall history of bare appearances to vulgar senses ... Tome 8 should contain Man's *desiderata* upon these things ... Tome 9 should proceede Analytically to seeke out ye meanes for this end ...' And the final tome would present 'The particular praecepts of searching in the right way, to get a true knowledge of all things.'[21] Here, once again, was the notion that mathematics could be used as the master-model for the reduction and recombination of all knowledge about all things – a hugely ambitious pansophical project, the detailed working-out of which was never actually attempted.

The second scheme, composed in late July, was somewhat more feasible – though still enormously ambitious. Written in the form of a letter to Hartlib,

---

[20] BL, MS Add. 4428, fo. 191^v (written on the back of a letter from Ithamar to Pell, which is addressed to him 'at Mr Hartlibs house in Dukes place'). This 'Mr Reinalds' (possibly a cousin of Ithamar's) has not been identified.
[21] BL, MS Add. 4409, fo. 252^v.

it was printed by him a few months later in broadsheet format (one version in English, another in Latin for continental readers), without Pell's name and without any title; later printings would entitle it *An Idea of Mathematics* or *Idea mathesews*.[22] This text contained, as its opening sentence put it, 'the summe of what I have heretofore written or spoken to you [*sc.* Hartlib] concerning *the advancement of the Mathematickes*'. The Baconian resonance of that last phrase was not accidental; writing in Latin to Mersenne, Pell would refer to it as an epistle 'de Augmentis Matheseos'.[23] At the same time, the use in this text of the term 'Idea' probably had self-consciously Comenian overtones: the 'axioms' in the *Conatuum ... praeludia*, published by Hartlib in the previous year, had set out a Neoplatonist metaphysics in which all created things were made in accordance with the 'ideas' of them in God's mind, and 'all things which are made, are made in accordance with ideas, whether they be works of God, or of nature, or of art.'[24] Indeed, there was something of a fashion in Comenian circles for the use of this term: during the same year, 1638, Hübner was writing his 'Idea politicae', and Jan Jonston was preparing an 'Idea historiae verè universalis'.[25] And a note by Pell on a different subject (the improvement of human sight by the use of lenses, etc.), written on 27 May 1638, suggests that 'idea' and 'conatus' ('endeavour') were being used almost as technical terms, with the former as the ideal of perfection which the improver formed first in his mind: 'Our Scope is a melioration of nature ... we should chuse ye greatest Ideas but after a sufficient endeavour rest content with *approximation...*'[26] What Pell was offering in his *Idea of Mathematics*, therefore, was first of all an ideal – the perfection of mathematics – and then a set of recommended means or 'endeavours' towards achieving it.

---

[22] For the English broadsheet see below, n. 27. Two exempla of the Latin broadsheet have recently been discovered: one by Christoph Meinel among the papers of Joachim Jungius (SUBH, MS Pe 5, fo. 43), the other by Eddie Mizzi (see below, at n. 34). The later English printing is in J. Dury, *The Reformed Librarie-Keeper with a Supplement to the Reformed-School* (London, 1650), pp. 33–46; the survival of a page-proof (pp. 45–6) among Pell's papers (BL, MS Add. 4408, fos. 233–4) shows that Pell, who was then in the Netherlands, was involved in its production. The later Latin printing (which closely follows the 1638 Latin text, being taken from a copy of that broadsheet supplied by Theodore Haak) is in R. Hooke, *Philosophical Collections*, no. 5, Feb. 1682 (London, 1682), pp. 127–34. The omega in '*mathesews*' indicates the Greek genitive of 'mathesis'. Minor differences between the 1638 (English), 1650, and 1682 texts are recorded in J. Bernhardt, 'Une Lettre-programme pour "l'avancement des mathématiques" au XVIIe siècle: l' "idée générale des mathématiques" de John Pell', *Revue d'histoire des sciences et de leurs applications*, 24 (1971), pp. 309–16.

[23] *MC* viii, p. 623.

[24] Comenius, *Conatuum ... praeludia*, pp. 31–5, axioms VI–XVI; here p. 31, axiom VII ('Omnia ergo, quae fiunt, ad *Ideas* fiunt, sive sint opera *Dei*, sive *Naturae*, sive *Artis*'). On the Neoplatonist nature of Comenius's theory see Schaller, *Die Pädagogik des Comenius*, pp. 24–9.

[25] For Hübner's work see above, ch. 2, n. 112; for Jonston's see his letter to Hartlib, 18/28 Oct. 1638, HP 44/1/21 (printed in Hitchens, Matuszewski, and Young, eds., *Letters of Jonston*, p. 134). Cf. also the comments on the Comenian character of Pell's term 'Idea' in J. T. Young, '"To Leave No Problem Unsolved": The New Mathematics as a Model for Pansophy', *Acta comeniana*, 12 (1997), pp. 85–95, here p. 87.

[26] BL, MS Add. 4407, fo. 151^r (note entitled 'To better our sight', dated '9950').

Pell's proposals consisted of a plan for a public library-cum-institute, and a promised set of publications to encourage and assist mathematical research. The library was to contain all existing mathematical books, 'and one instrument of every sort that hath beene invented'; it would also have a revenue sufficient for buying a copy of every new mathematical book published anywhere in the world. The library keeper, a person 'of great judgement', would examine the contents of all new books, correct errors in them, approve inventions, correspond about new publications with other such experts abroad, 'keepe a Catalog of all such workmen as are able and fit to be imployed in making Mathematicall instruments', and 'give testimoniall, after examination, to all sorts of practisers, as Pilots, Masters, Landmeters, Accomptants, &c.'

As for the publications Pell had in mind, the first of these was a general work entitled 'Consiliarius Mathematicus', which would perform several functions. It would describe the 'extent' and 'profit' of mathematics; it would list mathematicians and (in chronological order) mathematical publications; it would give advice 'directing a student to the *best* bookes in every kinde; In what *order*, and *how* to reade them'; and it would include a 'Paraenesis' (exhortation), addressed 'To all those who have *understanding* to estimate the worth of these studies, and *wealth* wherewith to purchase to themselves lasting honour by the wise dispensing of it, to take more notice of this sort of students'. Pell also proposed to write three treatises. One, entitled 'Pandectae Mathematicae', would summarize all mathematical discoveries, attributing each one to its earliest known author (to eliminate duplication and plagiarism): as he noted, this would in effect 'bring that *great Library* into farre lesse roome'. Another, entitled 'Comes Mathematicus', would contain 'the usefullest Tables and the Precepts for their use'. And the third, 'Mathematicus αὐτάρκης' would enable a mathematician to solve any problem, 'though he be utterly destitute of bookes or instruments'.

Although his habitual modesty forbade even placing his name on this proposal, there can be little doubt as to who, in Pell's mind, might have been a suitable candidate for employment as the 'keeper' of his proposed library. And at the end of his text he not only made an open plea for patronage – 'As for this present *Idea*, I am so farre from counting it *meerely-impossible*, that I see not why it might not be performed by one man, without any assistants, provided that he were neither *distracted* with cares for his maintenance, nor *diverted* by other employments' – but also hinted that some parts of this project were already accomplished. 'For mine owne part, the consideration of the *incomparable excellency, unstained pleasure, unvaluable profitablenesse, and undoubted possibility* of this whole designe, hath prevailed so farre with me, that notwithstanding all the discouragements that I have met withall, I have done more towards it than bare *Idea*.'[27]

---

[27] All these quotations are from the 1638 English broadsheet, of which one copy survives: BL, pressmark 528.n.20 (5*); this is printed (with minor inaccuracies) in P. J. Wallis, 'An Early Mathematical Manifesto – John Pell's *Idea of Mathematics*', *The Durham Research Review*, 5, no. 18 (Apr. 1967), pp. 139–48 (text on pp. 141–7).

This was both one of the briefest and in some ways the most significant of Pell's handful of published works – a Baconian–Comenian plan for the systematic ordering of existing knowledge, and for the facilitation of further advances. It has attracted more attention than any of his other writings; it has also been the subject of more confusion, at least where its dating is concerned. Some scholars have assigned it to 1630, identifying it with the 'Method' referred to in one of Hartlib's letters to Dury of that year; as we have seen, that reference was most probably to a pedagogical method of a quite different kind.[28] Others have given the date as 1634, in accordance with a mock-up titlepage surviving among Hartlib's papers: 'Mr. Pells / Idaea of Mathematiques / Written to S. H. /Anno 1634. or m.'[29] But the formulation '1634 or m.', meaning probably 'or more' or 'or *magis*' [greater], suggests that this titlepage was drawn up at such a distance in time (perhaps in connection with the 1650 publication) that Hartlib could not remember the date with any certainty. It is true that one of the proposed treatises, 'Comes Mathematicus', was described by Pell in a letter to Hartlib of 1635; however, the ambitious list of the 'uses' of his mathematical tables given in that description suggests that when Pell wrote those words the work was at a very embryonic stage – perhaps little more than a gleam in his eye.[30]

There is in fact abundant evidence to show when the *Idea of Mathematics* was written and printed. In one of his later notes Pell referred to 'my Idea printed 1638 Octob.'; and in a letter to Hartlib of 12 October 1642 he referred to 'my letter to you which you caused to be printed just this time 4 yeares'.[31] In a letter to Mersenne in November 1639 he confirmed that the text had been printed in both English and Latin ('for the sake of foreigners') at Hartlib's own expense.[32] And in the draft of another letter to Hartlib in August 1655, he commented: 'You printed it about 3 months after I wrote it.'[33] The issue is conclusively settled by a recently discovered copy of the Latin broadsheet, bearing the annotation (in Pell's own hand) 'Written in London, 23 July 1638, printed in London, 23 October 1638'.[34]

---

[28] See above, ch. 2, at nn. 26–7. The identification made by Turnbull is accepted also by P. J. Wallis in his entry on Pell in the *DSB*.

[29] HP 14/1/6A. This date is accepted by Braddick and Greengrass: 'Letters of Sir Cheney Culpeper', p. 153, n. 5.

[30] See above, ch. 2, at n. 149. Similarly, the idea of a 'Mathematicus αὐτάρκης' included in his *Idea* may have had its origin in the little treatise 'Trigonometria logistica Αυτάρκης' which Pell had sketched at roughly the same time (see ch. 2, n. 148).

[31] BL, MS Add. 4399, fo. 1^r; MS Add. 4280, fo. 206^v.

[32] *MC* viii, p. 623 ('propter exteros').

[33] BL, MS Add. 4364, fo. 139^r.

[34] 'Scriptum Londini Julij 23° 1638 Impressum [*ditto sign*] Octob. 23° 1638'. The broadsheet, discovered by Eddie Mizzi, is in the Bodleian Library: it was bound in at the end of an exemplum of J. H. Rahn, *An Introduction to Algebra*, tr. T. Brancker, pressmark Savile K.13 (which, according to a note on its titlepage, was donated to the Bodleian by Sir Christopher Wren in 1673), and has now been separately catalogued as pressmark Vet. A2 b1. On Pell's involvement in the production of this book, see below, ch. 7, at nn. 7–34.

While Pell had no doubt come under constant encouragement from Hartlib to write such a prospectus for use in his quest for employment, there was also a more particular reason for composing this text in July 1638. At some time in the previous few weeks, Hartlib had received a not very dissimilar proposal from an Austrian nobleman, Baron Johann Ludwig Wolzogen, who had recently visited Comenius in Leszno and had offered him his help in developing a new 'didactic' of mathematics.[35] A copy of Wolzogen's proposal was sent to Hartlib from Germany on 13/23 May 1638; it is not known exactly when he received it. Entitled 'Magnum opus mathematicum', it was a plan for a series of six volumes ('tomi'): the first would be a 'system' of algebra, the second a treatise on theoretical geometry, the third on applied geometry, the fourth an account of mathematical instruments, the fifth a book of trigonometrical tables and the sixth a set of tables of logarithms.[36] This may have prompted Pell to draw up, as a first response, his own more ambitious scheme of twelve 'tomes'; but some elements of Wolzogen's proposal, such as his volumes of tables, had much closer counterparts in the *Idea*. That Hartlib regarded Wolzogen's plan and Pell's *Idea* as similar in nature is indicated by the way in which he referred to both of them in his letter of 10 August 1638 to Johann Adolf Tassius. Having commented that Wolzogen's 'outline of his endeavours' ['Conatuum'] seemed to be a very ambitious project, he continued: 'Instead, I am sending you (enclosed) another Idea of mathematical endeavours ['Conatuum Mathematicor.'] by another author'.[37]

That letter by Hartlib is the first reference to his transmission of copies – at this stage, evidently, manuscript copies – of the Latin text. Pell's letter to Goad, written three days earlier, shows that the English text had already been circulated, or, at least, shown to interested parties. Having described the difficulty of finding a patron, he wrote: 'Yet there are some y^t have seene the delineation of my Idea of ye perfecting, reforming, advancing & facilitating ye whole study from ye first seedes of pure Mathematickes to their highest & noblest, as well as the meanest & most ordinary applications with my designes to y^t end.' (Those who had looked at his proposals were convinced, he insisted, 'y^t my designes are such as would repay y^{eir} patron with as lasting honour, as Poesy did Maecenas'.)[38] And one other piece of evidence completes the story, even though

---

[35] See Blekastad, *Comenius*, pp. 247–8; C. de Waard, 'Wiskundige bijdragen tot de pansophie van Comenius', *Euclides: tijdschrift voor de didactiek der exacte vakken*, 25 (1949–50), pp. 278–89, here p. 281. Wolzogen (c.1599–1661), Baron von Tarenfeldt and Freiherr von Neuhäusel, had become a Socinian and moved to Poland; he is best known for the critique of Descartes's *Meditationes* which he published in 1656 (see L. Chmaj, 'Wolzogen przeciw kartezjuszowi', in his *Bracia polscy: ludzie, idee, wpływy* (Warsaw, 1957), pp. 409–63, esp. the biographical details on pp. 423–4).

[36] BL, MS Sloane 652, fo. 190 (printed in de Waard, 'Wiskundige bijdragen', pp. 288–9).

[37] SUBH, MS Sup. ep. 100, fos. 60–3 (Hartlib to Tassius, 10 [/20] Aug. 1638), here fo. 63^r: 'Des H. Wolzogen Deliniation seiner Conatuum ... ein grosses weitaussehendes Werck. An dessen stat schicke ich hierbey eine andere Idäam Conatuum Mathematicor. eines anderen Authoris'.

[38] BL, MS Add. 4415, fo. 14^v.

it is itself undated: in his letter to Bisterfeld, Joachim Hübner wrote that Pell was writing 'a detailed exposition of his mathematical endeavours ['Conatuum Mathematicorum'] in English, and as soon as it is finished, I shall translate it into Latin and give it to you'.[39]

Once the broadsheets were printed in October, copies began to be circulated both by Hartlib and by Haak, who had returned from Holland at the end of the previous month.[40] Little is known about the distribution of the English version; presumably many copies were given to people in London with whom Hartlib was in personal contact (a transaction which, therefore, would not leave documentary evidence in the form of correspondence). A later letter from Sir Cheney Culpeper to Hartlib referred to Pell and what he 'vndertakes in his printed paper which I received from you': no doubt the *Idea* had been pressed into the hands of all those – such as Sir Cheney – who could be regarded as Pell's potential employers or patrons.[41]

More evidence is available about the distribution of the Latin text on the Continent of Europe. One of the first recipients, suitably enough, was Baron Wolzogen; he wrote what Pell described as 'an applauding letter in Latine to mee', and (as Wolzogen told Hartlib seventeen years later, in a letter complaining about Pell's failure to produce what was promised in the *Idea*) promptly abandoned work on his own project.[42] Another was Abraham von Franckenberg in Silesia, an enthusiastic Comenian and follower of the German theosophist Jakob Böhme, who discussed the *Idea* in a letter to Hartlib of [11/] 21 December 1638.[43] Hartlib also sent the *Idea* to Amsterdam, to a friend of Dury's who would soon become 'Hartlib's principal agent in the Netherlands': the Dutch-

---

[39] BL, MS Sloane 417, fo. 201ᵛ: 'ein aussfurliche detricationem seiner Conatuum Mathematicorum in Englisch, welche so bald sie fertig, ich ... in lateynisch transferiren Vnd dem Hn Zu schenken will'.

[40] On the date of Haak's return see L. W. Forster, 'Aus der Korrespondenz G. R. Weckherlins', *Jahrbuch der deutschen Schillergesellschaft*, 4 (1960), pp. 182–97; here p. 183.

[41] Braddick and Greengrass, eds., 'Letters of Sir Cheney Culpeper', p. 153 (29 Sept. 1641).

[42] BL, MSS Add. 4364, fo. 139ʳ (Pell to Hartlib, draft, early Aug. 1655); Add. 4408, fo. 263ʳ (Wolzogen to Hartlib, May 1655). For Wolzogen's complaint see below, ch. 5, at n. 126.

[43] BL, MS Add. 4416, fo. 38ʳ contains an extract from a letter of this date, over the initials 'A. F.', including the comment: 'You say that some people doubt whether Mr P. will produce those things, in the field of mathematics, which he promised in the letter you recently published' ('Dubitare quosdam ais, an praestiturus sit Dominus P. in Mathematicis quae nuper publicata a Te ejus epistola promisit'). Abraham von Franckenberg is the only person with those initials known to have corresponded with Hartlib: see Turnbull, *Hartlib, Dury, Comenius*, p. 431, and von Franckenberg, *Briefwechsel*, pp. 195–202, 219–20. (This extract, however, is not included in Telle's edition.) On von Franckenberg see Blekastad, *Comenius*, pp. 177, 379–80, and S. Rusterholz, 'Abraham von Franckenberg', in H. Holzhey *et al.*, eds., *Grundriss der Geschichte der Philosophie, begründet von Friedrich Ueberweg: Die Philosophie des 17. Jahrhunderts*, Bd iv, *Das Heilige Römische Reich, deutscher Nation, Nord- und Ostmitteleuropa*, 2 vols. (Basel, 2001), i, pp. 85–95.

German alchemist Johann Moriaen.[44] Some copies eventually reached Paris: in 1639 the young German scholar Johann Friedrich Gronovius, who had got to know Hartlib, Pell, and Haak in London, delivered the *Idea* to the astronomer Ismaël Boulliau and to the Jesuit scholar Denis Petau.[45] Other copies were sent to Leiden soon after publication, presumably by Haak. The Professor of Theology there, the Huguenot scholar André Rivet, had one by December 1638, when he summarized its contents in one of his letters to Mersenne.[46] Another was sent to the physician and Arabist Johann Elichmann (with whom Haak probably lodged during his stay in Leiden), who in turn forwarded it, via his friend Cornelis van Hogelande, to Descartes.[47] The French philosopher's eventual response was, though cool in tone, broadly sympathetic: 'I remember that I found nothing in it with which I would greatly disagree'. Descartes preferred to emphasize the difference between the 'history' and the 'science' of mathematics; he approved of Pell's plans to formalize the principles of the latter in his 'Mathematicus αὐτάρκης', but was less interested in the task of collecting all the materials of the former.[48] And Mersenne's reaction, when replying to the account of Pell's

---

[44] See Young, *Faith, Medical Alchemy and Natural Philosophy*, pp. 38 (quotation), 115–16 ('Moriaen received a copy soon after the work's publication in 1638'). Moriaen probably received multiple copies, for distribution: see his own account of the gift of a copy, below, n. 72.

[45] P. Dibon and F. Waquet, *Johannes Fredericus Gronovius: pèlerin de la République des Lettres* (Geneva, 1984), pp. 85–6 (Gronovius to Hübner, [2/] 12 Oct. 1639). Both Dibon and Waquet here, and the editors of *MC* annotating another letter from Gronovius (viii, p. 645), identify this 'Petavius' as Alexandre Petau, the owner of an important library; but Gronovius does also refer elsewhere to Denis Petau, the Jesuit theologian and chronologist. At this time, Denis would have been best known for his edition of an important collection of Greek astronomical texts, *Uranologion* (Paris, 1630). Boulliau was certainly Pell's choice of recipient: as Hartlib noted earlier that year, Boulliau's book 'Philolaus sive dissertatio de vero Systemate Mundi 1639. seemes to bee a most excellent booke to Pell' (HP 30/4/12B (Ephemerides, 1639, part 2)). Astronomy would seem to link Boulliau to Denis Petau, not to Alexandre.

[46] Rivet's letter does not survive, but Mersenne's reply (dated 20 Dec. 1638), referring to Rivet's comments on Pell, is printed in *MC* viii, pp. 238–46. The suggestion by de Waard that Rivet, like Mersenne, had not seen the actual text of the *Idea* ('Wiskundige bijdragen', p. 283) seems unwarranted.

[47] On Haak's sending of the *Idea* – 'together with some Comenian papers' ('una cum schedis quibusdam Comenianis') – to Elichmann, and his forwarding of both to Descartes, see R. Hooke, *Philosophical Collections*, no. 5 (Feb. 1682) (London, 1682), p. 144. On Elichmann (1601–39) see W. M. C. Juynboll, *Zeventiende-eeuwsche beoefenaars van het arabisch in Nederland* (Utrecht, 1931), pp. 191–4; J. Fück, *Die arabischen Studien in Europa bis in den Anfang des 20. Jahrhunderts* (Leipzig, 1955), p. 91; Toomer, *Eastern Wisedome*, pp. 51–2. It was apparently Elichmann's regular practice to have visiting students as lodgers and give them tuition, especially in Arabic: Isaac Vossius stayed with him for several months in 1637 (F. F. Blok, *Isaac Vossius and his Circle: His Life until his Farewell to Queen Christina of Sweden, 1618–1655* (Groningen, 2000), p. 63, n. 4), and Christian Ravius from late 1637 to the summer of 1638 (Toomer, *Eastern Wisedome*, p. 83).

[48] Descartes to van Hogelande, 8 Feb. 1640, printed in Hooke, *Philosophical Collections*, no. 5 (Feb. 1682) (London, 1682), pp. 144–5, and A&T iii, pp. 721–4 (p. 722: 'memini nihil me in eâ reperisse à quo multùm dissentirem'). Descartes's judgement seems to have been promptly transmitted by van Hogelande to Haak, who sent a copy of it to Mersenne in April or early May 1640: see *MC* ix, p. 305.

work given to him by Rivet, was fundamentally the same: 'As for Mr Pell's treatise, it is not at all necessary to make the sort of library he imagines, since he envisages a man who would be able to solve all sorts of problems in that field. It would be much better to get away from all the books, and have this man, whom he envisages, put everything into two or three volumes, according to his method.'[49]

The distribution of the *Idea* did thus help to attract interest in Pell on the part of some continental thinkers. Johann Moriaen in Amsterdam, who had strong mathematical interests of his own (he claimed to be able to teach people in a quarter of an hour to calculate any power of any number), bombarded Hartlib from the end of 1638 onwards with enquiries about Pell's work.[50] He had made a particular study of Viète's algebra, and was aware of the power of this new method of 'logistica speciosa', 'calculating with letters'. Viète, famously, had described the aim of his mathematics as 'nullum non problema solvere', 'to leave no problem unsolved'. Writing to Hartlib in December 1638, Moriaen enquired 'whether Pell's logistica extends as far as Viète's "Nullum non problema solvere"'.[51] The response given by Pell, in a letter to Hartlib drafted on 3/13 January 1639, indicated that he was now preparing, among other things, a revised version of Harriot's book, which he regarded as superior to Viète: Harriot taught, he said, 'more fully & farre more cleerly' than the French mathematician, but his book was hard to follow because of both the quantity of misprints and the inadequate explanation of the 'grounds' of its 'praecepts'. 'Wherefore,' Pell went on, 'I have I thinke so [>illustrated] & [>perfected] y^t Treatise, y^t it may be easily [>conceived], & so perfectly kept in mind y^t he y^t hath once understood it, shall never neede to see it againe.'[52] But Pell also made it clear that he was developing new methods of his own which would make possible a whole range of mathematical discoveries:

> There lyeth in ye Abysse of wisedome an untouched masse of mathematicall verities, some of which I had, it may be, brought to light ere y^s, if it had beene possible for me to have enjoyed meanes & Leisure both at once. [>yet] I may perhaps ere long tell ye world of Theoremes discovered in Geometry concerning ye Section of Angles

---

[49] *MC* viii, p. 240: 'Quant au discours du Sr Pell, il n'est point necessaire de faire de Biblioteque telle qu'il se l'imagine, puisqu'il suppose un homme qui puisse soudre toutes sortes de problemes en cette matiere. Il vaudroit bien mieux quitter tous les livres et que cet homme imaginaire comprist dans 2 ou 3 volumes le tout par sa methode.'

[50] For Moriaen's claim see HP 37/13B.

[51] HP 37/166A, Moriaen to Hartlib, [17/] 27 Dec. 1638: 'ob sich Mons. Pellii Logistica so weit erstrecke als des Vietae Nullum non problema solvere'.

[52] BL, MS Add. 4419, fo. 136^v. Moriaen responded, in a later letter to Hartlib (HP 37/7A-8B, [4/] 14 Feb. 1639): 'I should very much like to know whether Mr Pell's commentaries on Harriot are printed or available' ('Ob Domini Pellii Commentaten in Harriotum gedruckt oder zu bekommen sein möchte Ich wol wiszen'). The answer was evidently 'no'.

[>&] which all my antecessors saw not, no [>not] those y^t boasted most of their mirificall arts & invented abilities.[53]

Moriaen was so impressed by this that he promised, in his next letter to Hartlib, that he would send some money to support Pell's work.[54]

Another reader of the *Idea* who took a growing interest in Pell was the French Minim friar, scientist, and tireless 'intelligencer' – in many ways, indeed, Hartlib's counterpart in Paris – Marin Mersenne. When he first heard about Pell's *Idea* from Rivet in December 1638, he had no direct contact of his own with Hartlib or with any member of Hartlib's circle in England. The connection was made only in the autumn of 1639, thanks to a young Dutch–German (from a Palatine family), Philip Ernst Vegelin. On a visit to London during the first half of that year, Vegelin had got to know Hartlib, Haak, and Pell; in October he returned to Paris (where he had previously made the acquaintance of Mersenne) and gave the friar various items entrusted to him in London, including Comenius's *Pansophiae prodromus*, Pell's *Idea*, and a letter from Haak.[55] Commenting on the *Idea* in his reply to Haak on [22 October/] 1 November 1639, Mersenne gave a somewhat toned-down version of his earlier response. 'His plan is praiseworthy. But instead of making, as he proposes, a huge collection of all mathematical writings, it would be better to select a dozen of the best ones in each area...' And if, he added, Pell himself was really able to simplify and codify all existing mathematics in the best possible order, then the book Pell produced would suffice on its own, 'without bothering about the others.'[56] These remarks prompted Pell to write to Mersenne, on 21 November/1 December, offering a long and detailed defence of his scheme.[57] Mersenne's reply was emollient and complimentary; and in a letter written soon afterwards to Haak he asked him to let Pell know that 'I shall do everything I can here to contribute to his plan'.[58] A prolonged twin-track correspondence followed, with Mersenne writing to Pell in Latin and to Haak in French (sometimes enclosing materials which he asked Haak to pass on to Pell). In his letters to Pell, Mersenne became increasingly positive about the proposals

---

[53] BL, MS Add. 4419, fo. 136^r.

[54] HP 37/5A (Moriaen to Hartlib, [21/] 31 Jan. 1639).

[55] On Vegelin see N. Malcolm, 'Six Unknown Letters from Vegelin to Mersenne', *The Seventeenth Century*, 16 (2001), pp. 95–122; on his transmission of 'Comeniana et Pelliana' to Mersenne see *MC* viii, p. 689. There is some evidence that Pell kept up a correspondence with Vegelin in subsequent years: in a letter to Athanasius Kircher of 5 Dec. 1653, Vegelin wrote that he had been encouraged 'ten years ago' by both 'Father Marin Mersenne and the English mathematician Pell' to get in contact with him (Pontificia Università Gregoriana, Rome, MS 557, fo. 307^r: 'vor 10 jahren ... Patris Marini Mersenni und des Englischen Mathematici Pellen').

[56] *MC* viii, pp. 581 ('son dessein est louable. Mais au lieu du grand ramas qu'il propose de tous ceux qui ont escrit des Mathématiques, il vaudroit mieux faire le choix d'une douzaine des meilleurs en chacque partie'), 583 ('devroit estre retenu tout seul, sans se soucier des autres').

[57] Ibid., viii, pp. 623–32.

[58] Ibid., viii pp. 685, 723 ('vous pourrez encore luy confirmer que tout ce que je pourray icy faire pour contribuer à son dessein').

set out in the *Idea*, suggesting ways of organizing the mathematical library, and enthusing about the treatises Pell planned to write.[59] But he also discussed many other topics with the young English mathematician, asking his opinion about the physics of percussion and Descartes's theory of refraction, sending him a query about the type of curve known as a 'cycloid' or 'roulette', offering examples of numerical 'magic squares', and discussing at length current work on longitude and the declination of the magnetic needle. In his replies, Pell discussed some of his own interests (in particular, his work on magnetic declination), and mentioned in passing that he had prepared an almost complete English translation of Descartes's *Discours de la méthode* and *Essais* – a translation which, like so many of Pell's other undertakings, never saw the light of day.[60]

It is clear, then, that thanks to the efforts of Hartlib and his friends, Pell's work did receive a truly international distribution. As Pell later remarked in an undated note, 'almost all Christendome hath seene my undertaking. England, Ireland, Germany, Netherlands, France, Italy, Spaine, Poland, I thinke also Denmarke, Sweden, Bohemia & Hungaria, sure I am some of those countryes have seene it, my name is tost up & downe among all sorts of censors, able & unable ...'[61] This may have been gratifying in itself (assuming that his 'name' was known to the readers of his anonymous publication), but the chances of eliciting an offer of employment on the Continent must have seemed even slimmer than they were in his native land. Nevertheless, the publication of the *Idea* did stimulate, directly or indirectly, two attempts by well-wishers to find European patronage or employment for Pell during the following year.

One of these was an initiative taken by a German student of Arabic, Christian Raue or Ravius, who was two years younger than Pell. Ravius, who had studied in Leiden in 1637–8, made a brief visit to England at the end of 1638 (seeking both employment and financial support for a proposed trip to the Levant); thereafter he spent some time in Paris in 1639, where he got to know Mersenne,

---

[59] Ibid., ix, pp. 19–20, 49–50.

[60] Ibid., ix, e.g. pp. 50–4, 58–63, 168–71, 182–4, 228–33, 311–14. Pell's letter to Mersenne of [14/] 24 Jan. 1640 stated that he had 'translated most of it into my native tongue' (p. 59: 'ea maximam partem in vernaculam meam transtuli'); in his draft of this letter he wrote that 'almost all of the great Descartes's book has been translated into English, either in my hand or by my dictation' (p. 64: 'Magni illius Escartesij liber fere totus, mea manu vel dictamine Anglicus factus est'). Charles Adam assumed that it was only the 'Géométrie (one of the *Essais*) that Pell translated ('Descartes et ses correspondants anglais', *Revue de littérature comparée*, 17 (1937), pp. 437–60; here p. 444); but the draft refers specifically to the whole 'book'.

[61] BL, MS Add. 4423, fo. 376r.

before travelling to Constantinople in the autumn.[62] Christian Ravius had probably arrived in London with some sort of introduction to Hartlib: his brother Johann was a devoted Comenian, and, what is more, Christian had stayed with Elichmann in Leiden, possibly overlapping there with Haak.[63] However the connection was made, it is clear that Ravius became a passionate enthusiast for the project outlined in the *Idea*, and a personal admirer of Pell. During his stay in Paris, he published a pamphlet entitled *Dissertatio mathematica proponens novum summumque totius matheseos inventum problemata omnia certa methodo inveniendi & inventa solvendi* ('A mathematical dissertation, putting forward a new discovery, the most important in the whole field of mathematics, of how to find all problems by a certain method, and how to solve them, once found'). The pamphlet was in the form of an appeal to Ravius's own patron, Gerbrandt Anslo, an Anabaptist merchant in Amsterdam who had strong Orientalist interests; it ended by thanking Anslo for supporting Ravius's studies (and his visits to England and France), and asking him to sponsor his trip to Constantinople.[64] But the main part of the text was a hymn of praise to Pell, and an appeal to Anslo to extend his patronage to him too. Pell was, he said, 'a man commended not only for the purity of his morals, but also because of his supreme intelligence in mathematics, in which he equals the greatest mathematicians of our age'; were it not for his youth 'and his straitened circumstances', he would already have become universally famous for his 'discovery', which hugely simplified the solving of all problems.[65] Ravius had certainly seen Pell's *Idea*; from these comments it would appear that he had also heard Pell talk about his work on algebraic method – making, perhaps, some of the same claims that filled Moriaen with such eager expectation. No offers of support for Pell, from Anslo or anyone else, are known to have followed from Ravius's appeal; but the fact that he went to the trouble of having this pamphlet printed – and therefore, presumably, distributed to others as well – shows how strongly his imagination must have been gripped by Pell's proposals.

---

[62] See Toomer, *Eastern Wisedome*, pp. 83–4. Ravius is not mentioned in Mersenne's correspondence in 1639, but he stated himself that he had frequent discussions with Mersenne and Girard Desargues in Paris during that year: Apollonius, *Conicarum sectionum libri V, VI & VII*, ed. and tr. C. Ravius (Kiel, 1669), sig. *7ᵛ. However, although Ravius thus represents a link between the Hartlib circle and Mersenne, there is no evidence that he brought about any direct contact between them, as Vegelin did a few months later. (Unlike Ravius, Vegelin had known Mersenne before he went to London; no doubt he told Hartlib and Haak about Mersenne, which is why, on his return to Paris, he brought a letter to Mersenne from Haak.)

[63] On Johann see Tautz, *Die Bibliothekare*, pp. 18–43; Blekastad, *Comenius*, pp. 225, 376, 389–90; Turnbull, *Hartlib, Dury, Comenius*, pp. 363–4. On Elichmann see above, at n. 47.

[64] C. Ravius, *Dissertatio mathematica* (Paris, 1639), p. 8. A rare (apparently unique) copy of this pamphlet is BL, pressmark 529 g. 13 (7). On Anslo, who studied Hebrew under Menasseh ben Israel and had Turkish, Persian, and Arabic MSS in his library, see *NNBW* ix, cols. 26–7. No contact between Anslo and Pell is known to have followed.

[65] Ravius, *Dissertatio mathematica*, pp. 6 ('Vir non modo ob vitae integritatem commendatus sed ob summum eius in Mathesi ingenium summis nostri seculi Mathematicis par' ... 'nisi iuventus ... & res angustior domi'), 7.

The second initiative was equally unsuccessful in the short term – though the seed it planted would in fact come to fruition five years later. On [7/] 17 August 1639 the astronomer Martinus Hortensius (Maarten van den Hove), who had taught mathematics at the Amsterdam 'Athenaeum' (the forerunner of the University of Amsterdam), died, aged only 34.[66] In the words of Haak's later memoir: 'Mr. John Morian, a very learned and expert gentleman, gave me notice that Hortensius... was deceased, wishing that our friend Mr. Pell might succeed. Sir William Boswell, his majestie's ambassador in Holland, being here then, I conferred with him about it, who promised all his assistance.'[67] Boswell was the ideal person to intercede in such an affair: not only was he a skilled diplomat, he also had serious intellectual interests of his own, as a literary executor of Bacon, a former secretary of Edward Herbert, and an active participant in the 'Republic of Letters'. He had recently developed close and cordial relations with the Hartlib circle: the connection of Hartlib and Haak with the Palatine cause was one natural link (Boswell was intimately involved in Palatine affairs, as the court-in-exile of the Elector Palatine was in The Hague), and another was their mutual friendship with Georg Rudolf Weckherlin, the German poet (also a pro-Palatine activist) who was Charles I's Under-Secretary for the German, Latin, and French Tongues.[68] During a prolonged visit to England in 1639, Boswell had had frequent contacts with Hartlib – he is a constant presence in the 'Ephemerides' for the second and third quarters of that year – and had also met Pell on several occasions. (Hartlib's laconic note of Pell's account of one conversation reads: 'Sir W. Boswell hase composed diverse Treatises of Chronology, Geometry, etc. Pell'.)[69] So when news of Hortensius's death reached England, Boswell was quick to act: on 23 August [/2 September] 1639 he wrote to G. J. Vossius, the intellectual *doyen* of the Amsterdam Athenaeum, recommending Pell – 'who, I am absolutely certain, would maintain the dignity of that lecture-hall and attract

---

[66] C. de Waard, 'Martinus Hortensius', in *NNBW* i, cols. 1160–4; H. Brugmans, J. H. Scholte, and P. Kleintjes, eds., *Gedenkboek van het Athenaeum en de Universiteit van Amsterdam, 1632–1932* (Amsterdam, 1932), pp. 225–8.

[67] Aubrey, *'Brief Lives'*, ii, p. 130.

[68] Barnett notes that Haak, during his stay in Holland in 1638, received letters from Weckherlin, 'sometimes sent under cover to Sir William Boswell... with whom Haak seems to have enjoyed a long-lasting and close acquaintance' (*Theodore Haak*, p. 32). On Weckherlin see L. W. Forster, *Georg Rudolf Weckherlin: zur Kenntnis seines Lebens in England*, Basler Studien zur deutsche Sprache und Literatur, 2 (Basel, 1944).

[69] HP 30/4/14A (Ephemerides, 1639, part 2). Boswell took a personal interest in Pell: on 19 Mar. 1640 he sent him, from The Hague, a copy of a recently published mathematical tract by Johan Stampioen, *Solutie op sijn eygen problema* (The Hague, 1640) (see Pell's note on this in BL, MS Add. 4408, fo. 55^r). The tract was a defence of his earlier treatise, *Algebra ofte nieuwe stel-regel* (The Hague, 1639); among Pell's books there is a copy of that treatise, inscribed 'For Mr Pell at his House in the Bowling Alley in Westm^r: From Sr William Boswell. Hagh 5/15 Oct. 1640' (Busby Library, Westminster School, pressmark I C 32).

the highest praise'.⁷⁰ Moriaen was tireless in his encouragement of this plan; he wrote to Hartlib on 22 September that 'Mr William Boswell will make an effort to look after Mr Pell, one way or another', and suggested that Boswell should write to one of the Burgemeesters of Amsterdam, Albert Burgh, with a recommendation.⁷¹ The following week, he noted that Burgh was an enthusiast for 'pansophy', and that he would be keen to invite Pell if he knew that he was the author of the *Idea*, 'which he received from me a long time ago', adding that 'I can think of no one more suitable to tell him that than Mr Boswell'.⁷² Three months later, Moriaen was still actively lobbying on Pell's behalf: he told Hartlib that he had arranged for a copy of the *Idea* to be sent to Burgh, and that he had then visited Burgh and 'recommended Mr Pell's character and abilities'.⁷³ But these efforts remained, for the time being, unrewarded, and Hortensius's chair would be left vacant until 1644.

Meanwhile, the search for patrons and employers in England seems to have continued. One mysterious comment in a letter from Moriaen to Hartlib, written in April 1639, suggests that some promise had recently been made to Pell: 'It gladdens my heart to hear that Mr Pell has found two years' support'.⁷⁴ But it must be assumed that the promise, whoever it came from, was not kept; no regular connection with any employer emerges from Pell's own papers from this period, which offer instead a picture of sporadic contacts with a variety of potential patrons. How Pell supported himself financially is quite unclear; he may have been mainly dependent on the generosity of Hartlib, whose private income was liberally spent on such causes. (Many years later, Hartlib would write that 'as long as I have lived in England, I have spent yearly out of my own betwixt £3 or 400 sterl. a year.')⁷⁵ And the cost of Pell's upkeep was rising: Ithamar and the children had joined him in London by the end of 1639, and two more children were born in the following year (Elizabeth on 7 January, and William

---

[70] Bodl., MS Rawl. Letters 83, fo. 172ʳ ('quem certo certiús scio summâ cum laude dignitatem istius theatri sustenturum'). Boswell's acquaintance with Vossius went back to 1624–6, when he had repeatedly written, on behalf of his then employer, John Williams, Bishop of Lincoln, inviting Vossius to accept a Chair at Cambridge: see P. Colomesius, ed., *Gerardi Joan. Vossii et clarorum virorum ad eum epistolae*, 2 vols. (London, 1690), ii, pp. 35, 39–40, 42–3.

[71] HP 37/40B ('herr William Boswel sich bemuhen will H Pell auff einen oder andern weg zue versorgen'). On Albert Coenratsz. Burgh (1593–1647), who held several high offices in Amsterdam and also served as a diplomat in Moscow and Copenhagen, see J. E. Elias, *De vroedschap van Amsterdam, 1578–1795*, 2 vols. (Haarlem, 1903–05), i, pp. 327–8. Moriaen refers here to Burgh as 'Cunradi', latinizing his patronymic.

[72] HP 37/41A, Moriaen to Hartlib, [19/] 29 Sept. 1639 ('dessen Idaea Mathematica Er vorlängst von mir empfangen'; 'darzu ich niemand bequemer als H. Boswel erdencken kan'). He also noted that Burgh had wanted to invite 'die Architectos Pansophiae' to Holland.

[73] HP 37/50A, Moriaen to Hartlib, [16/] 26 Dec. 1639 ('H Pells persohn vnd qualitet commendirt').

[74] HP 37/21B, Moriaen to Hartlib, [9/] 19 Apr. 1639 ('Es ist mir eine herzliche freude zue hören das H Pelleus vff 2 Iahr vnderhalt gefunden hatt').

[75] BL, MS Add. 6269, fo. 29ʳ, Hartlib to John Worthington, 3 Aug. 1660.

on 21 December).[76]

Pell's desire to be supported as an 'Artist', not a jobbing mathematics-teacher, has already been noted. He did in fact receive one offer designed to enable him to remain master of his own studies; but unfortunately another of Pell's scruples came into play in this case, and he refused it. Theodore Haak, as he explained in his later memoir, recommended Pell to John Williams, Bishop of Lincoln, 'who became very desirous to see the man', and invited both Pell and Haak to dinner.[77] (Haak had had earlier contacts with Williams through his work for the 'Palatine collections'; and, as we have seen, Williams was probably the benefactor who had offered the use of his country house to Hartlib in 1632.)[78] Williams was so impressed by Pell that he offered him a benefice on the spot; whereupon 'Mr. Pell thankd his lordship, saying he was not capacitate for that, as being no divine and having made the mathematics his main studie, for the great publick need and usefullnesse therof.' The Bishop accepted this answer, and asked Pell to 'befriend him with his visits', promising that he would always receive a hearty welcome; nevertheless, 'Mr. Pell who was no courtier came there no more.'[79] (If Pell had other reasons for declining this offer, they are hard to penetrate. As several of his manuscripts from the 1630s show, he was not lacking in religious fervour; in the absence of any evidence of objections on his part to the ceremonies or hierarchy of the Church of England, the safest conclusion must be that his reasons were as he stated them.)[80]

So the quest for a secular patron was resumed. Among the people cultivated by Pell during this period, several names stand out. His contacts with Sir Thomas Aylesbury continued intermittently, and he also had some dealings with a Mr Stanier, one of Aylesbury's clerks.[81] Another contact, made in early 1639, was with Robert Long, an official who had previously served as personal secretary to the Lord Treasurer and was appointed Receiver of Recusant revenues in that

---

[76] A. M. Burke, *Memorials of St. Margaret's Church Westminster, Comprising the Parish Registers, 1539–1660 and the Churchwardens' Accounts, 1460–1603* (London, 1914), pp. 163, 169. As these parish records show, Pell was no longer staying at Hartlib's house (in Duke's Place, Aldgate).

[77] Aubrey, *'Brief Lives'*, ii, p. 129.

[78] Barnett, *Theodore Haak*, p. 32; above, ch. 2, n. 36.

[79] Aubrey, *'Brief Lives'*, ii, pp. 129–30.

[80] Two examples have been given above (ch. 2, at nn. 137–8); one more may be added here. In a note dated 6 July 1638 Pell wrote: 'if we follow these *implanted dictates* of *naturall reason* carefully we shall goe a great way, beyond all y^t ever yet went by ye light: But we might come to an infinitely greater height & light of truth if we had *revelations* or divine light which will bring us to y^e *Sanctum* of wisedome whereas ye Philosopher can not possibly get further y^n ye upper end of y^e *Atrium*. But when shall I enter into ye Sanctum Sanctorum! O light! Oh light! I pant for y^{ee} as ye embossed Hart! Be rationall in *working*: faithfull in *praying*, diligent & constant to ye end in *both*...' (BL, MS Add. 4409, fo. 253^r).

[81] HP 30/4/2A (Ephemerides, 1639, part 1, preceding a note dated 14 Jan.): Hartlib noted (from Pell) that Stanier had mathematical interests, and possessed a valuable collection of books.

year.[82] As Hartlib noted: 'Robert Long, a former secretary, lives independently. He has a library, worth more than £500, of all kinds of fine books. Pell has recently got to know him.'[83] Long had a special interest in optics: Hartlib also recorded the information (again, from Pell) that Descartes was 'judged by Mr Long to have stolen much out of Antonius de Dominis Optica', and later in the year Pell would note that Long sent him 'his Tubus Opticus or perspective trunke' (i.e. telescope).[84] Long, in turn, passed Pell on to another person with scientific interests: as Hartlib noted in his first entry on Long, he 'will bring him acquainted with one Harris or Harrington a lawer and great Mathematician'.[85] The introduction was duly made, and not long afterwards Hartlib recorded details of 'Harringhton' supplied by Pell: 'A great lover of the Hebrew language. 2. of Mathematics. 3. of Ches play. Taught his children of 12 years the deepest Algebras, and that himself.'[86] Notes from the summer of 1639 among Pell's papers show that he had continuing contacts with this man: on 25 May Harrington proposed an equation to him, and on 29 June he 'saw a paper at Mr Harringtons chamber with this title. Questions for Mr Warner.'[87] Neither Pell nor Hartlib gave any further details about Harrington's identity, but enough evidence is available to make a plausible identification: he was probably John Harington or Harrington of Kelston, Somerset (1589–1654), who had studied law at Lincoln's Inn and was a member of a prominent Puritan gentry family.[88]

Finally, one other potential patron developed a connection with Pell during this period: the senior official Sir William Beecher, who was Clerk of the Coun-

---

[82] Long became the Queen's Surveyor-General in 1639 or 1640; he was later knighted, and became Secretary to the Prince of Wales (the future Charles II): see Aylmer, *The King's Servants*, pp. 139, 321, 368.

[83] HP 30/4/2A (Ephemerides, 1639, part 1, preceding a note dated 14 Jan.): 'Robert Long ein gewesener Secret man lebet fur sich selbst. Hat ein Bibliothek vber 500 lib von allerhand statlichen buchern. Pell ist newlich mit ihm bekant worden.'

[84] HP 30/4/3B (Ephemerides, 1639, part 1); BL, MS Add. 4474, fo. 76^r (note dated 14 Oct. 1639). On the common suspicion that Descartes's 'Dioptrique' contained ideas borrowed without acknowledgement from Marc'Antonio de Dominis, see A. Ziggelaar, 'Die Erklärung des Regenbogens durch Marcantonio de Dominis, 1611: zum Optikunterricht am Ende des 16. Jahrhunderts', *Centaurus*, 23 (1979), pp. 21–50.

[85] HP 30/4/2A (Ephemerides, 1639, part 1, preceding a note dated 14 Jan.).

[86] HP 30/4/3A (Ephemerides, 1639, part 1).

[87] BL, MS Add. 4415, fo. 83. (Pell transcribed the problems addressed to Warner, and produced his own solutions to them.)

[88] The key evidence is contained in a manuscript of the poems of Wyatt and Surrey (BL, MS Egerton 2711) which belonged to John Harington, and which he filled up with notes and writings of his own: these include many pages of mathematical diagrams and calculations (e.g. fos. 4^v, 9–26, 63^r, 66^r), and some passages in Hebrew (e.g. fos. 6^v, 26^r). (Also included is a draft letter to John Selden, asking for the loan of a book from which to learn Arabic: fo. 94^r.) On Harington see I. Grimble, *The Harington Family* (London, 1957), pp. 178–81 (and p. 192, noting that his eldest surviving son John was born in 1627: if we substitute 'child' for 'children', this also fits the report of algebra lessons at the age of twelve); for further biographical details see also the Introduction to M. F. Stieg, ed., *The Diary of John Harington, M.P., 1646–53*, Somerset Record Society, lxxiv (Old Woking, 1977).

cil.[89] They seem to have been introduced in the autumn of 1639, and to have discussed the controversial claims made by Descartes in his 'Dioptrique' on behalf of hyperbolic lenses – controversial, that is, because it was widely believed that it would be impossible to cut them to precisely the right shape. In a note dated 23 October 1639, Pell wrote that 'Sr William Beecher sent to Monsieur Briot' and commissioned him to make triangles of 'ye clearest glasse' to Pell's designs; a contemporaneous note by Hartlib recorded that 'One Bryot a french-man in the Tower ... hase vndertaken to procure the making of those Optical Cartusion-glasses [sc. 'Cartesian-glasses'] according to Mr Pels ordering.'[90] Early in the following year, Hartlib learned from Pell that Beecher had written 'an accurat Treatise concerning the encreasing imbasing etc of monys fit for all government of State'; this, together with the connection with Briot (who was 'in the Tower' because he worked as a coin-engraver for the Mint, which was located there), suggests that Beecher had close links with Sir Thomas Aylesbury, the Master of the Mint.[91] Although Beecher spent parts of subsequent years in France, he seems to have kept up a connection with Pell: they corresponded in 1643, when Beecher sent Pell a copy of a newly published work on algebra by Jacques de Billy.[92]

A few other names of possible supporters of Pell crop up in passing during the period 1639–42. Sir Justinian Isham (a Comenian and a supporter of Hartlib) tried to engage the interest of Sir Christopher Hatton in Pell's cause in the autumn of 1639; two years later Pell gave some tuition in mathematics to Sir Cheney Culpeper, having been recommended to him by Hartlib; and in early 1642 Pell delivered a copy of his 'Progymnasma analytico-logisticum' (his tabulation of equations) to Sir Anthony Ashley Cooper, the future patron

---

[89] Beecher, who had previously served as an envoy to France, was knighted in 1623, appointed Clerk of the Council in 1623, and retired from that post in Jan. 1641, both for reasons of age and health, and because, 'if not a practising Roman Catholic, he was certainly a suspect' (see G. M. Bell, *A Handlist of British Diplomatic Representatives, 1509–1688* (London, 1990), pp. 104–5; W. A. Shaw, *The Knights of England*, 2 vols. (London, 1906), ii, p. 180; *CSPD 1619–23*, p. 480; *CSPD 1640–1*, p. 433; Aylmer, *The King's Servants*, p. 232 (quotation)).

[90] BL, MS Add. 4474, fo.77^r; HP 30/4/25A (Ephemerides, 1639, part 3). On the Huguenot coin-engraver Nicolas Briot (1579 or 1580–1646) see F. Mazerolle, 'Nicolas Briot: tailleur général des monnaies (1606–1625)', *Revue belge de numismatique*, 60 (1904), pp. 191–203, 295–314, and Webster, *Great Instauration*, pp. 405–6. The 'triangles' referred to here accord with the instructions given by Descartes in the tenth 'discours' of his 'Dioptrique', in which he explains how glass should first be made in triangular sections and measured for its refractive index, then cut to the shape of a hyperbolic lens (A&T vi, pp. 211–27).

[91] HP 30/4/42A (Ephemerides, 1640, part 1).

[92] See *CSPD 1641–3*, p. 426 (Beecher in Normandy, 1641, 1642); BL, MS Add. 4416, fos. 159, 166 (letter, 11 Oct. 1643, accompanying one by Pell, from a cousin of Beecher's, also William: this junior cousin was possibly being given mathematics tuition by Pell); below, Letter 16, at n. 11 (Pell's receipt in 1643 of J. de Billy, *Nova geometriae clavis algebra* (Paris, 1643)).

of John Locke.[93] But those figures seem to have played only the most peripheral role in Pell's life, compared with the attentions, sporadic thought they may have been, of Aylesbury, Long, Harington, and Beecher. This group of potential – or, intermittently, actual – patrons had, it should be emphasized, no particular cohesive character in ideological terms. Aylesbury and Long would be staunch Royalists; Beecher played no active political role after his retirement, but was unlikely to have Parliamentarian sympathies, being suspected of crypto-Catholicism; whereas Harington became a supporter of the Parliamentary cause and was a hard-line Presbyterian. (Harington's daughter would marry a nephew of 'King' Pym; Aylesbury's granddaughter, the daughter of Edward Hyde, would be the wife of King James II.) But most or all of them did share some characteristics. They had strong but private interests in pure and applied mathematics; most of them held administrative posts (a fact which, taken together with Pell's encounter with Gualter Frost outside the Council chamber, may suggest that Hartlib was using a well-placed friend such as Weckherlin to make introductions); and it seems that all of them – not only Aylesbury and Harington, but also Beecher and Long – were acquainted with the scientist and mathematician Walter Warner.[94]

This elderly scientist (born in c.1560) had evidently exerted a special fascination on Pell for some time before they met. His chief claim on Pell's attention was the fact that he had been a friend and colleague of Thomas Harriot: both men had been under the patronage of Henry Percy, the ninth Earl of Northumberland, and it was Warner who had edited and published Harriot's treatise on algebra, *Artis analyticae praxis*, in 1631, ten years after Harriot's death. However, although Warner shared some of Harriot's mathematical interests, he was not a particularly skilled mathematician himself, and his presentation of Harriot's arguments was marred in places by sheer incomprehension. Warner's own interests lay mainly in metaphysics, physics, physiology, and psychology; but he never

---

[93] Osborn Collection, Beinecke Library, Yale University, MS 16792 (Isham to Hartlib, 2 Sept. 1639; transcription in HP); Braddick and Greengrass, 'Letters of Sir Cheney Culpeper', pp. 153–4; BL, MS Add. 4398, fo. 188 (notes on 'Progymnasma').

[94] The evidence to suggest Beecher's acquaintance with Warner is both indirect (his presumed acquaintance with Aylesbury (see above, at n. 91), and the fact that Pell's first meeting with Warner took place during the period of his work on Beecher's lenses and is recorded in the same set of notes as Pell's account of that work (BL, MS Add. 4474, fo. 77^r)) and direct: an inventory of Warner's manuscripts drawn up by John Collins in 1667 includes 'A bundle intituled "Sir William Beccher"' (BL, MS Add. 4394, fo. 106^r). The evidence of Long's acquaintance with Warner is a note by Hartlib of information from Pell, deriving apparently from a conversation with Long: 'Warner hase many choice Exercitations in Mathematik's ... Warner is devicing a new kind of Logarithms' (HP 30/4/2A, Ephemerides, 1639, part 1).

PART I: THE LIFE OF JOHN PELL   83

published his own writings on these topics, voluminous though they were.[95] At the behest of his patron in the late 1620s and 1630s, Sir Thomas Aylesbury, he also wrote about metal alloys (in connection with Aylesbury's work as Master of the Mint), and conducted research on refraction.[96] This last topic was a special interest of Aylesbury's, an interest that had doubtless been instilled in him by the experimental work of Harriot, who was the first scientist to formulate the sine law of refraction.[97]

Pell had probably first heard about Warner from Hartlib, who, as early as 1635, had recorded information he had received from Henry Gellibrand: 'Mr Warner hase all Hariots MS. and is setting some of them forth.'[98] Working on his revision of Harriot's book in early 1639, Pell found to some extent that he was stepping into Warner's shoes, trying to complete what Warner had left unfinished. As Hartlib noted in the spring of that year: 'Hee [sc. Pell] hase finished those Problemes of Hariot which Warner should have perfected. Sir Thomas Alesbury promised to let them have Hariots papers but hee did solve them without them.'[99] At roughly the same time Pell also heard, apparently from Robert Long, that Warner had 'many choice Exercitations in Mathematik's'.[100] But Warner was away from London (probably at Aylesbury's country residence near Wind-

---

[95] The only full-length study of Warner concentrates on his work on physiology and psychology: J. L. M. Prins, *Walter Warner (ca. 1557–1643) and his Notes on Animal Organisms* (Utrecht, 1992). (Warner's date of birth is unknown; Prins's estimate is based on the fact that he proceeded BA at Oxford in 1578. S. Clucas suggests 1562: 'Walter Warner', in A. Pyle, ed., *The Dictionary of Seventeenth-Century British Philosophers*, 2 vols. (London, 2000), ii, pp. 858–62.) On his connections with Harriot and Northumberland see J. W. Shirley, 'The Scientific Experiments of Sir Walter Ralegh, the Wizard Earl, and the Three Magi in the Tower, 1603–1617', *Ambix*, 4 (1949), pp. 52–66, and J. Jacquot, 'Harriot, Hill, Warner and the New Philosophy', in J. W. Shirley, ed., *Thomas Harriot, Renaissance Scientist* (Oxford, 1974), pp. 107–28. On his physics and metaphysics see also S. Clucas, 'Corpuscular Matter Theory in the Northumberland Circle', in C. Lüthy, J. E. Murdoch, and W. R. Newman, eds., *Late Medieval and Early Modern Corpuscular Matter Theories* (Leiden, 2001), pp. 181–207. The extent of Warner's incomprehension of Harriot's algebra is demonstrated in J. A. Stedall, 'Rob'd of Glories: The Posthumous Misfortunes of Thomas Harriot and His Algebra', *Archive for the History of Exact Sciences*, 54 (2000), pp. 455–97, esp. pp. 469–71.

[96] See BL, MSS Harl. 6754, fos. 2–27 (treatise on analysis of bullion), fos. 28–39 (treatise on money and exchange); Harl. 6755, fos. 15–18^r (Sir Charles Cavendish's notes on Warner's treatise 'of the commixture of Metalls for the Mint'); Harl. 6756 ('De loco imaginis'). A treatise on refraction by Warner was published by Mersenne in his *Universae geometriae mixtaeque mathematicae synopsis* (Paris, 1644), pp. 549–66.

[97] On Harriot's work on refraction see Shirley, *Thomas Harriot*, pp. 381–5. In 1606 Harriot corresponded about refraction with Kepler, sending him tables of observations of the angle of refraction from air to various liquids and solids (crystal and glass): J. Kepler, *Gesammelte Werke*, ed. M. Caspar and W. von Dyck (Munich, 1938–), xv, pp. 365–6. A table of observations of refraction from air to glass, made by Warner and Aylesbury in July 1627, is in BL, MS Add. 4395, fo. 99^r.

[98] HP 29/3/41A (Ephemerides, 1635, part 4).

[99] HP 30/4/9B (Ephemerides, 1639, part 2). The other person (or persons) indicated by the word 'them' has not been identified.

[100] HP 30/4/2A (Ephemerides, 1639, part 1).

sor) until the late autumn: only then was Hartlib able to record that 'Varner is come againe to towne. Hee hase made ready a great Mathematical Worke which hee is to shew to Pell.'[101] The first account of a meeting with Warner in Pell's own notes (unfortunately, on a sheet of paper badly damaged by damp) comes from November 1639: that this was probably his first encounter with the man is suggested by the fact that he did not yet know his Christian name. 'Nov. 15 Mr W[*blank*] Warner told me that at Sr Thomas Ailesburies y^e [*page torn*] Table of refractions [*paper damaged*] ... That he had calculated his table of Analogicalls [*paper damaged*] extending it to 100 000 continuall proportionalls ... Nov. 20 He shewed me 10 thousand of y^m done.'[102] This was the 'great Mathematical Worke' referred to by Hartlib: a table of antilogarithms. As John Aubrey would later write (on the basis of information received from Pell), 'These ... before Mr. John Pell grew acquainted with Mr. Warner, were ten thousand, and at Mr. Warner's request were by Mr. Pell's hands, or direction, made a hundred thousand'; Pell's now fragmentary note shows that the idea of extending the table to 100 000 entries was discussed at their very first meeting.[103]

As a dogged compiler of mathematical tables, Warner was clearly a man after Pell's own heart. But there were plenty of other interests they had in common, and it seems that they were in frequent contact from now on. Hartlib soon recorded that Warner had given Pell 'some Hydraulical Papers'; he also heard from Pell that Warner had commissioned the manufacture of some hugely expensive 'Optical glasses', which he hoped could be used as burning glasses at a great distance.[104] Writing to Mersenne on [14/] 24 January 1640, Pell mentioned that he had got to know Warner during the previous two months, and often went to visit him; he said that, in view of Warner's old age and near-decrepitude, he was constantly offering to help him complete his works, and that if only Warner would accept his assistance, many 'recondite and completely new things, both in physics and almost the whole field of mathematics' would be vouchsafed to

---

[101] HP 30/4/26A (Ephemerides, 1639, part 3). Aubrey recorded that Warner 'vsually spent his sumer in Windesor park, and was welcom, for he was harmless and quet' (*'Brief Lives'*, ii, p. 16).

[102] BL, MS Add. 4474, fo. 77^r.

[103] Aubrey, *'Brief Lives'*, ii, pp. 292–3.

[104] HP 30/4/27B (Ephemerides, 1639, part 3); 30/4/28B (Ephemerides, 1639, part 4). The burning glasses were intended to have a military application, making it possible to destroy ships at a distance. Warner had tried to interest the Earl of Newcastle in this project in 1635, drawing the sceptical response from Hobbes: 'when it is demonstrated how ye glasse must be made to burne a mile of, if the glasse must be so bigge as cannot be made, the art will be no more worth, then the art of making ordinary burning glasses' (Hobbes to Newcastle, 15/25 Aug. 1635, in T. Hobbes, *The Correspondence*, ed. N. Malcolm, 2 vols. (Oxford, 1994) i, p. 29). A fragment of a work by Warner on hydraulics is in BL, MS Add. 4394, fo. 131. Pell's papers also include a set of notes in Warner's hand on mechanics (MS Add. 4421, fos. 71–2), omitted from Jan Prins's survey of Warner's MSS.

the world.[105] The topic in which Pell seems to have shown most interest was the measurement, mathematical analysis, and theoretical explanation of refraction. This was an essential matter for anyone engaged in the making of lenses (the task for which Pell had recently been employed by Sir William Beecher); it also raised larger questions about the physical nature of light itself, questions which Descartes had tried to answer in his 'Dioptrique'. An important note among Pell's papers (undated, but probably from the earliest phase of his acquaintance with Warner) records the following conversation:

> Mr Warner sayes y^t he had of Mr Harriot this proportion As the sine of one angle of incidence to ye Sine of its 1 refracted angle, found by experience So the sine of any angle of incidence upon ye same superficies to ye sine of its refracted angle, to be found by supputation. But that he never saw him demonstrate it. But said he upon [>some] occasion I did thoroughly demonstrate y^s proportion, but it was a long processe. I told him De Cartes in his Dioptric ... had ye same proportion.[106]

This is followed by detailed notes of the methods used by Warner when he had measured refraction in different media at Sir Thomas Aylesbury's house. Investigating the nature of refraction now became one of Pell's chief concerns.

It is in this context that a plausible account can be given of how, when, and why Pell was put in touch with the man who would become, for some years at least, the most important of his patrons and supporters: Sir Charles Cavendish. What appears to have happened is that Pell developed, in November and December 1639, a strong desire to collaborate with Walter Warner on the design of lenses and the experimental investigation of refraction, but that Warner was then turned aside from such work by his own patron, Sir Thomas Aylesbury. As Hartlib noted (recording information he had received from Pell), 'Hee [*sc.* Warner] is willed by Thomas Alesbury to absolve first the great Worke of Numeri Analogici before hee medle with his optical glasses.'[107] Most likely it was Warner who then proposed Sir Charles Cavendish as a suitable sponsor of Pell's optical researches, and made the introduction: Warner had corresponded with Sir Charles about optical matters in the mid-1630s, and would have been well

---

[105] *MC* ix, pp. 62–3 ('recondita et plane nova in Physicis et omni fere Mathesi'). 'Mathesis', translated here as 'mathematics', included (and here, no doubt, mainly referred to) applied or 'mixed' mathematics. There is no evidence of Warner making any significant advances in pure mathematics. One of Pell's earliest dated notes of his contacts with Warner suggests that Warner overestimated his own originality in that field: 'Mr Warner gave me these two Axiomes, as of his owne invention. [*marginal note:* But I find y^m in B. Pitiscus Lib. 2. 43. 44. 45]' (BL, MS Add. 4422, fo. 276^r, dated 'Decemb. 1639').

[106] BL, MS Add. 4407, fo. 183^r.

[107] HP 30/4/29B (Ephemerides, 1639, part 4).

aware of the seriousness of his interests in this field.[108] The Hartlib circle had had no contact with Sir Charles (or his brother, the Earl of Newcastle), and the first mention of Sir Charles's name in Hartlib's 'Ephemerides' comes in an account of people with mathematical interests, apparently given by Robert Long to Pell in a conversation in January 1639: 'Item Chaplain of my Lord Newcastle who lives now with Sir Charles Cavendish'.[109] This was a reference to the mathematician Robert Payne, who had served as Newcastle's chaplain since 1630; when Newcastle took up his appointment at Court as governor to Prince Charles in May 1638, Payne seems to have remained for much of the rest of that year with Sir Charles in Nottinghamshire. But in January 1639 he took up a canonry at his old Oxford college, Christ Church.[110] Sir Charles, whose mathematical interests – and competence – were much greater than those of his brother, must have felt the loss of such a mathematically skilled employee and companion. This, then, was the gap that was filled by John Pell in the early months of 1640.

In attracting the attention of Sir Charles Cavendish (who will be referred to hereafter simply as 'Cavendish'), Pell enjoyed an extraordinary stroke of good fortune. Not only was his new patron rich, and the brother of one of the most prominent men in England; he was also a man of unusual modesty, affability, and good nature. In the words of Edward Hyde, Earl of Clarendon: 'the gentleness of his disposition, the humility and meekness of his nature, and the vivacity of his wit was admirable. He was so modest, that he could hardly be prevailed with to enlarge himself on subjects he understood better than other men.'[111] Born in c.1595, Cavendish was the younger brother of William (who became Earl, Marquess, and eventually Duke of Newcastle); their father, Sir Charles, was a younger brother of the first Earl of Devonshire, and the youngest son of Bess of Hardwick by her second husband. Cavendish was of stunted growth, and apparently suffered from a degree of physical deformity; how great that deformity was is not at all clear, and a modern writer is probably exaggerating when he calls him 'a puny mis-shapen dwarf'.[112] Aubrey described him as 'a little, weake, crooked man ... nature not having adapted him for the court or camp'.[113] But he was clearly not grossly handicapped: during his adult life he served as a Member of Parliament and took part in his brother's military campaigns, charging with

---

[108] See BL, MSS Add. 4407, fos. 186–7 (Cavendish to Warner, 2 May 1636); Add. 4444, fos. 93–4 (Cavendish to Warner, 2 Sept. [1636]); Add. 4395, fos. 116–17 (Warner to Cavendish, draft, undated [Oct. 1636]). It is of course also possible that Sir Thomas Aylesbury played some part in the introduction of Pell to Cavendish; Aylesbury was an old friend of Cavendish's mathematical mentor, William Oughtred (see the letter from Aylesbury to Christopher Wren, 10 Apr. 1649, in C. Wren, *Parentalia: Or, Memoirs of the Family of the Wrens* (London, 1750), p. 184).

[109] HP 30/4/2A (Ephemerides, 1639, part 1, preceding a note dated 14 Jan.).

[110] See N. Malcolm, *Aspects of Hobbes* (Oxford, 2002), pp. 86–96.

[111] E. Hyde, *The Life of Edward Earl of Clarendon*, 3 vols. (Oxford, 1827), i, p. 293.

[112] A. S. Turberville, *A History of Welbeck Abbey and its Owners*, 2 vols. (London, 1938), i, p. 47.

[113] Aubrey, *'Brief Lives'*, i, p. 153.

the Royalist cavalry at Marston Moor.

Nothing is known about Cavendish's early life and education. He probably studied under tutors at home, and it has been suggested that his interest in mathematics was stimulated by a member of his father's household, Dr Gissop (or Jessup), who had been an assistant to the scientist William Gilbert.[114] Cavendish's brother spent some time at St John's College, Cambridge, but it is not known whether Cavendish joined him there. In her memoir of her husband's life, Margaret Cavendish (the second wife of William) merely wrote that while their father gave up trying to make William study, 'his other son Charles, in whom he found a greater love and inclination to learning, he encouraged as much that way, as possibly he could'.[115] In 1612 Cavendish travelled with his brother in the group of young gentlemen accompanying Sir Henry Wotton on a diplomatic mission to Savoy in 1612, passing through France on the way there and returning via Basel and Cologne. (Wotton himself, referring to the members of his entourage, mentioned only the elder of the two brothers, but the well-informed news-writer John Chamberlain recorded that Wotton's 'companie of note' included the 'two sonnes of Sir Charles Candish'.)[116] What is clear is that for most of his life Cavendish would remain very closely attached to his brother William; although he had his own estates (Wellingor in Lincolnshire and Slingsby in Yorkshire), he lived mainly with his brother, either at Welbeck Abbey in north Nottinghamshire or at William's town house in Clerkenwell. Cavendish was knighted at Welbeck in 1619, and served as MP for Nottingham in 1624 and 1628.

How and when did Cavendish's special interest in mathematics develop? It may be that he was taught as a child by Gissop, and it is also possible that the 'Mathematician' who features in Ben Jonson's masque of 1620, *An Entertainment at the Blackfriars* (written for the christening of another Charles Cavendish, the second son of the second Earl of Devonshire) was intended to represent him.[117]

---

[114] S. Hutton, 'Sir Charles Cavendish', in A. Pyle, ed., *The Dictionary of Seventeenth-Century British Philosophers*, 2 vols. (London, 2000), i, pp. 165–6.

[115] M. Cavendish, *The Life of the thrice Noble, High and Puissant Prince William Cavendishe, Duke, Marquess, and Earl of Newcastle* (London, 1915), p. 133. There is no trace of Cavendish's attendance at St John's in the official records of the College or the University; but, equally, there is no record of his brother's. (The oft-cited reference to a William Cavendish receiving an MA in 1608 relates in fact to their cousin, the future second Earl of Devonshire.) Nevertheless, a letter written to William by the Fellows of St John's in 1641 confirms that he did attend the college: St John's College Archives, C 7.15, 'The College Letter Book', p. 399. (I am very grateful to Dr Lucy Worsley for giving me this evidence, which is presented in her 'The Architectural Patronage of William Cavendish, Duke of Newcastle, 1593–1676' (University of Sussex D.Phil. dissertation, 2001), p. 98.)

[116] L. P. Smith, *The Life and Letters of Sir Henry Wotton*, 2 vols. (Oxford, 1907), ii, p. 2; N. E. McClure, ed., *The Letters of John Chamberlain*, 2 vols. (Memoirs of the American Philosophical Society, xii, parts 1, 2) (Philadelphia, 1939), i, p. 339.

[117] B. Jonson, *Works*, ed. C. H. Herford, P. Simpson, and E. Simpson, 11 vols. (Oxford, 1925–52), vii, pp. 776–7. However, the main concern of the 'Mathematician' in the masque is with casting a nativity, which does not appear ever to have been one of Cavendish's interests.

But the first definite evidence of his mathematical concerns dates from 1631. It was in that year that William Oughtred produced his treatise on algebra, the *Clavis*, remarking in his dedicatory epistle that he was encouraged to publish it by Cavendish; in the same year the French mathematician Claude Mydorge published his *Prodromi catoptricorum* with a dedication to Cavendish, calling him 'extremely skilled in all mathematics, and a very dear friend to me'; and at the end of that year Cavendish gave a copy of the recently printed treatise by Harriot, *Artis analyticae praxis*, to his brother's chaplain, Robert Payne.[118] In view of the connection with Mydorge, and with another French mathematician, François Derand, who corresponded with Cavendish in 1634 and 1637, it has been suggested that Cavendish had spent some time studying mathematics in Paris; but there is no evidence to support this claim, and the impression given by Cavendish's letters to Pell after his arrival in Paris in 1645 is not that of someone renewing a range of old acquaintances in that city.[119] Derand was probably put in touch with Cavendish by Mydorge, whom he knew; how Mydorge's connection with Cavendish had first come about is not known, but it may well have been through a mutual friend such as the French courtier Charles du Bosc, who lived for several years in the household of Cavendish's cousin, the Earl of Devonshire.[120] As for Cavendish's tuition in mathematics, it seems much more likely that this took place in London, at the hands of William Oughtred. This well-known mathematician had left his fellowship at King's College, Cambridge, in 1605, and had moved to a rectory near Guildford; he was contacted there in 1626 by the Earl of Arundel, who extended his patronage to him. In the words of Oughtred's later autobiographical account, Arundel 'was pleased to send for me: and afterward at London to appoint mee a Chamber in his owne house: where, at such times,

---

[118] W. Oughtred, *Clavis*, sig. A2^v ('ut nobilissimo Dn. Carolo Canditio ... satisfecerem'); C. Mydorge, *Prodromi catoptricorum et dioptricorum, sive conicorum operis ... libri duo* (Paris, 1631), sig. *4^r ('totius Matheseos summè perito, nobisque amicissimo'); Bodl., pressmark Savile O 9, T. Harriot, *Artis analyticae praxis*, inscribed by Payne: 'Ex dono nobilissimi Equitis Caroli Cauendysshe Decemb. 18 1631'. A letter from Mydorge was also forwarded to Cavendish in June 1631: see BL, MS Add. 70499, fo. 145^r (Richard Andrews to the Earl of Newcastle, 22 June 1631).

[119] The claim is made by Jean Jacquot: 'Sir Charles Cavendish and his Learned Friends', *Annals of Science*, 3 (1952), pp. 13–27, 175–91; here p. 15. (Sarah Hutton adds that Cavendish had got to know Mersenne in France 'in or before 1630' ('Sir Charles Cavendish', p. 165); but there is no evidence of contacts between Cavendish and Mersenne before 1640.) Jacquot's argument is based on a phrase in Derand's letter to Cavendish of [22 July/] 1 Aug. 1637: 'depuis le tems que j'ai eu l'honneur de vous pratiquer' (Cambridge University Library [hereafter: CUL], MS Add. 9597/13/1/99^r, printed in S. J. Rigaud, ed., *Correspondence of Scientific Men of the Seventeenth Century*, 2 vols. (Oxford, 1841), i, p. 28). Jacquot takes 'pratiquer' here to have its modern meaning, 'to frequent, associate with'; but in early modern French it could simply mean 'to seek favourable attention from, seek the help of' (see E. Huguet, *Dictionnaire de la langue française du seizième siècle*, 7 vols. (Paris, 1925–67), vi, p. 131).

[120] For contacts between Derand and Mydorge in 1630 see *MC* ii, p. 550; on du Bosc see Hobbes, *Correspondence*, ii, pp. 795–7. Mersenne's first surviving letter to Cavendish refers to du Bosc, in a way that seems to imply acquaintance between him and Cavendish: *MC* ix, pp. 315–16.

and in such manner as it seemed him good to imploy me ... I have been most ready to present myself in all humble and affectionate service'. Oughtred taught Arundel's son, William Howard; and he was also 'much frequented', during his sojourns at Arundel House, 'both by Natives and Strangers, for my resolution and instruction in many difficult poynts of Art'.[121] Cavendish was a close relative of the Earl of Arundel: Cavendish's father's half-brother, Gilbert Talbot, was the father of Arundel's wife (and Cavendish's brother William was Gilbert Talbot's executor). So it seems very likely that, during his stays in London in the late 1620s – for example, when attending Parliament in 1628 – Cavendish was a frequent visitor to Arundel House, and was given tuition there by William Oughtred. There is even some evidence of the form that some of that tuition took: as Pell later noted, Cavendish had a copy of Viète's *De aequationum resolutione*, with substantial corrections and annotations in Oughtred's hand.[122]

The choice of the mathematician Robert Payne as chaplain to the Earl of Newcastle (an appointment made in 1630) probably also reflected Cavendish's special interests in that field; indeed, it may have been Oughtred who first suggested Payne's name to the Cavendish brothers.[123] Payne played a key role in the intellectual life of Cavendish and his brother throughout the 1630s: he translated scientific works from Italian (including Galileo's *Della scienza mecanica*) for Newcastle, composed treatises of his own, and participated in discussion and correspondence with other scientists and philosophers. The first sign of Cavendish's contacts with Walter Warner is a letter from Warner to Payne of 17 October 1634; the exchanges that followed were concerned principally with Warner's work on optics (especially refraction), but Payne and Cavendish also took an interest in some of Warner's other concerns, including geometry, naval architecture, and metallurgy.[124] At the same time, there was a constant exchange of ideas with the philosopher (and mathematician, and optical scientist) Thomas Hobbes, who was an employee of the other Cavendish family, the Earls of Devonshire. Cavendish may have known Hobbes from a very early age (Hobbes was first employed by that other branch of the Cavendishes in 1608); the first evidence of particular contacts between them comes from 1624, when the future second Earl of Devonshire wrote to an acquaintance about his cousin William Cavendish (the

---

[121] W. Oughtred, *To the English Gentrie ... The Just Apologie of Wil: Oughtred, against the Slanderous Insimulations of Richard Delamain, in a Pamphlet called Grammeologia* (n.p., n.d. [1631]), sig. B1. The dedicatory epistle of *Clavis* is addressed to William Howard. Frances Willmoth discusses Oughtred's relationship with Arundel and comments: 'The room provided for Oughtred at Arundel House, in the Strand, gave him a convenient London base, where he probably spent several weeks of each year and certainly received pupils other than the Howards' (*Sir Jonas Moore: Practical Mathematics and Restoration Science* (Woodbridge, 1993), p. 47).

[122] BL, MS Add. 4423, fos. 146–151^r.

[123] Payne had probably known Edmund Gunter (who became a friend of Oughtred's) at Christ Church: he possessed a manuscript copied from Gunter's autograph original (Chatsworth, MS Hobbes B3). He also had a manuscript copy of a treatise by Oughtred on the construction of sun-dials (Chatsworth, MS Hobbes B2). See Malcolm, *Aspects of Hobbes*, p. 87.

[124] See Malcolm, *Aspects of Hobbes*, pp. 88–9.

future Earl of Newcastle) and added: 'Hobs shall put me in mind to write to W^{ms} brother.'[125] Hobbes was an occasional visitor to Welbeck; he would later refer, for example, to conversations he had had there with Cavendish and his brother about his 'doctrine of the nature and production of light, sound, and all phantasms and ideas'.[126] However, modern writers have given a somewhat misleading impression when they have referred to a 'Welbeck circle' or even a 'Welbeck academy' consisting of the Cavendishes, Payne, Hobbes, Warner, and Pell.[127] Neither Warner nor Pell ever visited Welbeck; Pell's connection with Cavendish began only after Payne had left Newcastle's service; there is no evidence that Pell had any contacts with Payne; and while it seems likely that Pell would have had some contacts with Hobbes, via Cavendish, in 1640, direct evidence of this is also lacking.

The first dated reference in Pell's papers to his dealings with Cavendish is a note referring to two mathematical problems, transcribed from a book by the sixteenth-century mathematician Guillaume Gosselin, and solved by Pell: it is annotated 'thus I titled ye paper which I gave Sr Charles Candish May 6. 1640'.[128] From this it would appear that Pell had been giving Cavendish tuition in mathematics. Another early sign of their contacts is a brief memorandum by Pell, referring to Warner's work on refraction: '1640 About ye beginning of June Sr Charles Candish going out of towne left in my hands a manuscript in [>a small] folio Of 18 leaves, or 9 sheetes of paper. With this title Mr Warners Tract for ye table of refraction transcribed by Hunt. Smithson. It begins thus. Problema, ad tabulas refractionum ex observatis construendas sequenti processu apodictico solvendum ['The problem of constructing tables of refraction from observations is to be solved by the following strictly demonstrative method'].'[129] And it is probably not a coincidence that the first direct contacts between Cavendish and Mersenne appear to have been made in the spring of 1640; Mersenne was already aware of Cavendish's intellectual interests, but it was no doubt through the good

---

[125] BL, MS Add. 70499, fo. 118^r (William Cavendish to Henry Bates, June 1624).

[126] Hobbes, *Correspondence*, i, p. 108; cf. his reference in his English Optical MS, written in 1645–6, to 'that wch about 16 yeares since I affirmed to your Lo^{pp} at Welbeck, that Light is a fancy in the minde, caused by motion in the braine' (BL, MS Harl. 3360, fo. 3^r).

[127] See Kargon, *Atomism in England*, pp. 41–2; Jacquot, 'Sir Charles Cavendish and his Learned Friends'; H. W. Jones, 'Der Kreis von Welbeck', in J.-P. Schobinger, ed., *Grundriss der Geschichte der Philosophie, begründet von Friedrich Ueberweg: Die Philosophie des 17. Jahrhunderts*, Bd iii, *Die Philosophie des 17. Jahrhunderts: England*, 2 vols. (Basel, 1988), i, pp. 186–91. For a nuanced and critical analysis of the concept of a 'Cavendish circle' see T. Raylor, 'Newcastle's Ghosts: Robert Payne, Ben Jonson, and the "Cavendish Circle"', in C. J. Summers and E.-L. Pebworth, eds., *Literary Circles and Cultural Communities in Renaissance England* (Columbia, Mo., 2000), pp. 92–114.

[128] BL, MS Add. 4431, fo. 270^v. The last numeral of '1640' is obscured by the binding, but the date is also contained in the text itself.

[129] BL, MS Add. 4417, fo. 39^r; Pell's copy of the first two pages of this treatise is on fo. 38.

PART I: THE LIFE OF JOHN PELL   91

offices of John Pell that he was enabled to write to him directly.[130]

With Cavendish's encouragement, Pell kept up his research in optics during the next two years. Scattered among his papers are many notes on the measurement of refraction in different media – often comparing the method recommended by Descartes with that used by Warner.[131] In the spring of 1641 Cavendish commissioned a new attempt to follow the instructions given by Descartes for the manufacture of a telescope using two hyperbolic lenses: making notes on 17 April of that year, Pell discussed methods of 'finding ye refraction by experience', described the glass triangles he had ordered from the instrument-maker Richard Reeve on 13 April, and also referred to 'ye Moulds which I gave to Monsieur Briot October 24 1639' – thus confirming that Cavendish was renewing the earlier project which, for whatever reason, Sir William Beecher had abandoned.[132] In the following month Pell made a detailed comparison between the measurements of refraction in glass and water reported by Descartes and Warner: he annotated his copy of this text 'This I gave to Sr Charles Candish at his going away in May 1641.'[133] During the summer, while Richard Reeve made painfully slow progress towards the manufacture of the hyperbolic lenses (as reported by Pell in his letters to Cavendish in Lincolnshire), Pell continued with his own researches, taking notes on the design of telescopes in June, making yet further comparisons between Descartes and Warner in August, and accumulating more observations of angles of refraction in September; when Cavendish briefly revisited London in October, Pell wrote up his findings and gave a copy to him.[134] The correspondence between Pell and Cavendish during the winter of 1641–2 shows what further difficulties their lens-making project encountered; although Reeve did deliver one lens to Pell at the end of January, it seems that all the glass triangles of which the refractive index had been so carefully measured had now been used up, so that they were obliged to proceed with ordinary,

---

[130] The first surviving letter from Mersenne to Cavendish (which was clearly not the first in the sequence) is dated [2/] 12 May 1640: *MC* ix, pp. 315–17. Mersenne had received a text emanating from 'Mr de Kandeise' from Edward Herbert in July or August 1639: *MC* viii, p. 476. The 1636 re-issue of Mersenne's *Harmonicorum libri* included a dedicatory epistle to Cavendish by the publisher, Guillaume Baudry; this has sometimes been described, incorrectly, as a dedication by Mersenne himself. (The confusion arises because Baudry substituted it for a dedicatory epistle (to de Montmor) by Mersenne, but made no change to the preface, which immediately followed the dedicatory epistle: this begins 'Praefatio ad eundem' ('Preface, addressed to the same person') and is signed 'Mersenne'.)

[131] See for example the notes on refraction in different media in BL, MS Add. 4431, fo. 285^v, or the notes comparing Descartes and Warner on this subject in BL, MS Add. 4423, fo. 390^r.

[132] BL, MS Add. 4428, fo. 178; for the earlier project (and the significance of the 'triangles'), see above, n. 90.

[133] BL, MS Add. 4415, fo. 290^v.

[134] Ibid., fos. 290^v (2 June), 295^v–7^r (5 Aug.), 292^v (6 Sept.), 282^r (2 Oct.). For Reeve's progress see Letters 2 and 4a, below.

unmeasured, glass.¹³⁵ Pell was still taking notes on the design of lenses during the following two months, and observing angles of refraction in April; but the lack of any records of observations made with the Cartesian telescope suggests that the work on it was never completed.¹³⁶

As those early letters between Pell and Cavendish show, they had other interests in common besides optics. In the first surviving letter from the sequence, written in June 1641, Cavendish thanked Pell for his 'waie of ordering aequations' (probably a reference to Pell's method of presenting the working-out of an algebraic problem in three columns) and urged him to make progress with his 'intended analyticall worcke' (meaning, perhaps, the work on the improvement of algebra which was so eagerly expected by Moriaen).¹³⁷ Pell's own comment on these projects, in a draft of a letter written probably in August, struck a familiar note: 'my owne analyticall worke stands still Of which that crotchet of mine of ordering & numbring ye aequations is but one peculiarity I have many others which might have beene published ere this if I could have beene acquainted with leisure sufficient for thoughts of y^t natu[re].'¹³⁸ When they met, they talked about mathematics; in his letters, Cavendish also asked Pell to obtain scientific books for him, and sought his opinion of the mathematical materials sent to him by Mersenne.¹³⁹ And in particular Cavendish seems to have encouraged Pell's idea of extending Warner's table of antilogarithms: he asked about this in June, and received the reply (in the following month) that 'I have begun Mr Warners worke & calculated ye first 4 thousands ... if I could set a scribe I would hope besides my other businesses to have ye whole hundred thousand ready for ye presse by Christmas.'¹⁴⁰ Among Pell's papers there is a set of notes detailing – with meticulous, well-nigh obsessive attention to detail – the progress of this project and what might be called the logistics of the operation. It began in June 1641 with his acquisition of special sheets of 'calculatory paper', which he bought from a Mr T. Browne at $2d$ per sheet and carried to Joachim Hübner's study in Hartlib's house. By November Pell had an assistant calculating antilogarithms for him 'according to my precepts in writing & instruction viva voce'; two months later Pell informed Cavendish that 40 000 were ready, half calculated by him and half by 'one whom I instructed'. On 11 May 1642 he noted that he had completed 74 950 out of the requisite 100 000: 'There remaines 25000 viz to 99950 for y^e last 50 are done. These, I reckon will cost me 225 houres...'¹⁴¹ And in mid-August,

---

[135] BL, MS Add. 4415, fo. 271ʳ: Pell noted delivery of the glass on 30 Jan., and the trials he made on 31 Jan. ('it seemed to gather y^e sun beames into a circle of $\frac{3}{4}$ of an inch diameter upon a plane 82 inches behind it; it selfe being 4 inches broad'). See also Cavendish's comments in Letter 10.
[136] BL, MS Add. 4415, fos. 273ʳ (2 Feb.), 274ʳ (10 Mar.), 281ʳ, 292ʳ (13 Apr.).
[137] Letter 1.
[138] Letter 4b.
[139] Letters 6–9.
[140] Letter 2.
[141] BL, MS Add. 4365, fos. 39ᵛ–36ʳ (retrograde); Letter 9.

just as the Civil War was about to start, he was able to tell a friend that he had 'finished' the work 'a month agoe', adding ominously that 'these times [are scarce *deleted* >must be a little] quieter before it goe to ye presse'.[142] Like so many of the other major projects on which he laboured, this work would remain unpublished.[143]

Although Pell was presumably well rewarded by Cavendish for his services, his references in his letters to his lack of leisure and the demands of his 'other businesses' give the impression that he was not being supported on a full-time basis by his new patron. How else he earned his living is not clear; possibly he was – despite his earlier reluctance – giving some occasional lessons in mathematics. He did have contacts with various members of the London community of 'mathematical practitioners', such as John Leake, John Marr, and Henry Bond; the T. Browne from whom he bought his 'calculatory paper' was also a mathematical instrument maker, whose 'serpentine scale' (a logarithmic scale arranged in a spiral, as a sort of non-sliding slide rule) particularly interested Pell.[144] Of these practitioners, Leake seems to have been the one with whom he had the closest relations; in the summer of 1645 he would write to Leake from Amsterdam, asking for his help in gathering support for Pell in his controversy with Longomontanus.[145] And a significant piece of evidence suggests that Leake was one of the first people to whom Pell explained his new three-column system for presenting the working-out of mathematical problems. In a letter to John Collins of 1666 or 1667 Pell would write: 'I hope amongst my papers in Essex, there is yet safe a faire coppy of a question so wrought with numbred steps & marginall references, which I shewed to Mr John Leake at my house in Westminster, 24 yeares agoe. If there be any yet alive that saw any thing of that kinde, before

---

[142] BL, MS Add. 4418, fo. 89^r (Pell to an unnamed correspondent, possibly Hübner, Haak, or Vegelin, 17 Aug. 1642; printed in *MC* xi, p. 245).

[143] Unpublished, at least, in Pell's lifetime and with his name: Prins notes a rumour that the work was eventually used by the mathematician James Dodson, who published it under his own name as *The Anti-logarithmic Canon, Being a Table of Numbers, Consisting of Eleven Places of Figures, Corresponding to all Logarithms under 100,000* (London, 1742) (Prins, *Walter Warner*, p. 20). However, two considerations make this accusation implausible: the fact that Dodson drew attention to the history of the Warner–Pell antilogarithms in his Introduction (p. ix), and the fact that his antilogarithms were given to a greater number of decimal places. On renewed attempts by Pell to prepare his material for publication in the 1680s, see below, ch. 6, at nn. 114–16.

[144] See Taylor, *Mathematical Practitioners*, pp. 203–4 (Marr), 207–8 (Bond), 210 (Browne), 236–7 (Leake). Pell's note of a geometrical problem proposed to him by Leake on 17 Feb. 1641 is in BL, MS Add. 4407, fo. 143^r. Pell described an encounter with Marr (in 1638) in his letter to Mersenne of 24 Jan. 1640: *MC* ix, p. 60. In that letter he said that he knew Bond by face, but that Bond did not know him; however, by June 1641 'Mr Bond' was offering to supply him with calculatory paper (BL, MS Add. 4365, fo. 38^r). Pell's papers include a printed sheet giving the design of Thomas Browne's serpentine scale, and a copy of an accompanying text, in Pell's hand: BL, MS Add. 4407, fos. 59–62. On the serpentine scale see N. Bion, *The Construction and Principal Uses of Mathematical Instruments*, tr. E. Stone, 2nd edn. (London, 1758), p. 16; Cajori, *History of the Slide Rule*, pp. 15–16; Taylor, *Mathematical Practitioners*, pp. 346–7.

[145] BL, MS Add. 4280, fo. 97 (Pell to Leake, 7 Aug. 1645).

that time, I doubt they are not of your acquaintance.'[146]

Pell's own activities were evidently dominated by his mathematical projects; but he also had (or made) time for other intellectual pursuits. By the spring of 1641 he was hard at work learning Arabic: he studied standard works by Erpenius and Giggeius, took notes on the trilingual publication by Haak's friend Johannes Elichmann, *Tabula Cebetis graece, arabice, latine* (Leiden, 1640), compared different Latin translations of the first Sura of the Koran, and wrote out a complete Latin translation of the 'Sura of the Cow'.[147] The principle of 'nulla dies sine linea' dominated Pell's life; characteristic evidence of this comes in the form of a little memorandum he drew up, probably in the spring of 1640, listing forty separate uses for the '40 sheetes of browne paper' he had recently bought. 'Arabicke' was no. 40 on the list (preceded by 'Dutch' and 'Hebrew Aramic [*sic*]'); other numbered entries included 'For Library papers', 'For occasionall thoughts, miscellanea', 'Steganography', and two of the treatises promised in his *Idea*, the 'Comes mathematicus' and the 'Mathematicus αὐτάρκης'.[148]

The fourth item on this list – and the one that makes it possible to date the memorandum – was given as 'One for Harrisonian Index'. This was a reference to a kind of filing-system invented by a certain Thomas Harrison, an Oxford graduate and rector of a parish in Northamptonshire, who hoped to facilitate the extraction of units of knowledge or argument from the works of all previously published authors.[149] Such a project was close to the heart of pansophists such as Hartlib, who saw the systematization of existing knowledge as essential to the task of extending and perfecting it. As Hartlib enthused in his diary-notebook in the spring of 1640, 'Harrisons booke-Invention is nothing else but an excellent and the compleatest Art that ever yet hase beene devised of a commodious and perfect art or slight of excerpendi ['making excerpts'] ... Hee aimes by it to gather 1. all the Authors. 2. their Notions or Axioms. 3. Argumentorum or their whole discurses...' The method, involving 'slices of paper ... which can bee removed and transposed as one pleases' could be used to create 'a most Real and judicious Catalogue Materiarum out of all Authors'.[150] A letter from Hübner to Gronovius, written in March 1640, tells the story of this otherwise

---

[146] BL, MS Add. 4278, fo. 131ʳ. This is headed 'Intended in mine of October 13 to Mʳ Collins But left out'; the year is not given, but the set of letters to Collins of which this forms part dates from 1666–9, and Pell's departure to Holland in Dec. 1643 limits the possible years to 1666–7.

[147] BL, MS Add. 4377, fos. 1ʳ (note dated 10982, i.e. 24 Mar. 1641); 2–3ʳ (Elichmann); 7ʳ (Erpenius), 18 (A. Giggeius, *Thesaurus linguae arabicae*); 20ʳ (translations, 1ˢᵗ Sura); 51–2 (Sura of the Cow).

[148] BL, MS Add. 4423, fo. 354.

[149] Harrison has not previously been identified. He was born in 1595, matriculated at St John's College, Oxford, in 1610, and became Rector of Crick in 1635: see C. J. Robinson, *A Register of the Scholars Admitted into Merchant Taylors' School, from A.D. 1562 to 1874*, 2 vols. (Lewes, 1882), i, p. 49; Foster, *Alumni oxonienses*, ii, p. 662; A. G. Matthews, *Walker Revised* (Oxford, 1948), p. 279.

[150] HP 30/4/46 (Ephemerides, 1640, part 2).

obscure inventor. Apparently he had spent nearly twenty years perfecting his system, and in 1638 he obtained an audience with the King, who discussed it for two hours with him. Soon thereafter, however, he was sent to gaol for slandering a judge (the judge had ruled against the Crown in the Ship Money case, and Harrison had called him a 'traitor'). Hübner and Haak went to visit him in prison, and on a second visit they brought Hartlib and Pell: the four of them were then granted a full explanation of the system Harrison had devised.[151] Hübner's letter merely stated that it would take too long to explain, assuring Gronovius that Harrison would publish it once he was set free; no such explanation of his system was ever issued in print by Harrison, and the few modern scholars who have discussed it have depended entirely on Hartlib's brief references to it.[152] However, a copy of Harrison's account of his invention does survive in manuscript; although his name is not given, the identification is confirmed both by the nature of the invention and by the fact that the account ends with a passage celebrating Comenius's 'pansophist' project, describing Hartlib as a friend, and praising his 'recently' published book, *Conatuum praeludia*.[153] In essence, Harrison's 'ark of studies' was a cabinet with an alphabetical filing system, in which excerpts from different sources were written on small pieces of paper: those pieces were then stuck on hooks attached to metal plates that bore alphabeticized subject-headings. Traditionally, when scholars had written excerpts from books, they had merely copied them into other books of their own; the novel idea here was that each excerpt or topic was a unit that could be organized and reorganized in different ways. And, as Harrison emphasized, this device could be used – unlike the personal and private commonplace-books of traditional scholars – for a collective enterprise in the accumulation and processing of knowledge.[154]

As this episode helps to show, Pell was still very much bound up with Samuel Hartlib and his inner circle of friends, sharing not only their company but also

---

[151] BL, MS Sloane 639, fos. 85–86ʳ (Hübner to Gronovius, 19/29 March 1640). Haak also mentioned Harrison's scheme to Mersenne, who wrote back asking for more information in his letter of [2/] 12 May 1640: *MC* ix, p. 305. For details of Harrison's slander of the judge, Sir Richard Hutton, see Anon., *The Tryal of John Hampden, Esq. ... in the Great Case of Ship-Money* (London, 1719), pp. 235–7; K. Sharpe, *The Personal Rule of Charles I* (New Haven, Conn., 1992), pp. 195, 716–17, 728.

[152] See, for example, S. Clucas, 'In Search of "The True Logick": Methodological Eclecticism among the "Baconian Reformers"', in M. Greengrass, M. Leslie, and T. Raylor, eds., *Samuel Hartlib and Universal Reformation: Studies in Intellectual Communication* (Cambridge, 1994), pp. 51–74, esp. pp. 65–6.

[153] BL, MS Add. 41846, fos. 194–205, entitled 'Arca studiorum' (fo. 204ʳ: 'nuper'). This copy of Harrison's text is in a bound volume of Sir Kenelm Digby's papers. It appears that copies of this text were circulated by Hartlib: one copy was sent to Tassius in Hamburg, and was later published by Vincent Placcius in his *De arte excerpendi, vom gelehrten Buchhalten, liber singularis* (Stockholm and Hamburg, 1689), pp. 124–49. Placcius did not know the name of the author of this text, and merely expressed doubts as to whether it was by Tassius himself (p. 122). He admired the system, and suggested some improvements to it; for this, Placcius has himself been credited by Bernhard Fabian with the invention of the card-index filing system (*Der Gelehrte als Leser: über Bücher und Bibliotheken* (Hildesheim, 1998), p. 23).

[154] BL, MS Add. 41846, fo. 203ᵛ.

their pansophist enthusiasms. The point would hardly need emphasizing, were it not for the fact that a reader whose knowledge of Pell was confined to Pell's correspondence with Cavendish would get almost no inkling of this side of Pell's mental life. While Cavendish was certainly enthusiastic about mathematics and the 'new philosophy', one may sense that Pell was aware that this patrician patron was not attuned to the projects, the ethos, or the ideology of pansophy, with all its utopian-millenarian *Schwärmerei*. (Pell did give Hartlib snippets of information he had received from Cavendish, and even showed him one of Cavendish's letters; but when writing to Cavendish he did not, so far as we know, mention Hartlib at all.)[155] A certain compartmentalizing seems to have taken place, as Pell adjusted to the differences between his patrons. One almost symbolic expression of that difference was the fact that Cavendish was planning, in or around 1640, to bring Descartes to England, while at the same time Hartlib was extending a similar invitation to Comenius.[156] Cavendish's plan never came to fruition; Hartlib's, on the other hand, did, and there is no sign whatsoever that any attempt was made to interest Cavendish in Comenius during the latter's stay in England.

Most of what is known about Comenius's visit to England comes from accounts written by Comenius himself. For a long time Hartlib had been urging him to come to London; then, in July 1641, he sent Comenius an urgent message, 'Come, come, come: it is for the glory of God: deliberate no longer with flesh and blood!' What exactly had prompted this sudden plea is not clear, but it was probably connected with discussions in Parliament in June about the confiscation of Church funds and the reform of higher education: Hartlib may have thought that the time was ripe for setting up a state-funded Baconian-pansophical college. Comenius was somehow given the impression that he had been summoned by special command of Parliament, but this was not true; at most, Hartlib had activated those friends and contacts he had in both the Lords (John Williams, Bishop of Lincoln, Lord Mandeville, Lord Brooke) and the Commons (Pym, Culpeper) to support Comenius and his projects.[157] Arriving in London in late September 1641, Comenius was greeted by Hartlib, Dury (who had recently returned from Holland), Hübner, Haak, and Pell. Writing to his friends in Leszno in the middle of October, he commented on the 'new and wonderful invention' of Thomas Harrison; noted that a translation of his *Mutterschule* had been 'pre-

---

[155] Information about Hobbes, derived from Cavendish by Pell, appears in HP 30/4/76B (Ephemerides, 1641). A phrase from Cavendish's letter to Pell of 18 Dec. 1641 (Letter 7) was copied down by Hartlib: HP 13/345 (printed in Braddick and Greengrass, eds., 'Letters of Sir Cheney Culpeper', p. 154).

[156] All that is known of Cavendish's plan to invite Descartes (and Mydorge) is contained in A. Baillet, *La Vie de Monsieur Descartes*, 2 vols. (Paris, 1691), ii, pp. 67–8.

[157] See Young, *Comenius in England*, pp. 38–41, 52. John Selden has sometimes been included in lists of Hartlib's supporters in Parliament; but, as Prof. Toomer has pointed out to me, this claim rests on very slender evidence (a single account by Comenius of an encounter with Selden: ibid., p. 46), and there are otherwise no reasons to think that Selden was an enthusiast for Comenian pansophy.

pared here before my arrival, but it has not yet been printed' (this was presumably the translation which Pell had revised in 1635); and mentioned that he had been taken by Hartlib and Dury to dine with Bishop John Williams.[158] During subsequent weeks, Hartlib and Dury lobbied hard for the adoption of their educational and irenicist schemes, and for a while it seemed as if Parliament would allot them both funds and a building (perhaps the Savoy Hospital in London, or Chelsea College) for their proposed new institute of universal learning. To stimulate further interest in such schemes, several tracts and proposals were written and circulated by Hartlib, Comenius, and Dury, including one suggesting that a special series of graded school textbooks should be written by them, Hübner, and Pell.[159] However, the rapid deterioration in the political situation during the winter of 1641–2 undermined all these plans, and although Comenius remained in London throughout the first half of 1642, nothing further was achieved. He left on 21 June, just two months before the beginning of the Civil War.

The outbreak of war must have placed new strains on Pell: now his friends and patrons were to be not merely compartmentalized, but kept apart as mutual enemies. The Hartlib circle was overwhelmingly Parliamentarian in its sympathies; Cavendish, on the other hand, was the brother of one of the leading commanders of Royalist forces. Pell's correspondence with him seems to have broken off in 1642 (the last surviving letter from Cavendish in England dates from February); however, intriguing evidence exists of a secret visit by Cavendish to London some time in the first three months of 1643. In January of that year the astronomer and Capuchin friar Anton Maria Schyrle von Rheita wrote and circulated a letter announcing his discovery of what he thought were new satellites of Jupiter, Saturn, and Mars. The French philosopher Pierre Gassendi wrote a brief refutation; the two texts were published together in Paris in April 1643, and republished (with further additions) in Louvain later that year.[160] On 7/17 September 1644 Pell would write to Cavendish, in response to a query about Rheita's work:

> Sr William Boswell sent me ye friers Epistle, [I received it ye day y^t you went out of London *deleted*] the next morning after I received it I brought it to your Lodging, where I learned you were gone out of [*words deleted*] London ye day before. I returned [a pretty large judgement of it *altered to* to Sr William a large letter] shewing many grounds of suspicion y^t it was a meere fable & there was no such matter ... A coppy of my letter was sent to a frend in France who thereupon sent me Gassendus his censure.[161]

---

[158] Ibid., pp. 64–7.

[159] Webster, *Hartlib and the Advancement of Learning*, pp. 35–8. On Comenius's tract, *Via lucis*, written in late 1641 but printed only in 1668 (in Amsterdam), see Blekastad, *Comenius*, pp. 317–26.

[160] See A. Rheita ['Reita'], *Novem stellae circa Iovem, circa Saturnum sex, circa Martem nonnullae, A P. Antonio Reita detectae & satellitibus adiudicatae* (Louvain, 1643), esp. pp. 3–8 (Rheita, letter to Puteanus), 12–59 (Gassendi, 'Iudicium').

[161] Letter 19.

The details given here by Pell suggest that this episode took place before the publication of Gassendi's refutation in April 1643: Sir William Boswell, a man of keen scientific interests and wide-ranging intellectual contacts, would surely not have sent Pell a mere manuscript copy of Rheita's original letter if the printed version, with Gassendi's objections, had already been available at the time.[162] It follows, then, that Cavendish was in London at some time between January and April 1643 – despite the fact that, on 2 February 1643, his brother the Earl of Newcastle was impeached in the House of Commons and expressly excluded, as one of the most dangerous enemies of the state, from any future amnesty.[163] Pell's comments refer to Cavendish's 'Lodgings' in London; no doubt it would have been too risky for him to stay in his brother's large town house in Clerkenwell, and indeed he may have been travelling *incognito*. The reasons for such a visit to London can only be guessed at; but they may well have included a desire to rescue items of value (especially personal papers, books, and manuscripts) from his brother's house before the entire contents of the building were seized by the Parliamentary authorities. This was probably the occasion on which, as their later correspondence makes clear, so many of Cavendish's 'bookes & papers' were passed to Pell for safe-keeping.[164]

With his richest patron rendered virtually incapable of supporting him, and with all hopes of state-funded support for Hartlib's educational projects obscured, for the time being, by the fog of war, Pell's prospects must have seemed as bleak as ever. But he still had his admirers, both in England and abroad; and thanks to the efforts of one of these, Johann Moriaen in Amsterdam, he would finally obtain proper employment as a mathematical 'Artist'. The first step in what became Moriaen's victorious campaign was taken in September/October 1642, after he had been visited by one of the Elzevier brothers, the famous publishers in Leiden, who had long been planning an edition of the collected works of Viète: they were looking for manuscript items by Viète, and also inquiring about mathematical experts who might be able to annotate his works. Moriaen wrote

---

[162] Confirmation of this point can be found in a list compiled by Pell of books he owned, with, in the case of works he had received as gifts, the names of the donors: this contains the entry 'Novem stellae circa Jovem visae & de eisdem Petri Gassendi Judicium. Parisijs 1643. Mr Hart[*page torn*]' (BL, MS Add. 4394, fo. 113^r). He thus received a copy of the Paris edition not from Boswell but from Hartlib, and it must seem unlikely that Hartlib, who was in close contact with him, would have given him a copy if he had possessed one already. A copy of Pell's letter to Boswell criticizing Rheita's claims was passed (probably by Boswell) to the learned Princess Elizabeth at the Palatine court in The Hague; she then showed it to other individuals, including Descartes's friend Andreas Colvius, who wrote to her on [30 May/] 9 June 1643: 'The Capuchin Rheita must reply to the objections put forward by the learned Pell, otherwise we shall have to regard him as a fantasist or a cheat' ('Il faut que le cappucin de Rheita responde aux difficultés que met en avant le docte de Pell, ou autrement nous le devons tenir pour fantastique ou imposteur': C. L. Thijssen-Schoute, *Nederlands cartesianisme*, 2nd edn. (Utrecht, 1989), p. 566). This provides a *terminus ad quem* for Pell's letter.

[163] G. Trease, *Portrait of a Cavalier: William Cavendish, First Duke of Newcastle* (London, 1979), p. 104.

[164] See Letters 90, 91.

PART I: THE LIFE OF JOHN PELL   99

to Hartlib on [22 September/] 2 October, describing this meeting and saying that he had given Elzevier Pell's name.[165] A week later he wrote again, obviously worried that Hartlib and the self-effacing Pell had failed to get the point: he urged Pell to volunteer his services for the project, insisting that 'Here Mr Pell has an opportunity to become known, and to set the furtherance of his career on a good foundation.'[166] At Hartlib's insistence, Pell then wrote a long letter (addressed to Hartlib, but evidently intended to be translated and forwarded to Amsterdam) in which he discussed Viète's works (mentioning in passing a manuscript he possessed of a section of one of Viète's unpublished texts, the 'Harmonicon coeleste'), suggested the names of competent commentators (including Fermat, Roberval, Tassius, Oughtred, Descartes, Golius, and 'Mr Warner, if his great age did not hinder him'), and ended with a characteristic reference – combining modesty and ambition – to what he could himself contribute to this field of mathematics:

> Lastly Mr Morian addes, 'or if he have anything else tending to the art it selfe'. I know not whether the printers desire this or not, because this might encrease the price of the booke and yet displease many lovers of Vieta: But perhaps it is onely his owne motion both because he hath seene my letter to you [*sc.* the *Idea*] which you caused to be printed just this time 4 yeares; And also because he knowes that Analysis is not the onely way to finde truths nor Algebra the onely field wherein to exercise Analysis nor is Vieta's way the only Algebra, nor hath Vieta perfectly taught all that belongs to his way. As perhaps I should have cleerely shown ere this time, had I not beene hindred by the continuance of innumerable diversions...[167]

It appears that Moriaen circulated this text as widely as he could. (Three years later, when Pell met the mathematician and orientalist Jacob Golius in Amsterdam, he reported their conversation in a letter to Cavendish as follows: 'I asked

---

[165] HP 37/112B–113A. The Elzeviers' project had first been announced in 1639, when they issued a broadsheet appealing for unpublished manuscripts: see the photograph in *MC* vii, opposite p. 109. The edition was eventually undertaken by F. van Schooten: F. Viète ['Vieta'], *Opera mathematica* (Leiden, 1646). This was not a complete edition of Viète's writings, and a further volume was planned; but the poor sales of this first volume precluded any further publication (see the letter from Nicolaus Heinsius to Carlo Dati in G. Targioni-Tozzetti, *Notizie degli aggrandamenti delle scienze fisiche accaduti in Toscana nel corso di anni LX. del secolo XVII*, 3 vols. (Florence, 1780), i, pp. 500–1).

[166] HP 37/114A, Moriaen to Hartlib, [29 Sept./] 9 Oct. 1642: 'Hie hat nun herr Pell [>gelegenheit] sich bekand zue machen vnd seiner promotion ein gut fundament zue legen.'

[167] BL, MS Add. 4280, fo. 206. This letter is printed, in a modernized and slightly inaccurate form, in Vaughan, *Protectorate*, ii, pp. 347–54 (and thence reprinted in *MC* xi, pp. 307–12, with the incorrect statement that the manuscript is unknown). The fragment of the 'Harmonicon coeleste' owned by Pell is BL, MS Add. 4417, fos. 26–37; on the rather uncertain history of this manuscript see N. Malcolm, 'Hobbes, the Latin Optical Manuscript, and the Parisian Scribe', *English Manuscript Studies*, ed. P. Beal and J. Griffith, 13 (2003), forthcoming.

him whence he came to have so great an opinion of [my abilities *deleted* >me.] He told me he had seene letters of my writing & specimina inough & saith he he y^t hath read your counsell to ye Elzevirs for ye printing of Vieta's workes, needes no other argument of your abilities.')[168] By thus establishing Pell's credentials as an expert in the field of algebra, Moriaen was preparing the way for another attempt at Hortensius's chair of mathematics – which had still not been filled – at the Amsterdam Athenaeum. The opportunity came in mid-October 1643, when Moriaen had another meeting with Albert Burgh, the senior civic official who served as one of the Athenaeum's 'Curators' or governors. Burgh told him that he was keen to see Pell elected to the post, but that it would be difficult to arrange unless Pell first came to Amsterdam to make himself known.[169] And so it was that in December 1643 Pell travelled to the Netherlands to seek his fortune. He would not return to England for almost nine years.

Although Pell had the friendship and guidance of Moriaen to look forward to, this was in some ways a peculiarly lonely expedition: his wife and children all remained in London. Admittedly, we know very little about the life that Ithamar led, even when she was with her husband. To the fifth and sixth children born in 1640 (one of whom would die in 1643), a seventh was added (Bathshua) in October 1642; the eighth (John) was born in February 1644, just two months after Pell's departure.[170] Ithamar did live not far from her sister Bathsua Makin, in Westminster, and it might be supposed that she benefited from Bathsua's support. But relations between them had in fact been very badly strained. In 1640 Francis Rogers, the second husband of their other sister Mespira, took Richard Makin (Bathsua's husband) to court; Ithamar was summoned as a witness, and testified on Rogers's behalf:

> That she doth know that the dep[onen]^t Richard Makins was about a yeare agone a prisoner in the Kings Bench & that the now p[lainti]^{ff} Francis Rogers was bound for his true Imprisonment for a while & for that the s[ai]d p[lainti]^{ff} would not soe stand bound anie longer he the s[ai]d dep[onen]t putt the bond of 40^{li} in this Inter[pellation] menconed [>in suit att the Common law] against the s[ai]d p[lainti]^{ff} Francis Rogers & caused him to be arrested thereupon.[171]

---

[168] Letter 35.

[169] HP 37/116A, Moriaen to Hartlib, 15 Oct. 1643. Haak later wrote that Boswell was also involved in this attempt to arrange Pell's appointment to Hortensius's chair (Aubrey, *'Brief Lives'*, ii, p. 130); but Haak's memory seems to have run together the two separate attempts, and there is no mention of Boswell's involvement this time in Moriaen's letters.

[170] Burke, *Memorials of St Margaret's*, pp. 180 (Bathshua), 188 (John), 603 (burial of William, 21 Mar. 1643). The fourth child, James, was buried on 23 Apr. 1641 (ibid., p. 596); his baptism is not recorded in the records of St Margaret's. He must have been born in Sussex between late 1635 and early 1639.

[171] PRO, C24/654 (Town depositions, Chancery), deposition by Ithamar Pell, 14 Nov. 1640.

The details of the case are rather obscure, but what is all too clear is the strong hostility to Makin on Rogers's side of the case: the second witness described Makin as 'generally reputed or knowne to be a man of corrupt & unconscionable condicion', and the third called him 'a very hard conscience & a badd dealing man'.[172] It seems that relations between Ithamar and Bathsua were patched up eventually: the choice of Bathsua's name for the Pells' seventh child in October 1642 may have been a symbolic gesture of reconciliation. Pell's occasional references to Bathsua Makin in his later letters to Ithamar would, all the same, be suspicious and apprehensive. But if any real sense of conflict lingered on, it was Ithamar, left on her own in London, who would have to deal with it.

---

[172] Ibid., depositions by John Guillim and Thomas Meade. Guillim also commented on an earlier dispute between Makin and Robert Wood, the Keeper of the King's Cormorants: to gain the reversion of Wood's office Makin had promised him money and 'did enter into covenants wth the s[ai]d Wood to performe diverse other thinges concerning the same wch the s[ai]d Makins in noe sort p[er]formed'; as a result, Makin had been 'banished the Court'. This puts a very different complexion on that dispute, which was hitherto known about only from Makin's subsequent petition in which he presented himself as the wronged party (cited, and accepted, in Teague, *Bathsua Makin*, p. 48).

4

Amsterdam, Breda, 1643–1652

John Pell had little idea of what awaited him in Amsterdam; apparently he thought his trip was to be merely a brief visit, for the purpose of being interviewed. As he wrote to Cavendish in August 1644: 'I came over hither in December last, not bringing any of my bookes or papers with me: Nor hardly clothes, for the worst I had, seemed good enough to weare o'shipboard. I then thought not to stay heere above a fortnight. Nor did I imagine that, before I gat out of this towne againe, I should be persuaded to clime the Cathedra...'[1] No doubt Moriaen took the necessary steps to introduce him to Albert Burgh, and to others with influence over the Amsterdam Athenaeum. But the authorities of the Athenaeum did not meet to consider the matter until April; how Pell spent his time during the early months of 1644 is not known. Moriaen was absorbed in his alchemical studies, and was closely attached to the controversial German Paracelsian chemist Johann Rudolf Glauber, who had moved into a house with a large laboratory in Amsterdam in 1643.[2] So it may be that Pell was made a participant in Moriaen's chemical and metallurgical studies there – though these would never become a major interest of Pell himself.

Finally, on 27 April 1644 the Burgemeesters of Amsterdam agreed to appoint Pell to the Chair of Mathematics at the Athenaeum, of a period of one year in the first instance, and at the salary enjoyed by the previous incumbent, Hortensius (a salary which, at 600 guilders per annum, was much lower than that of the other professors).[3] Whether there was any competition for the post is not known; news of an impending appointment seems to have got out only at the last moment, and the name of an alternative candidate from Leiden reached one of the Athenaeum's staff only after Pell's appointment was made.[4] At some time in May he made his inaugural oration: a brief 'Extract of a Letter from Mr Pell' among Hartlib's

---

[1] Letter 14. In his petition to the Council of State in 1652, Pell would write that he was 'called' in 1643 'from London to Amsterdam, there to professe publickly the Mathematicks' (BL, MS Lansdowne 751, fos. 71^r, 72^r; draft in BL, MS Lansdowne 754, fo. 465^r); this phrasing seems to involve a slightly misleading elision of the sequence of events.

[2] Young, *Faith, Medical Alchemy and Natural Philosophy*, pp. 186–9.

[3] D. J. van Lennep, *Illustris Amstelodamensium Athenaei memorabilia* (Amsterdam, 1832), pp. 110, 115; Brugmans, Scholte, and Kleintjes, eds., *Gedenkboek van het Athenaeum*, pp. 227, 233. The others were paid 2500 guilders (Vossius) and 1500 guilders (Barlaeus): F. F. Blok, *Caspar Barlaeus: From the Correspondence of a Melancholic* (Assen, 1976), p. 11.

[4] C. Barlaeus, *Epistolarum liber* (Amsterdam, 1667), p. 913, Barlaeus to F. Spanheim, from Amsterdam, 28 June 1644 ('Quem mihi commendasti Mathematicum, eodem me momento accessit, quo accedebat clarissimus Pellius, jam à D.D. Curatoribus ad professionem vocatus').

papers summarizes, in the third person, his own account of the event:

> Hee went in the streets of Amsterdam betweene 2 of the Burgimasters to the Publick [Audotory?] which hee found filled with the Magistrats Ministers Phis[ic]ians Sevillians Professors and all sort of Learnd Men Iewish Rabins and all expecting his coming to whome he made his first oratione in Latine & was brought backe to the same house where he put on his [goune?] and there having drunke a cup of wine or 2 with the Burgimaster theiy left him having first taken order with the master of the House to speake to the Ministers and Elders of the English Church to appoint him his place there.[5]

And by [23 May/] 2 June Moriaen was able to report, in a letter to Hartlib:

> By now he has not only performed his first and second inaugural orations in the presence of two Burgomasters, three Councillors, the two Governors of the Athenaeum, Professors, Doctors and ministers ... to great approbation, but today he has also delivered his Introduction or prefatory remarks to his future lectures, and in all this he has done more than was expected of him. Tomorrow, God willing, he will begin to expound Diophantus.[6]

And so Pell's new life as a Professor of Mathematics began. For the next two years (until he moved to a similar post at Breda) he would deliver a public lecture, in Latin, each week-day morning during term-time.[7]

Although Pell enjoyed a professorial title, the institution he had joined was something much less than a university. When the City Corporation of Amsterdam first discussed the creation of the Athenaeum in 1629, the reason it gave was that boys who completed their studies at the city's Latin schools were unprepared for university tuition: in particular, they needed a better grounding in philosophy, before they could embark on courses in the higher faculties (law, medicine, theology) of the universities. The Athenaeum was thus set up as an unusually modest type of 'École Illustre', with a pre-university propaedeutic role

---

[5] HP 43/74A. The phrase about Pell's 'place' in the English Church refers, apparently, only to his acceptance as a member of the congregation there. This was a 'Reformed' church, Calvinist and Presbyterian.

[6] HP 37/117A: 'Nunmehr hatt Er nicht allein Orationem Inauguralem primam et secundam coram 2 Consulibus tribus consularibus, 2 scholarchis Professoribus Doctoribus ministris ... magna cum laude gehalten sondern auch heute sienen Introitum oder Proeomium ad lectiones futuras gethan vnd in diesem allem mehr gethan als man von ihm erwarttet. Morgen geliebts Gott wird Er den diophantum beginnen zue erklären.' Moriaen also noted that Haak had witnessed these events: Haak was briefly in the Netherlands, on his return from a diplomatic mission to Denmark (see Barnett, *Theodore Haak*, p. 69).

[7] See Pell's comment that he was required to 'reade ... 5 houres in a weeke' (Letter 11). Part of the text of one lecture survives in a fair copy among Pell's papers: BL, MS Add. 4421, fos. 103–4.

to play: when it opened in 1632, it had just two professors, one in philosophy and the other in history. Only gradually thereafter would it extend into other areas of teaching covered by the universities; it would not itself acquire the official status of a university until 1877.[8]

The Professor of Philosophy was Caspar Barlaeus (Caspar van Baerle) (1584–1648), a brilliant humanist who was regarded as one of the finest neo-Latin poets in Europe.[9] His lectures were on Aristotle and on classical philosophy in general. The Professor of History was Gerardus Joannes Vossius (Geeraard Jan Voskens) (1577–1649), a polymathic classical scholar, philologist, and theologian: his initial course of history lectures began with the creation of the world and ended with Abraham.[10] He also lectured on Greek literature, and, in later years, patristics. These two eminent scholars were old friends, having taught together at the Statencollege (a seminary for Calvinist ministers) in Leiden. Barlaeus was a Remonstrant (Arminian), and Vossius had come under suspicion from hardline Calvinists because of his close links with the Remonstrant intellectual Hugo Grotius; both Barlaeus and Vossius had been dismissed from the Statencollege during the clampdown that followed the Synod of Dort in 1619.[11] The establishment of the Athenaeum in Amsterdam was opposed by the authorities at Leiden not only out of fear of a nascent rival university, but also because they thought it would become the Remonstrants' intellectual stronghold; Amsterdam had recently introduced a policy of tolerance towards the Remonstrants, and for a while Vossius hoped to recruit Grotius himself as the Athenaeum's third professor.[12]

By the time Pell joined it, the Athenaeum had extended its teaching into two new areas: mathematics and law. The Professor of Law, Joannes Cabeljau (Cabillavius) (1600/01–1652) was installed in 1640. An advocate at the Court of Holland at The Hague, he was probably chosen to give instruction in practical as well as theoretical jurisprudence; but he was not a success as a teacher, and

---

[8] See Brugmans, Scholte, and Kleintjes, eds., *Gedenkboek*, pp. 25–7; Dibon, *La Philosophie néerlandaise*, pp. 220–2; C. L. Heesakkers, 'Foundation and Early Development of the Athenaeum Illustre at Amsterdam', *Lias: Sources and Documents relating to the Early Modern History of Ideas*, 9 (1982), pp. 3–18.

[9] See the biography by Blok, *Caspar Barlaeus*, and K. van der Horst, 'A "Vita Casparis Barlaei" written by himself', *Lias: Sources and Documents relating to the Early Modern History of Ideas*, 9 (1982), pp. 57–83.

[10] The best modern study of Vossius is C. S. M. Rademaker, *Leven en werk van Gerardus Joannes Vossius (1577–1649)* (Hilversum, 1999): on the history lectures, see p. 192.

[11] Ibid., pp. 94–6, 116–24.

[12] Brugmans, Scholte, and Kleintjes, eds., *Gedenkboek*, p. 25; Dibon, *La Philosophie néerlandaise*, pp. 223–5; Rademaker, *Leven en werk van Vossius*, p. 196. A Remonstrant seminary was opened in Amsterdam in 1634: Rosalie Colie has mistakenly claimed that this was the same institution as the Athenaeum (*Light and Enlightenment: A Study of the Cambridge Platonists and the Dutch Arminians* (Cambridge, 1957), pp. 19–20).

was dismissed in 1646.[13] Pell's predecessor, Hortensius, who began lecturing in mathematics in 1634 and was given the title 'Professor' in the following year, was not a good lecturer either, and sometimes had only two auditors in the lecture-hall; the only popular course he gave was on navigation.[14] This last detail was a reflection of the fact that not only were many of the students at the Athenaeum the sons of merchants, but the lectures were open to the general public: burghers of Amsterdam made up the majority of the audience in the early years, and it was with good reason that Barlaeus's inaugural oration, *Mercator sapiens*, was about the role of moral philosophy in commerce.[15]

In these circumstances, it must be wondered whether Pell was not pitching his tuition at too high a level when he chose to lecture on the ancient Greek arithmetician Diophantus. It is not known what other courses he gave during his two years at the Athenaeum; Pell's papers contain almost nothing relating to his work there, apart from a fragment of the text of one lecture, and the Athenaeum's own records from this period do not survive.[16] One piece of evidence suggests that he also gave some tuition in astronomy (which had been Hortensius's special field): many years later, the minutes of a meeting of the Royal Society would record that Pell possessed 'a wooden hoop made by his direction in the year 1645, of twenty inches broad, representing the ecliptic, and at Amsterdam by Mr. Bleau [*sic*] lined with as much of the paper of Mr. Bleau's largest globes, as comprehends all the fixed stars within 10 deg. of the ecliptic on either side' – an object which he had designed partly in order 'to help young observators'.[17] (Astronomy was evidently an interest he kept up during this period: his papers include details of an eclipse of the moon observed by him in Amsterdam in February 1645, as well as notes on 'Phaenomena lunaria', dated December 1646.)[18] In addition to the public lectures, private tuition was also given to individual students by the professors at the Athenaeum. It is possible that Pell gave lessons on this basis to no fewer than three people who would later become well-known

---

[13] *NNBW* iv, col. 391; Heesakkers, 'Foundation and Early Development', p. 13.

[14] Rademaker, *Leven en werk van Vossius*, p. 196.

[15] Ibid., p. 192; Blok, *Caspar Barlaeus*, p. 18. On the *Mercator sapiens* see Dibon, *La Philosophie néerlandaise*, pp. 227–31.

[16] For the fragment see above, n. 7.

[17] Birch, *History*, i, p. 510.

[18] BL, MS Add. 4431, fos. 125^r, 371.

mathematicians: Johann Hudde, Nicolaus Mercator, and Gerard Kinckhuysen.[19] But the evidence for his tuition of them is largely circumstantial or indirect; otherwise, none of his private pupils is known by name, apart from one young Englishman who happened to pass through Amsterdam and had lessons with him for only five weeks.[20]

Nor is it possible to tell what kind of social life Pell enjoyed with his colleagues. Vossius and Barlaeus were liberal-minded and sociable men, with a wide range of acquaintances; in the 1630s they had been members of the 'Muiderkring', a circle of intellectuals, poets, and musicians (including the eminent statesman Constantijn Huygens, and the poetess Maria Tesselschade) that regularly met at Muiden castle, the residence of the poet Pieter Hooft.[21] When Barlaeus's son married a rich heiress in July 1644, Barlaeus told Constantijn Huygens in his letter of invitation that the wedding party would include 'knights, lawyers, governors of the Indies ... and professors too'; it would be pleasing to think that this last category included his new English colleague.[22] Vossius seems to have admired Pell's talents and befriended him. Writing to John Selden in January 1645, he mentioned that he had received news from London 'from my very close colleague Pell'; in a letter to John Sadler (a supporter of Hartlib in London) in September 1646, he referred to the recent departure for Breda of 'John Pell, our mutual friend', and commented regretfully that 'I would prefer to enjoy the friendship of that most learned man here in Amsterdam'; and in a book published posthumously in 1650 he wrote that Pell had been 'my dearest colleague', and mentioned that he himself had often attended Pell's brilliant lectures

---

[19] In a note headed 'Intended to be sent April 17 [1667] to Mr Collins but left out' Pell wrote of Hudde: 'He was one of my hearers sometimes at Amsterdam. Two and twenty yeares agoe He accompanied Mr Golius to my Lodging...' (BL, MS Add. 4278, fo. 131^r; on this meeting with Golius see below, at n. 81). That Pell gave Hudde some private tuition is suggested both by the fact that Hudde was known to Pell by name, and by the fact that he went with him to his 'Lodging'. On Hudde, who was 17 years old at the time of this meeting, see the entry in *NNBW* i, cols. 1172–6 (where, however, this period of study at the Athenaeum is not mentioned). The evidence for the tuition of Mercator (who was born in c.1619, in Holstein, and may thus be described as Danish or German) is that in July 1666 Pell wrote to Collins that he had known Mercator 'above 22 yeares' (BL, MS Add. 4278, fo. 118^v). Given Pell's habitual accuracy in chronological matters, this is strong evidence that he got to know Mercator in the first half of 1644; but whether Mercator was a private pupil or merely an auditor is not known. Reasons for connecting Kinckhuysen with Pell in the period 1643–6 (drawn from the internal evidence of Kinckhuysen's later published work) are presented in J. A. Stedall, *A Discourse Concerning Algebra: English Algebra to 1685* (Oxford, 2002), p. 147.

[20] On this pupil (Godolphin) see Letter 50.

[21] See H. W. van Tricht, *Het leven van P. C. Hooft* (The Hague, 1980), pp. 171–94; P. van der Sluijs, 'Constantijn Huygens en de Muiderkring', in H. Bots, ed., *Constantijn Huygens: zijn plaats in geleerd Europa* (Amsterdam, 1973), pp. 188–309.

[22] Blok, *Caspar Barlaeus*, p. 102 ('equites, advocatos, Indiarum profectos ... etiam professores').

on Diophantus.[23]

One thing at least is clear: Pell was not closeted in a rigidly Calvinist social world. Vossius and Barlaeus had contacts with Remonstrants, Catholics, and Jews (both were friends of Menasseh ben Israel).[24] The wider circle of Pell's contacts, generated by Moriaen, Hartlib, and Dury, would have consisted mainly of people who supported Dury's irenicist project – and these were a very varied group, ranging from liberal-minded Calvinist ministers such as Godefroid Hotton and Johann Rulice to the millenarian Peter Serrarius, a member of the non-ecclesial 'Collegiant' movement.[25] Pell also had some contact with the Socinian Baron Wolzogen, who spent some time in the mid-1640s in Amsterdam; years later Pell would tell Hartlib about 'A curious and accommodatious Weaving-Instrument ... which the Swedish Baron Wolzogens wife did vse in Holland to get her subsistence by'.[26] And soon after Pell's move to Breda, Wolzogen wrote to him from Amsterdam sending greetings to Stanisław Lubieniecki: from this it appears that the young Lubieniecki, who would later become famous for his writings on astronomy as well as Socinian history, went to visit Pell on his journey from Poland (via Elbing and Amsterdam) to France.[27] In 1646, just before his move to Breda, Pell had the pleasure of renewing his friendship with Christian Ravius, who stayed in Amsterdam for several months.[28] And it was probably during his time in Amsterdam that he made the acquaintance of the Dutch algebraist Jan Stampioen, who gave tutorials in mathematics both to Princess Elizabeth of the Palatinate, and to the young sons (Christiaan and Constantijn)

---

[23] Colomesius, ed., *Gerardi Vossii epistolae*, i, pp. 437 (to Selden, [23 Dec. 1644/] 2 Jan. 1645: 'ex conjunctissimo Collega Pellio' – the news being of successful efforts in Parliament to get Vossius paid an emolument), 451 (to Sadler, [6/] 16 Sept. 1646: 'Clarissimo viro, communi amico, Joanni Pellio ... Mallem hic doctissimi viri amicitiâ frui'); G. J. Vossius, *De quatuor artibus popularibus, de philologia, et scientiis mathematicis* (Amsterdam, 1650), 3rd part ('De universae matheseos natura & constitutione'), separate pagination, p. 38 ('collegâ carissimo').

[24] Blok, *Caspar Barlaeus*, p. 53; Rademaker, *Leven en werk van Vossius*, pp. 231–2.

[25] On Hotton (1596–1656), minister of the Walloon Church in Amsterdam, and Rulice (Rulicius, Rülz) (1602–66), minister of the German Reformed Church there, see the many references in Turnbull, *Hartlib, Dury, Comenius*, and Blekastad, *Comenius*. Moriaen's lobbying campaign on Pell's behalf in 1639 had been conducted jointly with Rulice: see HP 37/50A (Moriaen to Hartlib, [16/] 26 Dec. 1639). On Serrarius, who would later enjoy close relations with both Comenius and Spinoza, see E. G. E. van der Wall, *De mystieke chiliast Petrus Serrarius (1600–1669) en zijn wereld* (Leiden, 1987).

[26] HP 28/2/20A (Ephemerides, 1651, part 2). Wolzogen was 'Swedish' in the sense that he later lived in Stockholm under the protection of Louis de Geer (who had major trading, manufacturing, and mining interests in Sweden).

[27] BL, MS Add. 4279, fos. 309–10 (Wolzogen to Pell, 8 Sept. 1646); J. Tazbir, *Stanisław Lubieniecki: przywódca ariańskiej emigracji* (Warsaw, 1961), pp. 41–3.

[28] Ravius had moved to Amsterdam from Utrecht; see his letter of 23 May 1646 to Johannes Coccejus, in J. Coccejus, *Opera ἀνέκδοτα theologica et philologica*, 2 vols. (Amsterdam, 1706), ii, p. 666. Pell had evidently been in contact with Ravius in 1644, receiving his manuscript of Apollonius from him by Oct. of that year (see Letter 23), though whether Ravius had spent time with him then in Amsterdam is less clear.

of Constantijn Huygens.[29]

Fortunately there is one major source of information about Pell's life during this period: his correspondence with Cavendish. Contact between them was resumed in July 1644, soon after Cavendish and his brother had fled from England (in the aftermath of the battle of Marston Moor) and settled in Hamburg. Pell hastened to offer to help Cavendish 'If it please you to commaund me any service'; Cavendish did ask for assistance in obtaining newly published books, and would occasionally request favours such as tuition in mathematics for the sons of his friends.[30] But the nature of their relationship had definitely changed. This was not only because of Cavendish's straitened finances ('You will excuse me', he wrote in December, 'that I requite not your favoures for the present as I desire I can onelie yet acknowledge them, but shall heereafter by the grace of God endeavoure to requite them to my power').[31] Pell's status had been transformed: no longer an obscure mathematical practitioner or technician in Cavendish's part-time private employment, he now occupied a Chair at a famous institution. Even in his first letters to Cavendish in the summer of 1644, a new tone of intellectual self-confidence is audible, as he surveys the contemporary world of mathematical research; indeed, he could not resist the temptation to cite the passage in praise of himself which Mersenne included in his recently published work, *Universae geometriae synopsis*.[32] From this time onwards, although a sense of the difference in their social status was never absent from Pell's letters, his correspondence with Cavendish had much more of the character of a dialogue between two participants in the Republic of Letters, as they exchanged news and opinions about recent publications, current research, and their other encounters with learned men. For his own benefit as well as his correspondent's, Pell encouraged Cavendish to make contact in Hamburg with the scientist and philosopher Joachim Jungius, whose writings – especially his published treatise on logic –

---

[29] On Stampioen see Letter 86, n. 3, and *NNBW*, ii, cols. 1358–60; he lived in The Hague from 1638 to 1645, then in Leiden, so he met Pell most probably on a visit to Amsterdam. The evidence of their acquaintance is in a text written by John Collins in 1670: 'A very learned Algebraist [*sc.* Pell] affirmeth that he hath for above 30 yeares used to solve Aequations by tables, and that he gave unto Stampioen a Dutch Algebraist, all the rootes of an affected Aequation of the 6th power thus found' (Worcester College, Oxford, MS 64, unfoliated, 1st item in final section; cf. a similar statement in another 'discourse' by Collins: Royal Society, MS Classified Papers, XXIV (13)). Pell was familiar with some of Stampioen's work before he went to the Netherlands, having received two of his treatises from Boswell in 1640: see above, ch. 3, n. 69.

[30] Letter 11 (quotation).

[31] Letter 27.

[32] Letter 16, at n. 25.

he enormously admired.³³ (Pell had previously had some contact, via Hartlib, with Jungius's colleague Tassius, but does not seem to have communicated with Jungius directly at that time.) Cavendish had his own links with Paris (he had resumed correspondence with Mersenne by September 1644), and was able to pass on information about intellectual life there; in the spring of the following year he and his brother settled in the French capital, which meant that he became able to supply regular news of such luminaries as Gassendi, Roberval, and Hobbes. And Pell, in return, could report meetings with famous intellectuals such as Descartes and Golius, and offer details of recent or impending publications in both the United Provinces and the Spanish Netherlands.

Something that did not change was Cavendish's constant urging that Pell should publish his mathematical work. 'I praye you let me knowe whether Mr Warners Analogicks be printed', he wrote in his first letter from Hamburg; 'I hope you proceed in your intended worcke of Analyticks which if you doe not, I beseech you doe, & finish it with all conuenient speed.'³⁴ Pell's reply to the first query was not encouraging; after Warner's death (in March 1643) his manuscripts had been inherited by a relative, a merchant who had since gone bankrupt, and Pell feared that the papers had now been dispersed and lost. (In fact a large part of the table of antilogarithms, together with other papers, survived, and would later pass, as we shall see, to the divine and mathematician Herbert Thorndike.) As for Pell's promised work on algebra, his manuscript draft was with all his other papers in London, but he was now thinking seriously of re-composing and publishing it in Amsterdam. In passing, he gave one of his rare explanations of his reasons for non-publication, an explanation grounded not so much on modesty as on perfectionism: 'I have thought nothing elaborate [*sc.* 'worked-out'] enough to be printed, till it were so complete, that no man could better it and did therefore so long keepe my name out of the presse.'³⁵ Perhaps such reasoning helps to explain his continuing failure to publish this work. Another part of the explanation must be that he was working on too many large-scale projects at once. As he informed Cavendish in August 1644, he was now planning 'a New Edition of Diophantus wherein I would correct ye translation & make new Illustrations after my manner'; its publication had already been arranged with

---

³³Letter 16: 'having seene something of Jungius his writing wherein he seemed to me to make [>a truer &] better use of his reason ... y^n any other sonne of Adam, All other writers must pardon me if I professe to expect more solidity in Jungius his writings than in any other man now living.' Hartlib had received a copy of the 2nd edn. of Jungius's *Logica hamburgensis* (Hamburg, 1638; 1st publ. Hamburg, 1635), and had lent it to Pell: some of Pell's notes on it, dated '10025' [18 Aug. 1638] are in BL, MS Add. 4420, fo. 23^r. Hartlib recorded the following comment by Pell in early 1639: 'Cartes, Bisterfeld Comenius begin their philosophating a priori. But they will finde themselves deceived. Iungius goes more warily and does a posteriori not caring so much to teach as first to find out the truth that may not bee gain-sayed' (HP 30/4/3A (Ephemerides, 1639, part 1)). See also Clucas, 'In Search of "The True Logick"', pp. 62–3.

³⁴Letter 12.
³⁵Letter 14.

the famous Amsterdam publishing house of Johan Blaeu.[36] Two months later Pell mentioned another task of even greater complexity: he was also proposing to produce a new translation – from an Arabic manuscript, which his friend and admirer Christian Ravius had allowed him to borrow – of books 5, 6, and 7 of the *Conics* of Apollonius. The enthusiasm of his publisher was encouraging: 'Dr Blaeu [>tells me he] is ready to print Diophantus Apollonius or what I will & how I will'; but even so he needed 'some time to thinke how to satisfy ye Expectation of [>my] publishing Diophantus & Apollonius [this yeere *deleted*] within this halfe yeere'.[37] Neither work would in fact see the light of day, either within that half year or in any year thereafter.

But perhaps the greatest single impediment to the completion of any or all of these projects was Pell's increasing absorption in a wearying feud with the elderly Danish mathematician Longomontanus.[38] Born in 1562, Christian Sørensen Lomberg or Langberg – hence 'Longomontanus' – had worked in his youth as an assistant to Tycho Brahe, both in Denmark and in Bohemia (with Kepler). He had been appointed Professor of Mathematics at Copenhagen University in 1607, and had then been given a Chair in Astronomy (his real specialism) in 1621; he published a famous summary of Tychonic astronomical theory, *Astronomia danica*, in 1622, and supervised the building of the University's observatory in 1637–42. He also served for many years as the University's Rector Magnificus.[39] His consuming passion as a mathematician was squaring the circle: as he proudly stated in a book published in 1643, he had been working on this topic for 36 years.[40] By early 1644 he had published no fewer than eleven separate 'cyclometrical' treatises.[41] But it was his twelfth publication on that subject, a little book entitled *Rotundi in plano, seu circuli, absoluta mensura, duobus libellis comprehensa*, published by Blaeu in Amsterdam in the summer of 1644, that attracted

---

[36] Letter 14.

[37] Letter 23.

[38] A brief account of this dispute was given by E. J. Dijksterhuis, 'John Pell in zijn strijd over de rectificatie van den cirkel: bijdrage tot het jubileum van de Universiteit van Amsterdam', *Euclides: tijdschrift voor de didactiek der exacte vakken*, 8 (1931–2), pp. 286–96; this is largely superseded by the detailed account in van Maanen, 'The Refutation'.

[39] There is no full-length study of Longomontanus, and his personal archive has not survived. See the entry in the *Dansk biografisk leksikon*, 3rd edn, 16 vols. (Copenhagen, 1979–84), ix, pp. 109–10; see also H. F. Rørdam, *Kjøbenhavns Universitets historie fra 1537 til 1621*, 4 vols. (Copenhagen, 1868–74), iii, pp. 146, 304, 336–8, 477; H. F. Rørdam, 'Aktstykker til Universitets historie i tidsrummet 1621-60', *Dansk magazin, indeholdende bidrag til den danske histories og det dansk sprogs oplysning*, 5th ser., vol. 1 (1887–9), pp. 36–72, 133–57, 198–222, 332–55, and vol. 2 (1889–92), pp. 1–28, 126–52, 217–42, 320–49, esp. vol. 1, pp. 59–71 and vol. 2, p. 26; and J. R. Christianson, *On Tycho's Island: Tycho Brahe and his Assistants, 1570-1601* (Cambridge, 2000), pp. 131–2, 205, 313–19.

[40] C. S. Longomontanus, *Problema quod ... absolutam circuli mensuram praestat* (Copenhagen, 1643), sig. A4^r.

[41] These are listed in C. S. Longomontanus, *Ἐνέργεια proportionis sesquitertiae pro certis figuris circulo adscriptis, & per numeros determinatis, data circuli diametro* (Copenhagen, 1644), sigs. B3^v–B4^v.

Pell's surprisingly polemical attentions.

The origins of the dispute were recounted by Pell several times: in a letter to Cavendish in August 1644, in a letter to John Leake one year later, and in the major anti-Longomontanus text which he published in 1647.[42] What had happened one day in July was that on a visit to Blaeu's shop, to discuss his planned edition of Diophantus, Pell saw Longomontanus's book listed in Blaeu's latest catalogue, and asked if he could inspect it. He quickly realized that its argument was fallacious, and he told Blaeu that he could write a refutation of it on a single piece of paper. Pell then proceeded to do so, and persuaded Blaeu to print it on one leaf, in the same format as the book itself: copies of this 'refutatiuncula' or 'little refutation', as it became known, printed with the page-numbers '73' and '74', were then added by Blaeu to the book itself (which ended on p. 72).[43] That Blaeu was willing to treat one of his most distinguished authors in this way – Longomontanus's famous *Astronomia danica* was a Blaeu publication – is rather surprising; it must attest, at least, to the extraordinarily high regard in which Pell was held. But no less puzzling is the motivation of Pell himself, who was certainly not a polemicist or publicity-lover by nature. In each of his three accounts, he gave the same explanation: 'Since that book had been printed after my arrival in this city, and by the same printer to whom I was resolved to deliver, as soon as possible, my own Diophantus, I was very much afraid that perhaps other mathematicians might think that that book by Longomontanus had been printed, in this town and in that printing-house, *with my approval.*'[44] This was, on the face of it, a strange piece of reasoning. The mere fact of being appointed to the Chair of Mathematics at the Athenaeum did not give Pell any responsibility for licensing, formally or informally, mathematical publications in Amsterdam, and nobody would have supposed that it did. Nor would any informed person have thought that there was anything unusual about a large publishing house such as Blaeu's issuing different works in the same field by different authors without attempting to coordinate their contents. A better insight into Pell's underlying motives can be obtained from one passing phrase in his letter to Cavendish: he wrote that when he looked at Longomontanus's book, he 'found it dedicated to one of our Burgomeesters'.[45] The dedication was in fact to Albert Burgh, whom Longomontanus had met when he was on his

---

[42] Letter 14; BL, MS Add. 4280, fo. 97 (Pell to Leake, [28 July/] 7 Aug. 1645, printed, with some inaccuracies, in Halliwell, ed., *Collection of Letters*, pp. 89–90); J. Pell, *Controversiae de vera circuli mensura ... pars prima* (Amsterdam, 1647), pp. 1–4.

[43] 'Dr Blaeu ... intends to adjoine My refutation to every one of Longomontanus his bookes y^t he shall sell heereafter': BL, MS Add. 4280, fo. 204^r (Pell to William Petty, undated [Aug. 1644]).

[44] Pell, *Controversiae pars prima*, p. 4 ('Quum enim post meum adventum in hanc urbem, iste liber impressus fuisset, & ab eo Typographo, cui ego quamprimùm Diophantum meum excudendum tradere statueram: omnino metuebam ne forte alii Mathematici putarent, istos Longomontani libellos, in hâc urbe & in illo Typographéo, *me probante*, fuisse impressos'). Pell refers to the book as 'libelli' in the plural, in view of its title.

[45] Letter 14.

recent diplomatic mission to Denmark.[46] This, surely, was the real reason for Pell's determination to refute Longomontanus in public: he owed his own recent appointment to the efforts of Burgh, and wanted to demonstrate to Burgh (who, as a dedicatee, could not fail to take an interest in the dispute) that he had chosen well – that, indeed, he had chosen someone whose skills far outstripped those of one of the best-known professors of mathematics in northern Europe.

Pell's desire to impress Burgh is easily understandable; his desire to offend Longomontanus, less so. The text of his two-page 'refutatiuncula' was brusque and dismissive, rejecting the Danish mathematician's conclusions as 'absurdum' and 'absurdissimum', and emphasizing that it had taken Pell only three hours' work, using 'a handful of multiplications and divisions', to overturn Longomontanus's 'incredible labours over so many years'.[47] And as soon as the first copies of the 'refutatiuncula' were printed (on 18 August 1644), he sent one to Longomontanus with a covering letter that could scarcely have been thought conducive to a friendly settlement of the affair: 'At all events', it concluded, 'I see no reason why I should regard you as so stupid that you would be incapable of understanding the strength of my refutation, or so stubborn that you would refuse to yield to manifest truth. Therefore I have no doubt that you will examine this refutation of mine immediately, and publish a retraction as soon as possible.'[48] Not surprisingly, Longomontanus's response was a mixture of defiance and contempt. Having received the 'refutatiuncula' at the beginning of September, he wrote a short pamphlet in which he reprinted both it and Pell's covering letter. He added some dismissive comments in which he claimed (quite falsely) that Pell's argument depended on the use of unreliable figures in the printed tables of tangents, and ended with a letter of his own to Pell (dated 25 September), in which he tried to link Pell's arguments to those of the Jesuit mathematician Paul Guldin, with whom Longomontanus had previously quarreled. For good measure, the titlepage contained comic verses in Latin, ridiculing both Guldin and Pell.[49]

---

[46] Longomontanus, *Rotundi in plano*, sigs. *2ᵛ–*3ʳ; Longomontanus also mentioned that he had already given a copy of his demonstration to Burgh's son.

[47] The only known copy of the 'refutatiuncula' is in BL, MS Add. 4280, fo. 205 (given in photo-reproduction in van Maanen, 'The Refutation', pp. 324–5); here, fo. 205ᵛ ('post pauculas multiplicationes & divisiones, tot annorum incredibiles Longomontani labores prorsus periisse videbis').

[48] Pell, *Controversiae pars prima*, p. 17 ('equidem causam nullam video, cur te aut tam stupidum putem, ut ejus vim capere non possis; aut tam pertinacem, ut manifestae veritati cedere non velis. Ideoque mihi non est dubium, quin hanc nostram ἔλεγξιν statim examinaturus, & quàm primum παλινῳδίαν editurus sis'). This letter is dated here 'XVI Kalend. Septemb.' (17 Aug.), though Pell himself states (p. 15) that he collected the first copies from Blaeu on 18 Aug.; when Longomontanus reprinted this letter, he dated it 'XV Kal. Sept.' (18 Aug.) (C. S. Longomontanus, Ἐλέγξεως *Joannis Pellii contra Christianum S. Longomontanum de mensura circuli* Ἀνασκευὴ (Copenhagen, 1644), sig. π4ʳ). It is mis-dated in van Maanen, 'The Refutation', p. 328.

[49] Longomontanus, Ἐλέγξεως.

By the second week of November, a copy of this publication had reached Pell in Amsterdam. His reaction was expressed in a letter to Cavendish: 'But [>no doubt you desire to know what] Longomontanus [>hath] answered. All that can be expected from a peevish, obstinate, ignorant, infatuated old man. Verses in Pellium & Guldinum. Railings, frivolous impertinencies, nothing to ye purpose, any thing rather than nothing.'[50] The long, angry complaint which Pell poured out in this letter to Cavendish shows how easily his own pride had been wounded. It seems that, as someone with absolutely no previous experience of public controversy, he was both lacking in the ability to calibrate his own degree of offensiveness towards his opponent, and all too easily offended in return. Another sign of this was his treatment of his faithful friend and admirer Christian Ravius, who was now lecturing in oriental languages at Utrecht. Ravius had sent a copy of the 'refutatiuncula' to one of his acquaintances in Paris, the well-known mathematician Claude Hardy: in response, Hardy had written a letter criticizing Pell's argument. Ravius forwarded this, via a student who was travelling to Amsterdam; but only after a delay of more than a month did it reach Moriaen, who then showed it to Pell. Standing somewhat imperiously on his dignity, Pell wrote an angry letter of reproof to Ravius, saying that it was intolerable to think that a criticism of his work had been passed through the hands of 'adolescents' for weeks, before he saw it himself.[51] And when writing to Cavendish on 2/12 November Pell was still simmering with anger against Hardy: he described him as 'One y^t is famous ... for his blessed faculty of refuting errors by throwing heaps of new ones on top of them & so he stiffles y^m'.[52]

In the same letter, discussing Longomontanus's published reply, Pell also commented: 'I shall be constrained to reprint an answere, wherein I must be larger that I may be better understood.' Cavendish did his best to discourage Pell from devoting any more of his time to the matter: 'I hope you haue done with the waiewarde Dane', he wrote in late December, 'that you maye returne to Apollonius & Diophantes but especiallie to your owne analiticall worcke, which I extreamlie desire to see.'[53] For a while it seemed that Pell would follow his advice; at the end of January 1645 Cavendish was able to write that 'I am glad to heare you meane not to trouble your self with Longoman. more than as an appendix to some of your intended worckes'.[54] However, in his reply (of 8/18 February) to that letter, Pell gave the first sign that he was now planning a different approach: his strategy would be to make Longomontanus recognize 'that his demonstrations are generally counted paralogismes & his conclusions false,

---

[50] Letter 24.

[51] Pell's draft of his letter to Ravius is BL, MS Add. 4421, fos. 93–4. This is undated, but the letter from Hardy must have reached Pell before 2/12 Nov., when he quoted from it in Letter 24 (citing a passage which he also cited in his letter to Ravius).

[52] Letter 24.

[53] Letter 28.

[54] Letter 31.

in the judgement of all the able Mathematicians of Europe'.[55] Soon thereafter, Pell received a letter from Johann Ravius (the Comenian educationalist, brother of Christian) in Copenhagen, informing him that Longomontanus was on his death-bed.[56] This news eventually turned out to be false; accordingly, by mid-May, when Cavendish had moved to Paris and was in contact with several of the leading mathematicians there, Pell was now fully committed to his new strategy, and was urging Cavendish to collect demonstrations and statements of approval of Pell's argument, for publication.

Pell's justification of this procedure was that it was necessary to correct the imbalance in seniority between him and Longomontanus: 'And since he opposes his authority to mine, which overswayes exceedingly with those many that can [>number our yeares but] not weigh our reasons ... Therefore there is a necessity y^t I counterpoise [>that age] of his, with universality of consent of Mathematicians.'[57] Putting this in more personal terms, one might say that Pell was still smarting with resentment at the tone of Longomontanus's reply, which had treated him as a foolish and ignorant tyro. But at the same time, without imputing too cynical a calculation of self-interest, one might also note that this was an ideal opportunity for Pell to establish himself in the front rank of European mathematicians – first by circulating his 'refutatiuncula' among all the leading practitioners of advanced mathematics, and then by publishing their demonstrations in his support. The key point at issue was a theorem which Pell had made use of in his disproof of Longomontanus, but which he had merely stated, without demonstration. (Longomontanus had then claimed that it could not be demonstrated.)[58] Pell's idea was to invite other mathematicians to contribute different proofs of it; and as he was entirely certain, both of its demonstrability, and of the fact that there were many different ways of demonstrating it, this was, as he knew, a project that could not fail – so long as the right approaches were made to the right people. When the 'refutatiuncula' was printed, Pell had had it 'dispersed among all the mathematicians all over Europe', using whatever contacts he had.[59] (The young William Petty, who was then in Leiden, received a total of 21 copies for distribution to the leading mathematicians there, including Golius and Frans van Schooten; Petty also forwarded copies to England, to

---

[55] Letter 32.

[56] Pell, *Controversiae pars prima*, p. 42 (18/28 Jan. 1645). Ravius's name is not given, but the identification is made possible by Pell's reply (19 Feb./1 Mar.: pp. 43–5), which enclosed a letter from the unnamed correspondent's brother (i.e. Christian), and also referred to another letter from the same correspondent, dated 31 May 1645: that letter, signed by Ravius, survives among Pell's papers (BL, MS Add. 4280, fos. 8–9). Johann Ravius spent the years 1639–46 in Denmark, as a professor at the Ritterakademie in Sorö (Tautz, *Die Bibliothekare*, p. 21).

[57] Letter 35.

[58] For the theorem, see below, pp. 287–8.

[59] BL, MS Add. 4280, fo. 97^r (Pell to Leake, 7 Aug. 1645).

people such as William Oughtred and Thomas Barlow.)[60] All that was now required was another wave of letter-writing, text-circulating, and string-pulling.

The main burden of this labour fell on Cavendish: there was now an almost complete reversal of roles in their relationship, as Cavendish scurried round Paris performing this service for his former employee. Of the leading French mathematicians, one, Gilles Personne de Roberval, was quick to volunteer a demonstration of Pell's theorem, which he sent to Pell via Hobbes on [12/] 22 April; this took place just after Cavendish's arrival in Paris, and was no doubt prompted by some discussion by Cavendish of the dispute.[61] Pell then asked Cavendish to obtain demonstrations from Hobbes, Fermat, Mydorge, de Beaune, Boulliau, and de Billy – a task which, as he wrote rather unrealistically, Cavendish could perform 'with very little trouble to your selfe'. He also wanted Cavendish to ask Mersenne to make similar requests of the leading Italian mathematicians, such as Torricelli and Cavalieri; Mersenne was himself in Italy, but it seems that Cavendish was not able to contact him until after Mersenne's return to Paris in September.[62] Meanwhile Cavendish did obtain demonstrations from Hobbes and de Carcavi; he spoke to Mydorge, who promised to supply one; and he also attempted a demonstration of his own.[63] During the same period Pell also received a copy of Longomontanus's latest contribution to the debate, a short printed work entitled *Controversia inter Christianum Longomontanum & Johannem Pellium, de vera circuli mensura, ubi defectus canonis trigonometrici, sub initium ejusdem, ostenditur*. This booklet, written as if by a third party, was notably less polemical in tone; although it tried to maintain the argument that Pell depended on a defective table of tangents, it did also contain some cautiously positive statements about Pell's method. But those statements, commenting that the method 'deserves to be demonstrated more fully', merely strengthened Pell's determination

---

[60] Ibid., fo. 204^r (two undated letters from Pell to Petty [Aug. 1644]); BL, MS Add. 4279, fos. 184^r (Petty to Pell, from Leiden, 14/24 Aug. 1644, including the information that he has given two copies to Cornelis van Hogelande, 'who hath promised at his next writing to send one to DesCartes'), 185^r (Petty to Pell, from Leiden, [29 Aug./] 8 Sept. 1644). Exactly how and when Petty had been put in touch with Pell remains, like much else in this period of Petty's life, obscure. Possibly Petty had made contact with Pell, or with some mutual friend, when in London in 1639 (see E. Strauss, *Sir William Petty: Portrait of a Genius* (London, 1954), pp. 27–8).

[61] See Letter 35, referring to Hobbes's letter of [12/] 22 April and 'ye enclosed paper from Monsieur Robervall'. The demonstration by Roberval, printed in Pell, *Controversiae pars prima*, pp. 47–9, is dated [7/] 17 Apr. 1645.

[62] Letters 35, 40; *MC* xiii, p. 248.

[63] See Letters 37, 38, 39, 41. Hobbes's demonstration is in BL, MS Add. 4278, fo. 200^r; the demonstrations by Hobbes and de Carcavi are printed in Pell, *Controversiae pars prima*, pp. 50–1 and 51.

to proceed with his on-going accumulation of proofs.[64]

The French and Italian mathematicians mentioned above were not the only people Pell thought of approaching. Soon after he received Longomontanus's new publication, he wrote to John Leake in London, asking him to obtain demonstrations by various English mathematicians and statements of approval from the Gresham professors.[65] Among Pell's papers there is also a list of names, evidently representing actual and potential contributors to his planned volume: it begins with Roberval, de Carcavi, and Hobbes, continues with several others who did in fact contribute statements or demonstrations (le Pailleur, Cavendish, Tassius, Descartes, Cavalieri, Wolzogen, and Golius), but also has many other names that do not appear in the volume as eventually published. These include Moriaen, Sir William Boswell, Dr Samson Johnson (English chaplain to the Palatine court in The Hague), Mathias Pasor (Professor of Mathematics at Groningen), the German scientist Adam Olearius (who had been in contact with Pell, through an intermediary, since late 1644), and even the learned Princess Elizabeth (daughter of the Elector Palatine), who corresponded on mathematical matters with Descartes.[66]

Whether Pell did in fact approach those people is not known. There is, however, one poignant detail in this list: the name 'Wolzogen' was substituted for 'Jungius', which had originally been part of the list but had later been deleted. This deletion records one approach that signally failed, and it is clear that its failure must have caused Pell real personal and intellectual distress. Pell had asked for Tassius's approval of his demonstration as early as January 1645; but it was not until mid-May, when forwarding a letter from Cavendish, that he plucked up the courage to ask his intellectual hero, Joachim Jungius, to comment on his refutation of Longomontanus.[67] Jungius's reply was written on 6 June: he began by praising Pell's method in the 'refutatiuncula' as 'new and ingenious', but

---

[64] C. S. Longomontanus, *Controversia inter Christianum Longomontanum & Johannem Pellium* (Copenhagen, 1645), sig. A3^v: 'amplius ... demonstratu digna'. There is a copy of this rare work in the Kongelige Bibliothek, Copenhagen (bound with Longomontanus, $\mathit{Ἐλέγξεως}$: pressmark DA 1.–2. S 18 4° (18–271)). The work is dated 25 May 1645; Pell received a copy by the end of July, and wrote several pages of comment on it, dated 1 Aug. (printed in Pell, *Controversiae pars prima*, pp. 74–82). (See also his remarks in Letter 38.)

[65] BL, MS Add. 4280, fo. 97^r (Pell to Leake, [28 July/] 7 Aug. 1645: printed in Halliwell, ed., *Collection of Letters*, p. 89). No such demonstrations or statements, however, appeared in Pell's eventual publication, the *Controversiae pars prima*.

[66] BL, MS Add. 4421, fo. 92^r. Olearius (who was at the court of the Duke of Holstein, and is best known for his account of the mission undertaken on the Duke's behalf to Russia and Persia: *Vermehrte neue Beschreibung der Muscowitischen und Persischen Reyse* (Schleswig, 1656)) wrote to an intermediary on 2 Nov. 1644 that he wanted to correspond with Pell; on 1 May 1645 he apologized for not having written to Pell; and on 2 Aug. 1645 he promised to write to him in the next two weeks (BL, MS Add. 4421, fo. 101).

[67] For the approach to Tassius see Letter 30. Pell's letter (15 May 1645) is SUBH, MS Pe. 65, fos. 80–1; his draft is BL, MS Add. 4280, fo. 139. An extract from Pell's letter was later printed in J. Jungius, *Doxoscopiae physicae minores, sive isagoge physica doxoscopa*, ed. M. Fogel (Hamburg, 1662), sig. α3^r.

he went on, fatefully, to comment that it lacked 'αὐτάρκεια' or self-sufficiency.[68] This criticism, which was strangely similar to Longomontanus's claim that Pell depended on trigonometrical tables, troubled Pell deeply. In lieu of a reply, he sent Jungius copies of Longomontanus's original publication (*Rotundi in plano*) and his first published response to Pell, evidently hoping that Jungius would revise his opinion when he looked into the dispute more carefully.[69] By the last week of August, having received no further response, he grasped the nettle and wrote to Jungius, asking him to explain what he meant by his remark about the lack of self-sufficiency.[70] Jungius replied on 15 September; although he defended his criticism, he was clearly somewhat uneasy in this argument, as he sent further letters on 16 and 19 September (conceding some minor points), compiled detailed notes on Pell's argument (and on the theory of tangents, and on Aristotle's comments on αὐτάρκεια), and asked Tassius to check his reasoning.[71] Replying on 21 October, Pell stuck to his guns; he insisted that, if the comments made by Jungius in his first and second letters were to be included in his forthcoming publication, they would have to be modified, omitting the charge of non-self-sufficiency. This elicited a further letter from Jungius, also refusing to give way.[72] Thereafter, although Pell was happy to publish Tassius's defence of his theorem, his breach with Jungius seems to have been complete.[73]

In one other case, Pell's stubborn persistence had a more positive effect. When Cavendish finally informed Mersenne in early October 1645 of Pell's request that he solicit more demonstrations from Italian mathematicians, the French scientist was, as Cavendish reported, 'vnwilling to vndertake it, pretending that Roberuals demonstration onelie were sufficient'.[74] (Mersenne's argument – that if the theorem had been correctly proved once, all further demonstrations were superfluous – was one with which, in any other circumstances, Pell would surely

---

[68] Jungius's letter is BL, MS Add. 4279, fo. 75 ('nova & ingeniosa ... quamvis αὐτάρκεια istam non habeat': fo. 75^r); his draft is SUBH, MS Pe. 65, fos. 22^v–24^r.

[69] See Letter 38.

[70] Pell's letter (27 Aug. 1645) is SUBH, MS Pe. 65, fos. 82–6; his draft is BL, MS Add. 4280, fo. 96. An extract from Pell's letter was later printed in Jungius, *Doxoscopiae*, sig. α3.

[71] See BL, MS Add. 4279, fos. 76–8 (Jungius to Pell, 15 Sept. 1645), 79 (Jungius to Pell, 16 Sept.), 81 (Jungius to Pell, 19 Sept.) 87 (copy by Pell of Jungius's letter of 15 Sept.); SUBH, MS Pe. 65, fos. 13–20 (Jungius, draft of letter to Pell of 15 Sept.), 21, 37 (Tassius, undated note addressed to Jungius, saying he has checked his letter to Pell), 38–58 (Jungius, notes on Pell, tangents, and Aristotle).

[72] SUBH, MS Pe. 65, fos. 78–9 (Pell to Jungius, 21 Oct. 1645); Jungius's reply of 3 Nov. exists in draft (ibid., fos. 1–12) and in a copy by Pell (BL, MS Add. 4279, fos. 82–4).

[73] Tassius's statement of approval for Pell's theorem (printed in Pell, *Controversiae pars prima*, p. 55) was contained in his letter to Pell of 31 July 1645 (BL, MS Add. 4279, fos. 268–70^r), which Pell received on 17 Aug. Pell wrote to thank Tassius on 8 Oct., commenting pointedly that 'I am glad that you approve so strongly of my refutation, which Mr Jungius thinks is in some way deficient' (SUBH, MS Pe. 65, fo. 76^r: 'Gaudeo, tantoperè à te probari refutationem meam, in quâ D. JUNGIUS videtur aliquid desiderare' (copy by Pell in BL, MS Add. 4279, fo. 271)). For Pell's final comments on his disagreement with Jungius see Letter 48.

[74] Letter 40.

have been the first to agree.) Having lost contact with Pell in 1641, Mersenne was happy to have an opportunity to write to him again; he sent him a letter on 7 October, congratulating him on his appointment at Amsterdam, but also stating his opinion that Roberval's demonstration of his theorem was sufficient.[75] Pell's reply repeated his request; Mersenne responded by saying that he was willing to send letters to Italy if Pell insisted, but that, in view of the recent news of Longomontanus's death, he thought there was no point in pursuing the matter.[76] When this news turned out to be false, Pell was finally able to persuade Mersenne (via Cavendish) to comply with his request.[77] Copies of the 'refutatiuncula' were duly forwarded to Cavalieri and Torricelli, and in March 1646 the former did supply a demonstration for Pell to print.[78] Mersenne himself did not contribute a demonstration, but Pell was able to extract statements of approval of his refutation from both of Mersenne's recent letters to him, for inclusion in his book.[79]

Gradually Pell's intended publication was taking shape.[80] He now had, in addition to the other accumulated demonstrations, a valid proof by Cavendish himself; he was still expecting Mydorge's contribution; he was waiting (in vain) for demonstrations by Torricelli and Fermat; and he continued to cast around for other contributors, such as van Gutschoven in Louvain and Jacob Golius in Leiden. Golius, who was famous as a mathematician as well as an Arabist,

---

[75] BL, MS Add. 4278, fos. 217–18 (Mersenne to Pell, 7 Oct. 1645). This letter is omitted from *MC*; an extract is printed in van Maanen, 'The Refutation', p. 335, n. 33. Mersenne's previous letter to Pell was sent (via Haak) on [21 Apr./] 1 May 1641 (*MC* x, pp. 610–11); Pell received this (he marked it 'Accepi Julij 15/25': MS Add. 4279, fo. 141^r), but appears not to have replied. Several months later Mersenne complained to Vegelin that he had ceased to receive any letters from Pell: Rijksarchief in de Provincie Friesland, Leeuwarden, van Eysinga – Vegelin van Claerbergen papers, MS 67, second folder, 'Brieven van geleerde Lieden' (Mersenne to Vegelin, between Sept. and early Nov. 1642), printed in Malcolm, 'Six Unknown Letters from Mersenne', p. 105.

[76] *MC* xiii, pp. 497–9 (Pell to Mersenne, 18 Oct. 1645). Pell's copy of this letter (BL, MS Add. 4278, fo. 220^r) is accompanied by a copy of the enclosure sent with it, a refutation by Pell of another pretended quadrature on which Mersenne had asked him to comment; this refutation (fo. 221^r) is omitted from *MC*. Mersenne's reply, dated the Ides of Nov. (i.e. the 13th), is BL, MS Add. 4279, fo. 181; this is also omitted from *MC*. An extract is printed in van Maanen, 'The Refutation', p. 338, n. 42; van Maanen mis-dates the letter to before 8 Nov. (p. 337, n. 41), as it was sent enclosed in a letter from William Petty to Pell which bore that date (a date which was in fact Old Style, i.e. 8 [/18] Nov.).

[77] Letter 46.

[78] Pell, *Controversiae pars prima*, p. 60. Torricelli did not doubt the truth of Pell's theorem (though he thought the 'refutatiuncula' could have been made even shorter), but failed to send a demonstration: for his comments see *MC* xiv, p. 366.

[79] This material was printed in Pell, *Controversiae pars prima*, p. 55: the first passage there is from Mersenne's letter of 7 Oct. 1645 (BL, MS Add. 4278, fo. 217^r), and the second is from his letter of 13 Nov. 1645 (BL, MS Add. 4279, fo. 181^r). The editors of *MC* mistakenly assume that both passages were from a single letter dated late July or early Aug. 1645: *MC* xiii, p. 463.

[80] There are several drafts, mock-up pages, and fair copies of parts of this work among Pell's papers: see especially BL, MSS Add. 4280, fos. 93–4, 154; Add. 4400, fos. 1–2; Add. 4421, fo. 92^r; Add. 4430, fos. 78–81.

had visited Pell in Amsterdam in May 1645, on a rather ticklish mission: having heard of Pell's planned translation from the Arabic of books 5, 6, and 7 of Apollonius' *Conics*, he wanted to dissuade him from publishing it, as he himself had been working for many years on the same project. Golius did in fact have one good reason for asking Pell to give way to him: his own Arabic manuscript was a full translation of Apollonius' text, whereas the one used by Pell was merely an epitome or summary of it. (He did not, however, propose the obvious solution to the problem, which was that he should share his manuscript with Pell and produce a joint edition with him.) Pell's response had been both deferential and diplomatic, insisting that he would much rather see the work done by Golius, but not in fact renouncing the right to produce his own edition.[81] The awkwardness of the situation may have made Pell reluctant, at first, to ask Golius for a contribution to his anti-Longomontanus volume; but any such scruples were overcome by the arrival, some time in the spring of 1646, of a new booklet by Longomontanus, *Caput tertium libri primi de absoluta mensura rotundi plani*, which defended the original work attacked by Pell, and began with a dedicatory epistle addressed to Golius himself. 'Even though you may prefer to keep out of this dispute,' Longomontanus told him, 'nevertheless you cannot fail to bear witness to the truth.'[82] Pell now asked Golius (probably via his Arabic-speaking assistant, who visited Pell in May 1646) to write in his support; and Golius – who, in addition to his willingness to bear witness to the truth, had his own reasons for desiring Pell's good will – eventually complied.[83]

By this time (late May 1646) Pell was able to inform Cavendish that his book (a composite work, in which he reprinted some of the earlier materials of the controversy, added new comments of his own, and presented the demonstrations and statements of approval which he had received) was 'now in ye presse'. The printing – as always, in the case of works involving mathematical demonstrations – was slow, and had not been completed by the time Pell left Amsterdam for Breda in early August. His friend Baron Wolzogen stepped into the breach, writing to Pell on 8 September that 'I have corrected it with as much care as I can'.[84] On 21 October Pell wrote to Vossius that he thought the book was

---

[81] See Letters 35 and 60, and Toomer, *Eastern Wisedome*, pp. 184–5.

[82] C. S. Longomontanus, *Caput tertium* (Copenhagen, 1646), sig. A2^r ('Etsi autem fortè ab hac Controversia te integrum Conservare malles; tamen non poteris, quin veritati testimonium dabis'). This epistle is dated 2 Jan. 1646; Pell's first mention of it is in Letter 60 (11/21 May). The work is paginated 73–96, and was intended to be bound together with *Rotundi in plano*, as a continuation of it. It consists mainly of further propositions in support of that text, but ends with a short appendix on the inadequacy of trigonometric tables ('Appendicula de defectu canonis trigonometricis', pp. 92–6), directed implicitly against Pell's 'refutatiuncula' and containing one explicit reference to him (p. 95). There is a copy of this rare work in the Kongelige Bibliothek, Copenhagen (bound with Longomontanus, *Rotundi in plano*: pressmark DA 1.–2. S 18 4° (18–270 4° b).)

[83] See Letter 60; Golius's contribution is in Pell, *Controversiae pars prima*, pp. 95–6.

[84] BL, MS Add. 4279, fo. 309^r (Wolzogen to Pell): 'correxj eâ quâ potuj diligentiâ'.

almost ready, and that it lacked only the first and last gatherings.[85] Four weeks later he told Cavendish:

> I was forced to leave my booke in ye press imperfect nor doe I well know how much of it is since printed, A day or two after I came thence [*sc.* from Amsterdam] I heare Mr Golius came and left his censure of our controversy with ye Printer, desiring that it might be adjoined at least as an appendix ... I hope shortly I shall send you ye whole booke Or at least so much of it as I thinke fit to put to ye presse so inconveniently at y^t distance. I desire as soone as may be to employ our new printer heere [*sc.* in Breda] upon that [>greater] & perhaps farre better part which remaines.[86]

Pell's lack of involvement in the project at this crucial stage caused considerable difficulties for his printer, Johan Blaeu. The text, as finally printed, ended with the word 'FINIS' on p. 88; on p. 89 Blaeu added a note to the reader, in which he explained that Pell had promised more material. 'But he, having hurriedly moved away from this city, left only this incomplete account; nor, for the same reason, did it receive any corrections by him when it was in the press. No doubt he will be able to publish the rest of it more easily where he is now living.'[87] Following this, on pp. 90–6, were the last-minute contributions of Mydorge and Golius – the latter undated, the former consisting of extracts from letters written in June and August 1646. And yet the production of this volume was not completed until the summer of 1647: Pell himself received his first copy in June or early July of that year.[88]

Oddly, therefore, after all Pell's efforts, the final product thus had an air of *bricolage* about it – and, what was even more odd, a piece of *bricolage* abandoned by its creator. The delivery of letters between Breda and Amsterdam would normally take only a couple of days; the excuse given by Blaeu for Pell's non-communication does not really stand up, and it must be suspected that some deeper psychological cause was at work here. Having negotiated quite skilfully and (with the exception of his dispiriting exchange with Jungius) successfully the task of soliciting support from a wide range of European mathematicians, Pell seems almost to have lost interest in his project. It may be that, once his

---

[85] BL, MS Harl. 7012, fo. 229^r, Pell to Vossius, 21 Oct. 1646 ('Libellum meum ad umbilicam typis Blavianis perductum puto, ita ut duae tantùm phylirae, prima scilicet et postrema, desint').

[86] Letter 67.

[87] Pell, *Controversiae pars prima*, p. 89: 'At ille ... ex hac urbe festinatò migravit, hîc solam Narrationem relinquens à nobis inchoatam, ideoque nec ipsius correctione, inter excudendum emendatam: *reliqua* procul dubio, ibi libentius editurus, ubi ipse habitat.'

[88] Letter 70. Pell's copy, in which he corrected some of the marginal references, is in the Busby Library, Westminster School, pressmark I B 29. I know of no evidence to support the claim made by Dijksterhuis that this Latin edition of 1647 was preceded by an English-language edition in 1646 ('John Pell in zijn strijd', p. 295, n. 10).

wounded pride and self-esteem – which had so motivated him in the search for such support – were satisfied by the receipt of the demonstrations he had craved, the intellectual interest of presenting multiple proofs of the same theorem began to seem less compelling. Or it may be that his energies were now concentrated on preparing 'that [>greater] & perhaps farre better part which remaines' – a part in which he may have planned to set out some mathematical advances of his own in this field, instead of merely contesting someone else's errors. But that part, despite the promise of the title of the printed work (*Controversiae ... pars prima*: 'the first part ... of the controversy') appears not to have been written; the death of Longomontanus in October 1647 brought the entire dispute to its natural conclusion, and no further work on this subject was ever consigned either to Blaeu or to the Breda printing-house.[89]

Pell's reasons for moving to Breda are not at all clear. His friend Johann Moriaen thought he had been shabbily treated by the Amsterdam Athenaeum: writing to Hartlib in the autumn of 1647 about the possible employment of Christian Ravius there, he commented that 'once they have people here they pay them no attention; were they prepared to work for nothing, or indeed to pay for the privilege, they would be men fit for this city.'[90] Pell's starting salary of 600 guilders had indeed been low; but in early 1646 this had been raised to 900 guilders, and the salary offered at Breda was only 100 guilders more than that.[91] Possibly some additional promise was made at the outset that the authorities at Breda would pay the costs of bringing his family to the Netherlands (a payment that was, as we shall see, made after long delays). Perhaps the mere fact that he was sought out and personally invited by Prince Frederik Hendrik of Orange to accept a Chair at this new institution touched his sense of self-esteem. Or it may be that he felt that such a request, from such a source, simply could not be refused. That at least is the impression given by his somewhat grudging comments to Cavendish (which, however, may not have set out all his thinking on the matter): 'Thus the world [>goes on] to give me diversions enough, nor can I see when I shall enjoy [y^t which I doe so extreamely desire, *deleted*] leisure to draw the picture of my owne soule for the use of postcrity ... For without doubt something I could say, not meerely unworthy the reading. But this of Breda will robbe me of all my time...'[92]

---

[89] Nor – perhaps for the same reason – did Pell ever publish another work for which some notes survive, entitled 'Ioannis Pellii Apologia pro Francisco Vieta contra calumnias Christiani Severini Longomontani' ('John Pell's apology for François Viète, against the calumnies of C. S. Longomontanus'): BL, MS Add. 4397, fos. 21–24^r. (These notes date from before Pell's move to Breda: he refers to himself on fo. 22^r as 'Mathematicus Amstelodamensis'.)

[90] HP 37/123A, Moriaen to Hartlib, [14/] 24 Oct. 1647 ('wan man die leuthe hatt so achtet man Ihrer nicht wan sie vmbsonst arbeiten oder geltt zuegeben wolten das weren männer fur diese Statt'), cited and translated in Young, *Faith, Medical Alchemy and Natural Philosophy*, pp. 37, 64.

[91] Letters 53 ('This city hath lately multiplied me by 3/2') and 63.

[92] Letter 63.

The École Illustre or Illustre School at Breda (also known as the 'Collegium Auriacum' or College of Orange, although, strictly speaking, that was the name of the main building, a former nunnery, in which many of the students lived) was a completely new institution, set up to consolidate Protestant culture in a city which had been under Spanish rule from 1625 until its recapture by Frederik Hendrik in 1637. Its pupils were drawn from the merchant and gentry families of Breda and its province (plus a smattering of Calvinists from further afield, including several Poles); education at the Illustre School was designed to prepare them either for ministry in the Reformed Church, or for administrative and military service. The moving spirit in its foundation was a Calvinist divine, Lodewijk Gerard van Renesse, who had served as a military chaplain before becoming Pastor of Breda. Three eminent 'Curators' or governors were appointed: Constantijn Huygens, André Rivet, and Johan van Kerckhove, heer van Heenvliet. The elderly Rivet, who had served for many years as Professor of Theology at Leiden, moved to Breda in order to oversee the project, and van Renesse was appointed Rector Magnificus and Regent.[93]

The entire process of setting up this institution was managed with remarkable speed. Invitations were sent out to its prospective professors in May and June; the statutes were drafted in early June, and sent to be printed at the end of August; and the Illustre School was inaugurated in September.[94] Pell's invitation, which offered him the post of Professor of Philosophy and Mathematics, was sent on 1 June; at the beginning of July he travelled to Breda, where the offer was confirmed by the Prince of Orange in person.[95] Unlike most of the other invitees, he did not accept there and then, so his name was not included in the formal letter of appointment issued on 9 July. Probably the main sticking-point for him was the requirement that he lecture in philosophy too ('so extreamely', as he told Cavendish, 'doe I abhorre the senselesse wranglings of vulgar philosophy'); this

---

[93] See D. Langedijk, ' "De Illustre School ende Collegium Auriacum" te Breda', *Taxandria: tijdschrift voor noordbrabantsche geschiedenis en volkskunde*, ser. 5, year 1 (1934), pp. 257–70, 290–300, 328–36; ser. 5, year 2 (1935), pp. 28–39, 72–98, 128–35; G. van Alphen, 'De Illustre School te Breda en haar boekerij', *Tijdschrift voor geschiedenis*, 64 (1951), pp. 277–314; Sassen, 'Levensberichten'; F. F. X. Cerutti *et al.*, *Geschiedenis van Breda*, revd edn., 3 vols. (Schiedam and Breda, 1977–90), ii, pp. 65–7, 218–21.

[94] Algemeen Rijksarchief, The Hague, ref. 1.08.11 (Nassause Domeinsraad, 1581–1811), MS 7989 [formerly 1061], fos. 77^r (invitations, May and June), 80–90 (draft statutes, 7 June), 110–27 (printed statutes, 28 Aug.).

[95] Ibid., fo. 77^r ('Joannes Pellius geroepen tot Professor Philosophiae et mathem. . . . 1 Iunij'); Letter 63.

stipulation was eventually dropped.[96] (There is some evidence that negotiations over this were carried out on his behalf by Sir William Boswell and Dr Samson Johnson.)[97] The final version of the statutes, dated 28 August, set out his duties as follows:

> The Professor of Mathematics shall combine the study of ancient authors with that of modern ones, the demonstration of ancient truths with the searching out of new truths, and theory with practice. He shall mainly direct everything towards the aim of usefulness for his auditors, so that they will in due course become competent to make use of it in military or nautical matters; and he shall be free to show the applications and use of mathematical speculations in all other arts and sciences.[98]

There were seven professors in all. Van Renesse, in addition to his other duties, taught theology and church history.[99] The Professor of Law was a German, Johann Heinrich Dauber, who had previously taught at Marburg and at the Huguenot academy at Sedan.[100] The Professor of Medicine – another Ger-

---

[96] See Langedijk, ' "De Illustre School" ', (1934), pp. 290–2 (9 July letter); Letter 63; J. Pell, 'Oratio Inauguralis', in *Inauguratio illustris scholae ac illustris collegii auriaci, à celsissimo potentissimoque Arausionensium principe, Frederico Henrico in urbe Breda erectorum, cum orationibus solemnibus ipsâ inaugurationis die & seqq. aliquot habitis* (Breda, 1647), pp. 168–83, here p. 170: 'For when I had prepared for this day a lecture on the study of philosophy, after the arrival of my sixth colleague I understood that that I had been freed from that burden, and that I was destined to teach only mathematics' ('Quum enim in hunc diem parâssem orationem *De Philosophiae studio*; post *sexti* Collegae adventum, intellexi me isto munere levatum & *soli Mathematum professioni* destinatum'). The chronology of this is, however, unclear: Pell implies that the change was made shortly before the inauguration, but the appointment of a separate Professor of Philosophy was set out in the first draft of the statutes in early June. Possibly Pell's philosophy-teaching duties would have been only in the field of 'natural philosophy'; the Professor of Philosophy, on the other hand, was required by the statutes to teach only logic and ethics.

[97] BL, MS Lansdowne 751, fo. 55^r, Boswell to Pell, from The Hague, 5 Aug. 1646: 'Since my last I haue r[eceive]^d th^s enclosed from Dr Jhonson, wherin you may see his sense of yor busines at Breda: of wch I haue not yet had any speech wth Mijn H. de Willem...' (David de Wilhem, Constantijn Huygens's brother-in-law, was a senior counsellor to the Prince of Orange.)

[98] *Reglement ende ordonnantie by Syne Hoocheyt gestatueert voor Rector Magnificus, Professoren, Regens, Sub-Regens, collegisten, ende buyten-studenten van de Illustre Schole ende Collegium Auriacum tot Breda* (The Hague, 1646) (copy in Algemeen Rijksarchief, The Hague, ref. 1.08.11 (Nassause Domeinsraad, 1581–1811), MS 7989 [formerly 1061], fos. 110–27), sig. B3^r (= fo. 116^r): 'De Professor inde Mathesis, sal by de oude Autheurs, de nieuwe; by de Demonstratie vande oude waerheden het onderzoeck van nieuwe by de Theorie de Practijcke voegen, ende alles voornaemlijck richten tot nuttigheydt der Toehoorders die hun inden Oorlogh ofte Scheeps-vaert metter tijdt ghenegen zijn te laten ghebruycken, en sal hem vry staen de Applicatien, ende ghebruyck vande Mathematische speculatien in alle andere Konsten ende wetenschappen aen te wijsen.'

[99] *NNBW* ii, cols. 1193–5; Sassen, 'Levensberichten', pp. 124–8; Constantijn Huygens, *De briefwisseling*, ed. J. A. Worp, 6 vols. (The Hague, 1911–17), iii, p. 7 (n.); Cerutti *et al.*, *Geschiedenis*, ii, p. 220.

[100] Sassen, 'Levensberichten', pp. 128–31; see also Menk, *Die Hohe Schule Herborn*, p. 299.

man, from Königsberg in East Prussia – was Albert Kieper (Kyperus), who had proceeded MD at Leiden and lectured there; his earliest publication, a student disputation at Königsberg, had defended Galileo's cosmology, but his most famous book, *Institutiones physicae* (2 vols., Leiden, 1645–6), was a somewhat old-fashioned exposition of Aristotelian theory.[101] Johannes Philemon, a little-known figure from Palatine territory (probably from Bohemia), was Professor of History and Politics, and also served as Librarian; he had studied theology and law at Leiden, but nothing is known of his career between then and his appointment at Breda. Despite his relative obscurity, he was offended by the fact that Pell was placed above him in the order of seniority of the professors, and, just before the inauguration ceremony, told people that he was thinking of resigning in protest; but it seems that his bluff was called, as he neither resigned nor received promotion.[102] The Professor of Philosophy was a young man (b. 1617), originally from Utrecht, Hendrik Born (Bornius), who was interested in Cartesianism and had corresponded with Gassendi; he proceeded MA at Leiden on 27 September 1646 (in the week after the opening ceremonies at Breda), and began teaching only at the start of the following year.[103] But perhaps the most attractive character in this list was the Professor of Botany and Greek, Johan Brosterhuysen, who was to become Pell's closest friend at Breda. Born in Leiden (in 1596), he had studied botany there (and, probably, logic under Caspar Barlaeus). He was befriended by Constantijn Huygens, with whom he corresponded regularly from the mid-1620s onwards; as a lover of poetry and music, he was an active member of the 'Muiderkring' in the 1630s. He had a good knowledge of English: he discussed Bacon's English writings with Huygens, and made translations (never published) of Francis Godwin's *The Man in the Moone* and Sir Henry Wotton's *The Elements of Architecture*. Earlier ambitions to be put in charge of the new botanical–medicinal gardens at Utrecht (1630) or Amsterdam (1632) had come to nothing; but at Breda he was able to set up such a 'hortus

---

[101] Sassen, 'Levensberichten', pp. 131–4; Thijssen-Schoute, *Nederlands cartesianisme*, pp. 99–100; F. Sassen, *Het wijsgerig onderwijs aan de Illustre School te Breda (1646-1669)*, Mededelingen der Koninklijke Nederlandse Akademie van Wetenschappen, afd. letterkunde, n.s., vol. 25, no. 7 (Amsterdam, 1962), pp. 43–60 (on the *Institutiones physicae*). Sassen appears to have been unaware of Kieper's disputation, which was published under the name of his professor at Königsberg: A. Linemannus, *Disputatio ordinaria continens controversias physico-mathematicas quam publicae ventilationis subjicit M. Albertus Linemannus, respondente Alberto Kieper Regiom: Prusso* (n.p. [Königsberg], 1636); the reverse titlepage bears a dedication of the whole work, by Kieper. See esp. sig. A2^v for the praise of Galileo.

[102] Sassen, 'Levensberichten', pp. 139–40; van Alphen, 'De Illustre School', pp. 289–90. Philemon's complaint was reported by van Renesse in a letter to Constantijn Huygens of 12 Sept.: Constantijn Huygens, *De briefwisseling*, iv, p. 348.

[103] *NNBW* iii, cols. 147–9; Sassen, 'Levensberichten', pp. 140–4, and *Het wisgerige onderwijs*, pp. 16–32; Dibon, *La Philosophie néerlandaise*, p. 122.

medicus' himself, and also to encourage music-making among the students.[104]

A printed broadside advertised the inauguration of the Illustre School, informing the inhabitants of Breda that the proceedings would begin on 16 September with an address by Rivet, and that each professor in turn would then deliver his inaugural oration, 'with suitable ceremonies'.[105] By chance, an English artist who was then staying in Rotterdam, Edward Norgate, was invited to attend by the young Lord Stanhope (whose widowed mother had settled in the Netherlands and had married a brother of the Curator Johan van Kerckhove).[106] The description Norgate wrote of the whole occasion, in a long letter to a patron in England, is an extraordinarily vivid account of those 'suitable ceremonies', which began in the Groote Kerk in Breda. Distinguished by a sharp sense of humour as well as a marvellously painterly eye for detail, it deserves to be quoted at length.

> Bee pleased therefore to imagine, the great Church-Pulpitt, and pewes richlie sett out and adorned with Arras, the Towne in their holyday clothes, the horse-troopes, and foote companies making a goodlie Guard from the Castle to the Church, between whom the seven Professours were orderly ledd and attended severally one between two the principall Magistrates of the City ... In this Order and Equipage they were on Munday the 16th of this moneth brought to ye great Church, Ushered with blew officers in Liveries, and blew Musick, besides a formall Bedell, upon whose Ebon staffe, was percht a silver statue of Minerva, a graven image it was, but sure no Papisticall Pallas, for shee went to Church, and Ushered her new M^{rs} into the Chancell, where they were entertained with a varietie of excellent Musick, with voices and other instruments to the Organ: In this interim came the Princesse of Orange, a great Patronesse of Arts and friend to this soe great and good a worke, a goodlie Lady and of a noble aspect and presense, richly besett wth Orientall Pearle, and flaming Diamonds ... an Antheme was sung to ye Organ with verses and Chorus, during which time, the Professours were brought to their places neare the Pulpitt, which was ascended by Dr Rivett, who made a grave and learned Oration in Latine ... This speech ended they had another bout at Musick, and then went up Honsman their Secretarie, who reade ye booke of statutes ... This done the Princesse departed, and all the Professours, Curatores with many a Countrie Parson, and Predicant,

---

[104] *NNBW* ii, cols. 256–7; W. H. van Seters, 'Prof. Johannes Brosterhuysen (1596–1650), stichter en opziener van de Medicinale Hof te Breda', *Jaarboek van de geschied- en oudheidkundige kring van stad en land van Breda "De Oranjeboom"'*, 6 (1953), pp. 106–51; Dibon, *Regards sur la Hollande*, pp. 199–201.

[105] *Waer-schouwinghe uyt Breda* (copy in Algemeen Rijksarchief, The Hague, ref. 1.08.11 (Nassause Domeinsraad, 1581–1811), MS 7989 [formerly 1061], fo. 40: 'met de behoorlijcke Solenniteyten').

[106] Langedijk, ' "De Illustre School"' (1934), p. 293.

> returned to the Castle, guarded and attended as before: In the great Hall they were entertained with a Royall feast. The rest of that day was spent in Collations, Musick, fireworkes, and Drinke.

The next day, speeches were resumed in the chapel of the 'Collegium Auriacum' (the former nunnery building). The first oration was by the Rector Magnificus, van Renesse:

> It was as Long, Leane and drie as the Speaker, among other passages he saide, there needed no greater miracle to shew Gods good liking of this good worke then the fair weather, which instantly changed before his speech was done, for it presently began to rain, and so continued till night: to sweeten this was hear'd a service of very good Voices to a portative Organ; which done an affected frenchified high Germane [Dauber], mount's being for the Civill Law, himselfe of a small Volume, as was his discretion, his discourse in folio, enterlarded with a great deale of Hebrew, wherein he told ye Princesses and Ladies, to their great Edification, what were the laws among the Jewes, concerning Circumcision (a strange impertinence their persons and sex considered) Yet the Princesse with admirable patience satt him out, and to shew she was in her way, as Victorious and invincible as her husband in his, shee returned in ye after-noone, and Letts the Prussian Kiperus doe his worst, whose businesse being Physicke, the strong-hearted fellow had ye Conscience, to hold these poore Ladies three long houres, wth comparing ye Princesse to ye Moone, and the Ladies attending to lesser Constellations, but not a word of his own occupation. This dayes worke ended, the next beginnes wth Brosterhusius, who the Princesses sett, and ye Usuall preface of Musick done, beginn's to express the many and various uses of Plant's, herbes, trees &c. for Nutriment, Medicament, Navigation &c, and how in America, shipping, tackling, and Anchors too, were all made of the Coco tree, and other trees; This was the honestest man appeared as yet, and of Laudable Brevitie, and was succeeded by our Countrieman Mr Pell for the Mathematiques, who with admirable elocution, and most gracefull Deliverie, deservd and gott soe great Commendations, of all the auditory, as it was a measuring cast, whether his skill in his Art, or Oratorie Exceeded: Hee was little above halfe an hour, and yet found time to tell us soe much concerning the use and dignity of that studdy, the Antiquity it had, and Patronage it found from great Kings and Princes wth ample answeres to the usuall objections of those that deterr men from the study of the Mathematiques, as sufficiently shewed him a full M^r in his profession, an honour to his Countrie, and worthy the favour of so great a Prince, his conclusion was very respective, and Excellently

well worded, wth particular thankes to ye Princesse, mentioning the Late Royall Marriage, and his double obligation thereupon: A Complement to his Colleagues, and students ended ye Oration, and this forenoones worke.

The proceedings came to an end that afternoon, when Philemon had given his inaugural oration on history: 'he was an honest man, and made it short and good. Bornius for Logick (an Excellent scholler) ne're appeared to speake, being reserv'd against the Prince of Orange his coming.'[107]

Norgate's judgement on Pell's oration may be confirmed by consulting the printed text – probably an expanded version, as its fifteen pages would take well over half an hour to deliver – which was published in the following year. Elegant but not too florid in style, it extols the pleasures of mathematical studies, rejecting the complaint that they are either useless or too difficult; it also includes special praise of Viète, gives a brief account of how Pell's reputation was first established by the publication of his *Idea*, and pledges that Pell's own advances in mathematics will fulfil the promise made in Viète's famous phrase, 'nullum non problema solvere'.[108] Not for the first time, Pell seems to have been torn here by the conflicting forces of modesty and ambition: writing to Cavendish in July 1647, when the volume containing this oration was about to be published, he referred to this pledge and commented that 'This with some other passages [>of] y^t oration which I was forced to print in spite of my teeth, may perhaps raise me a [>hideous storme] of envy & contradiction.'[109] But no such storm appears to have broken.

Once again, Pell settled into the routine of a mathematics professor. An official broadside described the lecture-courses at the Illustre School, which began on 2 October 1646: each professor would give a daily one-hour lecture, with Pell's at two o'clock in the afternoon. 'In this winter term', it announced, 'he will teach only the elements of all those things which are of any relevance to the study and practice of mathematics. On the first of April next, he will begin to explain how

---

[107] BL, MS Lansdowne 1238, fos. 11–12^r, Norgate to Sir Abraham Williams, from Breda, 22 Sept. 1646 (copy). A somewhat inaccurate transcription of this letter is printed in Langedijk, ' "De Illustre School" ' (1935), pp. 128–32. The 'Honsman their secretarie' mentioned by Norgate was Johann Jacob Hausmann. The 'Late Royall Marriage' referred to by Pell was between Princess Mary (daughter of Charles I) and William II of Orange.

[108] Pell, 'Oratio Inauguralis', esp pp. 173 (Viète), 180 (phrase), 181 (*Idea*). An indication of the trouble Pell took over this oration is given by the quantity of drafts and preparatory notes for it among his papers: see BL, MSS Add. 4407, fos. 172–3, 175; Add. 4381, fos. 39^v–65; Add. 4431, fos. 155–6, 165–70, 410–12. Pell's copy of the printed work, bearing marginal annotations in his hand, is in the Busby Library, Westminster School, pressmark L B 2.

[109] Letter 70.

the ancients and moderns have built, or should build, on those foundations.'[110] As in the case of his courses at the Athenaeum, details of the contents of his lectures at Breda have not survived. If his second course concentrated (like his opening lectures at Amsterdam) on the interpretation of ancient mathematicians such as Diophantus, it may be doubted whether many of his young auditors were up to the task. Certainly that is the impression given in early 1648 in a letter from Rivet to his fellow-Curators Huygens and van Kerckhove; discussing the various professors, he noted that 'Mr Pell has very few students – not because he does not turn up to give his lectures, but because few take up this study, and it seems that he is too abstract for beginners.'[111] One student who would certainly have been capable of following Pell's lectures, however, was Constantijn Huygens's son Christiaan, who arrived at the Illustre School in November 1646 and stayed there until August 1649. Christiaan Huygens was encouraged to concentrate on the study of law (no doubt because his father wanted to prepare him for public life); he lodged with Professor Dauber, and received a personal tutorial from him every day.[112] But he did also attend other lectures: a letter from van Renesse to Constantijn Huygens in January 1647 stated that Christiaan would go to hear Dauber and Bornius in the morning and Pell at two o'clock in the afternoon, adding that he would receive private tuition in law or mathematics.[113] Since Christiaan was already corresponding on advanced mathematical problems with both van Schooten in Leiden and Mersenne in Paris, and since the latter specifically recommended to him 'Mr Pell, who is a very excellent algebraist', it is hard to believe that he never discussed his mathematical interests with Pell.[114] But although it has commonly been supposed that he was Pell's pupil, Huygens himself denied it in later life: scribbling in the margin of his copy of Adrien Baillet's life of Descartes (1691), he wrote that 'I neither studied with,

---

[110] *Series lectionum publicarum, in Illustri Schola Auriaco-Bredana, inchoandarum secundo die Octobris 1646* (broadside, n.p., n.d. [Breda, 1646]; a copy in the Lincoln Cathedral Archives, pressmark Bs. 5, is reproduced in Stedall, *A Discourse Concerning Algebra*, p. 133): 'hiberno semestri docebit tantum elementa eorum omnium, quae ad studium usumque Matheseos ullo modo spectare videbuntur. Quibus fundamentis quomodo veteres & recentiores superstruxerint, vel superstruere debuerint, futuris Kal. April. ostendere incipiet'.

[111] Constantijn Huygens, *De briefwisseling*, iv, p. 452 (4 Feb. 1648): 'Mons. Pellius a fort peu d'escholiers, non qu'il ne se presente pour faire ses leçons, mais peu embrassent cette estude, et semble qu'l soit trop speculatif pour ceux qui commencent'.

[112] Leiden University Library, MS Huygens 37, letters from Dauber to Constantijn Huygens, no. 4 (27 Nov. 1646).

[113] Christiaan Huygens, *Oeuvres complètes*, 22 vols. (The Hague, 1888–1950), i, p. 61 (16 Jan. 1647).

[114] Ibid., p. 45 (Mersenne to Huygens, 8 Dec. 1646: 'Mr Pell tresexcellent Analyste').

nor learned anything from, Mr Pell.'[115] Possibly Huygens was exaggerating here; or it may be that he attended some of the lectures in Pell's first course, found them too elementary, and was thereby discouraged from making further contacts with him. If so, it must seem peculiarly unfortunate that Pell's lectures were too advanced for most of his students, yet too simple for the one alumnus of Breda who would later become famous throughout Europe for his work in the physical and mathematical sciences.

Pell's personal and social life in Breda was probably more enjoyable than it had been in Amsterdam, for several reasons. First, and most importantly, his long-suffering wife and children were finally able to join him. When exactly the move took place is not known, but it must have been between July 1647 and March 1648.[116] (After a delay of more than three years, the costs of the move would eventually be reimbursed by the authorities of the Illustre School.)[117] Pell's intellectual life, too, must have benefited from this development, as Ithamar brought with her the manuscripts he had left in London: in April 1649 he was looking again at mathematical 'papers' which he had composed in 1640, and in early 1650 Pell was returning books and manuscripts to Cavendish which had evidently been left in Pell's safe-keeping in London before his departure for Amsterdam.[118]

A second reason why Pell may have taken more pleasure in life at Breda is that there was a significant English community there. Two long-term residents in the Netherlands, the half-Dutch John Ogle and his brother-in-law William Swann, lived in Breda; Swann was a captain in the Dutch army, and Ogle's father had been governor of Utrecht (1610–18). Ogle's sister Utricia (Swann's

---

[115] H.-J. Hess, 'Bücher aus dem Besitz von Christiaan Huygens (1629–1695) in der Niedersächsischen Landesbibliothek Hannover', *Studia leibnitiana*, 12 (1980), pp. 1–51, here p. 29 ('Je n'ay point etudie ni point appris de M. Pell).' For a statement of the common assumption see Thijssen-Schoute, *Nederlands cartesianisme*, p. 670: 'l'on peut présumer que les cours de mathématiques de John Pell furent pleins de séduction pour Christiaan Huygens.' Lisa Jardine similarly claims that Christiaan Huygens received a 'rigorous training' from Pell: *The Curious Life of Robert Hooke: The Man who Measured London* (London, 2003), p. 61. Nevertheless, that Huygens did visit Pell's lodgings 'sometimes' ('quelques fois') is confirmed by him in a letter to Mersenne of 20 Apr. 1648: *MC* xvi, p. 240.

[116] See Letter 73 (1/11 Mar. 1648), where Pell refers to his previous letter to Cavendish of 1/11 July 1647 (Letter 70) and comments: 'Since my wife & family came hither, I am removed from ye house where I first lodged, so y^t that addresse may be left out'. The move must have taken place after 25 July 1647: Pell's daughter Katherine was buried in Westminster on that day (Burke, ed., *Memorials of St Margaret's*, p. 616).

[117] Algemeen Rijksarchief, The Hague, ref. 1.08.11 (Nassause Domeinsraad, 1581–1811), MS 7989 [formerly 1061], fos. 182^r, (resolution by the Curators, 11 Apr. 1651, to pay the professors the costs of 'moving house and travel with their families and furniture' (''t verhuysen en verreysen met haere familien en Meubelen')), 188^r (resolution by the Curators, 17 Apr. 1651, to pay Pell 600 guilders 'for his travel and other costs incurred by him for his transportation to Breda' ('ouer [>reys en andere] oncosten by hem gehadt op syn Transport naer Breda')). Ferdinand Sassen writes that Pell was promised 500 guilders for this purpose at the outset ('Levensberichten', p. 137); I have not seen any evidence of such a promise.

[118] See Letters 92, 93, 96.

wife) was a talented musician, admired and befriended by both Caspar Barlaeus and Constantijn Huygens.[119] Royal connections also played a part here: Utricia was a lady-in-waiting to the new Princess of Orange, Mary Stuart. Breda was one of the residences of the Princes of Orange; it was while staying as their guest at Breda that the young Charles II learned of his father's execution. Indeed, Breda became something of a centre for Royalists in exile (Charles II remained there throughout the first half of 1649), and several of the young Englishmen who came to Pell for tuition were in this category. Another frequent visitor to Breda was Pell's friend Dr Samson Johnson (who had been preacher to Charles's aunt Elizabeth at the Palatine court in The Hague); he would later take over responsibility for the 'hortus medicus' at Breda which Johan Brosterhuysen had founded.

The third positive factor in Pell's life there was Brosterhuysen himself. Pell seems to have developed an unusually close friendship with this talented man, who was both an Anglophile and a keen musician. Brosterhuysen's private life was not above reproach: Rivet was shocked by the fact that he had taken an attractive 35-year-old widow to live with him as his housekeeper, and insisted that he either dismiss her or marry her. (He took the latter course: a brief diary of events compiled by Pell for 1648–9 includes the entry 'March 12. 22 [1648] Brosterhuysen married. Laetare ['Rejoice']'.)[120] Rivet also complained, in a letter to his two fellow-Curators, that Brosterhuysen 'socializes with the English, who are morally lax, smoking tobacco with them, which is his principal pastime'.[121] Brosterhuysen may have shared his musical interests with Pell, whose papers include music by William Byrd (notated in 'numerall prick-song') and a page of notes on 'Thomas Campions rules of counterpoint, after my fashion expressed'.[122] He also shared some of his own earlier correspondence: among Pell's manuscripts are two letters to Brosterhuysen from 1625 and 1637, the latter being a note written by Pell's predecessor at Amsterdam, Hortensius, commenting on the news that Descartes was complaining about his criticism of the newly published

---

[119] See van Seters, 'Prof. Johannes Brosterhuysen', p. 139; T. van Kempen-Stijgers and P. Rietbergen, 'Constantijn Huygens en Engeland', in H. Bots, ed., *Constantijn Huygens: zijn plaats in geleerd Europa* (Amsterdam, 1973), pp. 77–141, esp. p. 107. On 21 Jan. 1647 John Ogle wrote from Breda to Constantijn Huygens that 'frequenting the company of Dr Johnson, Mr Pell and other learned men should almost turn me into a philosopher' ('De omgang met Dr. Johnson, den heer Pell en andere geleerden zouden mij bijna tot een wijsgeer maken': Constantijn Huygens, *De briefwisseling*, iv, p. 376).

[120] Van Seters, 'Prof. Johannes Brosterhuysen', pp. 137, 140; Busby Library, Westminster School, loose sheet inserted at end of Busby, 'Accounts and Memoranda'.

[121] Van Seters, 'Prof. Johannes Brosterhuysen', p. 138 ('s'accomode avec les Anglois qui sont sans discipline, soufflant avec eux le Tabac qui est son principal exercise').

[122] BL, MSS Add. 4418, fo. 225^r; Add. 4423, fo. 379. W. H. van Seters writes that Pell had a good singing voice ('Prof. Johannes Brosterhuysen', p. 138), but does not give his evidence for this claim.

Essais.[123] The clearest evidence of the degree of intimacy between Pell and Brosterhuysen is Pell's own account (in a letter to Cavendish) of his reaction to his Dutch friend's death, which took place after a brief illness on 10 September 1650: 'Your letter found me standing by the death-bed of one of my Colleagues, whom I loved so well as few men doe their brethren. Which I expressed so immoderately, that my Physician sent me from his grave directly to my bed, from whence some thought I should not have risen.'[124]

About Pell's intellectual life in Breda, however, the surviving evidence is very inadequate. The strangest feature of his six years there is his complete failure to publish any mathematical work – despite his long-standing promises to Blaeu concerning Apollonius and Diophantus (William Petty had mentioned, in a letter to Pell from Paris in November 1645, 'yor Diophantus which was ready for ye presse before my departure from you'), and despite his renewed promise, made in his inaugural oration, to set out a new, all-powerful algebraic method.[125] Ten months after his arrival at Breda, he received from Blaeu, as we have seen, his first printed copies of his response to Longomontanus. A set of names among his notes constitutes what is clearly a distribution list for copies of this work: it includes the contributors (Cavendish, Hobbes, Roberval, etc.), people in Amsterdam (Moriaen, Barlaeus, Vossius, Hotton, Rulice, and others), in Leiden (Golius, the philologist Salmasius, the Cartesian theologian Abraham Heidanus), in England (Hartlib, who got four copies, Sadler, Haak, Busby, Pell's sister-in-law Makin and his old Sussex patron Dr Burton), in Groningen (Pasor and the theologian Hendrik Alting), in or around The Hague (Boswell, Descartes,

---

[123] BL, MSS Add. 4279, fo. 8 (Hendrik van Diest to Brosterhuysen, 2 July 1625); Add. 4280, fo. 19^r (Hortensius to Brosterhuysen, 25 July 1637). Pell's papers also contain a few notes on botany by Brosterhuysen: MS Add. 4280, fos. 288–90. On van Diest, Professor of Theology and Hebrew at Deventer, see *NNBW* iv, cols. 504–5. The letter from Hortensius is a document of some interest where the early reception of the *Essais* is concerned. He writes that 'Your Excellency can easily judge how unjustified M. Descartes's indignation is; I have never intended to write against him, and I am not aware of having said anything against the 'Dioptrique', which I had neither seen nor read – though it is true that I may have let slip the remark (in my cups, when everything one says should be treated as confidential) that I feared that his theory might be well-founded but not his practice, given that M. Descartes himself had not hitherto performed any experiments in this matter' ('VE can nu wel oordelen hoe ongefondeert de indignatie van M^r. Descartes is alsoo mij noijt inde gedachten is gekomen tegens hem te schrijven, oock en Weet ick niet iets tegens de Dioptrique gesegt te hebben die ick noch gesien noch gelesen en hadde, ten waere mij mocht ontvallen syn, inter pocula quando omnia oportet esse tuta, dat ick vreesde de Theorie goedt fondt sijn en niet de praxis, om dat M^r. voorteijdt selfs geen experimenten en heeft daer op gedaen').

[124] Letter 102. The burial took place in the Groote Kerk on 14 Sept.; Pell had been chosen to give the funeral address, but in the event this was given by Kieper instead (van Seters, 'Prof. Johannes Brosterhuysen', pp. 138, 142). During the following year, Hartlib noted that Pell was planning to write a funeral oration commemorating Brosterhuysen, in which 'many distinguished things' about his life would be recorded: HP 28/2/21B (Ephemerides, 1651, part 2: 'multa insignia'). This work was apparently never written.

[125] BL, MS Add. 4279, fo. 183^r (Petty to Pell, 8 [/18] Nov. 1645).

the two Curators Huygens and van Kerckhove); his colleagues in Breda; and others in Utrecht or elsewhere (Wolzogen, Lubieniecki, Samson Johnson, the mathematician Grégoire de Saint-Vincent, the philosopher Henricus Regius – the last two names with annotations indicating that Pell had personally given them copies).[126] The impression given by such a list is of an author keen to spread his fame as a mathematician as widely as possible; and yet Pell seems to have done virtually nothing thereafter, in the five years that remained to him at Breda, to satisfy whatever appetite for his work that far-ranging distribution had aroused. The only sign that he still had ambitions to publish is his remark in October 1651 that he had bought two reams of paper in order to make copies of 'my owne papers' that would be 'legible enough for a Printer'; how far he got with this project is not known, but it was certainly not as far as publication.[127]

Of course, Pell did not lack on-going contacts with mathematicians and philosophers – as that distribution list of 1647 helps to show. He kept up the connection he had established with de Saint-Vincent; he corresponded with Boswell on mathematical and scientific matters; he was visited in Breda by the mathematician Frans van Schooten in August 1649, by the Cartesian mathematician and physiologist Gerard van Gutschoven in the following month, and by the Catholic philosopher Thomas White in February 1650; and he continued to discuss mathematical problems and new publications of all kinds in his correspondence with Cavendish (who left Paris in the summer of 1648 and moved first to Rotterdam, then to Antwerp, from where he made visits to Pell in Breda in August 1649 and May 1650).[128] There is also some evidence relating to the private tuition Pell gave at Breda: a letter from one pupil, Sir Charles Lloyd, shows that Pell was using algebraic problems derived from a Dutch book by Sibrant Hansz. Cardinael, and that he was also discussing the work of Harriot with his

---

[126] BL, MS Add. 4416, fo. 102^r (with other names, not reproduced here). The inclusion of de Saint-Vincent, and of colleagues at Breda, shows that this must be a distribution list for the *Controversiae pars prima*, not the 'refutatiuncula'.

[127] Letter 115.

[128] See BL MS Add. 4431, fo. 408 (letter from de Saint-Vincent to Pell, from Ghent, 19 July 1647, thanking him for the *Controversiae pars prima*), and BL, MS Add. 4423, fo. 347^r (Pell, draft of reply [datable to 19 Aug. 1647]); BL, MS Lansdowne 751, fos. 58–9 (Boswell to Pell, Jan. 1648); Letter 89 (van Schooten visit); Busby Library, Westminster School, loose sheet inserted at end of Busby, 'Accounts and Memoranda', entries for Aug. 4–7 (visit by Cavendish), 13 Sept. (visit by 'Gutischovius'); BL, MS Add. 4413, fo. 246^r ('Jan. 24 [/] Febr. 3 1650 Thursday in ye Evening Mr Thomas White (Richardi Albij frater) was with me'); Letter 99 (referring to a second visit by Cavendish). The Busby Library has Pell's copies of Thomas White's *De mundo dialogi* (Paris, 1642) (pressmark H E 4) and Richard White's *Hemisphaerium dissectum* (Rome, 1648) (pressmark I D 11).

pupils.[129] These included a 'Mr Becher', probably the cousin of Sir William Beecher whom he had previously taught in London: Pell's summary diary for 1648–9 includes two records of payments from him in the spring of 1648.[130] And they also included a young nobleman from Cheshire, William Brereton (1631–80), later third Baron Brereton, who would remain a loyal friend and admirer, eventually becoming Pell's most important patron in Restoration England.[131]

One unusually interesting document written by Pell at Breda does survive, however, setting out his views on an issue of wider intellectual interest: the teaching of Cartesianism. On 1 July 1651 Count Lodewijk Hendrik of Nassau sent a letter to the authorities of the Illustre School at Breda, asking for their opinion on the advisability of permitting the teaching of Cartesian philosophy; he was trying to decide whether to ban it from his own university, Herborn, and was seeking advice on this matter from the universities and institutions of higher education in Breda, Harderwijk, Groningen, and Leiden.[132] This was an increasingly contentious issue: with the support of traditional Aristotelians, the hard-line Calvinists (whose biblical literalism made them hostile both to Descartes's rationalism, and to his support for the new science) were campaigning fiercely against Cartesianism, and had already obtained a decree from the Curators of Leiden University forbidding the invocation of Descartes's name in formal disputations there.[133] The wording of Lodewijk Hendrik's inquiry left little doubt that he was essentially hostile and suspicious towards this new philosophical vogue.

---

[129] BL, MS Add. 4408, fo. 88^r, Sir Charles Lloyd to Pell, from Dunghen, Dec. 1651: 'I haue sent yo^u Sybrant Hans his odd question resolued after my fashion ... If yo^u please to send me something to doe I shall endeor [sic] it, if it be not to much presumption I should desir yor Harriot for one day to Correct myn by it shalbe safly sent back.' Pell's note of his reply summarizes the problem (fos. 88^r, 90^r); cf. also his note of a problem from 'Sybrand Hansz at the end of his hundred Geometricall questions printed in Low Dutch at Amsterdam' in MS Add. 4398, fo. 180^r. (Aubrey's copy of this, evidently derived from Pell, is in Bodl., MS Aubrey 10, fo. 55^r.) Sir Charles Lloyd had served as quarter-master general in the Royalist army, and governor of Devizes, being knighted for his services in Nov. 1644 (Shaw, *Knights of England*, ii, p. 218). On Sijbrandt Hansz. Cardinael, a Mennonite schoolmaster and land-surveyor who had taught mathematics in the vernacular at the private 'Duytsche Academie', see H. F. Wijnman, 'De amsterdamsche rekenmeester Sybrandt Hansz Cardinael', *Het boek*, n.s., 22 (1933–4), pp. 73–94, and Thijssen-Schoute, *Nederlands cartesianisme*, pp. 87–8. The work cited by Pell was his 'Hondert geometrische questien met hare solutien', printed as an appendix (separately paginated) to his *Practijck des landmetens* (Amsterdam, 1614); on this work see Wijnman, *op. cit.*, pp. 77–9.

[130] Busby Library, Westminster School, loose sheet inserted at end of Busby, 'Accounts and Memoranda', entries for 18/28 March 1648 ('6 Riders of Mr Becher') and 29 March/9 April ('30 Rijxdalers of Mr Becher'). For the earlier tuition see ch. 3, n. 92.

[131] Ibid., entry for 1/11 Oct. 1649 ('17 Ducatons ... of Mr Brereton'); cf. the references to Brereton in Letters 90, 91, 99. Pell's papers include a note of an algebraic problem he gave Brereton at Breda in 1649: BL, MS Add. 4413, fo. 52^r.

[132] See J. Bohatec, *Die cartesianische Scholastik* (Leipzig, 1912), pp. 151–8; Thijssen-Schoute, *Nederlands cartesianisme*, pp. 463–5; Dibon, *La Philosophie néerlandaise*, pp. 188–91.

[133] Thijssen-Schoute, *Nederlands cartesianisme*, p. 102 (decree of 20 May 1647). A version of this ban was re-enacted in 1656: see P. C. Molhuysen, ed., *Bronnen tot de geschiedenis der Leidsche Universiteit*, 7 vols. (The Hague, 1913–24), iii, pp. 109–12.

Some of his respondents were happy to fortify his suspicions: the reply from Leiden emphasized that the teaching of philosophy should be kept within the bounds of Aristotelian orthodoxy, and that from Harderwijk declared that while Descartes was an excellent mathematician and a tolerable natural scientist, his metaphysics had been correctly condemned by theologians as 'pestiferous'.[134] Those institutions that had a more tolerant attitude to Cartesianism must have regarded framing a reply to the Count's inquiry as a peculiarly delicate task. In the case of Breda, it seems that various professors were asked for their opinions, or for written contributions to a reply. One of these was Pell – whose own strong interest in Descartes had first been aroused by the *Essais* (especially the 'Géométrie' and the 'Dioptrique') published in 1637, and had since been stimulated by Cavendish's enthusiasm for Descartes's work. The text Pell now wrote, entitled 'Responsionis Capita' ('Chief points of a reply'), survives among his papers; it is a cautiously phrased but nonetheless resourceful and robust defence of the 'liberty of philosophizing'.[135]

Pell began by pointing out that it was quite normal to study texts by authors with whose fundamental beliefs one disagreed: pagans, Muslims, Jews, and Papists. He observed that professors at Breda had the freedom both to disagree with famous philosophers, and to borrow good arguments from bad philosophers; and he suggested that Descartes should be treated on the same basis. He included some critical remarks directed against any kind of narrow and dogmatic Cartesianism, ingeniously pointing out that Descartes himself would have disapproved of it: 'For since Descartes dealt with such a small part of those matters which philosophers usually teach, anyone who insisted that only his works should be taught in the schools, to the exclusion of all others, would be – in the judgement of Descartes himself – ridiculous.' A marginal note gave the reference for that judgement of Descartes: 'Method pag 70'; by thus invoking Descartes's own opinions, Pell was going against the letter as well as the spirit of the rules then in force at Leiden.[136] He then added some more practical arguments against any policy of banning Cartesianism: Descartes's books were on sale and widely available, and if a ban were introduced at a university his doctrines would still get taught, though under other names. And he concluded that, in the opinion of the professors at Breda, it had been the wish of their founder Frederik Hendrik 'to keep the necessary liberty of philosophizing in good repair, and not to ban

---

[134] The replies are printed (from the originals in the Staatsarchiv, Wiesbaden) in Bohatec, *Die cartesianische Scholastik*, pp. 153–4 (Harderwijk), 154–5 (Leiden).

[135] BL, MS Add. 4474, fos. 9–11. Pell's copy of the letter from Lodewijk Hendrik is on fos. 5–6.

[136] Ibid., fos. 9^v (pagans, etc.), 10^r: 'Quum enim tam exiguam partem eorum quae Philosophi docere solent, tractaverit Cartesius; qui omnibus alijs rejectis, illum solum in Scholas recipiendum contenderent, ij, vel ipso Cartesio judice, forent ridiculi.' The reference is to a passage in the *Discours de la méthode* (A&T vi, pp. 69–70), where, however, Descartes's argument proceeds on a different basis: he argues against taking on trust any philosophical school or tradition claiming to represent the views of its founder, on the grounds that the founder's teaching will always have been distorted by his followers.

– invidiously, but also in vain – the name of any individual'.[137] The official response eventually submitted by Breda was broadly in line with Pell's position: it declared that while some things in Descartes's writings were 'excellent, and agreeing with the truth of nature', other things were wrong, or at least more clever than true, and it proposed that his writings should be studied like those of any other modern philosopher, though not to the neglect or contempt of the ancients (especially Aristotle).[138] But the fact that none of Pell's material was incorporated in this formal reply does leave the impression that he was somewhat side-lined in the deliberations of the Breda professoriate.

One other piece of evidence hints lightly at the same conclusion. When a new Rector had to be appointed in September 1650 (van Renesse having given up that post in 1649, and Dauber having replaced him for only a year), Rivet wrote a rather down-beat letter on the subject to Constantijn Huygens's brother-in-law, David de Wilhem. 'I do not think His Highness needs to submit to the order of precedence of the professors, since some of them are completely incapable of it.' Brosterhuysen was 'useless' (and, in any case, on his death-bed); Kieper was leaving (to take up a Chair in medicine at Leiden); Bornius was already serving as Sub-regent; so Rivet's choice was Philemon, 'for lack of anyone else'. Only then did he add: 'There remains Pell, who has good qualities, but I do not know if he would be so suitable.'[139] The Professor of Mathematics was thus relegated to an afterthought.

By this stage, Pell himself had already entertained some thoughts of leaving Breda. In March 1650 his friend Moriaen had suggested to Hartlib that Pell might be invited to replace the recently deceased Descartes as philosophy tutor to the Queen Christina of Sweden, 'if he is recommended to the Queen'; it is possible that Pell had hinted, in one of his letters to Moriaen from Breda, that he might be interested in more rewarding employment elsewhere.[140] Three months later, the senate of the University of Utrecht met to discuss finding a replacement for their Professor of Philosophy, Physics, and Mathematics, Jacob Ravensberger, who had died in April: various possible candidates were mentioned, 'but especially

---

[137] BL, MS Add. 4474, fo. 10 (fo. 10ᵛ: 'necessariam philosophandi libertatem sartam tectam conservare nec cujusquam nomen invidiosè sed frustra proscribere').

[138] Bohatec, *Die cartesianische Scholastik*, pp. 155–8 (p. 156: 'praeclara ac naturae veritati convenientia').

[139] Leiden University Library, MS BPL 293.1, Rivet letters to de Wilhem, no. 54 (9 Sept. 1650): 'Or je ne croy nullem[ent] que son Altesse se doibve assubjetir a l'ordre des Professeurs, Car il y en a qui ne sont nullem[ent] capables de cela. Brosterhuyse suivroit mais outre que ce nous est une piece inutile, il est à present fort abbatu de langueur, & apprehende de mourir. Nous perdons, comme vous savéz, Mons. Kipper qui eust esté capable de cela. En defaut, je donne ma voix a Monsieur Philemon ... Resteroit Pellius qui a de belles parties, mais je ne scay s'il y seroit si propre...' Philemon was in fact chosen, and served as Rector until his death in 1652.

[140] HP 37/146B, Moriaen to Hartlib, [15/] 25 March 1650 ('wan Er der Königinne commendirt wird').

the most famous mathematician Mr Pell, Professor at Breda'.[141] On 12 August, having considered and rejected the other candidates, they resolved that two delegates should visit Pell and make him a formal offer of the job, at a salary of 1000 guilders a year. This was the same as Pell's Breda salary – although, when he reported the matter to Cavendish, he wrote that he had informed the delegates from Utrecht that 'the Prince [of Orange] gave me more than they offered me'.[142] Reading between the lines of Pell's account, one may conclude that he was hoping they would do two things: raise the salary, and negotiate on his behalf with the Prince of Orange (whose permission was needed for him to vacate his post at Breda). By late September the two delegates had been to The Hague to take soundings on the possibility of getting such a release (obtaining, according to Pell's letter to Cavendish, a very negative response). But on 6 November 1650 the Prince of Orange died suddenly of smallpox; although Pell thought this might make it easier for Utrecht to get the permission it wanted, the authorities there took the opposite view, concluding at their meeting ten days later that 'this state of affairs is not conducive to the resumption of the invitation to Mr Pell'.[143] Only in early August of the following year did they renew their approach: again, two delegates were sent to speak to Pell, but on 18 August they announced that their invitation had been – for reasons which they did not elucidate – unsuccessful.[144]

According to Pell's own retrospective account, what finally obliged him to give up his post in Breda was the onset of the Anglo-Dutch war in 1652. After his return to England he sent a petition to the Council of State (in October 1652), in which he declared that 'your Petitioner ... of late having perceived an appearance of warre betweene this Common wealth and the United Provinces; and not being willing to serve any forraine State, longer than it is in amity with this Commonwealth, thought fit in the beginning of June last past to come hither, to see whether or no, he might be serviceable to his native Country'.[145] But although there had been tensions between the two states for many months, the hostilities did not break out until mid-May, when the British and Dutch admirals Blake and van Tromp had their famous clash off the Downs. Pell's thoughts about returning to England were certainly in train before that event – indeed, they may have constituted the real reason for his refusal of Utrecht's

---

[141] G. W. Kernkamp, ed., *Acta et decreta senatus: vroedschapsresolutiën en andere bescheiden betreffende de Utrechtsche Academie*, 3 vols. (Utrecht, 1936–40), i, p. 254, 25 June 1650 ('praecipue autem celeberrimum mathematicum D. Pellium, professorem Bredanum').

[142] Ibid., pp. 256–7; Letter 104.

[143] Kernkamp, ed., *Acta et decreta senatus*, i, pp. 257–8 (p. 258: 'bij dese constitutie van tijden niet dient te hervatten 't beroepen van D. Pellius'); Letter 104.

[144] Kernkamp, ed., *Acta et decreta senatus*, i, pp. 268–9 (p. 269: 'de niet gesuccedeerde beroepinge D. Pellii'). In June 1652 they finally appointed Johannes de Bruyn (making a considerable saving, as his salary was only 400 guilders): ibid., i, p. 273; G. J. Loncq, *Historische schets der Utrechtsche Hoogeschool tot hare verheffing in 1815* (Utrecht, 1886), p. 81.

[145] BL, MS Lansdowne 751, fo. 71^r.

offer in August 1651. On 5/15 March 1652 William Brereton wrote to Pell from London, thanking him for a letter written by Pell on 2 February, and commenting:

> I was glad to see, that you thought it not amiss to come hither your selfe; for (by what I find) there can be no such advantageous conditions offered you at this distance, as when you are heer, you may propose for your selfe, & (for ought I see) have them granted. Which I the rather beleeve, because some powerfull men profess themselves really bent to do you good; And it shall be my task to put them in mind of their promises; till your selfe be here to doe it, which Mr Hartlib judges your best course.[146]

It was thanks to Hartlib's efforts at spreading his name that Pell had gone to the Netherlands in the first place; and it may well have been thanks to Hartlib's encouragement of new expectations that he finally made the decision to go back to London. (Although very little correspondence between Pell and Hartlib survives from Pell's period in the Netherlands, it is clear that they had remained in contact: in the spring of 1650 Hartlib passed on to the young Robert Boyle information he had received in recent letters from Breda, and in March 1652 Pell wrote to Hartlib about Hobbes's criticism of Descartes's optical theories.)[147]

On [25 May/] 4 June 1652 Pell returned to the library at Breda the books he had borrowed from it; soon thereafter he travelled to England.[148] Whether this trip too was intended (like his initial voyage to Amsterdam) as a merely exploratory venture is not clear. The fact that he left his wife and children in Breda might support that explanation; but, on the other hand, it seems that he simply flunked the task of obtaining his leave from the authorities at The Hague. At the end of the year his friend Samson Johnson would write to him from Breda that 'I have not bin at ye Hagh otherwise I had spoken to ye Curators, They must express their dislike of your absence to your wife first wch they have not don, soe that I believe they will consider you will come if you had a pass, & cannot take it ill.'[149] It appears that Pell exploited the ambiguity of the situation, encouraging the Curators to think that he was eager to return, and that he was prevented from doing so only by the state of war. Some such claim seems to have been accepted by them, and in late August 1653 the Princess of Orange signed a resolution 'to pay the aforementioned petitioner [Pell] his salary or maintenance up to 1 August' – fourteen months after he had vacated his position. Only now

---

[146] BL, MS Add. 4278, fo. 104^r.

[147] Boyle, *Correspondence*, i, p. 89 (Boyle to Hartlib, 1 May 1650: 'thanks for the exact intelligence you are pleased to oblige me with from *Utopia* and *Breda*'); HP 8/4A-B (extract of a letter from Pell [to Hartlib], 18/28 Mar. 1652, on Descartes's 'hyperbolical glasses': 'I have heard that Mr Hobbes intends to shew that Des-Cartes made that Conclusion before hee had sufficiently considered his premises').

[148] BL, MS Add. 4408, fo. 92^r (copy by Pell of certificate of return of books, over initials of van Renesse and Bornius); MS Lansdowne 751, fo. 71^r ('the beginning of June').

[149] BL, MS Add. 4398, fo. 136^r, Johnson to Pell, [21/] 31 Dec. 1652. (The letter is not signed, but the writer is identifiable as Johnson on internal grounds.)

was Pell formally dismissed from his Chair.[150] Pell would not visit Breda again; when he next set foot in the Netherlands it would be in a very different capacity – and in conditions of great secrecy – as a special envoy in the diplomatic service of the English Republic.

---

[150] Algemeen Rijksarchief, The Hague, ref. 1.08.11 (Nassause Domeinsraad, 1581–1811), MS 7989 [formerly 1061], fo. 219^r (resolution of 25 Aug. 1653: 'aenden voorn. Suppliant syne gagie oft Tractement te betaelen totten 1. Augusty ... ende dimitteren denselven ... van den dienst ... als professor Philosophiae et matheseos in de Illustre Schole tot Breda').

5

London, Zurich, 1652–1658

After his return from Breda, John Pell spent nearly two years in England. The evidence for this part of his life is, unfortunately, extremely fragmentary, offering very little in the way of letters or datable personal papers. Contacts between him and Cavendish no doubt continued; but since they were both now in London (Cavendish having arrived there in November 1651, in order to negotiate with the authorities over his estates), such contacts did not take the form of correspondence.[1] Sadly, therefore, nothing more is known of Pell's relationship with Cavendish in the few years that remained before the latter's death in February 1654. But perhaps it was in this period that he witnessed the evidence for spontaneous generation which he later reported to the Royal Society: 'Mr. Pell relating, that Sir Charles Cavendish had kept a box of viper-powder, which being opened and found extremely stinking, had store of little moving creatures in it, like mites of cheese.'[2]

Pell's wife and children remained in Breda until October 1653: Ithamar presumably collected the final payment of Pell's salary that was agreed in late August, and it may be that some slowness in the issuing of that payment was the cause of this further delay. (There may have been other practical problems too: his 'bookes & papers' had still not been delivered from Breda by March 1654.)[3] Pell did have some other contacts with family members: in July 1652 he made a visit to Sussex, probably to see his uncle James, who was now rector of the parish of Horstead Keynes.[4] And in January 1653 he received a letter from

---

[1] Cavendish had travelled to London with his sister-in-law, Margaret, Marchioness of Newcastle, and had taken lodgings with her in Covent Garden: K. Jones, *A Glorious Fame: The Life of Margaret Cavendish, Duchess of Newcastle, 1623–1673* (London, 1988), pp. 75–6. In his letter to Pell from London of 5 [/15] Mar. 1652, Brereton wrote that he had had several meetings with Cavendish and Hobbes (who had arrived there at the end of Dec. 1651): BL, MS Add. 4278, fo. 104^r.

[2] Birch, *History*, i, p. 266 (24 June 1663).

[3] In a petition to Cromwell in Jan. 1654 Pell stated that his family 'arrived in England in October last past': BL, MS Lansdowne 751, fo. 85^r. On the final payment see ch. 4, n. 150. In another petition, written after his return from Switzerland in 1658, he wrote that Cromwell had sent for him in March 1654 'before his bookes & papers were come over to him out of Flanders': MS Lansdowne 754, fo. 465^r. In view of the paucity of surviving manuscripts from the Amsterdam–Breda period of his life, it may be suspected that some of those papers never did 'come over'.

[4] BL, MS Lansdowne 751, fo. 68^r (Brereton to Pell, 13 July 1652, referring to 'your appointed time of going into Sussex', and asking 'how long you think your absence from London is likely to be?'); R. W. Blencowe, 'Extracts from the Journal and Account Book of the Rev. Giles Moore, Rector of Horstead Keynes, Sussex, from the year 1655 to 1679', *Sussex Archaeological Collections*, 1 (1848), pp. 65–127, here p. 66 (noting James Pell's death on 10 Aug. 1655).

Bathsua Makin which indicates not only that relations between them were amicable, but also that he had recently been discussing astronomical matters with her: 'Most learned Brother, I pray send me a few lines of the position of the late Comet, out of the 3 papers you shewed me, that were sent you from beyond sea . . .'[5]

Where Pell's intellectual contacts more generally are concerned, the evidence is frustratingly sparse. It was apparently during this period that he made the acquaintance of the polymathic lawyer, historian, and Biblical scholar John Selden. Pell is known to have had several conversations with him in 1653, snippets of which were then passed on to Hartlib: details of a specialist orchard 'for all manner of Cherries' in Kent, for example, or the story of how Selden, 'having drunk himself drunk in his younger years', suffered from a swelling in his throat which was cured by smoking a pipe of tobacco with amber in it.[6] Selden's influence may well have been responsible for one intriguing offer which was recorded in a memorandum Pell later compiled on this period in his life: '1653 Biblia Polyglotta proposed to Mr Pell'.[7] This was a reference to the colossal scholarly enterprise undertaken by Brian Walton (with the encouragement of Selden, among others), which involved printing, in parallel, the biblical texts in up to nine different languages and scripts; Walton had obtained the approval of the Council of State in 1652, and the first of six large folio volumes was printed in the autumn of 1653.[8] The precise nature of the invitation to Pell is not known. It seems unlikely that he was asked to work on the Arabic text, as his knowledge of the language must have been inferior to that of the other Arabists who contributed to the project. However, one small piece of evidence suggests that he was asked to give the kind of general assistance which Samuel Clarke was later appointed to provide. In July 1653 the Oxford scholar Gerard Langbaine wrote to Selden that 'I understand ... from Mr Clark that in pursuance of y^r directions he has prevayled with Mr Pell to declare himself as to that employm^t for which he was designed'; it thus appears that Clarke took over a role which had previously been offered to Pell,

---

[5] BL, MS Add. 4279, fo. 103^r. This letter is dated 19 Dec., and annotated by Pell: 'January 23'. The year is not given, but can be established by comparison with BL, MSS Add. 4398, fo. 136^v (letter from Samson Johnson to Pell, from Breda, [21/] 31 Dec. 1652, describing a comet observed there from 7/17 to 20/30 Dec.), and Add. 4365, fo. 130^r (notes by Pell on observations of this comet, including a summary of Johnson's report, and details of observations made by Oughtred and 'Dr Martin'). For other observations of this comet (which was visible in Europe in Dec. 1652 and Jan. 1653), by Hevelius in Danzig and Richard White in Rome, see G. W. Kronk, *Cometography: A Catalogue of Comets*, i (Cambridge, 1999), pp. 346–7.

[6] HP 28/2/64B (Ephemerides, May–Sept. 1653), annotated 'Mr Pell ex Seldeno'; HP 28/2/77A (Ephemerides, Sept.–Dec. 1653), annotated 'Mr Pell from Mr Seldens owne mouth'.

[7] BL, MS Lansdowne 754, fo. 465^r.

[8] *Biblia sacra polyglotta*, ed. B. Walton, 6 vols. (London, 1653–7). On this project see Toomer, *Eastern Wisedome*, pp. 202–10 (with evidence of Selden's early interest in it: p. 203, n. 264), and P. N. Miller, 'The "Antiquarianization" of Biblical Scholarship and the London Polyglot Bible (1653–57)', *Journal of the History of Ideas*, 62 (2001), pp. 463–82. On Selden's support for it see also H. J. Todd, ed., *Memoirs of the Life and Writings of the Right Rev. Brian Walton*, 2 vols. (London, 1821), i, p. 45.

but which Pell had eventually turned down.⁹

One other acquaintance that appears to have been made during this period, probably through Hartlib, was with the young Christopher Wren, who was then at Oxford; Pell's later recollection was that he had got to know him in 1652 or 1653.¹⁰ Wren is known to have visited Hartlib in London in 1653, and two years later Hartlib would include a description of Wren's transparent beehive in his little book about apiary, *The Reformed Common-wealth of Bees*; Wren's main interests at this time were mathematical, so it may be assumed that his conversation with Pell was not confined to the discussion of bee-keeping.¹¹ And another contact made originally through Hartlib was evidently renewed at the same time: in early 1653 Pell was once again in touch with Sir Justinian Isham, who, as Hartlib noted, 'hath gotten all the MS Mathematical of Warner and Tin as himself shewed them Mr Pell'.¹²

Until the return of his family, Pell was staying at Hartlib's house in Charing Cross; so it is not surprising that he once again became a participant in some of Hartlib's multifarious plans and projects.¹³ A typical note by Hartlib, written soon after Pell's arrival in London, records: 'A great defect of skilful Potters and potter-work especially in England, which caused Palissy to write a tract of it[;] something Mr Pell hath suggested of it in his observations vpon my legacy'.¹⁴ This indicates that Pell had written a set of comments on one of Hartlib's publications (a work consisting mainly of materials by Robert Child and Cressy Dymock), *Samuel Hartlib his Legacie: Or, an Enlargement of the Discourse of Husbandry used in Brabant and Flaunders, wherein are Bequeathed to the Common-Wealth of England more Outlandish and Domestick Experiments and*

---

⁹Bodl., MS Selden supra 109, fo. 416ʳ (Langbaine to Selden, 14 July 1653). I am indebted to Prof. Toomer for this piece of evidence, and for the interpretation of it which is offered here. From the date of this letter it would appear that the key factor in Pell's decision was the agreement a few weeks earlier (on 24 June) by the Council of State to appoint him to a 'mathematical lectureship': see below, at n. 36. On Clarke see Toomer, *Eastern Wisedome*, pp. 156, 207, 226–9.

¹⁰BL, MS Add. 4278, fo. 118ᵛ (Pell to Collins, 4 July 1666: 'I have known ... D[r] C[hristopher] W[ren] neere 14 yeeres').

¹¹See G. H. Turnbull, 'Samuel Hartlib's Influence on the Early History of the Royal Society', *Notes and Records of the Royal Society of London*, 10 (1952–3), pp. 101–30, esp. pp. 108–15; J. A. Bennett, *The Mathematical Science of Christopher Wren* (Cambridge, 1982), pp. 14–19; S. Hartlib, *The Reformed Common-wealth of Bees, Presented in Several Letters and Observations to Samuel Hartlib* (London, 1655), pp. 50–1 (Wren to Hartlib, 25 Feb. 1654 [/1655]).

¹²HP 28/2/49A (Ephemerides, Jan.–Mar. 1653). 'Tin' was Charles Thynne, who had acquired some (but certainly not 'all') of Warner's MSS for Isham. Some of these Warner MSS survive among Isham's papers: Northamptonshire Records Office, IL 3422, bundle VI, fos. 1–23. See the comments in Stedall, 'Rob'd of Glories', p. 478.

¹³For letters addressed to Pell at Hartlib's house, see BL, MSS Lansdowne 751, fo. 68ʳ (13 July 1652), and Add. 4279, fo. 275ʳ (23 Dec. 1652).

¹⁴HP 28/2/31A (Ephemerides, Jan.–Oct. 1652). The reference is to the treatise *Discours admirables, de la nature des eaux ... des pierres, des terres, du feu et des emaux* (Paris, 1580), by Bernard Palissy, who worked as a potter at the French court in the late 16ᵗʰ century.

Secrets in Reference to Universall Husbandry (London, 1651; 2nd edn. 1652).[15] Unfortunately, however, Pell's text does not survive. Another jotting by Hartlib noted the information, supplied by Pell, that the famous surgeon Moleyns had 'above 400. choice Churgical Experiments and cures ... But is loath to publish them': stimulating the publication of such material was the sort of task that Hartlib was constantly undertaking.[16] And one other note records Pell's involvement in yet another typically Hartlibian project: 'Du Bosse Treatise of Graving and Painting should bee Englisht and Hollar make his Additions vnto it, besides other Observations which Mr Pell hath gathered in.'[17]

There is no reason to doubt the genuineness of Pell's interest in these matters; after all, he had been an enthusiastic Hartlibian in the late 1630s and 1640s, keen to contribute to the perfection of human knowledge across the whole range of theoretical sciences and practical arts. But there was, of course, one science to which he was particularly devoted: mathematics. Here, as in the previous six years at Breda, there is simply no sign, in the period 1652–4, of his trying to finish either the editions or the original treatises which he had promised to produce. His only attempt to prepare something for publication – an attempt which, once again, proved abortive – concerned the table of antilogarithms which he had prepared (on the basis of Warner's earlier work) in 1641–2. Contrary to Pell's fears, these papers had not been used to line pie-bottoms; they had been saved by Warner's relative Nathaniel Tovey, and passed to the divine and mathematician Herbert Thorndike (whom, as we have seen, Pell may have known at Cambridge). On 23 December 1652 Thorndike handed over the table of antilogarithms, together with various other Warner manuscripts, to Pell, in the hope that he would prepare them for publication.[18] Three weeks later Pell informed Thorndike that 'I feared I must change my resolution of putting out Mr Warners writings, because they were so incomplete.' He then obtained Thorndike's agreement that he should publish only the table of antilogarithms, with a 'discourse' of his own 'concerning the construction & use of it', but 'Adjoining nothing of Mr Warners, unless I did very well approve it & conceive y^t its imperfection

---

[15] The 2nd edn. included new material on Ireland. See Turnbull, *Hartlib, Dury, Comenius*, pp. 97–8, and Webster, *Great Instauration*, pp. 432, 473 (where it is described as 'one of the most important agricultural writings of the century').

[16] HP 28/2/68A (Ephemerides, May–Sept. 1653) ('Mollins'). This refers to the surgeon James (or Jacobus) Moleyns (or Moulins), who was licensed in 1627 to administer internal medicines in surgical cases (Munk *et al.*, eds., *Roll of the Royal College of Physicians*, i, p. 180); his prescriptions (pre- and post-operative) for Pepys's gall-stone operation of March 1658 are in BL, MS Sloane 1536, fo. 56.

[17] HP 28/2/52B (Ephemerides, Jan.–Mar. 1653). The reference is to the treatise by Abraham Bosse, *Sentimens sur la distinction des diverses manières de peinture, dessein & graveure, & des originaux d'avec leurs copies* (Paris, 1649). Material from this treatise was in fact 'Englisht' by John Evelyn, and deposited by him at the Royal Society before 1662: see J. Evelyn, *Sculptura, with the Unpublished Second Part*, ed. C. F. Bell (Oxford, 1906).

[18] BL, MS Add. 4279, fo. 275^r (Pell's copy of letter from Thorndike to Pell, 23 Dec. 1653; printed in Halliwell, *Collection of Letters*, p. 94).

might not bring some discouragement to his reputation'.[19] Soon afterwards, however, he discovered that one of the ten bundles of antilogarithms (from 29 750 to 39 750) was missing from what Thorndike had sent him.[20] It seems that this obstacle was sufficient to prevent the project from advancing any further at this stage.

There is, however, one other Hartlibian project which probably dates from this period, and of which Pell seems to have been the author: a hugely ambitious scheme for a 'General College' of mathematics, medicine, and other arts and sciences. Various plans for colleges, Baconian societies, and other institutions of research and higher education were put forward – mostly by Hartlib and his associates – in the late 1640s and early 1650s; they have been subjected to close scrutiny by modern historians, who have long recognized that such schemes formed part of the background to the creation of the Royal Society in 1660.[21] And yet the scheme of which Pell was probably the author does not appear to have received any attention, even though it was not only one of the most ambitious, but also one of the most carefully worked-out. It deserves, therefore, some detailed consideration here.

The College was to have a 'council' or 'senate' consisting of eleven men distinguished for their piety and erudition; and this council would be served by ten secretaries, who would correspond on its behalf with different parts of the world (one with France; one with Germany; one with Italy; one with the Netherlands; one with Scandinavia and Russia; one with Poland, Hungary, Transylvania, and the Ottoman Empire; one with Spain and Portugal; one with the East and West Indies; and two with England). The General College itself would be divided into eleven ordinary colleges or 'sub-colleges' (with one member of the council presiding over each of these), as follows: one college responsible for medicine, with 19 members; one for mathematics and military architecture (15 members); one for the education of children (17 members); one for commerce (15 members); one for agriculture (15 members); one for baking and brewing (13 members); one for the improvement of 'horse-riding, coach-driving, swimming, navigation, machines, aqueducts, clocks, etc.' (15 members); one for weaving and cloth-making (13 members); one for building and carpentry (11 members); one for metallurgy and glass-making (15 members); and one for music, painting, gold-smithing and typography (11 members).[22] Each member of the council would receive a salary of £400 per annum (plus 64s per month for food); the secretaries would get £300; members of the college of medicine would be paid £300, those of the college of mathematics £250, and the others various lower amounts, going down to £180.

---

[19] BL, MS Add. 4279, fo. 276^r (Pell, memorandum).

[20] Ibid., fo. 277^r (Pell, later note).

[21] See especially J. J. O'Brien, 'Commonwealth Schemes for the Advancement of Learning', *British Journal of Educational Studies*, 16 (1968), pp. 30–42; Webster, *Great Instauration*, pp. 221–39, 548–51.

[22] BL, MS Add. 4441, fo. 116 (fo. 116^v: 'pro Equitationibus, Aurigationibus, Natationibus, Navigationibus, Machinis, Aquaeductis, Horologiis, etc.')

Each of the colleges would, in addition, send six or eight men every year, 'at public expense', to travel the world, seeking out the arcana of their particular specialism.[23] The duties of the General College were carefully listed: apart from supervising and governing the eleven colleges, it would (among other things) take care 'to propagate and establish the Christian faith throughout the world', 'to give assistance in settling disputes wherever they may occur, especially among Christians', and 'to retain, in the safe-keeping of its archivists, fine descriptions, drawn up by the sub-colleges, of their experiments and tests of all things'.[24] And the special duties of the college of mathematics would be as follows: to inspect or supervise the college of navigation and machines; to correct common errors in mathematics in general; to improve geography and topography; to conduct research in optics; to conduct research in astronomy; and to carry out experiments with burning-glasses and other such devices.[25] Having listed the aims and duties of all the colleges, the proposal added: 'Many people will wonder why we do not intend to set up any college for theology or jurisprudence. They should know that we have nothing to do with controversial theology...' And it ended with a set of general principles, which included the rules 'We do not conduct any disputations about the Christian religion', 'We think it contrary to God's command to involve ourselves in any war, other than one undertaken for self-defence', and 'Each person is, by God's command, subject to the supreme magistrate.'[26]

The manuscript containing this proposal is in Pell's hand; but it does not contain a date, or any explicit statement of Pell's authorship. Neither of those things can be established with certainty. A date later than 1649 seems plausible, for two reasons. First, the division into eleven colleges appears to echo Hartlib's plan (drawn up at the end of the period 1647–9) for an eleven-college 'University of London': the contents of the Pell scheme are quite different from Hartlib's (which was a kind of omnium-gatherum of earlier Hartlibian plans for particular institutions, including a college of controversial theology), and the whole scheme might therefore be seen as a more strictly worked-out response or counterpart to Hartlib's.[27] Secondly, the emphasis in the final set of rules on renouncing war, and on accepting the 'supreme magistrate' as ordained by God, resonates with the concerns of the period after 1649, when, with the second civil war over and the King executed, the public doctrine of the new regime emphasized the need to

---

[23] Ibid., fos. 116^v–117^r (fo. 117^r: 'sumptibus publicis').

[24] Ibid., fo. 117^v ('Christianam fidem per universum Orbem propagare & stabilire'; 'Operam dare ut ubivis dissidia, Christianorum praecipue, componantur'; 'Omnium rerum Experientias et Tentamina à Collegijs inferioribus pulchrè descripta servare sub custodia Conservatorum').

[25] Ibid., fo. 118^r.

[26] Ibid., fos. 119^r ('Mirabuntur non pauci, quare pro *Theologia* et *Jurisprudentia* hîc nullum instituamus Collegium: Verum sciant illi, nobis cum Theologia disputatoria nihil esse rei'), 119^v ('Nos nullum ferre disputationem in Christianismo'; 'Nos contra mandatum Dei esse judicare se ullo bello involvere, nisi defensionis ergo suscepto'; 'Quemlibet ex mandato Dei Magistratui supremo subjectum esse').

[27] For Hartlib's plan see Webster, *Great Instauration*, p. 223 (noting that most of his eleven colleges 'may be related to projects announced by the Hartlib circle between 1647 and 1649').

submit to *de facto* rule. (This was the main line of argument used by defenders of the 'Engagement' – among them, John Dury – which, imposed on the entire adult male population from January 1650 onwards, required a statement of loyalty to the Commonwealth 'as now established'.)[28] As for the authorship of the text, it seems unlikely that it was by Hartlib (for the reasons just given), and although the emphasis on piety and Christian irenicism chimes with Dury's concerns, nothing like this detailed scheme can be found in Dury's own proposals for the reform of higher education, written in 1651 and 1653.[29] The text is in Pell's hand, and does not bear any attribution to another author – something which, when applicable, Pell was normally scrupulous in providing. The minute attention to numerical detail (giving the precise number of members for each college, and their carefully differentiated salaries) does seem characteristically Pellian. And the nature of the plan shares many features with the *Idea* of 1638: the accumulation of a public register of information, the duty of the mathematical experts to correct common errors, the cultivation of foreign correspondence, and the role of the mathematicians (emphasized in the *Idea*) in supervising practitioners of navigation and other such skills. If Pell is accepted as the most likely author of this text, then although it must remain possible that he wrote it while still in Breda, it is surely more probable that he composed it in London, once he was caught up again in Hartlib's infectious enthusiasms for educational and social reform.[30]

Many of the schemes produced by Hartlib and his friends may have been optimistic to the point of outright idealism; but their hopes were not completely unfounded. Hartlib had many friends and sympathizers in the upper reaches of the new regime – men such as his long-term supporter Francis Rous, or John Sadler's close relative Colonel Sydenham, both of whom were members of Cromwell's Council of State. Hartlib's 'Office of Address' (an agency for the gathering and dissemination of information) did receive offical encouragement from Parliament, and in 1657 some of his ideas would be embodied in the new

---

[28] On the Engagement see B. Worden, *The Rump Parliament, 1648–1653* (Cambridge, 1974), pp. 226–32; on the various *de facto* theories see the classic studies by Q. Skinner, 'Conquest and Consent: Hobbes and the Engagement Controversy', in his *Visions of Politics*, 3 vols. (Cambridge, 2002), iii, pp. 287–307, and J. Wallace, 'The Engagement Controversy, 1649–1652: An Annotated List of Pamphlets', *Bulletin of the New York Public Library*, 68, no. 6 (1964); *Destiny his Choice: The Loyalism of Andrew Marvell* (Cambridge, 1968) (esp. pp. 51–3, on Dury). Dury's role as a pro-Engagement publicist is discussed in Batten, *John Dury*, pp. 122–7.

[29] See Dury, *The Reformed School*, 2nd edn. (London, 1651), and the text 'Some Proposalls towards the Advancement of Learning' (HP 47/2), printed in Webster, *Hartlib and the Advancement of Learning*, pp. 165–92, esp. p. 191.

[30] I am very grateful to Dr Charles Webster for giving me his comments on this text: he agrees with the dating and authorship which I propose here, and suggests that Pell may have been putting forward a plan designed to rival the more educationally orientated proposals in William Petty's *The Advice of W. P. to Mr. Samuel Hartlib for the Advancement of some Particular Parts of Learning* (London, 1648). He also notes the influence on this text of the various 'Office of Address' plans that dated from 1646 onwards.

College established in Durham. So when William Brereton (with Hartlib's backing) had urged Pell to return to England on the strength of the 'promises' of 'some powerfull men' who were able to offer him 'advantageous conditions', there may have been some real basis for such enticement. Nevertheless, Pell's initial experience after his return to England in June 1652 was quite dispiriting. In October he sent a petition to the Council of State, complaining that he had 'in four months space not beene able to discover any encouragement to tarry heere any longer', and requesting a pass to return to his previous post at Breda.[31] On 29 October the Council of State decided that his petition should be referred to 'the Lord General' (Cromwell), who was Chancellor of Oxford, and to the Lord Chief Justice, Oliver St John, who was Chancellor of Cambridge, 'to consider what may be done for the peticion^r that may be an encouragem^t: unto him to reside in England'.[32] One significant piece of evidence indicates that Pell now had an interview with Cromwell: some time in November, Hartlib noted that 'There are 18. Millions of Acres in Ireland surveyed as my Lord General told Mr Pell.'[33] The significance of this is that it suggests that Pell was invited by Cromwell to go to Ireland, either to supervise the survey or to teach the relevant mathematical skills to surveyors at Trinity College, Dublin. (Members of Hartlib's circle did in fact fill many of the key roles here: his old friend Benjamin Worsley was the first Surveyor-General; Miles Symner, who was or became another Hartlibian, was appointed Professor of Mathematics at Trinity College specifically to teach surveying, and William Petty – who probably owed his acquaintance with Hartlib to an introduction from Pell – eventually master-minded the 'Down Survey', becoming rich in the process.)[34] But it seems that Pell turned down the invitation, either because he did not care for the peripatetic life of a surveyor, or because he disliked the idea of teaching only applied mathematics, or because he did not want to live in Ireland. On 16 November, accordingly, the Council of State decreed that Cromwell and St John, plus the senior official Bulstrode Whitelocke, should constitute a committee 'to consider of a fit encouragement to Mr Pell for his remaining in England, and to read Lectures upon the Mathematiques in some convenient place'.[35]

More than seven months passed before any more definite commitment was made. On 24 June 1653 the Council of State finally agreed that Pell should receive an annual salary of £200 'for being Mathematicall Lecturer' (with £50 paid

---

[31] BL, MS Lansdowne 751, fo. 71^r.

[32] Ibid., fo. 73^r.

[33] HP 28/2/40A (Ephemerides, Oct.–Dec. 1652).

[34] See T. A. Larcom, ed., *History of the Cromwellian Survey of Ireland A.D. 1655-6, commonly called 'the Down Survey'* (Dublin, 1851); Webster, *Great Instauration*, pp. 226, 435–44; T. C. Barnard, *Cromwellian Ireland: English Government and Reform in Ireland, 1649–1660*, 2nd edn. (Oxford, 2000), pp. 216–22, 224–6.

[35] BL, MS Lansdowne 751, fo. 74^r. 'Of which Order', Pell later wrote in an acerbic memorandum, 'Mr Pell Left coppies with the Lord chief-Justice, & the rest but nothing was done in it by them': MS Lansdowne 754, fo. 465^r.

in advance), and directed two officials 'to consider of a fitt place for the habitation of Mr Pell, & likewise wherein he may read his Mathematicall Lectures'.[36] Two months later it directed 'that the lodgings in James's House in possession of John Durie be assigned to Mr. Cradock and Mr. Pell, Mr. Cradock having the first choice': in 1649 Dury had been appointed Deputy Keeper (under Bulstrode Whitelocke) of St James's Library, the former royal library.[37] Pell must now have felt close to attaining the sort of position he had always wanted: independent status as a mathematician, with a good salary and comfortable, well-placed accommodation. But it was not to be. Apparently Mr Cradock (probably the Puritan divine Walter Cradock) exercised his option; and, what was worse, Pell's salary was not paid, despite repeated authorizations and directions for payment by the Council of State.[38] More long months of waiting followed. During this period Pell had some personal contact with one of the key members of the Council of State, Edward Mountagu, the future Earl of Sandwich. Mountagu had a serious interest in mathematics (he would become one of the first Fellows of the Royal Society), and within a few years Pell would be corresponding with him, sending him mathematical problems; but it seems that even this powerful patron, who served on the Committee for the Advancement of Learning, was unable to secure Pell's lodgings or his salary as mathematical lecturer.[39] In January 1654 (fifteen months after his original petition to the Council of State) Pell sent another petition, this time to Cromwell: 'whereas your Highness hath beene gra-

---

[36] BL, MS Lansdowne 751, fo. 79^{r} (also *CSPD 1652–3*, p. 437). All the evidence indicates that Pell's lectures were to be given in London. I know of no basis for W. C. Abbott's claim that Pell was 'kept in England, largely by Cromwell's interest, as a lecturer in Cambridge' (*The Writings and Speeches of Oliver Cromwell*, 4 vols. (Cambridge, Mass., 1937–47), iii, p. 232).

[37] *CSPD 1653–4*, p. 93; Batten, *John Dury*, pp. 127, 145. The appointment of Dury was made in 1649 (see R. Spalding, ed., *The Diary of Bulstrode Whitelocke, 1605–1675*, Records of Social and Economic History, n.s., xiii (Oxford, 1990), p. 243), and confirmed in 1650. It was in his capacity as librarian at St James's that Dury had published his *The Reformed Librarie-Keeper*, in which he reprinted as an appendix (in view of its suggestions about the maintenance of a mathematical library) the text of Pell's *Idea* (see above, ch. 3, n. 22). Dury eventually took up residence in St James's House himself: on 5 Jan. 1654 he wrote a letter from there to Whitelocke (see R. Spalding, *Contemporaries of Bulstrode Whitelocke, 1605–1675*, Records of Social and Economic History, n.s., xiv (Oxford, 1990), pp. 81–2).

[38] *CSPD 1653–4*, pp. 311, 345. In Mar. 1654 Walter Cradock was appointed commissioner for approving public preachers (see the entry in the *DNB*).

[39] On Mountagu see R. Ollard, *Cromwell's Earl: A Life of Edward Mountagu, 1st Earl of Sandwich* (London, 1994), esp. pp. 32 (Council of State, Committee), 155, 213 (interests in mathematics and astronomy). A letter from Samuel Morland to Pell in 1655 referred to their having met once before 'at Col. Montague's chamber at Whitehall' (Vaughan, *Protectorate*, i, p. 244); this must have happened after 7 July 1653 (when Mountagu was given lodgings in Whitehall: Ollard, *op. cit.*, p. 32) and before 2 Nov. 1653 (when Morland left London in Bulstrode Whitelocke's entourage: H. W. Dickinson, *Sir Samuel Morland, Diplomat and Inventor, 1625–1695* (Cambridge, 1970), p. 7). Pell seems to have been in frequent contact with Mountagu in the summer of 1654 (informing Ithamar on 26 Aug. that 'I cannot write this week to Colonel Montagu'); Mountagu wrote to him from on board ship in July 1656, and in Feb. 1657 Morland passed on Mountagu's thanks to Pell 'for the mathematical papers you formerly sent him' (Vaughan, *op. cit.*, ii, pp. 106, 391, 401).

ciously pleased to order your Petitioner a Salary and an house for his habitation: Notwithstanding which, your Petitioner is to seeke for both; by reason that Mr Gualter Frost (to whom a warrant from the Councell was directed for payment of one quarter) tells your petitioner that he hath no money to statisfy it withall.'[40] Cromwell did take notice of Pell's complaint: on 25 January he ordered Frost to pay the salary 'out of ye contingent moneys', and in late February officials were once again instructed 'to assign a lodging for Mr. Pell, the mathematical lecturer'.[41] But whether a house was found for him, and whether he ever gave any mathematical lectures, must seem very doubtful; for only a month later he was invited by Cromwell to take on a completely different employment, as special envoy to the Protestant cantons of Switzerland.

The immediate background to the choice of Pell for this task is easily explicable. He was not the person first selected for it: the choice originally fell on Theodore Haak, on the grounds that he was fluent in German, had many contacts throughout the German-speaking lands, and also had some experience of diplomacy, having served as a Parliamentary envoy to Denmark in 1643–4. (Since then he had performed various tasks, including translation work, for the Parliamentary authorities, and in 1650 he and Hartlib had been paid £50 each by the Council of State 'for many good services in their correspondence beyond the seas and to enable them to continue it'.)[42] But Haak had found his diplomatic work in Denmark deeply uncongenial; and so, in his own words, when the Secretary of State John Thurloe asked if he would undertake the Swiss embassy, he 'excused himself it and recommended Mr. Pell'.[43] The offer was made to Pell on 2 March 1654.[44]

It was Thurloe who, in his previous capacity as Clerk of the Council of State, had handled Pell's petition and the various decisions in his favour, so Pell's quest for employment would have been well known to him; Pell's fluency in 'Low Dutch' must have seemed close enough to a knowledge of 'High Dutch' (i.e. German); and it may be that Cromwell had been favourably impressed by him during their interview in the autumn of 1652. One of the great weaknesses of Parliamentary and Cromwellian diplomacy was that most of the experienced diplomats had taken the Royalist side; there was therefore much improvisation involving academics, Puritan divines, and others with no prior knowledge of such matters, so Pell's own lack of experience would not have been seen as a disqualification. As for his own reasons for accepting the offer, these may have comprised something more than mere frustration at the lack of progress with his mathematical lec-

---

[40] BL, MS Lansdowne 751, fo. 85^r. This Gualter Frost was the son of the Gualter Frost who had been Pell's 'old friend' (see ch. 3, n. 14).

[41] Ibid., fo. 85^r; *CSPD 1653–4*, p. 422.

[42] Barnett, *Theodore Haak*, pp. 53–70 (mission to Denmark), 92 (payment), 97–8 (choice for Switzerland).

[43] Aubrey, *'Brief Lives'*, ii, p. 131.

[44] The date is given in a chronology drawn up by Pell in 1658: BL, MS Lansdowne 754, fo. 481^v.

tureship. For the nature of the task that was now put to him (which involved strengthening friendly relations between England and the Swiss Protestants, and making contacts with other Protestant interests in France and the German lands) must have chimed with the religious and geopolitical ideals of Hartlib, Dury, and their other pro-Palatine friends from the first period of his acquaintance with them – building a grand, godly alliance in the service of a pan-Protestant foreign policy. Dury himself, on his return from Sweden in 1653, had stepped up his own campaign for a religious reconciliation of the main Protestant Churches (Lutheran, Calvinist, and Anglican); for this purpose he had obtained the official endorsement of various English ministers, both Presbyterian and Independent, and of leading academics at Oxford and Cambridge, and was seeking official authorization for a further round of his personal irenicist diplomacy in Europe.[45] No doubt his enthusiasm helped to sway Pell in favour of accepting the offer. Dury did get the authorization he sought from Cromwell, and the missions of the two men were in fact closely co-ordinated.[46]

The promotion of Protestant amity was not, however, the only reason for Pell's mission; the political background to it was more complicated than that. In the early months of 1654 serious negotiations were finally taking place to end the Anglo-Dutch war. The Protestant cantons of Switzerland had been worried by this conflict between the two most powerful Protestant states in Europe, and had sent an envoy, Johann Jakob Stockar (or Stokar), to England in 1653 with an offer of mediation. The offer was politely declined – not least because it was felt in London that the Swiss favoured Holland. (Being Presbyterian Calvinists, like the Dutch, they were known to have a distrustful attitude towards the English Independents. Some were also suspected of favouring the Royalist cause, thanks to their connections with the Palatinate and its partly Stuart ruling family.) Nevertheless, Stockar's diplomacy did lead to an improvement of relations between Switzerland and England, and stimulated a greater English interest in Swiss affairs: at his farewell audience with Cromwell on 25 January 1654, Cromwell suggested that 'the three republics, Switzerland, England, and Holland, should

---

[45] Batten, *John Dury*, pp. 146–8. A copy of the letter of recommendation from Oxford, Cambridge, and London is in the Staatsarchiv, Zurich [hereafter: SAZ], MS E II 457a, fos. 247–248^r: the recommenders included John Owen, John Wilkins, and Gerard Langbaine (Oxford); Lazarus Seaman, John Worthington, and Ralph Cudworth (Cambridge); and Edmund Calamy, Philip Nye, Stephen Marshal, Samuel Fisher, and Thomas Gataker (London).

[46] It thus appears that two separate developments came together: the invitation to Haak, transferred to Pell, and Dury's own project. The statement by one modern historian that 'Hartlib probably instigated the mission of Dury and Pell to the Swiss' (T. Venning, *Cromwellian Foreign Policy* (London, 1995), p. 26) seems unfounded. Even less plausible are the suggestions by Robert Pell that Pell may have been recommended for the post by Samuel Morland (then a very junior figure, whom Pell had met only once: see above, n. 39), or by his father-in-law Henry Reynolds (who had died in 1635): 'The Story of Dr John Pell', *Pelliana*, n.s., 1, no. 2 (Oct. 1963), pp. 1–48, here p. 18.

maintain a confidential correspondence among themselves'.[47] The drawing to a close of the Anglo-Dutch conflict also made possible, however, a more fundamental shift in English policy towards the major Catholic powers. Previously, the policy of the Parliamentary authorities had been deeply distrustful towards France (the power that had welcomed Henrietta Maria and the future Charles II), and had therefore cultivated good relations with France's great rival, Spain. As England had developed into a major military power, France had become more keen to placate it; yet a rapprochement had not been possible during the Anglo-Dutch conflict, when the seizure of French ships in privateering operations had brought France and England to the brink of war. Cromwell's personal preference was for an anti-Spanish foreign policy, which meant having good relations with France; but, realizing that he held more cards in his hand than the French did, he was able to lay down strict conditions when dealing with French negotiators in the first half of 1654. The most important of these were the requirement that France confirm all the liberties of its Protestant minority, and the demand that it expel the Stuarts from French soil. Throughout 1654, and until the late summer of 1655, the French were kept guessing about the long-term alignment of Cromwell's foreign policy. It was in his interests that they should feel uncertain and insecure: anything that weakened their geopolitical position was to be welcomed, as it would put them more in the position of supplicants in their negotiations with him. It was for this reason that a central feature of Cromwell's policy towards Switzerland was his desire to prevent the Swiss cantons from renewing their traditional 'league' with France – an arrangement under which they supplied the French king with significant numbers of Swiss troops. Cromwell thus wanted to weaken France – but only in order to facilitate the eventual formation of an advantageous alliance with it.[48] Whether this fundamental ambiguity in his policy was apparent to all the diplomats in his service is, however, far from clear.

On 30 March 1654 Pell received his official instructions from Cromwell. His duties were to assure the Protestant cantons of England's good will; to 'endeavour to shew the true Cause of the Warres which haue beene in England, Scotland and Ireland, and the necessity and Justice thereof on the Parlaments part'; to hinder any diplomatic efforts made there on behalf of 'Charles Stewart'; to explain the English position on the Anglo-Dutch war; to encourage the Swiss Protestants to send their sons to the English universities; and to send back to England 'frequent Accounts' of his dealings.[49] Alongside this official statement

---

[47] J. A. Balthasar, ed., 'Eidgenössiche Gesandtschaft an Cromwell im Jahr 1653', in his *Helvetia: Denkwürdigkeiten für die XII Freistaaten der schweizerischen Eidgenossenschaft*, i, Heft 4 (Zurich, 1823), pp. 561–98, here p. 586: 'wenn die drei Republiken Schweiz, England und Holland in vertraulichen Briefwechsel mit einander träten'.

[48] On these issues see Abbott, *Writings and Speeches of Cromwell*, iii, pp. 229–33; C. P. Korr, *Cromwell and the New Model Foreign Policy: England's Policy towards France, 1649–1658* (Berkeley, Calif., 1975); and Venning, *Cromwellian Foreign Policy*, pp. 39–50, 63–7.

[49] BL, MS Add. 4364, fos. 7–8^r.

of his duties, he was also given a set of secret instructions. These were directed mainly at the weakening of France: the aim of encouraging pan-Protestant cooperation was included here, but only, it seems, as part of a strategy of unnerving the French authorities by making them more fearful of a Huguenot rebellion. Thus the first item in Pell's secret list of duties was: 'Besides the matter of yor generall Instructions ... the mayne scope of yor. employment is, To settle a firme correspondence betweene vs, them [sc. the Swiss], and other protestants, aswell in France and Germany, as other neighbouringe Nations for ye support of the protestant Cause.' He was told that he could discuss this confidentially with the leaders of the Swiss Protestants. (It must surely have been expected that any such discussions would leak out, and reach the ears of the French.) The secret instructions also specified that, 'For ye better Management of this Worke, and obteyninge the desired Vnion, and Comunication of Counsell, and Interests', Pell was authorized to tell the Swiss that the Protestants in France 'thirst after a deliverance, and want nothing but Countenance to make them declare openly'. Another important duty was 'to vse all Diligence to hinder the renewinge of any League, betweene the French, and the Cantons'. Other items on this list advised that he should develop his own correspondence with Protestants in France and Germany (he was even authorized to go into France 'to conferre and Aduise with the protestants there, if you shall finde it for the good of the seruice'), and inform himself about 'the Affairs of the House of Austria, & what correspondencies they haue with Charles Stewart'. He was also instructed to communicate fully with Dury.[50]

Pell was equipped with a letter of accreditation to the Protestant cantons (dated 27 March), which described his mission only in the most general terms: he was to 'strengthen and increase' the 'friendship and affinity' between the two republics.[51] On the same day a letter of recommendation was written for Dury, addressed to the Protestant cantons; it did not describe him as an envoy, and explained that he had approached Cromwell with a request for approval for his own irenic project – approval which Cromwell was happy to give. On 28 March a more general letter was also issued on Dury's behalf, addressed to all kings, princes, and cities, asking that Dury be well treated and not subjected to any obstructions; again, there was no suggestion that he was acting in an

---

[50] Ibid., fos. 9–11^r.

[51] Zentralbibliothek, Zurich, MS B 102, fos. 149–150^r (fo. 149^v: 'Amicitiae, affinitatisq[ue] ... confirmendae et adaugendae'). There are multiple copies of this letter: one is in SAZ, MS E II 457b, fo. 25^r; one, in the Staatsarchiv, Basel, is printed in F. Holzach, 'Über die politischen Beziehungen der Schweiz zu Oliver Cromwell', *Basler Zeitschrift für Geschichte und Altertumskunde*, 4 (1905), pp. 182–245, and 5 (1906), pp. 1–58, here 4, p. 243; and one, in Bodl., MS Rawl. A 261, fo. 2, is printed in Abbott, *Writings and Speeches of Cromwell*, iii, p. 234. Pell's copy is BL, MS Lansdowne 751, fo. 101^r.

official capacity.⁵² Of course, as several details – the clause referring to Dury in Pell's secret instructions, the fact that Dury was paid by the government 'for his Imployment beyond the Seas', and the actual practice of the two men in the early months of their missions, as they sent back co-ordinated reports to London – clearly demonstrate, Dury was in reality serving the diplomacy, and the intelligence-gathering, of the English republic.⁵³ But it is quite misleading of the most recent study of Cromwell's foreign policy to state that 'John Dury was chosen for the 1654 Swiss embassy', and that 'The Oxford [sic] mathematician John Pell ... accompanied him.'⁵⁴

After accepting the offer of this post on 2 March, Pell had little more than a month in which to prepare himself for his mission. His papers include evidence of some hasty homework: he made a set of notes on a somewhat out-of-date book by Marc Lescarbot, *Le Tableau de la Suisse* (Paris, 1618).⁵⁵ On 21 March he hired a German servant, Marcus Föge, originally from Hamburg, whose previous employment in England had included work as a butler and a period spent as a trooper in Major-General Desborough's regiment; Föge would remain in Pell's service throughout his time in Switzerland.⁵⁶ On the following day Cromwell signed an order for the payment of £200 to Pell 'by way of aduance for his Journey into Swisserland'; curiously, he decreed in the following week that Pell should also continue to receive his salary as a mathematical lecturer, and ordered the payment of £50 for the three months just ended.⁵⁷

Pell's departure was delayed until the signing of the Anglo-Dutch Treaty of Westminster, which took place on 5 [/15] April; the next morning he and Dury went to Gravesend and boarded a frigate for Dunkirk, from where they travelled to The Hague.⁵⁸ Then they parted company for a while: Dury went to see his supporter Godefroid Hotton in Amsterdam (where he also met Menasseh ben Israel, and discussed his plans for the re-admission of the Jews to England), while

---

⁵²These two documents are printed in K. Brauer, *Die Unionstätigkeit John Duries unter dem Protektorat Cromwells: ein Beitrag zur Kirchengeschichte des siebzehnten Jahrhunderts* (Marburg, 1907), pp. 232–4. Copies of both are in SAZ, MS E II 457a, fos. 223ᵛ, 234ʳ. An original exemplum of the first, sealed and bearing the signature 'Oliuer P.', is in SAZ, MS E II 457b, fo. 334ʳ. The second is also printed in Abbott, *Writings and Speeches of Cromwell*, iii, pp. 236–7 (from the copy in Bodl., MS Rawl. A 328, p. 18).

⁵³Bodl., MS Rawl. A 328, pp. 13 (payment to Dury of £200, 22 Mar. 1654), 19 (quotation; £100, 18 July 1654), 147 (£100, 17 Nov. 1654).

⁵⁴Venning, *Cromwellian Foreign Policy*, p. 36. On the other hand, Holzach overstates Pell's role when he calls him 'Cromwell's agent-general for the Continent, based in Zurich' ('Über die politischen Beziehungen', 4, p. 218: 'Cromwells General-agent für den Kontinent mit Sitz in Zürich').

⁵⁵BL, MS Add. 4458, fos. 80–5.

⁵⁶BL, MSS Lansdowne 754, fo. 481ᵛ (chronology); Add. 4407, fos. 167ʳ (testimonials for Föge). Pell gave his name as 'Feege' or 'Veghe', but the man signed himself (fo. 163ᵛ) 'Föge'.

⁵⁷Bodl., MS Rawl. A 328, pp. 13 (quotation), 16 (£50, quarterly salary due on 25 Mar.).

⁵⁸Ibid., pp. 16–17 (the frigate 'Merlyn'); BL, MS Lansdowne 754, fo. 481ᵛ.

Pell went on to Utrecht and waited for him there.[59] Their movements – and, indeed, the fact of their mission – were at this stage subject to strict secrecy; the Dutch ambassador in London would be informed of Pell's embassy only in late June, long after Pell had passed through Dutch territory.[60] The reason for this emerges from a subsequent letter from Pell to Moriaen: the Orangist party in Holland strongly favoured the English Royalist cause, and there was a fear that a Cromwellian envoy might receive the same treatment as the ambassador sent to the Netherlands by the Rump Parliament in 1649, Isaac Dorislaus, who had been assassinated (with, as many thought, the tacit approval of the then Prince of Orange) by Royalist agents. 'I did not see Amsterdam, where you were then living', wrote Pell to Moriaen. 'In haste, and *incognito*, bearing Dorislaus in mind, I spent just a short time in hiding at The Hague, while I found a convenient way of travelling from there to Utrecht: there, however, I lay concealed, known only to one man, as I waited for Mr Dury's arrival. Nor did I identify myself to a single person (except your relative in Cologne) until we had reached Frankfurt.'[61] Thereafter, though the fear for his personal safety had receded, worries about the interception of his correspondence by Royalist agents continued. The rudimentary precautions taken against this included the use of pseudonyms: Pell signed himself with his manservant's name, 'Marcus Feege', Dury was 'John Robertson' (his father was Robert Dury), and Thurloe was addressed under the guise of a London merchant, 'Adrian Peters'.

---

[59] BL, MSS Lansdowne 745, fo. 2^r, Pell to Thurloe, from Utrecht, 20/30 Apr. 1654; Lansdowne 754, fo. 481^v; Add. 4365, fo. 227, Dury to Thurloe, from Amsterdam, undated (fo. 227^r: 'This day I set foreward towards Vtrecht where I shall find M^r. Pell againe, for we parted at the Hague'; fo. 227^v, commenting on Menasseh ben Israel's desire to come to England 'to sollicit a freedome for his Nation to liue there', and adding that 'I did once present to my Ld Protector from him an Hebrew booke of the Immortalitie of the Soule'). See also Brauer, *Die Unionstätigkeit John Duries*, pp. 16–17.

[60] J. Lindeboom, 'Johannes Duraeus en zijne werkzaamheid in dienst van Cromwell's politiek', *Nederlandsch archief voor kerkgeschiedenis*, n.s., 16 (1921), pp. 241–68, here p. 256 (n.). Henry Oldenburg, envoy of Bremen in London, did know of the mission by 7 [/17] Apr., but emphasized its secrecy in his report to Bremen of that date: H. Oldenburg, *The Correspondence*, ed. A. R. Hall and M. Boas Hall, 13 vols. (Madison, Wis., and London, 1965–86), i, pp. 27, 29. Cf. the comment by the Venetian ambassador in London on Cromwell's plan to send an envoy to Sweden in August 1654: 'This mission will probably be effected as suddenly and quietly as the one to the Swiss Protestant Cantons, to whom a person in the confidence of the Protector was despatched some time ago': *Calendar of State Papers ... in the Archives and Collections of Venice, 1653–4*, ed. A. B. Hinds (London, 1929), p. 251.

[61] BL, MS Lansdowne 751, fo. 256^r, Pell to unnamed correspondent (identifiable as Moriaen on internal grounds), 27 Jan. 1655: 'Neque enim ego vidi Amstelodamum, ubi tunc habitabas; sed festinans et incognitus, Dorislai memor, Hagae tantisper latitabam, dum commoditatem inde ad Ultrajectinos proficiscendi invenissem: Ultrajecti autem latebam, uni solo viro cognitus, D. Duraei expectans adventum. Nec ulli mortalium (si tuum affinem Coloniensem exceperis) nomen meum priùs indicaui, quàm Francofurtum venissemus.' Moriaen had served as Calvinist minister in Cologne, and had married the daughter, Odilia, of one of the elders of that church, Adam von Zeuel (see Young, *Faith, Medical Alchemy and Natural Philosophy*, p. 10). Robert Pell's speculation that Pell made secret contact with Charles II in Cologne ('The Story of Dr. John Pell', p. 19) seems quite unfounded.

On 18/28 May, six weeks after leaving London, Pell and Dury finally reached their destination, Zurich. Two days later they sent a joint report to Thurloe, informing him that all was well: 'yesterday some of the Chief by whom our way is to be made, hauing notice of our arriuall, visited us; & expressed much ioy to see us though they doe but guesse at our busines; wch wee see by all circumstances upon the place, must be managed with very great circumspection and secrety.'[62] A week later they wrote again, describing the honorific treatment they had received from the 'Burgemaster' (Bürgermeister, i.e. Mayor), and observing that 'Wee find by all manner of expressions y^t our coming hither is very acceptable.'[63] One great advantage they enjoyed was that Dury was already well known to one of the most influential people in Zurich, the 'Antistes' or overseer of the Calvinist Church there, Johann Jakob Ulrich, with whom he had corresponded since at least 1649.[64] Although Ulrich was somewhat conservative in his theological views (reacting with suspicion to the 'liberal' theology of the Saumur school), he was strongly in favour of the ostensible geopolitical aims of Pell's mission – building up a pan-Protestant bloc – and was therefore also sympathetic to Dury's project of effecting the reconciliation (or, at least, the mutual acceptance) of the Protestant Churches. Above all, he was opposed to any renewal of the 'league' between Switzerland and France.[65] Wisely, Pell cultivated his friendship from the outset; Ulrich would become his most valued ally among all the figures of authority in Switzerland with whom he had to deal.[66] Another person through whom some useful contacts may have been made was Jean-Baptiste Stouppe (or Stuppa, or Stoppa), a Swiss Protestant (originally from Zurich) who was one of the ministers at the French Protestant Church in London, and had recently been sent

---

[62] BL, MS Add 4364, fo. 115^r (Pell and Dury to Thurloe, 20 [/30] May 1654). This, like all the other joint letters, is in Dury's hand.

[63] Ibid., fo. 102^v (Pell and Dury to Thurloe, 27 May/6 June 1654). This letter is explicitly dated 27 May 'old stile'; since the Protestant cantons also used the Julian calendar, Pell would continue to use OS dating for his correspondence with England.

[64] SAZ, MS E II 457a, fos. 101–30 (copies of Dury's letters to Ulrich from London, 1649–50), 156–7 (Dury to Ulrich from London, 26 Jan. 1654). Possibly the contact had originally been made via Ulrich's first cousin once removed, Johann Ulrich, who was a member of Hartlib's circle in London in 1639 (Hübner's letter to Gronovius of 5/15 Dec. 1639 ends by sending greetings from Hartlib, Pell, Haak, and Ulrich: BL, MS Sloane 639, fo. 82^r). He studied under Christian Ravius in Amsterdam in 1646 (where he may have met Pell), and later travelled to England again (see Coccejus, *Opera ἀνέκδοτα*, ii, pp. 666–7, 687–8), apparently spending some time as a chaplain in the Parliamentary army, before returning to Zurich by 1651 to become minister and (in 1653) a professor at the 'Collegium Humanitatis' (see M. Godet et al., *Dictionnaire historique et biographique de la Suisse*, 8 vols. (Neuchatel, 1921–34), vi, p. 727).

[65] On Ulrich's biography and anti-French views see Godet *et al, Dictionnaire*, vi, p. 729; on his theology see Brauer, *Die Unionstätigkeit John Duries*, p. 18; see also the encomium of him, listing his writings on a wide range of subjects, in J. H. Hottinger, *Schola tigurinorum Carolina* (Zurich, 1664), p. 142.

[66] SAZ, MS E II 457c (1st file), fo. 45 (Dury to Ulrich, 29 May 1654) shows that by 29 May Dury had already introduced Pell to Ulrich; an accompanying note in Pell's hand requested another meeting with him.

on a secret mission by Cromwell to report on the conditions of the Protestants in France.[67] He arrived in Zurich less than a fortnight after Pell and Dury, and stayed there for one night. When he left, he took with him letters from them to both Thurloe and Cromwell. Pell would later write to Thurloe that he had heard that Stouppe travelled to England with one of Ulrich's sons: he described Ulrich as 'the principall Divine of this country' and 'a great hindrer of the League', and requested that his son be well treated.[68]

Pell's induction into the life and work of a professional diplomat was rapid. A special meeting of representatives of the Protestant cantons was arranged at Aarau on 13 June, to be addressed by him and Dury. Pell's speech on this occasion was brief and rather general, expressing the good will of the Lord Protector and emphasizing Cromwell's desire to bring together all the Protestants of Europe. He then added that his 'companion' would say more on that subject, and handed over to Dury, who spoke at greater length, outlining his own programme for Protestant union. Both men were well received, not only in the general session but also in subsequent private meetings with representatives of the other cantons.[69] Two weeks later, the Protestant cantons issued their formal response, expressing their approval of Dury's project and inviting him to visit each of them in turn, to discuss the matter in more detail. Dury promptly set off for Bern; he would spend the rest of the year touring the cantons, meeting in most cases with acceptance and encouragement, but also encountering opposition to his plans among the leading Calvinist ministers of Basel.[70]

Meanwhile Pell was left in Zurich, to get on with the other tasks associated with his mission. These included, as we have seen, gathering information about 'the House of Austria'; Pell was assiduous in transmitting to Thurloe what news he could glean from Vienna, and frequently added other items of news from the

---

[67] Stouppe's intriguing career (Calvinist minister; Cromwellian spy; lieutenant-colonel of a regiment in the service of Louis XIV; commentator on religion in Holland, with a special interest in Spinoza) has yet to receive the detailed treatment it merits. For a brief overview, see the essay by Elisabeth Labrousse, 'Jean-Baptiste Stouppe', in her *Conscience et conviction: études sur le XVIIe siècle* (Paris, 1996), pp. 60–8. See also SAZ, MS E II 457f, fos. 151 (Stouppe to Ulrich, from London, 14/24 Oct. 1652), 171 (Stouppe to Ulrich, from London, 17/27 Jan. 1653); 187a (Stouppe to Ulrich, from London, 2 [/12] Feb. 1653), 157 (copy of Stouppe's baptismal certificate); F. de Schickler, *Les Églises du refuge en Angleterre*, 3 vols. (Paris, 1892), ii, pp. 153–4, 170–3, 213–15, 231–4; K. O. Meinsma, *Spinoza en zijn kring: over Hollandse vrijgeesten* (The Hague, 1896), pp. 364–74.

[68] BL, MS Add. 4364, fos. 95^r (Pell to Thurloe, 5 Aug. 1654: quotation), 98^r (Pell and Dury to Thurloe, 3 June 1654).

[69] See SAZ, MS E II 457a, fos. 194 (Pell's speech; fo. 194^r: 'comes'), 219–220^r (Dury's speech); other copies are in MS E II 457b, fos. 111–27, 'Acta der Conferenz zu Araw Im Junio 1654': fos. 114 (Pell's speech), 119–23^r (Dury's speech). See also Brauer, *Die Unionstätigkeit John Duries*, pp. 27–8.

[70] Brauer, *Die Unionstätigkeit John Duries*, pp. 35–68; Batten, *John Dury*, pp. 152–5. The cautious response by the Calvinist Church in Basel, 'Judicium Theologorum, Pastorum, et Ministrorum Verbi Dei Basileensium' (18 Oct. 1654), is in SAZ, MS E II 457a, fos. 181–4.

Italian states and elsewhere.[71] He also acted on his instruction to encourage the Swiss to send their offspring to be educated in England, recommending, for example, that a bursary be given to another of Ulrich's sons.[72] (A third son was added to the list in 1656; meanwhile, unfortunately, the first of the Ulrich boys to arrive in London had run up huge debts, and Pell was obliged to pay £92 11*s* 7*d* out of his own pocket to secure his release from prison.)[73] The main purpose of his embassy, however, was to discourage any renewal of the 'League' between Switzerland and France. Here the fact that Pell was resident in Zurich was both a help and a hindrance. It helped, insofar as that canton was one of the most hostile to any new treaty of cooperation with France; he was thus treated as a close ally by the Zurich authorities, and made privy to their confidential dealings with the other cantons. But on the other hand the main threat to Pell's policy came from pro-French elements elsewhere (especially in Basel and Geneva); at such a distance from those places, he was unable to exert any direct influence over them. At most, he could orchestrate or galvanize the intra-Swiss diplomacy of the Zurich authorities: thus, for example, when Dury reported to him from Bern that the Council of that city had been told that Zurich had shifted in favour of renewing the League, Pell went immediately to 'one of those whom I most trust heere' (Ulrich), and asked him to 'get a letter written from hence', denying it.[74]

In the summer of 1654 Pell did become involved, for a while, in a scheme dreamt up by one of the senior officials in Zurich, which was intended to give England indirect leverage over both Switzerland and France. It concerned a valuable set of jewels, formerly the property of the French Queen Mother, which had been handed by France to the Swiss regiments in 1650 as surety for their arrears of pay. The subsequent failure of the French to make those payments – and thereby redeem the jewels – rankled in Switzerland, and was one of the main reasons for reluctance to renew the League (which had lapsed in 1651). There were fears in Zurich that if France did now redeem the jewels, this would remove the last obstacle to a renewal of the League by the Catholic cantons, and the

---

[71] Many of Pell's letters to Thurloe are printed in Vaughan, *Protectorate*: for Austrian news see, for example, i, pp. 29, 31, 47, 67 (and for other news: *passim*). An interesting exchange in the summer of 1654 shows how he had to learn the principles on which such information-gathering proceeded: when he told Thurloe that he was not filling his letters with news from elsewhere, because he knew that most such news would have passed from those places directly to Thurloe in much less time (i, p. 30: 15 July 1654), Thurloe instructed him to send the news anyway, since 'though I may hear it from other hands, yet yours will come as a confirmation, which is very useful in many cases' (i, p. 36: 4 Aug. 1654).

[72] Ibid., i, pp. 19–20 (1 July 1654). Another student, Johann Rudolf Suter, apparently a protégé of Ulrich, would travel to England in late 1656, where his education was supervised by Dury: SAZ, MS E II 457c (2nd file), fos. 115, 153, 156 (new foliation).

[73] Vaughan, *Protectorate*, ii, pp. 32, 139, 172, 183, 207, 215.

[74] BL, MS Lansdowne 745, fo. 41^v, Pell to Thurloe, 6 Jan. 1655 (Vaughan, *Protectorate*, i, p. 106).

consequent Catholic–Protestant split might even lead to a Swiss civil war.[75] The scheme that was now dreamt up was that the jewels would be bought from the Swiss by Cromwell – thereby leaving the French both embarrassed and (where the Swiss regiments were concerned) still in the wrong. Pell cautiously outlined this scheme in his letters to Thurloe, enclosing detailed drawings of the jewels. But the expected price of the jewels (estimated first at £60 000, then at £72 000) was forbiddingly high; and in any case, such a snub to France was now – in the late summer of 1654 – ruled out by the fact that serious negotiations were in train for a treaty of amity between the English and the French.[76] As relations between those two states blew hot and cold during the rest of the year, and as the Swiss continued to prevaricate over the renewal of the League, there was little further progress that Pell could make on the most important of his diplomatic tasks.

Early in 1655 a new issue arose, which would eventually absorb the energies of Pell and several other English diplomats, becoming a major focus of international interest: the fate of the Waldensians in the Duchy of Savoy, beyond the south-western borders of Switzerland. The Waldensians, who lived in villages in the mountains of Piedmont, to the west of Turin, had been heretics in the middle ages, but had subsequently been converted into Protestants by Calvinist ministers sent from Geneva in the mid-16th century.[77] They had been officially permitted by the Dukes of Savoy to practise their religion in a limited number of villages; over time, their population had spread far beyond that area. On 15/25 January 1655 an order was issued by a government minister in Turin, demanding that all those living outside the permitted area 'withdraw and depart' within three days; many were then forcibly driven from their homes, and destitute refugees began to flood into France and Switzerland.[78] By late February the authorities in Zurich were begging Pell to get Cromwell to intervene, with financial help for the victims and diplomatic or even military pressure against Savoy. Pell sent what further details he could obtain of the events in Piedmont; his reports were thus the first to notify London of this growing crisis.[79]

There was far worse to come. Sporadic resistance by Waldensians gave the government in Turin the pretext it needed to send military forces to the central Waldensian area; the policy now was (as the Regent, the dowager Duchess of

---

[75] Holzach, 'Über die politischen Beziehungen', 4, pp. 227–9.

[76] Vaughan, *Protectorate*, i, pp. 17, 25, 45–6, 58–9.

[77] See E. Cameron, *The Reformation of the Heretics: The Waldensians of the Alps, 1480–1580* (Oxford, 1984), pp. 157–229.

[78] See A. Molnar, A. Armand-Hugon, and V. Vinay, *Storia dei Valdesi*, 3 vols. (Turin, 1974–80), ii, pp. 73–77; for the text of the decree, see S. Morland, *The History of the Evangelical Churches of the Valleys of Piemont* (London, 1658), pp. 303–4.

[79] The key letters are BL, MS Lansdowne 745, fos. 59[v] (Pell to Thurloe, 24 Feb. 1655) and 60[v]–61[r] (Pell to Thurloe, 4 Mar. 1655). The former includes a news report from Geneva, dated 16 Feb.: 'A letter dated Febr. 1. out of the vallies of Piemont tells me of the sad condition of the reformed Churches there . . . ' (omitted from the printing of this letter in Vaughan, *Protectorate*, i, pp. 135–6); the latter is printed in Vaughan, *op. cit.*, i, pp. 137–9, but without its date.

Savoy, put it) 'to take a scythe to the root, and eliminate this evil [*sc.* Protestantism] once and for all'.[80] A body of troops entered the area in the third week of April and began killing, plundering, and burning. Pell was able to give London advance warning of these massacres: on 21 April he sent details of the new policy of the Savoyard authorities, which he had received, via Dury, from a well-informed minister in Geneva.[81] In his next letter, written nine days later, he enclosed a detailed account, drawn up by the city council of Zurich, of the killings that had now taken place.[82] More and more descriptions of the massacres were emerging, and were being transmitted hurriedly across the length and breadth of Protestant Europe. In the words of Samuel Morland's subsequent account:

> there then arrived Letters upon Letters, just like *Job*'s Messengers, one at the heels of another, with the sad and doleful Tidings of most strange and unheard of Cruelties ... Some of their Women were ravisht, and afterwards staked down to the ground through their Privities; others strangely forced, and then their Bellies rammed up with Stones and Rubbish: the Brains and Breasts of others sodden and eaten by their Murderers ... Others had their Flesh sliced from off their Bones, while they were yet alive, till such time as they were become meer Skeletons or Anatomies.[83]

Even if many of these details of the methods employed were exaggerated, it is clear that the killings were brutal, and on a large scale. Jean-Baptiste Stouppe, whose help was sought (via Pell) by the Zurich authorities, and who later published a collection of documents about the massacres, estimated that 6000 people died – either at the hands of the soldiers, or of cold and famine after being expelled from their homes.[84]

The English reaction was, on the face of it, robust: in Morland's words, 'The News of this *Massacre* no sooner came to his Highness ears, but he *arose like a Lion out of his place,* and by divers pathetical and quickening Letters, awoke the whole *Christian* World, and moved their hearts to pity and commiseration.'[85] Public opinion in England was indeed strongly engaged: there were days of public fasting and prayer and, over the next few months, a collection for the victims

---

[80] Molnar, Armand-Hugon, and Vinay, *Storia dei Valdesi*, ii, p. 81 ('di portare la falce alla radice, e di estirpare una volta questo male').

[81] Vaughan, *Protectorate*, i, p. 169.

[82] BL, MS Add. 4363, fo. 78, Pell to Thurloe, 30 Apr. 1655 (printed in Vaughan, *Protectorate*, i, pp. 178–9).

[83] Morland, *History*, sig. a1^v.

[84] J.-B. Stouppe, *A Collection of the Several Papers sent to his Highness the Lord Protector ... Concerning the Bloody and Barbarous Massacres ... Committed on many Thousands of Reformed, or Protestants Dwelling in the Vallies of Piedmont* (London, 1655), sig. A3^v. The request for Stouppe's help was transmitted via Pell with his letter of 30 April (above, n. 82).

[85] Morland, *History*, sig. a2^r. See the letters (drafted for Cromwell by Milton) in Milton, *Complete Prose Works*, v(2), pp. 684–710.

would raise the colossal sum of £38 097.[86] But although Cromwell's pronouncements gained him credit both at home and abroad, the issue had in fact arisen at a very awkward moment for him. As Thurloe drily observed in one of his letters to Pell, 'The peace with France is not yet concluded, nor do I believe that this action of Savoy will very much conduce to the promoting of it.'[87] The problem was not merely that any strong intervention in the affairs of Savoy might raise fears in French minds about a precedent for future English action in aid of a Huguenot rebellion. Worse than that, there were rumours (confirmed by Pell in his despatch of 12 May) that the French had known of the plan to extirpate the Waldensians, and that the forces carrying out the massacres had included units of French cavalry.[88] This was not a good time for cementing an Anglo-French alliance; nevertheless, since Cromwell was already committed to launching an attack on Spanish possessions in the Caribbean, an Anglo-French alliance was of greater geopolitical importance to him than the fate of a few thousand people in Savoy.

This underlying difficulty may explain the somewhat ineffectual nature of Cromwell's diplomacy over the Waldensians. First he despatched a very junior and inexperienced diplomat, the young Samuel Morland (previously a don at Magdalene College, Cambridge, where he had been Pepys's tutor), to Paris and Turin.[89] Morland achieved nothing there, and withdrew to Geneva, where he was joined by another English envoy, George Downing; Pell was instructed to go and meet them, to coordinate their efforts.[90] Cromwell then tried to encourage the Protestant cantons to go to war against Savoy; this suggestion, transmitted via Pell, was rejected as impractical. Even though he was asking the Swiss Protestants to stick their necks out in this way, Cromwell was also insisting on a unified Anglo-Dutch-Swiss diplomatic strategy, and was irked when the Swiss Protestants sent their own mission to Savoy. Meanwhile Louis XIV was also putting pressure on Savoy to come to a settlement with the Waldensians; fortunately for Cromwell, Louis also needed an Anglo-French treaty for the sake of his own anti-

---

[86] Abbott, *Writings and Speeches of Cromwell*, iii, p. 717 (fast days); Morland, *History*, p. 589 (collection, with a detailed breakdown of figures for the different parts of England and Wales).

[87] Vaughan, *Protectorate*, i, p. 175.

[88] Ibid., i, pp. 183–4.

[89] Morland's only previous experience of diplomacy was as a junior member of the entourage of Bulstrode Whitelocke on his embassy to Sweden in 1653 (see Spalding, ed., *Diary of Whitelocke*, p. 296, and Spalding, *Contemporaries of Whitelocke*, p. 205). On his career up to the mission to Turin, see Dickinson, *Sir Samuel Morland*, pp. 4–9.

[90] See Vaughan, *Protectorate*, i, pp. 228 (Thurloe to Pell, 27 July 1655, instructing him to go to Geneva), 268 (Thurloe to Pell, Morland, and Downing, 16 Sept. 1655, thanking them for a joint letter). On Downing, who had been a tutor at Harvard, a preacher in the Parliamentary army, and Cromwell's head of military intelligence, see J. Beresford, *The Godfather of Downing Street: Sir George Downing, 1623–1684* (London, 1925), esp. pp. 64–8 on this mission. While Pell developed a lasting friendship with Morland, he seems to have had no connections with Downing in later life.

Habsburg foreign policy. A settlement, brokered by the Swiss envoys, was finally agreed in late August; the Anglo-French treaty followed two months later.[91]

In all of this Pell played his part conscientiously, transmitting to London requests from the Waldensians for military help, urging the Swiss Protestants to take a stronger line, and explaining in detail to Thurloe the reasons for their reluctance to do so.[92] He also took pains to close the widening rift between the English and the Swiss Protestants: when Cromwell complained about the initiative taken by the Swiss diplomats in Savoy, Pell obtained and sent to London copies of their confidential instructions, showing that the cantons were not at fault, as they had ordered their envoys not to break ranks with England.[93] Nevertheless, some sense of estrangement between England and the Swiss Protestants did linger on thereafter; and the plain fact was that the securing of an Anglo-French treaty had greatly diminished the importance of Switzerland for English foreign policy. When hostilities broke out between Swiss Catholics and Protestants in the winter of 1655–6, both England and France made efforts to encourage a settlement; the English argument – enshrined in a new set of instructions to Pell, in February 1656 – was that the French should be encouraged to help the Protestant cantons, on the grounds that the Catholic ones 'adhere to Spayne'.[94] This was a very different approach from the grand anti-French strategy which had been the apparent *raison-d'être* of Pell's mission at the outset. Pell was now in a diplomatic backwater, and there was little more that he could accomplish before his eventual withdrawal in 1658.

An overall verdict on Pell's work as a diplomat must be that he carried out almost all the tasks given to him with intelligence, integrity, and extreme diligence; of all his official duties, none was more strictly adhered to than that of supplying 'frequent Accounts', as his voluminous correspondence with Thurloe makes plain.[95] To anyone who has followed his life up to this point, the striking thing is how quickly this very private individual acquired the confidence to fulfil such a public role. Part of the answer must be that, by good fortune, the place was peculiarly well-suited to the man. Pell would have been desperately ill-at-ease

---

[91] See Holzach, 'Über die politischen Beziehungen', 5, pp. 6–37; Venning, *Cromwellian Foreign Policy*, pp. 96–100; Vaughan, *Protectorate*, i, pp. 177–282.

[92] Pell's translation of the Waldensians' request (undated), which included also a request that Cromwell send Stouppe, 'as having great acqaintance there, and being very well knowne among them', is in BL, MS Add. 4365, fo. 239^r. For a detailed account of his lobbying of the Secretary of the canton of Zurich, and of the replies he received, see T. Birch, ed., *A Collection of the State Papers of John Thurloe*, 7 vols. (London, 1742), iii, pp. 431–2 (reprinted in Vaughan, *Protectorate*, pp. 177–83).

[93] Holzach, 'Über die politischen Beziehungen', 5, p. 36. Holzach comments that Pell thereby showed himself to be 'a thoroughly adroit diplomat and an excellent human being' ('als durchaus geschickten Diplomat und vortrefflichen Menschen').

[94] Birch, ed., *State Papers of Thurloe*, iv, p. 552.

[95] The only major failure to carry out his instructions concerns the task of cultivating correspondence with French (and German) Protestants. From the lack of complaint about this in Thurloe's letters, it seems that it was dropped from the agenda with at least the tacit consent of London.

PART I: THE LIFE OF JOHN PELL   161

as a diplomat at a status-obsessed princely court; but in Zurich he was dealing with non-aristocratic burghers, many of whom had both religious feelings and intellectual interests – a milieu quite similar to the world of Hartlib's supporters in London. The burghers and ministers of Zurich, in turn, respected him for his godliness and learning. And, lest the picture painted here become too dominated by sombre Calvinist piety, it should also be added that they went to some trouble to offer him recreation and social life as well: on 29 September 1654, for example, 'there came to my lodging 8 of them, whereof one was a Burgomaster, and tooke me along with them into their Barge. We had a large dinner and then rowed up the Lake that I might see the country. We spent that afternoone upon the water; save that we once went out to view a large Vineyard belonging to one of them. It was darke ere we got into the city.'[96]

Building on his knowledge of Dutch, Pell was able to acquire competence in German too. Naturally, his passive knowledge of the language was in advance of his active use: in February 1656, for instance, he had a meeting with the Bürgermeister of Zurich in which 'The BurgoM^r spake to me in High-dutch, I, without expecting [sc. waiting for] an Interpreter, answered in Latine'.[97] The peculiarities of Swiss German may have been an additional obstacle: one of his letters to Samuel Morland commented on the difference between the Swiss dialect and 'the Language of Luthers & Piscators Bibles'.[98] Although Pell's papers include two examples of documents translated by him into English, his own correspondence with German-speakers was conducted in French or Latin; what seems to be the only German-language item written by him among all his manuscripts is the brief testimonial he wrote for his valet, Marcus Föge, when dispensing with his services in 1658.[99]

There is, however, one German-language text that has sometimes been credited to Pell's pen: a translation of the anonymous, quasi-official defence of the Cromwellian regime, *A True State of the Case of the Commonwealth of England, Scotland, and Ireland, and in Reference to the Late Establish'd Government by a Lord Protector and a Parliament* (London, 1654), issued under the title *Gründtliche Beschreibung der neuen Regiments-Verfassung in dem gemeinen Wesen Engelland, Schott- und Irrland samt den zugehörigen Eyländern und andern Landschafften, unter dem Herren Protector und dem Parlament*

---

[96] BL, MS Add. 4365, fo. 22^r, Pell to Thurloe, 30 Sept. 1654.
[97] BL, MS Lansdowne 746, fo. 18^r: Pell to Thurloe, 27 Feb. [1656].
[98] BL, MS Lansdowne 747, fo. 85^r. On the German translation of the Bible by the Calvinist scholar Johannes Piscator (1546–1625) see H. Schlosser, *Die Piscatorbibel: ein Beitrag zur Geschichte der deutschen Bibelübersetzung* (Heidelberg, 1908).
[99] BL, MSS Lansdowne 752, fos. 136, 331–2 (translations); Add. 4407, fo. 167^r (testimonial, 31 Aug. 1658).

(Schaffhausen, 1657).[100] The printer of this translation, Johann Kaspar Suter, dedicated the book to Pell, calling him 'my highly honoured lord and great patron'. In his dedicatory epistle Suter described the book as 'a little work by one of my highly honoured and best patrons', and explained that he was dedicating it to Pell in accordance with the principle *suum cuique tribuere* ('to give to each his own').[101] Accordingly, Pell is credited as the translator in the catalogues of some modern libraries.[102] That Pell (whose official instructions required him to 'endeavour to shew the true Cause of the Warres which haue beene in England, Scotland and Ireland, and the necessity and Justice thereof on the Parlaments part') was involved in the production of this book is clear: proofs of the first eight pages are to be found among his papers.[103] That he was regarded as a 'great patron' by Suter is also explicable: it seems that Pell had obtained a bursary for one of Suter's sons to study in England.[104] But it is highly improbable that Pell was either the author of the text – there is no trace of any notes or drafts relating to it among his papers, and when he drew up a detailed list of his services to the state in 1658 he made no mention of it – or the translator.[105] If Pell was involved in any way in the work of translation, the final responsibility for it must surely have lain with a native speaker. Suter may simply have mistaken a translation commissioned by Pell for one performed by him; or, perhaps, he may have known the truth, and chosen to disregard it for the sake of a compliment. Nevertheless, if Pell was not even partly responsible for the translation, he can be presumed to have been the author of the brief Preface to the Reader in Suter's edition, in which specific comparisons were made between the new English con-

---

[100] The original text has been attributed to Marchamont Nedham (e.g. by C. H. Firth, *The Last Years of the Protectorate, 1656–1658*, 2 vols. (London, 1909), i, p. 156); the attribution is accepted by the anonymous editors of the modern reprint of this work (Exeter, 1978), and by Blair Worden, who also notes that one passage in the text is borrowed from Nedham's *Mercurius politicus* (*The Rump Parliament*, p. 362). However, apart from that passage (which amounts to only one sentence) there seems to be no basis for the attribution, apart from the similarity between this work's title and that of Nedhams's *The Case of the Commonwealth of England, Stated* (London, 1650), from which this work differs completely in both contents and style.

[101] Anon., *Gründtliche Beschreibung*, reverse of title page ('Joanni Pellio ... meinem Hochgeehrten Herren und grossen Patronen'); sigs. ¶1ʳ ('Wercklein von einem meiner Hochgeehrten und besten Patronen'), ¶1ᵛ ('Man soll *Suum cuiq; tribuere*').

[102] For example, the Library of Congress (pressmark DA 420. T 78 1657), and the Houghton Library, Harvard (pressmark *EC 65 A100 En657g). Other copies of this work are in London, Munich, Wolfenbüttel, and Zurich: see G. Berghaus, *Die Aufnahme der englischen Revolution in Deutschland, 1640–1669* (Wiesbaden, 1989), p. 266, item 333.

[103] BL, MS Add. 4279, fos. 234–41.

[104] See above, n. 72. Pell's original connection with Suter (and hence the reason for his use of a printer in Schaffhausen rather than Zurich) may also be guessed at: the Swiss envoy to England, Stokar, who was 'Secretary' of Schaffhausen, had arranged the printing there of Comenius's *Janua linguarum*. See BL, MS Add. 4364, fo. 143ʳ, Pell to Hartlib, 2/12 Oct. 1656: '[Comenius] Whose Janua is very lately printed with copper-plates at Schaffhausen, according to the coppy which you gave Mʳ Stockar'.

[105] BL, MS Lansdowne 754, fo. 465 (list of services).

stitution and those of Venice, the Netherlands, and Switzerland: in particular, he singled out the holding of 'fixed meetings of Parliament, as in the most laudable [Swiss] Confederation, and in the High and Mighty States-General of the United Netherlands'.[106]

This was not the only Swiss publication in which Pell became involved. Early in 1657 he took an interest in a newly completed Latin translation of Francis Potter's anti-Catholic treatise, *An Interpretation of the Number 666* (Oxford, 1642); the translator's grasp of English was apparently far from perfect, and Pell had to correct errors at a late stage, in the proofs.[107] But his involvement in Swiss cultural life more generally is difficult to assess, since very few of his personal papers (as opposed to his diplomatic papers, or his correspondence with England) have survived from this period. It is nevertheless clear that he was on friendly terms with some of Zurich's most prominent intellectuals: not only Ulrich, but also the theologian Johann Rudolf Stucki, and (before his departure for Heidelberg in the summer of 1655) the historian and oriental scholar Johann Heinrich Hottinger.[108] And Pell certainly had access to the well-stocked collections of the city library (of which Stucki was 'Bibliothecarius'): among his papers there is a fragment in his hand, dated 1654, listing all the mathematical books held there, in alphabetical order from A to J.[109]

One connection made with a Zurich intellectual would resonate in a later period of Pell's life: his friendship with the young mathematician Johann Heinrich Rahn. A member of a prominent Zurich family (his father had been a city councillor since 1611, and served as Bürgermeister from 1655 to 1669), Rahn had probably acquired his special interest in mathematics from his uncle, who was the engineer responsible for the city's fortifications. He himself was a 'Schützenmeister', which meant that he supervised shooting practice, and a 'Zeugherr', responsible for military supplies and artillery; he also served as a 'Zensor', passing materials

---

[106] Anon., *Gründtliche Beschreibung*, sigs. ¶3v–¶4v (sig. ¶4r: 'gewisse Land-Tags-Versamlungen, wie in Hochlöblicher Eydgenossschafft, und bey den Hochmögenden Herren Staaden der Vereinigten Niderlanden').

[107] BL, MS Add. 4398, fos. 133–4 (J. Schönauer to Ulrich, 14 Jan. 1657), 135 (proofs of pp. 201–2, sig. N1), 141 (Pell to J. R. Stucki, 11 Feb. 1657, saying that he and Schönauer have tried to correct an error, but that hundreds of copies have been sent off already). The title of the translation is given here as *Explicatio numeri bestiae*; I have not seen a copy of this publication, which was presumably a different translation from the one later issued at Amsterdam in 1677 under the title *Interpretatio numeri 666*.

[108] On Stucki see Hottinger, *Schola tigurinorum Carolina*, pp. 170–1 (commenting that he wrote more than he published, but listing 19 published disputations and a commentary on Peter Martyr), and R. Pfister, *Kirchengeschichte der Schweiz*, 3 vols. (Zurich, 1964–84), ii, p. 552. For letters from Stucki to Pell see BL, MSS Add. 4419, fos. 43, 47, 48, 112, 152; Add. 4421, fo. 1; Add. 4279, fos. 254–64. On Hottinger see Pfister, *op. cit.*, ii, pp. 552–3; Fück, *Die arabischen Studien*, pp. 91–2; H. Steiner, *Der Zürcher Professor Johann Heinrich Hottinger in Heidelberg, 1655–1661* (Zurich, 1886). For evidence of Hottinger's acquaintance with Pell see below, n. 115; Pell also corresponded with Hottinger from England in the period 1659–60 (see J. Worthington, *The Diary and Corrrespondence*, ed. J. Crossley, 3 vols., Chetham Society, vols. 13, 36, 94 (1847–86), i, pp. 172, 181, 198).

[109] BL, MS Add. 4407, fos. 42–3 ('Bibliotheca Civitatis Tigurinae 1654').

for publication.¹¹⁰ The first sign of his connection with Pell is a letter he sent, dated 4 November 1654, enclosing a short mathematical demonstration; in the letter, he thanked Pell for giving him a copy of his anti-Longomontanus compilation, the *Controversiae pars prima*.¹¹¹ What further contacts they had over the next two years is not known; but at some time in early 1657 Rahn began to receive regular weekly tutorials from Pell.¹¹² As John Aubrey would later write (summarizing what Pell had told him), 'Rhonius [the latinized version of 'Rahn'] was Dʳ Pell's scholar at Zurich, and came to him every friday night after he (J Pell) had writt his post-letteres.'¹¹³ A surviving letter from Rahn to Pell, dated 9 March 1657, confirms the details of Aubrey's account: it encloses the mathematical problem Pell has set him, and expresses Rahn's regret that he will not be able to come on the following Friday.¹¹⁴ Just eight days before that, Rahn had written to his friend Hottinger in Heidelberg, thanking him for sending a mathematical problem by Descartes (an extract from Descartes's correspondence with Princess Elizabeth of the Palatinate, which Hottinger had received from the Princess herself). Rahn passed on to Hottinger personal greetings from Pell, and added: 'I visit him every week, and experience no small signs of his special affection ... at each meeting we devote quite a long time entirely to mathematical recreations, in which field I regard him as incomparable.'¹¹⁵ Pell himself would refer to Rahn as his 'disciple', and would retain 'coppies of the most considerable papers that he wrought in my presence or that I gave him to transcribe'.¹¹⁶

In early 1658 Rahn was appointed 'Landvogt' (provincial governor) of Kyburg, a district at the northern edge of the canton of Zurich; writing to Pell from there on 3 March 1658, he remarked that his heavy administrative workload prevented him from spending any time on mathematics, but said that he had the consolation of remembering 'how many very delightful hours' they had

---

[110] See W. Schnyder-Spross, *Die Familie Rahn von Zürich* (Zurich, 1951), pp. 125–69 (father), 262–79 (son). See also R. Wolf, *Biographien zur Kulturgeschichte der Schweiz*, 4 vols. (Zurich, 1838–62), iv, pp. 55–66.

[111] BL, MS Add. 4423, fos. 50–1, referring to Pell's gift as 'that dispute about the measurement of the circle' ('Certamen illud cyclometricum').

[112] The date is uncertain. After the letter of 4 Nov. 1654, the next surviving evidence of contacts is Rahn's letter to Pell of 23 Jan. 1657 (BL, MS Add. 4279, fo. 212), referring to a copy of Viète. Rahn's account of his tuition by Pell in his letter to Hottinger of 1 Mar. 1657 (see below, at n. 115) presents it as recent news.

[113] Bodl., MS Aubrey 6, fo. 55ʳ.

[114] BL, MS Add. 4394, fo. 455.

[115] Zentralbibliothek, Zurich, MS F 71, fo. 235, Rahn to Hottinger, 1 Mar. 1657 (fo. 235ʳ: 'denne besuche Ich wochentlich; Vnd geffüre nit geringe Zeichen sonderer affection. Vnd ... wir bei Jeder Conferentz eine Zimmliche Zeit, alles Inn Mathematischen ergetzlichkeiten (Inn welchem studio Ich Ihnne für incomparabilem halte) Zuebringen'). Later in this letter Rahn suggested that the Princess might like to engage in mathematical correspondence with Pell; it seems that the suggestion was not taken up.

[116] BL, MSS Add. 4424, 260ʳ (note on Rahn: 'Pellij discipulus'); 4278, fo. 80ʳ, Pell to T. Brancker, 5 Mar. 1666 (quotation).

spent together.[117] Nevertheless, during the next year he somehow found time to compose the German-language textbook on algebra that would make him famous: *Teutsche Algebra, oder algebraische Rechenkunst, zusamt ihrem Gebrauch* (Zurich, 1659).[118] This work was heavily indebted to Pell, and it was a debt that Rahn was happy to acknowledge (though, probably in deference to Pell's own scruples, without mentioning his name): Rahn explained in his preface that 'in the solutions, and in the arithmetic too, I make use of a completely new method ... which I first learned from an eminent and very learned person.' This new method, he explained, consisted in 'a triple margin' – Pell's system of presenting the working-out of a problem in three columns. But that was not the only thing owed to Pell in this volume: Rahn also introduced Pell's division sign, and in a section on squaring the circle he presented the theorem Pell had defended in his *Controversiae pars prima*.[119] It has even been suggested by some writers (incorrectly, but not surprisingly) that the whole text had been plagiarized by Rahn from Pell.[120] Rahn's work, though indebted to Pell, was his own, but it would in turn occasion, nine years later, the only substantial mathematical publication that Pell ever produced: an English translation and adaptation of Rahn's book would contain nearly 100 pages of material reworked or added by Pell himself.[121]

As in the case of his six years at Breda, Pell's four years in Zurich yield no signs of any progress whatsoever on the various mathematical projects that he had once promised to complete. No doubt the heavy burden of his diplomatic duties is a major part of the explanation; but Pell's capacity for unremitting mental labour should not be underestimated, and it may be that many hundreds of pages of mathematical notes and tables have simply failed to survive from

---

[117] BL, MS Add. 4365, fos. 5–6, Rahn to Pell, 3 Mar. 1658 (fo. 5^r: 'quot horas dulcissimas').

[118] One comment by Pell might be taken to imply that Rahn had already composed part of his treatise before Pell taught him: writing to Collins on 17 Apr. 1667, he discussed Frans van Schooten's *Principia matheseos universalis* (Leiden, 1651) and noted that it was cited on p. 68 of Rahn's book, but added: 'Yet you may defend your conjecture, that Rhonius had written the beginning of his booke, before he had seene Van Schooten, which he had not heard of, till I told him of it' (BL, MS Add. 4278, fo. 122^r). But on the other hand it is not known how long the interval was between the time when Rahn was 'told of' the book, and the time when he first saw a copy of it.

[119] Rahn, *Teutsche Algebra*, sig. XX2^r ('In den Solutionen, und grad auch in der Arithmetic bediene ich mich einer ganz neuen manier ... die ich von einer hohen und sehr gelehrten Person erstmals erlehnet hab ... Dieser form bestehet in einem dreyfachen *Margine*'); pp. 8 (division sign), 187 (Pell's theorem).

[120] See Wolf, *Biographien zur Kulturgeschichte*, iv, p. 62, and Bom, *Het hooger onderwijs te Amsterdam*, p. 90, item 1234.

[121] See below, pp. 201–5. For a useful account of the contents of Rahn's book see G. Wertheim, 'Die Algebra des Johann Heinrich Rahn (1659) und die englische Übersetzung derselben', *Bibliotheca mathematica: Zeitschrift für Geschichte der mathematischen Wissenschaften*, ser. 3, no. 3 (1902), pp. 113–26.

this period.[122] One project about which he had apparently left some promises hanging in the air was the planned publication of the Warner–Pell tables of antilogarithms; here the patience of Warner's relative, Mr Tovey, was finally exhausted. In June 1655 Herbert Thorndike wrote to Pell: 'It was very much desired by Mr Tovey & his freindes for the remembrance of his Vncle, & the good of learning, to see Mr Warners Canon published by you, as the person y^t could giue most lustre to it. But having lost ye hope of seeing it done by you in reasonable time ... hee hath resolued to seeke what other means can be had, to bring it to light.'[123] Pell replied rather lamely in July that he had not returned the manuscript of the tables to Thorndike before his departure from England 'because I had then some reasons to expect a speedy returne'; a month later, he sent instructions to Ithamar to hand over to Thorndike 'the box, with ye broken cover, in my study, [which] hath Mr Warners papers in it'.[124]

That Pell was inwardly dissatisfied with his failure to produce his long-promised mathematical publications is revealed by a painful episode in his correspondence with Samuel Hartlib. In July 1655 Hartlib sent him a letter which began: 'Sir, these many months I have received not one substantial letter from you...', and ended with a postscript: 'I pray answer ... the adioyned Math: Paper'.[125] The paper enclosed with this letter was a text (which had been forwarded to Hartlib via Joachim Hübner in Berlin) by Baron Wolzogen, and it included a complaint about the non-appearance of the work promised in Pell's *Idea*:

> I myself began to attempt a sort of summary of writings on geometry several years ago, and had already made some progress in it. However, that wide-ranging promise of the 'Mathematicus αὐτάρκης', which was brought to me from England, interrupted my work, with the result that of those things which I had worked out (over many sleepless nights), some have perished, and others have been under attack for sixteen years (more or less) by moths and beetles. For that is indeed the length of time during

---

[122] A possible (and rare) exception is a set of notes on the problem of constructing a 14-sided polygon, taken from a text published in Munich in 1653: BL, MS Add. 4429, fos. 51^v–55^r. Some of Pell's mathematical papers from this period were taken in the early 1660s to his rectory at Fobbing, Essex (see below, ch. 7, at n. 18), and seem not to have been brought back to London after his death.

[123] BL, MS Add. 4364, fo. 167^r (Thorndike to Pell, 21 June 1655).

[124] Ibid., fo. 168 (Pell to Thorndike, 14/24 July 1655); BL, MS Add. 4280, fo. 215^r (Pell to Ithamar, 10 Aug. 1655). The reason for the delay emerges from Pell's letter to Hartlib of 14/24 July, in which his letter to Thorndike was enclosed: he comments that Thorndike is in danger of being 'seazed as a malcontent', and asks Hartlib to keep the enclosed letter until Thorndike is at liberty (BL, MS Add. 4364, fo. 136^r).

[125] BL, MS Add. 4408, fo. 262^r, Hartlib to Pell, 13 July [1655].

which the mountains laboured but failed to bring forth that English embryo.[126]

Hartlib may possibly still have hoped that Pell would produce the publications he had promised in his little broadsheet of 1638, and may have imagined that sending on this comment from Wolzogen would stimulate him to do so. All he achieved thereby, however, was to cut Pell's thin-skinned personal pride to the quick. Some time in early August, Pell drafted the following response:

> Sir. August ye second I received a letter from you, dated July 13. Wherein you tell me that in many months you had received not one *substantiall letter* from me. Till I know what you meane by a *substantiall letter*, I shall thinke it is not in my power to write such a one, & therefore cannot promise you to doe it heere after. If that [letter *deleted*] which I wrote to you above [>seventeene] yeares agoe were such, you have this weeke sent me a paper fit to deterre me from writing so hence-forward. You printed it about 3 months after I wrote it & gave away many coppies of it. I remember some liked it well. Our Austrian Baron then wrote an applauding letter in Latine to mee: But age hath turned his sweet wine into vinegar & now he sayes *parturiebant montes*. It made me thinke of ye Spanish proverb y^t sayes a man may eate their Kings oxe & pay for the bones a hundred yeares after. [>One may write] a letter to you & be called to account for it seventeene yeare after. I did never eate his bread and therefore I thinke I am not bound to give him account how I have spent my time ever since & why I have printed so little. If I be forced to write such an account you must not have a coppy of it, because it will reflect [>very] much upon his Lordship; & if by those hands y^t conveyed those passages of his letters to mee, my answer should returne to him, it might disquiet him so much y^t it might hinder him from publishing the booke which he now sayes he is fitting for ye presse. So that if heere-after a man should aske him why ye Polish plaines had been as barren as ye English mountaines, you & I must beare all ye blame.[127]

This outburst tells us much about Pell's pride, and about the rankling sense he must have had of his own unproductiveness; but, on the other hand, the fact that this draft was never committed to the post also tells us something about

---

[126] Ibid., fo. 263^r: 'Ego in Geometricis pandectis ante plurimos annos aliquid tentare cepi, & progressum iam aliquem feceram: sed largum illud Mathematici αὐτάρκεος promissum, ex Anglia ad nos allatum, opus ita interrupit, ut quae elaborata fuerunt non sine vigilijs, partim perierint, partim ab annis plus minùs 16. cùm blattis & tineis pugnent. A tanto verò tempore, non contigit Anglicanum illum foetum, quem Montes parturiebant videre.' This alludes to the proverb, 'the mountains have laboured, and have brought forth a mouse'. For Wolzogen's own project, see ch. 3, at nn. 35–6.

[127] BL, MS Add. 4364, fo. 139^r, Pell to Hartlib (draft, written between 2 and 11 Aug. 1655).

his self-control and his desire, in the end, not to hurt the feelings of an old and trusted friend. Instead, his next letter to Hartlib merely stated that 'I read over the extracts of the Barons letters that came in yours of July 13 and I began to pen something upon occasion of what he there writes so wooddenly & yet so confidently', and added that his other duties had then supervened, so that 'there was enough to take me off from thoughts of M[y] L[ord] Wolzogen.'[128]

Quite a full set survives of Pell's letters to Hartlib from Zurich – the copies Pell kept are methodically numbered from 1 to 51, though it is clear that the first of these (dated 24 March [1655]) was by no means the first he had sent his friend from Switzerland.[129] These letters cover a wide range of subject-matters, intellectual, political, and personal: not only was Hartlib as supportive as ever of his old protégé, but he was also engaged in the production of news-letters of his own, for which Pell's news from Zurich could furnish useful raw material.[130] At the same time, he was always eager for information that might be put to some other practical use: thus, for example, in March 1655 Pell sent him (via Dury) 'halfe a sheete of paper printed in High-Dutch concerning vines'.[131] From the evidence of these letters to Hartlib, it is clear that Pell was still keeping up his mathematical interests. In July 1655 he commented on Riccioli's *Almagestum novum*, calling it 'a disorderly piece of work'; in July 1657 he made some scathing remarks about the mathematician Luneschlos, and in December of that year he reported on a new publication by him, which had already received a refutation in print.[132] When told by Hartlib that the mathematician Nicolaus Mercator was travelling from England to Paris, he sent a list of queries for Mercator to make about Roberval – for instance, 'What Mr Roberval hath printed', and 'What pieces of his, he hath so communicated in Manuscript, as that coppies may be had for mony'.[133] And in one drafted letter (annotated 'Intended for Mr Hartlib, but not sent') he included a fascinating anecdote he had heard about Descartes:

> Sr Charles Cavendysh told me that at Paris one spake against ye new Algebra (as some call it) in the presence of Monsieur des Cartes. Who well perceived that it was done out of designe to provoke him to speake Wherefore he thought fit to reply something

---

[128] Ibid., fo. 156^v, Pell to Hartlib, 11 Aug. [1655].

[129] The numbered copies (together with some items sent by Hartlib) are in BL, MS Add. 4364, fos. 135–66; the first letter (fo. 135^r) includes a reference to a letter Pell had sent him, via Dury, the previous week. A less full set of Hartlib's letters to Pell from this period is in BL, MS Add. 4279, fos. 41–57.

[130] See for example BL, MS Add. 4364, fos. 49^r, 111^r (extracts, in the hand of one of Hartlib's copyists, from letters to Hartlib from Zurich, respectively 28 Apr. 1655 and 27 May 1654).

[131] Ibid., fo. 135^r, Pell to Hartlib, 24 Mar. [1655], referring to a letter sent on 17 Mar.

[132] Ibid., fos. 135^v (14 July [1655]), 150^r (9 July [1657]), 154^r (17 Dec. 1657).

[133] Ibid., fo. 146^r (9 Apr. 1657). By Aug. Mercator was back in England, without having supplied any answers to these questions. Returning to this subject in his letter to Hartlib of 27 Aug. (fo. 151^r), Pell remarked that Roberval used to publish some of his work in manuscript only, 'after the old fashion'; Hartlib's transcription of Pell's queries, including this passage, is HP 31/12/14A.

PART I: THE LIFE OF JOHN PELL 169

which he hoped all ye company would understand. S^r, said he, the advantage of those new calculations in Geometry, is most easily understood by comparing Midorge with Roberval. Mons Midorge hath a head *ten times* fitter for Geometry than Mons Robervals, I say *ten times*: And yet Roberval shall doe ten yea a hundred times more in Geometry than Midorge. Because Roberval is more exercised in that Logisticall way both for inquisition & demonstration & He hath more patience to pursue a tedious calculation when it is not his good fortune to take up the shortest way at the first setting out. Where-as Midorge keeps neerer to the old fashion; which oftentimes brings a man into a Labyrinth but gives him no clew to lead him out of it.

These words of Monsieur des Cartes were so much the more taken notice of: because it was well known y^t they were not spoken out of any affection to Roberval, who never had expressed any care to flatter Des Cartes, but rather a humour to crosse & disoblige him.[134]

Another of Pell's earlier concerns that resurfaced in this correspondence was the idea of a universal language, thanks to the interest taken by Hartlib in the work of George Dalgarno. In the mid-1650s this Aberdonian schoolmaster, who taught in Oxford, was trying to develop a new system of shorthand. It occurred to him that the symbols or characters on the page could be contracted still further if, instead of standing for letters or sounds, they stood directly for things or concepts. He then devised a system of symbols based on an overall categorization of reality, with a 'radical' symbol for the genus of a thing, an additional mark for a particular species, and surrounding marks to indicate opposites, privatives, or other conceptual qualifications. The result was what became known as a 'real character' – a term that would be popularized by John Wilkins, whose own *Essay towards a Real Character* (London, 1668) was stimulated by Dalgarno's work. And, what was more, once Dalgarno had organized his categorization of reality, he then found that it could also be expressed in a set of (arbitrarily chosen) sounds, so that the word for any complex concept would itself display, in its structure and the components out of which it was built, the nature of the concept itself. He had thus passed from shorthand to real character, and from real character to universal language.[135] In April 1657 Faustus Morstyn, a Pole studying at Oxford, wrote to Hartlib about Dalgarno's work; this was

---

[134] BL, MS Add. 4364, fo. 153 (undated; late Nov. or early Dec. 1657).

[135] See M. Greengrass, 'George Dalgarno', in A. Pyle, ed., *The Dictionary of Seventeenth-Century British Philosophers*, 2 vols. (London, 2000), i, pp. 233–4; V. Salmon, 'The Evolution of Dalgarno's "Ars Signorum"', in V. Salmon, *The Study of Language in 17th-Century England* (Amsterdam, 1979), pp. 157–75; Knowlson, *Universal Language Schemes*, pp. 72–107; Rossi, *Logic and the Art of Memory*, pp. 155–67; Lewis, 'John Wilkins's *Essay*', pp. 124–56. For a valuable contemporary account of Dalgarno's work (commenting also on Wilkins's relation to it) see R. Plot, *The Natural History of Oxford-shire* (Oxford, 1677), pp. 282–4.

followed by a letter from Dalgarno himself, giving a more detailed account of it. Remembering Pell's earlier efforts in this field, Hartlib forwarded both letters to him in Zurich.[136] Pell gave the matter his careful attention, but in June 1657 he replied, reasonably enough, that 'A very small specimen would better enable me to understand the designe, than twenty such letters can doe.'[137] In the following month Hartlib sent him both the 'specimen' he desired (a short extract from the Gospel of St John, ch. 16, in Dalgarno's real character) and a printed broadside, in Latin and English, announcing the invention.[138] Among Pell's papers there are a few pages of notes in which he tried out Dalgarno's symbols; but when the Scottish inventor finally contacted him directly at the end of the year, the verdict Pell gave him was unfavourable. The problem, in his opinion, was that the requirements of a shorthand system and those of a real character were not the same, and that Dalgarno's scheme was an awkward attempt to combine the two: 'I, indeed', as Pell put it in his polite but discouraging letter, 'do not want to hunt two hares at the same time.'[139]

Thanks to Hartlib, Pell was thus provided with new contacts, as well as being supplied with news of old friends. Other fresh contacts included Hartlib's assiduous correspondent from Herefordshire, John Beale, a man of wide intellectual interests who was also passionate about the improvement of agriculture.[140] The first reference to him in Pell's correspondence occurred in June 1657, when Pell mentioned 'the Hereford advice' which Hartlib had sent him.[141] More material

---

[136] BL, MS Add. 4377, fos. 147 (Morstyn to Hartlib, Apr. 11 1657, original, with endorsement in Hartlib's hand addressing it to Pell, and an annotation by Pell: 'Received May 2. 12. Z[u]r[ich] 1657'), 148 (Dalgarno to Hartlib, 20 Apr. 1657, scribal copy, with endorsement in Hartlib's hand addressing it to Pell, and an annotation by Pell: 'Zurich June 6. 1657').

[137] BL, MS Add. 4364, fo. 146^v (Pell to Hartlib, 11/21 June [1657]).

[138] BL, MSS Add. 4377, fos. 144 (broadside, in Latin (*Character Universalis*) and English (*A New Discovery of the Universal Character*), with endorsement in Hartlib's hand addressing it to Pell), 149^r (Dalgarno specimen, with letter from Hartlib to Pell, 3 July 1657); Add. 4364, fo. 150^r (Pell to Hartlib, 9/19 July 1657: 'The last week I received from you Mr Dalgarno's print, in Latine & English').

[139] BL, MS Add. 4377, fos. 150–2 (Dalgarno to Pell, undated, with annotation by Pell: 'Zurich Decemb 26 old style 1657'), 153–4 (Pell notes, using Dalgarno's symbols), 159^r (Pell, response to Dalgarno: 'Ego verò duos simul lepores venari nolo'). Dalgarno was not discouraged: the same MS contains his letter to Brereton of 17 Feb. 1658 (fo. 155), seeking his support, and a printed broadside dated 1 May 1658 (fo. 157), with the names of many academic commenders of his work (a list of names from which Pell was conspicuously absent). His treatise *Ars signorum, vulgo character universalis et lingua philosophica* was eventually published in 1661.

[140] On Beale (1608–83) see M. Stubbs, 'John Beale, Philosophical Gardener of Herefordshire. Part I: Prelude to the Royal Society (1608–63)', *Annals of Science*, 39 (1982), pp. 463–89, and 'John Beale, Philosophical Gardener of Herefordshire. Part II: The Improvement of Agriculture and Trade in the Royal Society (1663–83)', *Annals of Science*, 46 (1989), pp. 323–63; P. Goodchild, ' "No phantasticall Utopia, but a reall place" ': John Evelyn, John Beale and Backbury Hill, Herefordshire', *Garden History* 19 (1991), pp. 105–27; and M. Leslie, 'The Spiritual Husbandry of John Beale', in M. Leslie and T. Raylor, eds., *Culture and Cultivation in Early Modern England: Writing and the Land* (Leicester, 1992), pp. 151–72.

[141] BL, MS Add. 4364, fo. 150^r (Pell to Hartlib, 11/21 June [1657]: 'Concerning the Hereford advice, you desire me to enlarge, which I have not now time to doe').

from Beale was sent later in the year, although without revealing Beale's name; the reason for this was probably not any desire for secrecy on Beale's part, but merely the fact that parts of his letters to Hartlib were regularly included in Hartlib's scribally produced news-letters, under the bare heading 'Hereford'. In January 1657 Pell told Hartlib that he had received 'two Hereford-papers dated Octob. 22 & Novemb. 3 concerning L in T', adding that 'I hope your Hereford-friend will give you leave to tell me his name': Beale had millenarian interests too, and his papers here were comments on Comenius's compilation of prophecies about the fall of the Habsburgs, *Lux in tenebris*.[142] In response, Hartlib gave Pell not only Beale's name, but also directions for writing to him, and encouraged him to do so.[143] But some of Pell's communications with Beale continued to pass via Hartlib, so that the latter could recirculate them elsewhere: thus an account of the various types of wine, cider, and perry made in Switzerland (compiled by Pell primarily for Beale's benefit) was sent to Hartlib, who then included it in one of his letters to Robert Boyle.[144] (Pell was not yet known personally to Boyle, but the several references to him in Hartlib's letters to Boyle during this period would prepare the way for their personal acquaintance, which came about after Pell's return to England.) Another contact, facilitated by Hartlib, was with Hartlib's Cambridge friend Dr John Worthington: in November 1655 he asked for Pell's help in approaching the great Hebrew scholar Johannes Buxtorf. Pell did so, and forwarded letters from Buxtorf to Worthington both then and in 1658.[145]

Samuel Hartlib's letters sometimes supplied information about old friends – such as William Brereton, who sent a message apologizing for having been 'Pen-dead' for two and a half years.[146] Hartlib passed on news of Comenius, who suffered a grievous blow in May 1656 when the small town of Leszno – where he presided over the community of Bohemian exiles who had settled there – was sacked by Polish forces. Comenius had sent Hartlib an account of the losses he had undergone, listing unpublished manuscripts that had been destroyed: these included his 'refutations' of both Copernicus and Descartes. In his reply Pell could not resist the dry remark that 'of all his papers, there is none for whose losse I am lesse sorry'; but he did also go to the trouble of making copies of Comenius's account and distributing them in Zurich, in order to elicit donations

---

[142] Ibid., fo. 154^v (Pell to Hartlib, 7/17 Jan. 1658).

[143] BL, MS Add. 4279, fo. 49 (Hartlib to Pell, 4 Feb. 1658; printed in Vaughan, *Protectorate*, ii, p. 439).

[144] BL, MS Add. 4364, fos. 155^v–156^r (Pell to Hartlib, 11/21 Mar. 1658); Boyle, *Correspondence*, i, p. 261 (Hartlib to Boyle, 8 Apr. 1658).

[145] BL, MS Add. 4364, fos. 156^v (Pell to Hartlib, 1/11 Apr. 1658, with extract of letter from Buxtorf to Worthington), 165^r (Worthington to Hartlib, for Pell, 19 Nov. 1655), 166^r (copy of Buxtorf's reply).

[146] BL, MS Add. 4279, fo. 51^r (Hartlib to Pell, 18 Feb. 1658). Hartlib had previously reported that Brereton was 'transported or ravished' by Beale's writings: ibid., fo. 49^r (Hartlib to Pell, 4 Feb. 1658).

to Comenius and his flock.[147] A more direct act of charity was performed in the following year, when Pell learned (again, from Hartlib) that Moriaen, who had moved to Arnhem to supervise a dye-works, was now living in poverty. Pell had sent money to Moriaen (via Theodore Haak) before; he now asked that £10 be taken from the next quarterly payment of his salary and transmitted to his friend in the Netherlands. (Delays in the payment of his salary meant that the sum had to be advanced by Haak instead.)[148] And, last but not least, Pell's letters also show a real solicitude for Hartlib himself; when Hartlib gave details of the excruciating pains he suffered from a combination of gall-stones, bladder-stones, and piles, Pell did his best to supply helpful, and common-sensical, medical advice.[149]

Another old friend who was corresponding directly with Pell throughout this period was Theodore Haak; a handful of his letters have survived. They show that he frequently sent Pell books: in September 1655, for example, Haak sent 'Dr Gouge on ye Hebrews', and also mentioned that 'I was about to have bought Hobbes his book [*De corpore*], but Mr Brereton sends it you.'[150] In the same letter he informed Pell that the Oxford mathematician John Wallis was writing a refutation of Hobbes's book; and in June 1657 he sent 'a coppie of Hobbes against Wallis; if perhaps you should have gotten none by other frends'.[151] These are almost the first mentions of John Wallis in Pell's correspondence, and it is no doubt significant that they arose in connection with his dispute with Hobbes: although, as Pell would soon realize, Wallis was vastly superior to Hobbes in mathematical ability, Pell's starting-point was on the pro-Hobbes side of the divide, thanks to his own previous relations with Hobbes (who had contributed

---

[147] BL, MS Add. 4364, fo. 141^r (Pell to Hartlib, 17/27 July 1656); this letter is printed in Vaughan, *Protectorate*, ii, pp. 430–2, but with the intended recipient unidentified. On the sack of Leszno see Blekastad, *Comenius*, pp. 545–52.

[148] BL, MS Add. 4364, fo. 152^r (Pell to Hartlib, 19/29 Nov. [1657], commenting on the fact that Moriaen had failed to benefit financially from the death of his brother-in-law Peter van Zeuel). See also Hartlib's account in his letter to Boyle of 2 Feb. 1658: Boyle, *Correspondence*, i, p. 252. Haak's confirmation that he had made the payment is in MS Add. 24850, fo. 19^r (Haak to Pell, 28 Jan. 1658). The state of Moriaen's finances is hard to assess: see the comments on his activity as an entrepreneur in Young, *Faith, Medical Alchemy and Natural Philosophy*, pp. 58–62. It may be that he had borrowed heavily, on the strength of his expectations of an inheritance from his brother-in-law.

[149] For Hartlib's descriptions of his ailments see Vaughan, *Protectorate*, ii, pp. 435–7, and Boyle, *Correspondence*, i, p. 263. For Pell's advice see BL, MS Add. 4364, fo. 151^v (Pell to Hartlib, 15/25 Oct. [1657]). He noted Hartlib's strange insistence on going to stool three or four times a day, and suggested: 'Surely they doe better, who doe not regard the houre of the day, but the stimulations of Nature, whereby she craveth and urgeth Exoneration.'

[150] BL, MS Add. 24850, fo. 1^r (Haak to Pell, 14 Sept. 1655). The work by William Gouge (a leading Presbyterian divine and a supporter of Hartlib, who had died in 1653) was *A Learned and Very Useful Commentary on the Whole Epistle to the Hebrews* (London, 1655).

[151] BL, MS Add. 24850, fo. 12^r (Haak to Pell, 16 June 1657). Wallis's earlier refutation of Hobbes was his *Elenchus geometriae hobbianae* (Oxford, 1655); this work by Hobbes was ΣΤΙΓΜΑΙ... *Or, Markes of the Absurd Geometry, Rural Language, Scottish Church-Politicks and Barbarismes of John Wallis* (London, 1657).

to *Controversiae pars prima*) and the continuing friendship between Hobbes and Pell's pupil–patron, William Brereton.[152] The very first mention of Wallis comes in a fragment of a letter to Pell (copied by him, without identifying the author) of 27 April 1655, which was sent with a single printed leaf which Wallis had issued eleven days earlier, setting out a new method for squaring the circle. 'In ye meane time', the unidentified author wrote, 'accept of Dr Wallis Quadratura Circuli here adjoined, which I intreate you to handle soundly. For hee makes himselfe beleeve you will doe no great matters in Mathematicall studyes.' Pell's reply was cool and non-committal, expressing his uncertainty about the meaning of one of the technical terms used in Wallis's text.[153] Wallis's reason for rushing out this single leaf was in fact that he had heard of the impending publication of Hobbes's *De corpore* (which was almost ready at the printer's: its dedicatory epistle is dated 23 April), and was afraid that Hobbes's own circle-squaring chapter in that book might pre-empt him.[154] It seems very likely, then, that Pell's unidentified correspondent was someone with an interest in, or a special sympathy for, Hobbes. Haak is one plausible candidate, in view of the trouble he later took to keep Pell abreast of the Wallis–Hobbes dispute; Brereton is another; and a third – who is not otherwise known to have corresponded with Pell, but whose interest in this matter means that he cannot be ruled out – is Hobbes himself.

Two other themes recur in Haak's letters during these years: his concern at the constant delays in the payment of Pell's salary (what he called 'ye manifold disappointings about ye monie'), and his hopes that Pell would soon return to England.[155] In September 1657 he wrote that 'I did often of late wish for your companie heer, because of that stirre & pretty probable paines & designes of some ingenious persons heer to finde out & frame an universall character & language.'[156] And four months later he suggested that if Dr Goddard vacated

---

[152] The extent of Pell's direct personal acquaintance with Hobbes remains, however, uncertain. He may have met him through Cavendish in 1640, and again through Cavendish and/or Brereton in 1652–4, but direct evidence of such contacts has not survived. Siegmund Probst makes the unwarranted statement that Pell was, by the early 1660s, one of Hobbes's 'oldest acquaintances' ('ältesten Bekannten'), and wrongly claims that Pell had visited Paris with Cavendish in 1645: *Die mathematische Kontroverse zwischen Thomas Hobbes und John Wallis* (Hanover, 1997), pp. 112–13, n. 27.

[153] BL, MS Add. 4418, fo. 210^v (unidentified writer to Pell, 27 Apr. 1655; Pell's reply, 26 May 1655). The printed paper by Wallis was dated 16 Apr.; it was later reprinted in J. Wallis, *Arithmetica infinitorum* (Oxford, 1656), sig. Bb4. See the discussion of it in Stedall, *Discourse concerning Algebra*, pp. 165–7 (reproducing Wallis's text on p. 166, and giving Pell's reply on p. 167).

[154] See the account given by Wallis in a letter to Huygens: Huygens, *Oeuvres complètes*, i, pp. 336–7 (where the editors of Huygens mistakenly identify the whole dedicatory epistle to Oughtred of *Arithmetica infinitorum* as the preliminary publication referred to by Wallis). As he told Huygens, when he finally saw *De corpore* he realized he had nothing to fear; but he did find that some of Hobbes's comments on the rectification of the 'parabolaster' (*De corpore*, XVIII.2) were based on the same principles as the ones he was using in his work on the circle.

[155] BL, MS Add. 24850, fo. 5^r (Haak to Pell, 13 June 1656).

[156] Ibid., fo. 15^r (Haak to Pell, 24 Sept. 1657).

the Wardenship of Merton College, Oxford, in order to take up the Mastership of the Savoy in London, Pell should apply for the Wardenship himself. 'Perhaps it might not be remisse', he suggested, 'from these hints, to minde your great friends in time, of a convenient retiring place for your self.'[157]

One other person corresponding regularly with Pell was his wife, Ithamar: her letters were usually sent in the same cover as Hartlib's. Almost nothing from her side of the correspondence survives; those of Pell's letters that do are, unfortunately, dominated by his irritation at her failure to pass on the quarterly instalments of his salary in exact accordance with his instructions. His tone, in the words of one modern scholar, 'is consistently irascible, niggardly, and unpleasant' – though the facts presented in his letters do show that from time to time he had good reason to be irked by his wife's casual approach to financial matters.[158] Among other causes of irritation were Ithamar's grandiose plans for the impending marriage of their daughter Mary to a Captain Raven, which took place in November 1656: Pell warned against extravagance, while allowing as a dowry the not inconsiderable sum of £200.[159] (He was also displeased to learn, subsequently, that Mary was already pregnant at the time of the marriage.)[160] And another on-going problem concerned Ithamar's sister Bathsua, who was trying to extricate herself from the tenancy of a house (an arrangement in which, apparently, Pell had stood surety for her) and had begun to angle for free accommodation in Ithamar's home instead.[161] 'As for M[rs] M[akin],' he wrote in October 1654, 'you can hardly fall upon a foolisher resolution than to let her come to lodge, where you have power to keepe her out. Your best way will be to have nothing at all to doe with her, for feare of Had I wist.'[162] And again, two years later: 'I cannot but put you in minde of the great inconvenience which H[enry] R[eynolds] & his daughters found, in the boarding of R[ichard] M[akin] & his wife B[athsua]. And therefore I praise not your intention to give them two months board; you may be after troubled to get them out.'[163] Five months after that, his exasperation boiled over when he learned that his sister-in-law

---

[157] Ibid., fo. 19^r (Haak to Pell, 28 Jan. 1658).

[158] Teague, *Bathsua Makin*, p. 80. Many of Pell's letters are printed (from BL, MS Add. 4280) in Vaughan, *Protectorate*, ii, pp. 387–429. A fragment of a letter from Ithamar complains: 'it is a whole yeare that is due to you this Midsomer ... I am allready 20^{li} in depte ... I haue not one freinde where I cane be pleasured with 10^s therefore iudge how I am put to my shifts' (BL, MS Add. 4280, fo. 256^r).

[159] Vaughan, *Protectorate*, ii, pp. 412–19. The marriage, to 'Roger Raven, gent.', took place on 27 Nov. 1656: Burke, *Memorials of St Margaret's Church*, p. 370.

[160] BL, MS Add. 4280, fo. 220^r (Pell to Ithamar, replying to her letter of 9 July [1657] and commenting that it 'tells me of a forward Lad, borne, before his mother had beene married full 31. weekes'). Henry Raven was baptized on 1 July 1657 (Burke, *Memorials of St Margaret's Church*, p. 248), but died within the following year (see below, at n. 173).

[161] The details are obscure: see BL, MS Add. 4280, fo. 243 (Pell to Bathsua, 30 Apr. 1654; printed in Teague, *Bathsua Makin*, p. 82); Vaughan, *Protectorate*, ii, p. 391 (Pell to Ithamar, 26 Aug. 1654, about his personal liability to the landlord).

[162] BL, MS Add. 4280, fo. 241^r (Pell to Ithamar, 21 Oct. [1654]).

[163] Ibid., fo. 218^v (Pell to Ithamar, 2 Oct. 1656).

PART I: THE LIFE OF JOHN PELL   175

was encouraging Ithamar to ignore his instructions about his salary: 'I perceive Mrs Mak. hath found the way to you againe. Her counsel, added to your owne inclinations, will make you altogether unfit to meddle with any mony of mine. So that I shall be forced to seeke some other, who will dispose of it punctually, according to my order.'[164]

The delays and 'manifold disappointings' over the payment of Pell's salary were getting progressively worse: this may have been a reflection not only of general administrative (and fiscal) decay in the Cromwellian regime, but also of the reduced importance of Pell's embassy. Finally, on 6 May 1658, John Thurloe sent him the following brief message: 'Sir, His Highnesse hath thought fit, the face of things being much changed in those parts to recall you. I shall take care for the satisfying of your Arrears that you may come away with all convenience. You will please therefore upon this dispatch to dispose your selfe homeward.'[165] The promise made in that letter was not in fact kept: after his return to England, Pell would draw up a memorandum in which he calculated the total arrears owing to him as £1500 ('Without any consideration of Interest, though I pay interest to some of my Creditors').[166] But the instruction had to be obeyed, sooner or later. Cautiously, after receiving it on 22 May, Pell waited for the next two sets of offical despatches from England, in case it was countermanded.[167] He finally made his valedictory speech to the Zurich authorities on 23 June; they, to show their gratitude, presented him with a gold chain and a medallion.[168] He left Zurich on 26 June, stayed for two days with his friend Stucki at the nearby town of Baden, and then moved to Basel, where he spent just over two weeks, before continuing down the Rhine to the Netherlands.[169] With him came 'an industrious young man' from a prominent Zurich family, Balthasar Keller, to whom Pell had previously given tutorials in mathematics.[170] On 23 July [/2 Aug.] they reached Arnhem, where they spent three days with Moriaen: in his house, Pell saw again the Arabic manuscript of Apollonius which Ravius had once lent him (and which Ravius had later pawned to Moriaen). And it seems that Moriaen now received a substantial loan from Pell – one which would remain unpaid at the time of

---

[164] Ibid., fo. 214^r (Pell to Ithamar, 25 Mar. 1657, printed in Vaughan, *Protectorate*, ii, p. 406, but mis-dated there as 'Feb.' 1657).

[165] BL, MS Lansdowne 754, fo. 398^r; this was accompanied by an equally brief letter (of the same date, and giving the same reason for the decision) from Cromwell (fo. 400^r).

[166] Ibid., fo. 465^v (Pell, memorandum).

[167] Ibid., fo. 465^v (date of receipt); BL, MS Add. 4364, fo. 162^v (Pell to Hartlib, 27 May 1658).

[168] See BL, MS Add. 4280, fo. 67 (Pell's valedictory text, in which he referred once again to the fate of Dorislaus); Holzach, 'Über die politischen Beziehungen', 5, p. 48.

[169] Zentralbibliothek, Zurich, MS F 72, fo. 291^r (Stucki to Hottinger, 30 June 1658); Vaughan, *Protectorate*, ii, p. 495 (chronology of journey).

[170] BL, MS Add. 4414, fo. 10^r (Collins, draft preface to *An Introduction to Algebra*, quoting Pell's comment on Keller, whom he had taught how to make a table of prime numbers and the factors of composite odd numbers); Bodl., MS Aubrey 13, fo. 91^v (Pell to Haak, 20 June 1666, enquiring about Keller, and adding: 'Anno 1658 I brought him from Zurich to Flushing').

Moriaen's death, ten years later.[171] On 2 [/12] August Pell reached the port of Flushing (Vlissingen), from where he wrote to Hartlib.[172] Alerted by this to his imminent return, Ithamar then wrote the only letter to her husband that has survived in full from this period: 'My Dearest, I am glad to heere of your coming home for we all desire very much to see [>it] & for my bussinese at W[hite]H[all] it lies still there is noe monies for me ... your daughter Rauen hath a daughter 2 mounthes ould her Sone is dead the rest of our Children are all well.'[173] Pell boarded a ship on 11 [/21] August, transferred to Edward Mountagu's frigate, the Naseby, on the 12th [/13th], and on the following day, as his characteristically methodical later memorandum recorded, 'I came home about midnight'.[174]

---

[171] See Bodl., MS Aubrey 13, fos. 94^v (Pell to Haak, 17 Apr. 1668: 'Chr. Ravius had brought it out of Turky and pawned it to Mr Moriaen for some mony which He had not repaid when I last saw Mr Moriaen, that was in July 1658'), 95^r (Pell to Haak, 23 May 1668: 'I think you know that Mr Moriaen dyed in my debt'). Ravius had sent the MS to Moriaen, who was meant to arrange its sale to Claude Hardy, in 1651 (see Toomer, *Eastern Wisedome*, p. 238). It seems that Hardy may have reneged on the deal, and that Moriaen himself had then advanced the money Ravius needed. As Prof. Toomer has pointed out, Ravius would later claim that there were no debts for which this MS could be legitimately detained (Apollonius, *Conicarum sectionum*, ed. and tr. Ravius, sig. *6^v); but this appears to have been one of several matters in which his personal probity was open to question.

[172] Boyle, *Correspondence*, i, p. 287 (Hartlib to Boyle, 10 Aug. 1658, citing Pell's letter of 3/13 Aug.).

[173] BL, MS Add. 4280, fo. 46^r (Ithamar to Pell, 9 [/19] Aug. 1658).

[174] Vaughan, *Protectorate*, ii, pp. 495–6.

6

London, Essex, London, 1658–1665

John Pell's return to England in August 1658 coincided with the final illness of the Lord Protector. When George Fox, the Quaker, accosted Cromwell in Hampton Court Park in the middle of that month, he 'saw and felt a waft of death go forth against him; and he looked like a dead man'.[1] Pell had an interview with Thurloe on 17 August; but he was not able to report to Cromwell, who returned to Whitehall only to take to his deathbed, expiring on 3 September. Cromwell's son Richard was proclaimed Lord Protector on the following day, and Pell had an audience with him only five days later.[2] Apparently Pell was still regarded as a public servant: in late September he was granted fifteen yards of black cloth by the Council of State for his official mourning, and he also received a formal invitation to Cromwell's funeral (which was held on 23 November).[3] Thereafter he had two more meetings with Richard Cromwell, in December and January, at which he presented letters he had received from Switzerland; these, it seems, were the last of his official duties.[4] Unfortunately, in dispensing with his services, the government also unburdened itself of any sense of responsibility for his huge arrears of pay. A copy of Pell's final memorandum about this declares 'He hath not received a peny since he returned in ye end of August 1658', and is dated 'Easter eeve April 21. 1660'. Just a few days after that, the Convention Parliament met to agree on the restoration of Charles II: the debt owed to Pell by the English Republic was never to be paid.

Pell had a wife and several children to support. His eldest daughter, Mary, was now married, but her sister Judith (aged 23 at the time of his return from Switzerland) was not; and there were two other daughters (Elizabeth, aged 18; Bathsua, 15) and a son (John, 14). How Pell survived financially during this period is not known, though it may be assumed that he received some support from his generous pupil–patron, William Brereton. (He did try to recover from Ulrich the payment he had made on behalf of his errant son, forwarding an invoice for £112 14s 4d; both Haak and Stucki were pressed into service as

---

[1] Cited in A. Woolrych, *Britain in Revolution, 1625–1660* (Oxford, 2002) p. 696.

[2] BL, MS Lansdowne 754, fos. 465^v, 482^r (Pell, notes on chronology).

[3] Ibid., fo. 470 (invitation); Vaughan, *Protectorate*, ii, p. 341, and R. Sherwood, *The Court of Oliver Cromwell* (London, 1977), p. 46 (mourning). For a detailed description of Cromwell's lying-in-state and the funeral procession (in which, presumably, Pell took part), see H. Fletcher, *The Perfect Politician: Or, a Full View of the Life and Actions (Military and Civil) of O. Cromwel*, 2nd edn. (London, 1680), pp. 275–9.

[4] BL, MS Lansdowne 754, fo. 482^r.

intermediaries, but the money seems never to have been repaid.)[5] Pell was quick to resume his contacts with Edward Mountagu, and in September 1658 was evidently giving him regular lessons in mathematics; Mountagu's relative and protégé Samuel Pepys would later list Pell as one of the people who had taught Mountagu 'the whole course of the sciences Mathematical'.[6] Perhaps there were others who received such tuition, and rewarded him accordingly. Unfortunately the evidence of Pell's intellectual and personal life during this period is extremely slight; although he was living close to Hartlib and must have been in frequent contact with him, his name crops up in Hartlib's 'Ephemerides' for 1658–60 only twice (in connection with information about a type of goat's blood used for medicinal purposes in Switzerland and Savoy).[7] There are just a few references to Pell in Hartlib's letters to John Worthington: some mentions of the letters Pell was receiving from the oriental scholar Hottinger, and the news, in July 1659, that Pell had recently been introduced to the Cambridge Platonist Henry More when the latter was on a visit to London. 'Mr. Brereton brought yesterday Mr. Pell (who had a most earnest longing desire to discourse with him about his last book) to Mr. More's lodgings, and leaving them alone, they spent a good deal of time between them to their mutual satisfaction.'[8]

However, there is one important piece of evidence to show that Pell was actively taking part in discussions of mathematical matters with other specialists

---

[5] See the discussion of this in Barnett, *Theodore Haak*, pp. 122–3. Pell continued to pursue the debt with Ulrich, via Haak and a Swiss merchant, Tobias Zollicoffer, until Ulrich's death in 1668. On 19 May 1669 he wrote to Zollicoffer that he would not press Ulrich's widow for payment; but he still hoped to recover the money from the son, writing to Haak on 23 May 1669 that 'I think you know that Mr Moriaen dyed in my debt. I have the more need to endeavor that Heinrich doe not so' (Bodl., MS Aubrey 13, fo. 95).

[6] BL, MS Add. 4429, fos. 248^r (Mountagu to Pell, undated: 'S^r I haue Gone as farr as I can in y^t paper, be pleased to looke on it & perfect it & propose some other Thinges if you please'), 249^r (Pell to Mountagu, 7/17 Sept. 1658: 'My Lord, The inclosed is easier than that of yesterday; whereof your Lordship shall have an account tomorrow morning; or sooner, if it shall be required'); S. Pepys, *Private Correspondence and Miscellaneous Papers, 1679–1703*, ed. J. R. Tanner, 2 vols. (London, 1926), ii, p. 110 (Ward, Pell, Moore, and, 'as I remember', Wallis).

[7] HP 29/7/6A (Ephemerides, 1658); 29/8/6A (Ephemerides, 1659). (The first of these probably relates to correspondence, while Pell was still in Switzerland: cf. the reference to '[what] you have written conc the Helvetian Goates' in Hartlib's letter to Pell of 27 May 1658: BL, MS Add. 4279, fo. 47^r.) The editors of Oldenburg's *Correspondence* have made a claim which, if true, would help to explain the paucity of references to Pell during this period: annotating a letter from Hartlib to Oldenburg of 7 Mar. 1659, they write that by then Pell had 'evidently returned to Holland' (Oldenburg, *Correspondence*, i, p. 206, n. 10). This claim is based on references to Pell in the two passages Hartlib cited from letters he had received from an unnamed German correspondent (probably Moriaen) in the Netherlands, discussing a new pendulum clock and twice saying that he had seen a similar pendulum in Pell's house. However, both these references are qualified with the adverb 'etwan', meaning 'some time ago'; and the second passage concludes with a comment clearly implying that Pell is no longer available for consultation in the Netherlands (ibid., i, pp. 203–4).

[8] Worthington, *Diary and Correspondence*, i, p. 142 (quotation, 20 July 1659), 172, 181, 198 (Hottinger correspondence, Jan.–June 1660).

in the field. On 22 November 1658 (the day before Cromwell's funeral), the mathematician and instrument-maker Anthony Thompson sent him the following note:

> Mr Pell
> There is this day a meeting to bee in ye moore fields of some Mathematicall friends (as you know y^t costame hath beene) there will bee M^r. Rook & Mr Wrenn my Lord Brunkerd [*sic* – for 'Brouncker'] sir Pauel Neake [*sic* – for 'Paul Neale'] Dr Goddard Dr Scarburow [*sic* – for 'Scarborough'] &c. I had notice ye last night of your being in towne from some of ye Gentell men now named & of there desire to injoy your Company, their will bee no such number as you usually haue seene at such meetinges 12 is ye number invited, Sir I hope you will excuse ye short warming [*sic*] for it was shorte to mee
> yours to serue you
> Anthony Thompson
> [*addressed:*] Inquire for Mr Hartlib his house[9]

This document has long been recognized as a significant record of one of the gatherings that constituted the immediate precursors of the Royal Society. Rooke, Wren, and Goddard had previously been members of a group that met in Wadham College, Oxford, under the aegis of John Wilkins; Laurence Rooke had moved from Oxford to London to become Gresham's Professor of Astronomy in 1652, and Christopher Wren had joined him at Gresham's College (taking over his Chair, while Rooke transferred to the Chair of Geometry) in 1657. In his official history of the Royal Society, published in 1667, Thomas Sprat wrote that various members of the Oxford club moved to London 'about the year 1658', and that 'they usually met at *Gresham* College, at the *Wednesdays*, and *Thursdays* Lectures of Dr. *Wren*, and Mr. *Rook*: where there joyn'd with them ... The Lord Viscount *Brouncker*, the now Lord *Brereton*, Sir Paul *Neil* ... and divers other Gentlemen.'[10] By 1660 the custom was to meet in Rooke's rooms at Gresham College after his Wednesday lecture; and it was there, on 28 November, that a group of twelve people (chaired by Wilkins) decided to turn itself into the formal Society that would later receive a royal charter.[11] The gatherings described in Thompson's letter to Pell were evidently very similar: these meetings must have been connected with Gresham College (which was close to Moorfields); Rooke and Wren headed the list; the others named would all be early Fellows of the Royal Society; and the number twelve crops up in both stories. But minor differences are noticeable: this meeting was on a Monday, not a Wednesday; the

---

[9] BL, MS Add. 4279, fos. 273–4. On Thompson see Taylor, *Mathematical Practitioners*, pp. 220–1.

[10] T. Sprat, *The History of the Royal Society of London* (London, 1667), p. 57.

[11] Ibid., p. 58; M. Purver, *The Royal Society: Concept and Creation* (Cambridge, Mass., 1967), pp. 131–2.

fact that this note was written suggests that the previous meetings had not been so regular as to become a reliable weekly fixture; the emphasis on mathematics here makes this more of a specialist affair than the Royal Society; and the inclusion of Thompson, who was never made an F.R.S., probably indicates a greater openness to the ordinary 'mathematical practitioners' of London. Such a lack of precise correlation with the Royal Society as it eventually emerged is only to be expected. For the origins of the Royal Society are to be sought not simply in a canonical succession of clubs or associations (Oxford club – Moorfields meetings – Gresham College gatherings), but in the coalescence of a number of different groupings, networks, and activities; these included circles of mathematicians and physicians in London, as well as the network of correspondence and mutual enthusiasm that linked Samuel Hartlib with people such as Boyle, Oldenburg, Beale, and Pell.[12]

It was during this period that, thanks to Hartlib, a connection of some sort was established between Pell and Boyle. As we have seen, Hartlib had already been passing on to Boyle information he had obtained from Pell during the latter's stay in Switzerland; this practice continued after his return to England. Thus on 19 April 1659 Hartlib wrote to Boyle: 'I beg your promised communications about hatching of chickens. The enclosed paper on that argument I had from Mr *Pell*. It is pretty to observe, that this subject is counted worthy of academical considerations and trials.'[13] Whether Boyle and Pell had actually met by this stage is not clear; but Boyle was evidently well aware of Pell's interests and skills. Later that year, when he wanted an English translation made of a Dutch pamphlet, it was to Pell (via Hartlib) that he turned. This short work, by Lodewijk de Bils, was entitled *Kopye van zekere ampele acte van Jr. Louijs de Bils, Heere van Koppensdamme, Bonem, &c. rakende de wetenschap van de oprechte anatomie des menselijken lichaams* (Rotterdam, 1659); Pell's translation was published under the title *The Coppy of a Certain Large Act (Obligatory) of Yonker Louis de Bils, Lord of Koppensdamme, Bonen, &c. Touching the Skill of a Better Way of Anatomy of Man's Body. Printed (in Low Dutch) at Rotterdam* (London, 1659). Lodewijk de Bils was a land-owning gentleman with a particular enthusiasm for practical anatomy – that is, the preparation and dissection of corpses. Although he lacked formal medical training, he attracted the serious interest of leading academic anatomists, on account of 'the way, which he by long Experience and much practice hath found out for the *dissecting of a whole Body without spilling any blood, and for the Embalming it for whole Ages*'.[14] He issued a public promise (formalized by a notary in Rotterdam – hence the 'Act (Obligatory)' of the title) to reveal his methods, if he were paid the huge sum

---

[12] For a narrowly canonical interpretation, tracing the Royal Society exclusively to the Oxford club, see Purver, *Royal Society*. For an interpretation emphasizing the London connections and the role of Hartlib's circle, see Webster, *Great Instauration* (esp. pp. 91–3 on the significance of Thompson's letter).

[13] Boyle, *Correspondence*, i, pp. 339–40.

[14] L. de Bils, *Coppy of a Certain Large Act*, p. 8.

of 25 000 guilders (roughly £2300 sterling), and invited public subscriptions for that purpose.[15] Boyle was sent a copy of the pamphlet by a young French scientist, Pierre Guisony, who had visited him in England earlier in 1659 and then passed through the Netherlands on his way to Vienna.[16] It seems that Boyle was so enthused by de Bils's project that he promptly decided to circulate the appeal in England. The text was sent to Hartlib (with a dedicatory epistle addressed to him, signed 'R. B.') in mid-October, and Pell's very accurate translation was given to the printer a few days later. By 1 November Hartlib was able to tell Boyle that he had already despatched 25 copies to him in Oxford, and was now busy 'distributing to the rest of the gentlemen, whom you have named'. (Two weeks later he announced that he would also send 'ten or twenty copies' to New England, 'and as many to *Jamaica*'.)[17] The publication of this work did not lead, in the end, to the raising of any money for de Bils in England. But it may, in a minor way, have helped to draw attention to Pell, as someone of whom serious original work was expected: Boyle's dedicatory epistle included the explanation that the work had been translated 'at my request, by such a Person as you will readily think can translate very well, though he can better write things worthy to be translated; when I shall have told you that his name is Mr. *Pell*'.[18] And it may also be suspected that Pell's financial problems were temporarily alleviated by a generous reward for this service.

The Restoration of Charles II must have seemed to offer little hope of improvement in Pell's situation. His social and intellectual milieu in London, based as it was on the Hartlib circle, had been overwhelmingly Parliamentarian during the Civil War; the only prominent Royalist among his patrons, Sir Charles Cavendish, was long dead, and Pell never established any usable connection with his brother, the Marquess of Newcastle. A record as a Cromwellian diplomat was not, admittedly, a bar to a future career: Pell's former colleagues George Downing and Samuel Morland made almost seamless transitions to the service of the Crown. But they were ambitious young men, eager for preferment and power, while he was in his fiftieth year, hoping for nothing more than financial security and a tranquil life. By the autumn of 1660 he was in despair about his

---

[15] For a valuable contemporary account of de Bils and his project, see S. Sorbière, *Relations, lettres, et discours* (Paris, 1660), pp. 123–35 (noting that the faces of the deceased were astonishingly unchanged, but that the internal organs were so altered that they were hard to recognize, as different from their originals as raisins from grapes). See also H. J. Cook, 'Time's Bodies: Crafting the Preparation and Preservation of Naturalia', in P. H. Smith and P. Findlen, eds., *Merchants and Marvels: Commerce, Science, and Art in Early Modern Europe* (New York, 2002), pp. 223–47.

[16] Boyle refers, in his dedicatory epistle, to the 'French Gentleman' who had sent him the pamphlet in a letter from The Hague (de Bils, *Coppy of a Certain Large Act*, sigs. A3^v–A4^r), stating that he had recently been in England and had 'honoured me with severall of his visits'. He does not name him, but the details he gives make it possible to identify him with certainty as Guisony (on whom see Hobbes, *Correspondence*, ii, pp. 837–9). Guisony's letter does not survive, and is not mentioned by the editors of Boyle, *Correspondence*.

[17] Boyle, *Correspondence*, i, pp. 376–7, 379–80, 384.

[18] De Bils, *Coppy of a Certain Large Act*, sig. A2.

prospects: Hartlib reported to Worthington that 'Mr. Pell hath thoughts for America [where his younger brother Thomas had become a prosperous medical practitioner, merchant, and land-owner], nothing yet appearing to encourage his stay.'[19] Strangely enough, however, it was thanks to the Restoration (indirectly, at least) that financial security was eventually obtained for Pell, from a rather unexpected source: the restored hierarchy of the Church of England. On 31 March 1661 he was ordained a deacon; his ordination as a priest followed in June, and on the 16th of that month he was instituted Rector of Fobbing, a parish in the southern part of Essex.[20]

Accepting a Church living was a common enough expedient for an otherwise unsupported intellectual. In the words of one modern authority on this period, 'The support of appropriate kinds of scholarship was one of the Church's accepted objectives; and a blind eye was turned towards incumbents who ... left their pastoral duties to a curate, while they got on with their learned work.'[21] (However, such 'learned work' would most typically involve theology, church history, Biblical studies, or classical scholarship; mathematics might be regarded with less indulgence.)[22] John Aubrey – who would himself reject such a solution to his own financial problems, exclaiming 'fough the Cassock stinkes' – tells the story of Pell's complaint to the Archbishop of Canterbury about the unhealthiness of the Essex marshes: the Archbishop said 'I doe not intend that you shall live there' (to which Pell replied, 'No, but your Grace does intend that I shall die there').[23] Clearly, Pell had some friends in the hierarchy who felt that the Church should take him under her wing. The most likely candidate is George Morley, a man with strong scientific interests, who had visited Pell in Breda (probably receiving some tuition from him), and became Bishop of Worcester at the Restoration; Morley's close friends included Robert Sanderson, Bishop of Lincoln, who ordained Pell, and Gilbert Sheldon, Bishop of London, in whose diocese Fobbing lay.[24]

---

[19] Worthington, *Diary and Correspondence*, i, p. 230 (Hartlib to Worthington, Oct. 1660).

[20] These dates are supplied by Aubrey, *'Brief Lives'*, ii, p. 123, and the life of Pell in Birch, *History of the Royal Society*, iv, p. 446. Fobbing lies to the south of Basildon, just inland, to the north-west of Canvey Island.

[21] Sir Keith Thomas, 'The Life of Learning', *Proceedings of the British Academy*, 117 (2002), pp. 201–35; here p. 219.

[22] On the problematic nature of a clerical career for a mathematician or scientist during this period see M. Feingold, 'Science as a Calling? The Early Modern Dilemma', *Science in Context*, 15 (2002), pp. 79–119. A precedent for Pell's induction to the Church could be found in William Oughtred, who was vicar of Albury for fifty years; Aubrey recorded that 'he was a pittiful preacher; the reason was because he never studyed it, but bent all his thoughts on the mathematiques' (*'Brief Lives'*, ii, p. 111). However, as Feingold points out (*op. cit.*, pp. 101–3), Oughtred's position was fraught with tensions.

[23] Bodl., MS Ballard 14, fo. 98^{r} (Aubrey to Wood, 9 Apr. 1674); Aubrey, *'Brief Lives'*, ii, p. 124. In the MS, Pell has deleted the words 'but your Grace does intend that', perhaps for fear of seeming to commit archiepiscopal *lèse-majesté*.

[24] On Morley's scientific interests see J. C. Hayward, 'New Directions in Studies of the Falkland Circle', *The Seventeenth Century*, 2 (1987), pp. 19–48, esp. p. 42.

But if the Church's reasons for welcoming Pell are thus explicable, Pell's reasons for accepting an office in the Church – something he had firmly rejected when it was proposed to him by Bishop John Williams in 1638 – are harder to assess. His fundamental piety is not in doubt: as we have seen, he had devoted himself to the study of the Bible, and had long yearned for spiritual enlightenment – both for himself and for mankind in general. It is, rather, his attitude to institutional Christianity that is difficult to interpret. His personal contacts with churchmen had been mainly with 'godly' ministers (starting with Hartlib's friends William Speed and Philip Nye), most of whom were anti-Laudian and more or less anti-episcopalian; in London Hartlib's circle had close connections with the Calvinist 'foreign' Churches (especially the Dutch Church), and Pell had evidently felt at home in Calvinist Zurich. Yet to call Pell a 'Puritan' would be to mis-describe him, as he appears to have taken no interest whatsoever in disputes relating either to ceremonies, or to church-government. The disdain expressed for controversial theology in his plan for a 'Universal College' rings true to Pell's character; and the sincerity of his support for John Dury – who wanted all such disputes to be either resolved by broad agreement, or set aside as matters of indifference – cannot be questioned. Like many of the godly, both Presbyterian and Independent, he was alarmed by the proliferation of radical sects during the Interregnum: in March 1655 he wrote to Thurloe that 'I had one from Mr Hartlib; wherein he sent me the proclamation of Febr. 15, against the disturbers of Ministers &c. By which I perceived that many in England continue to convert Gods favour into wantonnesse & to turne their liberty of religious meetings into a licence of disturbing others'.[25] Some framework of order was needed for the worship of God; the form that framework took was perhaps unimportant in itself, and there was thus no reason to object to the form provided by the re-established Church of England.

One other factor must have played a part in Pell's decision: his long acquaintance with the Anglican cleric Dr Samson Johnson, who had become a close friend when they both lived in Breda. (A letter sent by Johnson to Pell from there, six months after Pell's return from Breda to England, shows that he was then helping to look after Pell's financial affairs, and that they were both in close touch with George Morley: 'For ye ten pounds to be left wth your wife ... if you please you may pay M^r morly in the meane time ten pounds & I shall not faile to procure soe much to pay her.')[26] Johnson had been rector of Fobbing from 1636 until 1645, when he was sequestered by the Parliamentary authorities; he

---

[25] BL, MS Lansdowne 745, fo. 62 (Pell to Thurloe, 11 [/21] Mar. 1655). Cf. his letter to Morland of 14/24 Jan. 1658, in which he referred to Quakers and Seekers 'venting all manner of extravagant opinions among the ignorant and simple', and speculated that many of the Quakers were Jesuit *agents provocateurs*: Vaughan, *Protectorate*, ii, pp. 309–10.

[26] BL, MS Add. 4398, fo. 136^r (Johnson [unsigned, but identifiable on internal grounds] to Pell, 31 Dec. 1652).

was restored in 1660, but died in May 1661.[27] It is surely possible that Pell was urged by Johnson himself to become his successor in the benefice.

With glebe land worth £4 a year and tithes estimated at £70, Fobbing was not a rich living, though it would have been quite sufficient for a priest willing to dispense with the services of a curate and live there himself.[28] Pell moved into the Rectory in June 1661: the parish register for June and July has several entries in his hand, recording burials at which he officiated. 'June 26. Thomas, an infant, the sonne of Thomas Reeve widower. July 12. A poore boy, a beggar, his name unknowen, buried at the Parish charge. July 23. Old Christopher Gowers, without a Coffin.'[29] For an intellectual with a Europe-wide reputation (and a former Ambassador), this was a strange come-down. Even though a curate would have to be paid nearly half the income of the benefice, Pell quickly installed one – the first of seven, who all succumbed to the unhealthy conditions of southern Essex and died within Pell's first ten years there – and returned to London.[30]

It seems, however, that Ithamar was left in Fobbing, and that the endemic sickness of the region (probably malaria) now claimed her too. In September she was brought back to London, where, on the 12th of that month, she died.[31] Her funeral took place at St Margaret's, Westminster, two days later.[32] A brief list,

---

[27] J. Foster, *Alumni oxonienses*; H. Smith, *The Ecclesiastical History of Essex under the Long Parliament and the Commonwealth* (Colchester, n.d. [1932]), p. 126; Matthews, *Walker Revised*, p. 156. The date of Johnson's death is supplied by Hartlib's letter to Worthington of 28 May, which says that he died 'last week': Worthington, *Diary and Correspondence*, i, p. 318.

[28] Smith, *Ecclesiastical History of Essex*, p. 239 (figures from the Parochial Inquisition of 1650). For a detailed study of the finances of an Essex clergyman during this period see A. Macfarlane, *The Family Life of Ralph Josselin, a Seventeenth-Century Clergyman: An Essay in Historical Anthropology* (New York, 1977), esp. pp. 33–67. Josselin's income from his living was only £60 (increased by voluntary contributions to £80); but he also had land of his own, and his total income in the 1660s was roughly £160 a year.

[29] Essex Record Office, Chelmsford [hereafter: ERO], microfiche D/P 414/1/2, Fobbing parish register, 1654–80.

[30] The name of the first curate is not known; the one recruited in 1662 was paid £30 a year (BL, MS Add. 4280, fo. 255^r, articles of agreement between Pell and William Adams). On the seven curates in ten years see Aubrey, *'Brief Lives'*, ii, p. 124. Pell's handwriting recurs in the Fobbing parish register in the months July–Nov. 1664 (ERO, microfiche D/P 414/1/2, entries for those dates); presumably that period was an interregnum between curates. Various notes for sermons among Pell's papers (e.g. BL, MSS Add. 4474, fos. 45–55; Add. 4431, *passim*) probably date from that period.

[31] Her last movements are not recorded, but are inferred here from other evidence. The funeral took place in Westminster; but a note by Pell, headed 'My wives debts paid 1661', includes, for 16 Sept. 1661, 'To Abigail her last maid ... 10s.' and 'To the same Abigail for her journey out of Essex ... 1s' (BL, MS Add. 4426, fo. 168^r). This suggests that Ithamar was brought from Essex only at the last moment. A note of expenses relating to her final illness and funeral, in the hand of one of her daughters, begins with an entry for 11 Sept.: MS Add. 4280, fo. 263^r. Aubrey also implies that Pell's wife died of an illness she contracted in Essex: *'Brief Lives'*, ii, p. 124.

[32] H. F. Westlake and L. E. Turner, eds., *The Register of St. Margaret's, Westminster, London, 1660–1675*, Publications of the Harleian Society, lxiv (London, 1934), p. 192.

drawn up by Pell, of those who attended the funeral shows how closely linked he still was to Samuel Hartlib and his circle: it includes 'Mr Freher [Hartlib's cousin] and his wife', 'Mr Clodius [Hartlib's son-in-law] and his wife', 'Mr Hartlib and his wife', 'Sr John Rode [Roth; also Hartlib's son-in-law] and his lady', and 'Mr Haak and his wife'.[33] But Hartlib – who was more or less the same age as Ithamar – was also in poor, and deteriorating, health. On 12 March 1662 Pell wrote to his son: 'I am now going out to accompany old Mr Hartlibs corps to the grave. He died last Munday morning early [10 March]. This day your mother hath beene dead just 26 weekes or halfe a yeare.'[34] Pell's world had sadly contracted; of his old close friends and supporters, only Haak and Brereton were left in England.

From the fragments that survive of Pell's correspondence with his son, it appears that John Pell Jr was now living in the Rectory at Fobbing, while his father and sisters remained in London. This teenager (his twentieth birthday would fall in February 1664) was thus acting, for most purposes, as his father's representative. His handwriting occurs in the Fobbing 'parish book': in March 1662 he signed his name as a witness to the statement that the churchwardens had accounted for all the money left to the parish in Samson Johnson's will, and there are entries in his hand for 1663, 1664, and 1665.[35] He does not appear to have inherited any of his father's intellectual abilities; in 1657 Pell had advised Ithamar to withdraw him from Westminster School, on the grounds that 'If he be not fit to get long lessons by heart, he will never be fit for that school', and the style and spelling of his letters give the impression that he was quite ill-educated.[36] The news he sent to his father was seldom good, where financial matters were concerned. In late February 1662, for example, he wrote that the curate had sickened and died: 'all his goods and cattells [*sic*] are seased upon for dept, so that the children (unlesse his Father provide for them) will fall to the parish.' And he added that 'Tuesday 18th by a great winde you are endamaged *5 pound or little lesse* it will cost to repair the chancell house and barne with tyles.'[37] Although his main role must have been to collect the tithes and forward them to Pell in London (the letter just quoted includes the announcement, 'Sir I heere send you a *tithe pig*'), he was constantly running up debts of his own. In the following month he wrote that 'I desired M^r M to furnish me with so much money as would [>pay] halfe the cessment [*sc.* 'assessment', for tax] he told me

---

[33] BL, MS Add. 4426, fo. 169^r. On Freher, Clodius, and Roth see Turnbull, *Hartlib, Dury and Comenius*, pp. 2, 4, 8.

[34] BL, MS Add. 4280, fo. 318^r.

[35] ERO, microfilm T/A 772, item 1, Fobbing 'parish book', 1630–1700, entry for 31 March 1662, and lists of 'nominated officers', 1663–5. John Pell Jr was presumably also the 'Pell' listed among the inhabitants of Fobbing in the Hearth Tax returns of July–Aug. 1662: ERO, microfiche SOG/12, p. 12 (assessed for 5s).

[36] Vaughan, *Protectorate*, ii, p. 407. The school is not named, but Pell makes a transparent reference to Richard Busby ('you must take care that you displease not Mr. B. by taking him away'), and refers to it as a 'great school' (p. 408).

[37] BL, MS Add. 4280, fo. 179^r (John Pell Jr to Pell, 26 Feb. [1662: the year of this and the following letters is not given, but can be deduced from internal evidence]).

he had but a little ... I have received 10^s of G Ford which I payd all away to those I was run in debt to'; a few days later he wrote to his sister Elizabeth, hoping that she would persuade their father to let him buy a horse for £2 10s.[38] He was, it seems, a spendthrift.

In the last week of February 1662 John Pell Jr informed his father that the rector of the nearby parish of Orsett (an elderly Scotsman, Dr John Michaelson) had died.[39] Another informant told Pell that 'Ozet is worth 300^{li} by ye yeare to him y^t lives upon the place and suffers not himselfe to be couzened in his tithes'; so he moved swiftly to obtain a document from the diocesan bishop, Gilbert Sheldon, who was patron of the living, conferring it on him.[40] Unfortunately for Pell, the news turned out to be false: Michaelson was in good health, and would remain in possession of the benefice until 1674. But Pell's prospects did improve in the summer of 1663, when his patron Gilbert Sheldon became Archbishop of Canterbury. One of Sheldon's last acts as Bishop of London was to give Pell another Essex living, that of Laindon with Basildon (which lay just to the north of Fobbing); once established in Lambeth Palace, Sheldon also made Pell one of his personal chaplains, and in October he awarded him a doctorate in Divinity.[41] No doubt Sheldon appreciated Pell's intellectual qualities in general, and his mathematical skills in particular; and in the following year Pell put those skills to work in his service, publishing a pamphlet on the dating of Easter. This short work, *Easter not Mis-timed: A Letter written out of the Country to a Friend in London, concerning Easter-Day* (London, 1664), explained the difference between the Julian and Gregorian calendars and the rules for the calculation of Easter laid down by the Book of Common Prayer; it was written to counter a popular belief that Easter Day in 1664 should have fallen on 3 April rather than 10 April. Characteristically, this was an anonymous publication; but its authorship was not concealed from Sheldon. (Indeed, it is possible that Sheldon actually commis-

---

[38] Ibid., fos. 144^r (John Pell Jr to Pell, annotated by Pell 'Received March 10 [1662]'), 147^r (John Pell Jr to Elizabeth Pell, 15 March [1662]: 'I have tried walking but that prevailes not with my ague. Therefore If my Father please I can buy a horse heere for 50^s...'). The address shows that Elizabeth was staying with her sister, Mary Raven.

[39] Ibid., fo. 177^r (Pell to John Pell Jr, 5 Mar. [1662], saying that he received three letters from him in the last week, the second of which contained this news). On Michaelson, who had suffered grievously after being sequestered in 1644, see Smith, *Ecclesiastical History of Essex*, pp. 174–9; Matthews, *Walker Revised*, pp. 158–9. Orsett lies roughly ten miles to the west of Fobbing.

[40] BL, MS Add. 4280, fos. 321^r (Sheldon document), 322^r (quotation). Pell also drew up a formal petition to the King for Orset, stating that he had been collated to it by Sheldon on 1 Mar. 1661[/2]: fo. 319^v. The estimate of £300 was excessive, but this was certainly a rich living: in the Parochial Inquisition of 1650 the tithe was estimated at £200 (Smith, *Ecclesiastical History of Essex*, p. 241).

[41] Pell was instituted Vicar of Laindon on 27 July 1663 (Birch, *History of Royal Society*, iv, p. 446), following the death of the incumbent, William Heywood; Sheldon became Archbishop of Canterbury on 31 Aug.; on the chaplaincy and the doctorate (awarded 7 Oct. 1663) see Aubrey, *'Brief Lives'*, ii, p. 124. This doctorate was a so-called 'Lambeth degree', given under the degree-awarding authority conferred on the Archbishop of Canterbury by a statute of Henry VIII.

PART I: THE LIFE OF JOHN PELL   187

sioned it.) The work received its imprimatur from another of the Archbishop's chaplains, Mark Frank, and the copy that survives from Sheldon's own library bears the manuscript annotation: 'By John Pell'.[42] How close Sheldon and Pell ever became is not clear; we know that Pell spent some time 'in attendance upon my Lord' immediately after the promotion to Canterbury, but there is no reference to Pell, or trace of his assistance, in Sheldon's letter-book of the 1660s.[43] In 1667, writing to the mathematician John Collins, Pell would make an acerbic comparison between his two patrons, William Brereton and Gilbert Sheldon: he asked Collins if he could find out what plans either of them had 'to see me in a better condition', and added that 'Some say there is this difference between them; that the Former [Brereton] hath Will & no Power, the Other [Sheldon] hath Power & no Will.'[44]

That comment seems positively ungrateful; for it was thanks to Sheldon's exercise of his power that Pell enjoyed a gross income of roughly £172 a year. The parsonage house at Laindon had glebe land yielding £15, and tithes worth £75; and the chapel of Basildon produced another £8.[45] As a non-resident, Pell had to supply a curate for this parish too: in the late 1670s his curate was John Nye, probably a grandson of Philip Nye (the Independent divine and supporter of Hartlib).[46] But the real drain on Pell's finances would come not from his curates' salaries, but from the profligacy of his own family members. Just as he had installed his son John at Fobbing, so too he placed his daughter Mary and her husband, Captain Roger Raven, in the parsonage at Laindon.[47] It was the debts accumulated by the Ravens (above all, by Pell's grandson, Miles Raven) that would cause Pell most trouble in later years. Pell's second daughter, Judith,

---

[42] CUL, pressmark Lib. 7. 66. 1.

[43] BL, MS Add. 4428, fo. 14^r (Pell to an unnamed correspondent [identifiable by comparison with MS Add. 4280, fo. 153^v, as Nicholas Gildredge], 3 Sept. 1663); Bodl., MS Add. C. 308 (Sheldon, letter-book).

[44] BL, MS Add. 4427, fo. 226^r (Pell to Collins, 8 May [1667]).

[45] Smith, *Ecclesiastical History of Essex*, p. 238 (again, these figures are from the Parochial Inquisition of 1650).

[46] See BL, MS Add. 4279, fo. 179 (John Nye to Pell, 1678). Philip Nye's son John, who had served as clerk to the Triers of Ministers (1654–60), conformed at the Restoration and was given the living of Quendon in north Essex, which he held from 1662 until his death in 1686 (Smith, *Ecclesiastical History of Essex*, p. 369; A. G. Matthews, *Calamy Revised* (Oxford, 1934), p. 369). The *DNB* entry on John Nye (which gives his date of death as 1688) records that he had two sons, Stephen and John. In 1681 a Steven Nye appears to have been acting as Pell's curate in Laindon (BL, MS Add. 4280, fo. 151^r, Steven Nye to Pell, 14 Sept. [the year is unspecified, but the letter is datable to 1681 on internal evidence]); this Steven and the previous curate John were presumably the sons of the John Nye of Quendon, though the date of Steven's letter here conflicts with the statement in the *DNB* entry on Stephen Nye (who became a prominent pro-Unitarian writer) that he obtained his own rectory in Hertfordshire in 1679.

[47] The date of this move is not apparent; it may have been made soon after Pell acquired Laindon, and had certainly happened before 19 Mar. 1669, when Roger and Mary Raven's daughter Sarah was baptized at Laindon (ERO, microfiche D/P 278/1/1, Laindon parish register, entry for that date).

had been married off to a Thomas Kirk, a linen-draper in Westminster; the date of the marriage is not recorded, but it must have taken place by 1660.[48] One other daughter – possibly Elizabeth, who had been living with the Ravens in Westminster – appears to have ended up homeless in London; when Pell was staying in Fobbing at the end of 1663, Haak wrote to tell him that his daughter was 'going around, here and there, like a lost sheep', and that a mutual friend, 'Frau Williams' (possibly Samson Johnson's widow, Lady Rebecca Williams) was so concerned about the girl that she planned 'to visit the Lord Archbishop in Lambeth this very afternoon, to complain to him in person about her needy state'.[49] It seems that the financial problems which would dog the last part of Pell's life were already snapping at his heels.

Notwithstanding all these problems, the early 1660s saw one very positive development in Pell's intellectual life: his involvement in the work of the Royal Society. The exact date at which his membership began is not known. He was not one of the twelve present at the founding meeting on 28 November 1660, nor was he in the list of forty suitable people which those founders then drew up; but he clearly belonged to the first wave of recruits, and the leading modern authority on the early Royal Society places him 19th in a chronological listing of members.[50] That his involvement was valued by the other members is indicated by the fact that he was exempted from paying the Society's dues. Although the Society had the blessing of Charles II, and became 'Royal' when it received its charter from him in 1662, it never obtained the sort of endowment from the state for which it hoped; its finances thus depended almost entirely on the admission fees paid by new members (10s at first, raised to £1 in 1661 and £2 in 1662) and the subscription fees (£2 12s a year) for which almost all members, in theory, were liable. A special exemption was made for Pell in August 1662, and confirmed in the following May; thereafter his 'arrears' were listed in the accounts, but were

---

[48] The existing set of parish registers (City of Westminster Archives Centre, parish registers of St Margaret's, Westminster, vols. 6 (marriages, 1653–8) and 7 (marriages, 1664–84)) lacks the relevant years. The baptism of John, son of Thomas and Judith Kirk, took place on 6 Jan. 1661: see Westlake and Turner, eds., *Register of St. Margaret's*, p. 1. A note by Pell on Thomas and Judith Kirk is in BL, MS Add. 4409, fo. 53[r]; by the time this note was written in 1666, they had two surviving sons, Thomas and William. Thomas Kirk or Kirke was born in Feb. 1636: see Burke, *Memorials of St Margaret's Church*, p. 150.

[49] BL, MS Add. 4299, fo. 1[r] (Haak to Pell, 30 Dec. 1663: 'umgieng wie ein verirret Schaf, bald hie bald da'; 'selbigen nachmittag nach den Herren Ertzbsch zu Lambeth zubesuchen, und ihm persönlich ihre noth zu klagen'). For references to Rebecca Williams see ERO, microfilm T/A 772, item 1 (Fobbing 'parish book'), entries for 1661–2, and Matthews, *Walker Revised*, p. 156. The other daughter, Bathshua, had been lodging with a relative, Nathaniel Brewster (possibly a grandson of Pell's uncle James) in Norfolk: see the letters from Brewster to Pell of 1662–3 in BL, MS Add. 4278, fos. 106–14, and Pell's note (fo. 113[r]) recording that he paid Brewster a total of £10 between May 1662 and July 1664. The last evidence of Bathshua's presence in London is a note in some of Pell's household accounts: 'Sept. 5. 1663 My daughter Bathshua tooke away halfe y[t] piece of $5\frac{1}{4}$ to make me a shirt' (BL, MS Add. 4426, fo. 112[r]).

[50] M. Hunter, *The Royal Society and its Fellows, 1660–1700: The Morphology of an Early Scientific Institution* (Chalfont St Giles, 1982), p. 162.

never demanded of him.[51]

Pell was, evidently, among friends. Among the fifty members who are thought to have joined within the first three months, there were all the other 'Mathematicall friends' mentioned in Anthony Thompson's letter of 1658: Rooke, Wren, Brouncker, Neile, Goddard, and Scarborough. Also included at this early stage were Robert Boyle, for whom Pell had translated de Bils's work; William Petty, who had performed services for Pell on the Continent in the mid-1640s and must have been helped, on his return to England, by an introduction to Hartlib; Pell's own patron William Brereton; and Pell's former pupil, Edward Mountagu, now Earl of Sandwich.[52] Hartlib himself was never proposed for membership; his prominence as a supporter of the Parliamentary and Cromwellian regimes may have been partly responsible for this, though there were in fact many members who had done well under Cromwell (including John Wilkins, Cromwell's brother-in-law). But several people who had corresponded enthusiastically with Hartlib on scientific matters did become Fellows: these included John Evelyn and (in 1663) John Beale. Another person who was close to Hartlib, not only corresponding with him in the late 1650s but also modelling his own activities as a news-gatherer and 'intelligencer' on his work, was the former diplomat from Bremen, Henry Oldenburg, who became Secretary of the Royal Society. At what stage his friendship with Pell first developed is not clear, but there is one scrap of evidence to suggest that he had made his acquaintance in Switzerland in 1658. A note in Pell's chronology of his time in Zurich records, for 23 May [/2 June] 1658, 'David Whitlaw came to me at Zurich'.[53] Whitelaw was a former manservant of John Dury's, who had entered the service of Richard Jones, the young nobleman whom Oldenburg was then accompanying as a travelling tutor.[54] On 11/21 May Oldenburg and Jones were in Geneva; Oldenburg's letter to Hartlib of that date said that he would write next from Frankfurt, implying that they were about to go there. But his next letter, written on the day of his arrival in Frankfurt, dates from 5/15 June.[55] The long interval between those letters suggests that their party had stopped somewhere on the way: if, as the note about Whitlaw suggests, that place was Zurich, then they are highly likely to have visited the person who was not only England's representative there, but also Hartlib's intimate friend.[56]

Pell's name first appears in the minutes of the Royal Society's meetings in

---

[51] Ibid., pp. 15, 88; Birch, *History*, i, p. 241.

[52] See the listing in Hunter, *Royal Society*, pp. 160–8.

[53] Vaughan, *Protectorate*, ii, p. 495.

[54] Ibid., ii, pp. 174, 195.

[55] Oldenburg, *Correspondence*, i, pp. 158, 160.

[56] That Oldenburg established other personal contacts in Zurich is clear from subsequent evidence: on 8 Oct. 1662 he read to the Royal Society a letter he had received from Zurich (ibid., pp. 474–5), and by Feb. 1663 Dury was able to assume, when writing to Ulrich in Zurich, that 'doubtless Mr Oldenburg will have written to you about English affairs' (SAZ, MS E II 457d, fo. 227^r: 'sans doute Monsr Oldenb: vous aura escrit les affaires d'Angleterre').

an entry for 24 December 1662, when it was recorded that 'Mr. Bruce [Alexander Bruce, future Earl of Kincardine] and Mr. Pell were desired to bring in the description of the several wind-mills and water-mills, which they had observed in Holland and other parts'.[57] (The absence of his name from previous records is of no special significance, as these minutes do not record attendance at the meetings; they provide only the names of those who spoke – or, as here, were called upon to perform particular tasks.) One week later Pell commented on the behaviour of differently shaped weights in 'pendulum-clocks'.[58] In the following month there was a discussion of 'the inquiries of the East-India voyages': Pell was asked 'to peruse the same inquiries, and to consider, what observable things might be added to them'.[59] In April and May 1663 he commented on the respiration of lampreys, and reported an observation he had made of the generation of flies from the bodies of dead silk-worms; in June he informed the Society that wheat was stored for up to eighty years in the granaries of Zurich. During that month he and three others were appointed 'to have the care of setting up the long glass tube for the Torricellian experiment with water'; two weeks later 'Dr Goddard and Mr Pell gave an account of their thoughts about the erecting of the long glass tubes.' (It was also at that meeting that Pell told his story about Sir Charles Cavendish's box of viper-powder.) Other matters discussed by Pell during 1663 and 1664 included vitrifying sand, and a simple device he had made for aiding observers of 'the fixed stars'.[60] On 11 January 1665 he remarked that there would be a lunar eclipse on the following Saturday, and suggested that it be observed; in March, April, and May of that year he made comments on brewing and fermentation, suggested measuring the refractive index of turpentine, proposed studying the growth of the roots of plants, put forward the idea of making 'experiments with May-dew', and contributed to a debate about measuring the angle of refraction in different 'liquors'.[61]

The continuity with Pell's earlier interests – both his investigations of refraction with Warner and Cavendish, and his collection of useful information for Hartlib – is apparent here; at the same time, these topics were fully in line with the concerns of the early Royal Society, which aimed not only at 'the promoting of physico-mathematical experimental learning' (to quote the resolution made at its founding meeting) but also at compiling practical information for a 'history of trades'. In the spring of 1664 the Society set up eight standing committees, each concerned with a different area of subject-matter or activity; Pell was put on the committee for collecting natural phenomena (but not, surprisingly, on the

---

[57] Birch, *History*, i, pp. 165–6.

[58] Ibid., i, p. 167 (cf. Moriaen's comment, cited above, in n. 7).

[59] Birch, *History*, p. 180. The compiling of such 'inquiries' was a task that had caught Pell's imagination in the 1630s: see above, ch. 2, nn. 85, 119.

[60] See below, at nn. 66–7 (sand); above, ch. 4, n. 17 (device). For other mentions of Pell in the minutes for 1663–4, see Birch, *History*, i, pp. 192, 193, 206, 258, 281, 296, 305, 307, 376, 390–1, 406–7, 416, 418, 442, 448–9.

[61] Ibid., i, pp. 218, 242, 253, 255, 266; ii, pp. 20, 27, 29, 42, 53.

astronomical and optical committee). There was also an agricultural committee, with a positively Hartlibian character to it: this was created in response to a suggestion by John Beale, and Hartlib's *Legacie* (the work on which Pell had written 'observations' in 1652) was discussed at its first meeting. Both Pell and Haak were invited to join it; Pell's attendance is recorded at one of its meetings (with Boyle, Evelyn, Wilkins, Hooke, and Oldenburg, among others), and at another William Brereton was asked 'to procure from Dr Pell, what he knows of the easy and cheap way used in Helvetia of drying of fruit'.[62]

Some of our knowledge of this committee system – which lasted only for a short time – comes in fact from a detailed list of the committees compiled by Pell.[63] This is just one of several signs that Pell was an enthusiastic member of the Royal Society, strongly interested in its activities and prepared to take trouble over his own contributions to them. A characteristic note among his papers, dated 5/15 February 1663, is reminiscent of his self-improving and self-admonishing little memoranda of the 1630s:

> 1. I am intreated to be one of the observers of the next Eclips of the Moone in London at Gresham colledge in London
> 2. I would prepare my selfe: that I may not be uselesse: or rather troublesome as a meer spectator
> 3. To this end I must [>fore]know the time, place, coadjutors & instruments more distinctly...

This is followed by detailed notes on how to calculate the exact time of the eclipse, and a further set of notes from Hevelius's *Selenographia*, including a list of 'Requisites for observing an Eclips of the Moone' which begins: '1. A furr'd cap & night-gowne'.[64] Pell had taken lodgings 'in a Grocers house next to Gresham Colledge', which may suggest that attendance at the Royal Society's meetings there was the main focus of his life in London.[65] It was from that address that he wrote a letter to 'My Lord' (probably Brouncker) in July 1663, enclosing a specimen of a special type of vitrifying sand, used in Sussex to line the bases of ovens, which Pell had talked about 'at our last meeting heere'; as he explained, 'When I had ended this Relation; Some of the members of this Society, then present, began to suggest, that this Sand might be capable of more noble

---

[62] M. Hunter, *Establishing the New Science: The Experience of the Early Royal Society* (Woodbridge, 1989), pp. 82–5, 105, 112. On the agricultural committee see also R. Lennard, 'English Agriculture under Charles II: The Evidence of the Royal Society's "Enquiries"', *Economic History Review*, 1st ser., 4 (1932–4), pp. 23–45, esp. pp. 23–7. Its minutes are in the Royal Society, MS Domestic V, items 63, 64. Pell's poor record of attendance as a member of this committee is largely explained by the fact that he was in Fobbing during most of its period of activity: see above, n. 30. For Pell's 'observations' on Hartlib's *Legacie* see above, ch. 5, n. 14.

[63] BL, MS Add. 4441, fo. 57 (Pell, '8 Committees of the Royall Society').

[64] BL, MS Add. 4424, fos. 106–107r.

[65] Pell gives this address in a letter to Nicholas Gildredge of July or Aug. 1663: BL, MS Add. 4280, fo. 153r. In this early period the Royal Society met at Gresham College during the term, and at the Temple during vacations.

uses ... There-upon, I received your Lordships command to endeavour to cause some of that sand to be brought hither: Which I was not able to doe till now: because of the Death of all my relations neere that place, & of the workeman himselfe'.[66] (In a letter written two weeks later to the person who had supplied the sand, Pell expressed the hope that 'perhaps some of our Curiosi heere, will finde other uses of that sort of sand, than for oven-bottoms'.)[67] And another area of Pell's interests was engaged in the following year, when the musical theorist John Birchensha (Birkenshaw) submitted proposals to the Royal Society for a new system of division of the octave; Pell was one of the three Fellows – the others being John Wallis and Nicolaus Mercator – who were asked to consider his theories. (Pell was not convinced, and wrote some critical notes on Birchensha's work in the following year.)[68]

Pell's eagerness to serve the Royal Society, and to be kept informed of its activities, is attested to by his correspondence with Theodore Haak in the winter months of 1663–4, when Pell was staying in Fobbing. On 24 November Haak forwarded to him the printed summons to the anniversary meeting of the Society on 30 November. (This was the annual gathering, held on a date close to that of the founding meeting, at which the officers and council for the next year were elected.) Pell replied that 'I need not tell you why I could not be there', but added: 'I pray you Let your Next tell me ... as much of that dayes action as Your leisure will permit [>you to describe].'[69] Haak then sent him the new printed list of Fellows, which Pell scrutinized with some care: in his next letter to Haak, he asked why Sir Justinian Isham's name had been omitted. (He also enclosed, with this letter, his copy of the Swiss translation into Latin of Francis Potter's book, with the explanation: 'You stood by, at Gr: Col: when I promised to send it to you for Mr Awbrey: To whom I pray you deliver it, & pray him to send it carefully to M^r Fr. Potter'.[70] This is the first sign of Pell's acquaintance with John Aubrey, which would become one of the most important friendships Pell enjoyed in the last part of his life.) Haak's next letter, written in mid-December, informed him that 'The Companie meetes usually in greater numb[er]s than they were wont to doe, & seem more eager for reall prosecutions. It is confidently hoped by yor

---

[66] BL, MS Add. 4279, fo. 35 (Pell to 'My Lord', 15 July 1663, adding that he had first observed this sand 'in Sussex, about 27 yeares before', i.e. in 1636). There is no mention of such a discussion in the minutes of the previous meeting (8 July); from this it appears either that the conversation took place outside the formal part of the meeting, or (as anyone studying the minutes from this period may reasonably suspect) that the minutes were far from comprehensive. The minutes of the meeting on 16 July record that Pell then presented the sand (and a letter from Gildredge), and that Pell and Goddard were asked 'to examine this sand, and to make some trials of it' (Birch, *History*, i, pp. 275–6).

[67] BL, MS Add. 4280, fo. 153^v (Pell to Gildredge, 28 July 1663).

[68] See P. Gouk, *Music, Science and Natural Magic in Seventeenth-Century England* (New Haven, Conn., 1999), p. 188; Pell's notes are in BL, MS Add. 4388, fos. 67–8, dated 9 June 1665.

[69] BL, MSS Add. 4443, fos. 3–4 (Haak to Pell, 24 Nov. 1663); Add 4365, fo. 16^r (Pell to Haak, 2 Dec. 1663).

[70] BL, MS Add. 4365, fo. 17^r (Pell to Haak, 9 Dec. 1663).

ffrends, you will present ye Societie with some speciall χειμήλιον ['stored-up treasure'] ere long, for yor & their credit'.[71] Isolated though he was in the rectory at Fobbing, Pell was able to contribute something: on 20 January he wrote an account of his observation of a solar eclipse that had occurred two days earlier, 'which', he informed Haak, 'I desire may be read at the next assembly R[egi]^{ae} S[ocieta]^{tis} in Gre-Coll. If you be there, you will be able to tell me what was said to it.'[72] A fortnight later he wrote again: 'I give you thanks for giving me account concerning my paper de Ecl. [solis]. Heerewith I send you a second; which needes not be read to the whole Gesellschaft: but may be shewed to some that most regard such things.'[73]

The Fellows of the Royal Society were not, of course, Pell's only contacts among mathematicians and other intellectuals during this period – though in some cases it may have been through the activities of the Society that he was put in touch with others outside the Society itself. An instance of such a process is indicated by a copy of a testimonial for the astronomer and mathematician Thomas Streete (who never became a Fellow): its declaration that Streete, 'by his long study and Industry, hath attained a greater Exactness in the discovery of the true Longitude by the Motion of the Moon then hath yet bene Extant', appears over the signatures of Brouncker, Wilkins, Pell, and Jonas Moore.[74] Streete had presented a copy of his *Astronomia carolina: A New Theory of the Coelestial Motion* (London, 1661) to the Society in February 1662, and at a meeting in November of that year a 'proposition about the longitude' by Streete had been read out, and referred to Brouncker and Sir Robert Moray.[75] It seems not only that Pell and Wilkins substituted for Moray, but also that the Earl of Sandwich took an interest: he added a testimonial of his own, and was probably responsible for the involvement of the mathematician Jonas Moore, who was

---

[71] Ibid., fo. 41^r (Haak to Pell, 16 Dec. 1663). Haak's next surviving letter to Pell, also containing some news of the Royal Society, is BL, MS Add. 4299, fos. 1–2 (30 Dec. 1663).

[72] BL, MS Add. 4365, fo. 14^r contains Pell's copy of this letter; his copy of the paper he sent is on fo. 13^v. Haak's reply, dated 27 Jan., informed Pell that he had delivered his 'anmerckung' (observation), and that the Society was grateful for it, as no one else had observed the eclipse: BL, MS Add. 4443, fo. 2^r. A.-G. Pingré notes that this eclipse (which he dates [17/] 27 Jan.) was not announced in any of the printed ephemerides; not knowing of Pell's observation, he writes that Hevelius was the only person to have observed it (*Annales célestes du dix-septième siècle*, ed. G. Bigourdan (Paris, 1901), p. 260). The paper sent by Pell is in the Royal Society, Classified Papers, VIII (1), no. 6, annotated 'read Jan: 27: 63'; this appears to be the only MS in Pell's hand now in the Royal Society's archives.

[73] BL, MS Add. 4365, fo. 14^r. 'Solis' ('of the sun') here replaces the astronomical symbol used by Pell; 'Gesellschaft' means 'society'.

[74] Bodl., MS Ashmole 423, fo. 236^r. On Streete see Taylor, *Mathematical Practitioners*, pp. 225–6.

[75] Birch, *History*, i, pp. 75, 124. Streete's book was later re-issued (London, 1710) in an edition prepared by Edmond Halley.

under Sandwich's patronage at the time.[76] In this way Pell had been brought into contact with two serious practitioners of astronomy and mathematics, Streete and Moore (the second of whom would eventually become a Fellow of the Royal Society in 1674). A tantalizing piece of evidence that dates probably from this period is an unfinished draft of a letter written by Pell to the specialist printer of mathematical books William Leybourn, which begins: 'you were standing by when some of your London Mathematicians made the motion to me to make use of your presses [>in the publishing some of my] Mathematicall [exercises *deleted* treatises *deleted*] exercises.' Those 'London Mathematicians' are not identified (and, needless to say, the 'exercises' were never printed).[77] Streete may have been one of them, and Pell's old friend John Leake another: both men were among a group of surveyors who would collaborate with Leybourn in producing a plan of the burnt area of the city immediately after the Great Fire of London.[78] And another likely candidate must be John Collins (F.R.S. 1667), with whom Pell would later develop a close friendship: the earliest evidence of contacts between them is a note of a mathematical problem, propounded by Collins on 4 March 1664 and solved by Pell ten days later.[79] No doubt Pell kept up many other connections too with other scientists and intellectuals: for example, when the Danish scholar Ole Borch (Olaus Borrichius) called on the Dutch inventor Caspar Kalthoff in London in May 1663, he encountered Pell at Kalthoff's lodgings.[80] There is also some evidence that he enjoyed close relations with the clergyman, physician, and educationalist Ezerel Tonge, who had shared many of Hartlib's enthusiasms in the 1650s.[81]

---

[76] On Moore and his relationship to Sandwich at this time, see Willmoth, *Sir Jonas Moore*, pp. 122–4. Sandwich also wrote his own testimonial for Streete, approving of his work on longitude: Bodl., MS Ashmole 423, fo. 237^r.

[77] BL, MS Add. 4408, fo. 388^r. The letter is not dated; it is addressed to Leybourn 'at his printing house', and Leybourn was active as a printer and bookseller from c.1651 to 1665 (see H. R. Plomer, *A Dictionary of the Booksellers and Printers who were at work in England, Scotland and Ireland from 1641 to 1667* (London, 1907), pp. 116–17). Conceivably, it could date from any time that Pell was in London between 1652 and 1665. However, later in the draft Pell commits a very uncharacteristic chronological mistake, giving 1644 (the date of his 'refutatiuncula') as the date of publication of his *Controversiae pars prima*: this suggests that he was writing at a considerable distance in time from that event. The draft indicates that Pell was preparing to send Leybourn his own demonstration of the theorem defended in that book, but that he became dissatisfied with his way of presenting it, and broke off the work.

[78] See M. A. R. Cooper, 'Robert Hooke, City Surveyor: An Assessment of his Work as Surveyor for the City of London in the Aftermath (1667–74) of the Great Fire' (City University, London, PhD dissertation, 1999), p. 131 (the others being John Jennings, William Marr, and Richard Shortgrave).

[79] BL, MS Add. 4422, fos. 159–162^r (Pell, notes).

[80] O. Borch, *Itinerarium, 1660–1665*, ed. H. D. Schepelern, 4 vols. (Copenhagen, 1983), iii, p. 22. For Pell's earlier interest in Kalthoff's perpetual motion machine, see above, ch. 3, n. 18.

[81] BL, MS Add. 4279, fo. 280 (Ezerel Tonge to Pell, 11 Oct. 1662, signed 'your most Affectionate friend'). On Tonge's work with Hartlib see Webster, *Great Instauration*, pp. 236–42, 259–60.

PART I: THE LIFE OF JOHN PELL   195

One leading intellectual who – to his own mounting irritation – was never invited to become a Fellow of the Royal Society was the philosopher (and mathematician) Thomas Hobbes.[82] Pell had been happy to accept a contribution from Hobbes to his *Controversiae pars prima*; but the philosopher's subsequent forays into circle-squaring and cube-doubling severely lowered Pell's opinion of his mathematical abilities. In the summer of 1661 Hobbes had published a critical response to Boyle's work on atmospheric pressure, making a mild criticism of the Royal Society as he did so: 'this new Academy abounds with most excellent ingenious men. But ingenuity is one thing and method is another. Here method is needed.'[83] Rashly, he included as an appendix to this work an erroneous demonstration of the duplication of the cube, which his arch-enemy John Wallis (who had become a Fellow of the Royal Society in March 1661) was quick to disprove. By the spring of 1662 Hobbes had made another version of his demonstration, and was preparing to have it printed as an appendix to his next book; although he must have had a growing sense that the Royal Society was a hostile body, he nevertheless still hoped to receive some recognition from it, and arranged for the dedicatory epistle of this book to be read out at a meeting of the Society on 19 March.[84] This was the context in which, a couple of weeks later, Pell had a chance encounter with Hobbes. The story of that meeting (as told here in Pell's own words) gives a good sense of the tact and scrupulousness with which Pell was capable of acting:

> Easter Munday. March 31. 1662
> This morning Mr Thomas Hobbes met me in the Strand, & led me back to Salisbury house, where he brought me into his chamber, and there shewed me his Construction of that Probleme, which he said he had solved, namely *The Doubling of a Cube*. He then told me, that Viscount Brouncker was writing against him. But, said he, I have written a Confirmation & Illustration of my Demonstration; and to morrow I intend to send it to the presse, that with the next opportunity I may send [>printed] coppies to transmarine Mathematicians, craving their censure of it. On this side of the sea, said he, I shall hope to have your approbation of it. I answered, that I was then busy, and could not perswade my selfe to pronounce of any such [thinge *deleted*] question, before I had very thoroughly

---

[82] On the complex issue of Hobbes's relationship to the Royal Society see Skinner, *Visions of Politics*, iii, pp. 324–45; S. Shapin and S. Schaffer, *Leviathan and the Air-Pump: Hobbes, Boyle and the Experimental Life* (Princeton, NJ, 1985); Malcolm, *Aspects of Hobbes*, pp. 317–35.

[83] Hobbes, *Dialogus physicus* (London, 1661), sig. π3^v (translation here from Shapin and Schaffer, *Leviathan and the Air-Pump*, p. 347).

[84] Birch, *History*, i, p. 78. Wallis's dismissive response to Hobbes was *Hobbius heautontimoroumenos, Or, A Consideration of Mr Hobbes his Dialogues* (Oxford, 1662). To 'duplicate the cube' means to obtain $2^{1/3}$ by ruler and compass construction (a task eventually proved impossible by the work of Galois in the early 19th century).

considered it, at leysure, in my owne chamber. Where-upon he gave me these two papers, bidding me take as much time as I pleased. Well, said I, if your work seeme true to mee, I shall not be afraid to tell the *world* so: But if I finde it false, you will be content that I tell you so [>But] privately, seeing you have onely thus privately desired my opinion of it. Yes, said he, I shall be content, and thanke you too. But I pray you, doe not dispute against my Construction, but shew me the fault of my Demonstration, if [there be any *deleted*] you finde any. Thus we then parted, I leaving him at Salisbury house, and returning home.[85]

True to his word, Pell did give Hobbes his response in private: on 17 April he visited him again in his chamber, where he defended Wallis's disproof of his earlier attempt, and on 5 May he delivered to him a copy of his own refutation of Hobbes's latest version.[86]

Hobbes was becoming more and more of an outsider in the intellectual life of Restoration London; Pell, it seems, was becoming ever more integrated into its premier institution. But there is one curious episode from this period that suggests that Pell had not entirely lost his old reserve – that strange compound of genuine modesty and a smouldering sense that his own worth was not being recognized. In January 1664 Haak transmitted to Pell a request by John Wilkins for 'a certification' or reference for Robert Hooke (the brilliant young scientist who had served as 'Curator' of the Royal Society, managing its practical demonstrations and experiments, since November 1662), in support of his application to become Gresham Professor of Geometry.[87] Pell's reply was as follows:

As for Mr Robert Hooke; I have knowen him many yeares: and doe think him a man very industrious, dexterous & ingenious; and more fit to succeed Mr Barrow in Gresham Colledge, than any Rivall that I can think on.

But yet I am not willing to certify so much under my hand in the forme of a Testimoniall to be shewen to The Electors: because

---

[85] BL, MS Add. 4425, fo. 238^r (Pell memorandum). Salisbury House, on the north side of the Strand near Covent Garden, was the London residence of Hobbes's patron the Earl of Devonshire during the early 1660s, rented by him from his brother-in-law, the third Earl of Salisbury. The text of Hobbes's demonstration, and a fragment addressed to Brouncker, both in Hobbes's hand, are on fos. 237^r and 237^v.

[86] Ibid., fos. 213^r (Pell, memorandum), 216^r (Pell, copy of refutation left with Hobbes).

[87] BL, MS Add. 4443, fo. 2^r (Haak to Pell, 27 Jan. 1664: 'ein Certificat'). Stephen Inwood discusses Hooke's application for this post (*The Man who Knew Too Much: The Strange and Inventive Life of Robert Hooke, 1635–1703* (London, 2002), p. 31), making the suggestion – which the evidence of Haak's letter tends to confirm – that it was 'probably with the Royal Society's support'.

> I am alltogether unknowen to them, & therefore must not expect that they should at all regard any Certificate of mine.[88]

A little later, Pell relented and wrote: 'I pray you let D[r] W[ilkins] see the nine last lines concerning M[r] R[obert] H[ooke] and if he think they will be usefull to him, let him cut them off and take them with him.'[89] Yet his initial reluctance is striking, nonetheless. It may have been true that Pell was not personally known to the electors to this post, who were merchants and other local dignitaries nominated by the Court of Common Council of the City of London; but it was their usual procedure to consider testimonials sent in by experts in the field, and it is very hard to believe that any other expert would have been held back by such a scruple.[90] Conceivably, Pell was irked that he had not himself been invited to apply for the job – though there is no sign that he was seeking employment of this kind, and a Gresham Professor was paid only £50 a year. More probably, he resented a certain lack of public acknowledgement of his stature as a mathematician, and – at a deeper level – resented himself for having caused that state of affairs by failing so consistently to publish. It is hard to escape the impression that such resentment, with no obvious target at which it could be expended, was sublimated willy-nilly into humility.

---

[88] BL, MS Add. 4365, fo. 13ʳ (Pell to Haak, 4 Feb. 1664). The most likely explanation of Pell's statement that he had known Hooke for 'many yeares' is that he had met him through Richard Busby, in whose house Hooke had lodged when he studied at Westminster School (probably in the years 1649–53).

[89] Ibid., fo. 14ʳ (Pell to Haak, also dated 4 Feb. 1664).

[90] For a full account of the procedure, see Cooper, 'Robert Hooke, City Surveyor', pp. 30–4 (giving a list of the twelve electors: pp. 34, 184 (n. 89)). As Cooper explains, Hooke was unsuccessful at this election, the post being awarded to Arthur Dacres, a physician at St Bartholomew's Hospital; but in the following year that appointment was set aside because of irregularities in the voting, and Hooke was given the job.

7
Cheshire, 1665–1669

'This day,' wrote Samuel Pepys on 7 June 1665, 'much against my Will, I did in Drury-lane see two or three houses marked with a red cross upon the doors, and "Lord have mercy upon us" writ there.'[1] He had observed some of the first signs of what quickly became London's worst-ever episode of bubonic plague, in which roughly one quarter of the city's population died. The Royal Society held its last meeting in London on 28 June; thereafter, most of its members fled the capital. Hooke, Wilkins, and Petty went to stay in a country house in Surrey; many others joined Boyle in Oxford; but John Pell had a different destination. On 6 July he left London and travelled north-west, to Brereton Hall, the handsome Elizabethan house in Cheshire that was the home of his patron William Brereton.[2]

Unlike the other Fellows escaping from London, Pell was making a long-term change of residence (he would not return until 1669), and doing so in response to a long-standing invitation. William Brereton had succeeded his father in April 1664, inheriting his title, his estates, and his manifold debts; and within a few weeks he had sent a letter to Pell, inviting him to come to Cheshire. 'I hope you will not refuse to come with me hither,' his message said, 'where if it be possible for me [>I will] fix *you*, as I hope you will see it may be for your quiet & advantage as well as my very great satisfaction it should be so ... I meane *you & yours*.'[3] There is no evidence that Pell took any members of his own family with him when he moved to Cheshire; it may be that he did not regard their company as conducive to his 'quiet & advantage'. But he did take (or was later joined by) Daniel Hartlib, the young son of Hartlib's brother Georg.[4] Daniel was not the only member of Hartlib's family to receive Brereton's hospitality: Frederick Clodius, the German alchemist who had married Hartlib's daughter, was also patronized by him. In September 1667 the physician and diarist John Ward would note the information that 'my Lrd Bruerton of Cheshire was Clodius his scholl[ar] also and y^t hee goes on very vigorously with Chemistry ... hee took Clodius downe with him into ye Country and setled something on him for life,

---

[1] Pepys, *Diary*, ed. R. Latham and W. Matthews, 11 vols. (London, 1970–83), vi, p. 120.

[2] The date of departure is supplied by a letter from Pell to Collins of 6 July 1668 (BL, MS Add. 4278, fo. 127^r), which says: 'Munday. July 6. Just 3 yeares since I last saw London.'

[3] BL, MS Add. 4280, fo. 102^r (Brereton to Pell, 2 May 1664). On Brereton Hall and the Brereton family see G. Ormerod, *The History of the County Palatine and City of Chester*, 2nd edn., 3 vols. (London, 1882), iii, pp. 85–7; G. E. C[okayne], *The Complete Peerage*, ed. V. Gibbs, G. H. White, and R. S. Lea, 12 vols. (London, 1912–59), ii, p. 301.

[4] BL, MS Add. 4278, fo. 65^r (note from Thomas Brancker to Daniel Hartlib at Brereton; undated, but probably 1667); on Daniel Hartlib see Turnbull, *Hartlib, Dury, Comenius*, p. 6.

but Clodius fell out with him and so left him and wanders; my Lrd keeps his wife and 2 Children'.[5]

How Pell spent most of his time at Brereton is not known. Possibly he participated in Brereton's 'vigorous' chemical researches. He may have taken an interest in the education of Brereton's son (who was born in 1659), but he was not employed to teach him. Two of his letters written from Brereton give the impression that he had duties to perform: in 1666 he wrote to Thomas Brancker that 'I am not free to dispose of my own time. I may be commanded to goe, to come, to doe this or that business', and two years later he told John Collins that 'I hope my freinds doe not think that all my time is spent in Mathematicks'.[6] But in both cases his remarks were made in the course of resisting suggestions that he complete and publish his own mathematical works – a context in which he was all too prone to exaggerate the obstacles that faced him. It was of course still necessary to supervise, from this great distance, the administration of his two parishes in Essex; and he was, in theory, still liable to be summoned to Lambeth Palace to wait on Archbishop Sheldon. Yet in most other ways this period must have been an unusually carefree one for Pell, as he enjoyed the benefits of free board and lodging, a sympathetic patron, and time in which to get on with his own work. Unfortunately the bulk of the manuscripts written by him during his stay at Brereton seem to have been left there, and have since disappeared. All that has survived is some of his correspondence; the evidence this supplies is valuable, though it must be borne in mind that the topics or activities that dominate the correspondence did not necessarily dominate, to the same extent, Pell's mental life.

The one topic that generated more letters than anything else was the preparation of an enlarged and improved English edition of Johann Heinrich Rahn's *Teutsche Algebra*. Some copies of Rahn's book had reached England (Pell was given one by Theodore Haak in 1660), prompting the thought that a translation of it would meet the need for an algebra primer that could replace the increasingly out-dated *Clavis* of William Oughtred.[7] Quite independently, two translations were undertaken. One was by Thomas Brancker, who, as he later explained, started to translate it for 'a friend' in 1662. At that time Brancker was

---

[5] Folger Shakespeare Library, Washington DC, MS V. a. 296 (John Ward, notebook), fo. 1^r (preceding an entry dated Sept. 1667 on fo. 2^r). The source of Ward's information was the chemical experimenter William Welden, who knew Clodius well.

[6] BL, MS Add. 4278, fos. 75^r (Pell to Brancker, 1 Sept. [1666]); 128^r (Pell to Collins, 28 Oct. 1668).

[7] Bodl., MS Aubrey 13, fo. 91^v (Pell to Haak, 13 June [1666]: 'I wish you could ... learne from Zurich, concerning Johan Heinrich Rahn ... You gave me his booke in November 1660'). The English admirers of Rahn's book were not aware that Rahn himself was translating it into Latin, completing his translation in 1667: Zurich, Zentralbibliothek, MS C 114a, 'Algebra Speciosa seu Introductio in Geometriam Universalem'. But, as Rahn explained in his preface (fo. iii^v), he had heard that distinguished men in England and Holland were working on the same topic, and therefore preferred to deposit his manuscript in a library instead of publishing it. It seems that Rahn had acquired some of Pell's modesty as well as his mathematics.

still a Fellow of Exeter College, Oxford, where he seems to have been a teacher of astronomy; but he was deprived of his fellowship for non-conformity in 1663, and found employment as tutor to the children of the widowed Lady Reynardson in the village of Tottenham, just to the north of London.[8] By May 1665 Brancker's translation was ready for the press; he found a suitable publisher, Moses Pitt, and obtained a license to print it. It was at this juncture that he was contacted by the mathematician John Collins: as Collins later explained, he had already commissioned another translation of Rahn's book, 'before I knew Mr Brancker, whose treating with Mr Pit to get his Translation printed, gave the first occasion to our acquaintance'.[9] Although Brancker and Pitt were on the point of printing the work, Collins persuaded them that they should not commit anything to the press without first consulting 'D[r] P[ell]' who was the man *from whom* Monsieur Rhonius had received the marginall style; and *of whom* he makes such honorable mention.' Towards the end of May, the three of them had a meeting with Pell in London, at which, Collins later wrote, 'I heard D[r] P[ell] give some cautions concerning the publishing of it.'[10] Brancker's own account of this meeting uses the same phrase: Pell gave him 'divers cautions concerning the Work'. And, in addition, Pell told him that 'he hoped to be at leisure, to review some of *Monsieur Rhonius* his Problemes, and to work them anew'.[11] Brancker and Pitt must have been gratified to think that this eminent algebraist, Rahn's own mentor, would now contribute to their volume; their pleasure would have been severely diminished had they known in advance of the paralysing slowness with which Pell would perform his part of the bargain.

---

[8] J. H. Rahn ['Rhonius'] and J. Pell, *An Introduction to Algebra, Translated out of the High-Dutch into English, by Thomas Brancker, M.A., Much Altered and Augmented by D.P.* (London, 1668), sig. A2^r, 'The Translator's Preface' ('a friend'); *DNB*, 'Thomas Brancker' (noting, incidentally, that Brancker's family name was originally spelt 'Brouncker': he and Viscount Brouncker may have been distantly related). The evidence of his astronomy-teaching is his printed broadside, *Doctrinae sphaericae adumbratio, unà cum usu globorum artificialium* (Oxford, 1662), clearly intended as an aid for students. BL, MS Add. 4278, fos. 35–6 (Brancker to Pell, 19 Dec. 1665), refers to 'ye dutys of y^s family' and tells Pell to address letters to him 'wth ye Lady Reynardson at Tottenham'); cf. *DNB*, 'Sir Abraham Reynardson' (former Lord Mayor of London, who died at Tottenham in Oct. 1661).

[9] BL, MS Add. 4414, fo. 10^r (Collins, 'The Publishers preface', intended for Rahn and Pell, *Introduction to Algebra*, but annotated by Pell: 'not sent, not printed'). A useful account of the preparation of the book is given in C. J. Scriba, 'John Pell's English Edition of J. H. Rahn's *Teutsche Algebra*', in R. S. Cohen, J. J. Stachel, and M. M. Wartofsky, eds., *For Dirk Struik: Scientific, Historical and Political Essays in Honor of Dirk J. Struik* (Dordrecht, 1974), pp. 261–74; but Scriba seems unaware of Collins's initially independent project.

[10] BL, MS Add. 4414, fo. 10^r. Collins specified that Pell encouraged Brancker 'to undertake the continuation of the *Table of incomposits* to 100,000. I, said he, first continued *that Table* for the more easy finding the Briggian Logarithms, when I had onely seen the first Chiliad of them.' While in Zurich he had instructed Balthasar Keller how to make such a table, which Keller then continued to 24 000; the table included in Rahn's book was a copy of the one Keller had made.

[11] Rahn and Pell, *Introduction to Algebra*, sig. A2.

PART I: THE LIFE OF JOHN PELL

The process of collaboration on this book began just five weeks before Pell's departure from London. On 1 June Brancker wrote to Pell from Tottenham:

I make bold to trouble you wth these lines attended wth a proof of y^e first sheet of Rhonius I have according to yor desire used your $\div$ & ω and $\therefore$ throughout y^e book having expunged my own former substitutions ... My design is y^e gratifying those many virtuosi y^t long for a view of y^s piece as raw as it is, and yet I will chearfully undertake any reasonable pains in compleating it, and gratefully accept of any direction; although my hopes at first were not to better ye treatise but my resolution not to wrong it.[12]

(It may be noted that it was thanks to the first of these concessions to Pell's personal system of notation that the division sign entered common usage in the English-speaking world.)[13] Two weeks later, Brancker informed Pell that he was at work on the third sheet of proofs. He also made some comments which show that Pell had been generous with his time and assistance: 'You were pleased ever since I had ye happines to be known to you to encourage me in these studys by yor freedome for wch I acknowledge my selfe ever obliged ... I beg pardon for this confidence, it ariseth from ... a sense of my happinesse of accesse to you ... I never before had soe much assistance a praeceptore vivo ['from a living teacher'] as from your selfe.'[14] Such remarks are worth noting, if only to counteract the impression of ever-increasing crotchetiness that arises from Pell's later correspondence about this project.

After Pell's departure for Cheshire, Brancker refrained from contacting him for several months, for fear of transmitting the plague. The correspondence resumed in December, by which time the printers had completed nine sheets in total.[15] (The book is a quarto, which means that each sheet contains eight pages of text.) This new material was sent to Pell for his comments and corrections; and by the end of February 1666 four more sheets were ready – which indicates that, given the difficulties of setting mathematical text, the compositors were making reasonable progress.[16] But at some time during that month, it had dawned on Pell that something more than a mere correction of, or tinkering with, the existing text was needed. The change of approach was signalled by his letter to

---

[12] BL, MS Add. 4278, fo. 31^r (Brancker to Pell, 1 June [1665]). The sign ω was for 'Evolution or Extraction of Roots out of Single whole Quantities', and the sign $\therefore$ was for 'therefore' (Rahn and Pell, *Introduction to Algebra*, pp. 9, 38).

[13] The division sign was first presented in Rahn's original text (*Teutsche Algebra*, p. 8: 'Das Haubtzeichen des Dividierens ist $\div$'); Brancker's reference to it as 'your $\div$' confirms that Rahn had derived it from Pell.

[14] BL, MS Add. 4278, fo. 33^r (Brancker to Pell, 15 June [1655]).

[15] Ibid., fos. 35–6 (Brancker to Pell, 19 Dec. 1665: 'It is now many months since I put you on y^e danger of perusing any thing from me out of these afflicted parts ... They have wrought off ye Book no further than K'). The printers were following the common practice of leaving sig. A – which would contain the prefatory material – till last.

[16] Ibid., fos. 39–40 (Brancker to Pell, Ash Wednesday [28 Feb.] 1666).

Brancker of 24 February, in which he wrote: 'I have adventured to send you this adjoined paper, which I desire may be printed in stead of Rhonij pag 110 & 111 & 4 lines on ye top of pag. 112 ... That probleme, which followes in Rhonij pag. 112. 113. 114. 115. 116., I would have wholly left out, although you should have nothing to put in its roome. But I hope I shall timely enough send you something more fit for that place.'[17] Writing again on 5 March, Pell enclosed 'what I desire may be printed in stead of the 112. 113. 114 115 & 116 pages of Rhonius', and added: 'I intend to examine & amend the rest, as my leisure &c will permit.' In any case, as he also explained in the draft of this letter, 'It would be much more easy for me to amend Rhonius, if I had my papers heere. For amongst them, I beleeve, I have coppies of the most considerable papers that he wrought in my presence or that I gave him to transcribe. But they were carried into Essex about 5 yeares agoe; Since which time I have not seene them.'[18]

At the same time, Pell did also try to set out the reasons for his decision to engage in a wholesale remodelling of Rahn's text: 'That translation can hardly be printed without taking notice of mee. There are many that have heard that M. Rhonius was my disciple: and, being desirous to see wherein my style differs from others, will hope to get a better view of it from his writings. Thence they will take false measures of it, if it appear in no better condition than a bare version & reimpression of that booke can bestow upon it.'[19] But – morally troubled, perhaps, by the self-regarding nature of this concern – he then omitted that passage from his letter. Writing again a month later, he took a different tack, commenting that 'I know not how my minde may alter: but for the present, I think it best not to name mee at all in the title or preface'; the complex interaction between his characteristic modesty, pride, and indecision then ensured that that passage too was omitted.[20]

Instead of simply printing off the manuscript he had, Moses Pitt was now required to wait for entirely new text to be produced piecemeal by Pell. (Worse than that, he had to begin by re-setting most of the sheet he had just completed.)[21] The strain was quickly apparent: in late March Brancker wrote to Pell that 'I shall persuade my Bookseller to keep your pace but yet he hints y^t he

---

[17] Ibid., fo. 71^v (Pell to Brancker, 24 Feb. 1666).

[18] Ibid., fos. 71^v (Pell to Brancker, 5 Mar. [1666]); 80^r (Pell to Brancker, 5 Mar. [1666], passage annotated 'designed ... but left out').

[19] Ibid., fo. 80^r (Pell to Brancker, 5 Mar. [1666], passage annotated 'designed ... but left out').

[20] Ibid., fo. 80^r (Pell to Brancker, 12 Apr. [1666], passage annotated 'designed ... but left out'). The date is probably meant to be the same as that of the letter of 11 Apr. (see below, n. 24). Indecision still reigned a month later: 'When you draw neere to the end of it, it will be time enough to consider what mention is to be made of mee' (fo. 74^r: Pell to Brancker, 21 May 1666).

[21] Brancker's letter of 28 Feb. 1666 thanked Pell for his letter of 24 Feb. and informed him that it had arrived 'just as ye sheet O was composed. I brought ye proof of it to Tottenham...' (ibid., fo. 40^r). In the book as finally printed, the main part of Pell's completely new material begins on p. 100, which is sig. O2^v.

PART I: THE LIFE OF JOHN PELL   203

hath been at above 20^{li} charge allready and would entreat convenient speed'.[22] Perhaps because of his desire for more convenient speed, Pitt now took to opening Pell's letters to Brancker (which were sent first to Pitt's shop in London, and then forwarded to Tottenham); when Pell was informed of this, by an apologetic Brancker, he took great offence, and began sending his letters to Brancker via John Collins instead.[23] He also started complaining to Brancker about the quality of Pitt's work. Of the way he set fractions, he wrote: 'To me, it is such an Eye-sore, that I would not employ that Printer, though he would give me his work & paper for nothing.'[24] Pell had the right to feel offended at the opening of his letters, but he was now over-reacting; and the excess in the reaction is easily attributable to the fact that he now felt himself to be under the pressure of Pitt's impatience for copy. This little *contretemps* did, however, have one beneficial effect: Collins, who had been following the progress of the work at a distance, became more closely involved, and a regular correspondence between him and Pell now developed.[25] Collins's assistance became indispensable when, in November 1666, Brancker accepted an invitation from Lord Brereton to come to Cheshire himself, to be tutor to his son.[26] As Brancker's last letter to Pell from London informed him, 'I am now sending T, corrected, to ye presse. Mr Collins undertakes it when I am gone to whom I pray be pleased to send V'.[27] By 4 December 1666, when Collins had taken over the London end of the operation, Pitt had printed sheet 'T' and sent it off to Cheshire: since the end of February (when he had completed his first version of sheet 'O') Pell's rate of production of the text had yielded only five-and-a-half sheets' worth of printed material in nine months.[28]

And so the work of preparation continued, with painful slowness. By early April 1667 only two more sheets had been completed, 'V' and 'X'.[29] In mid-

---

[22] BL, MS Add. 4278, fo. 41^v (Brancker to Pell, 23 Mar. 1666).

[23] Ibid., fo. 117^r (Pell to Collins, 5 May 1666). Brancker responded to Pell with emollience, compliments, and almost saintly patience: see his letters to Pell of 9 May, 21 June, 6 July, fos. 42–4, 44, 47–8, printed in Halliwell, *Collection of Letters*, pp. 97–101.

[24] BL, MS Add. 4278, fo. 73^r (Pell to Brancker, 11 Apr. 1666).

[25] A fairly full sequence of Pell's letters to Collins, from May 1666 to July 1669, is in ibid., fos. 117–29. The early part of Collins's side of the correspondence is missing up to Dec. 1667, with the exception of four letters: 28 Aug. 1666 (CUL, MS Add. 9597/13/1/80–81, printed in Rigaud, *Correspondence*, i, pp. 115–18), 4 Dec. 1666 (CUL, MS Add. 9597/13/1/86, printed in Rigaud, *Correspondence*, i, pp. 119–21), 9 Apr. 1667 (CUL, MS Add. 9597/13/1/89, printed in Rigaud, *Correspondence*, i, pp. 125–9), and 4 May 1667 (BL, MS Add. 4278, fo. 350). From Dec. 1667 a fairly full sequence survives up to Feb. 1669: ibid., fos. 326–47.

[26] BL, MS Add. 4278, fos. 78^v (Brereton to Brancker, 7 Nov. 1666); 56–7 (Brancker to Brereton, 13 Nov. 1666). Brereton, unlike Pell, had revisited London, and had evidently made Brancker's acquaintance there.

[27] Ibid., fo. 60^r (Brancker to Pell, 24 Nov. 1666).

[28] CUL, MS Add. 9597/13/1/86^r, printed in Rigaud, *Correspondence*, i, pp. 119 (Collins to Pell, 4 Dec. 1666).

[29] CUL, MS Add. 9597/13/1/89^r, printed in Rigaud, *Correspondence*, i, p. 125 (Collins to Pell, 9 Apr. 1667): 'I have M^r. Brankers [>(accompanied] with sheete X)'.

June Brancker wrote to Collins: 'You have staid so long for Z ... this sheet is now not unwelcome ... Yet I hope within a quarter of a year to send ye last sheet I shall let no time slip but be doing of my part assoon as more copy is delivered to me.'[30] Only three more sheets remained to be printed (apart from the 'Table of Incomposits' appended to the book, which seems to have been printed concurrently); and yet the completion of these, plus a small leaf of geometrical diagrams, took not a quarter of a year but eight long months.[31] In the end, Pell decided not to write an introduction or preface, and one written by Collins was, for unstated reasons, omitted; the work bore only a 'Translator's Preface', by Brancker, which referred to Pell under the initials 'D. I. P. [*sc.* Dr John Pell]'[32] The titlepage described the work as 'Much Altered and Augmented by D. P.'; and this was only fair, since Pell had in fact written just over half of the 198 pages of text. (As Brancker's preface explained, D. I. P.'s 'Alterations ... begin with *Probl. 24. pag. 100.* All from thence to the end is his Work. As also *pag. 79. 80. 81. 82.* which he sent last of all.' That last set of changes must have been particularly infuriating for Pitt, as it required altering the final leaf of sig. L and the first leaf of sig. M: those two leaves had to be cut out, and a new half-sheet was inserted when the book was bound.)[33] For Brancker, Collins, and Pitt, the whole sorry process must have felt like the extraction of blood from a stone; and yet, in the end, their patience and cajolery had achieved something of great value. They had succeeded in getting Pell to produce by far the most substantial mathematical text that he ever actually completed and published – and, what is more, a text that set out some of the most original features of his algebraic method.

In late May 1668, three years after his first meeting with Pell, Moses Pitt was at last able to send him a finished copy of the book. (In his accompanying note he added apologetically that 'I Wish it had bin Cleaner for In Removing In the fire Most of the sheets ware soild': Pitt had had a narrow escape during the Fire of London in 1666.)[34] Pell's reply was civil but not entirely uncritical: 'I

---

[30] BL, MS Add. 4278, fo. 63^r (Brancker to Collins [or possibly Pitt], 16 June 1667).

[31] Ibid., fo. 69^r (Pell to Brancker, 19 Feb. 1668: 'I have sent the page of schemes and the sheete Cc to Mr Collins ... They want nothing now, but a sheet or two, for Title, Preface, press-faults &c'). (Pell was by now corresponding with Brancker, as the latter was no longer at Brereton, having been installed as a minister in the nearby parish of Newchurch.) Some sign of the care with which Pell worked on this text is given by BL, MS Add. 4427, which consists mostly of his notes for it. MS Add. 4401, fos. 46^v–119 is a fair copy of material for it, including a mock-up of pp. 105–96 (all of which was written by Pell). MS Add. 4414, fos. 36–7, 47–50 are proof sheets corrected by Pell.

[32] Collins's preface, entitled 'The Publisher's preface' and dated 14 Apr. 1668, survives in a fair copy by Pell: BL, MS Add. 4414, fo. 10.

[33] Rahn and Pell, *Introduction to Algebra*, sig. A2^v (quotation); in the Savile Library copy in the Bodleian (pressmark Savile K. 13) the two cancel-stubs are clearly visible on either side of the inserted half-sheet.

[34] BL, MS Add. 4279, fo. 192^r (Pitt to Pell, undated; annotated by Pell 'I received this at Brereton June 1. 1668'). The letters sent by Pitt to Pell during the final stages of production of the book are in ibid., fos. 197–201.

have turned it over, and finde it no way imperfect; no sheet deficient, miss-folded or miss-placed. But ye great number of Press-faults keepes me from writing to you for coppies for my freinds, because I suppose they had rather tarry, till they may have coppies corrected by my hand throughout.' (His comments here were marked by two characteristic touches of modesty: he referred to the work simply as 'Mr Branckers new book', and said that this copy was 'more costly bound than I desired'.)[35] There were indeed numerous misprints: the prefatory materials included three pages listing 'Press-faults' in small type, and the fifty-page 'Table of Incomposits' – a list of all odd numbers from 1 to 99 999, identifying the prime numbers and giving the factors of the others – had an entire page of errata to itself.[36] Further errata would come to light later, including 145 more errors in the Table, spotted by John Wallis.[37] And yet such faults, tiresome though they were, could not obscure the vital merits of the work. These were fully apparent to the anonymous reviewer (none other than Collins himself) whose notice of the book was printed in the Royal Society's *Philosophical Transactions* on 18 May:

> *First*, as to the *Method* of this Book, it is *New*, such as contains much in a little, each distinct step of Ratiocination or Operation hath a distinct Line ...
>
> *Next*, as to the *Matter*, the Book consists of many excellent *Problems*; some whereof are such, as *Bachet* (that famous Commentator on *Diophantus*) either confesseth he did not attain, or at least left obscure: and others of them are such, as the celebrated *DesCartes* and *Van Schooten* have left doubtful, as not being by them thoroughly understood.[38]

Others were equally enthusiastic. Writing to Brancker on 3 June, Moses Pitt commented that 'I hope the Algebra will take well', and added: 'One of Dr Busby's scholars [at Westminster School] had one of me on Friday last and He told me the Doctor did highly extoll it and read it to them usually every night,

---

[35] Ibid., fo. 194^r (Pell to Pitt, 3 June 1668).

[36] Rahn and Pell, *Introduction to Algebra*, sigs. A3–A4^r, p. 198.

[37] BL, MS Add. 4278, fos. 340^r (Collins to Pell, 18 July 1668, sending errata: 'all the rest aboue, are the Errata as examined by Dr Wallis'); 81^r (Pell to Brancker, undated: 'Mr Collins in his letter of July 18th sent Dr Wallis his Catalog of 145 Errata in your Table'). Pell's own copy of the book is heavily corrected: Busby Library, Westminster School, pressmark I C 34.

[38] Anon., Review of Rahn and Pell, *Introduction to Algebra*, in *Philosophical Transactions of the Royal Society*, no. 35 (18 May 1668), pp. 688–90, here pp. 688–9. In a letter to Pell, Collins made a comment which might have been taken to imply that Oldenburg was the author, writing that the *Philosophical Transactions* 'give an account of your Booke but Mr Oldenburgh omitted to mention the errours committed by Rhonius' (BL, MS Add. 4278, fo. 336^r: Collins to Pell, 28 May 1668); however, Oldenburg was not a specialist in mathematics. The original text of Collins's review of the book is preserved as Royal Society MS Classified Papers XXIV (1), where it accompanies his letter to Oldenburg of 14 May 1668. (I am very grateful to Prof. Feingold for this reference.) 'Bachet' here is Claude Gaspar Bachet de Méziriac, who edited the Greek text of Diophantus, *Arithmeticorum libri sex* (Paris, 1621).

and all that have it doe like it well.'³⁹ The only substantial criticism that comes to light is that several early readers thought that the presentation of the basic assumptions and operations of algebra in the first part of the book was too compressed. John Collins noted this objection (which he privately shared) in a letter to Brancker in June 1668:

> In yours of May 29 you say the D^r [*sc.* Pell] is not so meanly Conceited of the Introductorie part as some it seemes will needs be ... I know none that Account the Introduction a bad one, but divers that thinke it might have been more plaine and ought to have been more large then [>it is. This] is the iudgm^t of diverse of the vertuosi and of some of the teachers of the mathematiques here, who all loue and honour the D^r and I hope I shall doe no lesse as long as I liue albeit I am of their mind, [>nor do] I endeavour to make others of the same opinion, but say [>to them] the D^r did not much concerne himselfe therein, but letts it come out as his Scholer left it.⁴⁰

When he wrote those words, Collins was already making plans – never to be realized – for an enlarged edition of the *Introduction to Algebra*, to contain not only further material by Pell but also a translation of a substantial part of a treatise by the Dutch mathematician Gerard Kinckhuysen. He reminded Brancker that he had recently asked him 'to incline the D^r to admit the first 7 sheets of the Introduction enlarged out of Kinckhuysen (which ... I haue now in my hands and vpon your answer am ready to send) to come out as your Translation, as soone as may be, the D^r taking what time he pleaseth to Supply the Defect at the beginning and to enlarge and compleate the Booke.'⁴¹ Although the wording is far from clear, this seems to refer to Kinckhuysen's general introduction to algebra, *Algebra, ofte stel-konst* (Haarlem, 1661) – a work which Collins had first encountered in 1667, and which, at his insistence, would later be translated into Latin by Nicolaus Mercator and annotated by Newton.⁴² Collins and Brancker had also entertained hopes of publishing a version of another work by Kinckhuysen, *De grondt der meet-konst* (Haarlem, 1660), a treatise on conics using the methods of analytical geometry. Collins obtained a copy of this book in 1666, and lent it to Brancker; by the spring of 1667 Brancker had not only translated this text into English, but also converted the working-out of its problems into

---

³⁹ BL, MS Add. 4398, fo. 196^r (Pitt to Brancker, 3 June 1668, extract).

⁴⁰ CUL, MS Add. 9597/13/1/42^r (Collins to Brancker, June 1668; printed in Rigaud, *Correspondence*, i, pp. 134–5).

⁴¹ Ibid. (printed in Rigaud, *Correspondence*, i, p. 136).

⁴² See CUL, MS Add. 9597/13/1/89 (Collins to Pell, 9 Apr. 1667, printed in Rigaud, *Correspondence*, i, p. 126); C. J. Scriba, 'Mercator's Kinckhuysen-Translation in the Bodleian Library at Oxford', *British Journal for the History of Science*, 2 (1964), pp. 45–58; Thijssen-Schoute, *Nederlands cartesianisme*, pp. 90–3; Westfall, *Never at Rest*, pp. 222–6.

Pell's tricolumnar method.[43] When Brancker sent the manuscript of his version to Collins at the end of that year, he warned: 'I pray take care that it be not printed, either as it is or wth any amendments without my own, but especially Dr Pells consent.'[44] Needless to say, the consent was not forthcoming.

Pell's reluctance to publish must have been peculiarly frustrating for John Collins, who was not only a highly competent mathematician, but also an enthusiastic 'intelligencer', always keen to circulate recent mathematical discoveries and to stimulate new ones. (He has even been described as the English Mersenne – though that title should more properly belong to Hartlib or Oldenburg.)[45] His correspondence with Pell is littered with hints, encouragements, and entreaties, designed to elicit more of Pell's unpublished work. The great prize, on which he had set his sights at the outset, was Pell's version of the *Arithmetic* of Diophantus, converted into modern algebraic notation and accompanied by many new solutions to its problems.[46] (Pell did eventually include a few examples of his treatment of problems from Book V of Diophantus in the *Introduction to Algebra*, commenting tantalizingly at one point: 'There might also *other* very different ways of solving this *Problema* 19. *V. Diophanti*, be added. But if all should be set down that might pertinently be written concerning the supplys of this Defect in the Manuscripts of *Diophantus*, it would make a large Treatise.')[47] As early as March 1666, one of Brancker's letters to Pell remarked that 'Mr Collins was hinting his desire y^t you would subjoin yor Diophantus wch he says lyes perfected by you. If you please I shall deal wth Mr Pits about it, or wth

---

[43] Brancker told Pell that Collins had lent him the book in his letter of 6 July 1666 (BL, MS Add. 4278, fo. 48^r, printed in Halliwell, *Collection*, p. 100); Collins expressed the hope that Brancker would lend him his translation in his letter of 9 Apr. 1667 (CUL, MS Add. 9597/13/1/89^v, printed in Rigaud, *Correspondence*, i, p. 127). Oldenburg later referred to it (in a note copied out by Pell) as 'translated into English, and put into Dr Pells method by Brancker' (BL, MS Add. 4407, fo. 118^r).

[44] CUL, MS Add. 9597/13/1/40 (Brancker to Collins, 28 Dec. 1667, printed in Rigaud, *Correspondence*, i, p. 131). Collins dutifully wrote to Pell on 6 Feb. 1668: 'Let not Mr Branker feare that I shall be Instrumentall or assenting to the Printing of Kinckhuysens Conicks without your and his leave' (BL, MS Add. 4278, fo. 331^r). The manuscript of Brancker's translation does not survive.

[45] *DNB*, 'John Collins'.

[46] On Pell's planned version of Diophantus see above, ch. 4, n. 36. Collins referred to this as 'Diophantus ... turned into specious algebra' in an undated letter [1671] to Francis Vernon, commenting that it was 'already done (but not printed) here by Mr. Kersey and Dr. Pell, as likewise by his scholar Rhonius' (Rigaud, *Correspondence*, i, p. 154). The mathematics teacher and surveyor John Kersey (d. 1677) had great difficulty in interesting publishers in his work (see Taylor, *Mathematical Practitioners*, p. 219); his general introduction to algebra, *The Elements of that Mathematical Art commonly called Algebra*, was published in London in 1673. Rahn's treatise 'Solutio problematum Diophanti Alexandrini' (Zentralbibliothek, Zurich, MS C 114b), written in 1667, was never published.

[47] Rahn and Pell, *Introduction to Algebra*, pp. 105–21 (problems from Diophantus V.18–19), 131 (quotation). Among Pell's papers there are a few notes, apparently written at Brereton, on Diophantus V.19: BL, MS Add. 4430, fo. 40^r.

any other person you choose to employ.'[48] Unfortunately this request coincided with the news that Pitt had been opening Pell's letters: Pell's reply included his outburst about the 'Eye-sore', and his declaration (partly cited above) that 'I would not employ that Printer, though he would give me his work & paper for nothing: So farre am I, from desiring you to treat with your Bookseller for Diophantus or anything else of mine.'[49] Two years later, just after the publication of the *Introduction to Algebra*, Collins raised the subject again: having quoted at length from a recent work by the mathematician Michelangelo Ricci, he added that 'what I mention about Riccio is to excite you to publish your Diophantus or an Additionall part to your Booke [>with Methods] which if done had been taken notice of by the King'.[50] And he returned once more to this topic early in the following year, when he learned that Fermat's son was preparing an edition of Fermat's annotations on Diophantus: 'Why should you not vouchsafe to hasten out yours, and prevent it?'[51]

Pell's Diophantus was not the only work that Collins hoped to elicit from him. In late 1666 he tried to persuade him to publish some of Warner's manuscripts, only to receive the response that 'I have not now leysure to say any thing concerning Mr Warners papers, save that I feare The times are not convenient for the publishing of them'.[52] One year later, when Collins had been elected to the Royal Society and had become an enthusiastic contributor to the *Philosophical Transactions*, he attempted to arrange for Oldenburg to print a short mathematical text by Pell, a set of critical comments on a 'Latine paper ... sent out of France' about biquadratic equations. Pell had offered those comments quite freely in a letter to Collins in May; but no sooner was the idea of publication mooted than Pell's usual scruples came into play. 'I pray you tell him [*sc.* Oldenburg], that I am not willing that he should print that which, 6 months agoe, I sent to you concerning new French inventions for Biquadratic Aequations. I would first see more of that French authors work in that kinde.'[53] In January 1669 Collins tried a different approach, invoking the interest of the courtier and

---

[48] BL, MS Add. 4278, fo. 41^v (Brancker to Pell, 23 Mar. 1666).

[49] Ibid., fo. 73^r (Pell to Brancker, 11 Apr. 1666).

[50] Ibid., fo. 336^v (Collins to Pell, 28 May 1668).

[51] Ibid., fo. 346^r (Collins to Pell, 14 Jan. 1669). The work by Fermat (consisting only of his marginal notes) was published as part of a re-edition of Bachet de Méziriac's Greek text: Diophantus, *Arithmeticorum libri sex*, ed. C. G. Bachet de Méziriac, with notes by P. Fermat, ed. S. Fermat (Toulouse, 1670).

[52] BL, MS Add. 4278, fo. 120^v (Pell to Collins, 26 Dec. 1666).

[53] Ibid., fo. 125^r (Pell to Collins, 16 Nov. [1667]). Pell refers to his letter to Collins of 22 May 1667 (fos. 123^v–124^r is the copy retained by Pell; the original is CUL, MS Add. 9597/13/1/219–219a, printed in Rigaud, *Correspondence*, i, pp. 132–3), in which he commented on 'the Latine paper, which you say was sent of out France, concerning Biquadratic Equations'. Collins's copy of that 'Latine paper' is CUL, MS Add. 9597/13/1/220, printed in Rigaud, *Correspondence*, i, pp. 133–4. Collins's copy of the relevant section of Pell's letter is Royal Society MS Classified Papers XXIV (3), entitled 'Dr Pells answer to the Latin Algebraick question sent out of France'. (I am very grateful to Prof. Feingold for this reference.)

politician Silius (or 'Silas') Titus, whom Pell had probably first got to know in Breda:

> I hereby signifie that Capt Titus is this day to be baloted into the Royall Societie, he speakes kindly of you, as willing to befreind you, and I much mistake, if your preferment be not designed by your freinds here, and if I may haue leaue to say it I thinke you should send up, for [>the use of] Capt Titus, some of your choice Algebraick Notions especially for the easy Resolution of high adfected Aequations in Numbers or Lines, it would be well accepted.[54]

Recognizing Pell's reluctance to appear before the general public in print, Collins was now trying to exploit the special opportunity offered by the Royal Society for communication, and the establishment of priority, without publication. As he added in the same letter, with regard to Pell's work on Diophantus, 'nor need you feare, to be deprived of the renowne of what you impart, for it may be Registred in the Societie, as Dr Wrens Theories of Motion were.'[55] Nevertheless, it seems that the request for Pell's 'Algebraick Notions', like the one for his Diophantus, was turned down. Even the invocation of Pell's own patron could not do the trick. When Collins once wrote to Pell that Brereton (who was then in London) 'intimates, as if you were willing to fit up somewhat more for the Presse and your owne Letters seeme to hint as much, I should be glad to be further informed about it', he received the definitive reply:

> You inquire concerning my fitting of somewhat more for the press. I hope my freinds doe not think that all my time is spent in Mathematicks; though I profess a desire to polish some of my rough draughts, that they may be thought not unworthy to be preserved by some understanding Reader. But whether I shall live to see them printed, I am not sollicitous. If in the meane time Men, of more leisure and skill, doe print better bookes of the same Arguments, I shall rejoice to see my selfe so prevented.[56]

Every incentive Collins could think of was thus deployed: the desire to be 'taken notice of by the King', the wish for 'preferment', the longing for 'renowne', even a simple willingness to gratify one's friends and benefactors. But the usually irresistible force of Collins's persuasion had met, in Pell, a truly immovable object.

Pell may have disliked being badgered for his unpublished work, but for many other reasons he was grateful for Collins's letters. He had not lost his appetite for other authors' publications: Collins was able to keep him informed about recent and forthcoming books, and often sent newly published works to him. Pell's

---

[54] BL, MS Add. 4278, fo. 346^r (Collins to Pell, 14 Jan. 1669). On Pell's earlier connection with Titus see Stedall, *Discourse concerning Algebra*, pp. 133, 141–2.

[55] BL, MS Add. 4278, fo. 346^r (Collins to Pell, 14 Jan. 1669).

[56] Ibid., fos. 344^r (Collins to Pell, 23 Oct. 1668); 128^r (Pell to Collins, 28 Oct. 1668).

letters are peppered with such requests: 'I long to see that new piece of Pascal; &, indeed, any thing of His doing'; 'I would know the bulk & price of Mr Hobbes his new piece against Euclid: though I cannot say, I long to see the booke it selfe'; 'I heare that Mr Evelyn hath put out his Gardiners Almanack ... If you can finde it, I pray you buy it and put it into the box'; 'When you see the third volume of Des Cartes his letters, I pray you take notice of the names of the persons to whom they are written and whether any of them be Mathematicall.'[57] Collins satsified most of these requests, sent copies of the *Philosophical Transactions*, and gave Pell information about other mathematicians and scientists, including Henry Oldenburg, John Wilkins, Nicolaus Mercator, James Gregory, and John Wallis.[58] He had met Wallis in Oxford in 1664 or 1665, and had kept up a correspondence with him ever since; his high opinion of Wallis's skills was apparently shared by Pell (who wrote to him in October 1668: 'I am glad to heare that so much of Doctor Wallis is under the Press'), and it seems to have been thanks to Collins that some sort of friendly relations were established between them.[59]

From his correspondence with Collins, and from other evidence, it is clear that Pell's interest in the activities of the Royal Society and its members continued unabated. Some of his connections with it were kept up via Brereton, who was an active participant in the Society on his visits to London. In 1666 Brereton gave him a copy of John Graunt's *Natural and Political Observations upon the Bills of Mortality*, which Pell read with great interest; in the following year he presented him with a copy of Sprat's *History*; and he also received letters from Oldenburg, and passed on their contents to him.[60] Despite his provincial seclusion, Pell cannot have been a completely forgotten figure at the Royal Society – though during the recurrence of the plague in the summer of 1666 he did have to rebut a rumour of his death, spread by Abraham Hill 'of our society'.[61] In one of his letters to Haak he sent (in addition to his personal good wishes to John Aubrey)

---

[57] Ibid., fos. 117^v (Pell to Collins, 10 May [1666]); 118^r (Pell to Collins, 9 June 1666); 120^r (Pell to Collins, 10 Nov. [1666]); 120^v (Pell to Collins, 26 Dec. [1666]).

[58] Ibid., fos. 326–50 (Collins to Pell, 1667–9).

[59] Rigaud, *Correspondence*, ii, pp. 458–609 (Collins–Wallis correspondence, 1665–77); BL, MS Add. 4278, fo. 127^v (Pell to Collins, 28 Oct. 1668: quotation). Among Pell's papers is an extract from a letter from Wallis (BL, MS Add. 4280, fos. 30–1), consisting of 'Animadversions' on Hobbes's *De principiis et ratiocinatione geometrarum* (London, 1666) – the book, published in the summer of 1666, referred to by Pell as 'against Euclid' (see above, at n. 57). This may have been sent to, or via, Collins. On the uncertain nature – and chronology – of Pell's relations with Wallis, see Stedall, *Discourse concerning Algebra*, pp. 141–53.

[60] Bodl., MS Aubrey 13, fo. 90^r (Pell to Haak, 26 May 1666: 'I pray you tell Captain Graunt that my Lord gave me one of his bookes and that I have read it over more than once'); Busby Library, Westminster School, pressmark H E 7, Sprat, *History*, inscribed by Pell 'William Lord Brereton gave this book to John Pell Septemb. IX. 1667' (and with some marginal comments in Pell's hand); BL, MS Add. 4278, fo. 127^r (Pell to Collins, 29 Aug. 1668, referring to Oldenburg's marriage: 'The Bride-groom hath written to my Lord'). Not many of Oldenburg's letters to Brereton have survived, but they were clearly in regular contact; cf. MS Add. 4280, fos. 43–4 (Oldenburg to Brereton, 22 Sept. 1668, printed in Oldenburg, *Correspondence*, v, pp. 60–1).

[61] Bodl., MS Aubrey 13, fos. 90^v–91^r (Pell to Haak, 4 June 1666).

'My service to all of the Royal Society', and asked for a copy of 'the new list of their names'.[62] When John Wilkins published his monumental *Essay towards a Real Character* in 1668, he went to the trouble of giving a copy (via Moses Pitt) to Pell; a few months later, remembering a remark by Collins that the King had liked the book so much that he wanted to make Wilkins a bishop at the next opportunity, Pell asked Collins to notify Wilkins's 'Freinds' of the sudden death of the Bishop of Chester.[63] Nor was Pell's name entirely absent from the formal business of the Society: in January 1669, when Oldenburg produced a letter from Huygens and 'some papers of his concerning motion', it was decided that they should be circulated to Fellows with specialist expertise – among them, Wallis, Wren, and Pell.[64] At the same time, while the Royal Society naturally features strongly in the two most informative sets of Pell's letters in this period (to John Collins F.R.S and Theodore Haak F.R.S.), he had contacts with others who were not members of it – such as his former diplomatic colleague Samuel Morland, who struck up a correspondence with him in 1666 over his own attempts to square the circle. (He asked Pell's advice on whether to publish his work on this topic; predictably enough, Pell tried to discourage him.)[65] The only aspect of Pell's intellectual life that appears to have shrunk significantly is his international correspondence: his news of Swiss, German, and Dutch matters – including Moriaen's death, in 1668 – seems to have come only via Haak (through whom, until the summer of 1669, he was still vainly attempting to recover some of the money owed to him by the Ulrich family in Zurich).[66] As for France: 'Since the death of Pere Mersenne & Sr Charles Cavendysh,' he told Collins, 'no letters have passed beweene any French Mathematicians & your serviteur J. P.'[67]

One other category of correspondence has not survived from this period, though it certainly existed: Pell's letters to and from the members of his family. The few traces of it that appear in his other letters are dispiriting. In January 1667 he told Collins that his son had written, begging him to arrange a loan of £20, and asked Collins if he would be willing to advance the money, 'upon my word & his bond'.[68] This was probably not the only such request Pell made

---

[62] Ibid., fo. 93^r (Pell to Haak, 26 Nov. [1666]).

[63] BL, MSS Add. 4279, fo. 193^r (Pitt to Pell, 28 May 1668); Add. 4278, fo. 127^r (Pell to Collins, 29 Aug. 1668). As Collins later explained, this news had in fact reached London before Pell's letter (fo. 342^r, Collins to Pell, 5 Sept. 1668). For the circumstances of Wilkins's promotion to the see of Chester, see B. J. Shapiro, *John Wilkins, 1614–1672: An Intellectual Biography* (Berkeley, Calif., 1969), p. 177.

[64] Birch, *History*, ii, p. 337. (A copy of Huygens's MS was duly given 'to the lord Brereton for Dr Pell' on 4 Feb.: p. 344.)

[65] BL, MS Add. 4279, fos. 158–66 (Morland to Pell, 3 Apr. 1666 – 4 June 1666, with a copy (fo. 160^v) of Pell's reply to the first letter, in which Pell notes a 'report' about Morland's calculating machine).

[66] Bodl., MS Aubrey 13, fos. 94^v (Pell to Haak, 17 Apr. 1668, responding to news of Moriaen's death); 94–5 (Pell to Haak, 24 Oct. 1667; Pell to Tobias Zollicoffer, via Haak, 19 May 1669).

[67] BL, MS Add. 4278, fo. 123^r (Pell to Collins, 29 Apr. 1667).

[68] Ibid., fo. 121^r (Pell to Collins, 28 Jan. [1667]).

– and not only for his son's benefit, if a scrap of paper in Collins's hand, 'Mrs Makins receipte for 3^l – 10^s –' is any indication.[69] By late April 1668 Collins was complaining to Pell, 'I never yet recd one Penny from the L[ord] B[rereton], and haue disbursed on his acc^t 47 – 12 – besides what I haue lent upon your request'.[70] The unfortunate truth was that Brereton had inherited mountainous debts; only a month later Collins – who, as a government accountant, was well placed for financial gossip – wrote that he had heard that Brereton's estate 'is more encombred and in Debt then it is worth'.[71] Later that summer, Brereton both sold off valuable estates to regain his solvency, and paid off all Pell's debts to Collins; but the root cause of those debts – Pell's profligate family – remained unchanged.[72] Early in 1669, Pell had to ask for Collins's help as an intermediary when he received a letter from a Mrs Sandbach or Sambach, whose son had been Pell's curate at Laindon, and who was now threatening legal action for the recovery of £27 (almost a whole year's salary) owed to him by John Pell Jr; Collins seems to have placated her for the time being, but the money had to be paid sooner or later.[73] Two years after that, Thomas Brancker, now settled by Brereton in the living of Tilston in Cheshire, admitted in a letter to Collins that he too had lent money to Pell's children – 'not without sufficient grounds to presume on speedy repayment, nor indeed will my great obligations to that eminent Person permit me to be backward in serving him as I shalbe abl, nor do I believe you have any reason to distrust his Integrity. But I hear of no payment yet.'[74] These were ominous signs of the problems that would darken Pell's last years.

---

[69] Ibid., fo. 330^r (Makin receipt, annotated by Pell 'Febr. 7. 1667/68').

[70] Ibid., fo. 335^r (Collins to Pell, 28 Apr. 1668).

[71] Ibid., fo. 336^r (Collins to Pell, 28 May 1668).

[72] Ibid., fos. 338^r (Collins to Pell, 6 June 1668); 340^r (Collins to Pell, 18 July 1668).

[73] Ibid., fos. 128^v (Pell to Collins, 15 Feb. 1669); 347^r (Collins to Pell, 18 Feb. 1669); 348^r (Collins to Pell, 22 Apr. 1669).

[74] CUL, MS Add. 9597/13/1/41^r (Brancker to Collins, printed in Rigaud, *Correspondence*, i, pp. 166–7).

8

London, 1669–1685

John Pell returned to London in the late summer of 1669. The reasons for the move are not entirely clear; there is certainly no sign that his relations with Lord Brereton had cooled. It seems that Brereton was now spending more time in the capital, and had brought his wife and child there, so that there was little point in leaving Pell on his own in Cheshire; but, on the other hand, he was presumably renting his accommodation in London, and may not have had room for Pell.[1] In mid-June Pell wrote to Haak from Brereton that he hoped to be in London by the end of July, 'if my son P. send me mony according to his promise'.[2] John Pell Jr may not have sent him much money, but he did provide some welcome news: on 5 July Pell wrote to John Collins that 'By my Sonnes last, I see I am to thank you ... for offering me a chamber in your house', and announced that he would set off in five days' time.[3] How long Pell stayed with Collins is not known, but it was evidently for many months: the first document indicating that he had ceased to live with him is a note mentioning that in September 1671 he had a 'chamber' in an inn on Pall Mall.[4] In the following year Collins would write to John Beale that 'he boarded long at my house' – adding, significantly, 'and I wish that he and his patron were out of my debt'.[5]

The problems caused by Pell's son were going from bad to worse. The threat of arraignment for debt had been averted (though Pell may still have been paying off that debt during his first months in London); but in September 1669 John Pell Jr was arrested for grievous bodily harm. According to the official indictment, he had assaulted William Gouldingham, a peaceful inhabitant of Great Burstead: 'he beat him with a stick, wounded him, and maltreated him to such an extent that his life was utterly despaired of, and committed other outrages against the

---

[1] The first sign of the impending move is given in Pell's letter to Collins of 23 Jan. [1669], in which he says that Brereton has written from London to Lady Brereton, suggesting that she might come there in two months' time; Pell adds that he might travel with her (BL, MS Add. 4278, fo. 128^r). In the following year, Brereton also made at least one trip to France: *CSPD 1670*, pp. 159, 553.

[2] Bodl., MS Aubrey 13, fo. 95^v (Pell to Haak, 19 June 1669). Pell calls his son 'my son P[ell]' to distinguish him from his sons-in-law, who could also be referred to as sons.

[3] BL, MS Add. 4278, fo. 129^r (Pell to Collins, 5 July 1669).

[4] BL, MS Add. 4424, fo. 78^r (Pell, notes on *Tabula numerorum quadratorum*).

[5] Rigaud, *Correspondence*, ii, p. 197 (Collins to Beale, 20 Aug. 1672). 'Boarded' here implies that Pell was paying for his board.

same William Gouldingham.'[6] Further details of the case are not recorded; if a fine was levied, it must have been a peculiarly unwelcome extra imposition on Pell's finances. It seems that, in an attempt to prevent Pell's son from getting into even more trouble in Essex (and, at the same time, to oblige him to earn some wages), a junior post was now arranged for him in the royal household: in the following year, a document would describe him as 'a Server in ordinary to the King'.[7] There were two categories of such officials: 'above stairs', and 'below stairs'. Those who held the most prestigious offices above stairs, such as the Grooms of the Bed-Chamber and Gentlemen Ushers of the Privy-Chamber, were members of the most prominent noble and gentry families of England, and John Pell Jr would have been completely out of place among them. He did have the status of 'gentleman', however, thanks to his father's ordination; and there were other, more junior, positions for which he might have qualified, such as that of a messenger. In a list of the 'below stairs' jobs one also finds a few reserved for gentlemen – for example, one such job in the cellar, and another in the buttery. So perhaps John Pell Jr had one of those.[8]

If he did, he held it only briefly. Some time in the winter of 1669–70, news arrived that would transform John Pell Jr's situation: Pell's brother Thomas, who had emigrated to America in 1635 and had become a rich landowner there, had died childless (probably in September 1669), and had left everything to his nephew.[9] His will contains one rather startling puzzle for any biographer of Pell: it seems to imply that Pell had married for a second time. 'I doe make my nephew, John Pell, living in ould England, the only sonne of my only brother John Pell, Doctor of Divinity, which he had by his first wife, my whole and sole

---

[6] ERO, MS Q/SR 422/5, indictment of 'Joh[ann]es Pell de Langdon' at Essex Quarter Sessions, 16 Sept. 1669: 'Will[ia]m Gouldingham' ... 'illum verb[er]avit vulnavit et maletractavit Ita q[uo]d de vita eius maxime desperabatur et al[ia] enormia eidem Will[iam]o ... intulit'. Gouldingham was one of Pell's parishioners in Laindon: his daughter Isabel was baptized there on 15 Apr. 1669 (ERO, microfiche D/P 278/1/1, entry for that date).

[7] Sir Robert Moray, letter to John Winthrop, 22 June 1670, in R. C. Winthrop, ed., *Correspondence of Hartlib, Haak, Oldenburg, and Others of the Founders of the Royal Society, with Governor Winthrop of Connecticut, 1661–1672* (Boston, 1878), pp. 44–5.

[8] See E. Chamberlayne, *Angliae notitia; Or, the Present State of England* (London, 1669), pp. 246–9 (below stairs), e.g. p. 246, '*In the Cellar*, a Sergeant, a Gentleman, Yeomen ... *In the Buttery*, A Gentleman, Yeomen, Grooms'; cf. pp. 249–58, listing the eminent 'Servants in Ordinary above Stairs', some of whom (e.g. Silius Titus and Sir Paul Neale) might have pulled strings in the royal household on Pell's behalf. For a slightly earlier and more detailed list of posts and their holders, see Anon., 'Select Documents XXXIX: A List of the Department of the Lord Chamberlain of the Household, Autumn, 1663', *Bulletin of the Institute for Historical Research*, 19 (1942–3), pp. 13–24. Robert Pell states that John Pell Jr was a server in ordinary from 1665, and was promoted to Groom of the Bedchamber in 1669 ('Sir John Pell, Second Lord of the Manor of Pelham', *Pelliana*, n.s., 1, no. 2 (1963), pp. 49–67, here p. 49); the latter claim is contradicted by Chamberlayne's list for that year, and he gives no evidence for the former.

[9] On Thomas Pell's estate (of 10 000 acres) in what became Westchester County, New York, see Tannenbaum, 'Thomas Pell', p. 264, and Bolton, *History of the County of Westchester*, ii, pp. 28–39. The probate valuation of the estate and chattels came to £1294 14s 4d (pp. 46–9).

heire' And again: 'in case my nephew ... deceased, and hath left no male issue, if my brother hath a sonne or sonnes by his last wife, he or they shall enjoy ye above said portion.'[10] If Pell had undergone a second marriage, it has left absolutely no trace in the documentary evidence; the most likely explanation of these phrases must be either that this was just a lawyer's safeguard against any claims that might arise from a possible remarriage, or that Thomas Pell, who seems to have had very little communication with his brother, had been misinformed by a third party. In fact there is no trace, among all of Pell's surviving papers, of any correspondence between the two brothers; and the news of Thomas's death seems to have come not from anyone close to him, but from John Winthrop, Governor of Connecticut, who was in contact with both Brereton and Boyle.[11] By early May 1670 John Pell Jr was ready to set sail for America, to claim his inheritance; the departure was delayed until late June, and may have been subject to further delays thereafter; but in September he finally arrived in Boston, equipped with letters to Winthrop from Brereton, Haak, and Sir Robert Moray, testifying to his identity.[12] There is some slight evidence that he travelled with, or was later joined by, one of his sisters.[13] At least one burden was thus lifted from Pell's shoulders; but his ne'er-do-well son-in-law Roger Raven remained in place at Laindon, where – together with his own young son, Miles Raven – he would soon run up debts that caused Pell more harm than anything John Pell Jr had inflicted on him.

Meanwhile Pell was re-introducing himself to the intellectual life of the capital. Records of his activities during these first two or three years in London are extremely scanty; once again, it seems that a whole section of his personal papers

---

[10] Bolton, *History of the County of Westchester*, ii, p. 44. It was presumably on the basis of this evidence that Agnes Clerke stated, in her *DNB* entry on Pell, that he remarried before 1669.

[11] Robert Pell's brief biography of his ancestor John Pell Jr states that 'a messenger from Lord William Brereton brought the news to Henry Reginolles [sc. Reynolds, Pell's father-in-law] ... that his grandson had inherited a vast fortune in America'; and that the King 'sent for John Pell [Jr] to hear about his good fortune and knighted him' ('Sir John Pell', pp. 49–51). Henry Reynolds had died in 1635, and no such knighthood is recorded, either in *CSPD 1670* or in the comprehensive listing by Shaw (*Knights of England*). The fact that John Pell Jr was occasionally referred to in subsequent American documents as 'Sir' John Pell probably reflects nothing more than local obsequiousness towards a major land-owner.

[12] See Oldenburg, *Correspondence*, vii, p. 8 (Oldenburg to Winthrop, 9 May 1670); Boyle, *Correspondence*, iv, pp. 180–1 (Boyle to Winthrop, 21 June 1670), 183–4 (Winthrop to Boyle, 27 Sept. 1670, reporting John Pell Jr's recent arrival); BL, MS Add. 4279, fo. 313 (Winthrop to Brereton, 11 Oct. 1670); Winthrop, ed., *Correspondence*, pp. 44–5 (Moray to Winthrop, 22 June 1670), 45–6 (Haak to Winthrop, 22 June 1670).

[13] In 1679 Pell sought permission from the Archbishop of Canterbury for three years' leave 'to goe to my children' (BL, MS Add. 4365, fo. 47^r, Pell to 'R. H.', 17 July 1679); this seems to imply that at least one of his daughters had joined his son. ('Children' might mean son and daughter-in-law in 17th-century usage, but John Pell Jr did not marry until 1684 or 1685: Bolton, *History of the County of Westchester*, ii, p. 60.) Cf. also a cryptic reference to 'Ulysses and Penelope' at 'Ithaca', meaning probably John Pell Jr and one of his sisters, in America: BL, MS Add. 4280, fo. 151^r (Steven Nye to Pell, *c*.1681).

has gone missing. It is clear that he renewed his friendship with Haak, and kept up his contacts with Brereton. In a letter to Brereton (then back in Cheshire) in September 1670 he presented Haak's greetings, gave news of Hartlib's old Cambridge friend John Worthington, passed on a message from Collins, and also conveyed some sad news about Henry Oldenburg, whose baby son had recently died.[14] The fact that Oldenburg's very young wife Dora Katherina (who had been only 14 years old at the time of their marriage in 1668) was John Dury's daughter must have created an extra sense of personal closeness between Oldenburg and Pell, who was one of Dury's oldest remaining friends in England. (Dury himself stayed on the Continent, still pursuing his irenicist projects, from 1661 until his death in 1680.) Later in the 1670s there are many signs of contacts between Pell and Oldenburg; in the period 1669–72, however, the evidence is sparse. In April 1670 Pell was able to use his position as chaplain to the Archbishop to perform a useful service for Oldenburg: he received an official license from Sheldon, addressed to the officers of His Majesty's Customs, authorizing him 'to veiw a Parcell of Bookes lately imported from Hamborough [*sc.* Hamburg] ... for the use of H. Oldenburg Esq'.[15] Acquaintance with Oldenburg was indeed useful for anyone who, like Pell, hungered after new publications. In the following summer Oldenburg told one of his correspondents that Pell was studying the new materials by Fermat that accompanied his recently published annotations on Diophantus; the book itself, which Fermat's son had sent to Oldenburg, may well have been on loan from him to Pell.[16] A brief letter from Oldenburg to Pell, from the summer of 1672, requests: 'Sir, I pray, send me Dr Willis's book de anima Brutorum, [if] you haue done wth it, as also the [>French] book of ye Quadrature of ye Circle and Perpetual motion.'[17] No doubt these were just two of many such borrowings by Pell.

Gradually, Oldenburg got into the habit of consulting Pell on matters of pure and applied mathematics. When, in May 1672, Oldenburg took an interest in the French inventor Cassegrain's design for a reflecting telescope with a convex

---

[14] BL, MS Add. 4412, fo. 323^r (Pell to Brereton, 6 Sept. [1670]). The year is not given, but can be inferred from internal evidence. Pell had evidently spoken to Worthington, who had returned to Hackney and London in August 1670 (Worthington, *Diary and Correspondence*, ii(2), p. 343). A strip has been torn from the section of the page containing the news about Oldenburg's son, which thus reads as follows: 'I beleeve you have [*page torn*] Mr Oldenburgs sonne died [a while since *deleted*] when it [*page torn*]en dayes old' – the incomplete numerical word could be 'seven', 'eleven', or anything from 'thirteen' to 'nineteen'. The existence of this son is apparently unknown to Oldenburg's modern biographer, who states that his first child (probably his daughter Sophia) was born in 1672: M. B. Hall, *Henry Oldenburg: Shaping the Royal Society* (Oxford, 2002), pp. 287–8.

[15] BL, MS Add. 4278, fo. 149^r (Sheldon, license).

[16] Oldenburg, *Correspondence*, viii, p. 127 (Oldenburg to Edward Bernard, 27 June 1671).

[17] BL, MS Add. 4424, fo. 27^r (undated, but datable by Pell's annotation and on internal grounds to 19 Aug. 1672 or a few days before). Oldenburg refers to T. Willis, *De anima brutorum* (London, 1672); I have not been able to identify the French book. This letter, like the one cited below at n. 32, is omitted from Oldenburg, *Correspondence*, where it is incorrectly stated (xiii, p. 411) that 'No letters from Oldenburg to Pell now survive.'

secondary mirror, and received a critique (or alleged improvement) of it from Nicolaus Mercator, it was to Pell that he turned for advice. In the words of Pell's subsequent jottings:

> May 28. 1672. I told Mr Oldenburg that if Monsieur Huygens should compare Mercators paper with Cassegrains Scheme, He would suspect that the paper was written by Mr Hobbes or some such doting bungler, not by Mercator.
>
> May 31. I met Mr Mercator at Mr Oldenburgs house. He defended his paper. I told him he had made much haste &c. He took it away with him: and the French Memoire also.[18]

Others, no doubt, also tried to make use of Pell's expertise from time to time. In October 1671, for example, Robert Boyle asked him for his opinion on 'a Chinese booke'. (Pell does not seem to have made much headway on it, but John Wallis was later able to deduce that it was an almanac, with signs representing days and months.)[19]

The one person who was most keen to tap Pell's mathematical expertise was his friend and sometime landlord John Collins. It is probably not imputing too Machiavellian a motivation to Collins to suggest that his underlying reason for inviting Pell into his home was the hope of finding out more about his mathematical methods. If this is correct, the months or years he spent with Pell under his roof must have been an extraordinarily frustrating time for him. His letters to other mathematicians during this period contain a litany of complaints about Pell's failure to impart his discoveries: he was, as Collins repeatedly said, 'incommunicative'.[20] Unable to learn the details of Pell's methods, Collins adopted a strategy of circulating to other mathematicians descriptions of the feats Pell claimed to be able to perform, with such hints as he could gather of how he performed them, in the hope that they could work out the missing steps in the argument. Thus in letters to Isaac Newton (in July 1670) and René-François de Sluse (in October 1670) he included almost identical passages, describing Pell's claim 'that he could most exactly limit any equation, shewing what the homogeneum must be to make any pair or pairs of roots gain or lose their possibility; and secondly, that out of that doctrine of limits he could fill up (with no great toil) columns containing all those ranks of roots both negative and positive'. All that he had been able to discover about the method Pell used was that it was achieved 'scandendo', by moving upwards. Significantly, he wrote that Pell had

---

[18] BL, MS Add. 4417, fo. 398^v (Pell notes). Oldenburg printed a translation of Cassegrain's proposal in the *Philosophical Transactions*, no. 83 (20 May 1672), pp. 4056–7. See also Oldenburg's discussion of this topic in his letter to Newton of 2 May 1672, and Newton's reply of 4 May: I. Newton, *Correspondence*, ed. H. W. Turnbull *et al.*, 7 vols. (Cambridge, 1959–77), i, pp. 150–1, 153–5.

[19] BL, MS Add. 4394, fo. 3^r (Pell, notes, 'Out of a Chinese booke lent me by Mr Robert Boyle Octob 29. 1671'); Boyle, *Correspondence*, iv, pp. 235–6 (Wallis to Boyle, 13 Dec. 1671).

[20] See for example Rigaud, *Correspondence*, i, pp. 149 (Collins to Sluse, Oct. 1670), 196 (Collins to Beale, 20 Aug. 1672); ii, p. 220 (Collins to Gregory, 25 Mar. 1671).

made this claim 'often ... to his familiars' – of whom Collins was one.[21] By March 1671 Collins was frustrated to find that neither the recent work of de Sluse nor that of James Gregory could supply the required method; as he wrote to John Wallis, 'Neither of them come up to what I have heard discoursed by Dr. Pell, to wit, that he finds the limits of high equations made by the multiplication of the known roots ascendendo'.[22] And just a few days later he wrote to James Gregory about another of Pell's feats in this field: 'Dr. Pell in discourse affirms that in a complete equation of the eighth power, between certain limits, the six intermediate terms may be all taken away; between other limits there can be but four of them taken away; and again between other limits but two of them taken away.' By this stage, Collins's patience had snapped. 'Dr. Pell communicates nothing', he fumed; 'he once refused me a proposition, and I am resolved never to move him more.'[23]

Possibly Pell's incommunicativeness was prompted not only by his habitual reserve, but also by some inkling of what Collins was up to – and, indeed, had been up to for several years. For as early as February 1667 Collins had written to Wallis giving what details he could of Pell's 'doctrine of limits' of equations, and introducing the information with the comment that 'in regard the Dr. is not likely in this treatise [the *Introduction to Algebra*] to publish something else I have heard him discourse of, I think fit to suggest it [to you]'.[24] Pell could be forgiven for disliking the prospect of having his most original ideas wormed out of him and then supplied privately to others, who might later present them to the public as their own. Paradoxically, therefore, this period of intimacy with such an active 'intelligencer' may have had the effect of making Pell even more reluctant to impart his discoveries to anyone. Collins's increasingly bitter remarks certainly suggest a growing secretiveness on Pell's part. By the summer of 1672 Collins was writing:

> As to Pappus ... Dr. Pell hath made notes, as I have heard him affirm, on that author; but to incite him to publish any thing seems to be as vain an endeavour, as to think of grasping the Italian Alps, in order to their removal. He hath been a man accounted incommunicable; the Society (not to mention myself) have found him so: had they not, possibly they might have recommended him to a pension from his Majesty of France, there being an intimation from the Royal Academy to allow two or three pensions to meet persons. As to his knowledge, I take him to be a very learned man,

---

[21] Ibid., i, 149 (Collins to Sluse, Oct. 1670); ii, 303–4 (Collins to Newton, 19 July 1670). Where Newton and Pell were concerned, Collins engaged in a two-way traffic: among Pell's papers is a transcript, in Collins's hand, of the mathematical section of Newton's letter to Collins of 10 Dec. 1672 (BL, MS Add. 4407, fo. 155^v; the original is printed in Newton, *Correspondence*, i, pp. 247–8).

[22] Rigaud, *Correspondence*, ii, p. 526 (Collins to Wallis, 21 Mar. 1671).

[23] Ibid., ii, pp. 219–20 (Collins to Gregory, 25 Mar. 1671).

[24] Ibid., ii, p. 472 (Collins to Wallis, 2 Feb. 1667).

more knowing in algebra, in some respects, (which I think I can guess at), than any other, and they in other respects than he; but as to other parts of the mathematics, I grossly mistake if divers of them do not parasangis bene multis ['by a great many parasangs': one parasang was roughly two miles] surpass him.[25]

And yet, for all his incommunicativeness, Pell had never lost the sense of a duty to serve the public with his skills. One sign of this was his publication of a 32-page volume of mathematical tables, *Tabula numerorum quadratorum decies millium ... A Table of Ten Thousand Square Numbers*, in the spring of 1672. As he explained in one of his private memoranda, he was prompted to undertake this work by the errors he found in the tables of 10 000 squares and cubes in Paul Guldin's *De centri gravitatis inventione* (Vienna, 1635). His minutely detailed note on the chronology of this enterprise, from his delivery of copy to Moses Pitt's printer to the appearance of the book, shows what a frustratingly slow experience it was – and, for a change, the slowness was not his fault. 'From September 13 to March 21 following, are 191 dayes, or 27 weeks, 2 dayes', he concluded. 'A long time for printing of 8 sheets.' Once again, the finished product was still riddled with misprints: he corrected some of the copies by hand, and his friend Sir Samuel Morland then helped to copy those corrections into the others.[26] Pell had decided, characteristically, that this should be an anonymous publication. It seems that his wishes were at first ignored by the printer, who, in his initial print-run, added the words 'By John Pell' in small print at the foot of the final page; most copies, however, have a corrected sheet, in which those words are omitted.[27] Nevertheless, Pell's responsibility for the work was not a secret. Oldenburg included a review of it in the issue of the *Philosophical Transactions* for 22 April 1672; when sending that issue to Huygens, he mentioned that it contained 'something by Mr Pell', and offered to send him a copy of the volume itself.[28]

At about this time Oldenburg seems to have begun taking more active steps to involve Pell in current intellectual life. (The fact that Pell was now living in

---

[25] Ibid., i, p. 197 (Collins to Beale, 20 Aug. 1672). Collins first told the story of this episode in a letter to Gregory of 6 May 1671: 'had not our Pell been a second Robervall, and no wayes obliging to [>his] acquaintance, or to ye R. Societie, they might, and probably would, have recommended him' (St Andrews University Library, MS 31009, fo. 26^v, printed in Turnbull, ed., *Gregory Tercentenary Volume*, p. 185). On Louis XIV's willingness to pay pensions to English writers, see J. J. Jusserand, *A French Ambassador at the Court of Charles the Second: Le Comte de Cominges, from the Unpublished Correspondence* (London, 1892), pp. 60–1.

[26] BL, MS Add. 4424, fos. 78^r, 80, 82^r (Pell, memorandum). Pell's papers contain one uncorrected copy: MS Add. 4394, fos. 89–98. The copy in the CUL (pressmark White a.5) has numerous pen-and-ink corrections, and its titlepage is annotated: 'The Press-faults, of this copy, were corrected by Doctor Pell.'

[27] A copy with the uncorrected sheet is Bodl., pressmark N.13 (3) (where the offending words have been deleted in ink); cf. Bodl., pressmark Ee.1 (2), which has the corrected sheet.

[28] Anon. (Oldenburg?), review of Pell, *Tabula numerorum quadratorum*, in *Philosophical Transactions of the Royal Society*, no. 82 (22 Apr. 1672), pp. 4050–2; Oldenburg, *Correspondence*, ix, p. 54 ('quelque chose faite par Monsieur Pell').

the same street as Oldenburg – Pall Mall – must have facilitated this; indeed, it may well have been a consequence of Pell's friendship with him.)[29] Oldenburg encouraged this process in two ways: by trying to involve him in Oldenburg's own activities as a correspondent with foreign mathematicians, and by getting Pell to resume his role as an active participant in the Royal Society. For several years, Oldenburg had been managing a complex mathematical dispute between Huygens and de Sluse about 'Alhazen's problem': as propounded by the eleventh-century Muslim optical scientist, this required constructing a mathematical solution to the problem of finding two points on a hemispherical concave mirror such that light falling on one point would be reflected to the other. Huygens first sent Oldenburg a solution in June 1669, and the dispute rumbled on for the next four years.[30] During the winter of 1672–3 Oldenburg passed to Pell all the letters relating to this problem, from both mathematicians, for his scrutiny. As Collins wrote to James Gregory in November: 'diverse other Letters have passed betweene Slusius and Hugens through Mr Oldenburgs hands, about Alhazens Probleme, and Dr Pell makes him beleiue he will digest and Print the whole with his owne thoughts thereupon, having to this Purpose all the Letters in his hands.'[31] At the end of December Pell returned all the Huygens letters; in mid-January he received a note from Oldenburg, 'Sir, I pray send me, for a day or two only, the letter of Monsr Slusius dated Decemb. 6. 1672. of wch I haue present occasion'; at the end of the month Pell returned all the letters by de Sluse; unfortunately, however, the verdict he gave on their labours was never published, and has not survived in any form.[32] Just as Pell was finishing his work on these materials, Oldenburg received another letter from de Sluse, proposing a new method for finding tangents of curves: in his reply, on 29 January, Oldenburg wrote that 'I have already shown your method to the most noble President of the Society [Brouncker] and also to Dr Pell, who ask for the demonstrations of

---

[29] See above, at n. 4; Oldenburg had a house in the middle of Pall Mall. How long Pell stayed in Pall Mall is not known with any accuracy; his next recorded address was a 'lodging in Jermin street' in July 1675 (BL, MS Add. 4416, fo. 54^r: Pell memorandum).

[30] See Oldenburg, *Correspondence*, vi, p. 43 (Huygens to Oldenburg, 16 June 1669), for the first solution, and vols. vi, vii, viii, ix, *passim*, for subsequent treatments by Huygens and de Sluse. For a useful summary of the dispute see Hall, *Henry Oldenburg*, pp. 185–7.

[31] St Andrews University Library, MS 31009, fo. 47, Collins to Gregory, 8 Nov. 1672 (printed with minor inaccuracies in Turnbull, ed., *Gregory Tercentenary Volume*, p. 247).

[32] BL, MSS Add. 4407, fo. 118^r (Pell, memorandum: 'Vltimo Decembris 1672, Hugenias, Vltimo [Februarij *deleted* >Januarij] 1672/3, Slusianas, chartas omnes de speculo sphaerico reddidi D. Henrico Oldenburg' ('I gave back to Mr Henry Oldenburg all the Huygens letters about the spherical mirror on 31 Dec. [1672], and all the de Sluse letters on 31 Jan. 1673')); Add. 4424, fo. 26^r (Oldenburg to Pell, 12 Jan. 1673; omitted from Oldenburg, *Correspondence*). Scribal copies of the Huygens and de Sluse letters, probably made at Pell's behest, are in BL, MS Add. 4400, fos. 49–72; a fair copy in Pell's hand of Huygens's letter of [21 June/] 1 July 1672 is in BL, MS Add. 4407, fos. 126^v–125 (retrograde); a scribal copy of de Sluse's letter of [12/] 22 Nov. 1670, lacking the diagrams, is in BL, MS Add. 4424, fos. 377–8.

the method.'[33] As this indicates, Oldenburg was not simply making use of Pell's expertise behind the scenes; on the contrary, he was eager that his name should be known and recognized.

A further opportunity for such recognition came just four days later, when the young mathematician, philosopher, and polymath Gottfried Wilhelm Leibniz, then visiting London, had an encounter with Pell. As he wrote in a letter addressed to Oldenburg (but evidently intended for wider circulation) on 3 February: 'When I was yesterday at the very illustrious Mr Boyle's, I met the famous Mr Pell, a notable mathematician, and the topic of numbers chanced to come up...' Leibniz had put forward an idea he had had for the creation of numerical series by means of what he called 'generative differences'; to his surprise, 'the famous Pell answered that this was already in print ... in the book by the meritorious Mouton, *De diametris apparentibus solis et lunae*'. Leibniz had never seen the book; 'for which reason, picking it up at Mr Oldenburg's ... I ran through it hastily, and found that what Pell had said was perfectly true.'[34] It was a sign of Pell's characteristic industry that he should have read this obscure astronomical work by Gabriel Mouton (*Observationes diametrorum solis et lunae apparentium* (Lyon, 1670)).

Leibniz thus acquired an instant respect for Pell; and he probably also heard, when discussing him with others, about Pell's rumoured discoveries in algebra. Just three weeks later he wrote to Oldenburg (this time from Paris): 'I very much wish to know whether anything has been set forth by those distinguished men among you ... Brouncker ... Wallis, Pell, Mercator, Gregory, or others concerning the reduction [of the degree] of equations.'[35] Oldenburg's next letter to Leibniz was drafted for him by Collins: in it, Collins referred to Pell's work (without giving his name), saying that there was 'a Learned Man' in England 'who hath thoughts of writing a Treatise *de canone Mathematico* or Table of sines, shewing what strange difficult Problemes and aequations may be solved

---

[33] Oldenburg, *Correspondence*, ix, pp. 386–96 (de Sluse to Oldenburg, [7/] 17 Jan. 1673), 427–30 (Oldenburg to de Sluse, 29 Jan. [/8 Feb.] 1673: quotation, p. 430).

[34] Ibid., ix, pp. 443–4 (Leibniz to Oldenburg, 3 Feb. 1673). This third-person reference to Oldenburg is one of several signs that the letter was intended for immediate circulation, to defend Leibniz against any possible charge of plagiarism. In fact this letter did later become the first item in a dossier of material relating to his dispute with Newton over the invention of the differential calculus. The dossier was published by the Royal Society as *Commercium epistolicum D. Johannis Collins, et aliorum de analysi promota: jussu Societatis Regiae in lucem editum* (London, 1712) (this letter: pp. 32–7), with a verdict appended to it: 'That when Mr. Leibnitz was the first time in London, he contended for the Invention of another Differential Method properly so call'd; and notwithstanding that he was shewn by Dr. *Pell* that it was *Mouton*'s Method, persisted in maintaining it to be his own Invention' (p. 121).

[35] Oldenburg, *Correspondence*, ix, p. 494 (Leibniz to Oldenburg, 26 Feb./8 Mar. 1673).

thereby'.[36] This was another topic on which Collins had long been trying, without success, to find out – or get others to replicate – Pell's methods: in November 1670 he had told Gregory that he had urged the mathematician Isaac Barrow 'to invent a method of solving all Aequations by the Canon of Sines or Log^{mes}, which Dr Pell pretends he hath done for many yeares'.[37] In the spring of 1675 Collins returned to this topic in another survey of current work, also intended for Leibniz. He referred first to Pell's alleged discovery of a method for finding the limits of equations and reducing them (the topic on which Collins had previously written to Newton, de Sluse, and Gregory), and then added: 'he also pretends to do wonders in Æquations in generall by ayd of a large Canon of Sines.'[38] This concept evidently caught Leibniz's interest, and in a later letter to Oldenburg he commented that 'To unravel equations by means of a table of logarithmic sines will be a most useful thing, if only the work of preparation be not so great that the benefit of the abridgement is lost'. This passage in Leibniz's letter was annotated in the margin by Oldenburg: 'Pell promises much about these things, but when.'[39] For at least a few months during 1675, it seemed that Pell's promises would finally be fulfilled. In June of that year Collins gave Gregory the following news about Pell: 'he upon the remoovall of Dr Barton [*sic* – for 'Barlow'] of Oxford to the Bishopric of Lincolne, hath obtained a graunt of a sine cura of 80^{pd} by the procurement of Sr Joseph Williamson and Col Titus, who haue engaged him to publish his [papers *page torn*] as soone as he can, he having

---

[36] Ibid., ix, p. 552 (Collins to Oldenburg for Leibniz, *c.*1 Apr. 1673). Among Pell's papers there is a large fair-copy table of sines (with their 'differentiae'): BL, MS Add. 4424, fos. 457–8. The earliest mention of this idea comes in a memorandum written by Collins in 1670 – apparently as a private record of what he had managed to winkle out of Pell on this subject – entitled 'A Discourse to prove that all Aequations may be solved by Tables'. It begins: 'A very learned Algebraist affirmeth that he hath for above 30 yeares used to solve Aequations by tables, and that he gave unto Stampioen a Dutch Algebraist, all the rootes of an adfected Aequation of the 6th power thus found, he indeed requireth a Canon of Sines, and each degree of an Arch to be divided into 1000 parts, and talkes of writing a Treatise of this Argumt:' (Worcester College, Oxford, MS 64 (unfoliated, in final section)). Another version of this statement (giving 'Dr Pell' instead of 'A very learned Algebraist') occurs in a Collins MS which is also undated, but which may have been sent by Collins as an enclosure to his letter to James Gregory of 15 Dec. 1670: St Andrews University Library, MS 31009, fo. 6^{r} (printed in Turnbull, ed., *Gregory Tercentenary Volume*, p. 142).

[37] St Andrews University Library, MS 31009, fo. 19^{r}, Collins to Gregory, 1 Nov. 1670 (printed in Turnbull, ed., *Gregory Tercentenary Volume*, p. 111).

[38] Oldenburg, *Correspondence*, xi, pp. 256–7 (Collins to Oldenburg for Leibniz, 10 Apr. 1675). Collins returned to Pell's work on equations in another letter prepared for Oldenburg for Leibniz: ibid., pp. 362–70 (Oldenburg to Leibniz, 24 June 1675).

[39] Ibid., xi, pp. 395–6 (Leibniz to Oldenburg, 2/12 July 1675); 'Pellius promittit multa de his, sed quando' (Royal Society, MS 81 (Commercium epistolicum), no. 26, 1st leaf, recto).

promised to read some of them to the Society.'[40] Possibly Collins's information about the sinecure – for which no other evidence emerges from anything in Pell's papers – was as inaccurate as his spelling of Thomas Barlow's name; nevertheless, the news of some sort of negotiation with Pell over the publication of his papers seems to have been authentic. Writing back to Leibniz in September, Oldenburg sounded optimistic. 'As for the resolution of equations by tables of sines and logarithms, our Pell has, I hear, promised to produce this. We very much hope that he will redeem his promise.'[41] That hope was never fulfilled; but if Pell had indeed made such a pledge, it must have been partly as a result of the stimulus of interest in his work which Oldenburg had helped to generate. (Leibniz did not forget the promise, and five years later a letter from Leibniz to Hooke was read out at a meeting of the Royal Society, 'inquiring concerning the undertakings of Dr Pell, and especially his way of resolving equations by a table of [sines]'.)[42]

Pell's involvement in the activities of the Royal Society during this period is not adequately documented, thanks to the failure of the minutes to list the members attending the meetings; Pell's name does not reappear in those minutes until 25 February 1675, when he was one of a handful of Fellows asked to examine the manuscript of a book about witchcraft by John Webster.[43] But there is clear evidence that he had been attending the Society's meetings before then. Only two weeks earlier (as Oldenburg mentioned in a letter to Martin Lister) he had been present at a meeting when 'very ingenious reflexions' were made in a lively discussion of the mechanics of avian flight, 'there being present many of our best Mathematicians and Mechanicians, and amongst ym ye Ld Brouncker, Sr Wm Petty, Dr Pell, Mr Hook'.[44] Among his papers there are the notices summoning him to meetings on 14 April and 12 November 1674.[45] Earlier indications of his closeness to the work of the Society are supplied by Robert Hooke's diary, which gives us glimpses of Pell socializing with Hooke and other Fellows (and/or

---

[40] St Andrews University Library, MS 31009, fo. 75^r, Collins to Gregory, 29 June 1675 (printed with some inaccuracies in Turnbull, ed., *Gregory Tercentenary Volume*, p. 310). That the missing word was 'papers' is indicated by Gregory's reply: 'I am glad that Dr. Pell is engaged to publish his papers' (Rigaud, *Correspondence*, ii, p. 268 (Gregory to Collins, 23 July 1675)). The prominent civil servant Sir Joseph Williamson (1633–1701) was an F.R.S. (1662) with strong interests in mathematics and the natural sciences; he served on the Council of the Royal Society from 1674 to 1690.

[41] Oldenburg, *Correspondence*, xi, p. 520 (Oldenburg to Leibniz, 30 Sept. [/10 Oct.] 1675).

[42] Birch, *History*, iv, p. 33; the uncomprehending minute-taker has written 'table of signs'.

[43] Ibid., iii, p. 192 (the other Fellows were William Petty and Daniel Milles). A memorandum among Pell's papers states: 'Friday. March 19. 1674/75. I began to read a Manuscript brought [>to] me (24 houres before) by Mr Oldenburg. The Title. The Displaying of supposed Witchcraft ... by John Webster' (BL, MS Add. 4255, fo. 39^r). On the chemist, physician and Puritan minister John Webster see A. G. Debus, *Science and Education in the Seventeenth Century: The Webster–Ward Debate* (London, 1970), esp. pp. 37–43. On the significance of his *The Displaying of Supposed Witchcraft* (London, 1677), see P. Elmer, *The Library of Dr. John Webster: The Making of a Seventeenth-Century Radical* (London, 1986), pp. 7–14.

[44] Oldenburg, *Correspondence*, xi, p. 190 (cf. Birch, *History*, iii, p. 181).

[45] BL, MS Add. 4424, fos. 381, 383.

employees) of the Society in 1673. A typically compressed entry for 27 January 1673 reads: 'Dr. Pell, Mr. Hill, Mr Haux, here, afterwards Mr Colwall' – the others named being Abraham Hill, Theodore Haak, and Daniel Colwall, all of them F.R.S. Hooke noted that on this occasion Pell told him about a method of printing discussed in one of the works of Giambattista della Porta, and also about techniques for grafting, and for propagating vines.[46] Another entry, for 3 March, must give the lie to any idea that Pell was generally unsociable or reclusive: 'Mr. Haux came back from Dr. Pell at 8 ... Dr. Pell here and dined at Dr. Goddards. DH [sc. 'Dined at Home']. At Lord Brounkers for Dr. Pell, Oldenburg there ... at Dr. Croons with Pell, Haux, Hill, Salisbury, Wise.'[47]

Hooke enjoyed quite a close friendship with Pell, interrupted only by a severe falling-out between them following Oldenburg's death in 1677. He seems to have involved Pell in his otherwise rather secretive work on his 'arithmetick engine' or calculating machine in the spring of 1673; Pell's name crops up repeatedly in his diary during these years; and in 1676, when his controversial lecture *Lampas* was published, he gave copies of it to just four people: his old mentor Dr Busby, his physician friend Dr Theodore Diodati, John Evelyn, and John Pell.[48] A few entries in Hooke's diary for March and April 1673 show that he even made an effort to solicit some sort of paid lectureship for Pell. On 16 March, in a conversation with Haak, he 'Spoke of Hospitall Lecture for Dr. Pell'; on 29 March he 'Told Pell of Wards Designe, he accepted it'; and then, on 12 April: 'At Scotland yard. Alderman Wards. Dr. Pell refused.'[49] Hooke was a governor of Christ's Hospital, the London school to which a special 'mathematical school' was attached later in 1673 to train boys to become navigators for the Royal Navy; so it might be supposed that Hooke's plan here was to instal Pell as headmaster of that new school.[50] But, on the other hand, schoolmastering would not have been described as a 'Lecture' or lectureship; and, in any case, the plans for the mathematical school were so embryonic at this stage that the nature of the job was as yet quite undefined. The school received its Letters Patent only in

---

[46] R. Hooke, *The Diary of Robert Hooke, 1672–1680*, ed. H. W. Robinson and W. Adams (London, 1935), p. 24.

[47] Ibid., p. 32. 'Dr Croon' was William Croone, F.R.S.; 'Salisbury' might be Oliver Salusbury (F.R.S. in 1681); Wise has not been identified.

[48] BL, MS Add. 4422, fo. 78^r contains a note by Pell on Hooke's 'engine for multiplying and dividing'; Inwood writes that Hooke 'worked with' Pell to produce this machine (*The Man who Knew Too Much*, p. 176); Hooke, *Diary*, e.g. pp. 21, 31, 33, 34, 97, 146–7, 149–52, 212, 253 (distribution of *Lampas*). (This Cutlerian Lecture by Hooke was controversial because it included an attack on Oldenburg for having failed to support him in his claims of priority against Huygens.) Hooke also bought 'Dr. Pells against Longomontanus' in 1674, and acquired a copy of 'Dr. Pells *Algebra*' in 1675: *Diary*, pp. 85, 152.

[49] Hooke, *Diary*, pp. 34, 36, 39.

[50] On Hooke's governorship see M. 'Espinasse, *Robert Hooke* (London, 1956), p. 96; on the mathematical school see E. H. Pearce, *Annals of Christ's Hospital*, 2nd edn. (London, 1908), pp. 98–126. This supposition is made by L. Murdin, *Under Newton's Shadow: Astronomical Practices in the Seventeenth Century* (Bristol, 1985), p. 47: 'Dr John Pell was the first nominee for the post.'

August 1673, and it was not until November that the duties of the headmaster – teaching, or supervising the boys, from 7 to 11 a.m. and from 1 to 5 p.m. – were laid down.[51] The job was then given immediately to Pell's old friend John Leake; but the reasons offered for his dismissal five years later (including the complaint by the governors, 'if he should be permitted to sitt privately in his Closett how could he Observe the misdemeanors of the Children?') make plain that his job would have been completely unsuited to Pell.[52] If this really was the employment suggested by Hooke in early 1673, Pell's instinct to say 'no' was, for once, fully justified.[53]

In 1674 Hooke served with Pell on a commission to examine a claimed solution to the problem of determining longitude. The solution had been put forward by Pell's old acquaintance Henry Bond, the mathematical practitioner in London, who was now well into his seventies.[54] Bond had taken a strong interest – as had Pell – in Gellibrand's work on magnetic declination in the 1630s, and had been hinting since the late 1640s that he had a method of determining position at sea, based on magnetic variation. In 1673 he published a broadsheet announcing this claim, whereupon Charles II appointed a commission under Sir Samuel Morland to examine Bond's work. This was not one of the Royal Society's own committees, but every one of its members apart from Morland (Pell, Hooke, Brouncker, Seth Ward, and Silius Titus) was a Fellow of the Society. Pell kept a detailed set of minutes of its activities, and one modern historian concludes that it was his 'diligence' that ensured that the commission did its work.[55] The members of the commission were generally sceptical about Bond's claims, but reluctant to reject them without further testing, for which special instruments would have to be made.[56] Perhaps some personal fondness for this elderly researcher also

---

[51] Guildhall Library, London, MS 12,873/1, Christ's Hospital, Royal Mathematical School, minute and memoranda book, unfoliated: Letters Patent, 19 Aug. 1673; minutes of meeting, 14 Nov. 1673 (duties). The salary was £50 a year, with the use of 'a convenient House in the Hospitall'.

[52] Ibid., minutes of meeting, 18 Nov. 1673 ('Mr John Leake ... was unanimously chosen'); minutes of meeting, 7 Dec. 1677 (complaints against Leake). Leake had presented himself at the meeting of 14 Nov., when the job-specification was drawn up; no other candidates were discussed or approached, and Pell is not mentioned anywhere in these minutes.

[53] It has been claimed that Pell did later serve the mathematical school in another capacity, being co-opted onto a committee in 1676 to discuss the topics on which the boys should be examined for their passing-out certificate: Willmoth, *Sir Jonas Moore*, p. 201. However, the source to which Willmoth refers makes no mention of Pell, listing the committee as 'Brouncker ... with Secry Pepys, S[r]. Xpher Wren, S[r] Jonas More, Esq[re]. Colwell and M[r]: Hooke': Pepys Library, Magdalene College, Cambridge, MS 2612, p. 229. (I am very grateful to Dr Richard Luckett and Mrs Aude Fitzsimons for their help in enabling me to check this.)

[54] See above, ch. 3, n. 144. Taylor gives his dates as *c.*1600–1678 (*Mathematical Practitioners*, p. 207).

[55] D. J. Bryden, 'Magnetic Inclinatory Needles: Approved by the Royal Society?', *Notes and Records of the Royal Society of London*, 47 (1993), pp. 17–31, here p. 17. Pell's notes (including a copy of Bond's 1673 broadsheet) are in BL, MS Add. 4393.

[56] Prof. Feingold has pointed out that a political consideration was also at work here: the King (or his favourites) did not wish to hear the truth.

came into play; Hooke's diary entry for the meeting of 16 April 1674 states that 'Dr. Pell examind Bond's theory. Found it ignorant and groundless and fals but resolved to speak favourably of it.'[57] In the end, only an interim report was made to the King in the following year, on the basis of which Bond was promised a single payment (not the annual payments he had requested) of £50.[58]

While this commission was still sitting, another claimant came forward with an alleged method of determining longitude – this time, based on making precise observations of the position of the moon in relation to various stars. The person making this claim was a little-known Frenchman called the sieur de Saint-Pierre; as he was a protégé of Charles II's mistress Louise de Keroualle, he succeed in obtaining an order from the King, at the end of 1674, that the commission should study his work too. Sir Christopher Wren was now appointed to join the commission, and the astronomer John Flamsteed was co-opted onto it in February 1675; the mathematician Sir Jonas Moore also became involved.[59] Once again Pell kept the minutes and managed the paperwork. Through him, Flamsteed sent Saint-Pierre two specimen observations, with details of the moon and the relevant stars, and a statement of the latitudes at which the observations were made, and asked him to work out the longitude; Saint-Pierre refused to do so, complaining that the observations had been fabricated.[60] Eventually, in April, Flamsteed handed in to Pell a damning judgement on Saint-Pierre's work: he suspected that it was plagiarized from the earlier work of Jean-Baptiste Morin, and pointed out that, even if the theory were correct, it would require measurements of such precision that they could never be made on the deck of a ship at sea. 'I marvel therefore', he concluded, 'at the stupid effrontery of the man.'[61]

Other duties, some less formal and some more so, were given to Pell by the Royal Society itself. His task of examining Webster's book has already been mentioned; in April 1676 he was similarly required, together with Hooke, Wren, Moore, Wallis, and others, to examine a treatise on mechanics submitted by a M. Joly of Dijon.[62] Pell had been elected to the Council of the Society in November (he is listed as present at a Council meeting in March), and on 18 May Brouncker

---

[57] Hooke, *Diary*, p. 97.

[58] Bryden, 'Magnetic Inclinatory Needles', pp. 19–20.

[59] BL, MS Add. 4393, fos. 89^r (appointment of Wren), 93^v (Pell, minutes: Moore). See also F. Baily, *An Account of the Revd. John Flamsteed* (London, 1835), pp. 37–8, 189, and E. G. Forbes, 'The Origins of the Greenwich Observatory', *Vistas in Astronomy*, 20 (1976), pp. 39–50, esp. pp. 40–1.

[60] CUL, MS RGO 1/50, fo. 262^r (Flamsteed memorandum, printed in J. Flamsteed, *The Correspondence of John Flamsteed, The First Astronomer Royal*, ed. E. G. Forbes, L. Murdin, and F. Willmoth (Bristol, 1995–), i, pp. 345–6).

[61] BL, MS Add. 4393, fo. 99 (Flamsteed to Pell, 26 Apr. 1675; printed with translation in Flamsteed, *Correspondence*, i, pp. 334–7 (here p. 337: quotation)).

[62] Birch, *History*, iii, p. 314; cf. Oldenburg, *Correspondence*, xii, pp. 195–7.

appointed him one of the Society's several Vice-Presidents.[63] Under the Society's rules, a meeting could take place only under the chairmanship of the President or someone formally deputed by him: hence, for convenience, the plurality of Vice-Presidents at any one time. But since this was the sole rationale of the office of Vice-President, the fact that Pell was appointed does suggest that he was known to be a habitual attender of the Society's meetings, even though his spoken interventions were comparatively rare.[64] Pell did not stay on the Council after 1676, but his continued involvement in some Council affairs is suggested by a note among his papers, dated 5 January '1677' [probably 1678], summoning him to 'a Committee of ye Council' of the Society, 'at Mr Boyls'.[65] Overall, it may be felt that the leading modern historian of the Royal Society has judged Pell a little too harshly when he has summed up his record as 'Active in early 1660s; slightly active thereafter'.[66]

Throughout this period, Pell had benefited from the constant stimulation of Henry Oldenburg's friendship. Many of the letters Oldenburg received from scientists at home and abroad are scattered among Pell's papers; some of these may have been acquired after Oldenburg's death, but most, probably, were passed to him for his information or to elicit his comments (for example, a letter from Wren of October 1673, or one from Huygens of November 1675.)[67] So it must have come as a cruel blow to Pell when Oldenburg died quite suddenly – possibly of a stroke – on 5 September 1677. Even worse, his young wife followed him to the grave only twelve days later, leaving two children aged just five and two-and-a-half. The ever-generous Robert Boyle paid for Oldenburg's funeral and took over immediate financial responsibility for the orphans. But there were also some awkward practical problems to be dealt with, arising from the fact that Oldenburg's will could not be found. His house contained many of the papers he had written or received in the course of his duties as Secretary of the Royal Society; that body naturally wished to recover them, but lacked the legal authority to search the premises. And one member of the Royal Society had a particular reason for wanting to make such a search: Robert Hooke, who had quarrelled bitterly with Oldenburg, was convinced that he would find evidence

---

[63] Birch, *History*, iii, pp. 309 (Council meeting, 6 Mar. 1676), 317 (Vice-Presidents: John Pearson, Moore, Pell, and Walter Needham). Brouncker's signed certification of Pell's Vice-Presidentship is in BL, MS Add. 4423, fo. 327^r.

[64] On the office see Hunter, *Royal Society*, pp. 78–9.

[65] BL, MS Add. 4426, fo. 166^r.

[66] Hunter, *Royal Society*, p. 163.

[67] BL, MSS Add. 4428, fo. 314^r (Wren); Add. 4299, fos. 92–3 (Huygens to Oldenburg, [13/] 23 Nov. 1675, printed in Oldenburg, *Correspondence*, xii, p. 54). The Wren letter is undated; reasons for dating it to c.8 Oct. 1673 are given in Oldenburg, *Correspondence*, where the text is printed only from the copy in the Royal Society, under the heading 'Wren to ? Oldenburg'. The doubt is removed by the MS, which is annotated: 'Dr Wren to Mr Oldenburg, drawn up according to his directions in ye paper adjoyned'. Several items of Oldenburg's correspondence among the Pell papers are omitted from the published edition; see N. Malcolm, *A Supplement to the Correspondence of Henry Oldenburg* (forthcoming).

that Oldenburg had been secretly passing details of Hooke's own discoveries and inventions to Huygens.

Eventually a Mrs Margaret Lowden – possibly a friend of Oldenburg's father-in-law, John Dury – was appointed 'administratrix' of the estate.[68] She was a woman of some social status, having inherited a large amount of London property from her father; but she must still have found it a little intimidating to be placed in a more or less conflictual relationship with the dignitaries of the Royal Society. Pell, putting his sense of obligation to his old friend Dury above his loyalty to the Society, gave her his support. Hooke's angry diary-entry for 7 November 1677 paints the scene: 'At ... Oldenburgs. The Books denyd, &c., and Dr. Pell noe freind to the Royall Society. Dr. Pell opend the seald paper from the key hole and unlocked the door. Mrs — and her Solicitor, Dr. Pell, I and H. Hunt [Henry Hunt, Hooke's assistant] enterd, we saw the things but she denyd delivery without paying money and giving Discharge.'[69] His next reference to Pell, on 21 December, reads simply: 'Pell a Dog.' Finally, on Christmas Eve, he was able to record that he had 'With much trouble retrievd the books out of Pells hands & Loudens &c.' Oldenburg's 'trunk' of papers was opened in the presence of Sir John Hoskins (F.R.S. and eminent lawyer), and during the next few days Hooke was able to search through the correspondence, hunting for evidence of intellectual treachery.[70] Pell had not in fact done anything to damage the interests of the Society; and his standing in the affair was formally recognized by the Society itself, which resolved on 2 January 'That care be taken to have the oaths of Dr. Pell and the administratrix made in chancery, that all the papers belonging to the Society had been delivered, and that they knew of none else.'[71] But it would take many months for Hooke's friendship with Pell to recover; as late as August 1678 he noted in his diary 'Haak fals in preferring Pell to Pett' – 'fals' here meaning, presumably, disloyal to Hooke.[72]

One person who had no reason to accuse Pell of disloyalty, on the other hand, was John Dury. A touching letter survives from the eighty-two-year-old Dury to Pell, written in Kassel in June 1678. 'I should be unthanckful', he told his old friend, 'if I did not acknowledge the paines wch you haue been pleased to take to let me know the concernes of my state in England which I seeke to b[r]ing to a settlement; for the education of my Grand-children, wch is now the Chief care I can take for them. Mrs Lowden doth much commend the favourable assistance wch you haue giuen her, for wch I owe you thancks, & beseech the Lord to requite

---

[68] On all these issues see Hall, *Henry Oldenburg*, pp. 209–305, 312–13.

[69] Hooke, *Diary*, p. 326.

[70] Ibid., pp. 335–7.

[71] Birch, *History*, iii, p. 369.

[72] Hooke, *Diary*, p. 370 (5 Aug. 1678). 'Pett' would seem to be Sir John Pett, F.R.S.; but there is a possibility that Hooke meant Sir John Pettus (also F.R.S.). At some stage, both Pettus and Pell prepared translations of a German book on mineralogy (see below, at n. 110); this might well have prompted some comparison of their merits as translators, which Haak was well qualified to judge.

it.'[73] This is the last known communication Pell received from Dury, and, indeed, almost the only letter he is known to have received from the Continent during this period – apart from one from his old admirer Christian Ravius in 1671, and a letter from his ex-pupil Johann Heinrich Rahn in 1675, thanking him for all the work he had done on the *Introduction to Algebra*.[74]

After the episode of Oldenburg's papers, Pell resumed his participation in the Royal Society – without, it seems, any ill-feelings on either side, apart from Hooke's lingering sense of personal grievance. When, in March 1678, his former publisher Moses Pitt put forward a proposal for a multi-volume atlas of the world and asked the Royal Society for its support, Pell was one of a group of Fellows (including Hooke, Wren, Haak, and Collins) who were deputed to consider the matter.[75] Pitt's plan for what he called 'The English Atlas' was immensely ambitious: a work of eleven folio volumes, on the same scale as the famous Dutch atlases by Blaeu and Janszoon which were by now becoming unobtainable. His correspondent and business partner in Holland, Steven Swart, had the plates of the Janszoon atlas, and Pitt's scheme depended mainly on reproducing these, with a different text; but the Royal Society, which decided to back the scheme, was eager for new and more accurate maps, despite the huge extra cost that their production would involve.[76] A committee was set up, consisting of Hooke, Wren, Pell, Isaac Vossius (the polymathic scholar and F.R.S., son of Pell's friend G. J. Vossius, now living in England), Dr Thomas Gale (High Master of St Paul's School, and recently elected F.R.S.), and Dr William Lloyd, the Dean of Bangor. Several of them (including Pell and Hooke) dined with Pitt on 30 April, and three days later Pitt issued a broadside prospectus for the edition; he listed them as 'Advisors', declared that the atlas would be 'a compleat Collection of Maps, Tables, or Delineations of the Heavens, Earth, or Seas', invited subscriptions, asked that anyone possessing rare or unpublished maps should lend them to him for copying, and promised that the work would be printed at Oxford 'upon as

---

[73] BL, MS Add. 4365, fo. 7^r (Dury to Pell, 28 May/7 June 1678).

[74] Ibid., fo. 59 (Ravius to Pell, from Berlin, 2 Jan. 1671, enclosing his printed bifolium *Synopsis chronologiae biblicae infallibilis*, fos. 57–8); BL, MS Add. 4398, fo. 143 (Rahn to Pell, from Zurich, 17 Apr. 1675, saying that he had only recently received a copy 'from a certain learned man ... thanks to your generosity' ('À studioso quodam ... ex munificentia tua')). Ravius had eventually given up all hope of Pell's translation of books V–VII of Apollonius's *Conics*, and had published his own (not very competent) translation in 1669. G. J. Toomer has expressed the suspicion that he simply passed off Pell's translation as his own (*Eastern Wisedome*, p. 186). However, comparison between the specimen of Pell's translation presented in Letter 23 and the same passage in Ravius's translation (Apollonius, *Conicarum sectionum*, ed. and tr. Ravius, p. 148) shows that Ravius's was a very different version.

[75] Birch, *History*, iii, p. 397 (28 Mar. 1678).

[76] See E. G. R. Taylor, '"The English Atlas" of Moses Pitt, 1680–83', *The Geographical Journal*, 95 (1940), pp. 292–9; L. Rostenberg, *The Library of Robert Hooke: The Scientific Book Trade of Restoration England* (Santa Monica, Calif., 1989), pp. 34–8. Pitt was probably stimulated by the various 'atlas' volumes – the most ambitious things of their kind hitherto produced in England – published by John Ogilby, who died in 1676: see K. S. van Eerde, *John Ogilby and the Taste of His Times* (London, 1976), pp. 95–143.

good, if not better Paper than *Bleau*'s.'[77] Further meetings or dinners followed at irregular intervals, with Pell, Hooke, Gale, and Lloyd being the most active members of the committee.[78] Pitt's plans were grandiose, fortified by the deal he had made in March 1678 to acquire (on a sub-lease from Dr John Fell) the most important printing privileges of Oxford University's own press; but he lacked the capital for such an undertaking as this. His first volume, promised for 24 March 1679, had to be postponed by a year at the last moment. In the end only one new map (of the polar regions) was actually drawn and engraved in England.[79] Four volumes were issued by 1680, and another two were 'almost finished' when, in 1683, two of Pitt's partners foreclosed on debts he owed them – whereupon the whole scheme foundered. (Further financial entanglements, arising from his attempts to develop property on land in London for which he was responsible as an executor, caused him to be thrown into prison for debt in 1689; there he wrote his *Cry of the Oppressed*, containing not only a statement of his own case, but also a detailed account of the abuses of the whole system of imprisonment for debt in England.)[80]

Involvement in the atlas project seems, by bringing Pell and Hooke into regular contact again, to have helped to effect a reconciliation between them; from September 1678 Pell's name recurs in lists of ordinary social gatherings attended by them both.[81] Some of these were dinners with Dr Richard Busby, the Headmaster of Westminster School, who was probably the person who had first introduced them. Twice in 1679 Pell and Hooke were at gatherings in Busby's house together with a master carpenter and a master bricklayer, and at one of these Hooke's diary entry records that they 'agreed about the new church'; so it seems possible that Pell was also consulted about Hooke's plans for the church he was building for Busby at Willen, near Newport Pagnell.[82] Busby himself, though a notorious disciplinarian where his pupils were concerned, was a kind and loyal friend, from whom Pell would greatly benefit in his final years. And this period also saw a growth in Pell's friendship with John Aubrey (who, as it happened, was also well acquainted with Busby). It was probably as a result of conversations with Aubrey that, in 1679, Pell wrote a short essay, 'Day-Fatality of Rome', on the subject of Roman beliefs in lucky and unlucky days. Long after Pell's death,

---

[77] Hooke, *Diary*, p. 356; BL, MS Add. 4394, fos. 405–6 (Pitt, broadside; another copy is in MS Add. 4421, fos. 326–7).

[78] Hooke, *Diary*, pp. 370, 374, 376, 440–2, 449. Rostenberg refers one of these dinners (*Library of Hooke*, p. 36, citing Hooke, *Diary*, p. 370), misreading Pell's name as 'John Fell'.

[79] See J. Johnson and S. Gibson, *Print and Privilege at Oxford to the Year 1700* (Oxford, 1946), pp. 79–81 (Oxford sub-lease); Taylor, '"The English Atlas"', p. 297 (polar map). The polar map was probably drawn by Hooke: see Inwood, *The Man who Knew Too Much*, p. 271.

[80] M. Pitt, *The Cry of the Oppressed, Being a True and Tragical Account of the Unparallel'd Sufferings of Multitude of Poor Imprisoned Debtors, in Most of the Gaols of England ... Together with the Case of the Publisher* (London, 1691) (quotation: p. 113).

[81] Hooke, *Diary*, p. 378 (26 Sept. 1678).

[82] Ibid., pp. 405 (29 Mar. 1679), 434 (30 Dec. 1679), with 'Bates' (carpenter) and Thomas Horn (builder); Inwood, *The Man who Knew Too Much*, pp. 302–3.

Aubrey would publish this in his own compilation of information about occult phenomena, 'accidents', and other such matters, entitled *Miscellanies* – thus contributing one further item to the otherwise slender corpus of Pell's published works.[83] Pell was thus far from friendless, even though this period did see the loss of his patron and admirer Lord Brereton. 'Never was there greater love between master and scholar', mourned Aubrey, 'then between Dr. Pell and this scholar of his, whose death March 17 1679/80 hath deprived this worthy doctor of an ingeniose companion and a usefull friend.'[84]

Pell was increasingly in need of loyal friendship, as his personal affairs were in a poor – and deteriorating – state. Discussing this period of his life in his notes on Pell, John Aubrey expostulated:

> Now by this time (1680), you doubt not but this great, learned man, famous both at home and abroad, haz obtained some considerable dignity in the church. You ought not in modestie to ghesse at lesse then a deanery – Why, truly, he is stak't to his poor preferment still! For though the parishes are large, yet (curates, etc., discharged) he cleares not above 3-score pound per annum (hardly fourscore), and lives in an obscure lodging, three stories high, in Jermyn Street, next to the signe of the Ship, wanting not only bookes but his proper MSS. which are many, as by and by will appeare. Many of them are at Brereton...[85]

Aubrey's figures here cannot be corroborated precisely, but may well have been correct. If the curates were paid £30 each, the remainder of Pell's gross income should have come to roughly £110. But there were other deductions to be made (taxes, and the setting aside of ten per cent for charitable uses), as well as occasional repairs and other forms of capital expenditure; and if the curates were not efficient in their collection of tithes, the net return to Pell might well have fallen below £80.[86] This, it must be said, was still a perfectly sufficient income for someone living on a modest scale in London. That Pell got into serious financial difficulties must be due to the fact that he was supporting other members of his family – whether willingly or (in the case of his son-in-law Roger Raven, who may have spent much of Pell's income before it reached Pell) unwillingly. In

---

[83] J. Aubrey, *Miscellanies* (London, 1696), pp. 22–6. The date of composition of Pell's text can be deduced from his statement that 'Some said that *Rome*'s Perdition should happen in the Year of Christ 1670. They have now been decryed Nine whole Years' (p. 22). It is noteworthy that, while Aubrey believed in the power of astral influences, the imagination, and other such occult forces (see M. Hunter, *John Aubrey and the Realm of Learning* (London, 1975), pp. 121–32), Pell's text is coolly critical, treating only the Roman beliefs in such matters and giving them no additional credence.

[84] Aubrey, *'Brief Lives'*, ii, p. 125.

[85] Ibid., ii, p. 124.

[86] No detailed accounts survive. There is one note among Pell's papers of the money 'Due to the King for the Tenths of Laindon', and of Fobbing, for 1680 and 1681, totalling £8 2s 8d: BL, MS Add. 4404, fo. 25^r.

Aubrey's words: 'his tenants and relations cousin'd him of the profits and kept him so indigent that he wanted necessarys, even paper and inke.'[87]

Those to whom Pell gave money probably included his sister-in-law, Bathsua Makin; details of this final period of her life are scanty, but the last surviving trace of her existence is a Latin poem she addressed to Robert Boyle in her eighty-first year (i.e., probably, in 1680), in which she urged him to continue 'giving aid, as you are accustomed to do, to pious widows'.[88] In the 1670s she had taught for a while at a school in Tottenham; it seems likely that she had gained this employment thanks to the intervention of Pell, who had heard about the plans to set up the school from Thomas Brancker in 1668.[89] Brancker (who had close connections with Tottenham, through his previous employment by Lady Reynardson) did not himself take up the offer to participate in this enterprise.[90] But by 1670 or 1671 the Comenian educationalist Mark Lewis – who happened to be also a friend of Pell's old acquaintance Ezerel Tongue – had founded a 'Gymnasium' at Tottenham, and by 1673 a girls' school had been attached to it, with Bathsua Makin as 'Governess'. Lewis was a prolific and repetitive author of tracts on educational method, and his words were echoed in other publications by one of his colleagues, Arthur Brett. Either Lewis or Brett was probably the author of an anonymous text presenting Lewis's methods, *An Essay to Revive the Antient Education of Gentlewomen* (London, 1673), which – unaccountably, in view of its author's explicit statement 'I am a Man' – has been commonly attributed to Bathsua Makin herself.[91]

With Pell's son transformed into a rich land-owner, it might have been supposed that Pell's own financial problems were over; but there is no sign of John Pell Jr sending any money to support his father. Aubrey records that Pell 'thought to have gone over to him'; this is confirmed by a letter Pell wrote in July 1679 to a friend, in which he said that 'Three weeks agoe I prayed A.B.C.

---

[87] Aubrey, *'Brief Lives'*, ii, p. 127.

[88] Boyle, *Correspondence*, p. 283 (Bathsua Makin to Boyle, undated).

[89] BL, MS Add. 4278, fo. 70^r (Brancker to Pell, 19 Feb. 1668): 'Yesterday I had a letter from Mr Pitts ... In it ... he tells me he is ordered by some friends of mine at Tottenham to represent to me their desires of my company there and purposes to encourage me in a school there.'

[90] See above, ch. 7, n. 8. Brancker did, however, accept the headmastership of Macclesfield Grammar School in 1675: *Victoria County History, County of Chester*, 3 vols. (London, 1979–87), iii, p. 238).

[91] Anon., *Essay to Revive the Antient Education*, p. 5 ('To the Reader'). The 'Postscript' to this text mentions that Bathsua Makin is 'Governess' of a new school for gentlewomen at Tottenham (p. 42). A collection of tracts by Mark Lewis, showing that he set up his school at Tottenham in 1670–1, is in BL, pressmark 622. d. 34; some of these are cited or referred to in the *Essay to Revive the Antient Education*, which, in addition, copies entire passages from them. The same collection also contains a tract by Arthur Brett, *A Model for a School for the Better Education of Youth* (n.p., n.d.), which praises Lewis's school and includes a list of features of its teaching (p. 7) identical with passages in the *Essay to Revive the Antient Education* (pp. 34, 43). On Lewis and Tongue see Aubrey's note in Bodl., MS Aubrey 10, fo. 13^r: 'M^r ... Lewis of Totenham (neer London, Schoolmaster) was Dr Tongues acquaintance'. See also Salmon, *Language and Society*, pp. 248–50.

[sc. the Archbishop of Canterbury] to give me leave to goe to my children &c and to be absent 3 yeares. He would not grant it.' Cryptically, he added: 'I think I may thank Don John for that repulse.'[92] From another jotting by Pell it is possible to deduce that 'Don John' was John Clotworthy, Viscount Massereene, the Irish landowner and politician who had keen scientific interests (he was an early F.R.S.) and had collaborated with William Petty on his innovative boat-building project; but both the nature of his relationship with Pell, and the reason for his opposing Pell's wishes in this instance, are utterly obscure.[93] Pell's correspondent cautiously replied: 'I wonder how Don John should have any interest with ABC but I am glad he was the occasion of stopping y^r Journey for I can not think but that he did it in hopes of good to you.'[94] The historian can share this unknown correspondent's wonder, but not, retrospectively, his hopes.

The crisis finally came in August 1680: on the 31st of that month Pell was arrested for debt, and on 6 or 7 September he was consigned to the King's Bench prison.[95] The details of the case – which does not seem to have been tried at the Court of King's Bench – are rather obscure.[96] A few references in subsequent correspondence suffice, however, to indicate what had happened. Roger Raven's son Miles, who was roughly 20 years old in 1680, had run up large debts, and had somehow persuaded Pell to stand surety for them; his creditors or 'assigns' had then taken action against Pell. In mid-September 1681, when yet more of Miles Raven's debts had to be paid, Pell dictated a message sent by a third party to one of his other relatives at Laindon, asking him to make a payment of £10 to 'the assignes of M Miles Raven. Who perhaps may give order to some body hee[re] to arrest the Doctor on the last of September, as He did it on the last of August in the yeare 80.'[97] The documents relating to this later episode of debt,

---

[92] BL, MS Add. 4365, fo. 47^r (Pell to 'R. H.', 1 July 1679).

[93] On Massereene see the entry in the *DNB*. The identification is made possible by Pell's note of his reply to 'R. H.'s' reply, in which he writes that 'I beleave Don John is now with his Aunt L. C. de R.' (BL, MS Add. 4365, fo. 46^r; cf. a reference to 'La Contessa di Ranelagh', fo. 46^v). Pell would have known Lady Ranelagh through both her brother Robert Boyle, and John Dury's wife, who, through her first marriage, had been Lady Ranelagh's sister-in-law. It may be guessed that the soubriquet 'Don John' probably related not to amorous affairs but to boating activities, by reference to Don John of Austria.

[94] BL, MS Add. 4365, fo. 45^v ('R. H.' to Pell, 17 July 1679). The identity of 'R. H.' is not known. This was not Robert Hooke; in his first letter Pell gave the date as 'July 1. just 11688 dayes since we first saw one another (11688 ÷ 32 = 365$\frac{1}{4}$)' (fo. 47^r). 'R. H.' was thus someone who had first met Pell in the summer of 1647, in Breda – perhaps as one of his pupils.

[95] BL, MSS Add. 4279, fo. 59^v (note dictated by Pell in reply to letter of 11 Sept. 1681 from George Hastings: 'on the last of August'); Add. 4280, fo. 151^r (Steven Nye to Pell, 14 Sept. [1681?]: 'such a Trick as MiRa shewed you on sept: 6. 80'); Aubrey, *'Brief Lives'*, ii, p. 126: 'He was cast into King's Bench prison for debt Sept. 7, 1680.'

[96] Pell's name is absent from the docket books for the Court of King's Bench for 1680 (PRO, IND 1/6066), and from the Plea Rolls of that court for Trinity 1680 (PRO, KB 27/2007 and 27/2008).

[97] BL, MS Add. 4279, fo. 58^r (note dictated by Pell in reply to letter of 11 Sept. 1681 from George Hastings, who was the husband of one of Pell's grand-daughters).

in 1681, give some sense of the conditions that had developed at Laindon, where both Roger and Miles Raven had spent their time living beyond their means and gambling. One of Pell's parishioners, Abraham Thresher of Billericay, wrote to Pell that 'yor sonn in Law Capt Rauen is endebted to me vppon acctt about six poundes. aboue all men I have ye least reason to loose for besides this mony if I should account what I haue lost wth Him & Myles at Play – vnfairly being dealt wth by them besides the loss of tyme – all wch would amount to a great sume.'[98] By this stage Roger Raven had fled to Dorset, to stay with his brother in Wimbourne; the brother then wrote to Pell on his behalf, asking him to take him back.[99] A letter to Pell from his curate Steven Nye summed up the situation:

> Abraham Thresher is an honest man & the debt he claimes is (I believe) due from Captain Raven, but you only can tel whether you are in a condition to pay other mens debts, or whether you [>yet] have not lost enough by CaRoRa [*sc.* Captain Roger Raven]? as to CaRoRa his going to Ithaca [*sc.* America], I think you have no Authority to send him thither, for you have given it to Ulisses & Penelope [*sc.* John Pell Jr and one of his sisters?], & if you send CaRoRa thither you wil be one day obliged (perhaps by such a Trick as MiRa [*sc.* Miles Raven] shewed you on sept: 6. 80) to pay for his board. CaRoRa has a son a Brother & a Sister in the West, & why so neer Relations should cast him on you, that have already four Children of CaRoRa to keep, I do not understand...
>
> S^r I advise you by al meanes to make use of my brother Boughton to put out those poor Children of CaRoRa to some honest Trades, els when your head is laid they wil know extreme want or to prevent it turn Rogues, your designe of sending them to N: England can advance them to no better preferment than being Cow-boyes, & the mony that wil send them thither wil bind them to good handy-craft trades.[100]

The lack of any reference here to Mary Raven, Pell's daughter, suggests that she was no longer alive; it seems that her young children – with the exception of Miles, who went to live with his father in Dorset – were entirely dependent on

---

[98] BL, MS Add. 4279, fo. 279^r (Abraham Thresher to Pell, 21 Sept. 1681).

[99] Ibid., fo. 209^r (William Raven to Pell, 24 Aug. 1681).

[100] BL, MS Add. 4280, fo. 151^r (Steven Nye to Pell, 14 Sept. [1681]). Pell's note of his reply to this letter includes the comment that 'I intend [>shortly] to write to the eldest of those birds of prey [*sc.* Roger Raven?] with a softer pen than any of them deserve' (fo. 152^r). Abraham Thresher may have been honest where this debt was concerned, but his moral character was not unblemished: in 1688 he would be arrested for riotous assault (ERO, MS Q/SR 460/37, indictment of 5 Sept. 1688, where he is described as a grocer, of Laindon). This perhaps gives some idea of the company the Ravens were keeping.

their grandfather for support.[101]

How long Pell spent in prison is not known; the conditions there, for any debtor who was so short of funds that he could not pay the 'chamber rent' of 8s per week, could be extremely harsh.[102] Presumably his friends rallied round and made whatever payments were necessary to secure his release. But it must be a sign of his straitened circumstances that even the third-storey lodgings in Jermyn St – which Aubrey had considered so demeaning – were now too expensive for him; he went to live instead with Stephen Boughton or Broughton, an ironmonger in Whitechapel (the 'brother Boughton' referred to in Steven Nye's letter).[103] In the spring of 1682 his conditions of life improved when he was invited by a Fellow of the Royal Society, the physician Daniel Whistler, to stay with him in the College of Physicians; this was a splendid new building designed by Robert Hooke, and fully completed only in 1679.[104] Aubrey's comment on Whistler's hospitality is uncharacteristically censorious: 'Dr Whistler invited D^r P[ell] to his house ... wch ye D^r likt & accepted of, loving good cheer & good liquour, which the other did also; where eating and drinking too much was the cause of shortning his daies.'[105] After prolonged poverty and a period of imprisonment, good cheer may have been just what Pell needed. A few months later, however, he fell 'extreme sick of a cold', and went to convalesce in the house of one of his

---

[101] I have not found any record, however, of Mary's death. Miles Raven married an Elizabeth Hopkins at St Vedast's in London on 11 Aug. 1687; he was then described as 'Miles Raven in Pudle towne, Dorsetshire' (W. A. Littledale, ed., *The Registers of St. Vedast, Foster Lane, and of St. Michael le Quern, London*, 2 vols., Publications of the Harleian Society, Registers, vols. 29–30 (London, 1902–3), ii, p. 31).

[102] Moses Pitt's account includes a detailed assessment of the costs (to both debtor and creditor) of imprisoning a debtor (*Cry of the Oppressed*, pp. 91–4), and notes that those who failed to pay 'chamber-rent' could be consigned to the dungeon (sig. A10^v). See also the lurid illustrations in Pitt's book, e.g. 'A Debtor Catching mice for his Sustenance', 'A Debtor Iron'd to a Wooden clog', and 'Debtors and Condem'd Criminals Log'd togeather' (facing pp. 3, 49, 62).

[103] William Raven's letter to Pell of 24 Aug. 1681 (BL, MS Add. 4279, fos. 209–10) is addressed to him 'att his lodging att the signe of the Leg an Iremongers shopp wthout Algate'; Abraham Thresher's letter of 21 Sept. 1681 (ibid., fo. 279) is addressed to him 'att Steuen Broughtons at the Legg in White Chappel'.

[104] Aubrey, *'Brief Lives'*, ii, p. 127 (dating the move to March 1681); a letter from Sir Samuel Morland to Pell of 13 May 1682 is addressed to him 'at Dr Whistler's house in the Phisitians Colledg in Warwick Lane' (BL, MS Add. 4279, fo. 150). On Hooke's Royal College of Physicians (his first major commission as an architect) see Inwood, *The Man who Knew Too Much*, pp. 131–2 and the plate facing p. 105; Cooper also notes that Whistler was responsible for supervising the design of the new anatomy theatre which it contained ('Robert Hooke, City Surveyor', p. 100). The building was destroyed by fire in 1876.

[105] Bodl., MS Aubrey 6, fo. 51^v (Aubrey, *'Brief Lives'*, ii, p. 128). Whistler, who became President of the College of Physicians in 1683 and died in 1684, would have been good company for Pell intellectually as well as socially: he had been Gresham's Professor of Geometry (1648–57), and was 'well skilled in the mathematics' (J. Ward, *The Lives of the Professors of Gresham College* (London, 1740), pp. 155–6).

grand-daughters and her husband, George Hastings, in the heart of Westminster; they later moved to Brownlow St, near Drury Lane, taking him with them, and he appears to have remained as their lodger until his death.[106]

Despite all his personal problems, Pell had picked up the threads of his normal life after being released from prison. He resumed his attendance at the Royal Society; among his papers are the printed notices summoning him to the anniversary meetings in 1681 and 1682.[107] His name crops up occasionally in the minutes of the Society: in May 1681, for example, he was one of four Fellows asked to advise on a proposal for a new geodetic survey of England and Wales, and in July, when the 'astronomico-chronological tables' of the German oriental scholar Matthias Wasmuth were discussed, it was recorded that 'Dr Pell ... who had formerly perused those tables, conceived them to be of little worth'.[108] A more striking entry in these minutes concerns a discussion of a classic German book on mining and metallurgy, Lazarus Ercker's *Beschreibung allerfürnemsten mineralischen Ertzt und Berckwercks Arten* (Prague, 1574). At a meeting on 9 March 1681 it was agreed that the work should be translated into English. The following week, however, Dr William Holder announced that the text had in fact already been translated by Sir John Pettus, though he had not yet found a publisher for it. Whereupon 'Dr. Pell mentioned, that he had translated the greatest part of Erker's book into English, but had not completed it, finding great difficulty to understand the mineral terms of art.'[109] Pell's problems are understandable; when Pettus's translation of this very technical work on the extraction and refining of metallic ores was finally published in 1683, it included a long appendix

---

[106] Ibid., ii, p. 127; a memorandum by Pell gives his address in Nov. 1684 as 'Mr Hastings his house in Brownlow street, neere Drury-Lane' (BL, MS Add. 4423, fo. 343[r]). Hastings signed a letter to Pell 'from youre Grand son George Hastings' (BL, MS Add. 4279, fo. 58[r], Hastings to Pell, 11 Sept. 1681); his wife was probably one of Thomas and Judith Kirk's daughters.

[107] BL, MSS Add. 4398, fo. 145[r] (for 30 Nov. 1681); Add. 4424, fo. 382[r] (for 22 Nov. 1682).

[108] Birch, *History*, iv, pp. 84 (4 May 1681), 96 (27 July 1681). On the survey proposal by John Adams, see E. G. R. Taylor, 'Robert Hooke and the Cartographic Projects of the Late Seventeenth Century', *The Geographical Journal*, 90 (1937), pp. 529–40 (noting that by 1684 Adams travelled 25 000 miles, triangulating and surveying, but that no sheet of his map ever appeared); the other Fellows appointed to advise were Hooke, Hill, and Hoskins. The work by M. Wasmuth was his *Idea astronomiae chronologiae restitutae* (Kiel, 1678).

[109] Birch, *History*, iv, pp. 72–3. On the Saxon-born Ercker (*c*.1530–1594), a mining supervisor, mint-master, and assayer, see P. R. Beierlein, *Lazarus Ercker: Bergmann, Hüttenmann und Münzmeister im 16. Jahrhundert* (Berlin, 1955), and P. O. Long, 'The Openness of Knowledge: An Ideal and its Context in 16th-Century Writings on Mining and Metallurgy', *Technology and Culture*, 32 (1991), pp. 318–55, esp. pp. 346–50.

in the form of a dictionary of technical terms.[110] It is noteworthy, however, that this major project by Pell, so casually mentioned in passing, has left no trace whatsoever among his papers; this fact must stand as a salutary reminder to any biographer of Pell that large sections of his personal archive have gone missing. One can only guess, therefore, at the number of other projects which, like this and like so many others, were neither completed nor published.

Pell himself published nothing in these final years of his life, but his friend Sir Samuel Morland did manage to extract some material from him, for publication in a book of his own. One of Morland's inventions was a new type of pump; he was now preparing a treatise on its use, together with a set of relevant mathematical tables (for the calculation of volumes of water in pipes, and so on). In May 1682 he wrote to Pell, asking a series of questions about the contents (in cubic inches, to several decimal places) of cylinders of different base diameters and a height of twelve inches. He added: 'And if your Leisure will permitt you I would beg a Table giving ye number of square Inches conteyned in ye Areas of all Circles, from 1 Inch diameter to a 100 Inches diameter.'[111] Three years later, when Morland's treatise was published (in French, since he was hoping that his invention would be taken up by Louis XIV), it included a table of the areas of circles of diameters between 1 and 100, with a note explaining that it was 'calculated by Mr Pell, whose name and great merit are known to the whole of the learned world'.[112] In 1697, two years after Morland's death, his nephew Joseph Morland edited a collection of his papers; in the Preface he explained that 'The following Tables I received from Sir *Samuel Morland*, amongst the rest of his Mathematical Papers, all of which Kind he was pleased to bestow on me not long before his Death'. The fifth table printed here was one 'Giving the true Content in Cubick Inches of Cylinders of different Diameters, from 1 to 12 Inclusive, and each of these Cylinders of a Foot or Twelve Inches in Height', with the cubic inches given

---

[110] Sir John Pettus, *Fleta Minor: The Laws of Art and Nature, in Knowing, Judging, Assaying, Fining, Refining and Inlarging the Bodies of Confin'd Metals* (London, 1683), second part (unpaginated). In his Preface to the volume (sig. B2^r), Pettus wrote that he had first 'caused *Eckern*'s Books to be Translated about Ten years since', and had then learned German in order to correct the translation, which he did 'with the help of a *German* here'; elsewhere (second part, sig. C1^v), he refers to 'my first Translation, about 14 years since'. Pettus was himself in prison for debt at the time of publication; but his courtly dedicatory epistle to the Warden of the Fleet prison (first part, sig. C1) suggests that he was receiving very favourable treatment. On this translation see also E. V. Armstrong and H. S. Lukens, 'Lazarus Ercker and his "Probierbuch": Sir John Pettus and his "Fleta Minor"', *Journal of Chemical Education*, 16 (1939), pp. 553–62.

[111] BL, MS Add. 4279, fo. 150^r (Morland to Pell, 13 May 1682).

[112] Sir Samuel Morland, *Elevation des eaux par toute sorte de machines reduite à la mesure, au poids, à la balance, par le moyen d'un nouveau piston, & corps de pompe, & d'un nouveau mouvement cyclo-elliptique* (Paris, 1685), pp. 61–2 ('Table des aires des cercles pour un diametre d'une grandeur donnée depuis (1) jusqu'à (100.)'), sig. H2^r ('calculée par *M. Pell*, dont le nom & le grand merite sont connus de tous les Sçavans'). On Morland's project see Dickinson, *Sir Samuel Morland*, pp. 81–7.

to nine decimal places.[113] In view of Morland's earlier request to Pell, it must seem overwhelmingly likely that this table was also Pell's work, and that either Morland or his nephew had simply forgotten to credit it to him.

In 1683 some efforts were made to persuade Pell to publish the table of antilogarithms which he had taken over from Walter Warner and completed in 1642. The manuscript of this table, 'elegant, and in a large folio', had passed from Herbert Thorndike to Richard Busby on Thorndike's death in 1672.[114] A note among Pell's papers, dated 22 September 1683, 'At Doctor Busbies', comments that 'I found that there are wanting 25 sheets namely all those that shewed the finding of *ten thousand* numbers belonging to the Logarithmes between 29750 and 39750' – precisely the problem which had thwarted his previous plan to commit this work to the press.[115] Pell's interest was sufficiently aroused, nevertheless, for him to sketch out a mock titlepage of the work as it might be published, describing it as 'The Resolution ... of the original analogical canon, to which is added, in a few pages, a specimen of a method, which should be sufficient for the intelligent mathematician to direct him in the practical use of it'.[116] That 'specimen of a method' was, perhaps, the key to the use of the table for which John Collins had searched in vain: in his letter to Oldenburg (for Tschirnhaus) in 1675 he had commented that 'I could not conceive but that this table was made properly for algebraical uses in resolving equations; what use it was intended for Dr. Pell is not free to disclose; none of his friends here can render him communicative'.[117] But if Collins's guess was right, the secret would go with Pell to his grave.[118]

Another aborted project of these final years also involved mathematical tables. At the end of the table of square numbers published in 1672, Pell had added a teasing half-promise: '*Having the two, three or four last figures of any Square number, to exhibit as many of the last figures of its side,* is a New Question: To which, the just Answers are manifold and not obvious. A particular account of

---

[113] Sir Samuel Morland, *Hydrostaticks: Or, Instructions Concerning Water-Works. Collected out of the Papers of Sir Samuel Morland*, ed. J. Morland (London, 1697), sig. A2^r (Preface), pp. 50–1 (table).

[114] Rigaud, *Correspondence*, i, p. 215 (Collins to Oldenburg, for Tschirnhaus, 30 Sept. 1675 (quotation); Collins had previously described it as still in Thorndike's possession (ii, p. 219, Collins to Gregory, 25 Mar. 1671). Anthony Wood noted that Thorndike's MSS all passed to Busby; Wood commented (using information probably provided by Aubrey) on the division of labour between Warner's original 10 000 entries and the rest by Pell, 'as the difference of hands will shew in the MS. if Dr Busby will communicate it' (*Athenae*, ii, col. 302).

[115] BL, MS Add. 4424, fo. 2^r. For the previous plan, see above, ch. 5, at nn. 18–19.

[116] BL, MS Add. 4424, fo. 3^r: 'Canonis Analogici Originalis ... RESOLUTIO Subjuncto paucis paginis artificij specimine quod ad praxim dirigendam Logistae intelligenti sufficiat'.

[117] Rigaud, *Correspondence*, i, p. 216 (Collins to Oldenburg, for Tschirnhaus, 30 Sept. 1675).

[118] John Wallis late recalled that Pell had asked him to 'promise to see the work finished, in case he should die before it were done', and that he had done so; but the printing of the tables had not begun by the time of Pell's death, 'And I fear, least by Dr *Busby*'s Death, the whole will be laid aside; especially as there is no one that will defray the Expences of the Edition' (cited in Dodson, *The Anti-logarithmic Canon*, p. ix). Busby died in 1695.

them is ready for the Press, when it shall be desired.'[119] A note in John Aubrey's 'brief life' of Pell refers to

> the last thing he wrote (wch he did at my earnest request) viz. *The Tables*, which are according to his promise in the last line of his printed Tables of Squares and Cubes (if desired) and wch Sir Cyrillus Wych (then President of the Royall Society) did License for the Press: there only wants a leafe or two for the Explanation of the Use of them, wch his death hath prevented. Sir Cyrillus Wych, only, knowes the Use of them.[120]

Aubrey's comment helps to fix the dating of this work: Sir Cyril Wyche was President of the Royal Society from November 1683 to November 1684.[121] But the text was never published, and the secret of its 'use' was not preserved.

That John Aubrey had come so close to success in persuading Pell to write something and publish it is testimony both to his persistence, and to the depth of his friendship with Pell in these last years. In some ways they were kindred spirits: Aubrey too got into increasingly severe financial difficulties, and, rejecting the solution of 'the Cassock', planned at one stage to emigrate to America or the West Indies.[122] He was also, like Pell, a man of wide intellectual interests (with a special passion for educational theory), who prepared many works that he never published. At the end of the 1670s he began writing the biographical sketches that would ensure his posthumous fame; and in the autumn of 1680, knowing that Pell had been acquainted with several of his subjects from the 1630s onwards, he turned to him for additional information. But it seems that Pell was then in particularly low spirits after his recent incarceration; as Aubrey later reported to Anthony Wood, 'I deposited my Minutes of lives in Dr Pells hands in Octob last (or Sept) expecting he would have made additions or amendments but (poor, disconsolate man!) I recieved it of him without any'.[123] Aubrey was not disheartened, however, and continued to cultivate Pell's acquaintance; indeed, there are signs that he even obtained some occasional tuition from him in mathematics – a subject in which Aubrey had long been interested, and in which he had taken lessons from Nicolaus Mercator. Among Aubrey's papers there is a copy of one of the problems Pell seems to have given to many of his students, the algebraic problem from the end of Sibrant Hansz. Cardinael's *Practijck des*

---

[119] Pell, *Tabula numerorum quadratorum*, p. 32.

[120] Bodl., MS Aubrey 6, fo. 51^{v} (Aubrey, *'Brief Lives'*, ii, p. 128: the title, italicized here, is given by Aubrey in larger script, represented as capitals in Clark's edition). Aubrey's imperfect recollection was that 'whereas some questions are capable of severall answers, by the help of these tables it might be discovered exactly how many, and no more, solutions, or answers, might be given'.

[121] On the politician and public servant Sir Cyril Wyche, see the entry in the *DNB*, and Hunter, *The Royal Society*, pp. 174–5.

[122] Hunter, *John Aubrey*, p. 33.

[123] Bodl., MS Wood F 39, fo. 351^{r} (Aubrey to Wood, 13 Jan. 1681).

landmetens.[124] Aubrey also persuaded him to compile a little booklet of seven leaves, containing all the mathematical problems set out in a famous Elizabethan mathematical work, Leonard and Thomas Digges's *Stratioticos* (London, 1579), together with Pell's own solutions to them.[125] And he evidently had many conversations with Pell about mathematical matters, fragments of which he later recorded; for example, 'Dr Pell affirmed to me, that Mr Oughtred's Demonstration of the Regula Falsi, is donne false: but he told me, Petiscus hath donne it true'.[126] Aubrey's comments on Pell make a welcome contrast to Collins's litany of complaints: he declared that there was 'no man more humble nor more communicative.'[127]

Many of these communicative conversations were related to the other major project on which Aubrey was engaged in the early 1680s, his draft treatise entitled 'Idea of Education of Young Gentlemen'. Aubrey was sympathetic to the new methods promoted by Comenius and the Hartlib circle; indeed, the very wording of his title may reflect, like that of Pell's *Idea of Mathematics*, a Comenian influence. He turned to Pell for advice on a range of subjects, including language-teaching: he noted his comments on learning Latin (which Pell thought should be more like learning a modern language, and less like 'slavery'), and recorded his advice that studying German (using Luther's Bible translation) would 'open ... a way for the ye understanding of ye old Saxon Lawes, &c: and ye old English: and Etymologies'.[128] But above all, it was mathematics that they discussed. Pell's emphasis, as always, was on method. In the draft of his 'Idea of Education', Aubrey recommended that children should 'ever use that [>excellent] method invented by Dr J. Pell, by steppes in the margent'; as he pointed out, 'After Dr Pell's way of solving questions, a man may steale a nappe, & fall to worke again afresh where he left off: which one cannot doe according to the Oughtredian method.'[129] And the virtue of this method was not confined to the solution of mathematical problems; Aubrey was convinced that it could also be used to solve 'Questions in the Civil Lawe', and in general he felt that 'Dr Pell's ratiotination by Syllogismes in his Solutions of Questions will teach Logick beyond all other waies.'[130] As he also noted: 'Dr Pell was wont to say, that in the Solution of Questions, the maine matter was, the *well-stating of them*; w^{ch}

---

[124] Worcester College, Oxford, MS 64, unfoliated, fourth item; Aubrey's copy of this is in Bodl., MS Aubrey 10, fo. 55^r. Cf. above, ch. 4, n. 129.

[125] Worcester College, Oxford, MS 63, unfoliated, second item. Aubrey commented elsewhere on 'Th. Digges's Stratioticos, which I have, which D^r. J Pell did me the favour to peruse, & hath solved the questions there after his owne way' (Bodl., MS Aubrey 10, fo. 101^r).

[126] Bodl., MS Aubrey 10, fo. 33^v. 'Petiscus' is the mathematician Bartholomaeus Pitiscus (1561–1613).

[127] Aubrey, *'Brief Lives'*, ii, p. 126.

[128] Bodl., MS Aubrey 10, fos. 26^r ('not to put them to the slavery to learn it all, which checks their spirits & keepes them back. We doe not learn modern languages after that fashion'); 104^r.

[129] Bodl., MS Aubrey 10, fos. 34^r, 33^r.

[130] Ibid., fos. 54^v, 55^r.

requires mother-witt, & Logick, as well as Algebra: for let the Question be but well-stated, it will worke almost of it selfe.'[131]

The last dated records of Pell's life arise from his friendship with John Aubrey. A leaf of paper summarizing the operations of algebra (dividing it into synthesis and analysis, dividing synthesis into logistical and syllogistical, and so on) is annotated by Aubrey, 'from Jo: Pell D.D. Septemb. 30 1684'.[132] Six weeks later a work on geometry by Thomas Baker was discussed at the Royal Society, and Haak was given a copy to convey to Pell; when this copy was found to have been mis-bound, it was sent back, and in April 1685, as Pell recorded in a somewhat querulous memorandum, 'Mr Aubry brought another'.[133] A note by Aubrey states that on 'Nov. 26' Pell 'fell into convulsion fitts which had almost killed him'; in the context, it is not clear whether this refers to 1684 or 1685.[134] Pell's health seems to have been good for most of his life; in his correspondence and personal papers there are no references to any of the chronic conditions (such as 'the stone') that plagued so many of his contemporaries. Two scraps of evidence do suggest, however, that in the last part of his life he was suffering from a heart condition. He copied out and translated an extract from Richard Lower's *Tractatus de corde* (London, 1669), referring to 'Wormes' between the heart and pericardium, 'which extremely hinder & disturb the naturall motion of the heart. For by gnawing it, they cause a Trembling of the heart, great anxiety & sadness, a pulse often ceasing, a pricking paine & swoonings.'[135] And among his papers there is also a set of notes on a prescription, from 'F. C.' (probably Hartlib's son-in-law Frederick Clodius), for 'liquor cordiacus'.[136] So perhaps Aubrey was referring more to physiology than to emotional distress when he wrote: 'He dyed of a broaken heart.'[137]

Pell's death occurred on 12 December 1685, in the afternoon, when he was visiting a friend in Dyot St, a short walk from his lodgings in Drury Lane. He had, as Aubrey indignantly recorded, 'not 6*d*. in his purse when he dyed'. The funeral took place at the local church of St Giles-in-the-Fields three days later. In early January Aubrey wrote to Anthony Wood that 'My deare [>& learned] friend D^r Jo: Pell dyed Decemb. 12 last, one of ye greatest scholars living: and dyed (to ye shame of ye Ecclesiasticks) so poore, that his buriall is not payd for'; the fee of £10 was eventually settled by Richard Busby and the Rector of St

---

[131] Worcester College, Oxford, MS 64, unfoliated, verso of blank leaf before fourth item.

[132] Worcester College, Oxford, MS 64, unfoliated, first item. Aubrey has noted at the end of this: 'Desire him to Pelliare Oughtredismos [*sc.* convert Oughtred's methods or symbols into his own style], wch would be of good use.'

[133] BL, MS Add. 4423, fo. 343^r. The work was Baker's *Clavis geometrica catholica* (London, 1684); Collins had put a proposal to the Royal Society in May 1682, recommending that it be printed (see Birch, *History*, iv, p. 155).

[134] Aubrey, *'Brief Lives'*, ii, p. 127.

[135] BL, MS Add. 4255, fo. 216^r (the passage is from p. 101 of Lower's book). He also noted the remedy Lower recommended, a herbal poultice, to be placed 'upon the Region of the Heart'.

[136] BL, MS Add. 4365, fos. 173–8.

[137] Aubrey, *'Brief Lives'*, ii, p. 128.

Giles.[138] Despite Aubrey's entreaties, Pell had never made a will; and although he had no money to his name, he did have a substantial collection of books, as well as his own personal archive. Two years later, the Consistory Court of London granted the administration of his estate to one of the people who had caused him most harm – his son-in-law Roger Raven. It was decreed that Raven should administer it 'during the absence, and for the use and benefit, of the said Charles and William Raven [two of Roger Raven's sons], who are now in the West Indies'.[139] The books and papers were purchased from him (for an unknown sum) by Richard Busby.[140] Whether Pell's profligate son-in-law did devote the proceeds to the 'use and benefit' of those two sons, or whether he spent them on 'gaming' with Miles instead, is not recorded.

A significant number of Pell's books have remained ever since in Westminster School, in what became known as the Busby Library. Many are identifiable by notes in his handwriting; some can be recognized by the binding, in a uniform style, which they later received. Altogether, this little-known set of books constitutes one of the most significant mathematical collections of early modern England, second only to the Savile Library in the Bodleian.[141] As for the manuscripts acquired by Busby from Roger Raven, they remained at Westminster School, mixed with some of Busby's own papers, in 'four large boxes', for the next seventy years; then, in 1755, Thomas Birch arranged for them to be transferred to the Royal Society.[142] They are now in the British Library, where they fill approximately fifty bound volumes. But these were not the only manuscripts Pell had left behind. Two months after his death, Robert Hooke reported to the Royal Society 'that the papers of the learned Dr. Pell ... were partly in custody of Dr. Busby, and the rest at Brereton in Cheshire'.[143] Even if the manuscripts at Brereton consisted only of items written during Pell's stay there, they might have amounted to a substantial quantity of material, representing four years' work in conditions of comparative tranquillity. Some of the most important items were mentioned by Aubrey: 'He hath written on the tenth booke of Euclid, which is in Cheshire at the lord Brereton's, and he hath also done the greatest part of Diophantus, which is there ... Also he hath donne the second booke of Euclid

---

[138] Ibid., ii, pp. 127–8; London Metropolitan Archives, microfilm X105/022, St Giles-in-the-Fields, parish register, burials, 1668–92: '15 Dec. 1685 John Pell D^r: D^{ty}: Dyott str.:'

[139] London Metropolitan Archives, microfilm X019/009 (Consistory Court of London, calendar of wills and administrations, 1670–1720), entry for 7 Apr. 1687: 'durante Absentia et ad usum et Beneficium dictorum Caroli et Gulielmi Raven modo in Indijs Occidentalibus'.

[140] Aubrey, 'Brief Lives', ii, p. 128.

[141] While this book was in its final stages of preparation, a substantial part of a catalogue of Pell's books, dated 1687 and numbering more than 700 items (perhaps half of them on 'mathematical' subjects, in the broad 17th-century use of that term), was discovered by Mr Eddie Smith among uncatalogued MSS at Westminster School. Mr Smith is now preparing an edition of this catalogue for publication.

[142] Birch, History, iv, p. 447.

[143] Ibid., iv, p. 458.

in one side of a large sheet of paper most clearly and ingeniously.'[144] But these papers were never recovered by Busby or the Royal Society, and at some later stage – perhaps after the extinction of the Brereton family line in 1722, or the dismemberment of the Brereton estates in 1817 – they appear to have been discarded or lost.[145] Another cache of Pell manuscripts was in Essex, where some of his papers (including all the problems he had set for Rahn in Zurich) were taken in 1661; these too have disappeared.[146] And if any other manuscripts had been left behind in other locations – Zurich itself, perhaps, or Breda – there is no trace of them today.

Nevertheless, as this study has tried to show, enough has survived in the papers rescued by Busby, and in evidence from other sources, to make it possible to enter into the mental world of this intriguingly awkward and intensely intelligent man – to appreciate the range of his interests, and to sense some of the connections between them. If intellectual history were to be constructed only on a narrowly conceived, strictly canonical basis, with significance depending entirely on the fact of having published influential texts, Pell would then appear to be a peculiarly insignificant figure. Indeed, on those terms his significance might be seen as essentially negative, as the historian's interest would have to focus on the potentially important works which Pell planned, drafted, or almost perfected, but which were never made available to the public: his 'discourse' on Gellibrand's theory of magnetic variation, his 'Manual of Logarithmicall Trigonometry', his 'Comes Mathematicus' and the other volumes promised in his *Idea of Mathematics*, his revised edition of Harriot, his great work on 'analytics', his table of antilogarithms, his new editions of Diophantus and Apollonius, his notes on Pappus, his edition of the letters of Huygens and de Sluse on Alhazen's problem, his treatise on equations and the 'Canon of Sines', and the 'Tables' he prepared for Aubrey – to say nothing of his translations of Comenius, Descartes, and Ercker, or the conversion by Brancker of Kinckhuysen into Pell's method, which he failed to sanction. The list does indeed make melancholy reading, and one does not have to be a narrowly positivist historian to regret the absence of these works. And yet, in the end, the story of Pell's life and work is the story of his presence – in the educational and scientific activities of the Hartlib circle, in the reception of Comenius and Descartes, in the world of mathematical practitioners in England and the teaching of mathematics in the Netherlands, in the Republic of Letters of mid-seventeenth-century Europe, in the international politics and Protestant irenicism of the 1650s, in the work of the Royal Society, and in the lives and thoughts of many pupils, patrons, and friends. Intellectual life depends on more than publications; it is made up of innumerable overlapping and

---

[144] Aubrey, *'Brief Lives'*, ii, pp. 125–6.

[145] For the later history of the family and estate see Ormerod, *History of the County of Chester*, iii, p. 86.

[146] See above, ch. 7, n. 18.

interacting presences of such kinds as these. Our understanding of seventeenth-century intellectual life can only become a little more complete, once it acknowledges the many and varied presences of John Pell.

Part II

The mathematics of John Pell

by Jacqueline Stedall

PART II: THE MATHEMATICS OF JOHN PELL

John Pell is known today, as he was in his lifetime, primarily as a mathematician. He immersed himself in mathematics for over fifty years, and for much of that time was renowned for his skill in the subject, both amongst his own countrymen and abroad. Any biography of Pell must therefore include an assessment of his mathematics, in the context of contemporary developments in the subject. One would say as much for any mathematician, but in Pell's case there are additional reasons for exploring his mathematical work in some detail. Through his peculiar modesty and unwillingness to publish he succeeded in keeping many aspects of his life and his friendships hidden from posterity, but in certain cases his mathematics betrays Pell's presence and character in a way that other contemporary evidence does not. The challenge for the biographer is that mathematics can be a daunting subject to those who are not familiar with it. Pell's mathematics, however, is not inherently difficult; to put it in perspective, the mathematics developed during Pell's lifetime even by the most advanced mathematicians in Europe is now well within the grasp of a student in the final years of school or the first years of university. In the essay that follows I have attempted to present the material in as accessible a way as possible, and for those prepared to persevere (or simply to omit the more difficult details) I hope that the study of Pell's mathematics will help to shed a little additional light not only on his life and acquaintances, but on his thinking, his motivation, and his unique and sometimes strange personality.

Pell's published mathematical works were few: the *Controversiae pars prima* in 1647, *An Introduction to Algebra* in 1668, and the *Table of 10000 square numbers* in 1672. These books, however, represent but a fraction of a lifetime's mathematical activity. The rest can be discovered only from Pell's correspondence and his unpublished papers and, fortunately, significant quantities of both have survived. The mathematical papers purchased by Richard Busby from Pell's Executors were eventually deposited in the British Museum and are now held in the British Library in London. They have been randomly bound into more than thirty substantial volumes, several of which contain as many as three or four hundred pages of assorted notes, tables, rough work, and jottings. Further papers are scattered through other volumes of the British Library's manuscript holdings. Many of Pell's scribblings were undated and untitled, and many folios in the modern volumes contain collections of tiny scraps or fragments, carefully preserved, but in no semblance of order. The papers span the full fifty years of Pell's working life but are both thematically and chronologically in complete disarray.[1]

A brief list of just a few of the items in two volumes (MS Add. 4415 and MS Add. 4416) will illustrate both the breadth of Pell's interests and the random nature of his papers. In MS Add. 4415 we have a note on Mercator's *Logarithmotechnia* of 1667; a note on finding the logarithm of the number 20 021; several sheets on mathematical progressions and difference tables (1638); several sheets of notes relating to Oughtred's *Clavis* of 1631; a piece headed 'The strife of Analytica and Synthetica'; another headed 'Stevin's rules for finding the roote of an equation'; another on 'The section of an angle into equal parts'; further calculations of logarithms (of tangents); calculations using the double-angle tangent formula; notes on Hérigone's *Cursus* of 1644; notes on Apollonius; notes on alligation as expounded by Gerhard de Neufville in 1624; a copy of Fermat's 'De contactibus sphaericis', probably composed in 1643; a list of mathematicians whose works were printed in folio; and a large table of calculations, for no stated purpose, based on the decimal expansion of π to 36 places. In the next volume, MS Add. 4416, we have a partial copy made in 1634 of some of Lansberg's tables; a letter from Thomas Strode dated 1672; calculations of sines, dated 1638; work on the ratio of the perimeter to the diameter of a circle, 1636; trigonometry, 1635; a note of books borrowed and returned in 1675; and a letter from John Collins to Thomas Brancker, dated 1667. The list gives a first impression of the nature and scope of Pell's mathematical interests, but selecting identifiable topics in this way makes his notes seem more coherent than they are. The items listed above are little more than occasional recognizable landmarks scattered amongst some 600 folios of unidentifiable rough working, repeated attempts at the same problem, incomplete fragments, scraps, and waste. It is clear from this list alone that it is not usually possible to date any of the papers, even approximately, from what lies next or close to them.

Even from a preliminary examination of the volumes and their contents, however, some general comments can be made immediately. First, there is abundant evidence of Pell's wide reading. There are numerous references and sometimes detailed notes on the work of authors ranging from the well-known to the obscure. Arranging them in alphabetical order (as Pell might have done) the list must include: Alexander Anderson, Apollonius, Archimedes, Bachet, Baker, Barrow, Bartholin, Beaugrand, de Billy, Briggs, Brouncker, Cardano, van Ceulen, Chauveau, Commandino, Coutereels, Dary, Dechales, Descartes, Diophantus, Euclid, Fermat, Foster, Frenicle de Bessy, Gellibrand, Gibson, Girard, Gosselin, James Gregory, Gunter, Harriot, Henrion, Hérigone, Huygens, Kersey, Kinckhuysen, Lansberg, Longomontanus, Luneschlos, Mengoli, Mercator, Metius, Mydorge, Neufville, Nuñez, Oughtred, Pappus, Pitiscus, Ptolemy, Ramus, Rahn, Recorde,

---

[1] Most of Pell's mathematical papers are in BL, MSS Add. 4397–404 and 4407–31. There is further relevant material in Pell's correspondence, especially with Cavendish, in MS Add. 4278–80, and in the mathematical papers of Cavendish, MSS Harl. 6001–2 and 6083, and of Warner, MSS Add. 4394–6. There is also related material scattered through other British Library manuscript volumes.

de Saint-Vincent, van Schooten the elder, van Schooten the younger, Stampioen, Stevin, Tacquet, Tapp, Viète, Wallis, Wassenaer, Wingate, and Xylander. In short, there were few classical or contemporary authors who escaped Pell's attention.

When it comes to Pell's own mathematics, two topics occupied him more than any others, and both of them for most of his life: the calculation of tables, and algebra. The first, the calculation of tables, dominated all others. Pell must have spent untold hours of his working life thinking about, calculating, or explaining the use of tables. The motto *Nulla dies sine linea*, 'No day without a line', of one of his early treatises could well be replaced by *Nulla dies sine tabula*, 'No day without a table'. Predominantly Pell's tables were of logarithms or antilogarithms, or of trigonometric quantities, but there are many others: of Pythagorean triples; of squares and sums of squares; of 'incomposits' or primes; of constant differences; and others related to the solution of particular problems. Some of them run to several consecutive pages neatly written, while others occupy just single pages or scraps. Some are large, and carefully ruled and copied, but others are squeezed onto the small octavo pages that Pell often used. Eventually Pell's interests in tables and in algebra came together, and he devised a method of solving equations by tables, though he could never be persuaded to publish it, and it can be reconstructed only from the hints left by him and others. More will be said of it later.

This brings us to another feature of Pell's mathematics that is obvious even to a casual observer. For all the prodigious time and effort Pell expended, one is left with an overriding sense of incompleteness. There are sheets that begin promisingly with neatly written date and heading but the sentences peter out into rough working or less; there are carefully designed title pages with no content behind them; repeated attempts at the same problem or question but no properly written up or published solution; ideas proposed but never carried out. Very little of the work Pell promised was ever brought to a state fit for publication. Major works on trigonometry and on algebra, tables of antilogarithms, editions of Apollonius and Diophantus, and many lesser texts, were promised but failed to materialize, to the endless frustration of Pell's friends and supporters.

In what follows I shall attempt to describe Pell's mathematics not so much by topic as by chronology, as far as it can be ascertained, in order to trace both the continuity and the changes in Pell's mathematical thinking, and to see how themes that preoccupied him as a young man were sometimes to re-emerge many years later. First, however, something should be said briefly about the mathematics known in England in the early years of the seventeenth century when Pell as a young man first became interested in the subject.

The modern theoretical study of polynomial equations stems from Girolamo Cardano's *Ars magna* of 1545, but Cardano's ideas were disseminated only gradually through the works of Raffaele Bombelli, Simon Stevin, and François Viète. Of these, the most important was Viète, who not only took up Cardano's methods for handling equations but had a new and profound vision of the power of algebraic methods. For Viète algebra was the key to uncovering the secrets of the Ancients, the 'analytic art' by which mathematical theorems could be both discovered and proved. His first book, his *Isagoge* or 'Introduction' to algebra, ends with a claim unprecedented in its boldness: *Nullum non problema solvere*, 'To leave no problem unsolved'.[2] Viète's work was taken up in England by Thomas Harriot, but Harriot's writings remained unpublished until ten years after his death, and Viète's own publications were hard to come by. It was not until 1631 that some of Viète's ideas became easily available to English readers through the publication of Harriot's posthumous *Artis analyticae praxis* and in a more elementary way through William Oughtred's *Clavis mathematicae*. Both of these texts were read by Pell during the 1630s, as were some of the writings of Viète himself (in manuscript copies), and the works of his predecessors, Cardano, Bombelli, and Stevin.

The other major influence on Pell was a different branch of mathematics developed closer to home, the logarithms of John Napier and Henry Briggs. Napier had published the first treatise on logarithms, *Mirifici logarithmorum canonis descriptio*, in 1614. Briggs immediately recognized the importance of Napier's invention and worked closely with him to develop it further, and in 1624, seven years after Napier's death, Briggs published his *Arithmetica logarithmica*, tables of logarithms for numbers from 1 to 20 000 and from 90 000 to 100 000, all calculated to 14 decimal places, together with detailed descriptions of his methods. The missing central portion, from 20 000 to 90 000, was completed by Adriaen Vlacq (though only to 10 decimal places) and published in the Netherlands in 1627.[3] Meanwhile, published tables of logarithms of trigonometric quantities began to proliferate: John Speidell's *New Logarithmes* of 1619, for example, and Edmund Gunter's *A Canon of Triangles: Or, A Table of Artificial Sines, Tangents, and Secants* in 1620. Gunter also devised a ruler marked with a logarithmic scale, which enabled multiplications to be carried out by adding appropriate lengths, the forerunner of the slide-rule, and Edmund Wingate took up the same idea in *The Construction and Use of the Line of Proportion*, published in 1628. The books of Gunter, Briggs, and Wingate were known to Pell during his Cambridge years. He corresponded with Briggs and appears to have known Wingate personally, the initial introduction perhaps having been made through Wingate's younger brother, Edward, who was an almost exact contemporary of Pell's at Trinity College, Cambridge.

Cambridge and Sussex, 1627–1635

In the papers that survive from his time as an undergraduate at Cambridge we see some of the long-term themes of Pell's life already beginning to develop. In these early years, Pell often created small booklets from folded sheets of paper,

---

[2] Viète, *Opera mathematica*, p. 12.

[3] A. Vlacq, *Het tweede deel van de nieuwe telkonst* (Gouda, 1627).

and filled them (or not) with topics that interested him. Several such booklets date from the years 1627 to 1629:

'Pro Latitudine 51^{gr}:0^m' (1627)
'A new almanack' (I. P. 1628)
'A prognostication for the yeare of our Lord God 1628' [title page only] (I. P. 1628)
'The description and use of the quadrant' (Jo. P. 1628)
'Notatu digniora' (1628)
'Tables for all kinds of Dialls' (1628)
'Tabula directoriae' (1628)
'Horologographia: the Art of Dialling' (1629)
'The construction and use of the proportioned line' [1629]
'Linea proportionata' [title page only] (R. FALE, Gent 1629)
'Imitatio Nepeira' [date unknown]
'Labyrinthus ingenii' (1629)

Note that only the earliest of the booklets carry Pell's initials, while one other bears a pseudonym but most show no name at all.

The first and earliest booklet in the list, 'Pro Latitudine 51^{gr}:0^m', has the subtitle 'Tables exactly calculated for the Latitude 51°0′ by Arthur Pollard, vicar of Eastdene and Frithston in Sussex. Serving to shew the houre of the day and night by the height of the Sun and starres. Ex dono authoris 1627', and is a copy in Pell's hand of tables giving the height of the sun at various times of day throughout the year.[4] Pell must have kept the tables by him for many years, for in 1637 he made a list inside the front cover of towns in England, and a few in France or further afield, which were at the correct latitude. Perhaps the tables were given to Pell to help him in the preparation of the almanack he planned for 1628. Pell's draft of 'A new Almanack', as far as it goes, contains lists of useful information for 1628: Holy Days and Ember Days according to both the Julian and Gregorian calendars and the date of the current year according to other systems of reckoning (those used, for example, by Ethiopians, Indians, Turks, Persians, or Egyptians); then a list of Oxford and Cambridge university terms and the signs of the zodiac; then Saints' Days and phases of the moon for each month, though only a few figures have actually been entered into the carefully ruled sheets.[5] The accompanying 'Prognostication for the yeare of our Lord God 1628' consists of notes explaining the meaning of terms and the basis of the calculations, the whole showing an easy proficiency in matters astronomical and calendrical.[6]

To make use of Pollard's tables to tell the time, one would have needed a method of measuring the height of the sun or stars, and it may have been for this purpose that Pell became interested in the use of the Quadrant and wrote

---
[4] BL, MS Add. 4397, fos. 47^v–65.
[5] BL, MS Add. 4403, fos. 120–34.
[6] BL, MS Add. 4403, fos. 135–45.

what was perhaps his first treatise: 'The Description and Use of the Quadrant'.[7] Dated 'Anno Dni 1628 May the 19th', the Preface promises two parts, one on the use of the Quadrant and plummet, the other on the use of the Quadrant and diopter. The text is incomplete (there is no second part) but nevertheless runs to sixteen pages of detailed and neatly written text, tables, and diagrams, ending with worked examples of useful possibilities: 'The year and suns place being known to find any day of the month', or 'Having ye altitude at Noone to find ye Latitude', or 'To find the time of Daybreake and shutting in twilight', etc. Pell's little treatise was almost certainly inspired by Gunter's *The Description and Use of the Sector Crosse-Staffe and other Instruments* published in 1624, where Gunter answered a similar set of questions: 'The houre of the day being given to find the altitude of the Sunne above the horizon', or 'To find the houre of the rising and setting of the Sun and thereby the length of the day and night', etc.[8] Many years later John Aubrey wrote that '[Gunter's] *Booke of the Quadrant, Sector and Crosse-Staffe* did open men's understandings and made young men in love with that Studie',[9] and it is likely that Pell was Aubrey's informant and had himself been one such young man.

Another use of Pollard's tables might have been in the construction of sundials, another topic on which Pell wrote in 1628 and 1629. We have the beginning of a small treatise entitled 'The best Way & manner of calculating of Tables for all kinds of Dialls & in all Countreyes whatsoever By Mr. Gunters Canon of Artificiall Sines and Tangents',[10] and a shorter but more finished treatise entitled 'Horologiographia The Art of Dialling', dated 1629.[11] The 'Horologiographia' is based on a standard work on the subject with exactly the same title, *Horologiographia: The Art of Dialling*, by Thomas Fale, from 1593. Pell's mock title page claims that the subject is: 'Of speciall use and delight not only for students of the Arts Mathematicall but also for divers Artificers, as also Architects, Surveyors of buildings and others', and described Fale's original book as now 'quite altered and changed into a most easy way By Redivivus Fale'. This is one of Pell's most complete and best written booklets from this period and it is clear that he imagined publishing it, for the mock title page ends with 'Printed by B.A. & T. Fawcet for Iohn Tap ... 1629', and we also have his letter to John Tapp on the subject, though ostensibly written on behalf of 'a freind' [see above, p. 22]. Pell's diffidence about revealing his name was to be characteristic of him

---

[7] BL, MS Add. 4401, fos. 3–14.

[8] E. Gunter, *The Description and Use of the Sector Crosse-Staffe and other Instruments. With a Canon of Artificiall Signes and Tangents* (London, 1624, reprinted 1636). The section on the use of the quadrant is pp. 230–63.

[9] Aubrey, *'Brief Lives'*, i, 276. Aubrey owned a copy of Gunter's *Description and Use of the Sector*, now in the library of Worcester College, Oxford, pressmark II T 5.

[10] BL, MS Add. 4387, fos. 7–26; 'Artificiall sines and tangents' are the *logarithms* of sines and tangents, and Pell later added the words 'Logarithmes of'. Gunter's *Canon of Triangles* was first published in 1620, but Pell may have used the tables in Gunter's *Description and Use of the Sector* of 1624.

[11] BL, MS Add. 4387, fos. 1–6.

throughout his life, yet at the same time he did not want to pass entirely unnoticed, and in the short Preface 'To the friendly reader' he deliberately drew attention to the mystery he was creating:

> I must desyre pardon in y^t I leave my reader in suspence, whither [*sc.* whether] I have assumed a name purposely not to be knowne, or whither my name be indeed ye same wth y^t in the Frontispiece or Title page of the booke.

If Pell had really wished to remain completely anonymous, he could simply have said nothing.

Throughout his life Pell would consider it useful to make concise versions of longer texts written by others, and in his student years Gunter and Fale were not the only authors he drew on for this purpose. In 1628 he prepared a booklet entitled 'Notatu Digniora ex B. Pitisci Trigonometria collecta', detailed notes on the third edition, printed at Frankfurt in 1612, of the *Trigonometria* of Bartholomaeus Pitiscus.[12] (The word 'trigonometry' was first defined in the opening sentence of the book: *Trigonometria est doctrina de dimensione Triangulorum*, 'trigonometry is the doctrine of the measuring of triangles'.) The *Trigonometria* was written in eleven Books, and Pell summarized the first four of them in a few pages containing tables of sines, tangents and secants, discussion of plane and spherical triangles, and a selection of problems in which trigonometry was applied to geodesy (surveying), geography, and astronomy. Pell's treatise is unfinished, ending abruptly after three lines of the second problem of astronomy, but it is very nearly complete, and he obviously had in mind that it might be printed, for the title page carries a suggested price of 9*d*. An English translation of Pitiscus, by Ralph Handson, had been sold through John Tapp in 1614 and this or similar books may have given Pell the idea of publishing his own small but useful treatises through Tapp.

From 1629 we have another 'improved' version of an existing book. Pell's small booklet 'The Construction and Use of the Proportioned Line' is clearly modelled on Wingate's *The Construction and Use of the Line of Proportion* of 1628.[13] Pell's title page (including the motto *Nulla dies sine linea*) is almost an exact copy of Wingate's, and Pell's four proposed chapters have headings nearly identical to Wingate's, except that Pell replaced 'Line of Proportion' with 'Proportioned Line'. In his Preface to the book, Pell wrote:

> Having met with a discourse of ye finding out of Logarithmes beyond the first Chiliad (wch are alwaies to be sold wth Mr. Gunters tables) written by Mr. Wingate under the name of the Line of Proportion: I have after the likenesse of it, invented another Scale, wch as it is like in the construction, so it differeth not much in the name.

---

[12] BL, MS Add. 4400, fos. 15–39.
[13] BL, MS Add. 4431, fos. 70–2.

Pell then set out the purpose of his scales:

1. Having a number to find his Logarithm
2. Having a logarithm to find his number.

Pell did not get very far; the text stops short after just four lines of Chapter 1. A second draft of the title page has the words 'Printed for ... ... 1629' but in this case there is nothing behind the title page except a verse: 'Upon the Arithmeticall Jewell'.[14]

There is another booklet which, from the handwriting and content appears to be from the same period, entitled 'Imitatio Nepeira'.[15] This too is unfinished but is nevertheless a substantial piece of work. The first part gives elementary properties of logarithms and a detailed description of Napier's Canon and its use; the second part is subtitled 'Of the excellent use of the marvelous canon of Logarithmes in Trigonometry' and has chapters on plane and spherical triangles; the third part has detailed descriptions of the tables of Briggs and Gunter as found in 'Mr. Wingate's booke'. The work is thorough and detailed but comes to an end in mid-sentence on the twenty-seventh page.

Pell's interests during his time at Cambridge are clear: calendrical and related astronomical matters; the theory and practical uses of trigonometry; and the theory and use of logarithms. On a lighter note, there are two other sets of tables compiled by him in his student years. The first, entitled 'Tabulae directoriae' and dated 'Anno Christi 1628 July 16' is a multiplication table up to 9×100, with instruction on how to make such a table (Pell explained that the numbers in each row increase in arithmetic progression), and how to use it to find square or cube roots or to carry out division.[16] Another booklet of a quite different kind, entitled 'Labyrinthus ingenii', is a game in which a player has to guess which of 60 words his opponent has chosen. It has some mathematical content, being based on the fact that numbers up to 63 can be written as sums of powers of two up to 2^5; Pell gave no mathematical analysis but neatly wrote out all sixty words in the six different arrangements required.[17]

Pell's early interest in trigonometrical and logarithmic tables is confirmed by a letter written to him by Henry Briggs, Savilian Professor of Geometry at Oxford, in October 1628.[18] Unfortunately, Pell's original letter to Briggs is lost, but Briggs's reply makes clear that Pell had asked him about the interpolation of tables of sines and the calculation of logarithms of fractions. Pell had also asked Briggs for his opinion of Wingate's *The Construction and Use of the Line of Proportion* (the book behind Pell's own treatment of 'the proportioned line'),

---

[14] BL, MS Add. 4431, fo. 73. There is also another draft of the Preface in BL, MS Add. 4426, fo. 96.

[15] BL, MS Add. 4407, fos. 65–79.

[16] BL, MS Add. 4397, fos. 1–4.

[17] BL, MS Add. 4431, fos. 1–8. The game is still in use today in various guises.

[18] Briggs to Pell, 25 October 1628: BL, MS Add. 4398, fo. 137, printed in Halliwell, *Collection of Letters*, pp. 55–7.

but Briggs replied that he had 'not so advisedly looked on it, that I may justly either except or approve all'.[19] It seems that Pell also asked Briggs about the missing chiliads (1000s) in Briggs's tables of logarithms (which were calculated only for numbers from 1 to 20 000 and from 90 000 to 100 000); Briggs replied that Vlacq had now filled in the missing central portion, though to four fewer decimal places than Briggs himself had used. Despite the availability of Vlacq's tables, there are numerous references in Pell's papers to the tables of Briggs, and suggestions as to how one might find logarithms of numbers in the missing chiliads from 20 000 to 90 000, and it is possible that some of his calculations were begun in the late 1620s.[20]

By 1630 we begin to see Pell also as a teacher. A simple exposition of logarithms is annotated: 'Dictated to George Schwarts at his going away Chichester 1630'. It explains how to use tables to find logarithms of whole numbers or fractions, how to find a number from its logarithm, and how to use a logarithmic scale marked on a 'graven Instrument' (with an explicit reference to Wingate).[21] A second section is annotated 'Given to William More at his going away from Chichester A° 1630' and explains how to extract square and cube roots by traditional methods and by logarithms.[22] These two pieces are in a booklet with the title 'Logarithmes. Extraction of rootes. Steganographia Trithemij', but the section on steganography was never added.

Pell appears to have been particularly active in 1634 and 1635. In April 1634 he wrote a small treatise entitled 'Eclipticus prognosta or The eclips Prognosticator or fforeknower of Eclipses', possibly what Hartlib was referring to when he wrote in the summer of 1634 of the 'Astronomical Histories, which [Pell] has perfected'.[23] In the same year Pell thought of copying and improving some of Philip Lansberg's tables: his draft title page describes them as 'The Everlasting tables of Heavenly Motions by Philip Lansberg 1632 Now turned from Latine into English & from the sexagesimall to the decimall subdivision Anno 1634'.[24] He also planned a book of astronomical tables for 1636 and began the calculations, but though several pages were ruled and prepared they are only partly filled.[25]

From the autumn of 1634 we also have 'The logarithmes for the Asscripts of a Circle'.[26] (The asscripts, or adscripts, of a circle are inscribed or circumscribed lines, that is, sines or tangents.) There are examples of ruled but blank pages, together with calculations of the number of pages needed, then there is a new heading: 'The Manual of Logarithmical Trigonometry Breifly containing the com-

---

[19] BL, MS Add. 4398, fo. 137.
[20] See, for example, BL, MSS Add. 4415, fo. 6; Add. 4425, fos. 330–1.
[21] BL, MS Add. 4430, fos. 22–9.
[22] BL, MS Add. 4430, fos. 30–1.
[23] BL, MS Add. 4397, fos. 12–20; see also HP 29/2/30B.
[24] BL, MS Add. 4416, fos. 1–3.
[25] BL, MS Add. 4431, fos. 52–65.
[26] BL, MS Add. 4431, fos. 30–5.

pend of both Mr. Briggs his great bookes viz his Logarithmicall Arithmeticke, & his British Trigonometry'. Clearly this was intended, like some of Pell's earlier booklets, as a useful compendium of larger works, in this case Briggs's *Arithmetica logarithmica* of 1624 and the *Trigonometria britannica* of 1633. From a letter that Pell drafted (apparently to Hartlib) in 1635 we know that he was in fact planning a substantial treatise on trigonometry, since he spoke of 'my new kinde of Trigonometry' and went on to say: 'though neither Golius nor I should finish that new trigonometry it may be done by any other mathematician'.[27] We also have, though undated, the proposed contents of a 'Trigonometria' in two books, the first on plane trigonometry, and some of its uses: making sundials, geodesics, measurement of altitude, and navigation; the second on spherical trigonometry, and its uses in astronomy, geography, and the construction of sundial tables.[28] On an adjacent sheet is the beginning of another or related plan, for a 'Trigonometria Logistica Αὐτάρκης', or, as the subtitle explains, the solution of plane triangles not by instruments but by computation.[29] Pell wrote the first line of the first problem, but no more. The 'Trigonometria' was to be one of his many unfulfilled aspirations.

The only treatise that Pell seems to have completed during this period, in 1635, was based on Gellibrand's *A Discourse Mathematicall on the Variation of the Magneticall Needle* (published early in 1635). Again we see Pell condensing and 'improving' someone else's work: he described his own role as 'making ye pith of [Gellibrand's] discourse ye foundation of mine'.[30] Fortunately, Gellibrand seemed to approve of Pell's treatise but, unfortunately, it is now lost.[31]

Sussex and London, 1635–1643

By the late 1630s the booklets of Pell's earlier years became fewer as his work was written more often on folio sheets, some of them precisely dated not only with a calendar date but with his own age in days. His subject matter also broadened in scope as he began to explore new topics.

He was still interested in logarithms, and an obvious method of constructing tables of logarithms is based on the fact that $\log ab = \log a + \log b$, so that $\log 1001$, for example, is easily found as $\log 7 + \log 11 + \log 13$. For this purpose, in May 1636, Pell created a table of prime factors for integers from 1001 to 1200, and noted that 'The use of this table is the easiest manner of construction of y^e tables of logarithmes exemplified in y^e 2 first centuries of y^e second Chiliad'. He found some useful short cuts in testing for divisibility by 9, 3, and 7, and by August that year he thought he had found a general method of testing for primes. First,

---

[27] BL, MS Add. 4425, fo. 11.
[28] BL, MS Add. 4409, fos. 354–5.
[29] 'Omnium triangulorum planorum rectilineorum solutio non per instrumenta sed per supputationem'; BL, MS Add. 4409, fos. 356^v–358.
[30] BL, MS Add. 4425, fo. 10.
[31] There are a few notes on the same subject in BL, MS Add. 4408, fo. 384.

he instructed, find the square root and test only for divisibility by primes less than that root, that is,

> ... divide ye number by 3, 7 and all ye incomposits betweene. But because this is tedious I have found quicker dispatch, thus ...

There follows a page of working out of a method based on sums of digits, but it was soon crossed out and abandoned.

In July of the same year, 1636, Pell was interested in another new topic: the problem of finding a good approximation for the ratio of the circumference of a circle to the diameter as a rational number.[32] Pell's approach was unusual because it involved no new method for finding the area of a circle; instead he took the decimal form of the number now called π (by then known to 40 decimal places) and tried to reduce it to a rational number that was more accurate than the well-known 22/7 but simple enough to be used in practical calculation. His method is described here, partly because it shows him pursuing an original train of thought, and partly because we shall see this work again in an unexpected context many years later.

Pell began by citing the values of the ratio found by Archimedes (22 to 7), van Ceulen (3. 14159 26434 89793 to 1), Lansberg (29 'cyphers', or places of decimals), and Briggs (40 cyphers). Pell wanted a ratio more precise than that of Archimedes, but more manageable than the lengthy decimal expansions found by the later mathematicians. He was not satisfied simply to cut the decimal expansion short (as he pointed out that Oughtred and Gunter had done) but wanted a ratio in its least terms. One possibility, for example, was the value of 355 to 113 discovered by Adriaen Metius (though Pell could not for the moment remember Metius's name and so left a space filled with a row of dots).[33]

Pell's first attempt was to find the highest common divisor of 3141592 and 1000000 by the Euclidean algorithm, thus reducing the fraction $\frac{3141592}{1000000}$ to $\frac{392699}{12500}$. This was still in much larger numbers than those found by Metius, so Pell tried another tack. He noted that the difference between Lansberg's value of 3.14159 26535 89793 ... and Archimedes' value of 3.14285 71428 57142 ... was 0.00126 44892 67349 ... , and the reciprocal of this difference he found by long division to be 790.83312593, or very nearly 791. Thus an improved value of Archimedes' ratio is $3\frac{1}{7} - \frac{1}{791}$ or $3\frac{16}{113}$, or $\frac{355}{113}$, as found by Metius. By working to greater degrees of accuracy (replacing 791 by 790.833, for example, or by 790.833126), Pell was able to find other and more accurate fractions with larger denominators. He was so pleased with his results that a few days later he wrote (still leaving Metius's name as a row of dots):

---

[32] BL, MS Add. 4416, fo. 31.

[33] The ratio 355 to 113 was discovered by Adriaen Anthonisz. Metius in 1584, and published by his son Adriaen in *Arithmetica et geometriae practica* (Franeker, 1611), p. 69.

Gross Proportion	I Kings 7.23	3	1
First correction	Archimedes	22	7
Second correction	.....	355	113
Third correction	John Pell A° 1636	13307632175	4235950883

Pell seems to have enjoyed this method because on another sheet, which happens to be in the same volume, we find him using it again to approximate 0.061773 72724 ... by the fraction $\frac{1}{16} - \frac{1}{1391} = \frac{1375}{22256}$, or better, $\frac{1375}{22256} - \frac{1}{135768}$ etc.[34]

Thus by the second half of the 1630s we see Pell beginning to extend his interests beyond the subjects that had preoccupied him as a student and in his early years of teaching. We can also detect another shift in his thinking during this period, as he began to develop a more general vision of mathematics, and to see it not only as a useful tool in the sciences, but as a system of knowledge in its own right. It is not easy to impose order on Pell's thoughts from the disorder of his papers, but it is, I believe, possible to discern something of the development of his mathematical thinking between 1635 and 1640, a period when he was perhaps at the height of his intellectual power.

In May 1635 Pell wrote a document in which he attempted to classify the subject matter and operations of arithmetic. It contains so many of the seeds of Pell's subsequent thought that it is quoted here at some length. It begins:[35]

> Logistica or the art of reckoning both in numbers and in kinds, to find out and determine the Equality, difference or proportion of any similar quantityes propounded. J. P. May 8. 1635

There follows a series of numbered points:

> 1. All determination of quantity is by comparing the propounded quantity with some other thing of ye same kind whose quantity is knowen, in the equality, difference or proportion, that they have to one another.
> 2. The equality or inequality of quantityes is seene and expressed by number, measure, weight or time.
> 3. Every figures quantity is either [1]a number or [2]a line or [3]an angle or [4]a superficies or [5]a solid or [6]weight or [7]some parallel or part of time.
> 4. Any of these 7 kinds may be expressed either numerically or speciously [in letters] or mixtly. as 18. $d + d$ or $d + 9$ or $2d$.
> 5. Every quantity is either real, positive, affirmed or more y^n nothing as 34 or +34, or $0 + 34$ for $+$ is y^e signe of affirmation; or imaginary, negative or lesse y^n nothing as -34 or $0 - 34$, that is 34 lesse y^n nothing as he y^t owes £50 and hath but £16 is worth £34 lesse y^n nothing.

---

[34] BL, MS Add. 4416, fo. 142.
[35] BL, MS Add. 4420, fos. 47–8.

PART II: THE MATHEMATICS OF JOHN PELL   259

> 6. The operations of Logistica are 6. Addition, Multiplication, Involution and their direct opposites Subtraction, Division or Application, Evolution or Extraction [of roots]. All which have their severall expressions in speech and writing.
> 7. Every quantity then, is either Aggregate as B or b (for $0 + B$), $B + C$. Or residuall as $c - b$, $-b$ (for $0 - b$). Or produced as $b * c$.

The piece ends, after 18 such points, with the words:

> Comparare is a most generall end of which an Exact Idea were a most excellent worke.

On the reverse of the same sheet Pell claimed that:

> All Mathematicall Problems are reducible to these two
>
> 1. To finde the habitudes of quantities to one another,
> 2. To finde and determine any further unknowne quantities by their habitudes as some [of] y^m are knowen.

Pell went on to explain what he meant by 'habitudes', a word he was to use often over the next few years:

> There is no other way to finde unknowen quantities nor to express them when they are found but this. How long is this piece of cloth? 7 yards.
> That is y^e habitude of y^e length of y^e cloth to y^e length of the yard is as 7 to 1 or 7 fold.
> How greate is this field? 2 acres and an halfe.
> Its habitude to an acre is as $2\frac{1}{2}$ to 1.
> Therefore those Problems and these only are insoluble in their owne nature which have not, a sufficient number of data to yield a discoverable habitude towards the quantity sought.

> Habitude) a more generall word than *proportion*, y^e Longest side of some rect[angled] triangle hath no proportion to y^e other side but it hath this habitude $a = \sqrt{q} : bb + cc$.

Two further quotations will perhaps help to give the full flavour of Pell's thought in this document:

> There are quality habitudes as well as of quantity but y^e Mathematician takes no notice of y^m.

> It is the work of the *Analyst* Having any proposition to find y^e habitudes of one of y^m to y^e other.
> It is *Synthesis* To put any jungibilia together whose habitudes are fit to produce y^e required end.

The sheet ends with the heading 'Progymnasma Synthetico-Logisticum' and a table. Down the side of the table are the quantities $0, a, c$, while across the top are the six possible operations of arithmetic: addition, subtraction, multiplication, division, involution (raising to a power), and evolution (extraction of roots), and the entries in the table show how to combine 0, a, c according to the headings:

	Addition to			Subtraction from			Multiplication by			Division[36] by			Involution[37]			Evolution[38]		
	0	a	c	0	a	c	0	a	c	0	a	c	0	a	c	0	a	c
0	0	a	c	0	a	c	0	0	0	0	0	0	0	0	0	0	0	0
a	a	$2a$	$c+a$	$-a$	0	$c-a$	0	aa	ac	0	1	a/c	0	$a\ominus a$	$c\ominus a$	0	$a\omega a$	$c\omega a$
c	c	$a+c$	$2c$	$-c$	$a-c$	0	0	ca	cc	0	c/a	1	0	$a\ominus c$	$c\ominus c$	0	$a\omega c$	$c\omega c$

Note that Pell already uses Harriot's ac for a times c (from the *Praxis*, 1631). He has not yet introduced the division sign ÷, but for involution and evolution he uses the spiral and the wavy line that he continued to use, and teach to others, for the rest of his life.

Many of the typical features of Pell's thought over the next few years are introduced in this piece of writing. He would go on to explore further: the classification of basic concepts; the proposal of an 'Exact Idea'; the notion of 'habitude'; the systematic working out of relationships of one quantity to another; the handling of problems that were 'insoluble' because of insufficient data; the relative properties and merits of Analysis and Synthesis; and the setting up of a Progymnasma, or foundation, of algebraic calculations.

The notion of habitude, in particular, was one that Pell used often at this time. In August 1636, for example, we find it again in a simple practical context:[39]

> Aug.4.1635. The habitudes of the sides of ye angles in a plaine rectangle triangle.
>
> What are the angles ... when ye proportions are 1, 1⌊5, 2⌊5 [sc. 1, 1.5, 2.5].

After a little working Pell wrote: 'Ans. ye angles are 36. 54. 90'. The following year, 1636, he went much further in his exploration of habitudes, now combining the six basic arithmetic operations in various ways to give new kinds of relationships. A large table of numbers and habitudes is headed:[40]

> Posoteticall and Logisticall habitudes, [1]their kinds, [2]signatures, [3]continuations, [4]homoschesy [or analogy] as well continued as interrupted, [5]Exegesis numerosa, [6]Use, [7]Sections.
>
> March 26 1636

---

[36] Pell takes $a/0$, $c/0$, and $0/0$ all to be 0.

[37] $c\ominus a$ is what we would now write as c^a. Pell supposed $a\ominus 0 = 0$, whereas it is now considered more natural to define $a^0 = 1$.

[38] $c\omega a$ is what we would now write as $c^{1/a}$. Pell supposed $a\omega 0 = 0$.

[39] BL, MS Add. 4416, fo. 34.

[40] BL, MS Add. 4415, fos. 29^v–30.

The table that followed was an attempt to make a systematic list of arithmetic relationships, and the progressions that arise from them. For example, simple addition, according to Pell, gave rise to the progression $d - 5x$, $d - 4x$, $d - 3x$, $d - 2x$, $d - x$, d, $d + x$, $d + 2x$, $d + 3x$, $d + 4x$, $d + 5x$, while simple multiplication gave the progression he wrote as $d(rrrrr$, $d(rrrr$, $d(rrr$, $d(rr$, $d(r$, d, $d*r$, $d*rr$, $d*rrr$, $d*rrrr$, $d*rrrrr$. Pell returned to this table two years later, in 1638, but in the meantime something new had entered his thinking.

Oughtred's *Clavis* was first published in 1631 under the title *Arithmeticae in numeris et speciebus institutio: quae tum logisticae, tum analyticae, atque adeo totius mathematicae, quasi clavis est*, a title that soon became abbreviated simply to *Clavis*. We do not know when Pell first read the book but the earliest indication of his interest in it is from August 1637. Oughtred (following Viète) used the letters A and E for unknown quantities, but also introduced Z for the sum $A + E$, X for the difference $A - E$, P for the product AE, R/S for the quotient A/E, $\mathbf{Z}$ for the sum of squares $A_q + E_q$, and $\mathbf{X}$ for the difference of squares $A_q - E_q$. Given any two of these quantities it is possible to find the others, and Oughtred had given a few examples at the end of Chapter XI and others in Chapter XVII. A few examples were not enough for Pell, who set himself the task of finding and tabulating every possible relationship between the eight quantities. He called the piece a 'Progymnasma', a preparatory exercise.[41]

> 1637 Aug 4 Aetatis 9654 inchoato Προγυμνασμα αναλυτικον.
> Datis duobus quibuslibet ex hic octo $A\ E\ Z\ X\ P\ R/S\ \mathbf{Z}\ \mathbf{X}$, reliqua sex invenire.
> [August 4 Aged 9654 days, the beginning of the analytic Progymnasma.
> Given any two of these eight ... to find the remaining six.]

A full-page table then follows with the headings A, E, Z, etc. across the top, and all possible pairings of the eight quantities down the left-hand margin (28 pairs). Over the next five days Pell filled in the six entries needed in each row, 168 entries in all, and the sheet ends:

> Laus Patri Luminum August 9

A week later he translated the table into English and made a few alterations, and on 17 August he made a fair copy, which he sent to Hartlib two days later.[42] In February 1638 Pell began an explanation of how he had compiled the table:[43]

> 9846 Febr 12 1637/8
> In the explication of my tabular Progymnasma finished 9666 Aug

---

[41] BL, MS Add. 4415, fos. 19–21. John Wallis too made a version of this table, though simpler than Pell's, when he first read the *Clavis* in 1647 or 1648. Wallis's table is written into two copies of the *Clavis* now in the Bodleian Library, pressmarks Savile Z 19 and Savile Z 24.

[42] BL, MS Add. 4398, fo. 170. In this copy Pell claimed to have found no fewer than 1892 equations because of the different ways of writing the same quantities.

[43] BL, MS Add. 4415, fos. 22–3.

16 1637 and coppied out the next day, and sent to Mr. Hartlib Aug.19 I promised, on greater leisure to show the whole order and manner of the making of that table and some further uses of it, besides ye numerall solution of those 168 Problemes. I thinke it not amisse to begin now, though I doe not yet abound in leisure.

Pell never did feel that he abounded in leisure. Nevertheless he wrote a fairly detailed explanation of his table, and his exposition ends with some alternative notation taken from Harriot.

The explanation for Hartlib was not the end of Pell's musings on the Progymnasma, for a few days later he was moved to write a further document that turns out to be crucial to the understanding of Pell's mathematics. It is not dated at the beginning, but the number 9853 appears part way through, suggesting that it was written around 19 February 1638, a few days before Pell's twenty-seventh birthday. It is entitled 'The Strife of Analytica and Synthetica for praeminence'. In Classical mathematics a synthetic proof began from known facts and worked towards a new conclusion. In the analytic approach, on the other hand, the mathematician began by assuming what he wanted to find or prove, and then worked 'backwards' to discover what truths it must be based on. As the Greeks and later mathematicians knew, the analytic process did not in itself constitute a proof, but in principle one could reverse the steps of the analysis to construct a rigorous and acceptable synthetic proof. Pell's discussion of the relationship between analysis and synthesis begins:[44]

> Synthetica taking it ill y^t Analytica should be called by Oughtred via artis, as though it selfe were error inertium, complained to Reason ...

After a little more preamble of this kind, Pell described some of the problems of the Progymnasma. Then he returned to Analytica and Synthetica:

> Being asked how they came by y^{se} rules they both answer y^y found y^m out by ye use of reason. Wherein then doe you differ, for all the while you say ye same things?

The difference between analysis and synthesis is not in the results, of course, but in the method, and Analytica explains that her way of approaching mathematical problems is through a 'general precept', while synthesis must rely on memory:

> Analytica answers. I never had such a question propounded me before, or if I had, I have forgotten what course I tooke to solve it. And so set my selfe to find out a generall precept for such questions. But Syntheticus did not so. But having heretofore found such a thing & being guilty to himselfe of the trouble which his groping way puts him to, when he sets himselfe to seeke any way: he wisely

---

[44] BL, MS Add. 4415, fos. 26–7.

laid it up in his Memory, for feare he should not have it, when he hath neede of it.

To which Syntheticus (now become male) has his reply:

Tush quoth Syntheticus, as if it were possible for you to doe any thing rationally without me; Or as if you were not more guilty of praetermitting [putting aside] many usefull truths than I was of producing such as are uselesse. For you contente your selfe if you find any way to solve a probleme, whereas I take such a course, y^t there can be no way that can scape me, different.

At one point the debate is written as a dialogue:

S. First I say that it is impossible for Analyticus to doe anything rationally without me.
A. No! See how I wrought this.

And here there follow yet more examples from the Progymnasma.

Finally the voice of Pell himself comes through. In a cramped note written sideways along the margin of the sheet he expresses his vision of a synthetic method, which will be imbued with a new logical clarity. Instead of analysis being a prerequisite for synthesis, this new kind of synthesis will shed light on analysis:

The Analysis of ye ancients nay of all my predecessors seemes to me but like ye groping of a man by darke, trusting his memory to tell him how to goe back in ye way which he came by day light. But ye way I aime at is first so nicely to observe all turnings from my setting out on my whole Syntheticall journey to ye top of problematicall daring as y^t I may know my way backe again without doubt or mistakes & yet to strike a light in analyticall regression & not to trust to groping. How much will this exceed ye former way!

Then Pell sets out a clear and important explanation of this new synthetic Method:

The great precept now of Synthetica is this (9853–)
Set downe all ye prime truths which are no conclusions (& therefore are naturall to us) in a due order, keepe y^s order constantly, then take ye first of these and apply it to ye second, then to ye third, & so to y^m all without any intermission, considering whether they agree in one terme, if y^{ey} doe not you can make no third out of y^m, if they doe, you may by one of ye 14 ways above, Set downe ye conclusion & so goe on. When you have done so, Begin with ye second, match y^t with ye first, second etc, & so combine all to all, till you have all ye Child-truths, y^n Begin againe & combine

every one of those Father & child truths with every one of ye child truths & so produce grand-child truths etc. Thus doing it shall not be possible to omit any one truth.

(In a table immediately above this, Pell has shown combinatorially that there are 14 ways of logically combining two three-term propositions to make a third.)

The 'great precept of Synthetica' is given in the first place as a mathematical method but Pell immediately links it to his ideas about language:[45]

> This was represented in ye seeming mad conceit of mine of Turris Babel, which should comprehend all ye possible words y^t all ye Sonnes of men could speake.

Pell went further, and began to write about his Method in an even broader way, as applicable to knowledge in general. It was clear, however, that any such method depended on the identification and ordering of the 'prime truths'. Again Pell went into dialogue:

> Q. How should we know these prime truths?
> A. 1. We must Analytically try every truth we heare & consider what makes us consent to it, for if we can give no reason of our consent, to us it seemes *a prime truth*.
>   2. Because these prime Truths come of necessity from our nature or Maker, therefore y^{ey} are in all men, wherefore this is a seconde signe of y^m, Universall consent &, from y^s second property, they are called Common Notions.
>   3. Consider whether any have gathered all or most or some of y^m already & so have saved you some Labour. ye Lord Herbert promiseth a Treatise of y^m.
> Q. In what order must we set y^m downe?
> A. It is all one to Synthetica how they stand, so they stand in one certaine order, for it is not possible for any Combination to escape her. As Though ye letters in ye Alphabet stand not in ye best order y^t may be, Yet y^t hinders not ye Syntheticus from finding all possible connections of y^m together.

And the piece ends:

> Ergo. It seemes y^t before we can do any thing more in y^s kind, we must first find out these prime truths.

Looking back to Pell's first writing on 'Logistica' in 1635, it is clear that over the course of three years he has moved from his first consideration of quantities and habitudes; first, to a more precise investigation of algebraic relationships in the Progymnasma and related tables, and from there to an investigation of

---

[45] For discussion of 'The strife of Analytica and Synthetica' in the context of Pell's ideas on universal language see Lewis, 'John Wilkins's Essay', pp. 18–23.

mathematical method. In 'The strife of Analytica and Synthetica' of February 1638, that mathematical method has in turn transformed itself into something like a general theory of knowledge based on a logical process of combining prime truths. In August 1638 Pell went further towards designing an algebra of knowledge, in which complex ideas could be built from 'simple notions':

> If we had all our simple notions set downe, we had as perfectly all the thoughts of men as if we had all our simple sounds, we have perfectly all ye words & speeches of men potentially.

Pell even went on to invent some special symbols for his algebra of knowledge, just as he had done for mathematical algebra: 'connecting notions is [×], removing notions is [_/]'.[46] Such ideas of universal language were later nurtured by a number of seventeenth-century thinkers, but were shared by few of Pell's English mathematical contemporaries, and certainly not as early as 1638.

An algebra of knowledge, however appealing in theory, was impossible in practice. Even if knowledge could be as logically structured as Pell suggested, and even if he or others could have identified the prime truths or simple notions on which it was supposed to be based, the number of possible combinations of truths and 'child-truths' would quickly run beyond human computational ability. If the Method could not be applied to all knowledge, however, it could at least be applied within mathematics. There is perhaps a parallel to be drawn here with Descartes, who offered a mathematical text, 'La Géométrie', as an illustration of *his* general theory of knowledge, his *Discours de la méthode*. By the end of 1639 Pell had translated the *Discours* and accompanying 'Essais',[47] but his own ideas about mathematical method had already begun to take form as early as 1635 when he had composed the 'Logistica' and when, in a draft letter to Hartlib in May of that year, he had spoken of:[48]

> A plaine and cleere delineation of the whole proceeding of invention in all kind of soluble problems in Mathematica, and which as it is more modest in undertaking than of Viète, Nullum non problema solvere, so I hope it will be more perfect in the performance.

Such thoughts must have been turned over, formed and reformed, in Pell's mind many times from about 1635 onwards. His unpublished manuscripts hold several hints, but his ideas were most fully worked out in early 1638, and were finally brought into the open later that year in one of Pell's few published pieces from this period, his *Idea of Mathematicks* of October 1638. The *Idea*, written originally as a letter to Hartlib in July 1638, brings together two major themes of Pell's life: first, his wish to provide the world with useful knowledge in a concise and easily available form, something he had been attempting to do since he was seventeen years old; second, his vision of a mathematical method that

---

[46] BL, MS Add. 4420, fo. 23.
[47] See above, Part I, ch. 3, n. 60.
[48] BL, MS Add. 4425, fo. 11.

would eventually enable the learner to be self-sufficient and therefore to manage without books altogether. Pell's plan was three-fold:

(1) To produce a *Consiliarius mathematicus*, which would direct the student to 'the *best* bookes in every kinde', and provide a chronological catalogue of mathematical writings, classified according to type.

(2) To establish a library containing those books, and to add to it new works printed either at home or abroad.

(3) To publish three new treatises: *Pandectae mathematicae*, an encyclopaedia of all mathematical results so far known; *Comes mathematicus*, a pocket-book containing the most useful tables and precepts; and finally, *Mathematicus αὐτάρκης*, showing how to resolve any problem from first principles even without the aid of books.

Although William Brereton later said that Pell produced 'a quire of papers' on the subject,[49] there is little evidence that Pell did much towards the preparation of the *Consiliarius* or the *Pandectae*. He did write some notes on a possible *Comes mathematicus*, but it is not clear whether before or after the *Idea*. Typically, Pell complained of lack of enthusiasm for the project on the part of those who might have funded it:[50]

> For a travailer *Comes mathematicus* is fittest which should contain some few fundamentall tables, ... in a mans study a book of Canons or tables ready calculated for all particular uses is a thing of very great use, and it argues but a dulnesse in our rich men y^t they doe not employ many in ye making such tables for their use, y^t being a worke of meere labour & therefore not requiring any ingenious or great Minds.

The proposed *Mathematicus αὐτάρκης* arose naturally from Pell's 'great precept' of February 1638 but, as we have seen, the concept had been taking shape as early as 1635 when Pell already believed that with an understanding of basic principles and methods any mathematician could reconstruct any piece of mathematics: 'So though neither Golius nor I should finish that new Trigonometry it may be done by any other mathematician.'[51] Thus a mathematician trained in Pell's method would be able to dispense with books and their imperfections, or as he put it in the *Idea*:

> Were the *Pandects* thus made and finished, I suppose it is manifest; that by their orderly, rationall and uniformed compleatnesse, ... they would spare after-students much labour and time that

---

[49] Rigaud, *Correspondence*, ii, p. 474.

[50] BL, MS Add. 4408, fo. 30; see also fo. 34, and MS Add. 4425, fo. 68. The last is dated 21 Feb. but without a year.

[51] BL, MS Add. 4425, fo. 11.

is now spent in seeking out of bookes, and disorderly reading them, and struggling with their cloudy expressions, unapt respresentations, different methods, confusions, tautologies, impertinences, falshoods by paralogismes and pseudographemes, uncertainties because of insufficient demonstrations, &c besides much cost also, now throwne away upon the multitude of bookes, the greater part whereof they had perhaps beene better never to have seene.

Was it was perhaps the inherent shortcomings of all mathematical texts that prevented Pell, the perfectionist, from writing them?

The *Idea* is important not so much as a practical project that was never carried out, as for the views Pell expressed in it as to how mathematics itself might progress by a process of ordered reasoning, views that are by now familiar from his earlier writings:

I should lay heavier lawes upon my selfe, than I have already mentioned; namely, *First* to lay downe such an exact *Method* or description of the process of Mans reason in inventions, that *afterwards* it should be imputed meerly to my negligence and disobedience to my owne lawes, (and not to their insufficiency) if, from my first grounds, seeds, or principles, I did not, in an orderly way, according to that praescribed Method, deduce not onely all that ever is to bee found in our Antecessor's writing, and whatsoever they may seeme to have thought on, but also all the Mathematicall inventions, Theoremes, Problemes and Precepts, that it is possible for the working wits of our successors to light upon and that in one certain, unchanged order, from the first seeds of Mathematics, to their highest and noblest applications.

Pell clearly echoed the hope first expressed by Viète that every problem in mathematics might be solved and every possible theorem discovered: *nullum non problema solvere*. But if Pell hoped, like Viète, to find a Method of solving all mathematical problems, he also differed from Viète in some important respects. Viète had believed that problems would be solved and discoveries made through the newly harnessed power of algebra, which he saw as the lost analysis or 'analytic art' of the Ancients. For Pell, on the other hand, the correct procedure was synthetic, a process of systematic logical deduction from first principles, from which the whole of mathematics past, present, and future, would fall out in an obvious and a well-defined order. Such a method was best written or displayed, of course, in the language of algebra, the language in which the relationships or habits between mathematical entities became immediately clear.

If it was too ambitious to hope that Pell's 'great precept' or Method could be applied to all human knowledge, or even to all of mathematics, it could certainly be applied with success to straightforward algebraic or geometric problems with a finite number of unknown quantities. Sometime in the very early 1640s Pell devised what Aubrey later called his 'excellent way or method of the marginall working in Algebra', a way of setting out his work in three columns.[52] The left-hand column carried a series of instructions while the wider right-hand column contained the calculations. A narrow central column held line numbers. Pell may have had the idea originally from the *Cursus mathematicus* of Pierre Hérigone, who frequently used two columns: a broad right-hand column for calculations, and a narrower left-hand column to give a brief reason for each step. The first five volumes of the *Cursus* were published between 1632 and 1637 (the sixth followed in 1644), and were perhaps made known to Pell through Mersenne who in 1639 suggested the *Cursus* as a text that might satisfy the requirements of Pell's *Idea*. Pell acquired his own copy, and from time to time borrowed some of Hérigone's notation: $\square AB$ for the square of a line AB, for example, rather than the AB_q introduced by Oughtred.[53]

At first sight Pell's three-column layout seems no more than a neat and tidy way of setting out a mathematical argument. But Pell (and both Wallis and Aubrey after him) saw it as more than that, and insisted on calling it a 'method'. On closer inspection Pell's three-column way of working is seen to be just that, for it follows very closely the 'great precept of Synthetica' that Pell outlined in February 1638. First one lists the unknown quantities on the left, and the known equations or relationships between them on the right. Then one works through the given equations in a systematic and logical way until the solution emerges. The Method became such a permanent and important feature of Pell's mathematics that a full example of it is given here.[54] This particular problem, of a kind familiar to anyone who has ever studied algebra, is about three amounts of money that must satisfy the three equations on the right.

---

[52] In the second half of the 1660s Pell wrote to Collins about 'a question so wrought with numbred steps & marginall references, which I shewed to Mr John Leake at my house in Westminster, 24 yeares agoe', suggesting that he demonstrated his three-column method to Leake in 1642 or 1643 before he left for Amsterdam; see above, Part I, ch. 3, n. 146. Pell added, 'If there be any yet alive that saw any thing of that kinde, before that time, I doubt they are not of your acquaintance'. As early as mid-1641 in a draft letter to Cavendish (Letter 4b) Pell had spoken of 'that crotchet of mine of numbering & ordering ye aequations', but Cavendish had died in 1654, and Collins in any case would not have known him.

[53] Pell's copy of Hérigone's *Cursus mathematicus*, i–v, with his inserts and annotations, is now in the Busby Library, Westminster School, pressmark I D 29.

[54] Rahn, *Teutsche Algebra*, p. 81; Rahn and Pell, *Introduction to Algebra*, pp. 71–2.

Prob. VIII. Three men have Money.

$a = ?$	1	$a - 100 = \dfrac{b+c}{4}$
$b = ?$	2	$b - 100 = \dfrac{a+c}{3}$
$c = ?$	3	$c - 100 = \dfrac{a+b}{2}$
$1 * \overline{4}$	4	$4a - 400 = b + c$
$2 * \overline{3}$	5	$3b - 300 = a + c$
$3 * \overline{2}$	6	$2c - 200 = a + b$
$4\pm$	7	$4a - b - c = 400$
$5\pm$	8	$-a + 3b - c = 300$
$6\pm$	9	$-a - b + 2c = 200$
$7 + 8 + 9$	10	$2a + b = 900$
$7 - 8$	11	$5a - 4b = 100$
$10 * 4$	12	$8a + 4b = 3600$
$11 + 12$	13	$13a = 3700$
$13 \div \overline{13}$	14	$A = \dfrac{3700}{13} =$ the sum of the first.
$14 * \overline{2}$	15	$2A = \dfrac{7400}{13}$
$10 - 15$	16	$B = \dfrac{4300}{13} =$ the sum of the second.
$14 + 16$	17	$A + B = \dfrac{8000}{13}$
$9 + 17$	18	$2C = \dfrac{10600}{13} =$ the sum of the third.
$18 \div \overline{2}$	19	$C = \dfrac{5300}{13}$

In its use of line numbers and its concise instructions, Pell's Method is remarkably close to a modern computer algorithm. The above example contains several features that were introduced by Pell as he gradually refined his Method, all of them intended to make the working clearer: the use of overlining (for example, at lines 4, 5, 6) to distinguish numbers used in the calculation from line numbers; the use of the symbol $*$ for multiplication (still used in computer-based mathematics); a new symbol $\div$ for division, to avoid the use of two-line fractions in the left-hand margin; and the use of lower case letters for unknown quantities, replaced by capitals as soon as their numerical values were established. The correct setting up of the equations at the beginning, as every student knows and as Pell recognized, is often the most difficult but also the most crucial part of the problem. As Aubrey later remarked: 'Dr Pell was wont to say, that in the Solution of Questions, the maine matter was, the well-stating of them; ... for let the question be but well-stated, it will worke almost of it selfe'.[55]

Pell's Method became the hallmark of his mathematics and of those who learned from him. It was not, and could not be, the key to all knowledge as Pell in his mid-twenties had hoped, but for the more limited range of mathematical problems to which it was suited it was a useful and systematic way of proceeding. Thus over a period of about five years the relationship between Pell's mathematics and his philosophy of knowledge turned full circle. It was algebra, Oughtred's in particular, that inspired him to a vision of mathematical reasoning as the basis of all knowledge, and to the setting out of his 'great precept of Synthetica', or Method. In the end, the Method could work only in the restricted field of straightforward mathematical problems, but there it gave a new impetus and meaning to algebra. In seventeenth-century mathematics it was unique.

In the course of 1638 Pell's study of 'habitudes', or relationships between quantities, led him also in other directions. In June that year he began an investigation of arithmetic progressions, in which the first differences are equal, and set himself the same sort of problems he had attempted in his Progymnasma: given any two of a list of quantities (first term, difference, sum of terms, number of terms, etc.) to find the others. In September he moved on to progressions in which the second differences are equal, and came to the conclusion that he now needed to know not two but three quantities in order to fix the rest. In Pell's mind this was the beginning of a longer project: 'Of the first sort I wrote Jun 27 Now I come to examine ye rest in order ...' and on 14 September he gathered his findings so far under the heading 'Logistica generalis'.[56]

It seems that he also intended to find a similar general scheme for mechanics (which Pell seems to have considered as a practical application of geometry, in the same way that 'logistica' was a practical aspect of arithmetic), for on another sheet he wrote the heading 'Mechanica generalis' and the date 'Oct 24'.[57] Some weeks later he returned to it and wrote: 'This title was written 40 dayes agoe, But nothing else'. Then in a caution that one wishes he had heeded more often, he added: 'if I let it lye thus some men may see it & understand nothing at all of my intent.' However he had little to add:

> The grand probleme of Logistica generalis is Datum quaesitum habitudine Logistica data afficere [to work out a given sought quantity given the arithmetic relationship]. So
> The grand probleme of Mechanica generalis is Datum quaesitum habitudine Mechanica data afficere [to work out a given sought

---

[55] Worcester College, Oxford, MS 64, first item.
[56] BL, MS Add. 4415, fos. 8, 9, 15.
[57] BL, MS Add. 4415, fo. 16.

quantity given the mechanical relationship].
To doe this I must know all Mechanicall habitudes & I know no man y^t has collected y^m.

The few abandoned notes that follow suggest that Pell intended to begin by exploring geometry as the ground of mechanics.[58] He began by defining a point in a completely practical (or mechanical) way as 'a pricke with a sharp pin, pen, etc.', a line as a stroke of the pen, and a figure as a meeting of two such strokes, but the work stopped abruptly. Listing the six basic operations or habitudes of arithmetic was easy, but listing the possible relationships of geometry was not. Pell suggested a few, including 'to copy', 'to continue', 'to break off', 'to measure', etc. but the task was ill-defined and it is not surprising that he quickly gave up.

Once before, probably as a result of queries from John Dury in 1631 [see above, Part I, ch. 2, n. 30], Pell had set out definitions of point, line, surface, and solid.[59] On that occasion he had defined a point in the traditional Euclidean way as 'that which hath neither length, Breadth nor thicknesse'. He went on to say:

1. Therefore a Point cannot be measured
2. Therefore a Point by moving will make a meere Length.
   Let this be called a line.

Pell's definition of a line as generated by a moving point was Aristotelian rather than Euclidean,[60] and similar to the definitions also proposed by Thomas Hobbes.[61] He distinguished between 'straight' and 'crooked' lines, the former being 'the shortest that can be drawne betweene its termes [ends]', and claimed that 'a straight Line must be the measure of the crooked ones' (unlike Descartes who claimed that the ratio between a curved line and a straight line could never be known).[62] As a moving point created a line, so 'A Line by moving will make a meere length & breadth. Call this a Superficies', and superficies, or surfaces, were again separated into 'flat' or 'crooked'. Finally, he said, 'The greatest finite quantity is a Solid'. Pell never put the same effort into defining the fundamentals of geometry, however, as he did into the fundamentals of arithmetic and algebra.

In 1639 the focus of Pell's mathematics was still algebra, though now no longer the Progymnasma or the manipulation of formulae that had preoccupied

---

[58] For further evidence of the link between 'geometrical' and 'mechanical' in Pell's mind, see BL, MS Add. 4429, fos. 281–6, where there is the beginning of a treatise entitled 'Geometricall problems, or mechanicall practise of mathematicall truths'. The treatise begins with basic geometric constructions (bisecting a line, erecting a perpendicular, etc.) but was never completed.

[59] BL, MS Add. 4429, fos. 40–1; see also fo. 29.

[60] See Euclid, *The Thirteen Books of the Elements*, ed. and tr. Sir Thomas Heath (Cambridge, 1908; reprinted New York, 1956), p. 159.

[61] See, for example, D. Jesseph, 'Of Analytics and Indivisibles: Hobbes on the Methods of Modern Mathematics', *Revue d'histoire des sciences*, 46 (1993), pp. 167–74.

[62] Descartes, 'La Géométrie', appended to his *Discours de la méthode*, p. 340.

him in 1638, but the classical theory of equations (quadratic, cubic, and quartic) propounded by Cardano, Bombelli, Stevin, Viète, and Harriot. Amongst Pell's papers there are notes on Cardano's *Ars magna* of 1545, Stevin's *L'Arithmétique aussi l'algebre* of 1585 (reprinted in the *Oeuvres* of 1634), Viète's *De aequationum recognitione* of 1615, and Harriot's *Praxis* of 1631.[63] The notes on Cardano, Stevin, and Viète are all now to be found in close proximity to material that Pell himself wrote in 1639.

As so often, Pell thought that he could usefully present the work of some earlier writers in more concise form. From Stevin's *L'Arithmétique*, for example, he summarized what he called 'Simon Stevins 10 rules of moulding equations' on a single sheet, and noted that these came from 'Lib 2. pag. 271.2.3.4.5.6.7.8.9, viz 9 pages in 8° [octavo] but expressed after my formes'.[64] The pagination cited here is from the original 1585 edition of *L'Arithmétique*. Pell also owned a copy of the 1634 edition published by Albert Girard, and in it he placed a sheet of paper dated Dec. 26 [1638?] on which he rewrote Stevin's Problems 79 to 81 and Theorems 1 to 6 from the end of Book II.[65] The sheet ends with the following note:

> And thus have I in 9 lines more cleerely expressed and more aptly demonstrated his 6 Theoremes y^n Stevin himselfe in 10 pages in Octavo. And a second worke would bring all y^s in much lesse roome with more light.

In a note dated 'Febr. 25. 1639' Pell made a similar remark about the treatment of algebra in the *Arithmeticae libri duo* of Petrus Ramus (reprinted for the fifth time in 1627). 'In that impression [1627]', wrote Pell, 'it takes up 27 pages in large 4to. I would frame ye same according to my Logistica in as little roome as might be'. He then proceeded to summarize the contents in two sides of a quarto sheet.[66] Such notes on the prolixity of other authors, and the possibility of reducing their work into a more concise form, are perhaps some of the first hints of the book on Analytics, or algebra, that Pell's friends came to expect of him.

Of all the writers on algebra that Pell read, it was Harriot who influenced him most. And Pell had recourse not only to the *Praxis* but to men who had known

---

[63] For Pell's notes on Cardano see BL, MS Add. 4413, fos. 196–7; on Stevin, BL, MS Add. 4413, fos. 194–5; on Viète, BL, MS Add. 4413, fos. 208–9; on Harriot, BL, MS Add. 4409, fo. 369. Pell's copies of the relevant books are now in the Busby Library, Westminster School: Cardano's *Ars magna*, I D 2, with Pell's notes and inserts; Stevin's *Oeuvres*, I F 10 with notes and inserts; Viète's *De aequationum recognitione* bound with a handwritten copy of *De potestatum resolutione*, I D 9. There is also a copy of Harriot's *Praxis*, I F 26, with a very few handwritten corrections, but it is not clear whether or not this was Pell's personal copy.

[64] BL, MS Add. 4413, fo. 194; see further notes also in fos. 194^v–195.

[65] Pell's copy of the 1634 edition of Stevin's *Oeuvres* is at Westminster School, I F 10. Problems 79–81 and Theorems 1–6 are on pages 90–101, and Pell's handwritten sheet is inserted between pages 90 and 91. The handwriting suggests that it was written in the late 1630s, and a date of 1638 would be consistent with Pell's known interest in equations at that period.

[66] BL, MS Add. 4397, fo. 38.

Harriot personally, for by January 1638 he had been introduced to Thomas Aylesbury, one of the executors of Harriot's will, who had overseen Walter Warner's editing of the *Praxis* and still held some of Harriot's mathematical papers.[67] Pell has left a detailed record of a conversation with Aylesbury in January 1638 (in the dialogue form he used again later that year in 'The strife of Analytica and Synthetica'). The conversation is quoted here at some length because it gives a unique glimpse of Pell at work, and shows that by the beginning of 1638 he was already thoroughly familiar with Harriot's notation and methods:[68]

> Jan 29 1638 Sr Thomas Ailsbury propounded me y^s probleme to solve extempore.
> A line being given to divide it by extreme and mean proportion. To divide it into two such unequal segments y^t y^e greater of y^m might be y^e mean proportionall between y^e lesser and y^e total. By y^s was meant, What is y^e habitude of y^e greater or lesser segment to y^e whole data linea?

This was the classic problem of dividing a line segment of length b into two portions a and $b - a$, with a greater than $b - a$ (see Fig. 1), and in such a way that a is the geometric mean, or 'mean proportional', between $b - a$ and b.

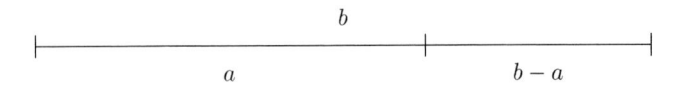

FIG. 1.

A seventeenth-century mathematician would have stated the relationship as a ratio, thus, $a : b - a = b : a$, or in Oughtred's notation, $a \, . \, b - a :: b \, . \, a$. In modern terminology, we would write $a^2 = b(b-a)$, which is a quadratic equation for a in terms of b. Pell (using the letter c instead of a) solved the quadratic equation correctly:

> After some analytical work, I gave him this: $c = \sqrt{_q} : bb + \dfrac{bb}{4} : -\dfrac{b}{2}$.

Aylesbury tried his hand at it too and, in passing, gave Pell an explanation of Harriot's choice of symbol for 'equals':

> Whereupon [Aylesbury] went to worke analytically and wrought thus. To finde
> $c = a$
> $d = b - a$

---

[67] According to Harriot's will, his manuscripts should have been kept by the Earl of Northumberland after the relevant material had been published, but Aylesbury retained the papers in the hope of publishing further extracts; see the letter from Aylesbury to Percy on the subject in BL, MS Add. 4396, fo. 90, and also Letter 114 from Cavendish to Pell.
[68] BL, MS Add. 4419, fo. 139.

A[ylesbury]. Mr. Harriot used y^e character $=$ for equall, not $y^s =$ which Warner and Oughtred use, because y^t in Vieta is used for a difference.

P[ell]. True, when it is unknown whether [which of the two] be greater, I remember it.

$\therefore b.a :: a.b - a$
$\therefore aa = bb - ba$
$\therefore aa + ba = bb$

Heere I interrupted him,

P. You are farre enough already S^r for a numerous exegesis.[69]

A. No.

P. Yes, I can exhibit a by y^t equation, for it is no other y^n Harriot's $aa + da = ff$.[70]

The equation $aa + da = ff$ is Problem 2 of the 'Exegetice numerosa', the part of the *Praxis* that deals with the numerical solution of equations, but Aylesbury was interested in an algebraic rather than numerical solution, and continued:

A. True. Let us goe on. Let b be equall to $2c$.

$b = 2c$

Heere I was not willing to stop him and say why doe you take $b = 2c$. But let him goe on. (More he might have sayd $4cc$ for bb, but he changed not that.)

A. $\therefore aa + 2ac = bb$
$\therefore$ (adde cc to both)
$\therefore aa + 2ca + cc = bb + cc$

Aylesbury had now 'completed the square', but (according to Pell) was somewhat confused about the next step until Pell set him right:

A. The former is a true square, therefore let y^e other be so, viz $= xx$, whose roote $x =$ to y^e roote of $aa + 2ca + cc$ which is $a - c$.
P. No S^r $a + c$.
A. No? are we mistaken? tis true.

$a + c = x$
$\therefore a = x - c$

and heere he left off as having found a.

Aylesbury may have been satisfied but Pell was not, and continued the working to the end, arriving at the answer he had first given.

---

[69]'Exegetice numerosa' is the title of the second part of Harriot's posthumous *Praxis*, which deals with the numerical solution of equations.

[70]Harriot, *Praxis*, pp. 119–21.

PART II: THE MATHEMATICS OF JOHN PELL   275

P. I pray S^r give me leave to draw downe this to y^e very termes given, viz to b, thus

P. $2c = b$
$\therefore c = b/2$
$\therefore cc = bb/4$
$\therefore bb + cc = bb + bb/4 = aa + 2ca + cc$
$\therefore \sqrt{}_q : bb + bb/4 := a + c$
$\therefore \sqrt{}_q : bb + bb/4 : -b/2 = a$

which is y^e very same equation with y^t which I found out above.
A. Well. But it needed not to be drawne downe so farre.
P. Yes, surely, to expresse y^e equation in y^e given termes.

A few days later, on 3 February, Pell returned to the problem, 'now my head is a little fitter for it', and referred both to the original source of the problem (Euclid II.11) and to Oughtred's treatment of it in the *Clavis* (Chapter 19.11).

Of particular interest in this conversation between Pell and Aylesbury is their method of solving the quadratic equation by 'completing the square'. Many years later, John Wallis, almost certainly acting on information from Pell, wrote that Harriot had 'a peculiar way of his own' for solving quadratic equations, and described exactly the method used (not quite correctly) by Aylesbury.[71] The method appears in Warner's papers also,[72] but it is not to be found in the *Praxis*, the only printed edition of Harriot's work. Pell's own method of solving the equation is not clear: the conversation only shows him following Aylesbury's. The general exchange, however, including the information about Harriot's equals sign, shows that Pell learned of Harriot's mathematics not only through the *Praxis* but also directly through discussions with Aylesbury, and later perhaps from Warner also, as well as from some of Harriot's original manuscripts still in Aylesbury's possession.

Pell evidently mastered Harriot's work much better than Warner had done. In 1639 Hartlib noted in his Ephemerides:[73]

> [Pell] hase finished those Problemes of Hariot which Warner should have perfected. Sir T. A. [Thomas Aylesbury] promised to let them

---

[71] Wallis described Harriot's method as follows:

> To each part of his Quadratick Equation, $aa \pm 2ba = \pm cc$; [Harriot] adds, the Square of half the Coefficient, bb, thereby making the Unknown part, a Compleat Square in Species equal to a Known Quantity.
> $aa \pm 2ba + bb = \pm cc + bb$
> And consequently, the Square Root of that, equal to the Square Root of this.
> $a \pm b = \sqrt{(\pm cc + bb)}$
> which being known; the value of a is known also.

J. Wallis, *A Treatise of Algebra* (London, 1685), p. 134.

[72] Northamptonshire Records Office, IL 3422, bundle VI, fo. 11.

[73] Samuel Hartlib, *Ephemerides* 1639, 30/4/9B.

have Harriot's papers but hee did solve them without them.

We cannot know for certain which problems Hartlib was referring to, but the most obvious candidates are the three problems that had defeated Warner in Section 3 of the *Praxis*. These problems, 19, 20, and 21, are concerned (as is the whole of Section 3) with the terms that disappear from a polynomial when certain conditions hold between the roots. For example, the equation $x^2+(a-b)x-ab = 0$ reduces to the simpler form $x^2 - a^2 = 0$ when $a = b$. For some biquadratic equations, Harriot had investigated the two independent conditions required for the simultaneous disappearance of two terms.[74] The biquadratic $(x-a)(x-b)(x-c)(x-d)$, for example, loses both the linear term and the cube term if $a+b+c+d = 0$ and $abc+abd+acd+bcd = 0$. Harriot's manipulations at this point were a little difficult to follow, because he made only the first condition explicit, while the second was buried in his working, which Warner was unable to disentangle. 'The reduction of these equations', wrote Warner, 'since they are delivered more obscurely in manuscript, must be referred to a better enquiry'.[75] Anyone who understood what Harriot was trying to show, however, would have had little difficulty in reconstructing his argument, even without recourse to the original papers, and it is likely that Pell was able to succeed where Warner had failed. If so, his ability must have seemed remarkable both to Aylesbury and to Hartlib.

It may have been such successes that led Pell, when he wrote about the *Praxis* to Hartlib in January 1639, to claim that he had 'illustrated and perfected that Treatise'. This comment in turn led Moriaen and perhaps others to expect from Pell a commentary on the *Praxis* (see above, Part I, ch. 2, n. 52), or even a revised version of it, but this was perhaps to read more into Pell's words than he intended. The full text of his letter shows that he was interested not so much in rewriting the *Praxis* as in rendering it unnecessary, a theme already familiar from the *Idea*. Pell wrote that Harriot's text was much clearer than Viète's but that:[76]

> ye printers faults being many & ye praecepts too concise, & their grounds not all shewed, y^t booke is not so usefull to learners as it might be, wherefore I have, I thinke, so illustrated it & perfected y^t Treatise y^t it may be easily conceived & so perfectly kept in mind y^t he y^t have once understood it, shall never neede to see it again, but may allwayes be able to write it againe completely though all copies of it were lost.

There is a small but important booklet amongst Pell's papers that suggests that he had indeed 'illustrated and perfected' some of the material from the

---

[74] BL, MS Add. 6783, fos. 172–4 (in reverse order) and fo. 204.
[75] Harriot, *Praxis*, p. 46.
[76] BL, MS Add. 4419, fo. 136.

PART II: THE MATHEMATICS OF JOHN PELL

Praxis in a systematic and concise form.[77] The booklet is undated but it is found amongst much other material on equations, all of which seems to have been written in 1638 or 1639. It is entitled 'Of Aequations' and begins by listing the different cases of quadratic and cubic equations that are to be treated. They are denoted simply by combinations of signs:[78]

Quadratic $+ - +$ & $+ + +$ pag. 2
or $- + -$ $- - -$
$+ + -$ & $+ - -$ pag. 1
or $- - +$ $- + +$

Cubicall (To avoide needlesse trouble I omit those which deny y^e highest power & affirme y^e homogeneum seeing y^s are y^e same with these y^t have all contrary signes)
$+ \,.\,.\, +$ & $+ \,.\,.\, -$ pag. 10
$+ \,.\, + +$ & $+ \,.\, + -$ 11   [etc.]

Pell noted that under each heading he would give examples of 'perfect', 'quasimperfect', 'imperfect', and 'impossible' equations, which he defined as follows:

> Perfecte which hath as many rootes (positive or privative, affirmative or negative) as the highest power in y^e Equation hath dimensions.
> A quadratic 2 A cubic 3 etc.
>
> Imperfecte which hath not so many rootes as dimensions, as A cubic having but one roote.
>
> Quasimperfecte, which is perfect but seemes to have a roote or two fewer than indeed it hath, because of their Equality as when y^e rootes of a Quadratic are 5, 5 or of a Cubic 5, 5, 5 or 5, 5, 6.
>
> Impossible which hath no roote at all neither positive not privative viz neither $=$ nor $>$ nor $<$ y^n 0.

All this is on the front cover in Pell's small neat writing. The text itself begins with the quadratic case $+ + -$ (Pell gives the example $+aa + 6a - 55 = 0$) or its equivalent $- - +$ ($-aa - 6a + 55 = 0$). The booklet does exactly what Pell claimed for it: each page is devoted to a particular type of equation, and lists perfect, quasimperfect and imperfect forms. Thus a typical page heading is '$+ - + -$ Or $- + - +$' with a list of forms and roots for each case. The influence of Harriot is everywhere obvious, not only in the notation but in Pell's system of classification, and in the way he handled pairs of equations with contrary

---

[77] BL, MS Add. 4413, fos. 198–224.
[78] The same method was used by Harriot, who similarly classified and counted the number of equations of each degree; see J. A. Stedall, *The Greate Invention of Algebra: Thomas Harriot's Treatise on Equations* (Oxford, 2003), pp. 175–7.

signs, including those that Harriot called 'secondary canonicals', in which the relationship between the roots causes one of the terms of the equation to vanish.[79]

The booklet has twelve numbered pages and appears to be complete, but seems to have been buried amongst Pell's papers ever since it was written. It is surrounded by many other notes on equations, mostly undated, but with two exceptions, dated 1 April 1639 and 2 August 1639.[80] The latter is simply a treatment of the equation $aaa - 147a = 286$. The earlier one is headed 'Of solving equations Aprill.1.1639', a title that leads one to expect a practical exposition, but in fact the subject matter is the definition and classification of equations, and how to reduce the infinite number of possible equations to a finite number of types. For biquadratics, for example, Pell found 46 types, 'as I have shewed in another paper above a year ago', and added 'Because men sticke at Cubes: I intend to try what may be done in y^m.' Many other pages are also devoted to various ways of classifying equations according to the number of terms, the number and signs of their roots, and so on. This was a subject that must have greatly appealed to Pell's desire to bring a substantial body of knowledge to order.

It seems that Pell, perhaps not surprisingly in view of his mastery of Harriot, soon came to be regarded as an expert on solving equations, and others besides Aylesbury presented him with problems that they found intractable. In May 1639, for example, John Harington asked him to solve the cubic equation that would now be written as $x^3 - 14x^2 + 64 = 0$. Harington presented it in old-fashioned cossist notation as $1C = 14Z - 64N$, but Pell rewrote it in Harriot's notation as $aaa = 14aa - 64$, and solved it by reference to the canonical equation given in the *Praxis*, page 12. For exercise he also solved it by Harriot's numerical method ('And so much the rather because they will be examples, in divers respects missing in Harriot'). For good measure he also added a method from Stevin, which he called 'A pretty processe!' (Thirty-five years later Pell's friends were still presenting him with difficult equations: In 1675 Michael Dary tried to solve $+yyyy + 8yyy - 24yy + 104y - 676 = 0$, but wrote that 'this soure crabb I can not deale with by no method'. Collins passed it on to Pell.[81])

All the equations discussed so far have been polynomials: quadratics, cubics, or biquadratics, but Pell also turned his hand to systems of linear equations with more than one unknown quantity. In May 1640, for example, he noted that he had sent Cavendish the solution to Problems 4 and 5 from the fourth book of Guillaume Gosselin's *De arte magna*. Problem 5, the last in Gosselin's collection of problems involving whole numbers, requires four numbers A, B, C, D, to satisfy the equations:[82]

---

[79] See, for example, BL, MS Add. 4413, fo. 204; Stedall, *Greate Invention*, 21, pp. 178–84.

[80] BL, MS Add. 4413, fos. 198, 199.

[81] BL, MS Add. 4425, fo. 57; printed (inaccurately) in Halliwell, *Collection of Letters*, 105.

[82] G. Gosselin, *De arte magna, seu de occulta parte numerorum, quae Algebra & Almucabala vulgo dicitur* (Paris, 1577), fos. 81^v–84.

$$1A + \tfrac{1}{2}B + \tfrac{1}{2}C + \tfrac{1}{2}D = 17$$
$$1B + \tfrac{1}{3}A + \tfrac{1}{3}C + \tfrac{1}{3}D = 12$$
$$1C + \tfrac{1}{4}A + \tfrac{1}{4}B + \tfrac{1}{4}D = 13$$
$$1D + \tfrac{1}{6}A + \tfrac{1}{6}B + \tfrac{1}{6}C = 13$$

(Problem 4 is of the same kind, but with only three unknown numbers.) Such problems were not common in sixteenth- and seventeenth-century texts, and the methods of solving them were *ad hoc*.[83] We do not know how Pell solved them for Cavendish on this occasion, but his three-column Method was well suited to such problems. The 'Three Men have Money' problem quoted above is of the same type, and we shall see others later.

In 1642, Pell tried a different, and for him somewhat novel, approach to solving equations, by means of geometrical construction. A carefully written sheet on the subject is dated 11486 (that is, from August 1642) and is headed 'Quadraticall Æquations solved by Delineation'.[84] (There is a first draft from the previous day, 11485, and another fair copy from the following day, 11487.) The text begins with some neatly drawn diagrams, after which Pell explained that: 'There are five sorts of Quadraticall Æquations, and they all have two rootes apiece, which are thus found'; he then gave instructions for solving 'The first sort, $aa = f$'; 'The second sort, $aa - da = f$'; 'The third sort, $aa + da = f$', and so on, all by straight line and compass constructions. He may have been inspired to this by his reading of Descartes' 'Géométrie', which gives the basis for such constructions, but it was an unusual subject for Pell who was by inclination an algebraist rather than a geometer, and nothing else on the subject survives amongst his manuscripts. The idea did appear again, however, in the work of his students and friends, and perhaps twenty years later Pell wrote: 'But *lines* may make a student of Algebra see a little more distinctly what is meant by *affirmative, negative & impossible* rootes; as also the necessity of admitting *equality* of explicatory rootes insome aequations'.

Before leaving Pell's algebra during this period there is one thing further that should be mentioned. In September 1642 Johann Moriaen proposed Pell's name to the Amsterdam publishers Elzevier as a possible editor for the collected works of Viète. Pell indeed knew Viète's work well, for at some point, probably in the late 1630s, he had written minutely detailed notes on it. Unfortunately the paper on which he wrote is now very badly damaged but enough remains to show that Pell used not only printed texts but also manuscript copies bound with them, all borrowed from an unnamed person. A note at the top of one page reads:

---

[83] There are similar examples, the first I have discovered in any European text, in Jean Borrell's *Logistica, quae et arithmetica vulgo dicitur in libros quinque digesta* (Lyon, 1559), p. 189.

[84] BL, MS Add. 4417, fos. 54–5, 58.

'Mr. Hartlib sent to me for Vieta & some [*words missing*] restore y^m to him of whom he had borrowed y^m for me'.[85] The owner was probably Aylesbury, and the books and manuscript copies perhaps those originally owned by Harriot.[86] Pell's response to Moriaen survives in a letter to Hartlib. He considered that the works of Viète needed 'illustration', 'confirmation', 'limitation', and 'vindication', together with a collation of the work of previous authors on the same topics; he put forward a number of names, though not his own, of those who might write the annotations, but ended by arguing that no notes at all would be better than bad ones; finally, he again held out some hope that he himself would improve on Viète:[87]

> Algebra is not the onely field wherein to exercise Analysis nor is Vieta's way the only Algebra, nor hath Vieta perfectly taught all that belongs to his way. As perhaps I should have cleerely shown ere this time, had I not been hindred by the continuance of innumerable diversions ... there lyes by me no small apparatus to this purpose, being the worke of all the spare minutes of many years.

Pell's letter was circulated to a number of recipients in the Netherlands, and helped to establish his reputation as an algebraist, yet closely read, the letter offers little but generalities and hopeful promises. Frans van Schooten was undoubtedly a more reliable choice as the editor of Viète's *Opera mathematica*, which was successfully published in 1646.

Not the least of the 'innumerable diversions' that hindered Pell during 1641 and 1642 was a project not at all connected with algebra, but a return to his early interest in mathematical tables. Walter Warner, one-time companion of Harriot and the eventual editor of the *Praxis*, had begun to work during the later 1630s on tables of antilogarithms, or as he called them 'analogics'.[88] Aubrey later wrote about Warner's tables but was puzzled as to their use:[89]

> Mr. Walter Warner made an Inverted Logarithmicall Table, i.e. whereas Briggs' table fills his Margin with Numbers encreasing by Unites, and over-against them setts their Logarithms, ... Mr. Warner (like a Dictionary of the Latine before the English) fills the

---

[85] BL, MS Add. 4474, fos. 78–9.

[86] Several of Viète's books were listed amongst Harriot's possessions at the time of his death: see BL, MS Add. 6789, fos. 448–50.

[87] BL, MS Add. 4280, fo. 206^v.

[88] John Wallis was later informed by Pell that the tables were probably begun by Harriot and then continued by Warner: see John Wallis, *Opera mathematica*, 3 vols. (Oxford, 1693–9), ii, p. 63. This may have been true; Harriot certainly experimented with the method of constant differences that Pell and Warner later used to interpolate their tables (see for example BL, MS Add. 6782, fos. 112–21).

[89] Aubrey, *'Brief Lives'*, ii, 292.

PART II: THE MATHEMATICS OF JOHN PELL 281

> Margin with Logarithmes encreasing by Unites, and setts to every one of them so many continuall meane proportionalls between one and 10 ... These, which, before Mr. John Pell grew acquainted with Mr. Warner, were ten thousand, and at Mr. Warner's request were by Mr. Pell's hands, or direction, made a hundred-thousand. *Quaere* Dr. Pell, what is the use of those Inverted Logarithmes?'

Aubrey's query was justified, for ordinary tables of logarithms could be used as tables of antilogarithms simply by reading them backwards. The rest of Aubrey's account is accurate; when Warner first met Pell he had already prepared a table of ten thousand antilogarithms but planned to extend it to one hundred thousand. Their initial meeting appears to have taken place on 15 November 1639 and, according to Pell's notes, Warner said:[90]

> That he had calculated his table of analogicalls an[*words missing*] speedy extending it to 100 000 continuall proportionalls.

Five days later Pell wrote that Warner 'shewed me 10 thousand of y^m done'. Pell rapidly became interested in the project. Warner estimated that the extended table would need '4 quire of paper', but Pell made his own calculation:[91]

> Mr. Warner says his table of 100 000 Analogicall numbers (being each of them 10 places; the first being 1 000 000 000; the last being 10 000 000 000) will take up 4 quires of paper, that is 100 sheets.
>
> Suppose it were 99.999| 999 999 9999|
>   100.000|1000 000 0000| ...

Pell went on to calculate that with such a layout, the work would require no more than 2 quires and 5 sheets of paper. Pell's estimate gives us a useful hint of the intended layout and shows that the antilogarithms were to be calculated to 10 decimal places. Some scraps of paper written by Pell over forty years later, however, tell us rather more about Warner's plans.

For any number α, its antilogarithm (working, as Warner did, in base 10) is 10^α. Since, for example, $10^{2.3} = 10^2 \times 10^{0.3}$, the antilogarithm of 2.3 is easy to find from the antilogarithm of 0.3, and so on. Thus all that is required is a table of antilogarithms of numbers between 0 and 1. According to notes written by Pell in 1683,[92] Warner's first table, or Canon, had only 10 subdivisions: 0.0, 0.1, 0.2, ..., 0.9, 1.0. His second canon had 100 subdivisions: 0.00, 0.01, 0.02, ...; his third canon had 1000 subdivisions, and his fourth, presumably the one he showed to Pell in 1639 had 10 000. His plan was therefore to extend it to the next level, that is, to create nine new entries between every existing entry in the fourth canon.

---

[90] BL, MS Add. 4474, fo. 77.
[91] BL, MS Add. 4426, fo. 209.
[92] BL, MS Add. 4424, fos. 1–3.

The mind reels at the labour involved, but there was one redeeming factor: the entries in Warner's fourth canon (according to Pell in 1683) were calculated to 13 decimal places, whereas those in the new fifth canon were to be to 10 decimal places only (as Pell's estimate of the paper required confirms).

Warner was already eighty years old and the task must have seemed daunting, and eventually Pell was engaged to help him. Pell was to receive remuneration for the work, and so kept an extraordinarily detailed record of the time he spent on each part of the task, down to the minutiae of folding and ruling of the paper.[93] He began:

1641

June
17 Thursday spent in seeking out some to rule their paper.
18 Friday spent in attending T. Browne for the businesse.

Pell spent the next day, Saturday, folding, pricking, sewing, and ruling some of the sheets himself, and calculated that Mr. Browne's share of the work would take 224 hours or 18 days and 8 hours. On Monday he began entering figures into the ruled columns:

21 Monday 1h[our] the marginall numbers of 7− columnes in tab-
ulis ipsis
2h _____ 13+ _____

whence I inferre y^t 1000 or 2000 columnes may be so numbered in 150 houres or 12 dayes $\frac{1}{2}$ at 12 h in y^e day.
3h the perpendiculars of black lead in 15 pages.

By Friday he was ready to begin the calculations themselves. His method was based on the fact that for antilogarithms of numbers between 0 and 1, the second differences increase very slowly, and over a limited range can be taken to be equal. Pell labelled his differences with the letters shown in the following scheme, so that a and e were the quantities he needed to calculate.

---

[93] BL, MS Add. 4365, fos. 36–9.

PART II: THE MATHEMATICS OF JOHN PELL   283

Antilogarithm	1st difference	2nd difference
$N + 10a + 45e$		
	$a + 9e$	
$N + 9a + 36e$		e
	$a + 8e$	
$N + 8a + 28e$		e
	$a + 7e$	
$N + 7a + 21e$		e
	$a + 6e$	
$N + 6a + 15e$		e
	$a + 5e$	
$N + 5a + 10e$		e
	$a + 4e$	
$N + 4a + 6e$		e
	$a + 3e$	
$N + 3a + 3e$		e
	$a + 2e$	
$N + 2a + 1e$		e
	$a + e$	
$N + a$		e
	a	
N		

Pell had every tenth entry, from which he could calculate the differences labelled f, h, and g in the table below. For the quantities $10a + 100e$ and $10a$, he used the labels 'upper $10a$' and 'lower $10a$', respectively.

Antilogarithm	1st difference	2nd difference
$A + 20a + 190e$		
	$f = 10a + 145e$	
	$=$ 'upper $10a$' $+ 45e$	
$A + 10a + 45e$		$g = 100e$
	$h = 10a + 45e$	
	$=$ 'lower $10a$' $+ 45e$	
A		

From the values of f (or h) and g, Pell could find a and e and therefore interpolate nine new antilogarithms between each existing pair.

Pell's record of his first day of calculation is as follows:

    25 Friday
    1h I drew y^e black lead perpendiculars in 5 calculatory pages
       I drew y^e Inky transposall lines in 5 calculatory pages

I transcribed into those 5 pages all y^e 50 analogicalls which lead y^e worke (y^{eir} first differences wanting but 10).

2h I calculated all y^e 2nd differences ($g = 100e$)
    subtracted their tenths ($10e$)
    bisected y^{eir} remainders $45e$
    subtracted those halfes from both f and h to find y^e upper and lower $10a$
    Tried y^m by adding g to y^e lower $10a$.
    Found and wrote downe a and e in every semi-columne of ye 500 in a little lesse y^n an houre.

3h I calculated 90 and y^e differences of y^e last 10 in one houre.
4h I calculated 80 and y^e differences of y^e other 20 in one houre.
5h I calculated a round century.
6h I calculated 90 and y^e differences of ye last 10 in one houre.
7h calcul [sic] all ye differences and all ye numbers save 12
8h transcribed 250 ⎱ 500 in 2 houres just
9h transcribed 250 ⎰ (see ye beginning of y^s day)

The first note about payment appears at the end to the following week. Pell mentioned no names, but from what follows later, it is clear that the person employing him was Warner himself:

3 July Saturday
He offered me 40^l [£40] for ye doing of y^m to fit y^m for ye presse.

On the following Monday, however, Aylesbury stepped in and made a better offer:

Monday July 5 Sr Thomas Ailsbury told me he would overrule him for 50^l so now y^t is 10s a sheete or 1000, y^t is 12d a hundred.

A few days later Pell had completed the first 2000 'analogicals' and noted that it had taken him 23 hours for the calculations, plus 12 hours for the transcribing. By 22 July he had completed the tables up to 4000, and reckoned he could do the next 4000 entries in 80 hours or 11 days, working for $7\frac{1}{4}$ hours per day. He exceeded his own expectations, for by 2 August he had completed the calculations up to 8000 in a mere 62 hours. By mid-August, however, he was complaining about the expense. Warner's name is now written in connection with the payments, but is then roughly crossed out leaving only 'Mr. W':

Aug.19

14 quire or 350 sheets 350^d or 29^{s}2^d. I have had but 20^s of Mr. W[arner] for it and have now thereof laid out 6^{s}6^d from 13^{s}6^d.

Perhaps it was the difficulties over payment that brought the work temporarily to a standstill. Warner asked for everything to be sent back to him, and on 27

August Pell delivered the original tables together with new tables completed as far as 8000, and his remaining stock of blank or ruled paper. At the beginning of September, however, he was persuaded to continue:

> September 8 He sent for me againe praying me to fill up ye sheetes, that is, to calculate and write faire 1750 more, viz to 9750.
>
> Sept 20 I began it (having layd away all other businesses) I sowed together ye tabular paper ...

In a further 29 hours he had reached the stipulated 9750, but the next sections of the table were evidently completed by two other calculators, a Mr Watts and a Mr Turner. Pell's record begins again in November:

> Nov 17 Coming home at night I found heere left 3 sheetes of 1200 calculated by Ed Watts (according to my precepts in writing and instructions viva voce) viz from 9750 where I left to 10950.
> 2) My 5th, 6th, 7th, 8th, Chiliads in the calculatory paper.
> 3) Mr. Warners own table cut off at 5100.
> 4) Calculatory sheetes having all the borrowed analogicalls (ten in a side, ten differences in a side) written into them viz from 10950 to 52350 that is 41400 Analogicalls.

Pell spotted some unwanted repetitions in the sheets done by Turner and sent him home to examine them again, and by December Turner and Watts had returned all their sheets. Meanwhile Pell himself had begun on the third myriad, from 20 000 onwards. When he first began the work he had hoped to have it ready for the press by Christmas,[94] but at the end of 1641 it was only about a quarter complete. Pell must have devoted most of his time to it at the beginning to 1642 because when his detailed record ended on 11 May he had completed calculations up to 74 950:

> May 11 I had finished 5200 viz from 69750 to 74950. There remaines 25000 viz to 99950 for ye last 50 are done. These, I reckon will cost me 225 houres viz 125 for ye perfecting ye differences and 100 for perfecting ye Analogics.

There follows a rough calculation that 225 hours would mean 45 to 56 days, depending on whether he worked 4 or 5 hours a day (rather less than the 7 or more hours a day he had optimistically but unrealistically planned when he first began). In fact he finished two months later, in mid-July,[95] but by then civil war was looming and Pell feared that 'these times must be a little quieter before it goe to ye presse'. For once, perhaps, external circumstances rather than Pell's own reluctance really did hinder the project.

---

[94] Pell to Cavendish, Letter 2.
[95] BL, MS Add. 4418, fo. 89.

The completed tables were returned to Warner who, according to Collins later, had paid just over £100 to Pell for their completion,[96] but Warner's death in 1643 and Pell's move to Amsterdam in December of that year brought to an end any immediate hope of publication. John Wallis in 1693 wrote, correctly, that the tables had been completed fifty years earlier, and added that even as late as the 1670s Pell still hoped to publish them.[97] Pell had turned down an opportunity to do so, however, in 1652, when Herbert Thorndike gave him some of Warner's papers, including the tables, specifically requesting that he would look them over with a view to publication. Unfortunately, Pell discovered that the tables were incomplete: the bundle containing antilogarithms for 29 750 to 39 750 was missing.[98] Once again the papers were put aside. In 1667 Thorndike gave some of Warner's papers to John Collins for safe-keeping, and the inventory includes 'The faire copy of a canon of 100,000 logarithmes',[99] but this and other items were later crossed out, presumably because they were returned to Thorndike or his Executors. After Thorndike's death in 1672, the papers passed to Richard Busby of Westminster School, and in 1683 Pell inspected them again.[100] Although the calculations for 29 750 to 39 750 were still missing, he did go as far as sketching a title page for publication of Warner's complete fourth canon, the one containing 10 000 entries: 'Canonis Analogici Originalis, Ex numeris proportionalibus 10 000', but as for so many of Pell's projected works, a title page is all there is to show. Pell's hours and hours of labour came to nothing: the fifth canon of antilogarithms has never been found.

Amsterdam, Breda, and Zurich, 1644–1658

Only a few months after Pell was installed as Professor of Mathematics at the Athenaeum Illustre in Amsterdam, the Danish astronomer Christian Severin Longomontanus published his attempted quadrature of the circle, *Rotundi in plano, seu circuli absoluta mensura*, and Pell, fearing, he said, that silence would be taken for approval, felt bound to refute it. Pell produced his counter-argument almost immediately in a single sheet, the 'refutatiuncula',[101] and the Amsterdam bookseller Willem Jansz Blaeu appended it to his unsold copies of the *Rotundi in plano*.

Longomontanus had found the ratio of the circumference of a circle to its

---

[96] Rigaud, *Correspondence*, i, p. 215, and ii, p. 218. The final version of the second letter is printed in Turnbull, ed., *Gregory Tercentenary Volume*, pp. 179–80.

[97] Wallis, *Opera mathematica*, ii, 63. See also Dodson, *The Antilogarithmic Canon*, p. ix.

[98] BL, MS Add. 4279, fos. 275–8. Thorndike's letter of 23 December is printed in Halliwell, *Collection of Letters*, p. 94.

[99] 'An inventorie of the papers of Mr. Warner', BL, MS Add. 4394, fo. 106, reprinted in Halliwell, *Collection of Letters*, p. 95.

[100] BL, MS Add. 4424, fos. 1–3.

[101] BL, MS Add. 4280, fo. 205, reproduced in van Maanen, 'The Refutation', pp. 324–5.

diameter as $\sqrt{18\,252:43}$, or approximately $3\,141\,859\,604\,427 : 1\,000\,000\,000\,000$.[102] This was significantly larger than the upper bound of $3.141\,592\,653\,897\,33$ calculated by van Ceulen, but Longomontanus claimed that van Ceulen's methods, based on repeated bisection of arcs and chords, were incorrect, because, he argued, one cannot find the weight of a piece of wire by dividing it into a large number of very small pieces and summing the weights of the tiny particles. He also rejected methods based on trigonometrical tables, on the grounds that the tables themselves were prone to accumulated error. In his refutation, Pell did use repeated bisection, to find the length of a circumscribed polygon of 256 sides, but avoided the use of trigonometrical tables by using a new formula for calculating tangents of double- (or half-) angles. For Pell the tangent of the angle θ was the length T in Fig. 2, and the tangent of the double angle, 2θ, was D, both angles being subtended at the centre of a circle of radius R. According to Pell, D could be calculated from the formula:

$$D = \frac{2RRT}{RR - TT}. \tag{1}$$

which, if we write $\tan\theta = T/R$ and $\tan 2\theta = D/R$, gives the more familiar modern form:

$$\tan 2\theta = \frac{2\tan\theta}{1 - \tan^2\theta}.$$

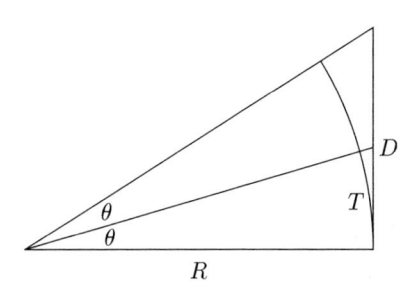

FIG. 2.

To find the perimeter of a circumscribing polygon of 256 sides Pell needed the tangent of $\frac{180}{256}°$ (or $\frac{45}{64}°$, or $0° \, 42\frac{3}{16}{}'$), so he needed to use the formula the other way round. Starting from $\tan 45° = 1$ he could find the tangent of the half-angle, $22° \, 30'$, then of $11° \, 15'$, etc. until after six bisections he arrived at the tangent of $0° \, 42\frac{3}{16}{}'$. In the 'refutatiuncula', Pell claimed that the value of $\tan 0° \, 42\frac{3}{16}{}'$ was less than $0.012\,272\,5$, and demonstrated (using formula (1) repeatedly) that any larger value led to the impossible conclusion that $\tan 45°$ is greater than 1. From

---

[102] Finding the length of the circumference was equivalent to finding the area of the circle, since it was well known that the area of a circle is the same as that of a right-angled triangle in which the two shorter sides are the lengths of the radius and circumference respectively (since $\frac{1}{2} \times r \times 2\pi r = \pi r^2$).

the upper bound of 0.01227 25 he was able to show that a circumscribing polygon with 256 sides around a circle of unit radius has length less than 2×3.14176, less than the circle itself according to the calculations of Longomontanus, who had estimated the circumference to be 2×3.14186. The contradiction immediately invalidated Longomontanus's result.[103] In the 'refutatiuncula' Pell did not say how he found the critical value of 0.01227 25 but his handwritten notes show that he calculated it, exactly as one would expect, by inverting formula (1) to give a quadratic equation for T, written by Pell as:[104]

$$TT + \frac{2RR}{D}T = RR.$$

The solution of this equation, again according to Pell, is:

$$T = \frac{RE - RR}{DD}$$

where $EE = RR + DD$, that is, from elementary trigonometry, $E = \sec 2\theta$.

Pell, of course, was interested only in the positive solution. In his notes we have his calculations, beginning from $\tan 45°$, of $\tan 22° 30'$, $11° 15'$, ... down to $\tan 0° \frac{675}{1024}'$ (or $\frac{45°}{2^{12}}$), whose tangent he calculated (slightly inaccurately) as less than 0.00019 (it is in fact 0.0001917).[105]

Pell was clearly proud of his double-angle formula, 'my fundamentall theorem' as he called it, yet he gave no proof of it, nor even his calculations for the critical value of 0.01227 25. Longomontanus, believing that the use of tangents implied the use of tables, refused to accept it and the affair dragged on for three more years as Pell attempted, using Cavendish as his intermediary, to gain support from some of the most renowned mathematicians in Europe: Hobbes, Fermat, Mydorge, de Beaune, de Billy, Torricelli, Cavalieri, Glorioso, and later Descartes. Pell requested from these men not only an endorsement of his methods but their own proofs of the 'fundamental theorem', and several of them obliged. Cavendish himself offered two proofs: the first (in August 1645) was hopelessly inept; a sounder proof, which Pell encouraged him to develop, followed in December 1645,[106] but Cavendish later admitted that it was not entirely his own work. Pell never did produce his own proof. The support that he, or rather Cavendish, had collected from the European mathematical community was eventually published in *Controversiae pars prima* (1647), and with the death of Longomontanus later the same year the controversy finally faded. Pell must have hoped it would enhance his reputation but it more likely had the opposite effect; his double-angle formula was a useful discovery but hardly ground-breaking, and the quadrature proposed by Longomontanus, even by contemporary standards, was simply not significant enough to justify the time and energy that Pell expended on it.

---

[103] For a detailed account of Pell's refutation see van Maanen, 'The Refutation'.
[104] BL, MS Add. 4418, fo. 228.
[105] BL, MS Add. 4417, fos. 265–6, 280–1.
[106] Van Maanen, 'The Refutation', pp. 333–5.

In the meantime, several rather more important projects that Pell was engaged on failed to come to fruition. The first of these was his book on 'analytics', or algebra. As we have seen, there are many signs that Pell thought about algebra intensively during 1638 and 1639, and by 1641 he had invented his three-column algebraic method. In 1642 he had briefly turned also to solving equations by geometrical construction, but although he read and even translated Descartes' 'Géométrie' Pell never really took up the new possibilities of analytic or algebraic geometry; his main interest was in the traditional subject matter of algebra, the study of equations. What would have been the final shape and content of an algebra text by Pell in the 1640s, we cannot know, for if begun it was never completed. Letter after letter from Cavendish from 1641 onwards refers to Pell's 'intended analyticall worcke', which Cavendish supposed would surpass any other book then in print, but he was to be disappointed, for Pell published nothing on algebra during Cavendish's lifetime.

Another major subject that began to occupy Pell in Amsterdam was the *Arithmetic* of Diophantus (c. AD 250). The six surviving books of the original thirteen of the *Arithmetic* made a profound impression on European mathematicians when they were rediscovered and translated in the sixteenth century. Raffaele Bombelli, who read the *Arithmetic* in manuscript, was the first to incorporate some of the problems into a printed text, his *L'algebra* of 1572. So later did other algebraists: Simon Stevin in *L'arithmétique* of 1585 and Viète in *Zeteticorum libri quinque* of 1593. Pell knew about Viète's *Zeteticorum* from the papers he had borrowed from Hartlib in the late 1630s, but he mostly drew on two later texts: Claude Gaspar Bachet's *Diophanti alexandrini arithmeticorum libri sex*, published in 1621, and Albert Girard's new and annotated edition of Stevin's *L'arithmétique* in the *Oeuvres*, published in 1634.[107] The tradition of treating the problems of Diophantus by the methods of contemporary algebra was therefore already well established and well known to Pell when he proposed in 1644 to produce 'a New edition of Diophantus wherein I would correct ye translation & make new Illustrations after my manner'.

Pell's 'manner' was his three-column Method, which by now he used for almost all his mathematical working. It was well suited to the problems of Diophantus, in which the starting point and the objective were always clearly stated, and in MS Add. 4419, in particular, there are numerous sheets of working on problems from the *Arithmetic*: Problems 1, 41, 45, 46 from Book IV;[108] Problems 18, 19, 32, 33 from Book V;[109] and Problems 10 and 24 from Book VI.[110]

---

[107] Pell's copy of Bachet's *Diophanti Alexandrini ... libri sex* is in the Busby Library, Westminster School, pressmark I F 21, with inserts at Problems IV.31, IV.36, and V.8, and frequent underlining of the text in red ink throughout. His copy of Girard's edition of Stevin's *Oeuvres* is also in the Busby Library, pressmark I F 10.

[108] BL, MS Add. 4419, fos. 222, 242, 243, 232^v, 237, 223.

[109] BL, MS Add. 4419, fos. 240, 247–8, 321. There are also many notes on Diophantus V.19 in BL, MSS Add. 4417, fos. 331–93, and Add. 4430, fos. 86–131.

[110] BL, MS Add. 4419, fos. 238, 249^v.

Although there are frequent references to both Bachet and Girard, all the problems are reworked by Pell in his own way, but the full edition that he promised was never published. Pell reported to Cavendish that he lost heart in the matter after his meeting with Descartes in 1646, but he may have abandoned it for quite other reasons. Aubrey suggested that Pell kept the work by him for many years but that it was left behind in Cheshire. It has never been found.[111]

It was probably his work on Diophantus that gave rise to Pell's special interest in indeterminate problems, in which the given conditions are too few to determine a unique solution. A problem of this kind has in general infinitely many solutions. If the problem is geometric the solutions may be described by a 'locus', that is, a line, curve, surface, or area containing the range of possible answers; if algebraic, arbitrary parameters may be introduced, giving rise to infinitely many solutions as the parameters vary. Diophantus, however, and also Pell, imposed the extra condition that solutions should be rational, in which case a problem might have only finitely many solutions, or even none at all.

Pell's three-column Method was easily adapted to indeterminate problems. He began, as usual, by listing all the unknown quantities on the left and the given conditions on the right, but since the conditions were fewer in number than the unknowns, the list on the right was shorter than the list on the left, so each absent condition was marked by the symbol $(*)$.[112] At a later stage each $(*)$ could be replaced by an arbitrary condition. The process, like so much of Pell's mathematics, is logical and systematic. There are several examples of it in *An Introduction to Algebra* where Pell explained it at length, but a much earlier example was given to Cavendish in Amsterdam in March 1645.[113] As an example of a 'determined' problem, Pell offered Cavendish a system of linear equations similar to those he had found in Gosselin's *De arte magna* in 1640:

$$3a - 4b + 5c = 2$$
$$5a + 3b - 2c = 58$$
$$7a - 5b + 4c = 51$$

He also, however, gave Cavendish an example of a problem not fully determined:

$$5a + 3b - 2c = 24$$
$$-2a + 4b + 3c = 51$$
$$(*)$$

---

[111] There is a handwritten version of the first six books of Diophantus in the Macclesfield Collection in Cambridge University Library, CUL, MS Add. 9597/5/1. The text is in Pell's three-column style, but is not in his handwriting, and the writer has not so far been identified.

[112] Asterisks were used by Descartes in 'La Géométrie' to stand for an absent term in a polynomial equation, and this may have given Pell the idea of using them in a similar way here.

[113] For Pell's copy of the problem see BL, MS Add. 4415, fo. 200 and Illustration 3, and for Cavendish's see BL, MS Harl. 6083, fo. 129. For a later reference to it see Letter 51.

PART II: THE MATHEMATICS OF JOHN PELL   291

In this second case, Pell imposed the conditions $a, b, c > 0$, and then investigated the upper limits for a, b, and c, which he wrote using Harriot's inequality signs, for example, '$0 < a$, $a < 15\frac{9}{11}$'.[114] Pell worked another example for Cavendish early in 1646, when Cavendish asked him to explain the following problem from Hérigone: 'Thirty people, men, women, children, have spent 30 sols, or 360 deniers, but in such a way that each man pays 5 sols, or 60 deniers, each woman 10 deniers, and each child three deniers. The question is, how many men, women and children were there?'[115] Pell began the solution by letting h, f, e stand for the numbers of *hommes*, *femmes*, *enfans*, with $(*)$ to denote the absent third condition:[116]

$$\begin{array}{rl} h = ? & \begin{vmatrix} 1 \\ 2 \\ 3 \end{vmatrix} \begin{array}{l} h + f + e = 30 \\ 60h + 10f + 3e = 360 \\ (*) \end{array} \\ f = ? & \\ e = ? & \end{array}$$

As with the problem he had given to Cavendish a year earlier, Pell investigated the limits on h, f, and e and so discovered the only possible integer solution: $h = 4$, $f = 6$, $e = 20$.

Another reference to indeterminate problems is to be found in one other seventeenth-century text, the *Algebra ofte stel-konst* (Haarlem, 1661) of Gerard Kinckhuysen:[117]

> But one has to pay attention, that for every Question one has as many Equations as one has introduced unknown Quantities, which one then transforms to a single [equation], in which only one sort of unknown quantity remains. If it then happens that one is unable to find as many equations as one has introduced unknown Quantities, even if one does everything that could possibly be done, it shows that there are not enough data given, and one may then for each of such unknown quantities for which one cannot find an Equation,

---

[114] This is a typical problem in 'linear programming', but Pell did not develop it in any systematic way.

[115] 'Trente personnes, hommes, femmes, enfans, ont depensé 30 sols, ou 360 deniers, en sorte neantmoins que chaque homme paye 5 sols, ou 60 deniers, chaque femme 10 deniers, & chaque enfant trois deniers: la demande est, combien il y avoit d'hommes, de femmes et d'enfans?'; Hérigone, *Cursus mathematicus*, vi, 92. See Letter 51.

[116] BL, MS Add. 4415, fos. 195–200.

[117] 'Maer men heeft hier op te letten, dat men op yder Questie, soo veel Verghelijckinghen moet vinden, als men onbekende Quantiteyten gesteldt heeft, die men dan tot soodanigh een eenighe brenght, in welck maer eenderley onbekende quantiteyten resteeren. Soo't dan ghebeurt, datter soo veel Verghelijckinghen niet te vinden en zijn, al is't dat men al doet, wat daer toe ghedaen kan worden, dat betoont datter te weynigh ghegheven is, ende men mach dan voor yder van soodanighe onbekende quantiteyten, daer men gheen Verghelijckingh toe vinden kan, naer dat het de natuer van de Questie vereyst, sulcken ghetal stellen, als men begheert, waer uyt volght, dat dierghelijcke Questien veel uyt-komsten konnen hebben': Kinckhuysen, *Algebra ofte stel-konst*, p. 96 (translated by Jan van Maanen). See also I. Newton, *The Mathematical Papers*, ed. D. T. Whiteside, 8 vols. (Cambridge, 1967–81), ii, p. 354, and Scriba, 'Mercator's Kinckhuysen-Translation', esp. p. 48.

according to the nature of the Question, put whatever number one likes, from which it follows, that such Questions have many solutions.

Discussion of indeterminate equations in seventeenth-century texts is rare, and Kinckhuysen's understanding is so similar to Pell's as to suggest that there may have been some communication between them on this subject. This would certainly not have been impossible because Kinckhuysen lived at Haarlem, only a few miles from Amsterdam.[118] He was eighteen years old when Pell arrived in the Netherlands, but had already published a treatise on quadrants (a subject familiar to Pell from his own student days), and he could well have studied under Pell at the Athenaeum Illustre. Further circumstantial evidence suggesting a link between Pell and Kinckhuysen will be discussed below.

If the *Arithmetic* of Diophantus proved for Pell, as it did for so many European mathematicians, a fruitful source of inspiration, another Greek text posed him a more difficult challenge. In 1644 Pell borrowed from Christian Ravius a manuscript copy of an epitome by 'Abd al-Malik al-Shirazi (*fl. c.*1150) of an Arabic version of Books V–VII of the *Conics* of Apollonius, and began to translate it into Latin. Books I–IV of the *Conics* had been translated into Latin in the sixteenth century by Johannes Baptista Memus in Venice in 1537, and by Federico Commandino in Bologna in 1566, but Books V–VII were extant only in Arabic, while Book VIII was lost altogether. There were, however, clues to the contents of Books V–VIII in the Lemmas given by Pappus in his *Collections* (*c.* AD 320), and it was not impossible for someone with an elementary knowledge of Arabic and a good understanding of the mathematics to make sense of the Arabic text without being an expert in the language (as Edmond Halley proved with great success some sixty years later). But while Pell was at work Jacob Golius (Gool) was also translating the same material from the earlier Arabic version made by Thābit ibn Qurra (836–901) and others, and unfortunately Pell was persuaded by Golius to put his own translation aside so that Golius could complete his. In the end neither was ever published, but we do have just one paragraph of Pell's translation that we can compare with the corresponding passage by Golius. According to Mersenne in 1644, Golius translated the first proposition of Book VII into Latin as follows:[119]

> Propositio 1. Si axis parabolae producatur ultra verticem, donec aequatur lateri recto, et a puncto quovis sectionis in axem ducatur perpendicularis, recta, sectionis punctum illud in sectione cum vertice connectens, poterit rectangulum contentum sub recta inter verticem et perpendicularis incidentiam interiecta, et tota ab hoc

---

[118] For biographical information on Kinckhuysen see C. M. P. M. Kempenaars, 'Some New Data on Gerard Kinckhuysen (*c.*1625–1666)', *Nieuw archief voor wiskunde*, 4 (1990), pp. 243–50.

[119] M. Mersenne, *Universae geometriae mixtaeque mathematicae synopsis* (Paris, 1644), p. 274.

incidentiae puncto per verticalem continuata.

If the axis of a parabola is produced beyond the vertex, until it equals the *latus rectum*, and from any point on the curve a perpendicular is dropped to the axis, [the square on] the line, connecting that point on the curve with the vertex, will be as the rectangle made by the line between the vertex and the foot of the constructed perpendicular, and the whole line drawn from that foot through the vertex.

In October 1644 Pell sent Cavendish his own translation of the same proposition:[120]

> Sit parabole, cujus axis *ad* qui producatur ad *c*, ita ut *ac* fiat aequalis lateri recto. Ex puncto *a*, ducatur linea recta *ab* secans parabolum ubilibet ut in puncto *b* unde demittatur perpendicularis *bd* in axem *da*. Dico Quadratum *ab* esse aequale rectangulo sub *cd* in *da*.

> Suppose there is a parabola, of which the axis *ad* is produced to *c*, so that *ac* is equal to the *latus rectum*. From point *a*, let there be drawn a straight line *ab* cutting the parabola anywhere, as in point *b*, whence let a perpendicular *bd* be dropped to the axis *da*. I say that the square of *ab* is equal to the rectangle [*sc.* product] of *cd* and *da*.

Pell's translation is, perhaps not surprisingly, more algebraic in style, and therefore clearer and more concise than that of Golius. Unless Mersenne has transcribed Golius' version incorrectly, Pell's is also more accurate. Pell claimed in his letter to Cavendish that he had translated the whole of Book V, and that he planned to complete Books VI and VII also, 'though', he complained, 'I have now very little or rather no Leisure'. Despite his usual complaint about lack of leisure, Pell completed the translation over the next few months, and in May 1645 was discussing the engraving of the diagrams when Golius dissuaded him from pursuing the matter further.[121] Unfortunately, as Pell had feared, Golius also failed to finish the task, so neither version was printed, and Pell's translation has never since been found. Amongst his surviving papers the only significant references to Apollonius are some notes on two propositions from Book I, on the definition of a parabola (I.11) and on the construction of tangents to a parabola (I.33) (based on the translation of Commandino).[122] Some of the notes are written in Pell's three-column Method and were therefore possibly written during the 1640s, but more than that it is impossible to say.

---

[120] BL, MS Add. 4280, fo. 109^v; see Letter 23 for Pell's diagram and his proof of the proposition.

[121] See Letter 35.

[122] BL, MSS Add. 4407, fos. 131–131^v; Add. 4413, fo. 401; Add. 4415, fos. 241–8.

Before leaving the subject of Pell's work on Apollonius, it may be noted that once again there is a possible connection with Kinckhuysen. In 1654 Kinckhuysen published a selection of the mathematical work of his teacher, Peter Wils, and amongst the contents is a treatment of Book V of Apollonius. Kinckhuysen's modern biographer, Kempenaars, was struck by the fact that Kinckhuysen was familiar with material then available only in manuscript, but did not investigate the matter further.[123] It is now possible to add that the only Apollonius manuscripts in the Netherlands before 1654 were the two used by Golius and Pell (plus a copy of the latter, also owned by Golius), and that Pell worked on his during 1644–5, precisely the period when Kinckhuysen was most likely to have met him.

In 1646, Pell was invited to become Professor of Mathematics at the new Illustre School in Breda and was to remain there until 1652. In his inaugural address he promised once again to publish his algebraic Method, and there are continuing references in his correspondence with Cavendish to the various editions he was supposed to be preparing for the press, but he published nothing further during his Breda years, and there is no sign of any new mathematical investigation. The evidence for this period is patchy, but one senses that Pell had lost the aspirations and creativity of his twenties, perhaps because he lacked the stimulus and encouragement from Hartlib and his friends, or simply because he was caught up in the pressures of teaching and supporting his family. Almost everything we know of Pell's mathematical activity during his years in Breda comes from a single source, his letters to Charles Cavendish. In the course of the correspondence, Cavendish repeatedly asked Pell for explanations of difficult or obscure points in the books he was reading, and Pell replied conscientiously, but there are none of the grand plans and schemes he had nurtured during the 1630s or even during his battle with Longomontanus. Instead Pell seems to have restricted himself to setting or answering questions that were moderately challenging but which required no original work on his part.

In 1649, for example, Cavendish on a visit to Breda asked Pell about Descartes' rule for factorizing a quartic, or biquadratic, equation as a product of two quadratics.[124] Descartes' had claimed that the equation:

$$+x^4 \,[\pm]\, pxx \,[\pm]\, qx \,[\pm]\, r = 0$$

could be replaced by two quadratic equations:

$$xx - yx + \frac{1}{2}yy \pm \frac{1}{2}p \mp \frac{q}{2y} = 0$$

and

$$xx + yx + \frac{1}{2}yy \pm \frac{1}{2}p \mp \frac{q}{2y} = 0$$

---

[123] Kempenaars, 'Gerard Kinckhuysen', p. 246.
[124] Descartes, 'La Géométrie', *Discours de la méthode*, pp. 383–5.

where y must satisfy the cubic equation (in y^2):
$$+y^6 [\pm] 2py^4 + (pp [\mp] 4r) yy - qq = 0.$$
Pell obliged Cavendish with a neatly written folio sheet with the method fully worked for each possible combination of signs, and Cavendish remarked that he liked Pell's 'waye of omitting no case'.[125] (And here yet again we have a suggestion of some communication with Kinckhuysen, for an almost identical explanation appeared in Kinckhuysen's *Algebra ofte stel-konst* in 1661.)

Cavendish also frequently asked Pell about cubic equations, and Pell sent him a brief one-sheet summary of the possible cases and the number of (real) roots in each case, something he had long since worked out in detail.[126] Pell claimed also, in 1649, to have a method of solving cubic equations by tables of sines. This method would have been based on the triple-angle sine formula:

$$\sin 3\theta = 3 \sin \theta - 4 \sin^3 \theta. \tag{2}$$

A cubic equation of the form:

$$3p^2 x - x^3 = 2q^3 \tag{3}$$

is easily converted to the trigonometrical form (2) using the substitution $x = 2p \sin \theta$. We then have $\sin 3\theta = q^3/p^3$, and hence real values of θ provided $|q| < |p|$. It is easy to reduce any cubic equation to the form (3) by removing the square term, so the method is more general than it first appears. There is an example of the method in volume VI of Hérigone's *Cursus*, published in 1644, which may be where Pell discovered it.[127] Pell also told Cavendish that Warner had spoken of using tables of logarithms for the same purpose, possibly tables of logarithms of sines. Taking logarithms of both sides of the equation $\sin 3\theta = q^3/p^3$ we have $\log \sin 3\theta = 3(\log q - \log p)$, and the right-hand side is very easy to calculate.[128]

We know of one further problem that was posed to Pell in Breda, this time not by Cavendish but by William Brereton. The problem is mentioned here, not because it is of any great interest in its own right, but because, like so much of the mathematics Pell did in the 1630s and 1640s, it was to re-emerge in a different guise many years later. Brereton's problem as originally posed was to find three numbers a, b, and c satisfying the equations:

$$aa + bc = 16$$
$$bb + ac = 17$$
$$cc + ab = 22.$$

It is not difficult to spot the obvious solution $a = 2$, $b = 3$, and $c = 4$, but Pell claimed that 'as triall of logisticall skill' he had changed 22 to 18, and had given

---
[125] Letter 88; BL, MS Harl. 6083, fos. 100^v–2.
[126] BL, MS Harl. 6083, fo. 152.
[127] Hérigone, *Cursus mathematicus*, vi, pp. 42–4.
[128] For the use of logarithms in solving equations see also below, p. 322–3.

Brereton the solution to this new and more difficult problem. Later, however, he was unable either to recall the answer or to reconstruct the working:[129]

> The manner of investigation I did not shew him. Neither do I now at all remember what course I tooke ... but I will heere endeavour to show a way ...

Pell's efforts were soon abandoned, but he returned to the problem on several occasions and we shall see more of it later.

Pell's mathematical career in the Netherlands is strangely disappointing in the light of expectations of him during his years in Sussex and London. It is even more so when seen against the background of the mathematical developments that were taking place in the 1640s in the Netherlands and elsewhere. By 1649 Frans van Schooten had not only edited Viète's *Opera mathematica* but had also produced the first Latin translation of Descartes' 'Géométrie', with a lengthy commentary of his own and notes by Florimond de Beaune. The *Geometria*, as it now became known, was without doubt the single most influential text in seventeenth-century mathematics in the Netherlands, and inspired such a wealth of new material that the second edition ten years later was extended to two volumes to accommodate it all. But Pell, despite having translated it into English for himself, paid little heed to the mathematics in it, apart from reworking some of the algebra, and seems not to have recognized the far-reaching possibilities that Descartes had opened up. Similarly, although he successfully persuaded Cavalieri to help him out with a proof of his double-angle tangent theorem, there is no sign that he was ever interested in Cavalieri's theory of indivisibles, either as it was presented in Cavalieri's own treatise, the *Geometria indivisibilibus* of 1635, or more simply in Torricelli's *Opera geometrica* of 1644. Yet the theory of indivisibles was no less significant than Descartes' algebraic geometry as one of the driving forces of seventeenth-century mathematics. Another important book read, but largely ignored, by Pell was Grégoire de Saint-Vincent's *Opus geometricum* of 1647, a book that took important steps towards the quadrature of the circle and hyperbola. Pell owned a copy, and mentioned it in correspondence with Cavendish,[130] but Christiaan Huygens as a student in Breda, puzzled as to whether de Saint-Vincent had really found the quadrature of the circle or not, found Pell uncommunicative on the subject. Huygens later wrote to Mersenne about de Saint-Vincent's book:[131]

> ... having never seen it other than sometimes at Mr. Pell's at Breda who never wanted to lend it to me, nor give me a definitive opinion about it, even though he had had it quite a long time.

---

[129] BL, MS Add. 4413, fo. 52.

[130] Letter 75. Pell's copy of de Saint-Vincent's *Opus geometricum* is at Westminster School, pressmark K F 6, with the words he quoted to Cavendish underlined in red on page 1226.

[131] '... n l'ayant jamais vue autrement que quelques fois chez Monsieur Pell à Breda qui me l'a jamais voulu prester, ny m'en dire une sentence definitive encore qu'il l'ayt eu assez long temps'; Huygens, *Oeuvres*, ii, p. 556, Letter 47b.

PART II: THE MATHEMATICS OF JOHN PELL

Pell's reluctance to discuss the *Opus geometricum* is perhaps an indication that he felt himself becoming somewhat alienated from the mainstream of European mathematics. The younger Pell who had corresponded so eagerly with Briggs and mastered everything in Harriot was beginning to be left behind. Although some of the most profound and important developments of seventeenth-century mathematics took place within Pell's circle of acquaintance, they seem to have passed him by. And while he failed to take advantage of the ideas that were circulating close at hand in the Netherlands he was also missing out on opportunities at home. The end of civil war in England brought about a major reorganization of the universities as new men acceptable to the new regime were put in place. John Wallis and Seth Ward became Oxford professors, but Pell had credentials as good as theirs and might well have been a potential candidate for those or other posts if he had been in England at the right time.

Pell returned to England in 1652 too late for such advancement, but through Hartlib and Haak quickly became reintegrated into the English mathematical community. Samuel Foster, who shared many of Pell's early interests, for instance in quadrants and dialling, and had been Gresham Professor of Astronomy since 1641, would have been known to Pell before he left London in 1643 and it is possible that he and Pell met again shortly before Foster died in 1652. It was probably also at this period that Pell first met or communicated with John Wallis, now Savilian Professor of Geometry at Oxford. The first definite evidence of interaction between them comes in an exchange to be found in Pell's papers, from April and May 1655, by which time he was living in Zürich. At the end of April he received from an unknown correspondent a leaflet, printed in mid-April, dedicated to Oughtred, and headed 'Circuli Quadraturam'.[132] The leaflet carried news of Wallis's forthcoming *Arithmetica infinitorum*, in which Wallis attempted a problem already long familiar to Pell, the quadrature of the circle. The letter that accompanied the leaflet suggested that Wallis was not impressed by Pell's mathematical abilities:[133]

> Apr 27. 1655
> In ye meantime accept of Dr Wallis Quadratura Circuli here adjoined, which I intreate you to handle soundly. For hee makes himselfe beleeve you will doe no great matters in Mathematicall studys.

Pell read the leaflet, picked up a typographical error, and questioned what Wallis might have meant by an *aequabilis curva*, or smooth curve. Wallis's intention was to construct a continuous curve between points with ordinates 1, 6, 30, 140, 630, ... at equally spaced intervals, but the leaflet showed only the curve itself and not

---

[132] See above, pp. 172–3. Of the three possible correspondents mentioned there (Haak, Brereton, and Hobbes) Haak was perhaps the most likely to be privy to Wallis's opinion of Pell.
[133] BL, MS Add. 4418, fo. 210^v.

the method of constructing it.[134] Wallis had proposed the problem to his Oxford colleagues Seth Ward, Lawrence Rooke, Richard Rawlinson, Robert Wood, and Christopher Wren in 1652, but Pell had never heard of it. He kept a copy of his reply but unfortunately without mentioning the name of his correspondent:[135]

> May 26. 1655
> Sir. I thanke you for yours of April 27 with that printed paper inscribed to Mr Oughtred. If his great age have not made him unwilling to looke upon things of that nature, perhaps he will make some reply. When it comes to your hand, I pray you to send it to mee. As also if the Author expresse himself more fully heere-after. Artists will not trouble themselves to make an enquiry concerning the *truth* of his new Theorems, till they be sure of the *sense* of it. They may soone find out the mysterie of continuing his numbers as farre as they desire and so may perceive that his Graver hath set 360 for 630. But out of that paper and those schemes, no man will be able to find what he means by *aequabilis curva*. He makes mention of a Probleme proposed by him, to many mathematicians, some years ago. Perhaps y^t problem joined with this printed paper, might help toward the finding of his meaning. I never saw that Problem, nor heard of it till now. But I should be glad to see it, especially if it have an intelligible Definition of *aequabilis curva*, in such a sense as he would have his Readers understand in his new Theorems.

Pell lived in Zürich from 1654 to 1658, and little is known of his mathematical activities during those years except that his pupil Johann Heinrich Rahn visited him weekly during 1657 (and perhaps earlier) and worked on problems that Pell set for him.[136] Such problems laid the foundations for the *Teutsche Algebra* that Rahn was to publish in Zürich in 1659. In his Preface, Rahn said that he learned his mathematics from an eminent and very learned person who would not allow him to reveal his name,[137] but it is abundantly clear from the contents of the book and from the style in which it was written that the learned person was Pell.[138] The *Teutsche Algebra* is therefore the best guide we have to Pell's mathematical preoccupations during the 1650s.

---

[134] Wallis's sequence may be written as $1 \times \frac{6}{1} \times \frac{10}{2} \times \frac{14}{3} \times \frac{18}{4} \times \ldots$; his problem was to find a rule that would allow him to interpolate intermediate values fulfilling certain conditions; see Wallis, *Arithmetica infinitorum*, Propositions 166–8.

[135] BL, MS Add. 4418, fo. 210^v.

[136] BL, MS Add. 4394, fo. 455.

[137] 'In den Solutionen/ und grad auch in der Arithmetic bediene ich mich einer ganz neuen manier/ die bey einichen Algebraischen Scribenten in offenem Trukk gebraucht worden/ und die ich von einer hohen und sehr gelehrten Person erstmals erlehrnet hab/ deren ich auch schuldiger maassen/ und zwaren zur bezeugung untertähnigen respects/ gar gern gedenken/ so sie es hette zulassen wollen'; Rahn, *Teutsche Algebra*, Preface.

[138] See also Malcolm, 'The Publications of John Pell', esp. pp. 286–7.

A book of 185 pages, the *Teutsche Algebra* opens with a lengthy exposition of the six basic operations of arithmetic (addition, subtraction, multiplication, division, involution, and evolution) applied in turn to whole numbers, fractions, algebraic fractions, and surds. Rahn's notation reveals Pell's hand immediately: ÷ for division, a spiral ↻ for involution, and a wavy ⍵ for evolution. As part of the treatment of surds there is also a table of primes up to 23 999. All this fills just over a quarter of the book. Then, after some instruction on equations and algebraic manipulation, comes the first worked problem: given any two of the quantities $a+b$, $a-b$, ab, a/b, $aa+bb$, $aa-bb$, find the other four. This, of course, is a simplified version of Pell's great Progymnasma of 1637. Several other problems follow, some geometrical, some purely algebraic ('Three Men have Money' is one of them), all worked by Pell's three-column Method, and including two that are indeterminate. There are no Diophantine problems but a few years later Rahn prepared a second book that treated Diophantus algebraically.[139]

At page 136 a new topic is introduced with the heading 'Stylus delineatorius'. the first subheading is 'Delineation der quadratischen' and explains how to construct ruler and compass solutions to:

1. $zz = az + bb$ und $-az + bb$
2. $zz = az - bb$

This is very similar to Pell's 'Quadraticall Æquations solved by Delineation' of 1642, but Rahn (or Pell) has now taken it further, because we also have 'Delineation der Biquadratischen Aequationen' and 'Delineation der Cubischen Aequationen'. As an example of the last, Rahn took the equation $zzz - 12z = 9$, but before he gave the construction he produced a table, of a kind not yet discussed but which appears several times in Pell's manuscripts. It shows how two of the roots of $zzz - 12z = N$ change from complex to real as N passes through 16:

z [1st root]	zz	$zz - 12$	$zzz - 12z\,[=N]$	[2nd root]	[3rd root]
5	25	13	65	$\frac{\sqrt{-27}-5}{2}$	$\frac{-\sqrt{-27}-5}{2}$
4	16	4	16	$0-2$	$-0-2$
3	9	-3	-9	$\frac{\sqrt{21}-3}{2}$	$\frac{-\sqrt{21}-3}{2}$

From page 167 onwards the book is concerned with bisection and trisection of angles, with a great deal of work on Pell's double-angle tangent formula (used now in both directions, for both doubling and halving), and the final section is a discussion of Pell's refutation of Longomontanus. Pell may not have allowed his

---

[139] Rigaud *Correspondence*, ii, p. 195; Turnbull, ed., *Gregory Tercentenary Volume*, p. 194; see also above, Part I, ch. 7, n. 46.

name to be printed, but his mathematics betrays his influence in almost every page of the book.

London and Cheshire, 1658–1669

Soon after his return to England in 1658, Pell once again renewed his acquaintance with English mathematicians. In November 1658 he was invited to a meeting with 'some Mathematicall friends' amongst whom were Lawrence Rooke, Christopher Wren, William Brouncker, Paul Neile, and Charles Scarborough, and the letter suggests that Pell had attended several such meetings before.[140] Another acquaintance must have been John Twysden, about whom little is known, but who in 1659 published a posthumous edition of some of Samuel Foster's papers, together with some problems of his own. His Preface sang the praises of Pell:[141]

> How far Mr. John Pell hath pierced into the depths of these Sciences you may from thence easily conjecture that he hath been able, and that in a way not troden by others, and within the compasse of one page, to overthrow the endeavours, and many years attempts of the famous Longomontanus touching the true measuring of a Circle.

Twysden's eulogy ends with a familiar refrain:

> He hath promised us other things and is fit to undertake far greater then this was.

Two of the problems in the section headed 'Certain Mathematical problems analytically resolved and effected By J. Twysden', are based on Pell's double-angle tangent formula.

About this time too, Pell must have renewed his acquaintance with John Wallis, whose opinion of Pell, it seems, had begun to improve, for by 1659 or very soon afterwards Wallis was using Pell's three-column Method. It is to be found in notes he made on Pascal's *Lettres de A. Detonville* of 1659, the publication in which Pascal dismissed Wallis's contribution to work on the cycloid, and which Wallis almost certainly read as soon as it was published. Wallis and Pell continued to discuss mathematics together in the early 1660s, for in 1662 William Brereton's problem surfaced again, and now Wallis too was working on it. Pell's papers contain page after page of attempts at the problem (in the version $aa + bc = 16$, $bb + ac = 17$, $cc + ab = 18$), but it was Wallis who came up with the eventual solution, and who wrote it out in Pell's three-column Method. Pell copied the solution and annotated it with the words: *Ex Iohannis Wallisii*

---

[140] BL, MS Add. 4279, fo. 273; printed in Vaughan, *Protectorate*, ii, pp. 478–9 and in Halliwell, *Collection of Letters*, pp. 95–6.
[141] S. Foster and J. Twysden, *Miscellanies or Mathematical Lucubrations* (London, 1659), Preface.

autographo exscripsi, Aprilis 14. 1663 ('I wrote this out from a copy in Wallis's hand, April 14 1663').[142]

Wallis's solution led to a sixth-degree equation in the unknown quantity a^2, but he replaced a^2 by $e^2/2$ and was able to reduce the equation to a quartic in e^2:

$$e^8 - 80e^6 + 1998e^4 - 14937e^2 + 5000 = 0.$$

Pell or Wallis or both went on to find the roots of this equation, and on another sheet in Pell's papers we have the process of solution, as far as fifteen decimal places.[143] A third person was also interested in it; in December 1662 Pell left notes about progress on the problem with Silius or Silas Titus, an Officer of the King's Bedchamber.[144] In 1649 and early 1650, then aged about twenty-six, Titus had spent time in Breda as a representative of the English Presbyterians in the protracted negotiations between Charles II and the Scottish Presbyterian Commissioners, and if he was at all interested in mathematics he would certainly have got to know the resident English mathematician, Pell. He had therefore probably been introduced to Brereton's problem when it first arose in 1649.[145]

Pell wrote over one hundred sheets on Brereton's problem,[146] and one must ask why he thought it worth the effort, but it was not unusual for him to expend such energy on questions of apparently limited interest. He calculated and wrote out, for example, several large tables for finding square numbers beginning or ending with given sets of four digits. He may have been inspired to try his hand at this problem in the year 1664, since he was particularly interested in that particular group of figures. He discovered no fewer than fifty-eight squares beginning with 1664 (the greatest being 16649773156) and others that ended with 1664 (the greatest being 9978411664).[147] Further tables that he promised in 1672 and produced for Aubrey in the last year of his life, 'Having the two, three or four *last* figures of any Square number, to exhibit as many of the last figures of its side', were almost certainly connected to the same work.

During this period Pell seems also to have maintained his interest in the difference methods he had used with Warner, and investigated their relevance to

---

[142] BL, MS Add. 4425, fo. 161.

[143] BL, MS Add. 4411, fos. 359–67. The solutions as calculated by Wallis and/or Pell were $a = 2.525\,513\,986\,744\,158$, $b = 2.969\,152\,768\,619\,848$, $c = 3.240\,580\,681\,617\,174$.

[144] BL, MS Add. 4425, fos. 367 and 368.

[145] The friendship between Pell and Titus continued for many years. Titus attempted to help Pell financially in 1675 (Turnbull, ed., *Gregory Tercentenary Volume*, p. 310) and there is a note in Pell's papers about newly printed pages of Dechales's *Cursus* that he borrowed from Titus in July 1675 and returned in May 1676 (BL, MS Add. 4416, fo. 54^r). Wallis, long before he met Pell, was aware of Titus and his activities in the Netherlands because in 1650 he had deciphered letters intercepted between Breda and London; Titus, then working assiduously for the return of Charles II, was mentioned in several of them and Wallis correctly discovered his name (Bodleian Library, MS e Mus 203, fos. 193, 195, 198, 200).

[146] BL, MSS Add. 4411, fos. 359–68; Add. 4412, fos. 36–202 *passim*; Add. 4413, fos. 52–3; Add. 4425, fos. 161–206 *passim*.

[147] BL, MS Add. 4418, fos. 112, 141, 150; see also MS Add. 4429, fos. 259–77.

trigonometric quantities and to logarithms. He calculated at least one table of eighth differences for sines,[148] and there are also difference calculations relating to the logarithms of sines and tangents appended by Joseph Moxon to Oughtred's *Trigonometria* in 1657.[149] There are also several examples of algebraic difference tables.[150] The one partially reproduced below is a particularly lengthy example displaying constant fourth differences, and continued for sixty lines from -20 to $+40$. The differences are arranged so that they are always positive, that is, the smaller number of any pair is subtracted from the larger. The pattern of differences in the second and third columns suggests that it was devised for tables of logarithms, since the numbers in the first column (representing logarithms) increase, while those in the second and third columns decrease.[151]

-20
.
.
.
-3 $A - 3b - 6c - 10d$
            $b + 3c + 6d$
-2 $A - 2b - 3c - 4d$ $c + 3d$
            $b + 2c + 3d$ d
-1 $A - b - c - d$ $c + 2d$
            $b + c + d$ d
0 $A \mp 0b - 0c - 0d$ $c + d$
            $b \pm 0c + 0d$ d
1 $A + b - 0c \mp 0d$ c
            $b - c + 0d$ d
2 $A + 2b - 0c + 0d$ $c - d$
            $b - 2c + d$ d
3 $A + 3b - 3c + d$ $c - 2d$
            $b - 3c + 3d$
4 $A + 4b - 6c + 4d$
.
.
.
40

---

[148] BL, MS Add. 4424, fo. 458.

[149] William Oughtred, *Trigonometria ... et logarithmorum pro sinubus et tangentibus* (London, 1657); BL, MS Add. 4415, fos. 126$^\text{v}$–131.

[150] BL, MS Add. 4415, fos. 87–91, 113 and 115.

[151] BL, MS Add. 4415, fo. 113. Log x is an increasing function of x, but the absolute value of any of its derivatives is a decreasing function.

One of Pell's calculations of logarithms using such a table is given in full below, for its intrinsic interest, and for what it says about Pell's willingness to engage in time-consuming calculations to an extraordinary degree of accuracy even when he was not being paid to do so.

In a note that may have been written soon after he met Collins, Pell noted the existence of copies of Briggs's *Arithmetica logarithmica* with an extra chiliad:[152]

> Mr. Collins hath the London edition of Henrici Briggij Arithmetica logarithmica printed 1624. At the end of the 100th Chiliad of Logarithmes that book hath 3 sheets more than any other book of the same edition.
> The first 5 leaves contain ... logarithms 100 000 to 101 000.
> The last leafe has ... tables of square roots to 200.
> But *in the book itselfe* I finde nothing added concerning any peculiar use of either of these accessory tables.
> I once heard that Mr. Briggs had written a short discourse shewing how to make that one chiliad serve for all. But I never saw any such discourse of his ...
> In the meane time why may not I see if I can supply this defect?

As Pell indicated, a few rare copies of Briggs's *Arithmetica logarithmica* contained an additional chiliad of logarithms from 100 000 to 101 000.[153] Pell also constructed some extra tables of his own, and explained how it was possible 'By the help of Mr. Briggs his 101st Chiliad & of my appendix to finde the Logarithme of any number less than 10000 000000 0000'. In the example following,[154] Pell wanted to find the logarithm of 317:

> 1st example. I seeke y^e logarithme of 317.
> $317 \div 101 = 3\frac{14}{101} = 3.138$ $317 \div 100 = 3.17$

Now Pell referred to his 'Appendix':

> In the Appendix I find 315 falling betweene these two quotients 3.138 and 3.170 wherefore $317 \div 315$ will fall betweene 100 and 101. It is 100.6349206349206349 etc. This in the Chiliad falls betweene 100634 and 100635. Wherefore my worke is to find the Logarithm of 10063492 and of 10063493. I seeke them thus.

---

[152] BL, MS Add, 4414, fos. 232–4.

[153] Collins's copy of Briggs's *Arithmetica logarithmica* with the additional 101st chiliad was held in the Library of the Earls of Macclesfield until June 2004 when it was sold by auction at Sotheby's of London. See sale catalogue *The Library of the Earls of Macclesfield removed from Shirburn Castle*, Part Two: Science A–C, p. 228. Some of the inserts mentioned there are in Collins's hand but there are none in Pell's. The only other known copy with the additional sheets is in the Library of Trinity College, Cambridge; on the flyleaf are the words: 'ab ipso authore emptus' ('bought from the author himself') with the additional information: 'p[re]tium 9s – 0d' ('price 9 shillings').

[154] BL, MS Add. 4413, fo. 230.

JOHN PELL (1611–1685)

Number	Logarithm	Difference 1st	Difference 2nd
1.0063400	0.00274 47353 6898		
		431 556 257	
1.0063500	0.00274 90509 3155		4288
		431 551 969	
1.0063600	0.00275 33664 5124		

The calculated figures are then compared with those in the following difference table (a small portion of what must have been a much longer table):

$$A$$
$$100b - 4950c$$
$$A + 100b - 4950c \qquad 10000c$$
$$100b - 14950c$$
$$A + 200b - 19900c$$

Equating corresponding entries in the two tables, the equations are easily solved to give (as Pell wrote them):

$C = 0000\,4288 \qquad B = 431\,5538\,7956 \qquad A = 0.00274\,47353\,6898$

Note Pell's usual technique of replacing lower case letters with capitals once he knew their value. Now he introduced a second table with finer gradations so that he could interpolate between A and $A + 100B - 4950C$:

$$A + 92B - 4186C$$
$$B - 92C$$
$$A + 93B - 4278C \qquad C$$
$$B - 93C$$
$$A + 94B - 4371C$$

Finally he substituted the calculated values of A, B, and C to give the required logarithms:

Number	Logarithm	Difference 1st	Difference 2nd
1.0063492	0.00274 87056 8812 2384		
		$B - 92C$	
		$= 431\,5499\,3460$	
1.0063493	0.00274 87488 4311 5844		4288
		$B - 93C$	
		$= 431\,5598\,9172$	
1.0063494	0.00274 87919 9909 6016		

'The very same', remarked Pell triumphantly, 'that you shall find in y^{e} 1st chiliad of H. Briggs'. This example was followed by two more, to find the logarithms of

6343 and 98179, respectively.[155]

Other problems that occupied Pell during the 1660s were proposed to him by his acquaintances, who clearly considered him an authority on a range of subjects. In 1663, for instance, at the request of Gilbert Sheldon, the Archbishop of Canterbury, he went to some trouble to confirm the date of Easter printed in the current Almanacs,[156] resulting in a brief anonymous publication, *Easter not Mis-Timed*, in 1664.[157] In May 1665 he tackled a different kind of problem that was perhaps typical of those brought to him: to find four numbers in continued proportion (that is, in geometric progression) such that eight times the first, plus four times the second, plus eight times the third, plus the last, makes sixteen. Pell noted that: 'Mr. John Collins brought me half a sheete wherein Mr Darie had begun the search of a solution to this problem',[158] and Dary, according to Pell, had written that 'in this pile of inferences I will abide by the censure of the learned Dr Pell and Mr. Kersey'. Pell obligingly found a solution: $\frac{2}{17}$, $\frac{8}{17}$, $\frac{32}{17}$, $\frac{128}{17}$. Another problem that Pell probably worked on during the 1660s was the *Problema austriacum*: to construct a 14-sided polygon on a given line,[159] raised by Collins in a letter to Pell in 1667.[160]

The lack of dates on the later papers makes it difficult to be certain of exactly what else Pell worked on during these years, but what comes over even more strongly than from his Breda years is a sense that Pell spent much of his energy on time-consuming but relatively unimportant problems. At the same time, he could be scathing about the efforts of others. When Nicolaus Mercator discovered the infinite series for $\log 1/(1+x)$ and published that and the quadrature of the hyperbola in his *Logarithmotechnia* of 1668, Pell wrote:[161]

> In his title page, he says his Logarithmotechnia had beene communicated in writing in August 1667. He says not *to whom*. Not to Dr. Wallis, I beleeve. I desire to know what *Hee* saith of it. Howsoever (as I wrote before) I look y^t some transmarine pen should fly at him. Englishmen, perhaps, will let him alone till he print the same *crudities* in English.

It was perhaps Pell's desire to know Wallis's views on Mercator's book that prompted Wallis to write an account of it for the *Philosophical Transactions of the Royal Society*.[162] Pell's note reveals, however, not only an ungenerous side of

---

[155] BL, MS Add. 4413, fos. 229–31.
[156] BL, MS Add. 4410, fo. 180; Pell's calculations are to be found throughout this volume.
[157] Malcolm, 'The Publications of John Pell', p. 286.
[158] BL, MS Add. 4417, fo. 59.
[159] BL, MS Add. 4429, fos. 51^v–56, 177–82.
[160] Rigaud, *Correspondence*, i, p. 128.
[161] BL, MS Add. 4415, fo. 2.
[162] J. Wallis, 'Logarithmotechnia Nicolai Mercatoris', *Philosophical Transactions*, 3 (1668), pp. 753–64.

his character, but an ignorance of what mattered most in contemporary mathematics. Mercator's book contained the first published example of an algebraic infinite series, and was an important step forward in the thirty-year struggle to find general methods of quadrature. Its publication prompted Newton, who had been working along similar lines, to reveal some of his own findings, and so brought into the open the full potential of the calculus, but none of this seemed to interest Pell. While Newton, in the winter of 1664, had discovered the general binomial theorem and laid the foundations of the theory of infinite series and the calculus, Pell had been preoccupied with listing square numbers. Pell's abilities were not Newton's, and one cannot criticize him for not achieving what was beyond him or most of his contemporaries, but someone as mathematically literate and experienced as he was should at least have recognized what was happening around him.

Nevertheless, it was during the 1660s that Pell did finally publish some of the algebra that Cavendish and others had so often pleaded for. *An Introduction to Algebra* was published in 1668. The book was based on Rahn's *Teutsche Algebra*, translated into English by Thomas Brancker, but also significantly changed, because Pell contributed much new material. Brancker started the translation in 1662, but Pell began to make improvements to the text only in 1665 after the first sheet had already been printed.[163] A draft of the preface in Brancker's handwriting dated 14 April 1668 and signed T.B. states that the book was given to him in 1662 by Francis Turner, then a student at New College, Oxford, who requested that Brancker should 'render it into English'.[164] The preface goes on to say that the translation was completed by 1665, and that only then did Brancker discover that Rahn's teacher was Pell:

> When it was thus gone to ye presse (and not before) I understood that that eminent person to whom Mr. Rhonius was so much beholden was ye reverend Dr Pell.

The preface was not very cleverly constructed, however, for near the beginning it had stated that Rahn's book

> was soon discovered to have had an English original. The method of his Processes by a double margin was enough to hint, even to diverse that understood not that language, that it was no Dutch artifice.

The 'method by a double margin' revealed not only English influence but more particularly Pell's hand in the matter. If Brancker was supposed to have recognized the English origins of the three-column Method, he must also have known that the Method was Pell's, and so cannot have been as surprised as he claimed in 1665.

---

[163] BL, MS Add. 4278, fo. 32.
[164] BL, MS Add. 4414, fo. 5.

There is another version of the story, in another draft preface, also dated 14 April 1668. This draft is in Pell's hand, and has been entered into a carefully ruled mock page, but it was marked by Pell in red ink as 'not sent, not printed'. It is headed 'The Publisher's Preface', and is purportedly by John Collins.[165] It states:

> I delivered a coppy of [Rahn's book] to a friend of mine, who understood the Language of it, better than the matter: He returned it to mee Englished Anno 1665; before I knew Mr. Brancker; whose treating with Mr. Pit to get his translation printed gave the first occasion of our acquaintance. I then told them, that ... they should do well to abstaine from putting it into the Press before they had shewed it to D.P. who was the man *from whom* Monsieur Rhonius had received the marginall style; and of whom he makes such honourable mention in the last page of his preface, adding that he would faine have expressed his name, but could not obtaine his leave so to doe.

It seems a little strange that two apparently different translations were completed in the same year: one at the request of Francis Turner by Thomas Brancker who claimed not to know whose work he was handling, even though the English influence was supposed to be plain; the other at the request of John Collins, by an unnamed translator whose work then sank without trace. The gaps and inconsistencies leave one feeling that neither preface discloses the whole truth. It also seems strange that Pell knew nothing of Brancker's translation until 1665; in 1662 Francis Turner was still at New College (he proceeded MA in 1663), but his father, Thomas Turner, had been with Charles I on the Isle of Wight, along with Silas Titus and Gilbert Sheldon, both of whom were active figures in Pell's life in 1662. It does seem from the later correspondence, however, that Pell genuinely knew nothing of Turner's and Brancker's scheme until 1665.

The preface eventually published in *An Introduction to Algebra* is a modified and much shortened version of the first of the above drafts, that written by Brancker. Under the heading 'The Translator's Preface' it reads:

> The copy which I have, was given me anno 1662, by a good friend who then told me he much desired to read it in some language that he understood; I then promised him to English it.

A margin note at this point has 'M.F.T' (Mr. Francis Turner).[166] The preface continues:

> A little after, I heard that there was at that time in London, a Person of Note very worthy to be made acquainted with my design,

---

[165] BL, MS Add. 4414, fo. 10.

[166] 'Francis Turner' is written in the margin of Aubrey's copy of *An Introduction to Algebra* in the library of Worcester College, Oxford, pressmark E.Z.2.

> before I made any farther progress in the Impression. Being admitted to speak with him, I found him not only able to direct me, but also very willing so to do, so far as his Leisure would permit.

Here the margin note has 'D.I.P', unmistakably Dr John Pell, and the preface goes on to note exactly which problems and pages Pell contributed.

The correspondence between Pell, Brancker, and Collins on the correcting and enlargement of the text after 1665 has been thoroughly documented elsewhere.[167] One suggestion, apparently from Collins,[168] was that some of the material from Kinckhuysen's *Algebra ofte stel-konst* should be included. The plan came to nothing, but given all the other circumstantial evidence that points to a connection between Pell and Kinckhuysen, the suggestion was perhaps not entirely fortuitous. Since *An Introduction to Algebra* was already based on work by one of Pell's pupils, namely Rahn, it would have made sense to include extracts from another, namely Kinckhuysen.

Pell certainly took more than a passing interest in Kinckhuysen's publications. It has generally escaped notice that Brancker, while he was working on *An Introduction to Algebra*, was also translating into English Kinckhuysen's *De grondt der meet-konst* of 1661, a treatment of conics by the methods of analytic geometry. Pell must have worked closely with Brancker on this project because between them they not only prepared the translation but also put the contents into Pell's three-column layout, as Oldenburg noted in 1672:[169]

> And Kinckhuysen's Anal ... nicks are translated into English and put into Pell's method by Brancker the publisher of Rhonius his Algebra.

There is a hole in the paper in the final sentence but the partly lost words may be read as 'Analytick Conicks', that is, Kinckhuysen's *Grondt der meet-konst*. As so often with Pell's projects, its demise is shrouded in mystery. In December 1667, Brancker sent the translations to Collins but asked him not to let it be printed 'without my own, but especially Dr Pell's consent'.[170] Collins wrote to Pell that he had received 'M. Brankers Translation of Kinckhuysen',[171] but in early February he wrote again:[172]

> Let not Mr Branker feare that I shall be Instrumentall or assenting to the Printing of Kinckhuysens Conicks without your and his leave.

---

[167] Scriba, 'John Pell's English Edition'.

[168] Rigaud, *Correspondence*, i, pp. 118, 126, 135–6. See also Scriba, 'Mercator's Kinckhuysen-Translation', pp. 50–3, and Whiteside, 'Kinckhuysen's Algebra and Newton's Observations upon it' in Newton, *Mathematical Papers*, ii, p. 279.

[169] BL, MS Add. 4407, fo. 118.

[170] Rigaud, *Correspondence*, i, p. 131.

[171] BL, MS Add. 4278, fo. 326.

[172] BL, MS Add. 4278, fo. 331.

We do not know the reasons for Brancker's, or more probably Pell's, reluctance to print but the translation was never published.

Brancker's translation of Rahn's *Teutsche Algebra*, however, came out in 1668, and the titlepage advised that it was 'Much Altered and Augmented by D. P.' The initials were a compromise that Pell struggled over. In a letter drafted (but not sent) to Brancker in 1666 he described his quandary about revealing his name:[173]

> I know not how my mind may alter but for the present, I think it best not to name mee at all in the title or preface: and yet you may be more ingenuous than Rahn was and not vent all for your owne devices. You may say, that the alterations and additions etc. were made by the advice of one of good reputation in those studies.

The letter displays the same ambivalence that we saw in Pell as the young man who wrote pseudonymously but at the same time drew attention to the fact. Pell's inner struggle was not easily resolved. His design for the title page of *An Introduction to Algebra* had the initials crossed out and then reinstated as Pell went through stages of indecision.[174] The initials were anyway not much of a disguise; few interested readers can have failed to surmise that D.P. was Doctor Pell, and Brancker's Preface described exactly which pages (79 to 82 and everything from page 100 onwards) Pell had added.

An Introduction to Algebra contains the most extensive pieces of mathematics that Pell ever published under his own name. The book opens in the standard way with an introduction to notation and to the basic rules of manipulation for simple and compound quantities, fractions, and surds. The material and its arrangement is similar but by no means identical to that at the opening of the *Teutsche Algebra*, so Brancker or Pell must have done some rewriting. Collins considered the opening pages inadequate but could not persuade Pell to change them, and protected Pell's reputation by claiming that the introduction was Rahn's, which was not quite true:[175]

> I know none that account the Introduction a bad one, but divers that think it might have been more plain, and ought to have been more large than it is. This is the judgement of divers of the virtuosi and of some teachers of the mathematics here, who all love and honour the Doctor [Pell]; and I hope I shall do no less as long as I live, albeit I am of their mind, nor do I endeavour to make others of the same opinion, but say to them the Doctor did not much concern himself therein, but lets it come out as his scholar left it;

After the introductory material, the main part of the book is entitled 'Resolution of Problemes. The Use of this Algebra'. Four short chapters set out some of

---

[173] BL, MS Add. 4278, fo. 80.
[174] BL, MS Add. 4414, fo. 2.
[175] Rigaud, *Correspondence*, i, pp. 134–5.

the basic theory of equations: that an equation has as many roots as the highest power, and as many 'divisors' or factors as it has roots (Chapter 1); that the second highest power may be removed by a suitable substitution, with examples for quadratics and a cubic (Chapter 2). Chapter 3 explains the standard rules for simplifying equations, and Chapter 4 demonstrates the rules using an example taken from van Schooten's *Principia matheseos universalis* (reissued with his second edition of the *Geometria* in 1659). There is also an unusual example of a quadratic solved, by completing the square, for *both* roots:

$$\tfrac{1}{4}zz - \tfrac{3}{2}Bz + \tfrac{9}{4}BB = \tfrac{9}{4}BB$$
$$\tfrac{1}{2}z - \tfrac{3}{2}B = \tfrac{3}{2}B, \quad (\text{also} = -\tfrac{3}{2}B)$$
$$\tfrac{1}{2}z = 3B, \quad (\text{also} = 0)$$

The fifth and final chapter begins at page 56 and ends at page 192, and so constitutes almost three-quarters of the book. It contains thirty-one problems, 'Arithmeticall and Geometricall', with no particular ordering except perhaps generally increasing difficulty. Problem 1, as in Rahn's original text, is: given any two of the quantities $a + b$, $a - b$, ab, a/b, $aa + bb$, $aa - bb$, find the other four. Problem 2 is to find the sides of a right-angled triangle in which one of the short sides is $3\sqrt{2} + 3$, and the sum of the hypotenuse and the other short side is $9\sqrt{2} + 9$. Further problems, up to Problem 11, are based either on elementary properties of circles and triangles, or on sets of two or three numbers that are required to satisfy certain conditions. Problem 11, for example, is:

> Three men [A, B, C] divide a sum. A gives both the others as much as they had before; B then gives both the others as much as now they have, and C doth the like, and at last each man had 8. How much had each man at first?

The first indeterminate problem is Problem 12. In a right-angled triangle with short sides b and c and hypotenuse h, we are given the difference D between the hypotenuse and one of the shorter sides, and the problem is to find all the three sides. Pell's solution begins:

$$\begin{array}{lll} h = ? & |1| & h = b + D \\ b = ? & |2| & hh = bb + cc \\ c = ? & |3| & (*) \end{array}$$

A note then refers the reader forward to Problem 15 where Pell explained (on one of the last pages to be added to the text):[176]

> this (*) ... shews *the defect of an Equation*. For in Probl. 12 and 13, and this fifteenth, the Margin shews that three Equations are required; but the Question affords not *more than two*. One Equation is wanting, to make up the number of *given* Equations, *equal*

---
[176] Rahn and Pell, *Introduction to Algebra*, p. 80.

to the number of *sought* Equations. For such *Equality* is necessary for the *limiting* of a question, that the answers may not be *innumerable*.

Problems 16 to 23 are geometrical, and up to this point the problems are very similar to those in the *Teutsche Algebra*. From Problem 24 onwards, however, the work is new. Problem 24 is a numerical problem:

Problem XXIV
To find two Numbers, either of which being subtracted from the square of their summe, will leav a remainder, that is a square Number.

This was a simplified version of what was to follow next, problems V.18 and V.19 from Diophantus (Pell's Problems 26 and 27):

Problem XXVI
To find three Numbers, which will make as many Cubes, by adding each of them to the Cube of their Sum (Dioph V.18).

Problem XXVII
To find three Numbers, which will leave as many Cubes after the Subtraction of each from the Cube of their Sum (Dioph V.19).

Pell devoted 20 pages to this last problem, working it out in different ways and commenting on the earlier solutions found by van Ceulen and Bachet.[177] Van Ceulen's solution was published by Frans van Schooten in 'Sectiones miscellaneae', the fifth book of his *Exercitationum mathematicarum* (Leiden 1657). A further problem from the 'Sectiones miscellaneae' fills most of the remaining sixty pages of *An Introduction to Algebra*: to find two isosceles triangles, equal in perimeter and area, with all their sides and perpendiculars rational multiples of each other. The book then ends with a table of 'Incomposits', or primes, calculated under Pell's instructions for numbers up to 23 999 by Balthasar Keller in Zürich, but now extended up to 100 000 by Brancker.

With Pell's new problems inserted, much of the original material of the *Teutsche Algebra* had to be omitted. The 'delineation of equations' disappeared, as did the table showing how the roots of a cubic equation changed according to the size of the constant term. All the material on Pell's double-angle tangent formula and the quadrature of the circle was also omitted. The 'delineation of equations' may not have found its way into *An Introduction to Algebra* but the ideas were not altogether lost. In 1684 Thomas Baker published a book called the *Geometrical Key*, which described at very great length how to solve quadratic, cubic, and biquadratic equations by geometrical construction, a great improvement, Baker thought, on the sparing hints given by Descartes in 1637. Pell's name is not mentioned, and it could be no more than coincidence that Baker's theme was so close to Pell's, except that Baker's book is written in Pell's three-column

---

[177] There are also many sheets on Diophantus V.19 in Pell's manuscripts: see n. 106.

Method, and we also know that Collins discussed some other pieces of Pell's mathematics with Baker in 1676.[178] The *Geometrical Key* was a long-winded and old-fashioned text, and its subject matter was of no practical use, but the Royal Society supported the publication and Pell was given a copy by Theodore Haak as soon as it appeared in 1684.[179]

We have little evidence of how *An Introduction to Algebra* itself was received except in a letter from the publisher, Moses Pitt, to Brancker in June 1668, in which he related that Richard Busby, headmaster of Westminster School, was pleased with it:[180]

> I have hopes the algebra will take well. One of Dr Busby's scholars had one of me on Friday last and He told me the Dr did highly extoll it and read it to them usually every night and all that have it do like it well.

London, 1669–1685

The year 1669 may be taken to mark the beginning of the final phase of Pell's mathematical career, and the one in which he put more difficulties than usual in the way of those who hoped then or now to understand his work. The best documentary evidence of his interests in this period is in the correspondence of John Collins, not with Pell himself, but with Wallis, James Gregory, Thomas Baker, Ehrenfried Walter von Tschirnhaus, and Leibniz, to all of whom, between 1667 and 1676, Collins tried to describe aspects of Pell's work.[181] From Collins we learn, for instance, that Pell knew how to calculate logarithms using Briggs's 101st chiliad, but that he was 'close fisted' on the matter.[182]

Pell also maintained or revived his long-standing interest in Diophantus, and made a careful study of Fermat's notes, written in the margins of Bachet's *Diophanti arithmeticorum* and published posthumously in 1670. Pell himself had studied Bachet assiduously in the 1640s, and his copy of the 1670 edition of Bachet's text with Fermat's notes is still in the Busby Library at Westminster School, with several inserts and numerous annotations in Pell's hand.[183] Fermat was a far superior mathematician to Pell, but they were alike in frustrating their contemporaries and posterity with unfulfilled promises. Collins wrote of Fermat

---

[178] Rigaud, *Correspondence*, ii, pp. 8–9, 10. In 1677 Robert Hooke recorded in his diary that he borrowed a solution to Brereton's problem from Baker in a London coffee house: see Hooke, *Diary*, p. 322.

[179] BL, MS Add. 4423, fo. 343.

[180] BL, MS Add. 4398, fo. 196.

[181] Many of the surviving drafts of Collins's letters are now in CUL, MS Add. 9597/13/5. Much of this correspondence was transcribed and published in Rigaud, *Correspondence*. Letters from Collins to Gregory were also published in Turnbull, ed., *Gregory Tercentenary Volume*.

[182] Turnbull, ed., *Gregory Tercentenary Volume*, pp. 153–4.

[183] Diophantus, *Arithmeticorum libri sex*, ed. C. G. Bachet de Méziriac, with notes by P. Fermat, ed. S. Fermat (Toulouse, 1670). Pell's copy is in the Busby Library, Westminster School, pressmark I F 22.

that 'when he comes to knotty problems he is frequently complaining that he wanted room, and therefore hath left the matter untouched.'[184] Pell generally complained of want of time rather than want of room, but the end result was the same. Pell informed Collins that he found Fermat's methods for indeterminate problems 'very deficient'.[185] As he had so often done in the past with other authors, he thought he could simplify and improve on Fermat's treatment, and told Collins that he was writing about twenty sheets which, together with Fermat's annotations, would conveniently replace the new edition of Bachet. It hardly need be said that no such sheets were ever published, but some of Pell's notes on Diophantine problems may date from this final effort to redeem the work he had begun thirty years earlier.

In 1669 Collins published an article in the *Philosophical Transactions* on the 'Resolution of Equations', and explained that it was part of a longer narrative 'touching on some late Improvements of Algebra in *England*, upon the occasion of its being alledged, that none at all were made since *Des Cartes*.'[186] Notes towards Collins's narrative survive in the British Library manuscript collection in a pair of folded folio sheets labelled on the outside: 'Of some improvements of Algebra in England',[187] and the text, under the same heading and in Collins's hand, is a discourse on methods of solving equations. The third paragraph reads: 'Concerning the Method of Progressional Differences long Scroles have been made by a learned person'. The 'learned person' is not named but the long scrolls of progressional differences are enough to identify him as Pell,[188] and much of the material that follows, including a specific reference to Brereton's problem, show that Pell was the source of most of the 'improvements of Algebra in England' that Collins had in mind. Collins's attempts at mathematical exposition often leave the reader feeling bewildered, and neither his notes nor his article in the *Philosophical Transactions* is an exception, but Pell's influence in the subject matter of both is clear. (Pell's methods of solving equations will be discussed in detail shortly.)

Collins was always hatching new plans for publishing mathematics, and between 1669 and 1672 he discussed with Wallis various schemes for new algebra texts.[189] Perhaps one of them was a book that would contain an account of 'improvements of Algebra in England', for in 1673 or thereabouts Wallis began to write just such a book, *A Treatise of Algebra, both Historical and Practical, Shewing the Original, Progress, and Advancement thereof, from Time to Time; and*

---

[184] Rigaud, *Correspondence*, ii, p. 196. One such 'knotty problem' became known as 'Fermat's Last Theorem' and was not finally proved until 1995.

[185] Rigaud, *Correspondence*, ii, pp. 196–7 and Turnbull, ed., *Gregory Tercentenary Volume*, pp. 194–5.

[186] J. Collins, 'An Account concerning the Resolution of Equations in Numbers', *Philosophical Transactions*, 4 (1669), pp. 929–34.

[187] BL, MS Add. 4474, fos. 1–4.

[188] For one such scroll see BL, MS Add. 4417, fo. 422 (and fo. 425^v).

[189] Rigaud, *Correspondence*, ii, pp. 515–6, 526, 552, 556.

by what Steps it hath Attained to the Heighth at which now it is. The manuscript was delivered to Collins in 1677,[190] and the book was eventually published in 1685. Although supposedly a general history of algebra, eighty-five of the one hundred chapters of Wallis's book are devoted to 'improvements of Algebra in England' at the hands of Oughtred, Harriot, Pell, and others.

The section ostensibly devoted to Pell is shorter than those given to Oughtred or Harriot, and only seven chapters (57 to 63) carry the running head 'Of Dr Pell's Algebra'. In Chapters 57 and 58, Wallis discussed indeterminate problems in a fairly general way, and then in Chapter 59 gave a long example of Pell's three-column Method taken verbatim from *An Introduction to Algebra* (Pell's Problem 29, an alternative solution to Diophantus V.19). In Chapter 60 Wallis turned to something different, Brereton's problem: $aa + bc = 16$, $bb + ac = 17$, $cc + ab = 18$, and devoted Chapters 60 to 63 to its solution. This was still under the heading 'Of Dr Pell's Algebra', but Wallis contrived to introduce the problem without mentioning Pell, and instead attributed it to Silas Titus. Given what we know from Pell's own papers: that the problem was put to him by Brereton in 1649, that he filled scores of sheets with attempts to solve it, and that he and Wallis worked together in 1662 and 1663 on the eventual solution, Wallis's failure to acknowledge him seems perverse. It was Pell himself, however, who erased his name from the record. Wallis had deposited his copy of the solution with Collins, who labelled it with the words: 'Dr Wallis his Resolution of an exercise upon a probleme put by Dr Pell'.[191] Some time later, Pell changed this wording, as he recorded on a slip of paper inserted into his own copy:[192]

> On the outside Mr. Collins had written thus
> Dr Wallis his Resolution of an exercise upon a probleme put by Dr Pell.
>
> I inserted seven words: so y^t now upon his paper it stands thus
> – put by Colonel Titus, who had received it from Dr Pell.

Wallis's introduction to the problem in *A Treatise of Algebra* follows Pell's revised wording; the problem, said Wallis, was 'proposed to my self, long since, by Colonel Silas Titus'.[193] Several lines later Wallis added 'The Process of which (because I understood from the Colonel, it was a Question Proposed by Dr Pell) I drew up in general Terms, (after Dr. Pell's Method, with which the Colonel was

---

[190] Rigaud, *Correspondence*, ii, pp. 606–7.

[191] Collins seems to have distributed the problem fairly widely. In 1672 James Gregory referred to it in a letter to Collins as 'that numerical problem which Dr Pell proposed to Dr Wallis'; Rigaud, *Correspondence*, ii, p. 229, and Turnbull, ed., *Gregory Tercentenary Volume*, pp. 212 (though Turnbull wrongly supposed (in his n. 8 there) that this may have been a reference to 'Pell's Equation'), 178. In October 1677 Robert Hooke recorded in his diary that he too borrowed a solution to the problem from Collins.

[192] BL, MS Add. 4411, fo. 361.

[193] Wallis, *A Treatise of Algebra*, p. 225.

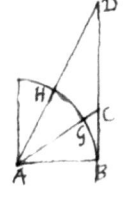

ILLUSTRATION 1. Letter 37, Cavendish to Pell (letter as sent), 27 July/6 August 1645 (BL, MS Add. 4278, fo. 210ʳ; by permission of the British Library).

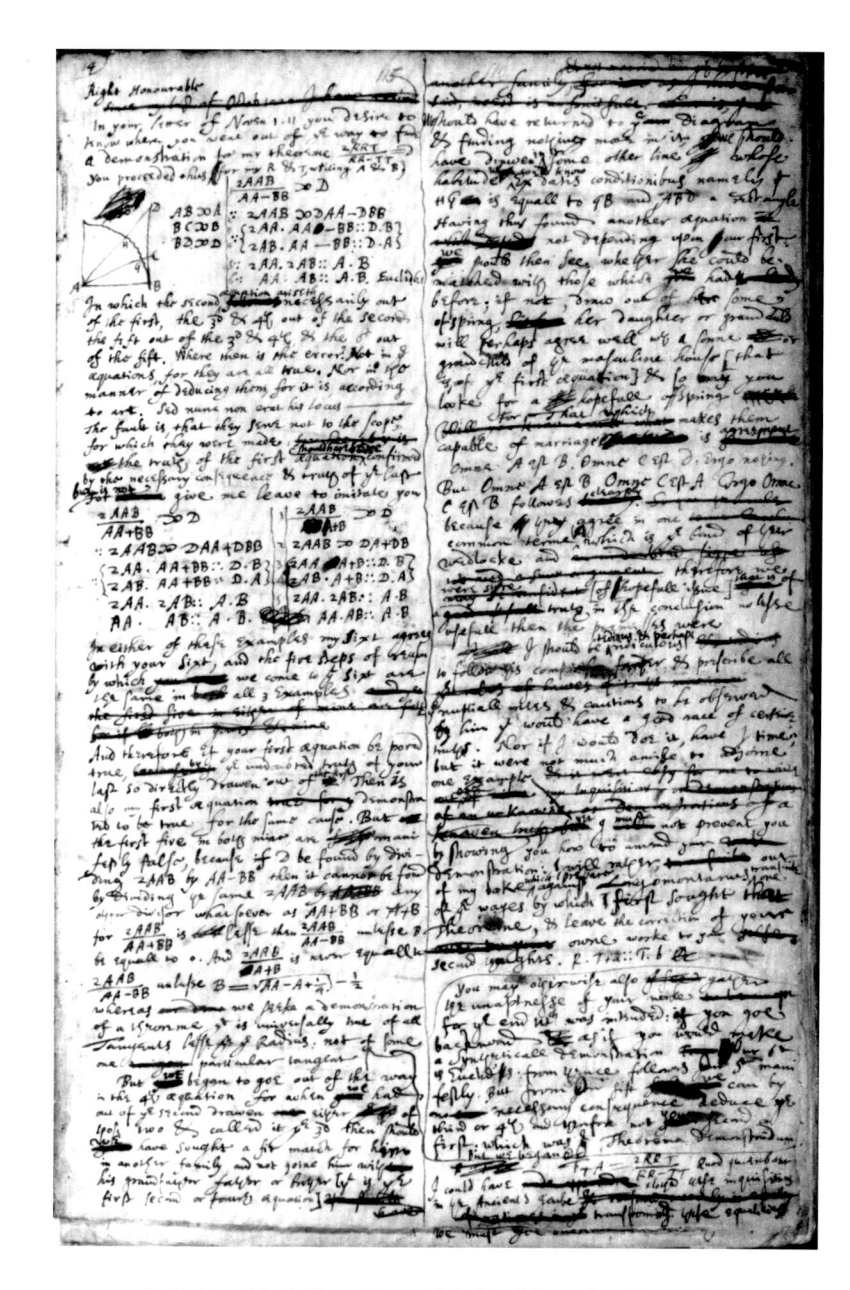

ILLUSTRATION 2. Letter 44, Pell to Cavendish (draft), 15/25 November 1645 (BL, MS Add. 4280, fo. 115ʳ; by permission of the British Library).

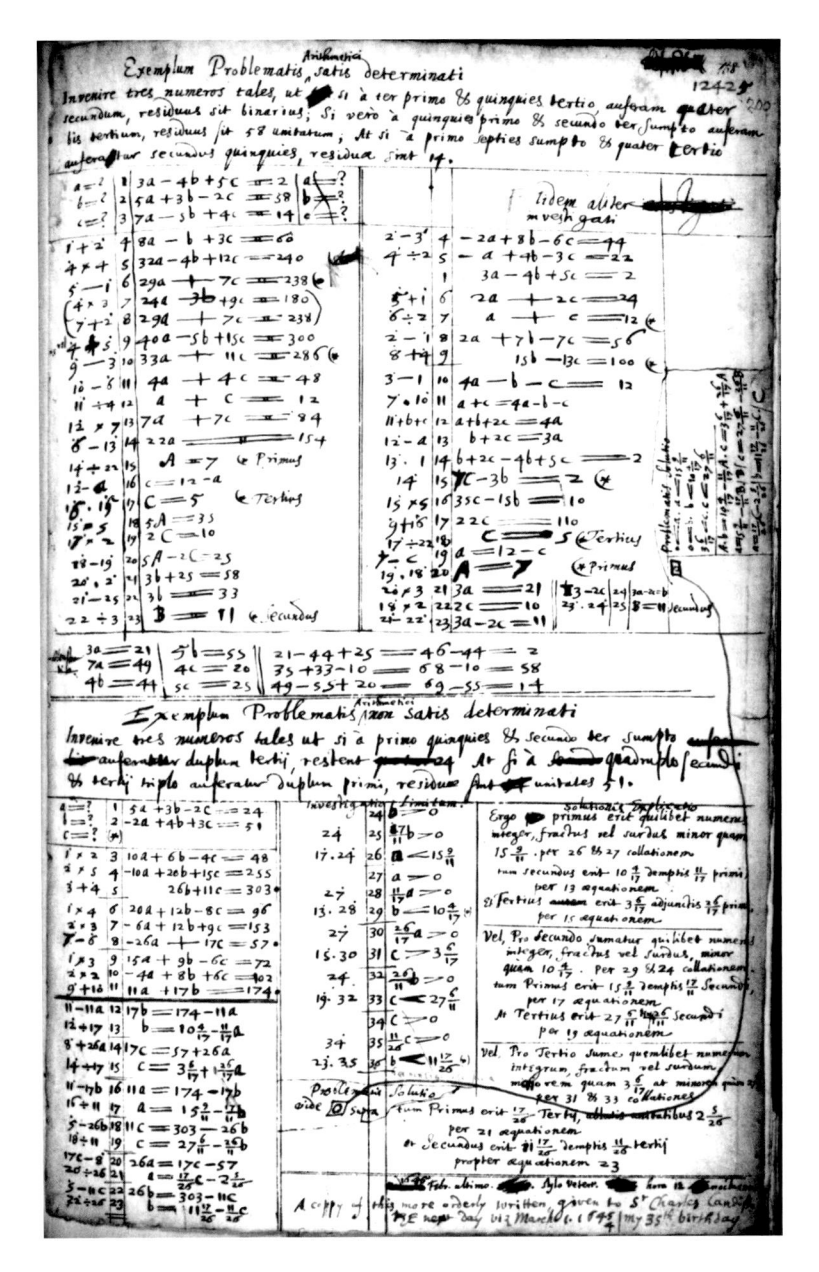

ILLUSTRATION 3. Pell's three-column method used to solve simultaneous equations. At the bottom right Pell has dated the sheet: '1645 Febr. ultimo. stylo veter. hora 12 nocturna', and has then added in red ink: 'A coppy of this more orderly written, given to Sr Charles Candish the next day viz. March 1. 1645/4 My 35th birthday'. At the top right, also in red ink, is Pell's age in days, 12425 (BL, MS Add. 4415, fo. 200^r; by permission of the British Library).

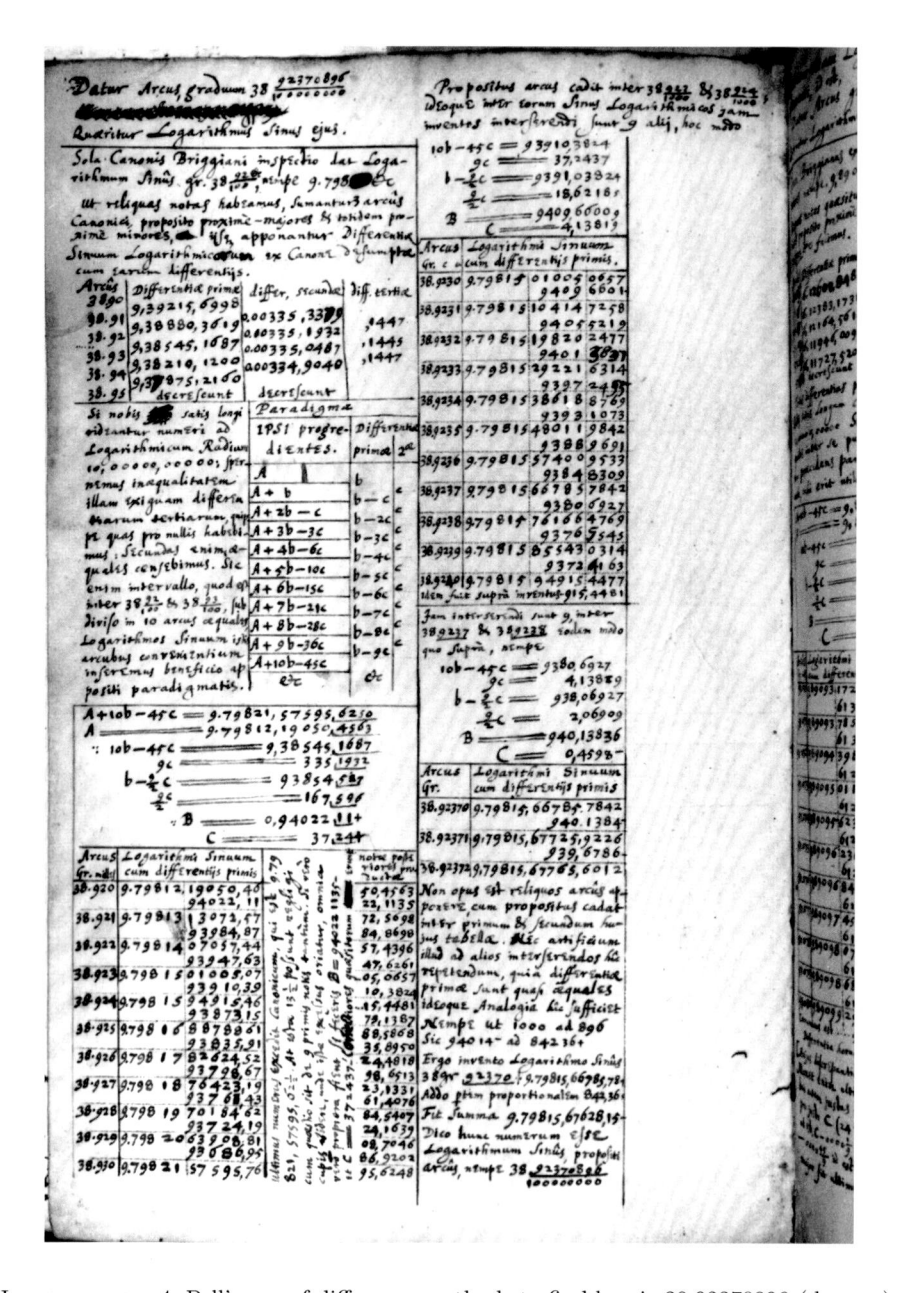

ILLUSTRATION 4. Pell's use of difference methods to find log sin 38.92370896 (degrees). The difference table headed $A + b$ in the first column has been cut out and patched from behind, so neatly that the repair is visible only from the back of the page (BL, MS Add. 4415, fo. 166^v; by permission of the British Library).

PART II: THE MATHEMATICS OF JOHN PELL   315

well acquainted)'. Wallis was evading the truth for he knew well enough that in 1662 Titus was little more than an interested bystander, and that the problem was posed and investigated at length by Pell. Nevertheless, he was willing to collude with Pell's desire to keep his name hidden.

This is not the only example of Wallis protecting Pell's anonymity. In Chapters 10 and 11 of *A Treatise of Algebra* Wallis explored at length the problem of approximating the ratio of the perimeter of a circle to the diameter by a fraction: $\frac{22}{7}$, $\frac{355}{113}$, and so on, exactly the problem that had exercised Pell in 1636. Wallis's introduction to the problem contains many echoes of Pell's. Where, for example, Pell wrote: 'we will enquire ... how [Metius] found this proportion in small numbers',[194] Wallis wrote: 'I find some have been wondering by what means Metius came to light upon those Numbers'.[195] And Wallis's method of working, using long division to ever greater degrees of accuracy was exactly the same as Pell's, though taken to much greater lengths. There can be little doubt that Wallis's treatment was based on Pell's, but once again he made no reference to him; instead he attributed the problem to Edward Davenant, whom he described in much the same terms as he described Titus, as 'very well skilled in the mathematicks, and a diligent proficient therein'.

It is possible to trace something of the role of Edward Davenant's interest in the problem through John Aubrey, who once learned a little mathematics from Davenant and in 1659 transcribed some of his algebra.[196] Towards the end of Aubrey's notes there appears the fraction $\frac{3456789}{9876543}$, and underneath it Aubrey has written:[197]

> Reduce this Fraction, to the nearest fraction that possibly can bee, whose Denominator shall bee only of three places, and the Numerator not to exceed three places. By this means (saith Doctor Davenant) [was *deleted*] is found out the Proportion of the Diameter to the Circumference.

The note suggests that Davenant had some inkling of the possibilities of the problem well before 1659, and indeed he may first have learned of it from Pell in 1636 or thereabouts, when his uncle, John Davenant, Bishop of Salisbury, was a friend and supporter of Hartlib and Dury.[198] In a later hand, under the same note, Aubrey has added:

> Dr. J. Pell hath donne this admirably well in one side of a sheet of paper, sc. ye backside of ye catalogue List of the Fellowes of the

---

[194] BL, MS Add. 4416, fo. 31.

[195] Wallis, *A Treatise of Algebra*, p. 36.

[196] See Aubrey, *'Brief Lives'*, i, p. 201. Aubrey's copy of Edward Davenant's algebra is now at Worcester College, Oxford, MS 64.

[197] Worcester College, Oxford, MS 64, fo. 62.

[198] We do not know Edward Davenant's date of birth, but he died in 1680 and so was more or less a contemporary of Pell's.

R. Society, and may be found amongst his papers. It was donne (I remember) at ye request of S^r Ch. Scarborough about A° D^n 1677.

The problem was in fact circulating amongst some members of the Royal Society a year earlier, in 1676, for on 19 August of that year Collins wrote about it to both Baker and Wallis, and mentioned Davenant by name. The problem as Collins presented it to them was exactly as Pell had posed it in 1636, to find the ratio of the diameter to the perimeter of a circle in its simplest terms.[199] Davenant had evidently grappled with this problem (as Titus had with Brereton's problem) and had found some solutions, but in a letter to Collins on 31 August Wallis expressed doubts as to whether Davenant's method was general enough. By the next day, 1 September, Wallis had checked his own papers on the subject (copies of which were already held by Collins), and confirmed that he had found 'hundreds' of solutions omitted by Davenant.[200]

Wallis's letter shows that he had worked on the problem in depth some time before it was publicly associated with Davenant in August 1676, and had already deposited his solutions with Collins, so the ascription to Davenant in *A Treatise of Algebra* was less than half the truth of the matter. The method Wallis applied, the accuracy and intricacy of the calculations, and the amount of labour expended on a problem that was of little obvious use, suggest that Pell collaborated with him (making it easy for Pell to perform the process 'admirably well' for Scarborough, Aubrey, and other bystanders in 1677). The calculations published in Chapters 10 and 11 of *A Treatise of Algebra* go to lengths that leave the modern reader awestruck, but to Pell or Wallis they probably represented no more than a few hours or days well spent.[201]

Once one is alerted to Pell's mathematics in *A Treatise of Algebra*, the examples multiply. Chapter 55, for example, contains a detailed explanation of Descartes' method of writing a biquadratic as a product of two quadratics. Wallis allowed his readers to suppose that this was his own treatment of the problem, but we know that the same work was done by Pell and presented to Cavendish in 1649. (It was also published by Kinckhuysen in 1661, though Wallis was unaware of Kinckhuysen's treatment when he wrote *A Treatise of Algebra*.)

Wallis's chapter on biquadratic equations follows immediately after twenty-five chapters on the work of Thomas Harriot. These chapters became both famous

---

[199] Rigaud, *Correspondence*, ii, pp. 8–9 and pp. 589–90.

[200] Rigaud, *Correspondence*, ii, pp. 589–90; the contents of Wallis's letter of 31 August can be surmised from the opening paragraph of his letter of 1 September.

[201] Wallis first published the work in an appendix entitled 'De rationum et fractionum reductione' in the second edition of the *Opera posthuma* of Jeremiah Horrocks (London, 1678, first edition 1673). For a further reference (naming neither Davenant nor Pell) see Rigaud, *Correspondence*, ii, p. 605.

For a modern analysis of the problem see D. Fowler, 'An Approximation Technique and its Use by Wallis and Taylor', *Archive for the History of Exact Sciences*, 41 (1990), pp. 189–233.

and controversial for Wallis's apparently exaggerated claims for Harriot and his repeated accusations that Descartes had made use of Harriot's work without acknowledgement. Pell's name was not mentioned anywhere in these chapters (except once in passing as the source of a story about Cavendish and Roberval), but eight years after Pell had died, Wallis's finally revealed how much he was indebted to him:[202]

> From [Pell's] own mouth I wrote down what I have said on this matter, and after it was written down, I showed it to him (to be examined, changed, or emended as he decided or preferred) before it was submitted to the press, and everything that was published was said with Pell's assent and approval.

Wallis, like Pell, understood Harriot's algebra well, and he was able to give a much more lucid account of it than Warner had done. In particular, he was able to fill the gaps that Warner had left in Problems 19, 20, and 21 of Section 3 of the *Praxis*, remarking:[203]

> Note, That in these three last Examples, Mr. Warner (the publisher) takes notice of some mistake in Mr. Harriot's Copy which he had: But would not adventure to restore it, but prints it as it was. It was, the omission of the latter of the two Suppositions in each Case: Which I have here supplied, according to Mr. Harriots mind.

The same problems, it was suggested above, were completed 'according to Mr. Harriots mind' by Pell for Aylesbury in 1639.

Even now it is not clear just how much more of *A Treatise of Algebra* was directly or indirectly influenced by Pell. Chapters 66 to 69, for example, immediately following the section on Pell's algebra, discuss the geometrical representation of complex numbers. We know that Pell investigated the way the roots of an equation changed as he systematically altered the coefficients, and in particular he was interested in the point at which real roots gave way to complex, or *vice versa*. Here, for example, is part of a table Pell drew up for the equation $xx - 6x + N = 0$, in which he let the 'homogeneum comparationis' or numerical term N range from 1 to 35:[204]

---

[202]'... & ex cujus ore descripsi quod hac de re dixi; eique postquam erat descriptum, ostendi, (examinandum, immutandum, emendandum pro arbitrio suo, siquid alias dictum malit) antequam prelo subjiceretur, totumque illud quod inde prodiit, assentiente & approbante Pellio dictum est.' Wallis, *Opera mathematica*, ii, 'De Harrioto addenda', sig. b1.
[203] Wallis, *A Treatise of Algebra*, p. 151.
[204] BL, MS Add. 4428, fo. 220.

$$xx - 6x + nu[\text{number}] = 0$$

nu	$x =$	$x =$
1	$3 + \sqrt{8}$	$3 - \sqrt{8}$
2	$3 + \sqrt{7}$	$3 - \sqrt{7}$
3	$3 + \sqrt{6}$	$3 - \sqrt{6}$
.		
8	$3 + \sqrt{1}$	$3 - \sqrt{1}$
9	3	3
10	$3 + \sqrt{-1}$	$3 - \sqrt{-1}$
11	$3 + \sqrt{-2}$	$3 - \sqrt{-2}$
.		
35	$3 + \sqrt{-26}$	$3 - \sqrt{-26}$

As N passes through 9, the roots change from real to complex, a change carefully underlined by Pell in his table. (A similar example for a cubic equation is to be found in Rahn's *Teutsche Algebra*: see above, pp. 299.[205]) Collins knew something of this idea and tried to describe it to Wallis:[206]

> I have seen two long scrolls, or tables of numbers, relating to a biquadratic equation, having all the powers extant;[207] in one, the resolvend or homogeneum comparationis [the numerical term] hath been kept constant, and the coefficient of any one term at a time altered successively in an arithmetical progression, till one or more of the roots have lost their possibility, and the said table hath given the roots very near. The other keeping the coefficients constant hath given the roots all along when the resolvend hath been an arithmetical progression. And such tables, [Pell] saith, are not very hard to make in relation to any equation, and that he disuseth the general method.

The table given above, for $xx - 6x + N = 0$, is an example of the second kind of table described by Collins, in which the numerical term increases in equal steps, or arithmetic progression. It shows how the roots become equal when $N = 9$, and 'lose their possibility' for greater values of N. Indeed, the imaginary part of the root becomes greater as N increases, and so is an indication of how far the equation is from being solvable (for real roots). Again Collins had some understanding of this:[208]

---

[205] Rahn, *Teutsche Algebra*, pp. 150–1.
[206] Rigaud, *Correspondence*, ii, pp. 472–3; see also ii, p. 247, para. 15.
[207] Possibly BL, MS Add. 4417, fos. 422 and 425[v]; see also Rigaud, *Correspondence*, ii, pp. 601–4.
[208] Rigaud, *Correspondence*, ii, p. 481.

These impossible roots, saith Dr. Pell, ought as well to be given in number as the negative and affirmative roots, their use being to shew how much the data must be mended to make the roots possible.

Wallis described complex roots in a similar way, as indicators of how far an equation was from being solvable, and his geometrical constructions in Chapters 66 to 69 showed such information visually. In the *Teutsche Algebra* too, the table of roots was linked to a geometrical construction, and we may recall Pell's belief that '*lines* may make a student of Algebra see a little more distinctly what is meant by *affirmative, negative & impossible* rootes'. At the end of Chapter 69 Wallis wrote:[209]

[The constructions] while declaring the case in Rigor to be impossible, shew the measure of the impossibility; which if removed, the case will become possible.

In other words Wallis, like Pell, was concerned with 'the measure of impossibility': how far an equation was from having real roots, and how much a given coefficient must be changed to render it solvable.

Wallis's discussion of geometrical representations of complex numbers is immediately followed by Chapter 70 which, by his own admission, does not fit comfortably into a textbook on algebra, and must have been a very late addition. It is entitled 'The geometrical construction of cubick and biquadratick equations' and reproduces results from Baker's *Geometrical Key* of 1684, a book that, as we have seen (above, p. 311), echoed themes that Pell had first explored in 1642.

Pell's influence may have extended also to the less mathematical and more historical parts of Wallis's text. In the opening chapters of *A Treatise of Algebra*, Chapters 1 to 5, Wallis drew heavily on the *De scientiis mathematicis* of Geeraard Johan Voskens (Vossius), professor of history at Amsterdam and a close colleague and friend of Pell's. We also know from Anthony Wood that Pell wrote a paper on the *Psammites*, or *Sand Reckoner* of Archimedes, and in Pell's papers we have the beginning of a translation into English of the *Psammites* together with a few calculations.[210] If there was more it is now lost, but in Chapter 6 of *A Treatise of Algebra* Wallis carefully explained the method used in the *Psammites* for describing large numbers. The subject has little to do with algebra in the strictest sense; Wallis justified its inclusion on the grounds that it was part of the history of the numeral system, without which algebra could not easily be managed, but it might well have been Pell who persuaded him to include it. In 1676, about the time he was writing *A Treatise of Algebra*, Wallis translated the *Psammites* into Latin, but whether or not Pell was involved in that project, we do not know.

---

[209] Wallis, *A Treatise of Algebra*, p. 272.
[210] BL, MS Add. 4411, fos. 114–5.

It seems that Wallis went on adding snippets of information from Pell even after the first draft was completed. When he sent the manuscript to Collins in 1677 he wrote: 'you may mind me also of the names of ancient algebraists of our own before Vieta. Such I have seen, but have forgot their names'.[211] Collins could not help much, for there were few English algebraists to consider, but one new writer did appear in Wallis's 'Additions and Emendations': John Tapp. His credentials as an algebraist were based only on *The Pathway to Knowledge* of 1613, a little known and not very good translation of Nicolaus Petri's *Practique om te Leeren Reekenen*, but he had been one of Pell's earliest contacts in the world of mathematics. Finally, in Chapter 2 of the Latin translation of *A Treatise of Algebra* in 1693 we find Wallis arguing that the name Sacrobosco was a Latinized form of 'holy bush', while in Aubrey's notes on Sacrobosco, we find: 'Dr Pell is positive that his name was Holybushe'.[212]

Pell's direct or indirect influence on *A Treatise of Algebra* can no longer be doubted. The chapters containing topics initiated, influenced, approved, or previously worked on by Pell constitute almost half the book, making it as significant as *An Introduction to Algebra* as a repository of Pell's work and ideas. In particular, *A Treatise of Algebra* recovers some of the work of Pell's early years: his questions about the ratio of the perimeter to the diameter of a circle from 1636; his familiarity with Harriot's algebra and, perhaps, his completion of Warner's unfinished problems from 1639; his interest in indeterminate equations from the early 1640s; his treatment of biquadratics from 1649; his work on Brereton's problem from 1649 and 1662; and his interest in complex roots as a measure of the 'impossibility' of polynomial equations from the 1650s onwards.

We cannot leave a discussion of Pell and Wallis without mentioning 'Pell's equation': $x^2 + Ny^2 = 1$, for integers N, x, y. The equation has not been mentioned until now because Pell never worked on it (except to rewrite Brouncker's solution from the *Commercium epistolicum* of 1658 in his own three-column Method).[213] It was Euler, many years later, who came across Brouncker's solution in the Latin edition of Wallis's *Treatise of Algebra* and erroneously replaced Brouncker's name by Pell's.[214] There is a double irony here: first, that Pell who was otherwise so reticent became universally known for an equation he never worked on; and second, that his name emerged only by mistake from a book in which it was otherwise so well hidden.

One more topic must be discussed in some detail because it came up so many times in Collins's correspondence between 1667 and 1676: Pell's claim that he had developed a method far superior to Viète's for solving equations.[215] Collins tried on many occasions to describe Pell's techniques, but he himself had only the

---

[211] Rigaud, *Correspondence*, ii, p. 607.
[212] Aubrey, *'Brief Lives'*, i, p. 408.
[213] BL, MS Add. 4414, fo. 254.
[214] See J. A. Stedall, 'Catching Proteus: The Collaborations of Wallis and Brouncker. II: Number Problems', *Notes and Records of the Royal Society*, 54 (2000), pp. 317–1, esp. p. 325.
[215] Rigaud, *Correspondence*, i, pp. 247–8.

most rudimentary grasp of them, and has muddied the waters by haphazardly describing elements of several different approaches.[216] Collins wrote more than once, for example, that Pell could solve any equation using tables of sines or logarithms, a claim that James Gregory, a much more skillful mathematician, gravely doubted.[217] The techniques that Pell did know and use seem to have been those that follow.

The first of Pell's techniques was the removal of intermediate terms (a method now known as 'Tschirnhaus transformation' following a paper on the subject by Tschirnhaus in the *Acta eruditorum* in 1683).[218] It is not clear how Pell performed his operations, and in the general case the method is neither easy nor helpful, except for quadratic and cubic equations where it leads to the usual well-known formulae.

A second approach was the method Pell had probably learned from Hérigone in 1644 for solving a cubic equation by converting it to the angle trisection equation, $3p^2x - x^3 = 2q^3$. Pell seems to have thought that he could solve higher degree equations by a similar method if only he had time to work it out, as Collins reported, again to Gregory, in 1671:[219]

> Being in discourse with D P, I [Collins] told him that Dulaurens had made such Aequations of the 3rd, 5th, 7th, 9th degree, as might be solved by finding
> 2 meanes or trisection
> 4 quinquisection
> 6 septisection } etc.
> 8 nonisection
>
> and asked whether all[220] Aequations of odde degrees might not be solved by the like methods, he said they might, and that he was in these Aequations always sure to find a root possible, that this Argument would be the subject of a treatise he long intended, to bear the title of Canon Mathematicus, that herein he should have to deall with the Chords of divers figures, besides the Circle, Ellipsis &c.

Viète and Briggs had written down the fifth, seventh, and ninth degree equations for quinquisection, septisection, and nonisection (cyclotomic equations) long before Dulaurens did so, but the point at issue here was whether *all* equations of

---

[216] See ibid., i, p. 247; ii, pp. 197, 215–6, 219–20, 243–8, 472–3, 526–7, 601–4.

[217] Ibid., ii, pp. 223, 230, and Turnbull, ed., *Gregory Tercentenary Volume*, pp. 187, 211.

[218] Rigaud, *Correspondence*, ii, pp. 219–20; see also i, pp. 213–5. For Tschirnhaus's method see E. W. von Tschirnhaus, 'Nova methodus auferendi omnes terminos intermedios ex data aequatione', *Acta eruditorum* (1683), pp. 204–7.

[219] Rigaud, *Correspondence*, ii, p. 198 and Turnbull, ed., *Gregory Tercentenary Volume*, pp. 195–6.

[220] Rigaud here has 'all' whereas Turnbull has 'also'; the sense of Collins's query seems to require the former reading.

those degrees could be solved by angular section equations. The answer (from Galois theory, much later) is that there exist quintic, septic, and nonic equations that cannot be reduced to cyclotomic equations. Pell seems to have had some inkling that he needed to go beyond simple circular functions, but it is doubtful that he got very far towards the full answer. Collins tried to send a fuller description of the method to Leibniz through Oldenburg, but knew that he was out of his depth, and begged Oldenburg 'to report it on account of your own knowledge, lest coming from me only it should seem incredible; and for me that am ignorant to give an idea of it, will be somewhat difficult'.[221]

The third method of solving equations was to plot a graph of the associated curve, which could then be used to estimate the number and nature of the roots for different values of the *homogeneum comparationis* or constant term. To use an example that Collins wrote out in 1669,[222] a few calculated values of $x^4 - 4x^3 - 19x^2 + 106x$ for integer values of x show that $x^4 - 4x^3 - 19x^2 + 106x = 120$ has four real roots (2, 3, 4, and -5), while $x^4 - 4x^3 - 19x^2 + 106x = 840$ has only two (7 and -6). If the values are plotted as ordinates against an x-axis, the graph can be used for establishing (by inspection) what Collins called the 'limits' of an equation, of which, he said, there were two kinds: the 'dioristic limits', or what we would now call the maximum and minimum values, and the 'base limits', or the least and greatest roots. Pell considered such limits important, for once he had them he was able to find solutions as accurately as he chose by use of logarithms:[223]

> Dr Pell asserts that in any Aequation, after he hath the Limits, viz where the Serpentine Curves for Aequations crosse the Base Line (when it so happens) and the greatest Ordinate or Homogeneum in the said Portion of the Serpentine Curve given that then he finds the Logarithmes of the rootes so precisely by Logarithmicall Operations, that the Logarithme found shall not erre an Unit in the last figure.

In a letter to von Tschirnhaus four years later Collins was able to be more explicit:[224]

---

[221] Rigaud, *Correspondence*, i, pp. 243–5.

[222] Collins sent this example to Barrow in June 1669 and to Gregory in 1670: see Turnbull, ed., *Gregory Tercentenary Volume*, pp. 109 and 113–5. A MS copy was preserved by Aubrey and is now in Worcester College, Oxford, MS 64 (amongst unpaginated loose papers at the end of the volume).

[223] Rigaud, *Correspondence*, ii, pp. 219–20 (see also p. 473), and Turnbull, ed., *Gregory Tercentenary Volume*, p. 180. Collins wrote further to Gregory about Pell's use of limits in 1675 (p. 298). Turnbull (in his n. 2 there) describes this as 'at last a statement of Pell's method', but it is even less clear than many of Collins's earlier attempts to explain Pell's work.

[224] Rigaud, *Correspondence*, i, pp. 216–7; see also related descriptions in vol. ii, pp. 219, 526–7.

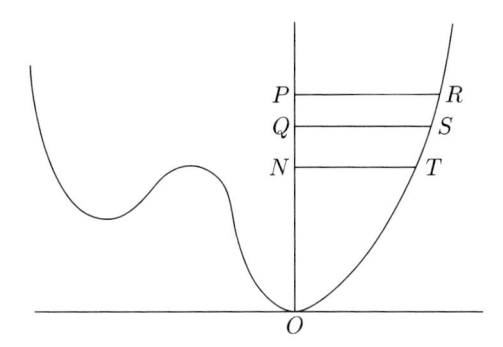

FIG. 3.

As the difference of the logarithms of ON and OP
Is to the difference of the logarithms of NT and PR;
So is the difference of the logarithms of ON and OQ
To the difference of the logarithms of NT and QS.

In modern terms Pell was finding an interpolated value OQ between ON and OP on the assumption that locally $y = x^r$ (so that $ON = NT^r$, $OQ = QS^r$, and $OP = PR^r$), and if the calculated value was not good enough it could be refined to the required degree of accuracy. This, thought Collins, was the true purpose of the tables of antilogarithms that Pell and Warner had taken such pains to calculate.[225]

As mentioned above, the graph could be used to estimate the 'dioristic limits', or maximum and minimum values, of a polynomial, but Pell also knew an algebraic method for finding such values. In 1682 Collins drafted a letter headed 'To describe the Locus of a Cubick Aequation',[226] and after several numbered points he wrote:

> 17. The Learned Dr Pell hath often asserted, that after the limits of an Aequation are once obtained then it is easy to find all the Rootes to any Resolvend offered.
>
> Now for instance (according to Hudden's method) in a biquadratick Aequation you must multiply all the termes beginning with the Highest and so in order by 4. 3. 2. 1 & the last terme or Resolvend by 0, whereby it is destroyed, and you come to a Cubick Aequation.

---

[225] Ibid., i, pp. 215–6; ii, pp. 197, 218–9.
[226] Collins's letter was published posthumously in *Philosophical Transactions*, 14 (1684), 575–82, from a copy sent to Wallis. A MS copy was kept by Aubrey and is now amongst his mathematical papers in Worcester College, Oxford, MS 64. The letter mentions 'divers of your surd canons' and so was probably intended for Baker, whom Collins described elsewhere as 'a learned analyst, and a person fit to labour in discovering canons for the surd roots of equations'; see Rigaud, *Correspondence*, i, p. 212.

Thus from the biquadratic equation $x^4 + ax^3 + bx^2 + cx + d = 0$ was derived the cubic $4x^3 + 3ax^2 + 2bx + c = 0$, and Collins went on to explain how the roots of this cubic, substituted back into the original equation, gave the maximum and minimum values. This was also the method he tried, less successfully, to describe to Wallis in 1671:[227]

> I have heard discoursed by Dr. Pell, to wit, that he finds the limits of high equations made by the multiplication of the known roots [sc. terms] ascendendo ...

Pell's method was not, as first appears, based on a knowledge of differential calculus, but on Hudde's Rule, first published in 1659 in van Schooten's *Geometria*, for finding maxima and minima from double roots.[228]

Pell's fourth and perhaps most typical approach is familiar from what we have seen of his methods already: the tabulation of certain values of a polynomial (as required also for sketching its graph), followed by interpolation by a method of constant differences.[229] A detailed example was written out by Collins in 1669 in a 'Narrative about Aequations', and once again it seems that he had learned much of what he knew from Pell.[230] In Collins's table (below) the cubic polynomial in the second column is the product of the linear and quadratic polynomials in the first and third. The quadratic (set equal to zero) is easily solved, thus giving the three possible roots of the cubic (set equal to zero). All this, of course, is similar to the method described ten years earlier by Rahn in the *Teutsche Algebra* in 1659 (see p. 299).

Root 1st	Cubic equation.		Root 1.	Root 2.	Root 3.
$a - 1$)	$a^3 - 15a^2 + 54a - 40$	$(a^2 - 14a + 40$	1	4	10
$a - 2$)	$a^3 - 15a^2 + 54a - 56$	$(a^2 - 13a + 28$	2	2.7251−	10.2749+
...					
...					
$a - 8$)	$a^3 - 15a^2 + 54a + 16$	$(a^2 - 7a - 2$	8	−0.2749+	+0.2749+
$a - 9$)	$a^3 - 15a^2 + 54a \pm 0$	$(a^2 - 6a \mp 0$	9	0	6
$a - 10$)	$a^3 - 15a^2 + 54a - 40$	$(a^2 - 5a + 4$	10	1	4
$a - 11$)	$a^3 - 15a^2 + 54a - 110$	$(a^2 - 4a + 10$	11	impossible	
$a - 12$)	$a^3 - 15a^2 + 54a - 216$	$(a^2 - 3a + 18$	12		
...					
...					
$a - 17$)	$a^3 - 15a^2 + 54a - 1496$	$(a^2 + 2a + 88$	17		

---

[227] Rigaud, *Correspondence*, ii, p. 526.

[228] J. Hudde, 'De maximis et minimis' in R. Descartes, *Geometria editio secunda*, ed. and tr. F. van Schooten, 2nd edn., 2 vols (Amsterdam 1659–61), i, pp. 507–16. Pell had met Hudde in Amsterdam in 1645 (see above, Part I, ch. 4, n. 19), so he might well have taken a special interest in his later work.

[229] See Rigaud, *Correspondence*, i, p. 247, and ii, pp. 472–3.

[230] Turnbull, ed., *Gregory Tercentenary Volume*, pp. 109, 116–17. In 1676 Collins sent the same material to Wallis: Rigaud, *Correspondence*, ii, pp. 601–3. A MS copy is preserved in Worcester College, Oxford, MS 64.

Collins also displayed columns of differences for the successive values of $a^3 - 15a^2 + 54a$ and of $a^2 - Na$:

```
   ...                              ...
  −16                               +2
         +16                                +2
   ∓0                               ∓0             +2
         +24                                +2
              +6                           +4
         +40                               +2
  +40                               −4
         +30                               +2
              +6                           +6
         +70                               +2
 +110                              −10
         +36                               +2
              +6                           +8
        +106                              
 +216                             −18
   ...                             ...
```

For anyone familiar with interpolation from constant differences (in which Pell was expert) it was possible to interpose intermediate values to any degree of accuracy, thus solving $a^3 - 15a^2 + 54a - N = 0$ for any value of N.

This is a numerical, not an algebraic, method of solving equations, and is not very practical in that it requires a fresh table for each equation, as well as considerable skill in interpolation. For Pell, so much at home with tables and interpolation of tables, the method would clearly have had great appeal. The 'learned person' whose work was described in Collins's 'Improvements of Algebra in England' was said to have devised:[231]

> 2. A Progressional table made with severall Columnes for the easy finding of the rootes without the ayd of the Generall method, and this part of the Doctrine ... seemes to be no other but a Generall method of interpolating such ranks whose third, fourth, fifth, sixth Differences are equall.

The 'Generall method' was the numerical method taught by Viète in *De numerosa potestatum resolutione* (and described by Harriot as the *via generalis*).[232] Pell may have first developed his alternative, interpolative, method in the 1640s in conjunction with Warner, from whom he learned much about difference methods, for he reported that Warner described Viète's method as 'work unfit for a Christian' (possibly a reference to its Islamic origins).[233] Pell surprised Leibniz in 1673 by telling him about examples of constant difference tables in a little-known book, the *Observationes diametrorum solis et lunae apparentium* of Gabriele Mouton, published at Lyon in 1670.[234] In a series of rather artificial

---
[231] BL, MS Add. 4474, fo. 1ᵛ.
[232] Stedall, *Greate Invention*, p. 201.
[233] Rigaud, *Correspondence*, i, pp. 247–8.
[234] Mouton, *Observationes*, pp. 368–96.

exercises Mouton applied constant difference methods to astronomical observations, but for Pell there could have been little new in what he wrote, for he himself had constructed and used constant difference tables quite as lengthy as those presented by Mouton.

It seems that Collins gleaned much from Pell, and frequently attempted to describe what he knew, but his accounts generally do more to confuse than to clarify. He always hoped, of course, that Pell might one day be persuaded to explain his methods for himself, but the truth of the matter was bluntly, but realistically, expressed in a letter to Gregory in 1670: 'Dr Pell communicates nothing'.[235]

By now we have seen numerous examples of work that Pell promised, but failed to deliver. There may, of course, have been completed treatises that are now lost. Aubrey listed several important items were left behind in Cheshire.[236] One was the 'greatest part' of Pell's edition of Diophantus. Another was Pell's exploration of the tenth book of Euclid's *Elements*, on incommensurable quantities, a book that exercised many mathematicians of the sixteenth and early seventeenth centuries. Pell is also supposed to have 'donne the second booke of Euclid in one side of a large sheet of paper most clearly and ingeniously'; this would have been an algebraic rendering of Book II, something that Harriot and Oughtred and no doubt many others tried their hand at as the new possibilities of algebraic geometry began to be understood. The other works mentioned by Aubrey have all been discussed elsewhere in this study, but there may well have been further documents that have simply disappeared without trace.

Any historian is used to dealing with incomplete evidence, but in Pell's case additional difficulties stem from his almost obsessional reluctance to reveal his name. His efforts to cover his traces, whether in anonymous publications, or in the prefaces to the *Teutsche Algebra* and *An Introduction to Algebra*, or in Wallis's *Treatise of Algebra*, now seem not only incomprehensible but somewhat inept. Nevertheless, to a large extent he succeeded, so much so that only now, three hundred years later, is a clearer picture of Pell and his mathematical interests finally beginning to emerge.

It is impossible not to conclude, and not to feel disappointed, that the promise of Pell's early years was never fulfilled. He was at his most enthusiastic and most creative between the ages of about seventeen and thirty, but from then on one senses a steady diminishing of his hopes and ambitions. Perhaps he became more realistic. But in many ways he also simply failed to move on. As a young man he was clearly excited by the new mathematics of the 1620s and 1630s: Briggs's logarithms, Oughtred's algebra, and Harriot's theory of equations, but he remained locked into those subjects that had first caught his imagination and never responded in the same way to the next wave of advances. Cavalieri's indivisibles, Descartes' algebraic geometry, Fermat's number theory, many explorations of

---

[235] Rigaud, *Correspondence*, ii, p. 220.
[236] Aubrey, *'Brief Lives'*, ii, p. 128.

quadrature and rectification, the possibilities of infinite series, and eventually the calculus itself, were all developed to levels of sophistication far beyond anything Pell achieved, and he was simply not part of that forward movement. Even Collins, who was loyal to Pell in many ways, and who considered him an expert in algebra, recognized his limitations:[237]

> Upon conference with the Doctor I do not find that he is knowing in the doctrine of infinite series; and although he grants they may be of good use, as to the theorems or rather habitudes thereby invented, yet as to the calculative or applicatory part, he says it may either be quite removed or exceedingly facilitated by his methods, which he will forbear to make common, till he first sees what will come out of Mr. Gregory's and Mr. Newton's ...

Time and again Pell thought that he could improve other people's work or present it in a form that he himself considered easier, more concise, more useful, or less expensive, but he added little of his own. Or else he responded to problems brought to him by others, and did so conscientiously and thoroughly, but without ever pushing much beyond the limits of what was already established. Even in those subjects in which he was most knowledgeable, he contributed little that was new. His reputation amongst his contemporaries was never in doubt, but it seems that it was based less on any specific achievement than on his general erudition, and his knowledge and understanding of the mathematics of others. He was not, by the standards of his or any other age, a great mathematician, but he was, nevertheless, a competent one. He may have lacked the vision of Harriot, the intuitions of Brouncker, the wisdom of Gregory, or the purposefulness of Wallis, but he was very much more skilled in mathematics than, for example, Warner, Cavendish, Aylesbury, Collins, or Dary, all of whom turned to him for help. It is hard to know how successful he was as a teacher. Huygens was cool in his appraisal of Pell, but Brereton and Titus became lifelong friends, while Rahn and Kinckhuysen, if the latter was indeed Pell's student, both went on to produce competent and worthwhile textbooks.

Pell was unique amongst seventeenth-century mathematicians in one area: his sense of the logical and deductive power of mathematics. First proclaimed in his private writings from around 1635 onwards, and publicly in his *Idea* in 1638, Pell's attempts both to use and create system and order are evident throughout his work, and the culmination was the three-column Method that became his hallmark. In his Method, Pell explored the possibilities of algebraic notation and symbolic reasoning to a much greater degree than any of his predecessors or contemporaries except perhaps Harriot, from whom he learned so much. The three men who influenced Pell and his mathematics most were Briggs (in logarithms and trigonometry), Harriot (in notation and theory of equations) and Hérigone (in the systematic approach to mathematical problems), but Pell also did much

---

[237] Rigaud, *Correspondence*, i, p. 248.

to develop his own idiosyncratic style, driven by a desire to make mathematics clear, logical, and comprehensible.

Pell was not one of the outstanding figures of seventeenth-century English mathematics but he remains an intriguing one. Mathematics was at the centre of his life for over fifty years, and for that reason alone he deserves the attention of historians of the subject. Pell was driven by his passion for mathematics until he died in 1685, but he had recognized his own obsession many years earlier, and we may leave him with this last word on the subject:[238]

> For there is no such εργοδιωχτης [taskmaster] as a mans owne Genius which at every little vacancy calls upon him with Pliny's *Poteras has horas non perdere* [you could not waste these hours], and is still setting him upon some particle or other of that study which he most affects, not permitting him to remember how uncertain it is who shall reape the fruit of all that Labour, whether it shall never be gainfull to himself or procure him *Viventi decus atque sentienti* [honour while he lives and feels]; Or whether he must not leave it to them that come after who perhaps will never understand or regard the effects of all that labour wherein he hath travailed.

---

[238] BL, MS Add. 4280, fo. 206^v.

Part III

The Pell–Cavendish correspondence

Textual introduction

That both sides of this correspondence with Sir Charles Cavendish survive in such a near-complete state is due to the methodical nature of John Pell, who preserved not only the letters he received from Cavendish but also his own drafts. Indeed, from the moment he resumed contact with Cavendish on the Continent in 1644, he systematically numbered his own drafts and also wrote numbers on Cavendish's letters. Pell's drafts are numbered from 1 to 47, and Cavendish's letters run from 1 to 59: these two sequences are used by van Maanen, in the form 'P1' and 'C1' respectively, in his tabulation of the entire correspondence, and are reproduced here, in the form 'van Maanen P1', at the head of each letter.[1] (For the correspondence before 1644 van Maanen has assigned letters: 'Pa', 'Pb', 'Ca', 'Cb', and so on.) Out of the two sequences numbered by Pell between 1644 and 1651, only three items are missing: P35, P36, and P37, which were written in the winter and spring of 1649–50. No trace of these letters has been found.

Pell's papers passed through the hands of the antiquary (and historian of the Royal Society) Thomas Birch in the mid-1750s; annotations in Birch's hands are to be seen on many of the letters. It was probably Birch who tried to collect all letters to and from Pell and arrange them in alphabetical order of writer. As a result, although many other letters remain scattered among Pell's papers, there are three bound volumes of his correspondence (BL, MSS Add. 4278–4280) in which the letters are placed roughly in alphabetical order of writer: Cavendish's letters are thus in MS Add. 4278, and Pell's drafts in MS Add. 4280. In a few cases, other copies made by Pell (of his own drafts, or of passages in Cavendish's letters) are found elsewhere among Pell's manuscripts, and some items sent by Pell are to be found in the much smaller surviving collection of Cavendish's papers. The listing given by van Maanen was confined to the letters and drafts contained in MSS Add. 4278 and 4280. Where duplicate material exists in other manuscripts, it is listed here in the head-note to the letter; the manuscript used as copy-text is the first item listed.

The letters from Cavendish pose no textual problems. The letters we have are the letters as sent; no prior drafts or retained copies survive. Cavendish wrote in a fairly large and regular italic hand; his spelling may occasionally give pause to modern readers ('on' for 'one', for example), but most will find that they adjust to it quickly. The only peculiarity of his writing is his use of the virgule or sloped stroke: /. This was traditionally used, in medieval manuscripts and sixteenth-century printed works (and printed works in German in the seventeenth century), to mark a slight caesura in the sense or construction – more or less as a comma.

---

[1] For the tabulation see van Maanen, 'The Refutation', pp. 348–52.

In Cavendish's writing, however, it marks a larger break, particularly a change of subject-matter. (Sometimes he uses a double virgule, //, for a stronger effect.) This comes closer to performing the function of a paragraph-break. Cavendish may indeed have employed it to some extent as a substitute for a paragraph-break (something he very seldom used), in order to fit more words onto the page. In some cases, however, a virgule comes at the end of each sentence in a row; if such a passage were broken into separate paragraphs, this would create an odd effect on the page. In this printing, therefore, the virgule is simply reproduced as such.

Pell's drafts are a little more problematic. His handwriting is generally legible: for formal writing he had developed an extremely neat, print-like script, and his informal hand can be read simply as a more cursive, cramped, and hasty version of it (with some vestiges of secretary-hand letter-forms, such as an 'h' below the line). But some of these drafts are very rough, with frequent crossings-out and interlineations; sometimes a whole section written elsewhere on the page is marked for insertion at a particular point. The main sequence of such rough drafts is to be found on a set of folio-sized pages (MS Add. 4280, fos. 106–35) divided by Pell into double columns. Not all the drafts are rough, however; in some cases (Letter 97, for example, on fo. 136^r), what is preserved is apparently a fair copy derived from some previous draft, and it has merely been lightly emended or added to in turn, in order to create the text from which the final version sent to Cavendish would presumably have been copied. (Similarly, the 'draft' of Letter 14 began as a fair copy of a letter dated 7/17 August 1644 but not sent; emendations were made and extra material added, to create the text of the letter of 10/20 August which was in fact sent.) In some cases we also have a fair copy made by Pell at the end of the writing process and retained by him; these are referred to in the head-notes as 'authorial copies'.

Where the drafts are concerned, the question naturally arises as to whether the draft that survives does in fact represent the text actually sent to Cavendish. Overall, the evidence suggests that while Pell might make major changes from one draft to the next, the last draft would correspond closely to the letter as copied out and sent. (In the case of one of his letters to Mersenne, we have a first draft, a second draft, and the letter itself: there are large differences between the first and second items, but none between the second and third.)[2] Letter 91 here survives in three forms: a rough draft, a second draft (in which changes marked on the rough draft are incorporated in the text), and an authorial fair-copy. Some changes were made between the second draft and the final version sent to Cavendish (which, we can presume, was recorded in the authorial copy). In the case of most of the letters, the lack of such authorial copies probably indicates that no further significant changes had been made, subsequent to the alterations present in the draft: this seems a reasonable assumption, but it is not entirely certain, as it depends on the further assumption that such copies,

---

[2] See *MC* ix, p. 57.

if made, would have survived. In one case, that of Letter 32, we do have the letter as sent, which has been preserved in a collection of Cavendish's papers: it corresponds very closely to the surviving draft. And in general it may be noted that whenever Cavendish responds to points raised in Pell's letters, his comments do correspond precisely to the material that is present in the final version of the draft as we have it.

For the modern reader, the fact that Pell's letters survive in draft adds an extra dimension of interest: it becomes possible to see his thought-processes in action, as he adds, deletes, and modifies his material. An edition may be either clear-text (presenting only a final version of the text on the page, and consigning all textual variants, etc., to an apparatus) or formulaic (using formulae in the text to mark the different status of material included there). An editor's choice between these two methods must depend largely on the nature of the text and the purposes for which it will be read. The editor of a poem by Milton or a treatise by Newton, for example, will normally want to present only the finished version of the text on the page: the aesthetic or intellectual unity of the work would be undermined by the intrusion of other material into the text itself. But a personal letter is not like a poem or a treatise; it seems a more private (and, often, provisional) thing, and the thought-processes of the writer may be a major focus of the modern reader's interest. For this reason a simplified method of formulaic transcription has been adopted here. In the case of the roughest drafts, this may create a somewhat rebarbative effect on the page; but readers will quickly become used to assimilating the material in square brackets and thereby following the flow of Pell's compositional process. For all the occasional awkwardness of this method of transcription, readers may appreciate the sense of immediacy it gives when, for example, they find Pell struggling to express his feelings about his enemy Longomontanus in Letter 44: 'an ignorant [dunce *deleted* bungler *deleted* >smatterer &] no such [cunning man *deleted* >Artist] as he would be thought to be'. It would be a pity if such vivid evidence of Pell's emotional state were to be buried in a textual apparatus where few readers, probably, would notice it.

The basic transcription aims to reproduce as accurately as possible the original text, altering or omitting only those kinds of detail that are of no importance to the study of its meaning. The original spelling is preserved, but long 's' and the ligatures 'æ' and 'œ' are normalized. Superscripts are preserved, but other sorts of contraction are expanded – silently if they are simple and unambiguous, but otherwise with the expansion placed within square brackets. The original punctuation is adhered to, even where this follows rules that are no longer recognized (for example, the use of a full point followed by a new clause which does not begin with a capital letter: this was a conventional way of signifying a caesura a little less strong than the ending of a sentence). Only in very rare cases (the missing second half of a pair of parentheses, for example) is punctuation emended, with the emendation presented in square brackets. Underlining is presented as italics; quotations are presented in single quotation marks, whether the text uses quotation marks or marginal lining. (In the latter case, the lining is recorded in a note.)

Editorial interventions, and the recording of information about the text in the text itself, are presented in square brackets. The most important of these are as follows. Where a deletion is legible, it is presented thus:

I have [sent *deleted*] delivered your letters

Given the frequency of deletions in the drafts, illegible deletions have not been recorded, and editorial judgement has been used in selecting which legible deletions to record: all significant material has (it is hoped) been presented, but if Pell has crossed out a few words merely because he has chosen to use those same words at a later point in the sentence, that deletion will not be recorded. Where Pell has added interlinear material, it is presented thus, using a sign reminiscent of a caret mark:

I have [>not yet] delivered your letters

Such material is placed where the original caret mark stands (if present) or should stand; but the original caret mark itself is not reproduced. Where an interlineation replaces material originally written on the line and then deleted, the deletion is presented first, and the two sets of square brackets are fused into one. Thus:

I have [already *deleted* >not yet] delivered your letters

Occasionally, given the nature of these drafts, bracketed material may appear within brackets. Thus an interlinear passage may itself contain a deletion:

I shall deliver your letter [>on [Monday *deleted*] Tuesday]

In general, such combinations will be self-explanatory. Equally self-explanatory are italicized statements about the text such as '[*page torn*]' or '[*blotted*]'. (Blotting, unlike deletion, is presumed to be unintentional.) This information may be combined with a conjectural restitution of the text, in the form:

I have [*page torn* deli]vered your letters

Also self-explanatory is material presented as follows:

I have [delivered *altered to* destroyed] your letters

Uncertain readings are placed in square brackets, with an italicized question-mark:

I have [never*?*] delivered your letters

Finally, a standardized order has been used for presenting the material in these letters. If the date or address is given at the head of the letter in the manuscript, it is silently transposed to fit this ordering: salutation; text of letter; valediction; place and date of writing; postscript; address; any annotation by Pell; any later annotation; enclosure.

The correspondence

Letter 1

26 June [/6 July] 1641

Cavendish to Pell, from Wellingor

BL MS Add. 4278, fos. 161–2 (original)
van Maanen Ca

Sr

I perceiue oure business of making the perspectiue glass,[1] proceeds not, & I know not well howe to help it, vnless there be some as good matter to make glass, in some other place to be bought, for it seemes, that, at Broadstreet,[2] will not be had; I am not willing to trouble Sr Robert Mansfeild[3] about it, though I thinke [you *deleted*] he would not denie me. Therefore if you or Mr Reaues[4] can finde fitting matter for vs somewhere els, ye should doe me a greate fauoure; Broadstreet I suppose will be the best place to make the glass, when ye

---

[1] As the further details supplied by Letters 2–10 show, Pell and Cavendish were trying to make a telescope in accordance with the design recommended by Descartes in the ninth 'discours' of his 'Dioptrique' (A&T vi, pp. 201–5), employing two hyperbolic lenses, one concave and the other convex. (See also the discussion in J. F. Scott, *The Scientific Work of René Descartes (1596–1650)* (London, n.d. [1952]), pp. 59–61.)

[2] Broad Street, in the City of London.

[3] Sir Robert Mansfield or Mansell (1577–1656), naval commander, was knighted at Cadiz in 1596, became Vice-Admiral of the Narrow Seas in 1603 and Vice-Admiral of England in 1618. He had obtained a monopoly of glass manufacture from James I; this was confirmed by Charles I in 'A Proclamation prohibiting the Importation of all sorts of Glass whatsoever made in foreign parts' (1635), which stated that 'our trusty and wellbeloved Sir *Robert Mansell* ... by his Industry and great Expences, hath perfected the Manufacture of all sorts of Glass with Sea Coal or Pit Coal' (T. Rymer, *Foedera*, 3rd edn., ed. G. Holmes, 10 vols. (The Hague, 1739–45), ix, p. 24).

[4] Richard Reeve or Reeves (active 1641–63; died 1666), a glass-grinder, who in early Restoration London would become the leading maker of telescopes, microscopes, and other scientific instruments. See M. Daumas, *Scientific Instruments of the Seventeenth and Eighteenth Centuries and their Makers*, ed. and tr. M. Holbrook (London, 1972), p. 67; A. D. C. Simpson, 'Richard Reeve – the "English Campani" – and the Origins of the London Telescope-Making Tradition', *Vistas in Astronomy*, 28 (1985), pp. 357–65; G. Clifton, 'The Spectaclemakers' Company and the Origins of the Optical Instrument-making Trade in London', in R. G. W. Anderson, J. A. Bennett, and W. F. Ryan, eds., *Making Instruments Count: Essays on Historical Scientific Instruments presented to Graham L'Estrange Turner* (Aldershot, 1993), pp. 341–64; G. Clifton, *Directory of British Scientific Instrument Makers, 1550–1851* (London, 1995), p. 229.

haue bought the stuff to make it of./ I shall write to Mr Reaues to giue vs his help heerin./ I must againe thanke you for your waie of ordering aequations,[5] & doe desire that you will proceed in your intended analyticall worcke,[6] as your occasions will [*one word deleted*] giue you leaue. I desire to knowe if M^r. Warners analogicall worck[7] goe on or not./ And so wishing you all hapiness I remaine

    Your assured freind to serue you
    Charles Cauendysshe

Wellingor[8] June 26 1641

[*postscript:*] If you knowe an easie and readie waie to measure the refraction in water, you should doe me a fauoure to let me knowe it; for I confess I knowe none.

[*addressed:*] To my worthie freind Mr John Pell, at his house in Westminster. giue this.

[*annotated by Pell:*] Received 5 July 10 ./. 14 dayes after.
    perspectiue glasse
My $\begin{cases} \text{ordering aequations} \\ \text{analyticall worke} \end{cases}$
    Warners analogicall tables
    refraction in water

[*cover annotated by Birch:*] 26 June 1644

[*cover annotated in another hand:*] Ch Cavendish

---

[5] Pell devised a method of working mathematical problems in three columns: the left-hand column carried a set of instructions and the corresponding calculations appeared in the right-hand column, while a narrow central column held line numbers, very much as in a modern computer programme. There are numerous examples of this layout or 'method' throughout Pell's mathematical work from the early 1640s onwards. It may have been inspired originally by Pierre Hérigone, who, in his *Cursus mathematicus* (the first five vols. of which were published in Paris, 1632–7), sometimes used two columns, one for the working itself and the other to give a brief explanation or reason for each step. Pell refers to the method in Letter 4b as 'that crotchet of mine of ordering and numbring ye equations'.

[6] François Viète (or Vieta) (1540–1603) supposed that algebra was the analytic tool by which classical geometrical theorems could be confirmed or completed, or new theorems discovered. Following him, for the greater part of the 17th century, algebra was known as 'analysis' or 'the analytic art'. Cavendish was thus encouraging Pell to write on one of the subjects Pell knew best, algebra. Letter after letter repeats this plea to Pell to bring out his book; but Pell published nothing on the subject until the *Introduction to Algebra* in 1668.

[7] The mathematician Walter Warner (1562–1643), friend, executor, and editor of Harriot, was preparing (with Pell's assistance) tables of antilogarithms. These, unfortunately, were never published; for their presumed fate see Pell's comments in Letter 14.

[8] Wellingor or Wellingore, Cavendish's seat in Lincolnshire (nine miles south of Lincoln).

Letter 2

19 [/29] July 1641

Pell to Cavendish, from London

BL MS 4278, fo. 165 (draft)
van Maanen Pa

Right honourable,
 Your letter dated June 26[1] came [>not] to my hands [>till] ye 10th of this month at which time Mr Reaves sonne[2] brought it to my house in St Peters streete[3] (for I am removed) with a piece of glasse made at Broadstreet[4] if we may beleeve ye spectacle-maker of whom Mr Reave bought it. I suppose others [have geven you account of those hindrances whereby we have meerely lost a whole quarter of a yeare. For so long it was since I gave Mr Reeve ye shapes of our triangles,[5] but they tell me that the glassehouse workemen in Broadstreet being overcome with y^e *deleted* >have sent you word that the workemen in Broadstreet being overcome with the] importunities of our men & some spectacle-makers, [>did at length] prepare a pot of choice mettall as they speake, but whether for want of skill or care, overfired y^r worke & brake their pot, then was all ye fat in ye fire, their Lady[6] so heated, that she will not heare of giving y^m leave to make a new adventure, so Mr Reeve found out this piece, but he having lost ye former patternes, I drew 4 more & went [*marginal note:* July 12] with your Goldsmith[7] to a cutter of gemmes, who promised to dispatch y^m speedily. y^e *last Saturday* [*marginal note:* July. 17] Mr Reeve went to his house & found a Journyman ready to polish y^m up, having no square or other direction, more than my draught, to examine y^m by. he caused him to lay y^m aside & sent me word of it, this morning [*marginal note:* July 19.] we have beene with the M^r workeman, he promiseth to finish y^m him selfe accurately by a square, which Mr Reeve is to send him, I am now come backe from y^{ence} to your goldsmiths shop heere to write y^s answere:

---

 [1] Letter 1.

 [2] See Letter 1, n. 4. Reeve's son, also named Richard, would continue his father's business from 1666 to 1668; he died in 1679 (Clifton, *Directory*, p. 229).

 [3] St Peter's Street, in Westminster.

 [4] See Letter 1, n. 2.

 [5] Triangular blocks of glass, as specified by Descartes in his instructions for making hyperbolic lenses: see his 'Dioptrique', discours 10 (A&T vi, pp. 211–13).

 [6] Unidentified; this may refer not to a person but to some aspect of the preparation of molten glass.

 [7] Among Pell's papers is a rough note (dated May 1641) giving measurements of angles of refraction in glass and water, headed 'To Sr Charles Cavendish [>leave y^m] at Mr Moslyes a goldsmith in Cheapside' (BL, MS Add. 4415, fo. 290^v). As Letter 4a shows, Cavendish transmitted letters and instructions via this goldsmith; cf. also Letter 7, n. 6.

Mr Reeve desires to be excused for not writing, having lately so overwrought his hand y^t he can not hold a pen. He sayes he hopes to have the [>perspective glass] finished by Michaelmas.[8]

I have begun Mr Warners worke[9] & calculated ye first 4 thousands by which I conjecture ye labour of ye whole, if I could set a scribe I would hope besides my other businesses to have ye whole hundred thousand[10] ready for ye presse by Christmas.

I had more to say, but I am so called upon to make an end for feare ye Carrier will be gone; y^t I [>have] scarce leave to adde y^t I am

Sr Your thankfull & humble servant.
John Pell

Cheapeside.[11]

[*addressed:*] For the right honourable Sr Charles Cavendish at his house in Wellingor neere Lincolne

---

[8] 29 Sept.

[9] See Letter 1, n. 7.

[10] By 1639 Warner had produced a table of 10 000 antilogarithms. He and Pell then set themselves the task of extending the table to 100 000 entries by interpolation (see Aubrey, '*Brief Lives*' ii, p. 292). Pell did not complete the calculations by Christmas, as he hoped, but only in July of the following year.

[11] Cheapside, a street running eastwards from St Paul's Cathedral, in the City of London.

Letter 3

24 July [/3 August] 1641

Cavendish to Pell, from Wellingor

BL MS 4278, fos. 163–4 (original)
van Maanen Cb

Worthie S^r
 I am glad to heare you haue got some glass, I hope it is good & fit for oure purpose, for I should be vnwilling that you & Mr Reaues[1] should bestowe your paines upon course glass./ When you haue tried what the refraction is in that glass, I desire to knowe it, & allso howe you like the glass./ I haue latelie receiued some propositions out of France.[2] some demonstrated & some not, but I will not diuert you from the business you haue in hand./ I am glad you haue begun the analogiques,[3] & hope allso that you proceed in your owne analiticall worcke./[4] And so wishing you all hapiness I remaine

 Your assured freind to serue you
 Charles Cauendysshe

Wellingor Julie 24 1641

[*addressed:*] To my verie worthie freind Mr John Pell giue this

[*cover annotated by Birch:*] 24 July 1641

---

 [1] See Letter 1, n. 4.

 [2] Possibly the tract entitled 'Elemens des indivisibles' which survives in a volume of Cavendish's papers (BL, MS Harl. 6083, fos. 279–302). William Oughtred may have been referring to this tract when he wrote to John Wallis in Aug. 1655: 'Full twenty years ago, the learned patron of sciences, Sir Charles Cavendish, shewed me a written paper sent out of France, in which were some very few excellent new theorems, wrought by the way, as I suppose, of Cavalieri' (Rigaud, *Correspondence*, i, pp. 87–8). Oughtred's 'twenty years ago' may have been an exaggeration, for in Oct. or Nov. 1645 he wrote to Robert Keylway about 'a geometrical-analytical art or practice found out by one Cavalieri, an Italian, of which about three years since I received information by a letter from Paris' (ibid., i, pp. 65–6), implying that he saw the paper some time during 1642. This date would be consistent with Cavendish mentioning the same paper to Pell in 1641.

 [3] See Letter 2, nn. 9, 10.
 [4] See Letter 1, n. 6.

Letter 4a

9 [/19] August 1641

Pell to Cavendish, from Westminster

BL MS Add. 4278, fo. 165^v (draft)
van Maanen Pb

Right honorable,

Yours of July 24[1] requires no answere till we have tried ye refraction which as yet we have not done: but your Goldsmith[2] calls upon me by y^s Carrier to give [>you] account how farre we are advanced. Mr Reeve[3] helped ye Workeman to a squire,[4] with which [>when] he had almost done the worke he made him lay y^m aside & begin anew upon a cleerer piece y^t he had after met withall this he dispatched & sent y^m me 4 dayes agoe: I went ye next morning to a Mathematicall joyner & left directions with him for ye making a reigle [>something like that which] des Cartes describes pag 137[5] but more convenient & so more costly. He promised y^t I should have it ye next weeke. In the meane time I wished ye Goldsmith to give order to ye Workeman to finish ye 4 former triangles also.[6] seeing he must be paid for his worke upon y^m though he should not finish y^m. When we have y^m we may try ye difference of [>those] 2 sorts of glasse and accordingly proceed: Mr Reeve longs to be at worke upon y^m. For want of a Scribe I make [no *deleted* >small] riddance of the Analyticalls Nor have I any leisure to perfect mine owne thoughts of a [>higher nature]. I have received from

---

[1] Letter 3.
[2] See Letter 2, n. 7.
[3] See Letter 1, n. 4.
[4] = square.
[5] Descartes's 'reigle' or ruler was a device for holding a triangular block of glass and enabling it to be cut into the shape of a hyperbolic lens: this is a reference to the discussion of it in his 'Dioptrique', discours 10 (A&T vi, pp. 211–13).
[6] A note by Pell, dated 5 [/15] Aug. 1641 begins: 'Of the 4 triangles of glasse y^s day brought me How to use y^m for ye measuring of refraction' (BL, MS Add. 4415, fo. 295^v). Pell thus had four new triangles, and four old ones in a different sort of glass.

Mersennus a supplement of Apollonius & a censure on des Cartes,[7] but cannot say I have [>yet] read a line in either of them: as reserving y^t labour till I am rid of some of those many diversions of different kinds with which I am now distracted: in ye midst of which I neverthe[less *page torn*] remaine

    Your humble servant
    John Pell.

St Peters streete Westminster
1641

August 9. 1641

---

[7] On [21 Apr./] 1 May 1641 Marin Mersenne had sent Pell (via Theodore Haak) two MSS: the 'Loca plana' by the mathematician Pierre de Fermat (1601–65), and a criticism of Descartes by another of Mersenne's friends (*MC* x, p. 610). The first of these was 'Apollonii Pergaei libri duo de locis planis restituti', a reconstruction of a lost work by Apollonius; it had been circulating in Mersenne's circle in Paris since its completion in 1637, and would be published in Fermat's posthumous *Varia opera* (Toulouse, 1679), pp. 12–43 (see M. S. Mahoney, *The Mathematical Career of Pierre de Fermat, 1601–1665*, 2nd edn. (Princeton, NJ, 1994), pp. 92–101, 416–17). The second MS, described as unidentifiable by the editors of *MC*, was probably the brief text on tangents by Jean de Beaugrand which survives in a volume of Cavendish's papers (BL, MS Harl. 6796, fos. 155–61); it begins: 'To give you a better idea of the defects of M. Descartes's way of finding straight lines that cut given curves at right-angles, I want to show you the device which Apollonius probably used to find the tangents of conic sections...' ('Pour te mieux faire connoistre les deffauts de la façon du S. des C. pour trouuer des lignes droictes qui coupent les courbes données a angles droicts, Je veux te monstrer l'artifice dont il est vray semblable que Apollonius s'est seruy pour trouuer les tangentes des sections coniques...': fo. 155^r). (On this work, composed probably in 1638–40, see C. de Waard, 'Un Écrit de Beaugrand sur la méthode des tangentes de Fermat à propos de celle de Descartes, *Bulletin des sciences mathématiques*, ser. 2, 17 (1918), pp. 157–77; de Waard's claim that it is in Hobbes's hand is corrected in Malcolm, 'Hobbes, the Latin Optical Manuscript, and the Parisian Scribe'. Pell's copy of this work is in BL, MS Add. 4407, fos. 38–9.)

Letter 4b

[August 1641]

Pell [to Cavendish?]

BL MS Add. 4278, fo. 175^r (draft, fragment)
Not in van Maanen.

But in the meane time, [>you easily beleeve that] my owne analyticall worke stands still Of which that crotchet of mine of ordering & numbring ye aequations is but one peculiarity I have many others which might have beene published ere this if I could have beene acquainted with leisure sufficient for thoughts of y^t natu[re *page torn*]

Pere Mersenne hath sent me [>in Manuscript] 2 bookes of Apollonius restored by a judge [>in] one of their Parliaments[1] And a censure of some Geometricall passages of de Cartes[2] but I have not yet reade y^m & doubt not but ere this he hath sent you a Copy of ye same & 't is likely in a fairer hand

---

This fragment is, evidently, closely connected with Letter 4a, repeating substantially the same information. Either it formed part of an alternative draft of that letter, written at more or less the same time; or, perhaps, it was part of a draft of a replacement letter, written when Pell thought that Letter 4a had not in fact reached Cavendish; or, possibly, it belonged to a draft of a letter to someone else. However, the last sentence shows that its intended recipient was also receiving letters from Mersenne, and that that person would have had a higher status in Mersenne's eyes. Cavendish satisfies both those criteria; the only other member of Pell's circle of acquaintances in regular contact with Mersenne was Haak, but it is known that Mersenne had sent these mathematical MSS to Pell via Haak in the first place (see *MC* x, p. 610).

[1] See Letter 4a, n. 7. Pierre de Fermat was a *conseiller* of the Parlement of Toulouse, and *commissaire aux requêtes du Palais* there; he also sat as a member of the 'Commission de la Chambre de l'Édit' (a court dealing with Catholic–Protestant disputes under the Edict of Nantes) in Castres.

[2] See Letter 4a, n. 7.

Letter 5

20 [/30] November 1641

Cavendish to Pell, from Wellingor

BL MS 4278, fos. 166–7 (original)
van Maanen Cc

Worthie S^r
    I hope Mr Reaues[1] is in a good forwardness with the conuex glass;[2] I dout not but you will trie all conclusions with it, which may conduce to informe you whether it be an hyperbole or no; as allso what proportion the diameter of the glass hath to the line of the contracted beames of the sun at the pointes of concourse./ as allso to obserue what aparences are made, the eye being placed in, before, or behinde, the pointe of [>con]course; and in the meane time before the concaue glass be made, to trie whether my concaue glass which you haue, will in anie sorte fit it. S^r I leaue the further scrutinie of this to your better consideration, & wishing you all hapiness remaine

    Your assured freind to serue you
    Charles Cauendysshe

Wellingor Nouemb: 20 1641

[*postscript:*] I praye you commend me to Mr Reaues when you see him./

[*addressed:*] To my verie worthie freind Mr John Pell. giue this

[*annotated by Pell:*] Received Decemb 2

[*cover annotated by Birch:*] 20 Nov. 1641

---

[1] See Letter 1, n. 4.
[2] On this, and the 'concaue glass' mentioned later in this letter, see Letter 1, n. 1.

Letter 6

13 [/23] December 1641

Pell to Cavendish, from London

BL MS Add. 4278, fo. 172^v (draft)
van Maanen Pc

Right honourable

As soone as I had received your letter dec. 2.[1] I went to Mr Reeve's[2] house but found neither [>him nor his sonne] within: I saw him [>not] since till this morning, he tells me that he [>accounts himselfe] in a great forwardnesse because he hath so perfectly fitted himselfe with all requisites for ye assured doing of it exactly according to ye patterne,[3] that as soone as the weather serves he can goe in hand with it

If I am not mistaken you told me that you had often in vaine enquired for Scheineri Ars nova delineandi[4] [>it] is now easy enough to be had [>heere] a thin booke in 4°

When we talked of demonstrating [>some of] ye precepts of [Arithmeticke *deleted* >Logistica], I doe not remember that you seemed to have seene Henischius the title is thus Georgij &c as above. [*note at head of page:* Georgij Henischij Arithmetica perfecta & demonstrata, doctrinam de numero triplici, vulgari, cossico & Astronomico nova methodo per propositiones explicatam continens, libris septem, Augustae Vindeolicorum 16[09 *concealed by binding*] in 4° pag. 410.][5] [>One] might guesse what manner of writer he is by this passage in his epistle Dedicatory Jam cui non displiceat &c [*note at head of page:* Jam cui non displiceat mos ille demonstrandi quo plerique Arithmetici delectantur, per nebulas id est per literas sine numeris & sine luce syllogismorum quae nihil est obscurius,

---

[1] This letter has not apparently survived.

[2] See Letter 1, n. 4.

[3] This presumably refers to the design of Descartes's hyperbolic convex lens: see Letter 1, n. 1.

[4] C. Scheiner, *Pantographice seu ars delineandi res quaslibet per parallelogrammum lineare seu cavum mechanicum mobile* (Rome, 1631). Christoph Scheiner (1573–1650) was a German Jesuit mathematician and scientist, specializing in optics and astronomy. The book referred to here (which bears a half-title, 'Ars noua delineandi' – hence Pell's reference to it under that title) described a device he had invented in 1603, the 'pantograph', by means of which a drawing could be reproduced in a reduced or enlarged form. (See A. von Braunmühl, *Christoph Scheiner als Mathematiker, Physiker und Astronom* (Bamberg, 1891), esp. pp. 2–5.)

[5] The date is 1609, and 'Augustae Vindelicorum' is Augsburg; the title is correctly given by Pell. Georg Henisch (1549–1618) was a humanist scholar who taught logic and mathematics at Augsburg.

nihil Heracliti libro similius]⁶ It may be you would find something in it worth ye price it will cost though it be extreamely short of what we desire. He might have beene able to have put more matter in lesse roome if he had not disaffected demonstration per literas⁷

This is all y^t in this Carriers haste I can thinke of at this time so that I must not forget in taking my leave to tell you y^t

I am S^r your most humble Servant
John Pell

Cheapeside dec. 13 1641

---

⁶'Indeed, who can fail to dislike that method of demonstration which charms most mathematicians – demonstrating by clouds of mist, that is, by letters without numbers, and without the light of syllogisms? Nothing is more obscure, or more like the book of Heraclitus' (Henischius, *Arithmetica perfecta*, epistle dedicatory, sig. 4^r). (The writings of the early Greek philosopher Heraclitus were proverbially obscure: see Aristotle, *Physics*, 185a–b, and Cicero, *De finibus*, II.5.15.)

⁷'By letters' (as used in algebra). The tension between the new symbolic algebra pioneered by Viète, Harriot, and Descartes, and more traditional geometric and rhetorical styles of proof, was to continue well into the 17th century. Both Fermat and Hobbes, for instance, were to eschew symbolism in mathematics, but Pell recognized the power of the new algebraic methods, and helped to pioneer the effective use of symbolism.

Letter 7

18 [/28] December 1641

Cavendish to Pell, from Wellingor

BL MS Add. 4278, fos. 168–9 (original)
van Maanen Cd

Worthie S^r

I thanke you for your letter of Dec. 13.[1] I am glad Mr Reeues[2] is so well fitted for oure worcke; when he hath done it, I dout not but you will make all such triålls as maye giue you satisfaction whether it be a true hyperbole or not; & then proceede to the making of the concaue glasse; if this fit it not, I shall still be in hope that a concaue on bothe sides will.[3] I haue not (to my remembrance) seene Henischius Arithmetick,[4] nor should [not *deleted*] desire, for his mislike of demonstration by letters; yet if you thinke there be anie thinge in him considerable, which is not in Vieta or de Cartes[5] I desire you will send it me, & Mr Moselei[6] will paye for it, as allso for Scheiners ars noua delineandi./[7] I confess I expect not an exact booke of analiticks till you perfect yours./[8] And so wishing you all hapiness I rest

   Your assured freind to serue you
   Charles Cauendysshe

Wellingor Decemb: 18 1641

[*addressed:*] To my verie worthie freind Mr John Pell. giue this

---

[1] Letter 6.

[2] See Letter 1, n. 4.

[3] See Letter 1, n. 1, Letter 5, and Letter 6.

[4] See Letter 6, n. 5.

[5] Viète planned an *Opus restitutae mathematicae analyseos, seu algebra nova* consisting of ten books on the scope and use of algebra. Eight of the books were eventually published, the most important being *In artem analyticam isagoge* (Tours, 1591), *Zeteticorum libri quinque* (Paris, 1593), and *De aequationum recognitione* (Paris, 1615). Descartes's algebra was published in the three books of 'La Géométrie', one of the 'essays' included in the *Discours de la méthode et les essais* (Leiden, 1637).

[6] This could possibly refer to Humphrey Moseley, a well-known London bookseller, active from 1630 until his death in 1661; however, previous references (see Letter 2, n. 7, and Letter 4a, n. 2) make it more likely that this is a reference to Cavendish's goldsmith.

[7] See Letter 6, n. 4.

[8] This sentence was copied by Samuel Hartlib on a letter to Hartlib from Sir Cheney Culpeper (which referred to Pell's mathematical skills) of 29 Oct. [/7 Nov.] 1641, with the note: 'Sir Charles Cavendish the 18 December 1641' (HP 13/345, printed in M. J. Braddick and M. Greengrass, eds., 'The Letters of Sir Cheney Culpeper', p. 154). Evidently Pell had shown Hartlib Cavendish's letter.

[*annotated by Pell:*] Received Jan: 4

[*cover annotated by Birch:*] 18 Decemb. 1641

Letter 8

8 [/18] January 1642

Cavendish to Pell, from Wellingor

BL MS Add. 4278, fos. 170–1 (original)
van Maanen Ce

Worthie S^r

I haue not much to write to you of; onelie I haue heard nothing of those bookes you writ me of./[1] I haue sent you heere inclosed what Mersennus latelie sent me./[2] I desire you will doe me the fauoure to write it oute & send it me for I confess his hand is an Arabicke character to me;[3] I praye you keepe his paper, till it please God wee meete./ I doute heere hath bin ill weather for Mr Reeues[4] to worcke in./ And so wishing you all hapiness I rest

    Your assured freind to serue you
    Charles Cauendysshe

Wellingor Jan: 8. 1641

[*addressed:*] To my verie worthie freind Mr John Pell. giue this.

[*annotated by Pell:*] Received Jan. 14.

[*cover annotated by Birch:*] 8 January 1641/2

---

[1] Probably the works by Henischius and Scheiner: see Letters 6, 7, and 9.

[2] This text has not survived; for a description of its contents see Letter 9.

[3] Mersenne's handwriting was tiny, cramped, and irregular: see the photograph in Malcolm, 'Six Unknown Letters from Mersenne to Vegelin', at p. 110.

[4] See Letter 1, n. 4.

Letter 9

17 [/27] January 1642

Pell to Cavendish, from London

BL MS Add. 4278, fo. 172^r (draft)
van Maanen Pd

Right honourable

 I having no occasions sufficient to draw me into London of late, your letter [>dated] decemb 18[1] lay in Cheapeside till Mr Reeve[2] sent it me January 4 I was to bee last Friday [>(Dec. 14)] at Mr Reeves house there I found the glasse[3] broken, he told me y^t as soone as the diamond touched it ye Cement let it goe because of ye weather & ye polished smoothnesse of y^e [>flat side of y^e] glasse: it served him so 3 times & in putting it on the 4th time ye fire brake ye glasse which chance he foreseeing had another ready but devised another Sort of Cement which [>holds it] as fast as can be wished, I saw it, it was so neere done y^t he promised ye next Wednesday (y^t is Jan. 19) to bring it to my house finished for me to examine[4] &c

 I went from his house into Cheapeside & there found [>another] letter of yours dated Jan. 8.[5] with a french demonstration,[6] [>of] which I [>heere send you not ye manuscript but a translation] how faithfully I have done it you will see when you come up: for ye french I keepe as you willed me. But I understand not why he should [>send you this as a Novelty or how he can] thinke y^t this Theoreme or its demonstration is new to you seeing ye Theoreme it selfe is to be found in most if not all ye bookes [>penned] of any such subiect within these 50 yeeres. And so is verie oft demonstrated Two of our English [>have done it viz] Mr Briggs in his Trigonom Britannica [>lib. 1] Perhaps this very demonstration & Mr Oughtred after another way in ye last page of his Clavis.[7]

 I have enquired for Scheiner[8] ye Coppies are all sold, they will not long be

---

[1] Letter 7.

[2] See Letter 1, n. 4.

[3] The hyperbolic convex lens: see Letters 6 and 7.

[4] A note by Pell records that Reeve delivered this lens to him on 30 Jan. [/9 Feb.] (BL, MS Add. 4415, fo. 271^r).

[5] Letter 8.

[6] The 'paper' in Mersenne's hand: see Letter 8.

[7] The theorem referred to here concerns the classical problem of trisecting an angle. Henry Briggs, in his *Trigonometria britannica, sive de doctrina triangulorum libri duo*, ed. H. Gellibrand (Gouda, 1633), devoted three chapters (4, 6, and 7) to trisection, quinquesection, and septisection of angles by trigonometric methods. William Oughtred discussed the same problem at the end (pp. 86–8) of his *Clavis*, a work written with direct encouragement from Cavendish.

[8] See Letters 6 and 7.

without more having so soone sold these. If [>I] had sent it, you could hardly have found a workeman in ye country to make ye instrument by which y^t new way of delineation is practised.[9] I have written a short description of ye fashion & use of it & have given order to a workman to make it, perhaps it will be ready by ye time you come to London. Till which time I thinke we were best deferre ye buying of Henischius,[10] [>also for] I can lend you one to read which if you like you may then buy one of y^m.

40 thousand analogies are calculated[11] 20 by my selfe & as many by one whom I instructed,[12] who will be able to dispatch y^e [>remaining 60,000 in short time] though I should not helpe him.

I take my leave & rest

Sir, your humble servant
J Pell

Jan. 17. 1641/42 Cheapeside

---

[9] The pantograph: see Letter 6, n. 4.
[10] See Letters 6 and 7.
[11] See Letter 1, n. 7.
[12] Pell's notes identify this person as 'Ed. Watts'; they also mention another assistant, called Turner (BL, MS Add. 4365, fo. 37^r).

Letter 10

5 [/15] February 1642

Cavendish to Pell, from Wellingor

BL MS Add. 4278, fos. 173–4 (original)
van Maanen Cf

Worthie S^r

I thanke you for your letter[1] & the transcript of Mersennus probleme;[2] if he hath read Mr Brigs or Mr Oughtred, I wonder he would send it;[3] but it maye be he hath found it the analyticall waye himself./ I am glad Mr Reaues[4] was in such forwardness when you writ; but I doute the glass (which I hope is nowe finished) is not of the same which you tried your refraction in; because m^r Reaues hath broken in his triall so much glass that I doute[5] there is none left of that which you tried your refraction in; but I hope fine glass differs so little in refraction that it will not doe vs much harme./ I hope you goe on with your owne Analyticall worcke as your occasions will permit you. I haue no more at this time to trouble you with, but remaine

   Your assured freind to serue you
   Charles Cauendysshe

Wellingor Feb: 5 1641

[addressed:] To my verie worthie Freind Mr John Pell giue this

[cover annotated by Birch:] 5 Febr. 1641/2

---

[1] Letter 9.
[2] See Letter 9, n. 6.
[3] See Letter 9, n. 7.
[4] See Letter 1, n. 4.
[5] In 17th-century usage, 'I doubt' meant 'I suspect', i.e., believe to be probably true.

Letter 11

20/30 July 1644

Pell to Cavendish, from Amsterdam

BL MS Add. 4280, fo. 107^v (draft)
van Maanen P1

Right Honourable,

The newes of your comming to Hamburg[1] is now a weeke old, in this city but I was none of ye first y^t beleeved it: Nor can I yet certainelye heare, whether you be come thither or not, which makes me write as one y^t suspects his letters shall not come to ye hands to which y^{ei} were intended. Yet can I not any longer hold my hands from writing to you were it but to let you understand y^t I have beene heere about 7 months, having left in Westminster my wife, children & y^t little all y^t I have besides, nor can I yet tell when I shall send for them hither. Yet heere I am Hortensii[2] successor that is Totius Matheseos, in Amstelodamensium [>illustri] Gymnasio, professor publicus.[3] [where I and my collegues Vossius,[4] Barleus[5] & Cabeliavius[6] must reade each of us 5 houres in a weeke *deleted*] I shall trouble you no further till I heare y^t this is come safe to your hands.

If it please you to commaund me any service your letter will surely find me if it be directed to me In t'oude Convoy op de zeedijk[7] in Amsterdam.

Where I continue

Sir
your humble & faithfull servant
John Pell

---

[1] Cavendish and his brother had arrived in Hamburg on [28 June/] 8 July (Trease, *Portrait of a Cavalier*, p. 142).

[2] Martinus Hortensius (1605–39) had been Professor of Mathematics at the Amsterdam Athenaeum, 1634–9; the post had remained vacant until Pell's appointment in Apr. 1644.

[3] 'Public Professor of the whole of Mathematics in the high school of the inhabitants of Amsterdam.'

[4] Gerardus Joannes Vossius (Geeraard Jan Voskens) (1577–1649), historian, philologist, and theologian, was Professor of History at the Amsterdam Athenaeum from 1632 until his death.

[5] Caspar Barlaeus (van Baerle) (1584–1648), poet, physician, and philosopher, was Professor of Philosophy at the Amsterdam Athenaeum from 1632 until his death.

[6] Jan Cabeljauw (Joannes Cabillavius) (1600/01–1652) was Professor of Law at the Amsterdam Athenaeum from 1640 until his dismissal in Jan. 1646.

[7] The Zeedijk is a street in north-central Amsterdam, running northwards from the Nieuwe Markt; the 'oude Convoy' was a former customs-house at the northern end of the street, where merchants had paid the duty known as 'convoy' or 'konvooi' (originally, a payment for protection by armed escorts at sea) (see C. Commelin, *Beschryvinge van Amsterdam* (Amsterdam, 1693), p. 216).

July 20/30

[*added by Birch:*] 1644.

[*postscript:*] DesCartes his new booke de principijs Philosophiae is come out heere within these 3 or 4 dayes in 4° there is adioined with it his Methode dioptrique & Meteores (that is to say all ye Old booke save ye Geometrie) translated into [French *deleted* >Latine][8] But there are some few coppies printed without y^t appendix for their sakes y^t have bought ye French & understand it well inough. [One of which I can send you. You *deleted* >One] may have of either sort in this towne at Louys Elzevirs shop.[9]

[*Later annotation by Pell:*] His answere dated [Aug *deleted* >July] 26.[10] came to me Aug. 1

---

[8] The 'Old booke' was the *Discours de la méthode et les essais*, the 'essais' there consisting of the 'Dioptrique', 'Météores', and 'Géométrie'. In 1644 the Amsterdam printer Louis Elzevier (see n. 9) published simultaneously Descartes's *Principia philosophiae* (the 'new booke') and a Latin translation of the 'Discours de la méthode', the 'Dioptrique', and the 'Météores (under the title *Specimina philosophiae*). These two publications were normally issued bound together (with the *Specimina* preceding the *Principia*).

[9] Louis Elzevier (d. 1670), a member of the numerous family of printers and book-sellers from Leiden, had established a printing house in Amsterdam in 1638. His shop was on a quayside street known as 'op 't Water' ('on the water'), which ran alongside the Damrak (see A. Willems, *Les Elzevier: histoire et annales typographiques* (Brussels, 1880), p. lxiii).

[10] Letter 12.

Letter 12

26 July [/5 August] 1644

Cavendish to Pell, from Hamburg

BL MS Add. 4278, fos. 176–7 (original)
van Maanen C1

Worthie S^r

I giue you manie thankes for your letter,[1] which I receiued yesterdaie, & since things goe not so well in england I am glad you are so well placed for the present, for I suppose it a place of good means as well as of honor; yet I hope by the grace of God wee shall one daie meet in england, & liue more happilie there than euer wee did. I thinke wee shall remaine a whyle in this towne & so be depriued of the happiness of your conuersation but by letter, which I desire you will be pleased sometimes to afford me at your best leasure./ I desire you will doe me the fauoure to send me one of De Cartes his new bookes De principiis philosophiae, without anie addition of his olde worckes,[2] except he hath either added or altered something in the matter; I desire you will let me knowe the price of it, & howe I maye with most conuenience returne monei to you, for I am likelie to trouble you for more bookes./ I praye you let me knowe whether M^r: Warners Analogicks[3] be printed; I hope you proceed in your intended worcke of Analiticks which if you doe not, I beseech you doe, & finish it with all conuenient speed; for I confesse I expect not anie absolute worcke in that kinde but from your self; I praye let me know what new inuentions are extant in the Mathematicks latelie, if anie. And so wishing you all happiness I rest

Your assured friend to serue you
Charles Cauendysshe

Hamburg. Julie 26. 1644

[*addressed:*] To my verie worthie friend M^r: John Pell, Publick Professor of the Mathematickes at Amsterdam. giue this in t'oude Conuoy op de Zeedijk Amsterdam

[*annotated by Pell:*] Received Aug. 1/11

[*cover annotated by Birch:*] 26 July 1644

---

[1] Letter 11.
[2] See Letter 11, n. 8.
[3] See Letter 1, n. 7.

Letter 13

8/18 August 1644

Cavendish to Pell, from Hamburg

BL MS Add. 4278, fos. 178–9 (original)
van Maanen C2

Worthie S^r:

doduting that my letter [>of late][1] came not to your handes makes me nowe trouble you with this; I giue you manie thankes for your letter & am most glad that you are well & in so good a condition; I doute not but you haue hearde [>the occasion] of oure comming ouer,[2] I take no pleasure to write it nor I suppose you to reade it; Gods will be done, & to that I humblie submit./ I desire you will be pleased to send me de Cartes de principiis philosophiae & none of his olde bookes except there be some addition or alteration./[3] I desire your opinion of the late discourse of newe stars, I see Gassendes doutes of it./[4] I desire to knowe if m^r: warners analogicks[5] be printed; & if there be anie newe bookes of Analiticks but I expect no greate aduancement of analitickes but by your self, therefore I beseech you proceede in your intended worcke./ I am likelie sometimes to trouble you for bookes, therfore I desire you will let me knowe howe I maye returne monei

---

[1] Letter 12.

[2] The defeat at Marston Moor on 2 [/12] July 1644 (for which Cavendish's brother bore a large share of the responsibility).

[3] See Letter 11, n. 8.

[4] The astronomer and Capuchin friar Anton Rheita (Anton-Maria Schyrle von Rheita, 1604–59) had observed, on [19/] 29 Dec. 1642 and [25 Dec. 1642/] 4 Jan. 1643, what he took to be five new moons of Jupiter. He announced the discovery in a letter to Erycius Puteanus (see Letter 19, n. 18) written on [27 Dec. 1642/] 6 Jan. 1643; copies of this were then circulated to other scientists. Pierre Gassendi wrote a letter to Gabriel Naudé on [25 Mar./] 4 Apr. 1643, arguing that the alleged moons were in fact fixed stars in the constellation of Aquarius; this was published soon thereafter, with Rheita's original letter, as *Novem stellae circa Jovem visae, et de eisdem P. Gassendi iudicium* (Paris, 1643). These texts were reprinted in Sept. with additional letters (one to Gassendi, three to Rheita) by the Cistercian scientist Juan Caramuel y Lobkowitz, supporting Rheita's claims: *Novem stellae circa Jovem, circa Saturnum sex, circa Martem non-nullae, a P. Antonio Reita detectae & satellitibus adiudicatae, de primis ... D. Petri Gassendi iudicium D. Ioannis Caramuel Lobkowitz eiusdem iudicii censura* (Louvain, 1643). (See also S. Débarbat and C. Wilson, 'The Galilean Satellites of Jupiter from Galileo to Cassini, Römer and Bradley', in R. Taton and C. Wilson, eds., *Planetary Astronomy from the Renaissance to the Rise of Astrophysics, Part A: Tycho Brahe to Newton* (Cambridge, 1989), pp. 144–57, esp. p. 148. On Rheita and his astronomical work see A. Thewes, *Oculus Enoch ... Ein Beitrag zur Entdeckungsgeschichte des Fernrohrs* (Oldenburg, 1983); on his biography more generally see A. Thewes, 'Ein Ordensbruder und Astronom aus dem Kapuzinerkloster Passau', *Ostbairische Grenzmarken: Passauer Jahrbuch für Geschichte, Kunst und Volkskunde*, 30 (1988), pp. 95–104.)

[5] See Letter 1, n. 7.

to you; I longe to see you, in the meane time I hope to conuerse by letters./ I remaine

   Your assured friend to serue you
   Charles Cauendysshe

Hamburg Aug: 8/18

[*addressed:*] To my verie worthie friend M^r: John Pell, publike Professor of the Mathematicks at Amsterdam; giue this in t'oude Conuoy op de Zeedijk in Amsterdam

[*cover annotated by Birch:*] 8/18 Aug. 1644

Letter 14

10/20 August 1644

Pell to Cavendish, from Amsterdam

BL MS Add. 4280, fo. 105 (fair copy of letter written 7/17 Aug., but not sent, and subsequently used as draft for letter of 10/20 Aug), and fo. 106 (draft of additional material for letter of 10/20 Aug.)

van Maanen P2

Right Honourable

[Six *deleted* >Nine] dayes agoe, was your answere[1] to mine, left at my lodging by a man that would have nothing for postage. Had I seene him, perhaps I might have learned some conveniency of sending backe to you againe, so that my letters might be less chargeable, than if they come thus single to you by the ordinary poste.

[*The following two paragraphs were written in the opposite order, then the order was reversed by listing them in the margin as 'a' and 'b'*]

[>As for des Cartes his new Booke] Though the titlepage of the other part, tell us that it is ab auctore perlecta varijsque in locis emendata, Yet I have given order to leave it out and binde his principia philosophiae apart for you.[2] Whereby halfe the price is abated. [I suppose I shall haue it to morrow. And then I shall soone finde some ship to send it by: for such opportunities betweene this towne and Hamburg are very frequent. *deleted*]

[I had sent you Des Cartes his new booke forthwith *deleted* >I had sent it as soone as it was bound], if your letter had given me direction how to convey it to you; Which seeing you have not done, I resolve to send it to Hamburg by the next ship, to be left for you either at Berthold Offermans, a booksellers;[3] Or else at some other freinds house which I shall signify by letter.

---

[1] Letter 12.

[2] The 'other part' was the *Specimina* (see Letter 11, n. 8); the phrase on its titlepage describes it as 'read through by the author and corrected in various places'.

[3] Barthold Offerman was a prominent bookseller in Hamburg, and the father-in-law of one of the leading printers there, Johann Naumann (see W. Kayser, *Hamburger Bücher, 1491–1850, aus der Hamburgensien-Sammlung der Staats- und Universitätsbibliothek Hamburg*, Mitteilungen der Staats- und Universitätsbibliothek Hamburg, vii (Hamburg, 1973), pp. 58, 62). He published the 1635 and 1638 editions of Jungius's *Logica hamburgensis* (see H. Kangro, *Joachim Jungius' Experimente und Gedanken zur Begrundung der Chemie als Wissenschaft: ein Beitrag zur Geistesgeschichte des 17. Jahrhunderts* (Wiesbaden, 1968), p. 366).

Des Cartes himselfe, is gone into France.[4] Monsieur Hardy[5] tells us, in a letter lately written, that Des Cartes met him in Paris & blamed him for offering so much mony to our Arabicke professor at Utrecht for his Arabicke manuscript of Apollonius.[6] Which Mr Hardy interprets as a signe of envy in Des Cartes as being unwilling that we should esteeme the ancients or admire any man but himselfe for the doctrine of lignes courbes.[7]

But I thinke France alone will afford me argument for a large letter and therefore I leave it till the next time.

Come we therefore to England. And first for Mr Warners Analogickes,[8] of which you [inquire *deleted* >desire to know] whether they be printed. You remember that his papers were given to his kinsman, a merchant in London, who sent his partner to bury the old man,[9] himselfe being hindred by a politicke gout, which made him keepe out of their sight that urged him to contribute to the Parliaments assistance, from which he was exceedingly averse. So he was looked upon as one that absented himselfe, out of Malignancy, and his partner managed the whole trade. Since my comming over, the English merchants heere, tell me that both he and his partner are broken and now they both keepe out of sight, not as Malignants but as Bankrupts. But this you may better inquire among our Hamburg merchants. In the meane time I am not a little afraid that all Mr Warners papers, and no small share of my labours therein, are seazed upon and most unmathematically divided betweene the Sequestrators and Creditors, who (being not able to ballance the account when there appeare so many numbers, and much troubled at the sight of so many crosses and circles in the superstitious Algebra and that blacke art of Geometry) Will no doubt determine once in their lives to become Figure-casters and so vote them all to be throwen into the fire; If some good body do not reprive them for pye-bottoms etc for which purposes, you know Analogicall numbers are incomparably apt, if they be accurately calculated.

I cannot tell you much better newes of my Analyticall speculations, of the finishing of which, you desire to heare. I came over hither in December last, not

---

[4] Descartes left the Netherlands in late June and was in Paris by [29 June/] 9 July; he visited family and friends in Brittany and Poitou, spent some of the autumn in Paris, and returned to the Netherlands in Nov. (see A&T iv, pp. 126–47).

[5] Claude Hardy (c.1598–1678), a lawyer and mathematician, was a member of Mersenne's circle and a supporter of Descartes in his quarrels with Fermat and Roberval. He knew Arabic, and had edited Euclid's *Data* (from an Arabic MS) in 1625.

[6] The German scholar Christian Raue (Ravius) (1613–77), who had studied Arabic under Golius at Leiden, had spent two years (1639–41) in Istanbul, where he acquired a large number of MSS, including this Arabic translation of Apollonius's *Conics* (on which see Letter 23, n. 3). In 1643-4 he had a temporary lectureship in Oriental languages at the University of Utrecht, but failed to acquire a formal position there; he later moved to Amsterdam (1645–6), then to England (1647), and eventually to Sweden (1650).

[7] 'Curved lines'.

[8] See Letter 1, n. 7.

[9] Warner died in Mar. 1643.

bringing any of my bookes or papers with me: Nor hardly clothes, for the worst I had, seemed good enough to weare o'shipboard. I then thought not to stay heere above a fortnight. Nor did I imagine that, before I gat out of this towne againe, I should be persuaded to clime the Cathedra and make inaugurall Orations and praelusions and afterward reade publikely 5 dayes in a weeke, an houre every day in Latine.[10] Which had I foreseene, I thinke y^t all the bookes and papers that I had, both yours and mine, should have come along with me, to enable me to doe those things the more easily. And yet I have no great minde to goe fetch them, nor to send for them. So long as they are there unstirred, they seeme to be safe. But the disasters of the whole Kingdome put me in minde of what Melancthon used so often to say, Non est tutum quieta movere.[11] What may happen to them in the remoovall, by Searchers,[12] pirats &c I am not willing to try. Yet so long as they are there, We can not count them out of danger, But should that befall Mr Warners papers & mine, which we feare, it would put me into an humour quite contrary to that in which I have hitherto beene. I have thought nothing elaborate enough to be printed, till it were so complete, that no man could better it and did therefore so long keepe my name out of the presse: But now I begin to count nothing safe enough till it be printed and therefore I have almost resolved to secure my thoughts, not by burying my papers in England, nor by fetching them hither but by publishing the same notions heere, that I have committed to paper there.

[*The final section of the 7 Aug. letter is struck through, as follows:*] [I had thought heere to have given you account of what I have now to hand: but being desirous by this poste to let you know that your letter came safe to my handes I am constrained to breake off heere, deferring the rest till my next. In the meane time I remaine

    Sir

    Your humble servant

    John Pell.

Amsterdam Aug. 7. *deleted*]

[*and is replaced by the following in the draft on fo. 106:*]

To this purpose I have in hand a New Edition of Diophantus wherein I would

---

[10] See Letter 11.

[11] 'It is not safe to disturb things that are at rest', a saying of the Reformation theologian and educationalist Philipp Melanchthon (1497–1560). This was apparently a version of the legal dictum 'stare decisis et non quieta movere', meaning to adhere to decided cases and not to unsettle things that are established.

[12] Customs officials.

correct ye translation & make new Illustrations after my manner.[13] And Dr Blaew[14] hath promised to print it: Going one day to his shop to speake with him about it I there found a Catalogue of bookes printed by him and amongst y^m I find C. S. Longomontani Astronomia Danica, item Ejusdem de verâ Circuli mensura libelli duo which I asked for & found it dedicated to one of our Burgomeesters & printed A° 1644.[15] I therefore feared least some should thinke it was printed by my approbation. I therefore conceived a kinde of necessity layd upon me to let all y^t know me understand it was not so. Besides I supposed it no way disagreeing with my profession to disabuse all those that might be misse-led by ye name of [page torn] ancient a Professor as he is. I then [page torn] caused a refutation to be printed[16] which I intend with ye next opportunity to send to Longomontanus himselfe. Whose challenging of all Mathematicians in his Epistle dedicatory[17] might serve for a sufficient excuse of what I have done You have heere one Coppy I shall send more with des Cartes his booke[18] which I intend to send by sea by the next opportunity which will be about 3 or 4 dayes hence. In the meane time I remaine Sir

---

[13] The *Arithmetic* of Diophantus (*fl.* AD 250) is the only known work of classical mathematics on arithmetical problems. Only six of the original thirteen books survived in the original Greek, but the problems made a profound impression on European mathematicians when they were rediscovered and translated in the 16th century, and they were to play a significant role in the later development of number theory. Some of the problems were treated by Raffaele Bombelli in *L'algebra* (Bologna, 1572), by Simon Stevin in *L'Arithmetique* (1585), and by Viète in *Zeteticorum libri quinque* (1593). The first six books were first translated into Latin by Xylander (Wilhelm Holzmann), and published in Basel in 1575, and then by Claude Gaspar Bachet de Méziriac in Paris in 1621. Albert Girard published a revised and augmented edition of Stevin's *L'Arithmetique* in his edition of Stevin's *Les Oeuvres mathematiques* (1634), in which he reprinted Stevin's simplified and adapted version of books 1–4 of Diophantus' work and added his own similar treatment of books 5 and 6. Pell's treatment of Diophantus, like his intended work on 'analytics', was referred to often in the subsequent correspondence but, like that work, was never published. For 'after my manner' see Letter 1, n. 5.

[14] Johan Blaeu (1596–1673), son of the printer and astronomer Willem Jansz. Blaeu (see Letter 102, n. 6), had a doctorate in law from Leiden University; he ran the family business (with his brother Cornelis) from 1638 until his death. From 1637 to 1667 the printing-house was on the Bloemgracht.

[15] Longomontanus, *Rotundi in plano*. The work was dedicated to Albert Coenratsz. Burgh (1592–1647), a member of the city council of Amsterdam and a trustee of the Amsterdam Athenaeum (see Elias, *De vroedschap van Amsterdam*, i, pp. 327–9). Longomontanus explained that he had got to know Burgh when Burgh was a member of a Dutch mission to Denmark in 1641 (*Rotundi in plano*, sigs. *2^v–3^r).

[16] This single-leaf refutation, paginated '73–4' (in order to be inserted at the end of Longomontanus's book, which finished on p. 72), became known as the 'refutatiuncula'.

[17] Longomontanus wrote that 'Indeed, I have had so much faith in my discovery that I have not hesitated to wager my fame and fortune against anyone – especially any of the famous mathematicians – who might take up his pen against it' ('Invento nostro equidem tantam fiduciam tribui, ut non veritus sim, de fama pariter & fortuna mea, cum quovis, praecipue vero inter praestantes Mathematicos, calamum contra stringente, periclitari': *Rotundi in plano*, sig. *3^r).

[18] See Letter 11, n. 8.

your humble servant
John Pell

Amsterdam Aug 10/20

[*postscript:*] Heere is a rumor that My Lord of Newcastle hath taken a house in ye Hagh[19] and y^t you are shortly to come thither.

Any merchant can tell you of Jean D'Orville[20] a merchant of Hamburg whose Correspondents heere are well knowen to me & will pay me whatsoever you shall there deliver to him for y^t purpose.

---

[19] William Cavendish, Earl, Marquess (from 1643), and later Duke of Newcastle (1593–1676), Cavendish's brother. The rumour was false.

[20] Unidentified. Presumably an ancestor of the scholar Jacques Philippe d'Orville (b. 1696), whose father was an Amsterdam merchant named Jean d'Orville (and thus, perhaps, the grandson of the Jean d'Orville referred to here). That one branch of the family was established in Amsterdam in the mid-17th century is suggested by a publication there by a Guillaume or Wilhelmus d'Orville: *De constantia in adversis* (Amsterdam, 1666).

Letter 15

10/20 August 1644

Pell to Cavendish, from Amsterdam

Right Honourable,

Heere have you des Cartes his [>new] booke without ye translation of the Old[1] thus bound it costs 3 gulden 6 stuyvers, that is, as I reckon, about 5^s 6^d English. Had it had ye rest, the price had beene double. I have [adjoyned 4 coppies of my great worck *deleted*] augmented ye weight of it by adding a sheete of paper of mine owne[2] [It may be you will hardly finde 3 more on whom you *deleted*] Jean Dorville[3] of Hamburg hath so many correspondents knowen to me in y^s towne y^t I shall certainely receive what mony so ever you deliver to him to be paid to me.

  your humble servant
  John Pell

Aug 10/20

[*postscript:*] Any merchant can tell you of d'Orville

---

[1] See Letter 11, n. 8.
[2] The 'refutatiuncula' (see Letter 14, n. 16).
[3] See Letter 14, n. 20.

Letter 16

14/24 August 1644

Pell to Cavendish, from Amsterdam

BL MS Add. 4280, fos. 106$^\mathrm{v}$–107$^\mathrm{r}$ (draft)
van Maanen P3bis

Right honourable,
 In My second of Aug 10/20[1] I would gladly have sent you y$^\mathrm{e}$ Shippers name y$^\mathrm{t}$ brings you des Cartes his booke[2] which made me waite till it was in a manner too late to send my letter to y$^\mathrm{e}$ poste I therefore sealed [>it] & sent it away in such poste haste y$^\mathrm{t}$ I forgat to put one of my refutations[3] into it which I therefore now send in this I have put up 4 more with des Cartes [>one] Of which I desire may be sent to Johannes Adolphus Tassius [>who is] Conrector Gymnasij Hamburgensis[4] And if you please another to D$^\mathrm{r}$ Joachimus Jungius who is Rector of y$^\mathrm{e}$ same.[5] [>Of] their abilities & worth I have so great an opinion that I make no question but you will give me thankes for endeavouring to make you know them If you first desire to know them by their writings ere you seeke to be acquainted with their persons, that may easily be done. now you are upon y$^\mathrm{e}$ place you may finde divers pieces of Jungius in print And with a little narrow inquiry you might come to see divers dictates & Theses of theirs in manuscript among the students.
 Tassius hath 2 yeeres agoe reprinted Jungij Geometriam Empiricam [>in 5 sheets of paper] but without diagrams.[6] [If in some of y$^\mathrm{m}$ you should desire y$^\mathrm{e}$ diags which Tassius will presently helpe you to, if you desire you *deleted* >If he have printed any thing of his owne it is more than I know] But if your leisure will permit, I pray omit not y$^\mathrm{e}$ opportunity of acquaintance with such a paire of solid heads as they are. I am told y$^\mathrm{t}$ Tassius is a very courteous, affable, open man. The other a little more reserved, but through Tassius you may come to see & know what you desire. Without doubt you have [*page torn* fo]rgotten y$^\mathrm{t}$

---

[1] Letter 15.
[2] See Letters 11, n. 8, and 15, n. 1.
[3] See Letter 15, at n. 2.
[4] Johann Adolf Tassius (1585–1654) had studied in Rostock, Heidelberg, and Tübingen, before becoming Professor of Mathematics (and deputy Rector) at the Gymnasium in Hamburg in 1629. He published his *Disputatio de rebus astronomicis et geographicis* there in 1635.
[5] Joachim Jungius (1587–1657) studied in Rostock, Giessen, and Padua; he became Professor of Logic and Natural Philosophy at the Gymnasium in Hamburg in 1629. He was best known in his lifetime for his logic textbook, *Logica hamburgensis*; several of his other works were published only after his death.
[6] Jungius's *Geometria empirica* was first published in Rostock in 1627; this 2$^\mathrm{nd}$ edn. was published in Hamburg in 1642.

I [>often] made mention of y^m to you at London and of Jungij Apodictica vel Apodidactica: /⁷ Added / withall y^t the most barbarous nations had some thing to worship & there are few men that have not some Idol, some man or woman whom they esteeme, and admire, above all [>ye rest of] mankind & Jungius is mine For I esteeme of men more or lesse as I find y^m more or lesse rationall and therefore having seene something of Jungius his writing wherein he seemed to me to make [>a truer &] better use of his reason & to manage y^t divine instrument of instruments with more dexterity & skill y^n any other sonne of Adam, All other writers must pardon me if I professe to expect more solidity in Jungius his writings than in any other man now living. If you finde but ye one halfe of what I [>imagine] to be in Jungius, you will never be able to relish any other moderne philosopher And [>perhaps] you will find enough in Tassius to make you take no further thought what becomes of Pellius or any [others *deleted*] of ye great promisers in France

I made no mention of them in my former letters because I thought it might be a goode beginning of acquaintance with y^m if you should send my refutation to Tassius & des Cartes to Jungius & desire their judgements of y^m but now this rumor of your comming to ye Hagh⁸ puts me in a great feare least you should come thence before you have any knowledge of those men. And therefore I doe now exceedingly blame my selfe y^t I did not in my first letter put you in mind of y^t opportunity.

Your letter inquires what new inventions are extant in ye Mathematickes. To which I have so much to answere y^t I see it will cost you ye postage of yet another letter.

There is an industrious Carmelite about my age in paris called Ludovicus à Sancto Carolo who will shortly give us a Catalog of all ye bookes printed in paris within these 2 yeeres:⁹ then no doubt we shall finde many that have escaped my inquiry Till then, you will accept of as much as I know

Jacobus de Billy [>a Jesuit] hath [>set] forth [>an Algebraicall] booke in 4° called Nova Geometriae Clavis,¹⁰ Sr William Beecher sent it me about this time

---

⁷No work by Jungius was published under such a name; Pell may perhaps be referring to unpublished writings on method, of which one fragment, 'Protonoeticae philosophiae sciagraphia' (CUL, MS Baker Oo. VI. 113, fos. 220–3; printed in Kangro, *Joachim Jungius' Experimente*, pp. 256–70), was in Hartlib's possession.

⁸See Letter 14, n. 19.

⁹Louis Jacob de Saint-Charles (Ludovicus à Sancto Carolo), *Bibliographia parisina, hoc est, catalogus omnium librorum Parisiis, annis 1643 & 1644 inclusivè excusorum* (Paris, 1645). Another such volume by this Carmelite friar and librarian (who was three years older than Pell: 1608–1670), covering the publications of 1645, was issued in 1646; further volumes appeared in 1649 and 1650.

¹⁰Jacques de Billy SJ (1602–79), *Nova geometriae clavis algebra* (Paris, 1643).

twelvemonth but I left it behind me in England.[11]

La Bosse the incomparable graver of paris printed last yeare in 8° the opuscula of ye unlearned great Geometer Desargues[12] Theon Smyrnaeus his illustrations of Platos mathematicall passages were never [>published] till 1642 when they were printed in Greeke & Latine cum notis Bullialdi in 4°.[13] the price about 4 guilders, bound.

Mersennus hath put forth 2 Tomes in large quarto, each of y^m about an inch & $\frac{1}{2}$ thicke they came out of y^e presse about 4 months agoe The former is called Mersenni Cogitata Physico-Mathematica In quibus tam naturae quam artis effectus admirandi certissimis demonstrationibus explicantur

The latter, Universae Geometriae Mixtaeque Mathematicae Synopsis, Et bini refractionum demonstratarum tractatus.[14] Therein hath he Epitomized Euclid, Ramus, Archimedes & with supplements ex Snellio Kepplero et Lucas Valerius Theodosius, Menelaus, Maurolycus, Autolycus Apollonius, Serenus, Mydorgius,

---

[11] William Beecher was knighted in 1622 (Shaw, *Knights*, ii, p. 180) and was appointed Clerk of the Council in 1623 (*CSPD 1619–23*, p. 480). He resigned this post in January 1641 (*CSPD 1640–1*, p. 433) and appears to have spent parts of 1641 and 1642 (and 1643: see below) in Normandy: a medical certificate of 1642 stated that he needed to drink the waters of Saint Paul, near Rouen, for the sake of his health, 'he having been much benefited thereby last year' (*CSPD 1641–3*, p. 426). Pell had been in contact with him in London in the autumn of 1639; together, they had embarked then on the project of making Cartesian hyperbolic lenses which was later taken over by Cavendish (see BL. MS Add. 4474, fo. 77^r (Pell, memorandum), and HP 30/4/25A (Ephemerides, 1639)). Shortly before his departure from London, in Oct. 1643, Pell was in touch both with Sir William and with his 'cousin', also called William Beecher: see BL MS Add. 4416, fos. 159, 166. Pell's copy of de Billy's book survives: Busby Library, Westminster School, pressmark I C 31. It bears an inscription from Sir William: 'to Mr Pell, at his howse in the bowling Alley in Westminster Roüen the 20th of Novemb. 1643', and an annotation by Pell: 'Received Nov. 27/Dec 7 1643'.

[12] A. Bosse, *La Maniere universelle de Mr Desargues ... pour poser l'essieu & placer les heures & autres choses aux cadrans au soleil* (Paris, 1643). Girard Desargues (1591–1661) was one of the most talented geometers of his age; details of his education are, however, entirely lacking (see R. Taton, *L'Oeuvre mathématique de G. Desargues* (Paris, 1951), pp. 12, 39). The engraver Abraham Bosse (not 'La Bosse' – this is perhaps a confusion with the name of the mathematician Guy de La Brosse) had studied perspective under Desargues.

[13] I. Boulliau, ed., *Theonis ... eorum, quae in mathematicis ad Platonis lectiones utilia sunt, expositio* (Paris, 1644): the date given by Pell is less inaccurate than it seems, as the work was in fact printed by Aug. 1643 (see H. J. M. Nellen, *Ismaël Boulliau (1605–1694): astronome, épistolier, nouvelliste et intermédiaire scientifique* (Amsterdam, 1994), p. 97(n.)). Ismaël Boulliau, a converted Huguenot, became a Catholic priest in c.1640. He had lived in Paris since 1633, where he had been befriended by Gassendi and had become an accomplished astronomer, making his name with his *Philolai, sive dissertationis de vero systemate mundo, libri IV* (Amsterdam, 1639).

[14] Both these works by Mersenne, *Cogitata physico-mathematica* and *Universae geometriae mixtaeque mathematicae synopsis*, were published in Paris in 1644.

Pappus, Vietae angulares sectiones,[15] & he hath concluded with 2 demonstrations of refraction by Warner and Hobbes[16]

In his preface before this Synopsis he tells us of ye Decanus Diniensis (he meanes Gassendus).[17] *cujus philosophia tam styli pulchritudine, quàm novarum observationum multitudine, rationumque subtilitate philomusos in tanti viri raptura sit admirationem, vix ut unus supersit, qui deinceps atomos rejiciat.*[18] Also of *Curiosa R. P. Niceronis perspectiva.*[19] also *Opus illud ingens Eximij Astronomi I. Boullialdi, quod jam praelum exercet, quod facilè superabit quidquid hactenus Astronomicum editum est, quodque nihil aliunde supponet, ut qui volumen illud habuerit, omnia generis istius* [obtineat *blotted*][20]

---

[15] The writers on geometry whose theories were presented by Mersenne were: Euclid (*fl.* 295 BC); Petrus Ramus (Pierre de La Ramée) (1515–72), author of *Geometriae libri septem et viginti* (Basel, 1569); Archimedes (c.287 BC – 212 BC); Willebrord Snel (1515–72), author of *Apollonius batavus* (Leiden, 1608), a restitution of the lost work by Apollonius on plane loci (see Letter 42, n. 10), and *Doctrina triangulorum* (Leiden, 1627); Johannes Kepler (1571–1630); Luca Valerio or Valeri (1552–1618), author of *De centro gravitatis solidorum libri tres* (Rome, 1604) and *Quadratura parabolae* (Rome, 1606); Theodosius of Bithynia (2nd century BC); Menelaus of Alexandria (*fl.* AD 100); Francesco Maurolico (1494–1575), author of many mathematical works, including a reconstruction of books 5 and 6 of Apollonius's *Conics*; Autolycus of Pitane (*fl.* 300 BC), author of *De sphaera quae movetur*; Apollonius of Perga (late 3rd – early 2nd century BC), author of the *Conics*; Serenus of Antinoupolis (4th century AD), author of a (lost) commentary on Apollonius and an extant book on the section of the cylinder; Claude Mydorge (on whom see Letter 34, n. 1), author of *Conicorum operis ... libri quatuor* (Paris, 1639); Pappus of Alexandria (*fl.* AD 300–350), author of the *Synagoge* or 'Collection', an influential compilation (made probably after his death) in eight books of eight originally different treatises and commentaries on different parts of the mathematical sciences; and François Viète or Vieta (1540–1603), whose *Ad angularium sectionum analyticen theoremata* was published in Paris in 1615, edited by Alexander Anderson.

[16] Mersenne printed, as books 6 and 7 of the 'Optica' in his *Universae geometriae synopsis*, a treatise on refraction by Walter Warner (pp. 549–66) and a treatise on optics by Thomas Hobbes (pp. 567–89).

[17] The philosopher Pierre Gassendi was Dean ('Decanus') of the church of Notre Dame du Bourg in Digne, in his native Provence.

[18] 'Whose work on philosophy will, as much by the beauty of its style as by the great quantity of its new observations and the subtlety of its reasoning, cause lovers of the muses to be gripped by admiration of such a great man, so that scarcely anyone will be left to reject atomism from then on' (quoting Mersenne, *Universae geometriae synopsis*, sig. 2+2^v). On this work see Letter 18, n. 4.

[19] Jean-François Niceron, *La Perspective curieuse* (Paris, 1638), a treatise on anamorphic perspective. Niceron (1613–46) was a friend and colleague of Mersenne, belonging to the same order of Minim friars (hence 'R[everendus]. P[ater].', 'reverend father'). Mersenne's comments are in *Universae geometriae synopsis*, sig. 2+2^v.

[20] Ismaël Boulliau, *Astronomia philolaica, opus novum, in quo motus planetarum per novam ac veram hypothesim demonstrantur* (Paris, 1645). Mersenne's comment here is: 'that huge work by the outstanding astronomer I. Boulliau, *which he is already putting to the press* [the emphasis is Pell's] – it will easily surpass whatever has been published on astronomy up till now, and it will presuppose no knowledge from other sources, so that anyone who possesses the book will have everything that pertains to that subject' (*Universae geometriae synopsis*, sig. 2+3^r). Mersenne had been expecting the appearance of this major work since at least Oct. 1643 (see *MC* xii, p. 339), but the work was not consigned to the printer until the end of 1644 (see Nellen, *Ismaël Boulliau*, p. 123).

he bids us learne Analysin Algebraicam Ex Vieta, Cartesio, Diophanto: nisi, saith he, malis eam industriâ Geometrae D. Chauveau adornatam & majori claritate facilitateque donatam Expectare[21]

He mentions Viri Laudatissimi Domini de Mets pulcherrimum & limatissimum opus de munitione seu fortificatione not yet published,[22] Arithmeticam, saith he, si his in rebus vir incomparabilis D. Freniclus suis novis repertis augere velit, omnium possit animos rapere in admirationem.[23]

From these Mathematicians he passeth to speak of admirable pieces in Divinity [>not yet published] by Petavius, Marandeus, Dinetus, Montarsius, Lacombeus.[24]

But before he passeth to those French divines he makes one step into England, Non desunt autem egregij viri apprimè tàm in Geometricis puris quàm mixtis versati qui scientias istas maximè promoveant quos inter Pellium Anglum, praeter eos quos hoc Opere laudavi, ad ea quae paravit in lucem edenda urgere,

---

[21] 'Algebraic analysis from Vieta, Descartes, and Diophantus – unless you would prefer to wait for it to be set out by the industry of the geometer M. Chauveau and presented with more clarity and facility' (Mersenne, *Universae geometriae synopsis*, sig. 2+3^r). Little is known of the mathematician Jean-Baptiste Chauveau, who was active in the late 1630s and early 1640s, and was an associate of Carcavi, Roberval, and Desargues. Apparently he wrote several treatises but published none of them (see P. Tannery, 'Sur le Mathématicien français Chauveau', *Bulletin des sciences mathématiques*, 19 (1895), pp. 34–7, and R. Taton, 'Desargues et le monde scientifique de son époque', in J. Dhombres and J. Sakarovitch, eds., *Desargues et son temps* (Paris, 1991), pp. 25–53, here pp. 45–6). A MS 'Traicté d'algèbre' by him survives amongst Cavendish's papers (BL, MS Harl. 6083, fos. 350–79; Pell's copy of this MS is in BL, MS Add. 4407, fos. 31–7).

[22] 'The extremely beautiful and elegant work on defence or fortification by that most praiseworthy man, M. de Mets' (*ibid.*, sig. 2+3^r). Adrianus Metius (1571–1635), Professor of Mathematics at Alkmaar, had published only a brief treatise on the application of geometry to the science of fortification: *Geometriae practicae pars quarta, continens munitionum delineandarum muniendarumque genuinam & propriam institutionem* (Franeker, 1625); this was also issued as part of his *Arithmeticae libri duo, et geometriae libri VI* (Leiden, 1626). The longer work referred to by Mersenne, however, remained unpublished and is not extant.

[23] 'As for arithmetical knowledge, if that incomparable man M. Frenicle is willing to augment it with the new discoveries he has made in these matters, he will be able to inspire the minds of all men with admiration.' Bernard Frenicle de Bessy (*c.*1605–75) was a skilled amateur mathematician and friend of Mersenne. He had a gift for performing large calculations, and his 'discoveries' arose out of such computations, but he showed no interest in the theoretical aspects of number theory.

[24] Denis Petau SJ (1583–1652), teacher of theology in Paris, was the author of many works, including *Uranologion* (Paris, 1630), a collection of Greek astronomical texts, and *Dogmata theologica*, 4 vols. (Paris, 1644–50); Léonard de Marandé, a well-known lawyer-turned-priest, was author of *Le Théologien français*, 3 vols. (Paris, 1641); Jacques Dinet SJ (1584–1653), Provincial of France (1639–42) and of Champagne (1643–7), confessor to Louis XIII in his final months, left some works in MS; Pierre Baudouin de Montarcis was author of *Traité des fondemens de la science générale et universelle* (Paris, 1651); Jean de Lacombe, a friar at the Minim convent in Blaye, corresponded with Mersenne (see *MC* ix, x, and xi) but does not appear to have published anything.

juuareque oporteat.[25] I had almost sent [>you] Mersennus with des Cartes but I had no commission so to do

Heere are also to be had bookes which you sought for in vaine else where as Cevallerij Geometria per indivisibilia Ejusdem Uranometricum directorium.[26] As also new bookes out of Germany & these countries but of these heereafter, God willing To whose protection I commend you & take my leave, continuing

Sir
Your humble servant
J. Pell

Amsterdam Aug 14/24

[*postscript:*] The Shippers name is Evert Everts[27] [that is Eberhardus Eberhardi filius *deleted*] He hath promised to [>deliver it] with his owne hand.

---

[25] 'However, other exceptional men are not lacking, especially ones equally skilled in pure and mixed geometry, who are developing those sciences to the greatest degree – among them, the Englishman Pell. We should urge and help him to publish those things which he has prepared, in addition to [alternative translation: 'even more than'] those other people whom I have praised in this work' (Mersenne, *Universae geometriae synopsis*, sig. 2+3^r).

[26] Pell refers to two major works by Bonaventura Cavalieri (c.1598–1647). The *Geometria indivisibilibus promota* (1635) was a seminal text that considered areas or volumes in terms of an infinite number of infinitely fine, or 'indivisible', lines or planes. The book was not easy to come by: in Oct. or Nov. 1645 Oughtred reported that he was 'very desirous to see' it, but that when he had tried to obtain a copy three years earlier, he (or perhaps Cavendish on his behalf) 'could not get it in France' (Rigaud, *Correspondence*, i, pp. 65–6). Five or six years later, John Wallis was unable to obtain it in England (Wallis, *Arithmetica infinitorum* (Oxford, 1656), Preface, sig. Aa2^v). The second work referred to by Pell is Cavalieri's *Directorium generale uranometricum* (Bologna, 1632), which discussed logarithms and trigonometry.

[27] Unidentified.

Letter 17

16/26 August 1644

Cavendish to Pell, from Hamburg

BL MS Add. 4278, fos. 180–1 (original)
van Maanen C3

Worthie S^r;

[>yesterdaye] I received yours of the 10/20 August,[1] for the which I giue you manie thankes; I heare nothing yet of our remouing from hence, if we should remoue in to Hollande I should be in hope to see you, & intended to see de Cartes, but you write he is gone to Paris. I desire your iudgment of de Cartes his new booke,[2] douteless he is an excellent man; I hope M^r. Hobbes & he will be acquainted[3] & by that meanes highlie esteeme one of an other./ I am sorie M^r: warners Analogicks are not printed[4] but I yet hope they maye, as alsso other worckes of that excellent olde man. I am glad you intend to secure your thoughts by publishing them, & that you are printing Diophantus with newe illustrations after your manner,[5] which I am exceeding greedie to see, but I doute you will not in this worcke teach vs the whole science of Analiticks with all that pertaines to it, which if you doe not nowe I beseech you doe heereafter for I suppose you intended such a worck, & if you finish it not, I doute wee shall haue no better Anali[>ti]cks than wee haue./ I haue not yet receiued your refutation of C. S. Longo: his quadr. of a circle.[6] but I shall inquire dilligentlie for it; for I longe to see where he goes out of the waye. I am well aquainted with M^r: Gascoine whoe was prouidore to our armie, he is an Ingenious man & hath shewed me howe perspectiues maye be much improued;[7] I onelie mislike his glass next the eye which he makes conuex on both sides, I tolde him it woulde make confused sight, if De Cartes his doctrin be true, but vpon triall it proued more distinct than I expected, yet I thinke a concaue on that side next the eye would doe better. his perspectiue did not multiplie more than myne as I thinke; but his speculation is most true, & this was one of his first trialls & not made to the

---

[1] Letter 14.

[2] The *Principia* (see Letter 11, n. 8).

[3] Hobbes had been living in Paris since Nov. 1640.

[4] See Letter 1, n. 7, and Letter 14, at n. 8.

[5] See Letter 14, at n. 13.

[6] See Letter 14, n. 16.

[7] William Gascoigne (*c.*1612–1644) was a talented amateur astronomer, a designer of telescopes ('perspectiues'), and the inventor of the micrometer (a device for measuring the images of planets and other astronomical objects). When Cavendish wrote these words he was evidently unaware that Gascoigne had been killed at the battle of Marston Moor on 2/12 July 1644.

manner of his first inuentions./ wee lodge neere S^t: Johns Church./[8] And so wishing you all happiness I remaine

Your assured friend to serue you
Charles Cauendysshe

Hamburg Aug: 16/26. 1644.

[*addressed:*] To my verie worthie friend M^r: John Pell, publike Professor of Mathematicks at Amsterdam, giue this In t'oude Conuoy op de Zeedijk in Amsterdam

[*cover annotated by Birch:*] 16/26 Aug. 1644

---

[8] The Johanniskirche was next to the 'Johanniskloster', a former Dominican convent which had been used as the 'Gymnasium' (the high school or academy at which Jungius and Tassius taught) since 1613. Neither building survives (the church was demolished in 1813, the convent was burnt down in 1842); their site is now occupied by the Rathausmarkt. (See W. H. Adelung ['Adelungk'], *Die annoch verhandene hamburgische Antiquitäten oder Alterthums-Bedächtnisse* (Hamburg, 1696), pp. 11, 31 (descriptions of school and church); E. H. Wichmann, *Heimatskunde: topographische, historische und statistische Beschreibung von Hamburg und der Vorstadt St. Georg* (Hamburg, 1863), pp. 51, 51–2 (n.).)

Letter 18

1/11 September 1644

Cavendish to Pell, from Hamburg

BL MS Add. 4278, fos. 324–5 (original)
van Maanen C4

Worthie S^r:

    I receiued this daie des Cartes his newe booke with 4 of your refutations;[1] to morrowe God willing I will giue or send des Cartes his booke & your refutations of Longo: to Doctor Jungius & Mr Tassius;[2] for I longe to be acquainted with them; I desire you will let me knowe what it was of Doctor Jungius that made you so esteeme of him./ Mersennus continues still his ciuilitie & kindness to me; for I receiued a letter (since I came hither) from him, wherein he writes that he sendes me [>his] two bookes;[3] which I expect daielie to receiue./ I am so in loue with the titles of those bookes you writ of, that I intend god willing heereafter to haue them all; but for the present onelie Gassendes Philosophie,[4] & D. Chauueau, which latter Mersennus writes teacheth Analitickes with more cleereness & facilitie than anie yet extant./[5] Longo: maye be much ashamed to be refuted in one smalle leafe of paper;/[6] I shall as soone as I maye returne [>to] you fiue poundes sterling by M^r: Jean D'oruille[7] of Hamburg; I desire your opinion of des Cartes his newe booke. And so wishing you all hapiness I remayne

    your assured friend to serue you
    Charles Cauendysshe

Hamburg: Septemb: 1/11 1644

[*addressed:*] To my verie worthie friend M^r: John Pell Pubicke [*sic*] Professor of the Mathematicks at Amsterdam; giue this in t'oude Conuoy op de zeedijk in Amsterdam

[*annotated by Pell:*] Received September 11/21

[*cover annotated by Birch:*] 1/11 Sept. 1644

---

[1] See Letter 15.
[2] See Letter 16, nn. 4, 5.
[3] See Letter 16, n. 14; this letter has not survived.
[4] The 'philosophia' or 'work on philosophy' by Gassendi, referred to by Mersenne (see Letter 16, n. 18), was not published for another five years: *Animadversiones in decimum librum Diogenis Laertii, qui est de vita, moribus, placitisque Epicuri*, 3 parts in 2 vols. (Lyon, 1649).
[5] Cavendish translates here the phrase quoted by Pell (see Letter 16, n. 21).
[6] Pell's 'refutatiuncula' was printed on a single leaf (see Letter 14, n. 16).
[7] See Letter 14, n. 20.

372   JOHN PELL (1611–1685)

Letter 19

7/17 September 1644

Pell to Cavendish, from Amsterdam

BL MS Add. 4280, fos. 107^r, 109^r (draft)
van Maanen P4

Right honourable
   My last was Aug 14/24.[1] Since which time I have received a second, and a 3^d from you[2] I sent ye letter by the poste but Des Cartes his booke with 4 of my refutations comes by sea by a Skipper whose name is Evert Everts,[3] he sayes he dwells op den Keerwedder in Aelricks hof in Hamborch.[4] But I hope you have both before this time. In my letter I commended a paire of learned men in Hamburg[5] more largely than is my custome to doe, but whither more than they deserve, by this time I suppose you know better than I. In the same I [>told] you what we expect out of France, I should now tell you what we must not expect from thence And first We must expect no more from *Beaugrand* [for which I am exceeding sorry *deleted*] he died Anno 1642[6] I expected [>from him almost] as much as he promised, & that [>you know was pretty large]. I could have beene well content that he should have saved me the labour of reducing [>into precept] the subtilties & close conueyances of the old Greekes & his later countrymen: That so I might have had nothing to doe but to adde what I had observed in ye processe of mine owne reason. But now I give over all [>hopes of seeing] any thing of that nature, of his doing. What noster Geometra (so Mersennus allwayes

---

   [1] Letter 16.

   [2] Pell refers to Letters 13 and 17; at the time of writing this letter he had not received Letter 18.

   [3] See Letter 15, and Letter 16, nn. 1–3.

   [4] The Kehrwieder is a quayside street on the southern side of the Binnenhaven in Hamburg; 'Aelricks hof' (Ulrichshof?), a house on the ('op den') Kehrwieder, has not been identified.

   [5] Jungius and Tassius: see Letter 16, nn. 4, 5.

   [6] Pell was misinformed: the mathematician Jean de Beaugrand (born *c.*1595), a pupil of Viète and friend of Mersenne, had died in Dec. 1640 (*MC* x, p. 519).

calls *Robervall*)[7] will doe I know not, I feare, not enough to make others thinke so [>highly] of him as men say he seemed to thinke of himselfe. Some give me the Character of him to be the most selfe-conceited, proud bubble in all [France *deleted* >Paris] except the minim frier Niceron.[8] But yet the man without doubt, hath [>very] good worth in him and his greatest fault is that he knowes it too well.

Nor must we expect any more from *Herigone*, he died ye last yeare, and perhaps you are not much sorry for it. It is true he promised not so much as Beaugrand but he performed more than he. His 5 Tomes I suppose you have seene long since. But perhaps not his 6^t conteining Algebrae Supplementum & Isagogen and some other things.[9] And though he have little or nothing which [>may] seeme new to you or mee, yet [>being] separated from all [>my] bookes & papers, [I begin to like this *deleted* >methinkes] such kinde of writers as Mersennus & Herigon are better than none at all. Let me adde one thing more [>which I expect not & that is] that ye Genius either of Gassendus or of Mr Hobbes should ever close with that of des Cartes [>though you seeme to hope y^t des Cartes & Mr Hobbes will [>grow] acquainted in Paris][10] I suppose you know that Gassendus hath written a large work (heere printed in Amsterdam in 4°) against des Cartes his Metaphysicall meditations.[11]

I perceive you have seene Gassendus his judgement of ye new Joviales.[12] S^r

---

[7]'Our geometer' was the soubriquet used by Mersenne in many of his works to refer to his friend Gilles Personne de Roberval (1602–75), who was Professor of Mathematics at the Collège Royal in Paris. See, for example, Mersenne, *L'Optique et la catoptrique*, in J. F. Niceron and M. Mersenne, *La Perspective curieuse du R. P. Niceron... avec l'Optique et la catoptrique du R. P. Mersenne*, ed. G. P. de Roberval (Paris, 1652), book 2, p. 98: 'the person who is called simply "our geometer" in several places in our writings' ('celuy qui en plusieurs lieux de nos oeuures, est nommé absolument nostre Geometre'). 'Geometry' should be interpreted here in the Platonic sense, as one of the four major branches of mathematics (along with arithmetic, astronomy, and music). At a time when arithmetic was of limited interest, and algebra as a distinct branch of mathematics was still in its infancy, 'geometry' could be interpreted as a general word for 'mathematics'.

[8]Such contemporary judgements on the character of Niceron (on whom see Letter 16, n. 19) have not survived; but Roberval was notorious for undervaluing the achievements of others, and/or accusing them of stealing their ideas from himself (see the comments in E. Walker, *A Study of the Traité des indivisibles of Gilles Persone de Roberval* (New York, 1932), p. 7).

[9]Pierre Hérigone (d. 1643) had worked as a teacher of mathematics in Paris. His *Cursus mathematicus nova, brevi ac clara methodo demonstratus* had been published between 1634 and 1642; the sixth volume was entitled *Cursi mathematici tomus sextus ac ultimus, sive supplementum*. It included a French translation of Euclid's *Elements* (1st pagination, pp. 1–280); a 'Supplementum algebrae' in Latin and French (2nd pagination, pp. 1–73); and an 'Isagoge de l'algebre' (2nd pagination, pp. 74–98), as well as sections on perspective, astronomy, chronology, and fortification.

[10]Cavendish expressed this hope in Letter 17.

[11]P. Gassendi, *Disquisitio metaphysica seu dubitationes et instantiae adversus Renati Cartesii metaphysicam et responsa* (Amsterdam, 1644).

[12]On the dispute between Gassendi and Rheita, discussed in this letter, see Letter 13, n. 4.

william Boswell[13] sent me ye friers Epistle, [I received it ye day y^t you went out of London *deleted*] the next morning after I received it I brought it to your Lodging, where I learned you were gone out of London ye day before. I returned [a pretty large judgement of it *altered to* to Sr William a large letter] shewing many grounds of suspicion y^t it was a meere fable & there was no such matter.[14] [>Yet suspecting my judgement because it was possible for a man to have art enough to invent a Telescop so much better than ordinary and yet not wit enough to make [>a probable] report of it] A coppy of my letter was sent to a friend in France who thereupon sent me Gassendus his censure: I thereupon examined y^t censure and found y^t it was impossible for [>Gassendus] fixed [>starres] to make such an appearance [>with Jupiter] as the Frier hath described, and therefore I pronounced Gassendus his censure to be a very [*page torn*] of wit and nothing more. Or if you will, such a fine discourse of nothing as that which he hath adjoined with it concerning a weight hanged up at the end of an extreame long line shewing that ye earth forsooke her center [of gravity *deleted*], and I know not [>how many other] learned fooleries.[15] But let us heare Mersennus, (whose experiments for ye most part are not worth the taking up, yet in this I beleeve him) (Phaenomena Ballistica Propos. 16.) Haerebat animus num forsan observatores decepti fuissent ob funes intortos, vel fila sive cannabina sive bombycina, quae (praeterquàm diutissime detorquentur dum suspensum plumbum in orbem agitur) omnibus aeris mutationibus sunt obnoxia; quapropter filo usus sum argenteo, per foramen chalybeum ducto, cujus observatio clarissimè docuit nullum in eo motum sive 6 sive centum horarum spatio fieri: mane siquidem in linea LB positus, in eadem pluribus diebus, pluribusque testibus, permansit.

Unde, saith he, concludendum quanta sit in observationibus adhibenda diligentia, priusquam illarum rationes & causae, vel utilitates quaerantur, nisi enim de facto satis constet, quid ulterius inquires? huic autem Phaenomeno falsò credito quidpiam simile contigisset in 5 novis planetis Jovialibus, quos non nemo 4 Medicaeis addebat & jam de novenario musarum numero hisce 9 planetis comparando viri docti cogitabant, nisi faelicissimus observator fidelissimusque, Gassendus hunc errorem abstersisset, epistola in lucem edita, quá demonstrat

---

[13] Sir William Boswell (d. 1649), British Ambassador at The Hague since 1633, a man of wide scholarly and scientific interests, and one of the literary executors of Bacon's estate. It was Boswell who had first proposed Pell as the replacement for Hortensius in the Amsterdam Athenaeum (Bodl. MS Rawl. Letters 83, fo. 172, Boswell to G. J. Vossius, 23 Aug. 1639).

[14] Pell's letter does not survive. A copy of it was passed to Princess Elizabeth at the Palatine court in The Hague, and circulated further by her: see Thijssen-Schoute, *Nederlands cartesianisme*, p. 566.

[15] This refers to the 'Postscriptum' appended by Gassendi to his letter about Rheita's claims, in which he reported experiments which appeared to demonstrate a twelve-hourly cycle of displacement of a pendulum from a north-south alignment (see P. Gassendi, *Opera omnia*, 6 vols. (Lyon, 1658), iv, pp. 520–22).

stellas pro planetis acceptas.[16]

But by Mersennus his leave, Gassendus had ill lucke in y^t whole booke; for that novelty y^t he writes in favour of, we see confuted by a silver wire and the other novelty against which he writes, proves, they say, a verity.

For first, as I said, Gassendus his conjecture of those fixeds being rigidly examined, [can *deleted* >will] by no meanes agree with the place of Jupiter at that time, [>no nor with Jupiters possible motion at any time] so as to make [>such] phaenomena as the frier hath described.

Secondly, Gassendus, denying the new [planets *deleted*] satellites which I care not for, seemes to beleeve that ye Frier hath so excellent a glasse (which is ye maine thing that I care for, let him give us such a glasse & we will make observations our selves) [Thirdly, it seemes beyond all belief y^t ye frier could *deleted*] And yet [>(wch seemes very strange to me)] he would have us beleeve y^t the friers glasse inverts all the visibles As if a man that had wit enough to [make *deleted* >invent] a glasse [>so much better than all the best of Italy] had not wit enough to foresee, nor eyes inough to discerne y^t all visibles would appeare inverted in his glasse

But further, Caramuel (whose comment on Trithemius I beleeve you have seene [>long agoe]) hath since y^t time put forth a booke entitled Mathematicus audax (about 4 shillings price) where he not onely beleeves this newes of y^e

---

[16]'Phaenomena ballistica' is the running-title of 'Ballistica et acontismologia', the sixth part of Mersenne's *Cogitata physico-mathematica*. In this passage (on p. 46) Mersenne is discussing, and dismissing, Gassendi's account of the apparent twelve-hourly cycle of displacement of the pendulum (see the previous note): 'I hesitated over the possibility that the observers were perhaps deceived by the fact that the ropes were twisted, or made of hempen or cotton thread, which (apart from the fact that over a very long time they become unwound, while the lead weight is subjected to circular motion) are sensitive to all changes in the state of the air. For that reason I used a silver thread, drawn through an aperture in steel; observation of this clearly showed that there was no motion in it, whether over a period of six hours or over a hundred. Indeed, if it was placed in the line LB in the morning, it remained in that same line for several days, as several people witnessed. From this we should draw some conclusion about how careful we need to be in making observations, before we make enquiries about the reasons or causes of them, or their utility. If the fact is not sufficiently well established, what is the point of pursuing any further enquiries? A similar case of a falsely credited phenomenon would have happened where the five new moons of Jupiter are concerned (which someone added to the four Medicean moons of Jupiter, so that learned men were already thinking of comparing the set of nine muses to the number of these nine moons), had not that most successful and faithful observer, Gassendi, cleared up this error by means of his published letter, in which he demonstrates that stars have been mistaken for moons.'

Capuchin of Collen but confidently averrs it against all gainsayers.[17]

And which is more [>whereas] the french [>gave out that ye Capuchin] was ready to hang himselfe for griefe when Gassendus had discovered his error. It proves no such matter: The same Capuchin a yeare after (VII Kal Decemb. 1643) writes againe to ye same Putean:[18] that he saw ye same starres about Jupiter in May June & July but not since: that for a month together 2 of the Galilean satellites could not be seene And other matters worth the reading. I [>intend] therefore the next time send you a coppy of his letter

[I caused inquiry to be made at Collen among the Capuchins whether this frier were there & if not, where he was. It was answered that he was at Augsburg in Germany. By which we perceived y^t ye workman which he speakes of in his letter[19] is ye same, whom some yeares agoe, we had heard y^t he had a most incomparable hand glasse *deleted*]

with some other passages from others, tending to the confirmation of this report.

In the meane time I remaine.

Your Honours most humble servant
J: P:

Amsterdam Sept 7/17

[*postscript:*]
In a loose paper

Whereas, in answere to your inquiry how you might with most conveniency returne mony to me, I wrote that John Dorvill,[20] a merchant of Hamburg, had many correspondents in this towne knowne to me, and I thinke all his kindred & Correspondents [>heere] also know me. Yet because there are so many of them, it you make any use of him, it will be good to send me a Line or two under his hand that I may know of which of them to demaund it & not inquire of them, that know nothing of it.

---

[17] Juan Caramuel y Lobkowitz (1606–82), the son of a Bohemian engineer resident in Spain, was a Cistercian monk and a prolific author of scientific and theological works. These included an edition of book 1 of Johannes Trithemius's *Steganographia* (a treatise on cryptography), with a commentary on each chapter: *Steganographia necnon claviculae ... Ioannis Trithemii genuina, facilis, dilucidaque declaratio* (Cologne, 1635). His *Mathesis audax, rationalem, naturalem, supernaturalem, divinamque sapientiam ... substruens exponensque* (Louvain 1644) included a letter from him to Gottfried Wendelin, dated [7/] 17 Sept. 1643 (pp. 60–4), in which he defended Rheita (referred to here by Pell as the 'Capuchin of Collen', *sc.* of Cologne). His defence of Rheita against Gassendi's criticisms was published in Louvain in 1643: see Letter 13, n. 4. For a valuable modern study of Caramuel y Lobkowitz see D. Pastine, *Juan Caramuel: probabilismo ed enciclopedia* (Florence, 1975).

[18] 'Putean' was Erycius Puteanus (1574–1646), philologist at the University of Louvain, and the recipient of the original letter from Rheita announcing his discoveries. The date of this later letter was [15/] 25 Nov. 1643.

[19] See Letter 21, n. 15.

[20] See Letter 14, n. 20.

Letter 20

c.13/23 September 1644

Cavendish to Pell, from Hamburg

BL MS Add. 4278, fos. 182–3 (original)
van Maanen C5

Worthie S^r:

manie thankes for yours of the 7/17 of Septemb:[1] I am sorie for Beaugrand & Herrigons deathes.[2] but I hope it will make you nowe seriously thinke of pollishing & publishing your former thoughts of Analiticks./ I neuer sawe Herrogons 6th: tome, nor I thinke his 5th: yet somewhat of Algebra in those tomes I haue, but nothing newe as I remember, or verie little./ from Roberuall & Fermat I expect much;/[3] Nicerons perspectiue I thinke I haue at London, & as I remember one hath manifestlie conuinced his booke of error;[4] but if I mistake & that you aproue of Nicerons Perspectiue, I desire you will send it me./ If there be anie more then is in that little booke allreadie extant, concerning the newe Iouiales,[5] you shall doe me a fauoure to send it me./ I am extreamelie taken with Des Cartes his newe booke;[6] yet I thinke Kercker the Jesuit, of the Loadestone,[7] hath preuented Descartes for they differ little as I remember; I confess I conceiue not howe the particulae striatae by theyr motion, can reduce a loadestone, or touched needle (formerlie moued from theire meridian,) to their meridian againe; or if they doe, the situation of the poles would be contrarie to DesCartes his description./[8] I

This letter is undated; Pell notes that he received it on 19/29 Sept. Cavendish's other letters from Hamburg typically took 8–10 days to reach Amsterdam (see Letters 18, 30, 31), but in one case (Letter 12) took only six. Presumably carriage in the opposite direction took a similar length of time. Since Cavendish notes the receipt of Letter 19, which was written on 7/17 Sept., this letter is dated here c.13/23 Sept., which allows six days in each direction.

[1] Letter 19.
[2] See Letter 19, nn. 6, 9.
[3] On Roberval see Letter 19, n. 7; on Fermat, Letter 4a, n. 7, and Letter 4b, n. 1.
[4] On Niceron's book see Letter 16, n. 19; the critic has not been identified.
[5] See Letter 13, n. 4, and Letter 19, at n. 12. The 'little booke' known to Cavendish was the volume published in September 1643, containing Rheita's original text, Gassendi's criticism, and Caramuel y Lobkowitz's defence of Rheita (see Pell's comment in Letter 21, at n. 11).
[6] The *Principia* (see Letter 11, n. 8).
[7] The German Jesuit scientist Athanasius Kircher (1602–80) had published a short treatise on magnetism, *Ars magnesia*, at Würzburg in 1631. Cavendish is, however, more likely to have known Kircher's much longer work on this topic, *Magnes, sive de arte magnetica* (Rome, 1641), of which an enlarged edition was published in Cologne in 1643.
[8] A significant section of part 4 of Descartes's *Principia* was devoted to an explanation of magnetism (A&T viii, pp. 275–311), which he attributed to the motion of streams of minute 'grooved particles' ('particulae striatae').

beleeue M^r: Hobbes will not like so much of Des Cartes newe booke as is the same with his metaphisickes,[9] but most of the rest I thinke he will./ Doctor Jungius[10] hath bin once with me, I like him extreamelie well, but I can not speake latin well nor readilie, which hindered me of diuers quaeres & besides I would not trouble him too much, at the first visit I finde him verie free & intend God willing to be better acquainted with him./ he approues of your confutation of Longo:,[11] though he seemed to thinke Long: would take exceptions that you vse tangents.[12] which he refuses to be tried by./ If Gassendes Philosophie[13] be extant I praye you send it me. I haue not yet seen M^r: Tassius,[14] but he sends me word he will come to me. They are commonlie full of business or els I would often visit them./ I meruaile you haue not receiued 5^l, which I paid to M^r: John D'oruill[15] longe since, who promised to doe it with all conuenient expedition; but I hope you haue receiued it before nowe./ I haue troubled you enough for once. I remaine

    Your assured friend to serue you
    Charles Cauendysshe

[*postscript:*] Doctor Jungius preferrs the analiticks of the ancients before vietas by letters[16] which he saies is more subiect to errors & mistakes, though more facile & quick of dispatch, but I conceiue not yet whye.

[*addressed:*] To my verie worthie friend M^r: John Pell, publike Professor of the Mathematicks at Amsterdam, In T'oude Conuoy op de Zeedijke in Amsterdam giue this.

[*cover annotated by Pell:*] Re.ud 19/29 Sept. 1644

---

   [9] Hobbes's hostility to Descartes's metaphysics had been publicly expressed in his 'Objections' to the *Meditationes* (Paris, 1641).
   [10] See Letter 16, n. 5.
   [11] The 'refutatiuncula' (see Letter 14, n. 16).
   [12] Pell's refutation of Longomontanus made use of the double-angle property of tangents, in modern notation, $\tan 2\theta = 2\tan\theta/(1-\tan^2\theta)$. Pell was the first mathematician to formulate and use this property.
   [13] See Letter 18, n. 4.
   [14] See Letter 16, n. 4.
   [15] See Letter 14, n. 20.
   [16] On Viète see Letter 16, n. 15. See also Letter 6, n. 7, on the tension between the methods of classical mathematics and the new symbolic style of Viète and his successors.

Letter 21

2/12 October 1644

Pell to Cavendish, from Amsterdam

BL MS Add. 4280, fo. 109 (draft)
van Maanen P5

Right Honourable

Since my last of Sept 7/17[1] I have received 2 from you the former dated Sept 1/11; the latter without date[2] In which you write that you have paid Mr John d'Orvill[3] 5 pounds for me [>It seemes] your messenger understood him not when he offered a bill of exchange and therefore M^r d'orvill knowing no other way sent it to his correspondent A man with whom I had lesse acquaintance than with any of the rest of the whole kindred: [>yet one that knew me well inough when he had it I was a little amused with it, yet at ye last it came to my hands: it is a bill [>(dated Septemb 2) appointing him to pay [>me] La valeur de Vingt et deux Rixdalers et 10/48 pour autant ici receu du Sr Charles Candish[4] &c which heere comes not [>full] to so much as 5 pounds sterling: But for so much as I have received (Banke Dalers 22 5/24) I shall be accountable to you, as you shall please to command it.[5]

I am glad that Mersennus his letters finde ye way to you.[6] henceforth I shall trouble you with no more french newes, I shall rather hope [>now] to have from you how the learned world goes there. But either he or I have [>not] written so plainely to you as we should have done, if you gathered either out of his letters or mine that Gassendus his philosophy[7] or Chauveau his Algebra[8] are in print: For if one leafe of either be printed it is more than I can heare, and I thought I had

---

[1] Letter 19.

[2] Letters 18 and 20.

[3] See Letter 14, n. 20.

[4] 'The value of 22 10/48 Reichsthalers, for the same amount received here from Sir Charles Cavendish'.

[5] The Reichsthaler (equal to 3 Marks or 8 Flemish schillings) was the standard unit of account in Hamburg. It had two values: as 'bank money' (the unit of account) and as current money in coinage, with a slight premium on the former (which is what Pell clearly refers to here). In 1643–5 the average rate of exchange for bank money was 37.32 Flemish schillings, i.e. 4.665 Reichsthalers, to £1 sterling (see J. J. McCusker, *Money and Exchange in Europe and America, 1600–1775* (Chapel Hill, NC, 1978), pp. 62, 70). At that rate, Pell would have received the equivalent of £4 15s $2\frac{1}{2}d$.

[6] This responds to Cavendish's comment on a letter from Mersenne in Letter 18; however, no letter from Mersenne to Cavendish has survived from 1644.

[7] See Letter 18, n. 4.

[8] See Letter 16, n. 21 (and Cavendish's request in Letter 18).

beene inquisitive enough. But that you understood me as though I commended Nicerons perspective is certainly my fault [>in writing something] in favour of Robervall but in such a manner y^t you understood it as spoken of Niceron.[9] As for Nicerons perspective I have not taken so much heed of it as to approve it or dislike it. But seeing you have bought it once already and Mersennus tells us he will shortly augment it,[10] I thinke it is better to expect the second edition than to buy this once againe. As for the Joviales I thought you had read Gassendus his judgement of y^m in the first [french *deleted* >Paris] Edition [>in quarto] but I perceive now y^t it [>was] in the Lovain Edition in small 12°.[11] Wherein you see what Caramuel thinkes of de Reita, and of Gassendus his censure, My refutation of Gassendus[12] seemed to me something more violently convincing than that of [de Reita *deleted* >Caramuel] but I have it not heere. In that booke you have nothing from de Reita [>since] Apr. 24 1643. I therefore send you heerewith a letter of his 8 months younger.[13] I transcribed it out of a coppy, which was with[out *deleted* >halfe] date and so false written y^t in some places I could not make sence of it. I therefore [>by] freinds caused inquiry to be made at Collen[14] among ye Capuchins where de Reita was: they answered he was [>gone to] Augspurg. I was then assured that this letter was also written thence and y^t his workeman is [>y^t Augustanus] who some yeares agoe we heard y^t he had an incomparable hand for glasse.[15] [>I have much to say but am constrained heere

---

[9] On Niceron see Letter 16, n. 19; on Roberval, Letter 19, n. 7. Pell's ambiguous statement (referring to Roberval, but following a reference to Niceron) is in Letter 19. For Cavendish's misinterpretation of it, see Letter 20.

[10] Mersenne made this comment in *Universae geometriae synopsis*, sig. 2+2^v; Niceron was revising his entire work while preparing a Latin translation of it, but did not complete that task by the time of his death in 1646.

[11] See Letter 13, n. 4, and Letter 20, n. 5.

[12] See Letter 19, n. 14.

[13] See Letter 19, at n. 18.

[14] Cologne.

[15] Rheita's 'workeman' from Augsburg (hence 'Augustanus') was Johann Wiesel (1583–1662), whom he described as 'citizen and optical practitioner of the famous city of Augsburg, from whom ... you can obtain superbly well-made telescopes, of every sort of length and appearance, at a suitable price' ('Ciuem & Opticum inclytae Vrbis Augustae Vindeliciae, à quo competenti pretio, cuiuscunque etiam speciei qualitatis, & longitudinis Telescopia perfectissimè quidem elaborata ... habere obtinereque poteris': *Oculus Enoch et Eliae, sive radius sidereomysticus: pars prima* (Antwerp, 1645), p. 339). Cavendish would soon receive confirmation of his identity: a letter from an anonymous informant in Augsburg, dated [10/] 20 November 1644, called him 'a verry excellent master in that arte, whose name is Wisell' (BL, MS Add. 4278, fo. 193^r). Wiesel was put in contact with Samuel Hartlib in the late 1640s, and corresponded with Christiaan Huygens in the 1650s; he was the father-in-law of the famous Italian telescope-maker Giuseppe Campani (see Daumas, *Scientific Instruments*, pp. 64, 88, and I. Kiel, 'Technology Transfer and Scientific Specialization: Johann Wiesel, Optician of Augsburg, and the Hartlib Circle', in M. Greengrass, M. Leslie, and T. Raylor, eds., *Samuel Hartlib and Universal Reformation: Studies in Intellectual Communication* (Cambridge, 1994), pp. 268–78). The fullest study is I. Kiel's monograph, *Augustanus opticus: Johann Wiesel (1583–1662) und 200 Jahre optisches Handwerk in Augsburg* (Berlin, 2000); Kiel prints the text of the letter from Cavendish's anonymous informant on p. 284.

to breake off which] I pray you to excuse & to beleeve that I am

   Your Honours Most humble servant
   John Pell

Amsterdam. Octob 2/12.

Letter 22

10/20 October 1644

Cavendish to Pell, from Hamburg

BL MS Add. 4278, fos. 184–5 (original)
van Maanen C6

Worthie S^r:

manie thankes for yours of Octob: 2/12./[1] I conceiue there might easilie be a mistake in the waie of returning that 5^l to you, but I am glad you nowe haue it; what difference there is betweene Banck Dallers & Rixdalers I knowe not, but I intended you should receiue to the value of 5^l sterling;[2] if it want anie considerable sum I praye let me knowe it, for though the whole sum be not much considerable, yet the cosenage is./ I receiued yesterdaie a letter from M^r: Hobbes, who had not seen de Cartes his new booke printed, but had reade some sheets of it in manuscript & seems to receiue little satisfaction from it, & saies a friend of his hath reade it through, & is of the same minde;[3] but by their leaues I esteeme it an excellent booke though I thinke Monsr: des Cartes is not infallible. Mersennus is gone towardes Roome;[4] Those bookes I desired, might I suppose haue nowe bin in print; for Mersennus mentioning them so manie months since, as worckes either printed or readie for the press, made me vpon that supposall desire them if extant./ M^r: Hobbes writes, Gassendes his philosophie[5] is not yet printed, but that he hath reade it, & that it is as big as Aristotles philosophie but much truer, & excellent Latin./ Though you discommend Niceron for a vaineglorious man, yet your naming of his booke commended it to me;[6] yet not knoweing certainelie whether I haue it or not, & expecting a second edition; for the present I desire it not./ I desire Gassendes his refutation of Des Cartes his metaphisicks printed in 4to.[7] & if there be anie other booke of philosophie or mathematickes latelie printed I desire it allso; especiallie your Diophantes[8] if extant, or anie thinge

---

[1] Letter 21.

[2] The difference between Reichsthalers and Bankthalers was not in fact at issue here; see Letter 21, n. 5.

[3] It was probably thanks to Mersenne that Hobbes was enabled to see part of the MS of Descartes's *Principia*; the identity of his 'friend' can only be guessed at (perhaps Roberval, or Gassendi).

[4] The precise date of Mersenne's departure from Paris is not known, but it must have been in late Sept. or the first days of Oct. He would return in late July 1645 (see *MC* xiii, p. 223(n.), and R. Lenoble, *Mersenne ou la naissance du mécanisme* (Paris, 1943), pp. 51–2).

[5] See Letter 18, n. 4.

[6] See Letter 19, n. 8.

[7] See Letter 19, n. 11.

[8] See Letter 14, n. 13.

els of yours./ I desire you will dispose of the remainder of that smalle sum I returned, to your owne vse./ manie thankes for the copie of Reitas letter;[9] I admire his glass & would gladlie buye such a glass, & get aquaintance with the friar & his worckman[10] by letter if I could./

Your assured friend to serue you
Charles Cauendysshe

Hamb: Oct. 10/20 1644.

[*addressed:*] To my verie worthie friend M^r: John Pell Publick Professor of the Mathematickes. [>in Amsterdam] In T'oud Conuoy op de Seedijk in Amsterdam.

[*cover annotated by Birch:*] 10/20 Oct. 1644

---

[9] See Letter 21, n. 13.
[10] See Letter 21, n. 15.

Letter 23

19/29 October 1644

Pell to Cavendish, from Amsterdam

BL MS Add. 4280, fo. 109^v (draft)
van Maanen P6

Right honourable

When I finished or rather brake off my last of Octob 2/12[1] I intended not to deferr my next till I had an answere from you, [but you have prevented me *deleted* >though it be now so fallen out I being hindred by] various diversions as you will easily beleeve of him y^t must reade publikely 5 dayes in a weeke, y^t hath none of his bookes or papers heere, y^t hath ye care of an [>absent] family none of the least though consisting of many little ones: besides unavoidable visits, letters: and some time to thinke how to satisfy ye Expectation of [>my] publishing Diophantus & Apollonius[2] [this yeere *deleted*] within this halfe yeere, I say Apollonius for [>at this instant there lyeth] under my hand Apollonius with 36 other authors which taken together make a bulke almost twice as great as Apollonius. All 37 in Arabicke.[3] You smile and beleeve not that I have Arabicke enough to teach any of these to speake Geometricall Latine in any reasonable

---

[1] Letter 21.

[2] For Pell's previous statement of his plan to publish an edition of Diophantus, see Letter 14, at n. 13; this is his first mention of an edition of Apollonius too. The *Conica* (*Conics*) of Apollonius of Perga (*c*.262–*c*.190 BC), a treatise on conic sections in eight books, was one of the most important texts of Greek geometry. The first MS of the Greek text to be brought to the West (in the 15th century) contained only books 1–4. A complete Latin translation of these books was published by J.-B. Memus (Giovanni Battista Memo) in Venice in 1537. A second (and greatly superior) Latin translation of books 1–4 was published by Federico Commandino in 1566 (see below, n. 6), and reprinted in Paris in 1626. (On these early versions of Apollonius see L. Maierù, *La teoria e l'uso delle coniche nel Cinquecento: intersezione fra la conoscenza dei testi di Apollonio e dei 'Veteres' e il senso delle loro traduzioni* (Caltanisetta, 1996).) From 1629 onwards, several Arabic MSS were brought to the West, containing versions of books 5–7. The most accurate of these was the one owned by Jacob Golius (see below, n. 9); a Latin translation of this version of books 5–7 was eventually published by Edmond Halley in Oxford in 1710. (See Apollonius, *Les Coniques*, ed. and tr. P. ver Eecke (Bruges, 1923), esp. pp. xliii–xlviii, and G. J. Toomer, 'Introduction' to Apollonius, *Conics, Books V to VII: The Arabic Translation of the Lost Greek Original in the Version of the Banū Mūsā*, ed. and tr. G. J. Toomer, 2 vols. (New York, 1990), i, esp. pp. xxii–xxiv.) Pell's work on Apollonius was never published.

[3] Pell had borrowed from Christian Ravius (see Letter 14, n. 6) an Arabic MS containing an epitome of books 5, 6, and 7 of Apollonius' *Conics* (now Bodl. MS Thurston 3), with some other items.

time.[4] And [when I have a printer *altered to* if I had done it] the drawing of those Conicall diagrams is no childrens play [>nor can we every where find] a Printer for such a crabbed & costly piece of worke. But some of these difficulties are over. Dr Blaeu [>tells me he] is ready to print Diophantus Apollonius or what I will & how I will, in Arabicke & Latine if I will: he hath good Arabicke types, a compositor exercised in [>setting them] & what not.[5] But I have in a manner resolved to make lesse adoe, namely to publish onely the 5^t, 6^t & 7th (not repeating Commandines 4) & those onely in Latine, adjoyning apt figures, some notes & Pappus his Lemmata upon the 5^t. 6^t 7^t & 8th booke also: though the Arabicke booke wanteth the 8th [6] [>all which will require about 240 diagrams] to this purpose

Heere lyes by me the [>whole] fift booke translated into Latine but not faire written for ye presse nor one figure yet drawen [>so well as I] intend they shall be. The sixt & seventh [>simul][7] are to ye 5^t as 23 to 18 therefore it is almost halfe translated [>And] though I have [>now] very little [>or rather no] Leisure yet in this postehaste let me give you a specimen out of this Abdil Melik of Schiraz in Persia[8] though he write Arabick (the manuscript is above 450 yeeres old)

If I misunderstood not one of your letters you have Mersennus his new [>booke] in quarto In his Synopsis pag 274 you have Golius his translation of

---

[4]It was not impossible for someone who had only a limited knowledge of Arabic but who understood mathematics to translate Apollonius (as Edmond Halley also did, with great skill).

[5]Blaeu (on whom see Letter 14, n. 14) was one of the few printers in Europe with Arabic type (see Juynboll, *Zeventiende-eeuwsche beoefenaars*, p. 245; R. Smitskamp, *Philologia orientalis: A Description of Books Illustrating the Study and Printing of Oriental Languages in 16th- and 17th-Century Europe* (Leiden, 1992), nos. 280h, 355a).

[6]Federico Commandino (1509–75) published a number of classical texts, including important editions of Euclid, Archimedes, Ptolemy, and Pappus. He issued a Latin translation of the first four books of Apollonius' *Conics* (Bologna, 1566). Books 5–7 survived in Arabic translations (see above, n. 2), while book 8 was lost altogether; but there were 'lemmata' or lemmas (subsidiary propositions) to some of the problems of books 5–8 in the *Collections* of Pappus of Alexandria (see Letter 16, n. 15).

[7]'Together'.

[8]The mathematician Abu 'l-Husain 'Abd al-Malik b. Muhammad al-Shirazi (*fl.* c.1150?) prepared an epitome or summary of an earlier (9th century) Arabic translation of books 1–7 of the *Conics* (see H. Suter, *Die Mathematiker und Astronomen der Araber und ihre Werke* (Leipzig, 1900), pp. 126, 158).

[>yᵉ beginnings of yᵉ 6ᵗ & 7ᵗʰ of] his Arabian Apollonius.⁹ You shall see the difference In this Abdil Maliks Arabicke translation of Apollonius [>which I have.] Initium septimi [>libri] ita se habet sit parabole, cujus axis ad qui producatur ad c, ita ut ac fiat aequalis lateri recto.

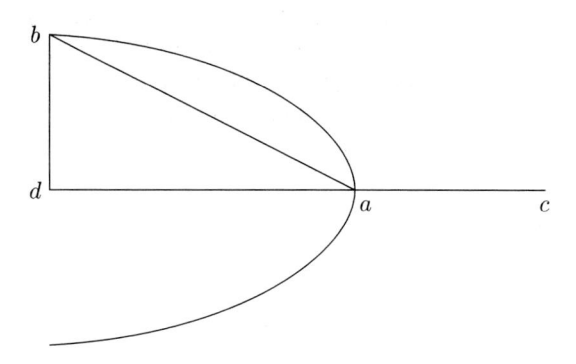

Ex puncto a, ducatur linea recta ab secans parabolam ubilibet ut in puncto b unde demittatur perpendicularis bd in axem da. Dico Quadratum ab esse aequale rectangulo sub cd in da

Est enim (per 11 primi) rectangulum sub ca in ad aequale quadrato bd Ergo rectangulum ca in ad cum quadrato ad, hoc est rectangulum cd in ad erit aequale aggregato quadratorum bd & da hoc est quadrato ba Ergo Quadratum ba est

---

⁹Mersenne, *Universae geometriae synopsis*, p. 274. Jacob Golius (Gool) (1596–1667) had studied Oriental languages at Leiden and became a Professor there in 1624. He possessed not only a copy of Ravius's MS but also his own MS of another, fuller, Arabic translation, by Thabit ibn Qurra and others, supervised by the Banū Mūsā (see J. J. Witkam, *Jacobus Golius (1596–1667) en zijn handschriften*, Oosters genootschap in Nederland, x (The Hague, 1980), esp. pp. 57–8; this MS is now Bodl. MS Arch. o. c. 3). Mersenne was the first to publicize the fact that Golius was planning to issue a Latin translation of books 5–7 of the *Conics*. The translation by Golius of the beginning of book 7, as quoted by Mersenne, was as follows: 'Si axis parabolae producatur ultra verticem, donec aequatur lateri recto, et a puncto quovis sectionis in axem ducatur perpendicularis, recta, sectionis punctum illud in sectione cum vertice connectens, poterit rectangulum contentum sub recta inter verticem et perpendicularis incidentiam interiecta, et tota ab hoc incidentiae puncto per verticalem continuata' ('If the axis of a parabola is produced beyond the vertex, until it equals the *latus rectum*, and from any point on the curve a perpendicular is dropped to the axis, [the square on] the line, connecting that point on the curve with the vertex, will be as the rectangle contained by the line between the vertex and the foot of the perpendicular, and the whole line drawn from that foot through the vertex').

aequale rectangulo *cd* in *da*. Quod erat demonstrandum.[10]

You will pardon this abrupt end what I had more to say, I hope I shall say in my next wherein I intend (amongst [>many] other things) to tell you by what shippers I send you Gassendus contra Cartesij meditationes Metaphysicas,[11] till when I might have deferred this, but that I am confident y^t you will thinke this worth ye postage and y^t you will accept [>heerin] ye service of

Your honours most humble servant
John Pell

Amsterdam Octob 19/29

---

[10]'The beginning of the seventh book goes as follows. Suppose there is a parabola, of which the axis *ad* is produced to *c*, so that *ac* is equal to the *latus rectum*. From point *a*, let there be drawn a straight line *ab* cutting the parabola anywhere, as in point *b*, whence let a perpendicular *bd* be dropped to the axis *da*. I say that the square of *ab* is equal to the rectangle [*sc.* product] of *cd* and *da*. For (by book 1, proposition 11) the rectangle of *ca* and *ad* is equal to the square of *bd*; therefore the rectangle of *ca* and *ad* with the square of *ad*, that is, the rectangle of *cd* and *ad* will be equal to the sum of the squares of *bd* and *da*, that is, to the square of *ba*. Therefore the square of *ba* is equal to the rectangle of *cd* and *da*. Q.E.D.' The letters italicized in this transcription are over-lined in the MS.

[11] See Letter 19, n. 11.

Letter 24

2/12 November 1644

Pell to Cavendish, from Amsterdam

BL MS Add. 4280, fo. 110 (draft)
van Maanen P7

Right honourable

    I have not yet a convenient shipper to send your booke[1] by, but I shall lay hold on ye first opportunity. I suppose you looke to heare more concerning Apollonius, but [>my Persian][2] must be content a little to tarry till I have dispatchd ye testy old Dane.[3] I thinke in [>one of] my former letters I told you y^t I had [>sent] to Longomontanus one coppy of my refutation with a [frendly *deleted*] letter advising him to publish a recantation ere he fell into the hands of others y^t would use him [>more harshly &] with lesse respect than I had done.[4] He hath sent me nothing written but 4 printed sheetes of paper, [>one against me[5] and] 3 against Guldin[6] a Jesuit (I thinke professor of Mathematickes at Vienna, in Austria) who [>it seemes wrote against him 9 yeeres agoe.][7] In that against me he hath reprinted [>not onely] my refutation [>but also] my letter y^t I wrote to him. To these he hath added not a [>recantation] as I desired, but a reply [>in scurvy language &] long inough to let the readers know that as yet he neither dislikes his owne bookes nor understands my discourse against them.[8] I shall be constrained to reprint an answere, wherein I must be larger that I may be better understood. Dr Jungius had a shrewd guesse, when he said

---

[1] Gassendi, *Disquisitio metaphysica*; see Letter 19, n. 11, and Letter 23, n. 11.

[2] Al-Shirazi, author of an Arabic summary of Apollonius: see Letter 23, n. 8.

[3] Longomontanus.

[4] The letter to Cavendish referred to here has not survived; but Pell's original intention to send a copy of his 'refutatiuncula' to Longomontanus was expressed in Letter 14. The text of the letter he sent to Longomontanus (dated 17 Aug. 1644) was later printed in Pell, *Controversiae pars prima*, pp. 17–18.

[5] Longomontanus, Ἐλέγξεως ... *In Pellium & Guldinum verae Cyclometriae adversantes*: this work consists of four leaves (one quarto gathering).

[6] Longomontanus, *Problema quod ... absolutam circuli mensura praestat*: this work consists of three quarto gatherings (A^4–C^4).

[7] Paul Guldin (1577–1643), a Jesuit of Swiss-German origin, had taught mathematics at Graz and Vienna. The work referred to here was his *De centro gravitatis*, 4 vols. (Vienna, 1635–41). Pell's '9 yeeres agoe' was evidently calculated from the date of vol. i (as given in Longomontanus, *Problema*, sig. B1^r); but Guldin's criticism of Longomontanus's previous circle-squaring efforts in fact appeared in vol. ii (1640), pp. 27–9.

[8] Longomontanus, Ἐλέγξεως (which includes criticism of Guldin too).

Longomontanus would take exceptions that I use tangents,[9] which he refuses to be tried by. Which when I read in your letter, I [thought he would prove a false prophet *deleted*] had and yet have too great an opinion of Dr Jungius to beleeve that he thinkes I should not have used Tangents For whosoever pronounces the circle greater than indeed it is (as Longomontanus hath done) such a one can not be confuted by any other possible Technicall medium than by circumscription of right lined figures. But the halfe sides of [>any] Circumscribed figure are Tangents of those arches which are intercepted betweene the right lines which connect ye center with ye middle & extremes of all the sides of ye circumscribed figure, I therefore understand Dr Youngs[10] meaning to be that Longomontanus [>would] find nothing to say for himselfe unlesse he fell a cavilling with *the name* of Tangents: For Longomontanus hath shewed himselfe in all his writings (this printed at Amsterdam [which I refuted, is *deleted* >was] the twelfth) so absurd, that men now beleeve he will not sticke to write any thing, be it never so monstrous if it seeme to have any thing in favour of his dreame. But Longomontanus refused not to be tried by Tangents but by the *table* of tangents[11] which is all that I charge him withall. (Trigonometrarum *Canones*, tanquam minimè accuratos, quamvis injustè, respuit)[12] And therefore to please him in y^t humour I appealed [>not] to ye authority of any Canon of tangents, but dealt with him as if no Canon of tangents had ever beene made. [>And therefore I concluded in that manner] (Demonstravi igitur idque sine tangentium Canone[13] etc).

But [>no doubt you desire to know what] Longomontanus [>hath] answered. All that can be expected from a peevish, obstinate, ignorant, infatuated old man. Verses in Pellium & Guldinum. Railings, frivolous impertinencies, nothing to ye purpose, any thing rather than nothing.[14]

For not at all understanding the force of my demonstration he tells his reader (but 'tis a lye) that I say Tangentes in Canone sunt veri[15] & those in my paper are precise and heereupon he declaims against ye imperfection of the Canon and

---

[9] On Jungius see Letter 16, n. 5; his 'shrewd guesse' was reported by Cavendish in Letter 20 (at n. 12).

[10] Jungius (Jung).

[11] Longomontanus objected to the use of trigonometrical tables on the grounds of inaccuracy (see van Maanen, 'Refutation', p. 321). Pell therefore took the lengths of his tangents not from tables but from his newly discovered double-angle formula (see Letter 20, n. 12).

[12] 'He rejects, albeit unjustly, the tables of the trigonometers, on the grounds that they are very inaccurate'; Pell quotes here from the prefatory paragraph of his 'refutatiuncula' (p. '73'), where he wrote that since Longomontanus rejected any method based on appeal to the trigonometrical tables, he would use a different method to refute him.

[13] 'Therefore I have demonstrated (and have done so without the table of tangents)...': again, Pell quotes from his 'refutatiuncula' (pp. '73–4').

[14] Longomontanus, Ἐλέγξεως, sig. π3^r.

[15] 'The tangents in the table are true' (Longomontanus, Ἐλέγξεως, sig. π3^r).

compares me with ye Spanish blunderer Alphonsus Molina.[16]

As for my Theoreme [>(Tangens cujuslibet &c)][17] upon which my whole Demonstration is grounded, the [>incomparable] Dunce calls it compendium quoddam ex Logarithmis, ut videtur, Neperi vel Briggij, caeteroquin Geometrice indemonstrabile.[18] As if hee knew what belongs to a Geometricall demonstration; a dotard that hath these 33 yeeres troubled ye world with silly Paralogismes in abundance but never [understood *deleted*] gave us one [>true] demonstration of his owne in his life.

I should doe you wrong to tell you it is demonstrable, or to send you the demonstration, if you please to examine it I know you [>can] in [>one] quarter of an houre find a demonstration for it

Some perhaps will thinke I might have done better and prevented such an extravagant answer if I had added ye Demonstration of that theoreme and had not used the *name of tangents* at all but had onely made mention of sides of ordinate polygons as the Ancients were wont to doe. To such I will answere, that I beleeved not that Longomontanus would be so absurd as to cavill at the *name* of tangents: and if any man will take the paines to new Language my refutation & in all places for *tangent* write halfe side of a right lined regular polygon circumscribed of so & so many sides I doubt not but he will find it to grow so unwieldy, cumbersome & cloudy, that he will begin to thinke it hard to mend y^t refutation for words or matter. And as for the demonstration of that Theoreme, I had many reasons inducing me to leave it out. It would require a Diagram; it would lengthen the refutation; true Mathematicians would acknowledge it [>true] without a demonstration, Or presently finde one without my helpe: Others would be prating against it, taking it to be meerly new, & so bewray their owne Bayardisme:[19] which is fallen out as I would wish, I hope it will discover more of these empty Mathematicasters. You see what Longomontanus sayes of it. Will you heere another learned man how he writes to [>one of my acquaintance] thanking him or me for ye coppies y^t were sent him?[20]

Gratias ages Er^mo Pellio, meo nomine, qui argutâ admodum usus est demonstratione & in quâ hoc requirendum ab adversario vereor, quod tangentes arcuum

---

[16] Juan Alfonso de Molina Cano, author of *Descubrimientos geometricos* (Antwerp, 1568); a translation of this work by Nicolaus Jansonius, *Nova reperta geometrica*, was published at Arnhem in 1620. (It was also criticized in Guldin, *De centro gravitatis*, ii, p. 30.)

[17] 'The tangent of any [arc of less than 45 degrees...]', the opening words of Pell's demonstration in his 'refutatiuncula'. Pell's 'Theoreme' was his double-angle tangent formula (see Letter 20, n. 12).

[18] 'A sort of abbreviated version, it seems, of the logarithms of Napier or Briggs, otherwise incapable of geometrical demonstration' (Longomontanus, Ἐλέγξεως, sig. π3^r).

[19] Foolhardy stupidity (from the proverbial phrase, 'as bold as blind Bayard', referring to a blind horse).

[20] The writer of the letter was Claude Hardy; the recipient, Pell's acquaintance, was Christian Ravius, who was then in Utrecht. Pell's draft of a reply to Ravius (complaining that Hardy's criticism had been made available to others before Pell saw it) is in BL, MS Add. 4421, fos. 93–4.

subduplorum commensurabiles videatur statuere semidiametro, atque adeo latus quadrati diagonio[21] (He takes me for a dull beast in ye meane while. But hee is a [cunning man *deleted* >rare Artist] For He addes) Alio calculo non adeo prolixo invenio secundùm ratiocinium Christiani Severini, perimetrum figurae XXXII laterum circumscriptae esse minorem circumferentiâ circuli.[22] Alas! [>What will become of] poore Longomontanus That is so angry because I say his circle is greater y^n a circumscribed polygon of 256 sides, Nay saith this man, y^n a Polygon of 32 sides. O, what a simple child doe these men take me to be! Is it so hard a matter to finde what ordinary Polygon is proximè majus plano proposito?[23] You would hardly guesse who this cunning man is y^t can [>see such absurdities on that paper of mine & so farre outdoe it]. I can tell you, it is no small foole; One y^t is famous for understanding Euclid in Greeke. & for his blessed faculty of refuting errors by throwing heapes of new ones on top of them & so he stiffles y^m. It is no lesse man than un Conseilleur du Roy au Chastelet de Paris, even Claudius Hardy.[24] [>Was not] that paper [>of mine] hatched at an unlucky houre to bring so much disgrace upon men that would faine be thought some body? One poore Theoreme turned loose into ye world without his demonstration baffles all the scioli[25] it meetes withall & discovers the hugeness of their ignorance & oultre cuidance.[26] [>But] what will these profound Geometers [>say] when [>in the treatise y^t I have in hand] they shall see the truth [>plentifully] backed with all manner of confirmation? Non putâram.[27] I have a great desire to see how M^sr Hardy will prove 10 864/100 to be [more *deleted* >lesse] than an unite. (for to that he must come, if he will confute Longomontanus with a regular circumscript of 32 sides) and therefore I have intreated y^t friend to write to M^sr Hardy for a coppy of this calculus non adeo prolixus,[28] which he tells me, he hath done, but

---

[21]'Give thanks on my behalf to the most learned Mr Pell, who has made use of a sufficiently subtle demonstration. In it, I suspect that this will be questioned by his opponent – that he seems to take the tangents of the half arcs as commensurable with the radius, and therefore the side of the square with the diagonal.' It had been known since classical times that the side of a square was not commensurable with (i.e., not a rational multiple of) the diagonal. Similarly, tangents (in the geometrical sense) subtending angles at the centre of a circle are in general incommensurable with the radius. If Pell had claimed that his tangents of half-angles were always commensurable with the radius, it would therefore have immediately invalidated his proof.

[22]'By another, not so prolix, calculation I find that, if one follows Christian Severinus's [*sc.* Longomontanus's] reasoning, the perimeter of a circumscribed figure of 32 sides will be less than the circumference of a circle.'

[23]'Is closest to being greater than the proposed plane figure [i.e., the circle]'. Pell used the double-angle tangent property in a carefully constructed argument to show that, according to the reasoning of Longomontanus, a circumscribing polygon of 256 sides would be less than the circle itself – an obvious contradiction. Hardy claimed that a simpler calculation showed that the same would be true even of a 32-sided polygon, a claim of which Pell was scornful.

[24]See Letter 14, n. 5.

[25]Sciolists, people with only a smattering of knowledge.

[26]'Presumptuousness'.

[27]'I would not have thought it.'

[28]'Not so prolix calculation' (using the phrase quoted at n. 22).

as yet he hath received no answere.

[>I pray, let me know] what Tassius[29] sayes to my refutation: and that, before you tell him y^t I am writing a full treatise of that argument.

As for D'orvill.[30] I received no more than was in his bill of exchange, Vingt & deux Rixdalers & 10/48, in Banke money, as all bills of exchange are to be paid.[31] Banke-dallers are onely the best sort of dallers whose intrinsicall value is truly worthy as much as they are heere current for & are therefore such as will be received in the Banke. You know how much you paid to him there, my merchant will tell you whither he [>allowed] you lesse for that mony then at that time was the [>ordinary] rate of exchange betweene Hamburg & Amsterdam where I continue your Honours most Humble servant John Pell

Nov 2/12 1644.

---

[29] See Letter 16, n. 4.
[30] See Letter 14, n. 20.
[31] See Letter 21, n. 5.

Letter 25

6/16 November 1644

Cavendish to Pell, from Hamburg

BL MS Add. 4278, fos. 186–7 (original)
van Maanen C7

Worthie S^r:

manye thankes for your letter,[1] wherin you write that you haue Apollonius 3 bookes of conicks in Arabick, more than wee had in the greeck & 36 authors more,[2] I hope some of those are of the mathematicks; howsoeuer I dout not but they are worthie the press./ I like extreamlie both the proposition & demonstration of Apollonius in your letter; & to my aprehension the expression of the same proposition in Mersennus his booke is perplexed & no demonstration translated./ I wonder Goleas hath not published it all this whyle;[3] yet being nowe in your handes I am not sorie he did not; for I assure my self wee shall nowe haue it with more aduantage than the loss of so much time./ Though I doute not but your explication of Diophantes[4] will put vs in to a more sure waye of Analiticks than formerlie; yet I suppose there is so much to be added & explaned concerning Analiticks that it will require a large volume; & I hope you continue your intention of publishing such a worcke; which I beseech you thinke seriouslie of to publish with all conuenient speede; for it is a worcke worthie of you;/ I haue sent to inquire at Auspurge of the famouse Reieta, [>&] to procure me one of his best sorte of glasses; but I haue yet no answeare./[5] If your occasions will permit you I shall take it as a fauoure if you will visite my Lo: widdrington[6] nowe in his passage towardes france./ And so wishing you all happinesse I remaine

    Your assured friend to serue you
    Charles Cauendysshe

Hamburg Nouemb: 6/16 1644.

[*addressed:*] To my verie worthie friend M^r: John Pell Publick Professor of the Mathematicks at Amsterdam. In T'oude Conuoy op de Zee dijke in Amsterdam

[*cover annotated by Birch:*] 6/16 Novr. 1644

---

[1] Letter 23.   [2] See Letter 23, n. 3.   [3] See Letter 23, nn. 9, 10.
[4] See Letter 14, n. 13.   [5] See Letter 21, n. 15.

[6] William Widdrington (1610–51), a zealous Royalist, had served under the Earl of Newcastle, who had appointed him President of his Council of War; he was created Baron Widdrington of Blankney in 1643. He travelled to Hamburg with Newcastle and Cavendish in July 1644, before entering the service of the Prince of Wales in Jersey and France. (See Cokayne, *Complete Peerage*, xii, part 2, pp. 625–7.)

Letter 26

30 November/10 December 1644

Pell to Cavendish, from Amsterdam

BL MS Add. 4280, fo. 111^r (draft)
van Maanen P8

Right honourable,

I had just finished the sealing up of Gassendus contra Cartesium[1] to send by an Amsterdam skipper that goes away hence to day towards Hamburg; when Sr William[2] (y^t brought me your letter[3] 12 dayes agoe) sent to know whither my letter were ready. I went to him & told him I had begun to write & to [>give] you [>notice] of ye booke which I was sending to you. he desired to carry it, so it now comes more certainely.

I should tell you the price but you have prevented me in your last save one.[4] When I had spoken of giving you an account, you then willed me to dispose of your mony to mine owne use. I confesse I am not in case to refuse any favour of that nature: But I pray you neverthelesse to continue to make use of my service in procuring of such bookes as may heere be had.

[*paragraph transposed from foot of letter:*] I would have sent you Theon Smyrnaeus upon Plato, but that I am assured by one whom I dare beleeve that Bullialdus hath found more pieces of the same author and hath promised to put them out with all speed, and for that cause I buy it not my selfe till I see what alteration the new edition will bring with it.[5]

I must make an end thus abruptly for Sir William is going away The rest therefore heereafter in the mean time I am

your Honours most humble servant
J. Pell

Nov. 30. Dec. 10.

---

[1] Gassendi, *Disquisitio metaphysica*; see Letter 19, n. 11.

[2] Sir William Carnaby, from a Northumberland gentry family, was an old friend of Cavendish and his brother; he had been knighted at Welbeck Abbey, together with Cavendish, on 10 [/20]August 1619 (Shaw, *Knights of England*, ii, p. 173). As Sheriff of Northumberland in 1636 he organized the collection of Ship Money for that county. A book of rentals (1634–42) for the Earl of Newcastle shows that Carnaby managed all the Earl's estates in Northumberland from 1634 to 1638: Nottingham University Library, MS Pl E 12/10/1/9/1. A staunch Royalist, he was with the forces under the Earl of Newcastle's command in early 1644, serving as 'Treasurer of the Army' (Cavendish, *Life of Newcastle*, p. 113). He was also closely associated with the Widdrington family. (See *CSPD 1636–7*, pp. 215, 318; *CSPD 1640*, p. 140; *CSPD 1644*, p. 10).

[3] Letter 25.   [4] Letter 22.

[5] On Boulliau's edition of Theon see Letter 16, n. 13; no other works by him were edited by Boulliau, however.

Letter 27

10/20 December 1644

Cavendish to Pell, from Hamburg

BL MS Add. 4278, fos. 188–9 (original)
van Maanen C8

Worthie S^r:

manie thankes for your letter & Gassendes his booke which I receiued by S^r: William Carnabye./[1] I am of your opinion that Gassendes & de Cartes [>are] of different dispositions; & I perceiue M^r: Hobbes ioines with Gassendes in his dislike of de Cartes his writings, for he vtterlie mislikes de Cartes his last [>newe] booke of philosophie[2] which by his leaue I highlie esteem of./ I am sorie the peeuish Dane Seuerin diuerts you from your better studies, for to my aprehension your refutation is full & plaine, & the proposition you builde on, so obuious, that I thinke I could demonstrate it in 1/4 of an houer./ It seems the word Tangent is odious to him,[3] but enough of him./ I haue not yet receiued answeare from Auspurge concerning Reitas glass,[4] but I perceiue Mr: Hobbes esteemes neither of his glass nor beleeus his discoueries, for he is ioined in a greate friendship with Gassendes. I writ to him Gassendes might be deceiued as he was about the varieing of the perpendiculare;[5] he excuses him what he can./ Mr: Tassius[6] is sick, but I perceiue hee is verie courteous. I haue not latelie seen Doctor Jungius,[7] but I haue a greate opinion of his abilities./ I desire to knowe if all Bonauentura Cauallieros worckes be to be had,[8] allso Sethus Caluisius worckes of musick,[9] for I sawe heere but some fragments which Mr: Tassius lent me./ You will excuse me that I requite not your fauoures for the present as I desire I can onelie yet acknowledge them, but shall heereafter by the grace of God endeauoure to requite them to my power./ And so wishing you all happinesse I remaine

---

[1] See Letter 26, nn. 1, 2.

[2] Descartes, *Principia* (see Letter 11, n. 8).

[3] See Pell's comments on Christian Severinus ('Seuerin') Longomontanus in Letter 24.

[4] See Letter 21, n. 15.

[5] See Letter 19, at nn. 15, 16.

[6] See Letter 16, n. 4.

[7] See Letter 16, n. 5.

[8] Bonauentura Cavalieri (see Letter 16, n. 26); for the works see Letter 29, nn. 4, 5.

[9] Seth Calvisius (Kalwitz) (1556–1615), music theorist, composer, astronomer, and chronologist, taught at Leipzig, where he became Cantor of the Thomaskirche. His main work of music theory was *Melopoeia sive melodiae condendae ratio, quam vulgo musicam poeticam vocant* (Erfurt, 1592); he also wrote a didactic work, *Compendium musicae pro incipientibus* (Leipzig, 1594), which was later issued in a simplified version, *Musicae artis praecepta nova et facillima* (Leipzig, 1612).

Your assured friend to serue you
Charles Cauendysshe

Hamburg: Decemb 10/20 1644

[*addressed:*] To my verie worthie friend M^r: John Pell Publike Profesor of the Mathematicks at Amsterdam. In T'oude Conuoy op de Zee dijke in Amsterdam. giue this

[*cover annotated by Birch:*] 10/20 Decemb. 1644

Letter 28

17/27 December 1644

Cavendish to Pell, from Hamburg

BL MS Add. 4278, fos. 190–1 (original)
van Maanen C9

Worthie S^r

I hope you haue receiued my letter which I sent last weeke by my brothers seruant;[1] I haue little to adde but this inclosed concerning Reietas glass;[2] I desire you will be pleased to inquire if he be at Antwerp, & if he be to inquire of him if one of those glasses maye be had both sooner & at an easier rate;/ I desire by your next to knowe if Bonauentura Caualliere, & Sethus Caluisius of musick be to be had./[3] I hope you haue done with the waiewarde Dane,[4] that you maye returne to Apollonius & Diophantes[5] but especiallie to your owne analiticall worcke, which I extreamlie desire to see./ Mr Hobbes puts me in hope of his Philosophie, which he writes he is nowe putting in order, but I feare that will take a longe time./[6] I confesse I expect much from him, your self, M^r: Doctor Jungius;[7] & M^r: De Cartes I hope hath not yet done./ I expect allso some rarities in Analiticks & Geometrie from Fermat & Roberuall;[8] so that I hope oure age will be famous in that kinde./ And so wishing you all happiness I remayne

Your assured friend to serue you
Charles Cauendysshe

Hamb: Decemb: 17/27. 1644:

[*addressed:*] To my verie worthie friend M^r: John Pell Publicke Professor of the Mathematicks at Amsterdam. In t'oude Conuoy op de Zee dijke in Amsterdam

[*cover annotated by Birch:*] 17/27 Decembr 1644

---

[1] Letter 27; the servant of the Marquess of Newcastle has not been identified.

[2] Cavendish enclosed a report from an anonymous correspondent in Augsburg, dated [10/] 20 Nov. 1644 (BL MS Add. 4278, fos. 193–4); see also Letter 21, n. 15.

[3] For these two requests see Letter 27, nn. 8, 9.

[4] Longomontanus.

[5] See Letter 14, n. 13, and Letter 23, n. 2.

[6] This refers to Hobbes's *De corpore*; Cavendish's fears were justified, as the work would not be finished until 1653. It was eventually published in London in 1655.

[7] See Letter 16, n. 5.

[8] On Fermat see Letter 4a, n. 7, and Letter 4b, n. 1; on Roberval, Letter 19, n. 7.

Letter 29

10/20 January 1645

Pell to Cavendish, from Amsterdam

BL, MS Add. 4280, fo. 111^r (draft)
van Maanen P9

Right honourable

    Since that which Sr William Carnaby brought,[1] I have received two more of yours Dated Dec 10/20 & 17/27.[2] By which I perceived y^t Gassendus was come safe to your hands. And y^t you approve of my refutation of Longomontanus as full & plaine. If other men were as able to apprehend it, I neede not write any more of it. Nor doe I intend [>to waste] much time in refuting his falshoods but if I print any thing more against him, the greater part shall be [>spent] in producing new truths or at least new applications of old truths. I cannot finde Calvisius his musicke[3] in this city, nor any of Bonaventura Cavalerio his Italian pieces of which I have 4 or 5 titles[4] But [>we have] 2 Latine pieces of his [>in 4° viz] Geometria indivisibilibus continuorum nova quadam ratione promota[5] [where I finde many things which I had thought no man had knowen but my selfe but I demonstrate y^m otherwise than he does *deleted*]. It will cost bound about 3 shillings [The other is precisely of the same price but in my account not of y^t same worth and indeed is no booke for an Englishman to buy, unless out of meere curiosity, it is *deleted* >The other] called Directorium generale Uranometricum is of the same price, but I suppose no Englishman neede learne of a frier of Millain how to use our Logarithmes.[6] I thanke you for your newes of Reita;[7] the same day that I

---

[1] Letter 25 (see nn. 2, 3).

[2] Letters 27, 28.

[3] See Letter 27, n. 9.

[4] Bonaventura Cavalieri had published the following works in Italian: *Specchio ustorio overo trattato delle settioni coniche* (Bologna, 1632); *Compendio delle regole dei triangoli colle loro dimostrazioni* (Bologna, 1638); *Centuria di varii problemi per dimostrare l'uso e la facilità dei logaritmi* (Bologna, 1639); *Nuova pratica astrologica di fare le direttioni secondo la via rationale* (Bologna, 1639); the last three works were also issued in a single volume, with the addition of *Annotazioni all'opera* (Bologna, 1639); and the *Nuova pratica astrologica* was supplemented by an *Appendice della nuova pratica astrologica* (Bologna, 1640).

[5] Cavalieri's major work on 'indivisibles', published in Bologna in 1635 (see Letter 16, n. 26).

[6] This work, on logarithms and trigonometry, was published in Bologna in 1632 (see Letter 16, n. 26); Pell refers to 'our' logarithms because of the pioneering work of Napier and Briggs. Cavalieri was born in Milan; he joined the 'Gesuati', or Apostolic Clerks of St Jerome, an order of Augustinian friars, becoming prior of the convent of S. Maria della Mascarella in Bologna in 1635.

[7] See Letter 28, n. 2.

received your letter, heere was a student lately come from Antwerpe who said, he had there talked with Reita but said he would not let me see his glasse onely he told me he would shortly publish a booke [>to] demonstrate that all this while we had beene besides the cushion[8] both for matter & forme of Telescops. I will not adventure to tell you my conjecture of both. But I hope shortly to finde leisure to write to him. In the meane time let me tell you some newes out of England.

8 Months agoe (an old beginning, you will thinke, for newes) I received a letter from London with this postscript.[9] *We have a Gentleman heere that makes exceeding good Telescops.* In my answere I desired to know this Gentlemans *name* and *how good* his Telescops were. [>He replyed thus] 'You will not beleeve what I could write of our new Telescops. Its nothing with us to looke backe & view the first atomes that begat the Chaos: Or to looke forward & blow away the last, & least, ashes that shall outlive this little point, which we call Earth. We laugh at those that onely see the things that are. And yet you cannot wonder at all this when you shall know that the Author is that Lo Generall of those ordinances that did atomize Pr. Ruperts whole Army, the other day, before Yorke.[10] In good earnest his glasses are very good; and discover 30 (or more) in the pleiades. Many new satellites also, not yet heard of. Himselfe gave me one of his best, (then by him), but I left it at Cambridge (after one howers tryall, onely upon Earth, in a cleere day,) & I cannot yet recover it againe from my friend. Yet of all discoveries possible through that Telescope, none would please me so well as to see you returning to us againe, &c.'

This answere I received July 20 and could therefore have sent it with my first letter to you. But I first desired to know his name, which you see is not heerein expressed at least for heads so ignorant how ye world goes as mine is. I therefore wrote to him againe to tell me his name. Which it seemes he forgat to doe. But at last Dec 20/30 I received a letter with this postscript. 'That Curious Artist

---

[8]'Away from the main purpose or argument; beside the mark' (*OED*, 'cushion', 10b).

[9]This letter has not survived; the identity of the writer is not known.

[10]General Sir Alexander Hamilton (see n. 11 below) commanded the Parliamentarian artillery at the battle of Marston Moor (2 July 1644).

is L. Gen^ll Hammond.[11] He makes those glasses of which I spake so much, so little.'

And this is all that I have heard of these new English Telescops, which it seemes are as good as Galilaeo's best, if not better.

[To the rest of your letter I hope I shall shortly find time to send you a long answer, in the meane interim I take my leave But now I *deleted*]

As soone as I heare any further certainety of those [Antwerpian *deleted*] dutch or these English glasses I shall not faile to acquaint you with it or in what else I may [>so] approve my selfe to be

your Honours most humble servant
John P.

Jan. 10/20

---

[11] This reference to 'Hammond', and the reference above to the artillery commander at the battle of Marston Moor (Hamilton), indicate a confusion between two different people: Lt-Gen. Thomas Hammond and Gen. Sir Alexander Hamilton. Little is known about Thomas Hammond (brother of the theologian Henry Hammond), who was rapidly promoted to the rank of Lieutenant-General of the Ordnance in the Parliamentarian army; he later became one of the Commissioners of the High Court of Justice, and a signatory to the death warrant for Charles I (see M. Noble, *The Lives of the English Regicides*, 2 vols. (London, 1798), i, pp. 277–8). His skill as an engineer was commented on in a report on the siege of Bridgewater in 1645, when 'Bridges upon Carts of 30. foot long, were provided by the skill of Major General Hammon' (H. Peters, *Mr Peters Report from the Army, to the Parliament, made Saturday the 26 of July 1645* (London, 1645), p. 1). Another report described him as 'a gentleman ... of a most dextrous and ripe invention' (cited in I. Gentles, *The New Model Army in England, Ireland and Scotland, 1645–1653* (Oxford, 1992), p. 69). He died some time after 1653 (Gentles, *op. cit.*, p. 554). Hammond was not present at the battle of Marston Moor, however; the Parliamentarian artillery commander there was Sir Alexander Hamilton, who was also an ingenious engineer and inventor. He had learned his trade as a gunner under Gustavus Adolphus, being employed at Örebro in 1630 'in making of Cannon and fireworkes'; in August 1641 King Charles, inspecting the Scottish Army at Newcastle, had been impressed by a new type of light ordnance invented by Hamilton (see P. Young, *Marston Moor, 1644: The Campaign and the Battle* (Moreton-in-Marsh, 1997), p. 70). Most probably he was the inventor referred to in the earlier letter to Pell, his name then being confused with Hammond's in the later one.

Letter 30

10/20 January 1645

Cavendish to Pell, from Hamburg

BL, MS Add. 4278, fos. 194–5 (original)
van Maanen C10

Worthie S^r.

it is so longe since I hearde from you that I doute my letters which [>I] last writ to you, came not to your handes. I sent you inclosed in one of them the answeare which I receaued from Auspurge concerning Reyetas glass; & allso my desire that you would be pleased to enquire of Reyeta, who I heare is at Antwerp, of what conditions one [>of] his glasses may be had.[1] I suppose some Merchant of your acquaintance will doe so much for your sake./ I haue latelie had some discourse with Doctor Jungius & M^r: Tassius;[2] douteless they are both verie learned men; I was at Doctor Jungius his house where he shewed me manie treatises of his owne in Manuscript; uid^{ct}: de locis planis,[3] de motu locali,[4] staticks, Hydrostaticks,[5] & some obseruations of Insects./[6] I haue as greate an opinion of his abilities as of anie mans. I finde he is not yet minded to print anie thinge, but I hope heereafter he will & I despaire not in the meane time but he will imparte somewhat to me./ I asked M^r: Tassius (as you desired me) his opinion of your refutation of the Dane who aproues of it, & that you doe it without helpe of the Table of Tangents./[7] I writ in some of my last letters to knowe if Caualieros worckes, & Sethus Caluisius of Musick were to be had./[8] S^r I haue no more at this time but wishing you all happiness remaine

---

[1] See Letter 28, at n. 2.

[2] See Letter 16, nn. 4, 5.

[3] This probably refers to the major work by Jungius, 'Apollonius saxonicus', a reconstruction of Apollonius's lost treatise *De locis planis*: see Letter 42, n. 10.

[4] A short treatise by Jungius was later published: *Phoranomica, id est de motu locali* (Hamburg, 1650). A scribal copy of this text survives among Cavendish's papers: BL, MS Harl. 6083, fos. 236–71.

[5] A collection of Jungius's papers on statics and hydrostatics (with some dated items from the 1640s) is preserved under the general title 'Statica natantium': SUBH, MS Wo. 22.

[6] A list of Jungius's MSS, drawn up after his death by Martin Fogel, included one entitled 'Meditationes de insectis et vermibus' (C. Meinel, ed., *Der handschriftliche Nachlass von Joachim Jungius in der Staats- und Universitätsbibliothek Hamburg* (Stuttgart, 1984), p. xxiii). Some of its contents were later printed in J. Jungius, *Historia vermium e mss. schedis ... a Johanne Vagetio communicata* (Hamburg, 1691): see C. Meinel, *Die Bibliothek des Joachim Jungius: ein Beitrag zur Historia litteraria der frühen Neuzeit* (Göttingen, 1992), p. 142.

[7] See Pell's comments in Letter 24.

[8] See Letters 27, 28.

Your assured friend to serue you
Charles Cauendysshe

Hamburg: Jan: 10/20 1644.

[*addressed:*] To my verie worthie friend M^r: John Pell, Publicke Professor of the Mathematicks at Amsterdame; In T'oude conuoy op de Zee dijk in Amsterdam

[*annotated by Pell:*] Received Jan. 18/28.

[*cover annotated by Birch:*] 10/20 Janua. 1644/5

Letter 31

21/31 January 1645

Cavendish to Pell, from Hamburg

BL, MS Add. 4278, fos. 201–2 (original)
van Maanen C11

Worthie S^r
    I giue you manie thankes for your letter of the 10/20 of this moneth;[1] I am glad to heare you meane not to trouble your self with Longoman: more than as an apendix to some of your intended worckes; which I confess I longe much to see whatsoeuer it be; especiallie if it be of Analytycks. I giue you allso manie thankes for your inquirie of Reieta[2] & hope shortelie to heare from you whether there be anie hopes to procure one of his best glasses. I am glad he will print somewhat, howe to make them./ Who that is that makes those [>excellent] glasses in England [>I knowe not.] I should haue guessed [>him] to haue bin Hammilton, who was Generall of the Artillerie to Leslei; but you write him Haman;[3] whom I haue not heard of; but I hope heereafter wee maye procure one of them./ I am sorie Caluisius can not be had;[4] I shall not yet trouble you for the other,[5] but returne you manie thankes for your inquirie of them./ I am now growne into some aquaintance with Doctor Jungius who is pleased to visit me commonlie twice a weeke, & to imparte to me some of his conceptions de motu locali;[6] wee are yet but in the definitions, which are verie well expressed; he hath allso imparted to me diuers Theorems on the same subiect but not demonstrated them; but I doute not but he can & will; after he hath laied foundation enough in definitions & axiomes to builde vpon;/ I am sorie I lost so much time before I was aquainted with him, but I shall indeuoure to redeeme it during my staie heere;/ And so hoping to heare shortelie from you, & wishing you all happiness I rest

    Your assured friend to serue you

---

[1] Letter 29.

[2] See Letter 29, final paragraph.

[3] On the Hamilton–Hammond confusion see Letter 29, n. 11. 'Leslei' here refers either to Alexander Leslie, first Earl of Leven (c.1580–1661), commander of the Scottish army which entered England in support of the Parliamentarian forces in January 1644; or, more probably (as Leven was commonly referred to by his title) to David Leslie, later first Baron Newark (d. 1682), who served under Leven as Major-General and played a key role at the battle of Marston Moor.

[4] See Letter 27, n. 9, and Letter 29, at n. 3.

[5] See Letter 27, n. 8, and Letter 29, n. 4.

[6] See Letter 30, n. 4.

Charles Cauendysshe

Hamburg Jan: 21/31 1644.

[*addressed:*] To my verie worthie friend M^r: John Pell Publicke Professor of the Mathematickes at Amsterdame In T'oude Conuoie op de Zee dijke in Amsterdam.

[*annotated by Pell:*] Received Jan 30/9 Febr.

[*cover annotated by Birch:*] 21/31 January 1644/5

Letter 32

8/18 February 1645

Pell to Cavendish, from Amsterdam

BL, MS Harl. 6796, fos. 295–6 (original); BL, MS Add. 4280, fos. 111^v–112^r (draft); BL, MS Harl. 6083, fo. 143 (copy, by Cavendish, of the central part of the letter, from '[Whereas we ought] to distinguish ... to '... that will not easily be untyed')

van Maanen P10

Right Honourable

To yours dated Jan 10/20 & 21/31[1] I have nothing to answere save that I hope shortly to give you an account concerning Reita; and that I feare not but you will finde, in both those Hamburgenses, enough to keepe you from blaming of me, for commending them to you: [I am glad Mr Tassius *deleted* >And since one of them *in draft*] And since one of them sees the force of my demonstration,[2] I hope he will not doe himselfe so much wrong as to refuse to say as much, under his hand, when I shall heereafter demaund it, if I finde no other meanes to make Longomontanus understand that his demonstrations are generally counted paralogismes & his conclusions false, in the judgement of all the able Mathematicians of Europe.

But I cannot want matter to write of, if I answere that passage in your Last (as almost in all yours) expressing a very great desire to see a large volume concerning Analytickes after my fashion. But I feare it will be long ere I shall finde leisure to finish such a volume for the presse. In the meane time perhaps some other man will spare me the labour. You know how largely Mersennus writes of their Chauveau,[3] and I suppose, you have heard as well as I, that, for certaine, Oughtredi Clavis Mathematicae is now in the presse at Cambridge,[4] with I know not what additions: I hope I shall have it as soone as it comes forth. I would gladly know what your great demonstrative men of Hamburg intend in y^t argument, for I heare that when they saw Oughtredi Clavem, they said it was

---

[1] Letters 30, 31.

[2] See Letter 30 ('Tassius ... aproues of it'); the two 'Hamburgenses' were Tassius and Jungius (on whom see Letter 16, nn. 4, 5).

[3] See Letter 16, n. 21.

[4] William Oughtred's classic work, *Clavis* (1631), had attracted the attention of Seth Ward, then a young Fellow at Cambridge, who visited Oughtred in 1642 and (as Oughtred put it) 'by a gentle violence induced me to publish again my former Tractate in a manner more moulded and perfected'. This led to the publication of an English version (*The Key of the Mathematicks new Forged and Filed* (London, 1647: the quotation just given is from the Preface, sig. B6^r) and a second edition of the Latin (London, 1648). Neither edition was printed or published in Cambridge, though it is possible that publication there may originally have been mooted by Ward.

not enough: I suppose they meant, *not enough* to deserve that title. Though I thinke I can guesse pretty neere, what those things are, which they expected under that title. For, whereas in one of yours, you write that 'Dr Jungius preferres the analytickes of the ancients, before Vieta's by letters, which, he sayes, is more subject to errors & mistakes, though more facile & quick of dispatch':[5] I doe thence gather that he is rather of the delineatory than of the supputatory tribe of Mathematicians.[6] For of these things he speakes but as most men doe, preferring the one before the other, some this, others that, according as their heads are fitted for delineation or supputation. Which difference appeares not onely in their discourse, but also in their practise; demonstrating Arithmeticall truths Geometrically & Geometricall truths Arithmetically: And as in demonstration, so doe they in investigation, especially in insuperable problemes, as of the Circles Quadrature or the Lawes of Caelestiall motions. For the Mathematicasters[7] are divided into the same tribes: A supputatory pate hopes to finde all things in his numerall concinnities:[8] whilest the delineatory noddle pursues his game through a wildernesse of lines & beleeves that all difficulties may be subdued by an army of ten thousand perpendiculars, five thousand parallels, & Circles innumerable. And to say the truth, you shall hardly finde a man that seemes distinctly to conceive the uses of those two faculties of his soule, or that speakes not confusedly of the meanes of searching of hidden truths. Whereas we ought to distinguish the questions that are proper to one sort, from those that are *both* wayes tractable: and to give infallible & manifest tokens enabling us to discerne whether the inquiry must be made in meere termes of delineation & mensuration (rectum, curvum, cavum, sinuosum, convexum; planum, longum, latum; infinitum, terminatum; cohaerens, continuum; directum, obliquum, acutum, obtusum; vertex, apex, latus, figura; directim-, deinceps-, alternatim situ; coalterna; inclinatio, tactio; parallela, perpendicularis; applicata, congrua &c)[9] Or whether in Logisticall termes (numeratio, additio, subtractio, multiplicatio, divisio, quotus, ratio, proportio; radix, potestas, involutio, evolutio, gradus parodici, coefficiens &c)[10] Or whether it may not be done either way: And if so;

---

[5] Pell cites here the postscript to Letter 20; the quotation (signalled by quotation marks here) is marked by a marginal line in the MS.

[6] Pell outlines what he sees as the difference between geometers, those who work with lines and figures, and arithmeticians, those who prefer to work with numbers. Geometry and arithmetic were both, according to Plato and all later writers, major branches of mathematics; classical geometry, however, was the dominant influence in early modern mathematics, replaced only gradually by the new methods made possible by algebra.

[7] = petty or inferior mathematicians.

[8] = harmonies, congruities.

[9] 'Straight, curved, concave, sinuous, convex; flat, long, broad; infinite, bounded; coherent, continuous; direct, oblique, acute, obtuse; vertex, apex, side, figure; directly, successively or alternately situated; coalternate; inclination, contact; parallel, perpendicular; applied, congruous, etc.'

[10] 'numeration, addition, subtraction, multiplication, division, quotient, ratio, proportion; root, power, raising to powers, extraction of roots, degree of a term, coefficient, etc.'

then whether both wayes be alike good, or for what reasons to preferre the one before the other.

Moreover, in *Diagraphicall mensurative*[11] speculation, to know in what *Locus* the imagination & understanding must meete, to joine their forces to goe a hunting after these lurking truths.

And in *Logisticall supputative* speculation, to determine whether the inquisition must proceede *directly*, or have somewhat in it of the *inverse*: & how to proceede in either sort: When to use *position* & what kinde, *Numerall* or *Abstract*.

If *numerall* (which unlesse by chance is allwayes *false*) how to manage that falshood, for the discovery of the truth: that is; When & how to use the *particular* rules of falshood taught by the Arabians & enlarged but not perfected by Gemma, Simon, Weber; as also that *universall regula falsae positionis*[12] practised by Diophantus, but reduced into precept by no man that I know. And who knowes whether it will please God to graunt me life & leisure to perfect them both?

If the positions must be *abstract* (which cannot easily be *false*) then whether the inquiry must be meere *Regression*, or whether it must proceed by *Rejection*: If *Rejection*, then whether *Abolition* or *Substitution*: and if by *Abolition*, how amongst those curious & artificiall praeparatories, to chuse the most convenient: if by *Substitution*, then whether terminorum propositorum or assumptorum,[13] or both; and in both Kindes, how to finde & make *apt substitutes* for the most convenient finding out & demonstrating of a *truth* or discovering & refuting a *falshood*.

Nor is this all: we should have precepts teaching us to discover when a probleme is *impossible* & when *possible*; & among possibles, when it is sufficiently conditionate, when too little, when too much:

If too much; how to correct it, if our selves be to propose it: or how to make the best advantage of that superfluity, when others propose it to us.

If too little; to know how much too little, what to doe in such a case, how to limit it sufficiently, if our selves be to propose it; Or if it be proposed to us, how fully to supply that defect, by assumption of aequations; how to know the utmost bounds of our liberty therein; When to assume numerall data, when abstract; When to answere such problemes in individuo determinato,[14] when indefinitely;

---

[11] 'Diagraphicall' is an adjective derived from 'diagram'; 'mensurative' is from 'mensuration'. These two terms refer back to Pell's earlier phrase, 'delineation & mensuration'.

[12] The 'rule of false position', in which a 'false' or incorrect answer is adjusted to give a 'true' or correct solution. The rule was taught in many textbooks of arithmetic. Those mentioned by Pell include R. Gemma (known as Gemma 'Frisius'), *Arithmeticae practicae methodus facilis* (Antwerp, 1540), and Stevin, *L'Arithmetique*. The identity of Weber here is uncertain; this may possibly be a reference to Johann Weber, *Gerechnet Rechenbüchlein auf Erfurtischen Wein- und Tranks-Kauff* (Erfurt, 1570; 2nd edn. Erfurt, 1583) (as cited in D. E. Smith, *Rara arithmetica: A Catalogue of the Arithmetics Written before the Year MDCI*, 2 vols. (Boston, 1908), ii, p. 338).

[13] 'Of proposed or assumed terms'.

[14] 'In a determined individual case'.

and in this latter, to know the whole artifice of limiting all the quaesita[15] within their precise bounds, whether it be à parte ante vel a parte post, or utrinque.[16] Or if one & the same quaesita quantitas[17] be diversly & alternately limited & free, to know how often it is thus limited & precisely where.

Nor are those directories to be omitted that serve to keepe men from running round; for want of which, very often, men after a long inquisition, finde themselves to have spent a great deale of ratiocination to no purpose, as concluding some truth proposed in the beginning, whereas they sought something meerely unknowen, like Watermen in a mist, who, in stead of crossing the Thames, after halfe an houres rowing, doe sometimes bring their fare to the same staires, where he tooke boate or at least some other place not farre off, but on the same side of the water.

Heereunto in Grand Problemes should be added the invaluable artifice of determining the number of cases or under-problemes, that is, How many & what they be; How to reject the unprofitable ones; How to make an orderly recension of them; How to finde their second order, that is, the best order of investigation of all their solutions, by making those knots to untie one another.

Likewise, the generall manner of transforming of all Problemes & Theoremes that are capable of such alteration, by stripping them of their Diagraphicall termes & clothing them in Logisticall termes & contra.

[And in all this, to *draft*] And all this is to *finde the habitudes* of quantities; But if the habitudes be given, then should we have an universall way to *examine* them and to finde media demonstrationis if they be true, or refutationis,[18] if they be false.

But because the *genuine* habitudes are oftentimes too troublesome, there should be added the art of finding the best *coactive* habitudes, and the whole misterie of making numeros solennes, eorumque Canonicas recensiones.[19]

As also Alexander Mathematicus, that is the great art of cutting all knots, that will not easily be untyed.

But I should forget my self if the paper did not admonish me to make an end. You have heere some of the heads of that multitude of thoughts, that I would willingly be delivered of; but it may be somebody else must bring them forth. We see there be great names & promises abroad, Logistica speciosa sive Algebra nova, docens nullum non problema solvere.[20] Idea universalis Matheseos

---

[15] 'Things sought'.

[16] 'As preconditions or as consequences, or both'.

[17] 'Sought quantity'.

[18] 'The means of demonstration' ... or 'of refutation'.

[19] 'Fixed numbers, and their recensions, established by rules'.

[20] F. Viète's promised *Algebra nova* (see Letter 7, n. 5) 'teaching how to leave no problem unsolved'. The phrase 'nullum non problema solvere' is the final sentence of the *Isagoge* (1591), p. 12.

seu primae Mathematicae.[21] Methodus practicae Arithmeticae universalis.[22] De resolutione & compositione Mathematicâ.[23] Clavis Mathematicae.[24] Artis Analyticae praxis.[25] Nova Geometriae clavis.[26] Encyclopaedia Mathematica[27] and what not? Must we not reade over all these once more and see if all these my desiderata be not there satisfied? For if they be not, there is little hope that they should be supplyed by a stupidus, stolidus, cani similis, temerarius petulans juvenis,[28] for those and the like titles hath Longomontanus bestowed upon

Your Honours most humble servant
John Pell.

Amsterdam. Febr 8/18 1644[/]5

[*addressed:*] A Monsieur
Monsieur Cavendysshe Chevalier Anglois à Hamburg. Recommandée a Monsr son frere Le Marquis de New Castel demeurant nahe bij St Johannis Kirche.[29]

---

[21] This is apparently a reference to Pell's own *Idea*.

[22] This is apparently a version of Gemma Frisius's title: see n. 12, above.

[23] M. Getaldus [Getaldić, Ghetaldi], *De resolutione et compositione mathematica libri quinque*, ed. A. Brogiotti (Rome, 1631).

[24] See n. 4, above. (Oughtred's book was usually referred to as *Clavis mathematicae*, or simply *Clavis*.)

[25] Harriot, *Artis analyticae praxis*.

[26] De Billy, *Nova geometriae clavis algebra* (see Letter 16, n. 10).

[27] *Encyclopaedia mathematica ad agones panegyricos anni 1640 ... in Claromontano Parisiensi Societatis Jesu Collegis* (Paris, 1640).

[28] 'Stupid, doltish, dog-like, rash, petulant youth': Pell quotes from Longomontanus, Ἐλέγξεως, sig. π1^v.

[29] 'M. Cavendish, English knight, at Hamburg. Directed to his brother, the Marquess of Newcastle, staying near St John's Church.'

Letter 33

26 March/5 April 1645

Cavendish to Pell, from Antwerp

BL, MS Add. 4278, fos. 203–4 (original)

Worthie S^r:

not to trouble you with more particulares of oure iournie than that at Rotterdam S^r: William Boswell[1] came to vs, whom I found to be that which I supposed him formerlie to be; a discreete ciuill gentleman; I perceiue he thinkes Monsr: de Cartes his last booke[2] to be full of fancie, though he esteemes much of him;/ Heere I mett with S^r: Kenelm Digbies booke[3] but had no time to reade it all; but it apeares to me to haue some things in it extraordinarie./ Wee met heere with the famous Cappuchin Rieta, his booke is nowe in the press heere allmost finished;[4] Hee solues the aparences of the planetes by excentricks without aequants or epicicles & yet not according to Copernicus System of the worlde; he teaches allso in this booke the making of his new Telescope;[5] his tube for the best glass was spoiled so that wee could not see it; but wee sawe an other [>made at his directions] but had not the oportunitie of looking at a conuenient obiect far distant; but as I guess it is not better than myne; it represents obiects euers;[6] I had not discourse enough with him alone to aske him manie quaeries. but douteless he is an excellent man & verie courteous, & I found him free & open in his discourse to me. he saies he can easilie & infalliblie finde the longitude; but he discouers not that in this booke./ he saies the Satellites of Jupiter are little sunns; & diuers other nouelties he hath obserued; manie of which he will not discouer in this booke./ I haue no more nowe to trouble you with but to desire that you will thinke of publishing some of your rarities especiallie in the Analitickes./ And so wishing you all happiness I remaine

  Your assured friend & seruant
  Charles Cauendysshe

---

[1] See Letter 19, n. 13.

[2] Descartes, *Principia*.

[3] Sir Kenelm Digby, *Two Treatises: In the one of which, the Nature of Bodies, in the other, the Nature of Mans Soule, is looked into* (Paris, 1644). This work had been published only a few weeks earlier: the prefatory materials include a note of approbation by the Catholic priest Henry Holden, dated 1 Mar. 'ab Incarnationis anno 1644', i.e. 1644/5 (sig. u2^r).

[4] A. M. Schyrleus de Rheita, *Oculus Enoch et Eliae sive radius sidereomysticus: pars prima* (Antwerp, 1645). This was shortly followed by the publication of *Oculi Enoch et Eliae pars altera sive theo-astronomia* (Antwerp, 1645). On Rheita see Letter 13, n. 4.

[5] For previous discussions of Rheita's improved telescope, and the lenses made for it, see Letters 19, 21–2, 27–9.

[6] 'Everse', inverted.

Antwerp March 26 olde style

[*postscript:*] wee are nowe goeing towardes Bruxells

[*addressed:*] To my verie worthie friend M^r: John Pell, Publike Professor of the Mathematickes at Amsterdam; In T'oude Conuoye op de zee dijke in Amsterdam giue this

[*annotated by Pell:*] Received 9/19 Aprill 1645

[*cover annotated by Birch:*] 26 March 1645 O.

Letter 34

1/11 May 1645

Cavendish to Pell, from Paris

BL, MS Add. 4278, fos. 205–6 (original)
van Maanen C13

Worthie S^r:
    after our parting from you, wee made no hast hither, but went little iournies & made some staie by the waye. since my comming hither I haue not yet made any newe aquaintances with the learned men heere, but my old aquaintance & friend M^r: Mydorge hath bin with me, who complaines of some diuersions which haue hindered his studies, so that I doute wee must yet awhyle longer expect the rest of his Conicks.[1] I perceiue he hath a verie greate opinion of M^r: de Cartes, M^r: Hobbes is so auers from a friendship with M^r: de Cartes that he would not see him when he was heere. M^r: Hobbes commends M^r: Roberual[2] extreamelie;/ I send you heere a problem solued & demonstrated by M^r: Hobbes;[3] The figures were made in haste by himself, onelie to express the demonstration to me; I thought to send you better figures, but when I had made them I liked them worse than these./ Hee cites one proposition demonstrated by himself in his intended worck; that is that the arithmeticall meane is greater than the geometricall meane between the same extreames;[4] it is obuious enough but he takes it by the waye in his philosophie, following out of his principles which are some of them peculiare to himself./ I doute it will be longe ere M^r: Hobbes publish anie thinge, so far as I haue reade I like verie well; he proceeds euerie daie somewhat but he hath a

---

[1] The mathematician and physician Claude Mydorge (1585–1647) had published the first two books of his 'Conics' fourteen years earlier (*Prodromi catoptricorum et dioptricorum: sive conicorum operis ... libri primus et secundus* (Paris, 1631), and the next two books eight years thereafter (*Prodromi catoptricorum et dioptricorum: sive conicorum operis ... libri quatuor priores* (Paris, 1639).) The first of these editions bore a dedication to Cavendish, with whom Mydorge corresponded in 1631 (see BL, MS Add. 70499, fo. 145^r, Richard Andrews to the Earl of Newcastle, 22 June [/2 July] 1631).

[2] See Letter 19, n. 7.

[3] This may refer to an item in Hobbes's hand, which is preserved in close proximity to this letter: a rough diagram relating to a hyperbola and a parabola (BL, MS Add. 4278, fo. 209^r). (Cf. the description given of it by Pell in Letter 38 as a 'Conicall demonstration'.) It was apparently accompanied by a text and one or more other 'figures'.

[4] The text of Hobbes's demonstration of this proposition is preserved in a set of notes made by Cavendish on Hobbes's draft chapters of *De corpore* in Paris: BL, MS Harl. 6083, fo. 203^r (draft of XIII.15; printed in T. Hobbes, *Critique du De mundo de Thomas White*, ed. J. Jacquot and H. W. Jones (Paris, 1973), p. 496). Hobbes presented a revised treatment of this topic in the published text (*De corpore*, XIII.26).

greate deale to doe./[5] I hope you aduance youre Analitick worck & Diophantes,[6] so that I hope you will be the first of my aquaintance that will publish anie thinge: I desire you will doe me the fauoure to send this inclosed letter to my worthie friend M^r: Doctor Jungius[7] at Hamburg./ And so wishing you all hapiness I rest

   Your assured friend to serue you
   Charles Cauendysshe

Paris Maie: 1/11 1645

[*addressed:*] To my verie worthie friend M^r: John Pell Publicke Professor of the Mathematickes at Amsterdam In t'oud Conuoy op de Zee dike in Amsterdam.

[*added to address in another hand:*] Amsterdam

[*annotated by Pell:*] Received May 10/20

[*cover annotated by Birch:*] 1 May 1645 O.S.

---

[5] This refers to Hobbes's work on the composition of *De corpore* (see Letter 28, n. 6).

[6] See Letter 14, n. 13.

[7] The letter to Jungius (on whom see Letter 16, n. 5), dated [1/] 11 May, which contains an important summary of Hobbes's theory of local motion, is SUBH, MS Sup. ep. 97 (= Pe. 1a) (printed in C. von Brockdorff, *Des Charles Cavendish Bericht für Joachim Jungius über die Grundzüge der Hobbes'schen Naturphilosophie*, Veröffentlichungen der Hobbes-Gesellschaft, iii (Kiel, 1934), pp. 2–4).

Letter 35

9/19 May 1645

Pell to Cavendish, from Amsterdam

BL, MS Add. 4280, fo. 112 (draft)
van Maanen P11

Right Honourable

I am in the first place to give you thankes for what I received from you since I saw you last viz. a letter from Antwerpe dated March 26 and a greeting from Paris in Mr Hobbes his letter dated Aprill 22.[1] To whom though I count my selfe much obliged for his letter & ye enclosed paper from Monsieur Robervall[2] yet I have not hitherto beene able to send him an answere, because he [forgot to adde *deleted*] adjoined [>no] manner of addresse for my letters either to his owne lodging or yours. And therefore I have beene constrained to trouble Sir William Boswell[3] for ye conveying of this: who I make no question, will direct it to y^m y^t will easily find you.

I heare no further newes of Reita's new booke which you write was almost finished 6 weekes agoe.[4] But our booksellers make no haste to send for new bookes. They have not yet Bullialdi Astronomiam Philolaicam[5] though I have often inquired for it, as also [>I have done] for all those Conicall french writers Mydorgius, Desargues. Paschalus [and I thinke *deleted*] Pallier.[6] Which they are so farre from having, that I cannot perceive that ever y^{ei} had y^m or saw them. They would otherwise be of some use to me now in ye edition of Apollonius for I have now the translation finished of those 3 latter bookes out of the Arabicke and d^r Blaew[7] hath a minde to print it and also to reprint Commandines edition of the first 4 bookes[8] with my notes [& my translation of it *deleted*] Of which when

---

[1] Cavendish's letter is Letter 33; Hobbes's letter has not survived.

[2] This may refer to the demonstration of Pell's theorem by Roberval (on whom see Letter 19, n. 7) printed in Pell, *Controversiae pars prima*, pp. 47–9.

[3] See Letter 19, n. 13.

[4] See Letter 33, n. 4.

[5] Boulliau, *Astronomia philolaica*. The text of this long-awaited work (see *MC* xii, p. 339) had been consigned to the printer at the end of 1644; the first copies were apparently distributed by the author just before his departure from Paris (for Italy) on [9/] 19 May 1645 (see Nellen, *Ismaël Boulliau*, p. 123, and *MC* xiii, p. 449).

[6] On Mydorge see Letter 34, n. 1; on Desargues see Letter 16, n. 12; Blaise Pascal (1623–62) had published his *Essai sur les coniques* in 1640; Jacques Le Pailleur (d. 1654) was a poet, amateur mathematician, and friend of Pascal and Mersenne.

[7] See Letter 14, n. 14.

[8] See Letter 23, n. 6.

[>M^r] Golius⁹ heard & y^t we began to talke with a graver about the diagrammes, he made a Journy from Leiden hither o' purpose to dissuade or deterre both Dr Blaew & me from y^t intention. He had beene heere divers times in ye city & inquired where I dwelt but never came to see me: till now he was in a manner constrained. [>6 dayes agoe] He brought his Arabicke 7 bookes of Apollonius with [>him], a goodly faire manuscript [and as he tells me *deleted*] given him [>18 yeares agoe] at Aleppo by a gentleman of this country, to whom he sayes he hath given a faithfull promise to publish it or otherwise he would put it into my hands & give me all the assistance y^t he could in ye publishing of it. Besides this promise, he hath beene at some cost to pay for the delineating of all the figures and indeed they are exceeding neately drawen [>in paper by his Collegues sonne] & some of them [>are also graven] He tells me he hath gotten a greeke coppy of that which Commandine translated and y^t he hath compared it with his manuscript which he finds to agree verbatim with the greeke in those 4 first bookes & therefore he is confident of the latter 3 that they doe in like manner give us not onely the sense, as my Epitomist, but the very phrase and style of Apollonius.¹⁰ I answered him y^t I Could say no other y^n what I had allwayes done, that it was more proper for him to publish it than for any other man as agreeing so well with both his professions, and therefore I had rather see it done by Mr Golius than by any man: but rather by my selfe, than by no man. As for my selfe, I told him, that I knew it was farre [better for *deleted*] easier for him to doe it in all respects, than for me. He replyed, that though he knew Ravius was unable to make [a tolerable translation *deleted*] an intelligible translation yet if it fell into my hands he was confident y^t I would make [something *deleted* >sence] of it [>& so supply all defects as to make it abundantly sufficient to quench the] thirst & satisfy the curiosity of Mathematicians, though it be impossible for me to restore ye very style & forme of reasoning which Apollonius useth for saith he I have a Coppy of Ravius¹¹ his Arabicke manuscript, which you use & know very well how bold that [>Persian] makes with Apollonius. I asked him whence he came to have so great an opinion of [my abilities *deleted* >me.] He told me he had seene letters of my writing & specimina inough & saith he he y^t hath read your counsell to ye Elzevirs for ye printing of Vieta's workes,¹² needes no other argument of your abilities. Doe you then, said I, approve of all y^t I there wrote? Yes, saith he, all & alltogether. But, said I, I thinke it was but lost labour, for I heare not that the Elzevirs goe in hand with it. Oh yes, said he, all the diagrams are graven and as soone as they can dispatch some of those things that they have

---

⁹See Letter 23, n. 9.

¹⁰On these Arabic texts see Letter 23, nn. 8, 9; see also Toomer, *Eastern Wisedome*, p. 184 (and pp. 239, 242 on the figures engraved for Golius).

¹¹See Letter 14, n. 6.

¹²Pell had written a long letter to Hartlib about this project on [2/] 12 Oct. 1642 (Pell's copy is in BL, MS Add. 4280, fos. 206–7); it appears that this was then circulated by Hartlib. The edition of Viète's works, prepared by Frans van Schooten, was published by Elzevier in Leiden in 1646.

now in hand, they will begin Vieta, for it is not for their profit to lose all that charge y^t y^{ei} have been at.

But what doe I trouble you with a dry tale of so much of our talke? The upshot is this. Mr Golius was never in so much feare [>of losing much] honour & I know not what profit which he might have [>secured] by publishing Apollonius as well 16 yeeres agoe as now.

His intention I perceive otherwise ran thus First to finish his Arabicke Lexicon,[13] of wch there are now perhaps 8 sheetes printed, [>Next] to get all his diagrams first graven & then to begin to translate into Latine not onely those 3 [other *deleted*] bookes of Apollonius, but also Barulcum Heronis[14] & some other things lost in ye Greeke but preserved in ye Arabicke & so come to his hands. and this he thought to doe at his best leisure, twenty yeeres hence [>perhaps] because he thought no man could prevent him. What haste he will now make I know not; but in ye meane time Apollonius hath ill lucke, if betweene [>us both] he doe not now in short time obtaine his liberty to fly abroad.

As for Longomontanus he is not dead, but recovered from his dangerous fit to perfect health nondum tamen ad sanam mentem.[15] And since he opposes his authority to mine, which overswayes exceedingly with those many that can [>number our yeares but] not weigh our reasons and [>therefore rather] beleeve y^t [>I was petulant & foolehardy to attacque him than that he is so miserably ignorant & so injuriously perverse as I say the] author of that reply is. Therefore there is a necessity y^t I counterpoise [>that age] of his, with universality of consent of Mathematicians, Wherein you may help me much, with very little trouble to your selfe. By obtaining the verdicts of as many as may be [>had] concerning Longomontanus his periphery (which is ye upshot of all the rest) and my refutation. But I shall desire them in Latine & subscribed by themselves as in France [>by] Mr Hobbes Monsieur Fermat, [>Midorge], de Beaulne Bulliald

---

[13] Golius's *Lexicon arabico-latinum* was eventually published in Leiden in 1653.

[14] Pappus (see Letter 16, n. 15) referred to a treatise by the Greek mathematician Heron (or 'Hero') of Alexandria entitled $Βαρουλκός$; Pappus gave the impression that this was a different work from Heron's treatise on mechanics, the $Μηχανικά$, to which he also referred, although the title of the former also suggested a mechanical subject-matter ('$βαρουλκός$' means 'that which draws up a heavy weight'). Neither treatise survived in Greek. However, an Arabic translation of the $Μηχανικά$ did survive, and the translator (a 9th-century Levantine Greek, Kosta ben Luka) referred to it as $Βαρουλκός$; this created some uncertainty as to whether there were two treatises, or one treatise with two names. Modern scholarship has concluded that the $Βαρουλκός$ was indeed a separate work, but that part of it was incorporated into the $Μηχανικά$. Golius possessed one of the rare manuscripts of the latter (then an unpublished and untranslated work), which, misled by the Arabic title, he described as the $Βαρουλκός$: it is now Leiden University Library, Cod. 51 Gol. (See Heron of Alexandria, *Opera quae supersunt omnia*, 5 vols. (Leipzig, 1899–1914), ii, *Mechanica et catoptrica*, ed. and tr. L. Nix and W. Schmidt, 'Einleitung', pp. xviii–xxiv, and Anon., *Catalogus librorum tam impressorum quam manuscriptorum bibliothecae publicae Universitatis Lugduno-Batavae* (Leiden, 1716), p. 454, item 1091.)

[15] 'Though not yet to sanity'.

[>P. de Billy &c]¹⁶ And by one passage in a letter to P. Mersenne you can set him a worke to procure the like from divers Italians as Tauricellus, Bonav Cavalieri, [>(and especially Camillus Gloriosus¹⁷ for y^t Longomontanus himselfe much desired)] It will be a singular pleasure to you to see what pretty variety of demonstrations these learned men will make of my fundamentall Theoreme (especially if they know y^t my Antagonist [>have] written y^t it is Geometricè indemonstrabile)¹⁸ and yet perhaps none of them will light on ye way, [>by which I sought it which I intend also to publish. But] if any are not at leisure to doe so much, I shall be contented with [>their] bare acknowledgement that I have laid a sure ground & built a just & complete [>refutation] sufficient to demonstrate ye falsity of y^t assertion which I undertooke to overthrow. And if any professed Mathematician refuse to doe thus much, all others will count him [>unworthy to have knowen the truth or] to finde anybody to backe him when he shall be worried by [>a crew of] prodigiously ignorant & [>extremely impudent scioli]¹⁹ in as grosse a manner, as I have beene: They will be the willinger to doe this if y^{ei} know y^t it appears in Longomontanus his 14 [>different] treatises wch he hath [>published at severall times] in y^s argument, amounting to 89 sheetes²⁰ that he beleeves y^t all Mathematicians are of his minde if they had ye manners to thanke him for his [>instruction but most men] are too ignorant to understand him, or

---

[16] On Fermat see Letter 4a, n. 7, and Letter 4b, n. 1; on Mydorge see Letter 34, n. 1; Florimond de Beaune (1601–52), lawyer, mathematician, and astronomer, wrote a commentary on Descartes's 'Géométrie' which was translated and appended to the first Latin edition of that work edited by Frans van Schooten (Leiden, 1649)); on Boulliau see Letter 16, n. 13, and above, n. 5; on de Billy see Letter 16, n. 10.

[17] Evangelista Torricelli (1608–47) was one of the leading mathematicians in Italy, author of *Opera geometrica*; on Cavalieri see Letter 16, n. 26; Giovanni Camillo Glorioso (or Gloriosi) (1572–1643: the fact of his death was evidently unknown to Pell), mathematician and astronomer, was the author of *Exercitationum mathematicarum decas prima in qua continentur varia et theoremata et problemata*, 2 vols. (Naples, 1627–35). On Glorioso see P. D. Napoletani, 'Galilei e due matematici napoletani: Luca Valerio e Giovanni Camillo Gloriosi', in F. Lomonaco and M. Torrini, eds., *Atti del convegno 'Galilei e Napoli', 12–14 aprile 1984* (Naples, 1987), pp. 159–90.

[18] 'Incapable of geometrical demonstration'.

[19] Sciolists, people with only a smattering of knowledge.

[20] The fact that Pell specifies the total number of sheets shows that he derived his information from a list published by Longomontanus at the end of his *'Ενέργεια proportionis sesquitertiae pro certis figuris circulo adscriptus, & per numeros determinatis, data circuli diametro* (Copenhagen, 1644), sigs. B3^v–B4, which gives the details of thirteen publications by Longomontanus on cyclometry, including the number of sheets (quarto gatherings) in each work. These publications are as follows: *Cyclometria ex lunulis reciproce demonstrata* (Copenhagen, 1612); *Cyclometria vere ac absolute in ipsa natura circuli* (Hamburg, 1627); *Fundamentum cyclometriae rationalis* (Copenhagen, 1628); *Geometrica quaesita tredecim* (Copenhagen, 1631); *Inventio quadraturae circuli* (Copenhagen, 1634); *Coronis problematica* (Copenhagen, 1637); *Problemata duo geometrica* (Copenhagen, 1638); *Literulae supplices ad S. R. Daniae* (Copenhagen, 1639); *Inventio theorematis nobilissimi* (Copenhagen, 1643); *Problema, quod ... absolutam circuli mensuram praestat* (Copenhagen, 1643); *Admiranda operatio trium numerorum* (Copenhagen, 1644); *Rotundi in plano* (Amsterdam, 1644); *'Ενέργεια proportionis sesquitertiae* (Copenhagen, 1644).

too envious to acknowledge so rare an invention to be reserved for denmarke [>invidiam, saith he ac malevolentiam in patriam nostram.]²¹

I published [>sayes hee] a challenge, Anno 1639 which unà cum Quadratura Circuli, ob eandem causam, in Germaniam, Italiam, Hispaniam, Galliam, Belgium [>&] Angliam transmitti curavi & simul literis meis nonnullos praestantes Mathematicos ad iudicia sua de hoc invento meo ferenda, eâ, quâ par erat, humanitate & liberaliter invitavi. Nec tamen in hunc usque diem, quisquam repertus est, qui se scripto suo publicato nobis opponeret, praeter unum Paulum Guldinum &c²² Yes now there is another hath taken upon him [who hopes to *deleted*] to be the Mathematicians champion and therefore he hopes they will owne him; in France & Italy by your procurement: in other parts by ye helpe of other freinds. I have received a letter from Hamburg from m^r Tassius,²³ and y^s gives me occasion of another long story but y^t my discreeter paper bids me make an end [>onely] gives me leave to adde y^t I am

Your Honours most humble servant,
John Pell.

May 9/19 1645

---

²¹'Envy and ill-will towards our native land': Pell quotes from Longomontanus, *Problema*, sig. A4^v.

²²'[... which,] together with the quadrature of the circle, and for the same reason, I arranged to be sent to Germany, Italy, Spain, France, the Netherlands, and England, with letters from me to several eminent mathematicians, inviting them to give me their judgement, freely and with suitable humanity, on this discovery of mine. And until this day no one has been found who opposed me in print, except for a certain Paul Guldin...'. Pell quotes from Longomontanus, *Problema*, sigs. A4^v–B1^r. The challenge, and the quadrature, were printed in Longomontanus, *Literulae supplices ad S. R. Daniae*. On Guldin see Letter 24, n. 7.

²³See Letter 16, n. 4.

Letter 36
27 June/7 July 1645
Cavendish to Pell, from Paris

BL, MS Add. 4278, fos. 196, 198 (original)
van Maanen C14

Worthie S^r
    I giue you manie thankes for your letter of Maye 9/19[1] which I receiued not longe since. I writ to you a letter from hence a greate whyle since in which was one inclosed to Doctor Jungius; & a proposition demonstrated by m^r: Hobbes:[2] [>all] which I hope you haue receiued./ I haue as you desire, procured not onelie the approbation but demonstration of your fundamentall proposition, by m^r: Hobbes his meanes;[3] Fermat[4] is not in this towne & Mersennus is on his waye hither;[5] so that I knowe not whither to write to him; but I doute not but more handes with demonstrations might be procured if you desire it; I heare some of them wonder you would not publish a demonstration of it your self: I tolde M^r: Hobbes I assured my self you coulde demonstrate it. & thought it more obuious than I perceiue they thinke it./ I am glad you are publishing of Appolonius & Diophantes[6] though if Goleas meane to publish his Appolonius, I could be content you forbore yours, his being the perfecter copie,[7] & he hauing bin at some cost about it, though I doute we shall be loosers by it; for I beleeue your notes would abundantlie recompence the defect of your coppie./ I haue not yet seen Bullialdes astronomie,[8] but m^r: Hobbes thinkes he hath not much aduanced oure knoweledge. Mydorgius hath pubblished nothing of late[9] nor anie of the rest more that I heare of./ I hope you proceed with your analytickes & that er longe you will publish it; for I expect not so much from anie man in that kinde as from your self. I suppose you will one daye aduance [>allso] the doctrine of conickes & of other difficulties [*page torn* of g]eometrie./ I knowe not [>when] wee remoue, nor whither; but when [*page torn* or wh]ere soeuer I remaine

---

[1] Letter 35.

[2] Letter 34 (on Jungius see Letter 16, n. 5; on Hobbes's proposition see Letter 34, n. 3).

[3] Hobbes's demonstration (BL, MS Add. 4278, fo. 200^r) was printed in Pell, *Controversiae pars prima*, pp. 50–1.

[4] On Fermat see Letter 4a, n. 7, and Letter 4b, n. 1; Pell did not in the end obtain a demonstration of his theorem from Fermat.

[5] See Letter 22, n. 4.

[6] See Letter 23, n. 2, and Letter 14, n. 13.

[7] See Letter 23, n. 9, and Letter 35, at n. 10.

[8] See Letter 35, n. 5.

[9] Mydorge published nothing after his *Prodromi*: see Letter 34, n. 1.

Your assured friend to serue you
Charles Cauendysshe

June 27 olde style 1645

[*addressed:*] To my verie worthie friend M[r]: John Pell Publicke Professor of the Mathematicks at Amsterdam. In t'oude conuoy op de Zee dijke in Amsterdame.

[*added to address in a different hand:*] Willm Boswell.[10]

[*cover annotated by Birch:*] 27 June 1645

---

[10]This letter appears to have been sent via Sir William Boswell, who was resident in The Hague (see Letter 19, n. 13).

Letter 37

27 July/6 August 1645

Cavendish to Pell, from Paris

BL, MS Add. 4278, fos. 210–11 (original)
van Maanen C15

Worthie S^r:

it is so longe since I hearde from you & our staie heer is not likelie to be longe, [>that] I nowe write, least heereafter I finde not so fit an oportunitie./ Mersennus is not yet come hither but wee expect him daielie:[1] M^r: Hobbes is gone to Roen[2] & I doute will not returne whylst wee are heere; he much commends Taricell,[3] but till Mersennus come I doute I shall not get him./ I haue hearde nothing of the Capuchin Reieta,[4] which I wonder at./ I hope you haue by nowe, printed your Diophantus,[5] & as you write, Apollonius Pergeus is likelie betweene Goleas & you to be printed;[6] which I should allso be glad of, but extreamelie more glad of your Analyticall worck; your grand Problem[7] to begin with, & soone after the rest./ I haue [>not] seene Gassendus nor Mydorge this longe time;/ Mydorge had of his owne accorde demonstrated your Fundamentall [problem *deleted*] Theorem,[8] but he supposed the angle DAC = ADB which I tolde him of; & desired he would doe me the fauoure as others had done, which he willinglie consented to; but I haue not since hearde of him. I confess it is more difficult to proue than I thought; yet not extreame, since I can easilie (if I mistake not) trace it Analyticallie to

---

[1] See Letter 22, n. 4.

[2] Two days later Hobbes wrote to Edmund Waller, from Rouen: 'I came hether to see my lord of Deuonshire' (*sc.* William Cavendish, third Earl of Devonshire, Hobbes's former pupil and employer): Hobbes, *Correspondence*, i, p. 124 ([29 July/] 8 Aug. 1645).

[3] See Letter 35, n. 17.

[4] See Letter 33, at n. 4, and Letter 35, at n. 4.

[5] See Letter 14, n. 13.

[6] Cavendish refers to the concluding sentence of Pell's account of his discussion with Golius in Letter 35.

[7] A reference to a problem that Pell took up from Oughtred's *Clavis*: given any two of the quantities a, b, $a+b$, $a-b$, ab, a/b, a^2+b^2, a^2-b^2, find the other six. In 1637 Pell drew up a large table containing all the solutions and called it an 'analytic Progymnasma' (an algebraic preparatory exercise); he described the work to Hartlib as the 'greate Probleme' (see BL, MS Add. 4415, fos. 19–21 and MS Add. 4398, fo. 170). It later appeared as the first problem both in Rahn's *Teutsche algebra* (1659) and in Rahn and Pell's *Introduction to algebra* (1668); from Cavendish's words we may assume that it would also have provided the opening to the 'Analyticall worck' that Pell was planning in the 1640s.

[8] Pell did eventually receive a demonstration of his theorem by Mydorge (on whom see Letter 34, n. 1), which he published at the end of his *Controversiae pars prima* (pp. 90–4).

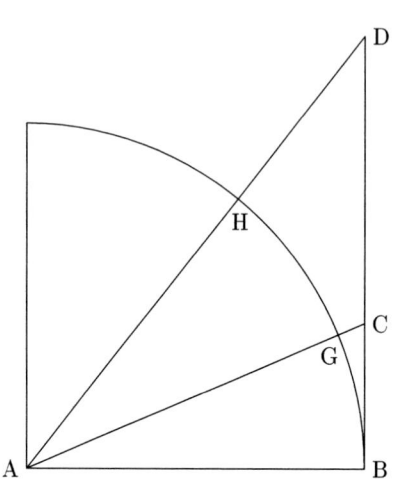

the 1, of the 6th: of Euclide; as in this paper[9] which I send to auoide writing it ouer againe in this letter, you maye see; I finde the regress harde;[10] & therefore desire your information therin./ S^r. I rest

> Your assured friend & seruant
> Charles Cauendysshe

Paris. Julie 27, olde stile

[*addressed:*] To my verie worthie Friend M^r: John Pell, Publicke Professor of the Mathematicks at Amsterdam, In T'oude conuoye op de zee dijke in Amsterdam.

[*added to address in another hand:*] Amsterdam.

[*annotated by Pell:*] Received Aug 9/19

[*cover annotated by Birch:*] 27 July 1645

---

[9] Cavendish's enclosed paper, with diagram and mathematical demonstration, is BL, MS Add. 4278, fo. 212; Pell recapitulates its argument in Letter 44.

[10] In his enclosed paper Cavendish argued from Pell's theorem to Euclid I, Proposition 6 (see Letter 44), a piece of mathematical 'analysis'. To give a satisfactory proof of Pell's theorem, however, he needed to provide the 'synthesis' by working in the other direction, from Euclid I.6 to Pell's theorem. This was the 'regress' that he found hard.

Letter 38

2/12 August 1645

Pell to Cavendish, from Amsterdam

BL MS Add. 4280, fo. 113 (draft); fo. 114^r (authorial fair copy of the mathematical demonstration)

van Maanen P12

Right Honourable

Since my last of May 9/19[1] I have received 2 of yours dated May 1/11 & June 27[2] [>I sent] your letter to Dr Jungius inclosed in a letter to him[3] wherein I offered to convey his [letters *deleted*] to you & desired his judgement of my paper. To the former He answered thus Nobilissimi domini Cavendissij literis humanissimis respondebo quamprimùm de Phoranomicis cogitare licuerit, inanes enim nolui mittere.[4] To the other, he answeres like one y^t approved it very well, It is Nova, ingeniosa, solida & necessariae firmitatis [>ratio][5] Of [>my] fundamentall praecept he sayes [>ego quidem] novi certum esse & in magni viri inventis fundatum, sed adversarius tuus Calculator potius trigonometricus est quàm Geometra.[6] And yet for all this [>he addes] a passage or two y^t makes me suspect y^t he does not fully reach ye bottome of my [>whole] demonstration.[7] Which will cost me another letter to draw him to explaine himselfe. In the meane time because he desired to see Longomontanus his booke I have sent it him [>with] his *first* reply. I say *first*, for He hath sent me now a second reply

---

[1] Letter 35.

[2] Letters 34, 36.

[3] For Cavendish's letter to Jungius (on whom see Letter 16, n. 5), see Letter 34, n. 7; Pell's letter, dated 15 Mar., is SUBH, MS Pe. 65, fos. 80–1.

[4] 'I shall reply to the very kind letter from the most noble Mr Cavendish as soon as I have a chance to think about the theory of local motion, for I did not want to send him something insubstantial.' Jungius's letter, dated [27 May/] 6 June 1645, is BL, MS Add. 4279, fo. 75.

[5] 'A new, ingenious and solid argument, of the necessary strength'. Pell has extracted these terms from the following sentence: 'Nova & ingeniosa est tua ... ratio, & quamvis αὐτάρκειαν istam non habet, quam ea quae per extractiones radicum fiunt, solida tamen est, & necessariae firmitatis' ('Your argument is new and ingenious, and although it does not have the self-sufficiency which those arguments that proceed by the extraction of roots have, nevertheless it is solid, and of the necessary strength': ibid., fo. 75^r).

[6] 'I, at any rate, knew that it was certain, and that it was founded on the inventions of a great man; but your opponent is rather a trigonometrical calculator than a geometer.'

[7] See the quotation in n. 5, above.

of 2 sheetes of paper printed Haunick 8 Calend Junij [>1645][8] upon occasion of y^t letter which I shewed you when you were heere, As soone as he had gotten a coppy of it he threatned it should be answered and now it is done, after [>his] fashion, But it is penned as by a third person & no name to it. This defender now sees, by ye helpe of my letter, that I rely not upon any of those things y^t he excepts against [>And therefore buckles himselfe to examine this new way of mine.] My fundamentall precept (which Longomontanus said was Geometricè indemonstrabile)[9] this defender begins to feare yt it is demonstrable Pulchra certè methodus, saith he, & à Pellio amplius, si possit, Geometricè, demonstratu digna &c[10] And in another place, Interim ea laudanda est, praecipuè modò Geometricè demonstrabilis; nam per ipsam, Canonem Trigonometricam construi posse video;[11] And neere the end Experiemur, si placet vel ad ipsius Pellij methodum, quantus tangens arcûs dupli &c[12]

But all ye wisedome in Regia danorum Academia Hauniensi[13] will not serve to demonstrate that poore Theoreme, and therefore, if y^{ei} may be procured, I desire to have [a few *deleted*] more demonstrations from severall hands [(& Sr Charles Cavendish for one) *deleted*] that these selfe-conceited Danes may be so much ye more confounded & cured of their oultrecuidance,[14] when y^{ei} shall see y^t so many single men have done it so variously [>whilest] all they put together can not doe it once. I could wish also to have a line or two (to this purpose, Longomontani Peripheriam majorem esse quàm oportuit, idque solide à Pellio demonstratum esse scimus)[15] subscribed by such professors & students as by demonstrating my fundamentall Theoreme can shew that they know what they say & doe not approve or dislike for company. I am afraid Sr Charles Cavendish will forget to send me his [>owne] approbation & demonstration, if I doe not expressly crave [>that favour] of him. As for my selfe, I smile when I see those of Copenhaven or Paris make a question whether I can demonstrate it my selfe: as if I were able to seeke it & find it & now knew not how to demonstrate it

---

[8] The original 'booke' was Longomontanus, *Rotundi in plano*: see Letter 14, n. 15. The 'first reply' was Longomontanus, Ἐλέγξεως; the second was the anonymous pamphlet (written by Longomontanus), *Controversia inter Christianum Longomontanum & Ioannem Pellium, de vera circuli mensura, ubi defectus canonis trigonometrici, sub initium ejusdem, ostenditur* (Copenhagen ['Haunick' here], 1645: reprinted in Pell, *Controversiae pars prima*, pp. 63–73). The second reply consisted of two quarto gatherings (A^4–B^4).

[9] 'Incapable of geometrical demonstration'.

[10] 'A beautiful method, indeed ... and one that deserves to be demonstrated by Pell more fully, geometrically, if possible...' (Longomontanus, *Controversia*, sig. A3^v).

[11] 'Meanwhile that method deserves praise, especially if it happens to be geometrically demonstrable; for, by means of it, I see that it will be possible to construct a trigonometrical table' (ibid., sig. B2^v).

[12] 'Let us try to find out, if we may, by Pell's method, how much the tangent of the double arc...' (ibid., sig. B3^r).

[13] 'The Royal Danish Academy, Copenhagen' (*sc.* Copenhagen University).

[14] 'Presumptuousness'.

[15] 'We know that Longomontanus's circumference exceeds the required length, and that Pell has solidly demonstrated that that is the case.'

but it had [>not] beene so great credit for [>those] of Paris to demonstrate it, if they had seene mine first, whereas now it proves y^m Capable to be judges of our Controversy [>I had resolved to] publish mine in ye method y^t I sought it [>unknowen] which is a step higher than barely to demonstrate a truth knowne. And [>if so er you come to see a] strange piece of worke y^t I have in hand de demonstratione[16] you will beleeve y^t I [>could] demonstrate y^t five hundred wayes, if neede be. So that I desire not their demonstrations, to helpe mee, but So much ye more to [>make knowen to the world y^{eir}] abilities & ye weaknesse of ye Haffnians.

This [wise *deleted* >learned] Haffnian defender further acknowledges that if the Tangent of 0 gr 42 3/16 be *equall* to 0.01227.25 then it is manifest y^t Longomontanus hath made ye part greater then the whole. [>A man in his wits would have added that the absurdity is] much greater, if y^t tangent be lesse y^n 122725 as I demonstrate it to be[17] But that he [>sayes not & strives to put face as] y^t without a Canon of tangents [>there is no finding of that] unlucky number 122725. So senselessely talke [>these men]

I have therefore [enlarged *deleted* >opened my former] demonstration that these blundering Danes may see ye force of it. I hope I shall within a few dayes get it printed: To which I would adjoine as many demonstrations of other men as I can get.[18] But [>if] your French men [>write not] their names [more] distinctly my [>Dutch] printer will miscall y^m. [>We reade] Presonenus de Roberval & De Carcavy [>Another] y^t makes me beleeve he knew y^m both, in Paris, sayes it must be Personerius & Petrus Carcavy.[19] I thank you for [communicating *deleted* >both] Mr Hobbes his demonstrations[20] Of which heereafter I shall say more. Onely thus much now That theoreme which is one of ye legs upon wch his Conicall demonstration stands (Medium Arithmeticum Geometrico majus est) [>may] as [>you say,] easily [>be] demonstrated [>yea variously though we] take it more narrowly not onely saying *it is greater* but shewing *how much greater*

---

[16]'About demonstration'. No such work was published, nor does any draft of it survive.

[17] $0° 42\frac{3}{16}' = \frac{1}{64} \times 45°$, and $\tan 0° 42\frac{3}{16}' = 0.012\,272\,46$. This value of $\tan 0° 42\frac{3}{16}'$ is crucial to Pell's refutation, as it enables him to show that, according to the reasoning of Longomontanus, a circumscribing polygon with 256 sides would be shorter than the circle itself – an obvious contradiction. Note that Pell's double-angle formula gives $\tan 2\theta$ in terms of $\tan \theta$. Pell had to invert the formula so that starting from $\tan 45° = 1$ he could find $\tan 22° 30'$, $\tan 11° 15'$, etc., by solving a quadratic equation at each stage.

[18]This plan was eventually fulfilled with the publication of Pell's *Controversiae pars prima*.

[19]On Gilles Personne de Roberval see Letter 19, n. 7; his demonstration of Pell's theorem was printed in Pell, *Controversiae pars prima*, pp. 47–9. Roberval latinized his name as 'Personerius'. The mathematician Pierre de Carcavi (*c.*1600–84), originally from Toulouse but resident in Paris since 1636, was a friend of Roberval and Fermat: his demonstration of Pell's theorem was printed in *ibid.*, p. 51.

[20]See Letter 34, n. 3, and Letter 36, n. 3.

For[21]

Quadratum medij Arithmetici est aequale ad quadrato medij Geometrici & quadrato semidifferentiae extremorum

A.b.C. A,C Extrema
A. β.C. b medium Arithmeticum β medium Geometricum
id est, $b = \frac{A+C}{2}$. $\beta = \frac{AC}{2}$
Utrum majus, b an β?

Sit $A = C + 2d$

$\therefore A + C = 2C + 2d$ $\quad|\quad \therefore AC = CC + 2Cd \quad|\quad \therefore 2d = A - C$
$\therefore \dfrac{A + C}{2} = C + d$ $\quad|\quad \therefore \beta\beta = CC + 2Cd \quad|\quad \therefore d = \dfrac{A - C}{2}$
$\therefore b = C + d$ $\qquad\qquad\qquad\qquad\qquad\qquad \therefore A > C$
$\therefore bb = CC + 2d + dd$ $\qquad\qquad\qquad\qquad \therefore d > 0$
$\qquad\qquad\qquad\qquad\qquad\qquad\qquad\qquad \therefore dd > 0$
$\qquad\qquad\qquad\qquad\qquad\qquad\qquad\qquad \therefore \beta\beta + dd > \beta\beta$

$\therefore bb = \beta\beta + dd$

$\therefore bb > \beta\beta$
$\therefore b > \beta$

I make no question but Monsieur Roberval[22] & others of Mr Hobbes his acquaintance have received Langreus first Map of y^e Moone[23] and therefore I shall not neede to say any more of it, [>My paper admonishes me to take my leave, but your favours oblige me to continue]

Your Honours most humble servant
[John Pell *deleted*]

Aug 2/12 1645

---

[21] The demonstration which follows begins: 'The square of the arithmetic mean is equal to the square of the geometric mean plus the square of half the difference between the extremes. A and C are the extremes; b is the arithmetic mean; β is the geometric mean ... which is greater, b or β?' When Pell writes '$\frac{AC}{2}$', he means the square root of AC. This is not Pell's usual notation; presumably he used it here to emphasize the analogy between the arithmetic mean and the geometric mean. He goes on to give a simple algebraic proof of the well-known fact that the arithmetic mean of two positive numbers is greater than the geometric mean. Note his use of capital letters for the fixed quantities A and C but lower case for b, d, and β.

[22] See Letter 19, n. 7.

[23] Michael Florent van Langren (c.1600–1675), a member of a Flemish globe- and map-making family, 'Royal Mathematician and Cosmographer' to Philip IV, had hit on the idea of using the phases of the moon as a celestial clock in order to solve the problem of determining longitude. By 1643 he had prepared 30 detailed drawings of different phases of the moon. When he heard of similar selenographic projects by other scientists (Hevelius and Caramuel), he printed his first map as a broadsheet in March 1645 (see E. A. Whitaker, 'Selenography in the Seventeenth Century', in R. Taton and C. Wilson, eds., *Planetary Astronomy from the Renaissance to the Rise of Astrophysics: Part A: Tycho Brahe to Newton* (Cambridge, 1989), pp. 119–43; here pp. 127–8).

Letter 39

20/30 September 1645

Cavendish to Pell, from Paris

BL, MS Add. 4278, fos. 213–14 (original)
van Maanen C16

Worthie S[r]
   it is longe since I hearde from you, as I thinke, & I perceiue not by your letters that you euer receiued anie of myne, though I haue writ diuers; I writ to Doctor Jungius longe since, which letter I inclosed in one of myne to you,[1] desiring you would conueie it to him, but I doute those letters neuer came to your handes; I hope the last & this, by S[r] Williame Boswels[2] fauoure will./ wee expect shortelie Gassendes his replie to a Jesuit who opposed Gallileos doctrine of motion;[3] Gassendes expresses things verie cleerelie; I am much beholding to him for his visits & freedom of discourse with me; yet I decline not from my opinion of Mons[r]: des Cartes abilities;/ I haue not seene Mydorge[4] a longe whyle, I meruaile he hath not yet giuen me a demonstration of your proposition; I beleeue he is nowe busie at his vintage. I hope you haue receiued M[r]: Hobs his demonstration, & an others[5] which I sent you longe since./ Mersennus is returned[6] & was with me; He hath brought the greatest rarities he could finde in Italie & doutles he was a diligent Inquisitor;/ Tauricells worckes all,[7] as I thinke, which are rare heere;/

---

  [1] See Letter 34, n. 7.

  [2] See Letter 19, n. 13.

  [3] P. Gassendi, *De proportione qua gravia accidentia accelerantur epistolae tres, quibus ad totidem epistolas R. P. Cazrei ... respondetur* (Paris, 1646), a response to criticisms of Galileo's theory of gravitational acceleration by Pierre Le Cazre SJ (1589–1661). On this dispute see P. Galluzzi, 'Gassendi and *l'affaire Galilée* of the Laws of Motion', in J. Renn, ed., *Galileo in Context* (Cambridge, 2001), pp. 239–75, esp. pp. 242–50, and C. R. Palmerino, 'Two Jesuit Responses to Galileo's Science of Motion: Honoré Fabri and Pierre Le Cazre', in M. Feingold, ed., *The New Science and Jesuit Science: Seventeenth-Century Perspectives* (Dordrecht, 2003), pp. 187–227, esp. pp. 204–14.

  [4] See Letter 34, n. 1.

  [5] The 'demonstration' referred to here is apparently Hobbes's demonstration of Pell's theorem (BL, MS Add. 4278, fo. 200[r], printed in Pell, *Controversiae pars prima*, pp. 50–1). The phrase 'an others' may refer to the demonstration of Pell's theorem by de Carcavi (see Letter 38, n. 19); it is less likely to refer to the one by Roberval, as that is specified later in this letter.

  [6] See Letter 22, n. 4.

  [7] Cavendish refers to Torricelli's *Opera geometrica*, his only published work at this time: see Letter 35, n. 17, and the enclosure to Letter 46.

A short treatise in manuscript de maximis et minimis;[8] and as I remember the 5th & 6th bookes of Apollonius Pergeus conickes, translated by Maurolicus in to Latin.[9] & this was all he spoke of to me./ If you woulde haue me to procure more demonstrations of your proposition I will endeuour it; but if you haue receiued those I sent you with that of Mons[r]: Roberualls[10] & your owne; I beleeue they are sufficient to stoppe the waywarde Danes mouthe. I hope to heare what oure admirable Doctor Jungius doeth; whether he publish anie thing yet. as allso the like of oure learned Tassius.[11] I am not less inquisitiue what you publish; I hope your Diophantus;[12] but especiallie your owne analitickes; which I confess I much thirst after; I hope at the least your grand problem[13] to be doeing with all till the other come./ And so wishing you all happiness I remaine

your assured friend & seruant
Charles Cauendysshe.

Paris Septemb: 20 olde stile. 1645.

[*addressed:*] To my verie worthie friend M[r]: John Pell publick Professor of the Mathematickes at Amsterdam. In T'oude Conuoy op de Zee dijke in Amsterdam

[*annotated by Pell:*] Received Octob 1/11

[*cover annotated by Birch:*] 20 Sept. 1645.

---

[8] Unidentified. This is unlikely to refer to the brief treatise by Fermat, 'Methodus ad disquirendam maximam et minimam', of which Pell's papers contain both a scribal copy (BL, MS Add. 4398, fos. 151–152[r]) and a copy in Pell's hand (BL, MS Add. 4407, fo. 105): Mersenne had had a copy of this text since at least as early as Jan. 1638 (see Mahoney, *Mathematical Career of Fermat*, p. 416). Possibly the reference is to Fermat's later text 'Ad methodum de maxima et minima appendix', of which two copies survive among Cavendish's own papers (BL, MS Harl. 6083, fos. 116[v], 330[v]). However, if (as Cavendish seems to imply here) Mersenne had acquired the 'short treatise' in Italy, it was probably not by Fermat (who lived in Toulouse – which was not on Mersenne's itinerary – and was, in any case, in direct epistolary contact with Mersenne). No likely text by an Italian mathematician on this subject (prior to Vincenzo Viviani's *De maximis et minimis* (Florence, 1659), and Michelangelo Ricci's *Exercitatio geometrica de maximis et minimis* (Rome, 1668)) has been identified. Prof. Feingold has suggested that Mersenne might have obtained some early version of part of Richard White's *Hemisphaerium dissectum: opus geometricum in quo obiter tractatur de maximis inscriptibilibus, & minimis circumscribentibus* (Rome, 1648).

[9] The mathematician Francesco Maurolico (1494–1575) had lectured on geometry in his home town, Messina. Among his manuscripts was a 'restitution' of the missing books 5 and 6 of Apollonius's *Conics*: this was eventually published as part of his *Emendatio et restitutio Conicorum Apollonii Pergaei* (Messina, 1654). It was widely believed, both before this publication and for some time thereafter, that Maurolico had translated the missing books into Latin; in fact he had reconstructed or invented them himself. (See G. Rossi, *Francesco Maurolico e il risorgimento filosofico e scientifico in Italia nel secolo XVI* (Messina, 1888), pp. 107–9, and Maierù, *La teoria e l'uso delle coniche*, pp. 83–97.)

[10] See Letter 38, n. 19; the phrase 'those I sent you' probably stands for the demonstrations by de Carcavi, Hobbes, and Le Pailleur (the last of which was printed in Pell, *Controversiae pars prima*, p. 54).

[11] On Jungius and Tassius see Letter 16, nn. 4, 5.

[12] See Letter 14, n. 13.   [13] See Letter 37, n. 7.

Letter 40

[27 September/7 October 1645]

Cavendish to Pell, from Paris

BL, MS Add. 4278, fos. 215–16 (original)
van Maanen C17

Worthie S^r.

I receiued yours of the 2/12 of Aug:[1] the other daie; the next daye I went to Pere Mersenne to desire him to procure more demonstrations of your Theorem; but I found him vnwilling to vndertake it, pretending that Roberuals demonstration[2] onelie were sufficient; & so tolde me he would write a letter to you to that effect, but I had rather he would haue procured some more demonstrations of your Theorem; but I could not force him to it; so I doute I shall procure you no more./ I doute not but you knowe Reietas booke is nowe finished & printed at Antwerp;[3] & wee expect to receiue one shortelie. Mersennus sent me this inclosed, this morning, wherein I perceiue he makes some quaeres to you.[4] I can not but in all my letters mention my desire of seeing something published by you, which I hope I [>shall] shortelie see. I must not forget to giue you particulare thankes for your neate & excellent demonstration that the Arithm: Medium etc:;/[5] I haue no more to trouble you with at this time but wishing you all happiness remaine

> Your assured friend to serue you
> Charles Cauendysshe

[*postscript:*] If wee remoue from hence directing your packet to M^r: Hobbes he will conuei it to me

---

This letter, which is undated, is dated here by reference to the letter from Mersenne (see n. 4), which is dated [27 Sept./] 7 Oct. 45 and annotated by Pell 'Accepi 5/15 Octob.'

[1] Letter 38.

[2] See Letter 38, n. 19.

[3] See Letter 33, n. 4.

[4] Mersenne's letter to Pell is BL, MS Add. 4278, fos. 217–18; this letter is not included in *MC*, but an extract from it is printed in van Maanen, 'The Refutation', pp. 129–30. Mersenne's 'quaere' is as follows: 'I also hope for a short refutation from your pen of another quadrature that someone here with us claims to be true. He wants the side of the square, inscribed in a circle, to be to a quarter of the circle as 9 is to 10' ('vt sperem etiam à tuo calamo breuem refutationem alterius quadraturae, quam esse veram aliquis apud nos contendit. Nempe vult latus quadrati circulo inscripti esse ad quadrantem circuli, ut 9 ad 10': fo. 217^r). Pell's reply to Mersenne, dated 8/18 Oct. 1645, is printed in *MC* xiii, pp. 497–9 (from the letter as received by Mersenne; Pell's fair copy is BL, MS Add. 4280, fo. 220, and his refutation of this quadrature – omitted in *MC* – is fo. 221^r).

[5] See Pell's demonstration in Letter 38, which was sent in response to the demonstration by Hobbes which accompanied Letter 34.

[*addressed:*] To my verie worthie friend M^r: John Pell, Publicke Professor of the Mathematicks at Amsterdam. In T'oude Conuoy op de zee dijke in Amsterdam.

[*added to address in another hand:*] Amsterdam

[*cover annotated by Birch:*] Sept. 1645

Letter 41

10/20 October 1645

Pell to Cavendish, from Amsterdam

BL MS Add. 4280, fo. 113^v (draft)
van Maanen P13

Right Honourable

    Since my last of Aug 2/12 I have receved 3 of yours. [>The 2 former] bearing date July 27. Sept 20.[1] By this last without date which came [>to my hands] 5 dayes agoe I perceive y^t you have received my last and y^t you like my demonstration [>de medio Arith &c][2] I confesse I like it well my selfe, who am none of y^e easiest to please: and I doubt not but you & Mr Hobbes will see y^e reason of every stroke & tittle in it. In y^t first of your 3 letters [>I saw] you had beene seeking y^e demonstration of my new Theoreme 2RRT/RR − TT = D,[3] but it seemes in an unlucky minute, for I have seene harder things of your doing; but in this, your bowle met with a rub as soone as it was out of your hand & turned it quite out of y^e way So y^t if a man write RR + TT or RRRR or TTTT or what he will in y^e place of RR − TT, your demonstration will prove y^m all It is [>so] universall y^t it will demonstrate ye true & ye false all alike. I am sorry y^t I have [>yet] no time to give you [>a compleate preservative] y^t you may never fall into ye like againe. But [>if you thinke good to attempt this againe,] make a rectangled triangle & bisect one of his acutes, [>which done,] you can not lightly draw [>out of any point a [>fift line] parallel or perpendicular which will not serve to helpe you to a medium demonstrationis,[4] especially now y^t you have seene 4 [>demonstrations] allready I hope my answere to Longomontanus shall shew varieties enough to make ye world see, y^t I needed no man to helpe me demonstrate my owne Theoreme. I have much to say of the whole art of demonstration & [>hope one day] to spoile their sport that thinke it a great matter to adde a second or third demonstration to a Theoreme.

    I thanke you for your relation of Mersennus his rarities,[5] I hope heereafter to have it more particularly from himselfe.

---

    [1] Letters 37, 39, and (the third, undated letter) 40.

    [2] See Letter 40, n. 5.

    [3] This is Pell's double-angle tangent theorem, now written as an algebraic formula (with $R/T = \tan\theta$ and $D/T = \tan 2\theta$). Pell responds here to the attempted demonstration sent by Cavendish (see Letter 37, n. 9); he explicates Cavendish's error more thoroughly in Letter 44.

    [4] 'Means of demonstration'.

    [5] See Letter 39, at nn. 7–9.

From Tassius I have received an approbation of my paper against Longomontanus, which you will shortly see in print.[6] I wrote againe to Jungius [>&] he hath answered me largely enough[7] so y^t I [>am more assured of his meaning than before] but I shall write to him againe to[day *deleted* >morrow] & let him know y^t I am not yet pleased & y^t there is more artifice in [>that paper of mine] y^n he is aware of. [>(In his answere he prayes me to tell him where you are & how you doe) But of my] answere to Longomontanus my readers [should *deleted* >will] be able to learne something more y^n that Longomontanus is an ignorant testy &c [>But our quarrell being tripartite I shall] meddle with y^t first part [>onely] in this booke & let ye other [>2] alone till 2 or 3 yeares hence.[8] That [>in the meane time] I may finde leisure to publish some of those observations & rules, which hitherto have enabled me to runne through, where some others sticke fast.

I shall trouble your Honour no further for y^s time, onely praying you to thinke upon this enclosed memoriall thereby further obliging

your Honours most humble servant
IP

Octob. 10/20 1645
Amsterdam

[*note added at foot of text:*]
1. To cause these inclosed to be delivered to Mersennus.[9]
2. To prescribe me a forme of directing letters to Mr Hobbes[10]
3. To send me Mr Robervalls name & Mr Carcavys name & title distinctly written.[11]

---

[6] The judgement by Tassius (on whom see Letter 16, n. 4) was printed in Pell, *Controversiae pars prima*, p. 55.

[7] Pell's fair copy of this letter from Jungius (on whom see Letter 16, n. 5), dated [5/] 15 Sept. 1645, is in BL, MS Add. 4279, fo. 87 (with another, very faint, copy on fos. 76–8); the original draft is in SUBH, MS Pe. 65, fos. 13–20.

[8] Pell's published response was thus entitled 'the first part', *Controversiae pars prima*; however, no further parts were published.

[9] Pell's letter to Mersenne of 8/18 Oct., which was accompanied by his refutation of the proposed quadrature (see Letter 40, n. 4) and some other 'papers' (see Letter 42).

[10] See Letter 43, n. 1.

[11] See Letter 38, n. 19, and Letter 43, n. 1.

Letter 42

1/11 November 1645

Cavendish to Pell, from Paris

BL MS Add. 4278, fos. 223–4 (original)
van Maanen C18

Worthie S^r

I giue you manie thankes for your letter & telling [>me] of my error;[1] but I should haue [>bin] glad if your leisure had permitted that you would haue shewed me where I went out of the waye; for though I beleeue I haue because you tell me so, yet I perceiue not wherein & therefore am bolde to trouble you againe with this inclosed paper./ I am glad to vnderstand by your letter to Mersennus that vieta is nowe printing[2] but should be gladder to heare that some of yours were printing. Mersennus hath your letter & papers to him & I hope he will doe that at your request which I could not get him to doe, which is to get you more testificates of the truth of your theorem.[3] I haue reade some of Tauricell & I esteeme him one of the most admirable men that euer I reade, so that I haue sent for his booke to Florence[4] for there is none heere to be bought. wee receiued Reietas booke yesterdaie & I haue looked ouer his manner of making perspectiue glasses[5] & I beleeue it is goode, & so I must for him, for he demonstrates nothing, nor medles with the speculatiue parte of opticks; I haue not had time to reade the rest, but I doute not but you haue the booke, & therefore no more of it; onelie this much, he infinitelie commends Gerardus Gutschonius a Doctor of phisick in the lowe countries, for a rare Mathematician, & his inuention for describing hyperboles fitting to the seuerall lengths of tubes

---

[1] Letter 41.

[2] See Letter 40, n. 4; the postscript to Pell's letter to Mersenne included the information that 'They say that Viète is in the press, at the Elzeviers, in Leiden' ('Ajunt Vietam apud Elzevirios, Leydae, sub praelo esse': *MC* xiii, p. 498). This refers to Frans van Schooten's edition of Viète, *Opera mathematica* (Leiden, 1646).

[3] Pell's letter to Mersenne enclosed letters to Torricelli and Glorioso (requesting their demonstrations of the truth of his theorem), which he asked Mersenne to forward (see *MC* xiii, pp. 497–8; cf. Letter 35, n. 17). The other 'papers' enclosed by Pell included his refutation of another attempt to square the circle (see Letter 40, n. 4).

[4] See Letter 39, n. 7.

[5] See Letter 33, n. 4.

of perspectiues;⁶ if he haue printed anie thing I praye you giue me notice of it by your next; he allso mentions that the quadrature of the circle is allso published, but names not the author; but when I was at Antwerp I hearde there was a Jesuit [>in those partes] publishing such a booke;⁷ if there be anie such I desire you will giue me notice of it by your next. And so wishing you all happiness I remayne,

 Your assured friend to serue you
 Charles Cauendysshe

Paris Nouemb: 1/11 1645.

[*postscript:*] M^r: Hobbes intends to publish as soon as he can a treatise of opticks; he hath done half of it, & M^r: Petit hath writ it faire;⁸ It is in english at my brothers request.

[*second postscript:*] M^r: Hobbes promised me to giue me all the names of Roberuall & Carcaruin & theyre titles;⁹ but not hauing yet done it, & the poste goeing awaye, I must doe it by the next./ My seruice to Doctor Jungius when you write to him; I hope he will shortelie publish his Apollonius Saxsonicus,¹⁰ Logick,¹¹ &

---

⁶Gerard van Gutschoven (1615–1668) had worked in his youth as an assistant to Descartes, before returning to his native town of Louvain to take a medical degree (in 1635). He became Professor of Mathematics (1640), and later Professor of Anatomy, Surgery, and Botany (1659) at Louvain University. His first publication, *Arithmeticae practicae regulae brevissimae*, appeared in 1654. Rheita's praise of him (as 'ingeniosissimus'), and of his 'machina' for making hyperbolic lenses, is in *Oculus Enoch et Eliae ... pars prima*, pp. 344, 345–7.

⁷Cavendish may have been thinking either (possibly) of Jean Charles della Faille, or (more probably) of Grégoire de Saint-Vincent: see Letter 44, n. 15.

⁸Hobbes's 'A Minute or First Draught of the Optiques'(BL, MS Harl. 3360): the copyist of this manuscript was William Petty (1623–87), the future scientist, economist, and Fellow of the Royal Society, who had travelled to Paris from the Netherlands in Sept. or Oct. 1645 with an introduction to Hobbes from Pell. Hobbes's treatise was not published, but a Latin adaptation of the second part of it was later published as the first part of his *Elementorum philosophiae sectio secunda de homine* (London, 1658).

⁹See Letter 38, at n. 19, and Letter 42, n. 1.

¹⁰A reconstruction of Apollonius's lost work *De locis planis*. The surviving MSS of Joachim Jungius (on whom see Letter 16, n. 5) include a copy of this major work in two volumes (of 503 and 520 fos.), prepared for the printer by his student and collaborator, Woldeck Weland (1614–1641): SUBH, MSS Suppl. 1 and Suppl. 2. This has been printed in a modern edition: B. Elsner, ed., *"Apollonius Saxonicus": die Restitution eines verlorenen Werkes des Apollonius von Perga durch Joachim Jungius, Woldeck Weland und Johannes Müller* (Göttingen, 1988). The titlepage of the manuscript (MS Suppl. 1, fo. 10^r) describes it as 'Apollonius Saxonicus seu Exsuscitata Apollonii Pergaei Analytica'. This title played on those of earlier works by Viète (*Apollonius Gallicus, seu exsuscitata Apollonis Pergaei περὶ ἐπαφῶν geometria* (Paris, 1600) – a reconstruction of the *De tactionibus*) and Snel (*Apollonius Batavus seu exsuscitata Apollonii Pergaei περὶ διωρισμένης τομῆς geometria* (Leiden, 1608) – a reconstruction of the *De sectione determinata*).

¹¹After the publication of his *Logica hamburgensis* in 1638 Jungius had written two additional treatises, 'Logica didactica' and 'Logicae hamburgensis praestantia' (SUBH, MSS Wo 30, 31): these have now been printed in J. Jungius, *Logicae hamburgensis additamenta*, ed. W. Risse (Göttingen, 1977), pp. 98–171 and 54–93, respectively.

Phoranomicks;[12] douteless he is an incomparable man.

[*addressed:*] To my verie worthie friend M^r: John Pell publicke Professor of the Mathematicks at Amsterdam In T'oude conuoy op de zee dijke in Amsterdam giue this

[*annotated by Pell:*] Received Nov 8/18

[*cover annotated by Birch:*] 1/11 Nov. 1645

---

[12]See Letter 30, n. 4.

Letter 43

8/18 November 1645

Cavendish to Pell, from Paris

BL, MS Add. 4278, fos. 227–9 (original)
van Maanen C19

Worthie S^r:

I send you heer inclosed the best satisfaction that I could get to your demandes./[1] I heare that Mersennus will indeuoure to get you more learned mens aprobation to your Theorem[2] which I am glad of./ He hath nowe Reietas booke but I haue not seene him since he had it. M^r: Hobbes likes it not, & I suppose the worse, because he commends hyperbolick glasses, aboue sphaericall, for perspectiues./[3] I longe to heare that you publish something, which I hope I shall shortelie doe. I haue nothing more to trouble you with at this time. And so wishing you all happiness I remaine

    Your assured friend to serue you
    Charles Cauendysshe

Paris Nouemb: 8/18. 1645

[*addressed:*] To my verie worthie friend M^r: John Pell Publick Professor of the Mathematickes in Amsterdam In T'oude conuoy op de Zee dijke At Amsterdam

[*annotated by Pell:*] Received Novemb. 16.26

[*cover annotated by Birch:*] 8/18 Nov. 1645.

---

[1] This apparently refers to a note in Hobbes's hand, enclosed with this letter (fo. 228^r), which states: 'Directions for your letters to m^r Hobbes. A Monsieur Monsieur Hobbes Gentilhomme Anglois au Cerf volant sur le Pont St Michael a Paris. Monsr Robervalls name and title Aegidius Personerius de Roberual Mathem. Scient. in Collegio Regio Franciae, Professor. Monsieur Carcauy his title (his christen name he being out of towne I cannot learne) Dominus Carcauius in supremo Galliae Consistorio Senator' ('To M. Hobbes, English gentleman, at the sign of the flying stag, on the Pont S. Michel, Paris'; 'Gilles Personne de Roberval, Professor of the mathematical sciences in the Royal College of France'; 'M. Carcavi, senator in the supreme court of France').

[2] See Letter 42, n. 3.

[3] Rheita's astronomical treatise *Oculus Enoch et Eliae ... pars prima* ended with a short section on the design of telescopes (pp. 336–56), in which he recommended hyperbolic lenses and specifically commended Descartes's writings on the subject (p. 339).

Letter 44

15/25 November 1645

Pell to Cavendish, from Amsterdam

BL, MS Add. 4280, fos. 115–116^r (draft)
van Maanen P14

Right Honourable

[Since my last of October 10.20[1] I have received *deleted*] In your letter of Novem 1.11[2] you desire to know where you were out of y^e way to finde a demonstration for my theoreme 2RRT/RR − TT = D You proceeded thus (for my R & T, writing A & B)

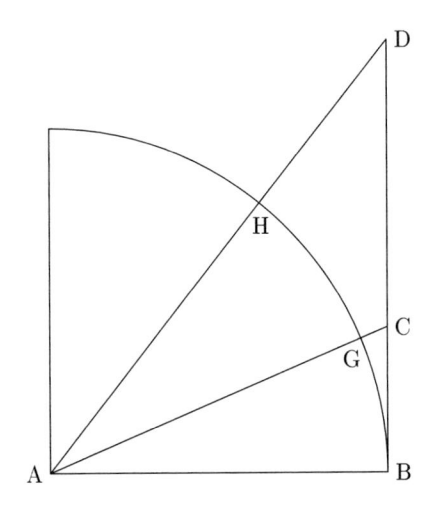

AB = A
BC = B
BD = D

1 $\dfrac{2AAB}{AA - BB} = D$

2 ∴ 2AAB = DAA − DBB
3 ∴ 2AA . AA − BB :: D . B
4 ∴ 2AB . AA − BB :: D . A
5 ∴ 2AA . 2AB :: A . B
6 ∴ AA . AB :: A . B Euclid 6.1

---

[1] Letter 41.
[2] Letter 42.

In which the second [>aequation ariseth] necessarily out of the first, the 3^d & 4th out of the second, the fift out of the 3^d & 4th & the 6^t out of the fift.[3] Where then is the error? Not in ye aequations, for they are all true. Nor in the manner of deducing them, for it is according to art. Sed nunc non erat his locus[4] – The fault is, that they serve not to the scope, for which they were made: the truth of the first aequation [>should have beene] confirmed by the necessary consequence & truth of y^e last [>but is not] For give me leave to imitate you

$$\frac{2AAB}{AA+BB} = D \qquad \Big| 1 \Big| \quad \frac{2AAB}{A+B} = D$$
$$\therefore 2AAB = DAA + DBB \quad \Big|2\Big| \quad 2AAB = DA + DB$$
$$\therefore 2AA \cdot AA+BB :: D \cdot B \quad \Big|3\Big| \quad 2AA \cdot A+B :: D \cdot B$$
$$\therefore 2AB \cdot AA+BB :: D \cdot A \quad \Big|4\Big| \quad 2AB \cdot A+B :: D \cdot A$$
$$2AA \cdot 2AB :: A \cdot B \quad \Big|5\Big| \quad 2AA \cdot 2AB :: A \cdot B$$
$$AA \cdot AB :: A \cdot B \quad \Big|6\Big| \quad AA \cdot AB :: A \cdot B$$

In either of these examples my sixt agrees with your sixt, and the five steps of reason by which we come to y^t sixt are the same in all 3 Examples[5] [and yet the first five in either of mine, are false for if both in yours & mine *deleted*]

And therefore if your first aequation be proved true, [>by] ye undoubted truth of your last so directly drawn out of [>the first] Then is also my first aequation demonstrated to be true for the same cause. But the first five in both mine, are manifestly false, because if D be found by dividing 2AAB by AA − BB then it cannot be found by dividing the same 2AAB by any other divisor whatsoever as

---

[3] This was Cavendish's first attempt (hopelessly inept) to prove Pell's double-angle theorem. In modern notation his argument runs as follows:

$AB = A$
$BC = B$
$BD = D$

$2AAB = DAA - DBB$
$2AA : (AA - BB) = D : B$
$2AB : (AA - BB) = D : A$
$2AA : 2BB = A : B$
$AA : AB = A : B$ (Euclid, VI.1)

Pell's explanation is then not difficult to follow (see van Maanen, 'The Refutation', pp. 331–3).

[4] 'But there was no place at this time for these.'

[5] In modern notation:

$\dfrac{2AAB}{AA+BB} = D$ $\qquad$ $\dfrac{2AAB}{A+B} = D$
$2AAB = DAA + DBB$ $\qquad$ $2AAB = DA + DB$
$2AA : (AA+BB) = D : B$ $\qquad$ $2AA : (A+B) = D : B$
$2AB : (AA+BB) = D : A$ $\qquad$ $2AB : (A+B) = D : A$
$2AA : 2AB = A : B$ $\qquad$ $2AA : 2AB = A : B$
$AA : AB = A : B$ $\qquad$ $AA : AB = A : B$

That is, two new and different premises (as well as Cavendish's) apparently lead to the same conclusion (Euclid, VI.1).

AA + BB or A + B for $2AAB/AA + BB$ is lesse then $2AAB/AA - BB$ unlesse B be equall to 0. And $2AAB/A + B$ is never equall to $2AAB/AA - BB$ unlesse $B = \sqrt{AA - A + \frac{1}{4}} - \frac{1}{2}$ whereas we seeke a demonstration of a theoreme y^t is universally true of all Tangents lesse y^n y^e Radius: not of some one particular tangent

[*next paragraph inserted from following column:*]

You may otherwise also gather the unaptnesse of your worke for y^e end wch was intended: if you goe backward as if you would make a syntheticall demonstration Our 6^t is Euclid's: from thence followes your 5^t manifestly. But from Our fift [>we] can by no necessary consequence deduce ye third or 4th and therefore not [>your] second or first: which was ye Theorema demonstrandum.[6]

But [you *deleted* >we] began to goe out of the way in the 4th aequation for when [you *deleted* >we] had out of ye second drawen either of those two & called it ye 3^d then should [you *deleted* >we] have sought a fit match for him in another family and not joine him with his grandfather father or brother (y^t is ye first second or fourth aequation) [you should have another family, & not married him wth his sister, for as I have often said, incest is unfruitfull. That is, you *deleted*] We should have returned to ye diagram & finding nothing more in it, we should have drawn [>in] some other line whose habitude [>we could know] ex datis conditionibus[7] namely y^t HG is equall to GB and ABD a rectangle Having thus found another aequation not depending upon our first: [you *deleted* >we] should then see whether she could be matched with those which [you *deleted* >we] had before; if not, draw out of her some ofspring, her daughter or grandchild will perhaps agree well wth a sonne or grandchild of the masculine house (that is, of ye first aequation) & so may you looke for a hopefull offspring [>For that which] makes them capable of marriage is [>agreement] Omne A est B.[8] Omne C est D, Ergo nothing. But Omne A est B Omne C est A Ergo Omne C est B followes [>clearly] because they agree in one common terme [>(A)] which is ye band of their wedlocke and therefore we [>were sure] (of hopefull issue) [>that is of] truth in the conclusion no lesse usefull then the premisses were

I should be [>tedious & perhaps] ridiculous to follow this comparison furder & prescribe all ye nuptiall rules & cautions to be observed by him y^t would have a good race of certaine truths. Nor if I would doe it, have I time, but it were not much amisse to adjoine one example yet I [>must] not prevent you by showing you how to amend your demonstration: I will rather, out of my booke [>which I

---

[6]'The theorem to be demonstrated'. A mathematical 'analysis' began from the stated theorem and proceeded by logical steps to a necessary and accepted truth. This did not in itself provide a proof of the theorem, since it is also possible to arrive at a true statement from a false or irrelevant theorem (as Cavendish has done once, and Pell twice). It should be possible, however, to construct a sound 'synthetic' demonstration by reversing the steps of the analysis. Cavendish had found the reversal difficult (see Letter 37, n. 10) and Pell explained that in this case it is in fact impossible.

[7]'From the given conditions'.

[8]'Every A is B' [etc.].

prepare] against Longomontanus, [>transcribe] one of ye wayes by which I first sought that Theoreme, & leave the correction of your owne worke to your second thoughts. R.T + a :: T.b &c

T + A + 2RRT/RR − TT Quod quaerebatur.[9] I could have clothed these inquisitions in the Ancients garbe transforming these equalities of products into equalities of rationes[10] & [>the worke] might have [>appeared y^e] more admirable but I had rather men should admire me lesse & thanke me more for the plaine dealing which I use with y^m for I doubt not but you see the steps of reason to be farre more obvious, than Analogies. Though some kinds of inquisitions & demonstrations runne smoothly inough in Analogies: Of which I thinke that booke will have also some examples.

But of this matter inough for once: I have other things in your letter to answere. I desire to know ye bulke price & argument of Tauricells booke[11] [>I have seene] Reitas booke[12] [>it] seemed to me [(for I have looked it over) *deleted*] a most monstrous Frierly composition: [>having] nothing, meerely nothing worthy of the hundred thousandth part of ye noise that he had made of it Gutschovius is not onely Dr of Physicke but also a Professor of Mathematicks, (I thinke) in Lovain.[13] I have given order to a frend to visite him & [>to] inquire what he hath published as also Langrenus[14] at Bruxells & to endeavour to finde out him y^t [>Reita saith hath] squared ye Circle. If it be a Jesuite & in Brabant I should thinke it to be Della Faille, (an ominous name) who hath been tampering at it a great while by the center of gravity.[15] I shall much desire to see what he sayes, as thinking my selfe able to [examine *deleted* >detect] & refute all pseudographers & paralogismes in that argument.

I wrote to Dr Jungius [& his colleague Tassius *deleted*] Octob 11.21 but he hath not yet answered me;[16] I suppose I have almost angred him, by praying

---

[9]'Which is what was sought.'

[10] Greek mathematics relied heavily on the concept of ratio (here 'rationes'), whereas 17th-century mathematicians from Descartes onwards began to work instead with absolute or dimensionless qualities, thus with products and quotients rather than ratios.

[11] See Letter 39, n. 7.

[12] See Letter 33, n. 4, and Letter 43, n. 3.

[13] See Letter 42, n. 6.

[14] See Letter 38, n. 23.

[15] Jean Charles della Faille (1597–1654), a Flemish Jesuit who taught mathematics at Louvain and Madrid. He had published a treatise on the centre of gravity of curve-bounded figures thirteen years earlier: *Theoremata de centro gravitatis partium circuli et ellipsis* (Antwerp, 1632). (On della Faille see O. van de Vyver, 'L'École de mathématiques des Jésuites de la province flandro-belge au XVIIe siècle', *Archivum historicum societatis iesu*, 49 (1980), pp. 265–78, esp. pp. 267–8.) However, later comments by Pell indicate that the circle-squarer was Grégoire de Saint-Vincent (see Letters 66 and 67).

[16] Pell's letter to Jungius (on whom see Letter 16, n. 5), dated 11 [/21] Oct. 1645, is in SUBH, MS Pe. 65, fos. 78–9. A fair copy of Jungius's reply, dated [24 Oct./] 3 Nov. 1645, is in ibid., fos. 1–12. (Pell's own fair copy of that letter is in BL, MS Add. 4279, fos. 82–4.) Pell did also send a letter to Tassius (see Letter 16, n. 4), dated 8 [/18] Oct. 1645: SUBH, MS Pe. 65, fos. 76–7.

him to alter something in the discourse which he sent me to insert in my booke against Longomontanus.

Two dayes after ye receit of yours I received another from Mr Petty inclosing a second from [>P.] Mersenne, wherein he tells me y^t he will send my papers to Tauricellus & Cavalero (for Galileius & Camillus are dead) but he must first know whether my intention be not altered by Longomontanus his death.[17] What reason he hath so to write, I know not, nor he neither, unlesse perhaps he dreamed so. Yet sure I should know it as soone as he: but to put ye matter out of doubt, 3 dayes agoe an acquaintance of mine received a long letter (I thinke written) I am sure underwritten manu propriâ Longomontani,[18] wherein he talkes like himselfe & tells my freind that he will print y^t third letter also against me, if he will give way that his also may be adjoyned My freind had written to him about 3 months agoe shewing him how [>supposing] ye Tangent of 45.00 [>to be 100000.00] I could demonstrate ye Tangent of 22.30 to be lesse y^n 41421,36[19] The old [dunce *deleted* >bungler] in his letter prayes him to shew him also how out of those two I would demonstrate Tangentem 11,15 esse minorem quam 19891,24 for saith he, Davus sum non Oedipus[20] that is an ignorant [dunce *deleted* bungler *deleted* >smatterer &] no such [cunning man *deleted* >Artist] as he would be thought to be. So I have caused my freind to chew this morsell small & to shew him this great piece of cunning: so that it may be [>in 18 months] he may come to understand my paper against which he hath [>all this while] railed & scribbled so learnedly. The answere we have sent towards danemarke today.

I doubt not but you are weary of reading this tedious letter & therefore you will easier beleeve y^t I am weary of writing it. Wherefore to save me ye labour of answering P. Mersenne & Mr Petty, I will presume to crave y^t favour of you to assure Mersennus y^t Longomontanus is not dead but Excepto quod sit ἀγεωμέτρητος Caetera sanus:[21] and therefore I hope he will without further delay write into Italy.

[I should commend Mr Petty to you, as a hopefull young man, who by countenance & counsell may become exceeding usefull but I perceive you know him & that as well as I shall not neede to commend Mr Petty to you *deleted*] I perceive you have beene pleased to take notice of Mr Petty I hope you finde him worthy of wise mens counsell & countenance.

---

[17] On Petty see Letter 42, n. 8 Petty's letter from Paris, dated 8 [/18] Nov. 1645, begins: 'Father Mersen his desire to convey this inclosed to you...' (BL. MS Add. 4279, fo. 172^r); Mersenne's letter, dated [3/] 13 Nov., is in ibid., fo. 181. Galileo died on [29 Dec. 1641/] 8 Jan. 1642; the mathematician Giovanni Camillo Glorioso (see Letter 35, n. 17) died in the following year. The news of Longomontanus's death was false.

[18] 'In Longomontanus's own hand'; 'underwritten' means 'signed'. The 'acquaintance' has not been identified.

[19] See Letter 38, n. 17.

[20] 'I am Davus [a simple fellow], not Oedipus [a solver of riddles]': a saying from Terence, *Andria*, I. ii. 23.

[21] 'Apart from the fact that he is incapable of doing geometry, he is otherwise healthy.'

You [>often] expresse a desire to see me in print, I desire it too. but [>finde my selfe too many wayes hindred by] my publike imployment & many interruptions in private by visits letters & besides ye particular discontents which arise sometimes when I thinke upon what conditions I live heere now 2 yeeres banished from all y^t is deare to me my bookes & papers my wife & children, not seeing when it will begin to be otherwise. [>Notwithstanding] as I may I shall allwayes strive to approve my selfe

Your Honours most humble servant
I. Pell.

November 15/25/. 1645.

[*postscript:*] I pray, forget not to send me in your next an addresse for my letters to Mr Hobbes[22] or [>to] command Mr Petty to doe it.

---

[22] Cavendish had already sent this with Letter 43 (see Letter 43, n. 1), but that letter reached Pell one day after he sent Letter 44.

Letter 45

6/16 December 1645

Cavendish to Pell, from Paris

BL, MS Add. 4278, fos. 230–1 (original)

Worthie S^r:

I heere send you an other demonstration of your Theorem; I was trieing to demonstrate it but I found I was in the waye of this demonstration of Monsr: Palliers[1] & so I desisted;/ I longe to heare from you & howe you proceed in your worckes, for I hope you will shortelie publish some of them; I desire allso to heare from Doctor Jungius[2] & what he publishes; I praye you when you writ to him present my seruice [>to] him. I thinke wee shall remayne heere this winter. And so longing to heare from you & wishing you all happiness I rest

Your assured friend to serue you
Charles Cauendysshe

Paris Decemb: 6/16. 1645

[*addressed:*] To my verie worthie friend M^r: John Pell Publick Professor of the Mathematickes at Amsterdam, In Toude Conuoye op de zee dijke in Amsterdam

[*annotated by Pell:*] Received dec 15/25

[*cover annotated by Birch:*] 6/16 Decemb. 1645

---

[1] On Le Pailleur see Letter 35, n. 6. His demonstration of Pell's theorem, forwarded here by Cavendish, was printed in Pell, *Controversiae pars prima*, p. 54.

[2] See Letter 16, n. 5.

Letter 46

12/22 December 1645

Cavendish to Pell, from Paris

BL, MS Add. 4278, fos. 232–3 (original), 236^r (enclosure, original)
van Maanen C21

Worthie S^r:

I giue you manie thankes for yours of the 15/25 of Nouemb:[1] which I receiued yesterdaye. I am glad you aproue of my deducing from your aequation to the 1 of the 6th. of Euclid; but for the regress as I remember I writ to you I found it so difficult that I left it to your better disquisition, supposing it might be found, but it seems it can not./ I should haue remembred that which I learned longe since, that though the analyticall thred be rightlie deduced, yet if the proposition taken in be not conuertible the regress or direct proofe will not proceede;[2] as if I would proue Omne B est A.[3] which suppose true, I ioine to it Omne C est B, which I knowe to be true. from these [>two] propositions there followes omne C est A. which being knowne to be true yet I can not proue Omne B est A by regress or directlie except the proposition taken in be conuertible vic^t: Omne B est C be allso true; & then it will followe (Omne C est A)

(Omne B est C)

(Omne B est A) so it seemes that this deduction of myne was in somewhat taken in not conuertible./ But I am glad I gaue you occasion to intimate howe to ioine aequations so that they maye be vsefull, which I hartilie thanke you for & for your two excellent [>concise] proofs of your theorem./ Mersennus remembers him to you & will according to your desire write by the next oportunitie to Tauricell & Caualiero./[4] I send you heere the titles of Tauricells worckes,[5] the price I knowe not for there is none heere to be solde./ I vnderstand by Sr: Williame Boswell,[6] that Monsr: de Cartes aproues of your refutation of Longomontanus etc: with high commendations. If de Cartes demonstrate your theorem I desire you will do me the fauoure to send it to me./[7] wee haue nothing newe heere./ By the next the opinion of our learned men heere

---

[1] Letter 44.

[2] See Letter 44, n. 6, on converting analytic reasoning into a sound synthetic proof.

[3] 'Every B is A' [etc.].

[4] See Letter 42, n. 3, and Letter 44, n. 17.

[5] See the enclosure to this letter.

[6] See Letter 19, n. 13.

[7] Descartes did not put forward a demonstration of his own, but he did contribute a statement of approval of Pell's demonstration, which was printed in Pell, *Controversiae pars prima*, p. 59 (see Letter 49, n. 3).

PART III: THE CORRESPONDENCE, LETTER 46   445

of Rietas booke;[8] Gassendes is nowe reading of it./ I desire you will continue your fauoure to me in inquiring of those bookes I writ of./ I haue wearied you enough for once./ And so wishing you all happiness I remayne

Your assured friend to serue you
Charles Cauendysshe

Paris Decemb: 12/22 [*altered from* 15/25] 1645

[*addressed:*] To my verie worthie friend Mr: John Pell Publicke Professor of the Mathematickes at Amsterdam In T'oude Conuoy op de Zee dijk in Amsterdam.

[*annotated by Pell:*] Received Jan 3/13

[*cover annotated by Birch:*] 12/24 [*sic*] Decemb. 1645.

[*Enclosure to Letter 46:*]

Opera Geometrica Euangelistae Toricellii. De Solidis Sphaeralibus / De Motu / De Dimensione Parabolae / De Solido Hyperbolico / cum Appendicibus De Cycloide et cochlea. / All in one volume in 4to: aboute the thickness of De Cartes his first french worckes. / printed at Florence 1644.[9]

I dare not shewe your [>owne] demonstrat. of your theorem[10] without your leaue. / I praye let me knowe by your next whether I maye.

---

[8]See Letter 33, n. 4.

[9]Cavendish lists the contents of Torricelli, *Opera geometrica*; the work by Descartes is the *Discours de la méthode*.

[10]This demonstration by Pell – the only one he is known to have written – does not survive.

Letter 47

17/27 December 1645

Cavendish to Pell, from Paris

BL, MS Add. 4278, fos. 235, 237 (original); fo. 234^r (enclosure, original)
van Maanen C22

Worthie S^r:

I send you heer a demonstration of myne of your Theorem;[1] that which I like most in it is that [>it] is not longe & that it depends not of anie thinge I haue seene in anie demonstration of the same Theorem; but yet till you haue examined it & aproued of the truth of it; I shall not be so fullie satisfied./ There is one Faber a Jesuit I thinke at Lions who hath printed a booke of mathematicall Theses[2] which he saies he can proue; there are manie rare propositions; & Mersennus tells me the first parte of those Theses are nowe printed & that wee shall presentlie haue them heere. when they come I shall informe you what they are & howe they are aproued on by oure Mathematicians heere./ My seruice to Doctor Jungius when you write to him; I longe to heare of the printing of his Logick, Appolonius Saxsonicus or ani thing els of his.[3] but much more of your Analytickes, Appolonius, & Diophantus.[4] And so longing to heare from you & wishing you all happiness I remaine

   Your assured friend to serue you
   Charles Cauendysshe

Paris Decemb: 17/27 1645.

[*addressed:*] To my verie worthie friend M^r: John Pell Publicke Professor of the Mathematickes at Amsterdam. In T'oude Conuoy op de Zee diike in Amsterdam.

[*annotated by Pell:*] Received Jan 3/13

[*cover annotated by Birch:*] 17/27 Decemb. 1645

---

[1] See the enclosure, below.

[2] Honoré Fabri SJ (1607–88) taught philosophy at Arles (1636–8) and mathematics and philosophy at Lyon (1639–46), before moving to Rome, where he spent the rest of his life. His *Theses de universa mathematica* was printed at Lyon at the end of 1645, with a titlepage dated '1646'.

[3] See Letter 42, nn. 10, 11.

[4] See Letter 23, n. 2, and Letter 14, n. 13.

PART III: THE CORRESPONDENCE, LETTER 47   447

[*Enclosure to Letter 47:*][5]

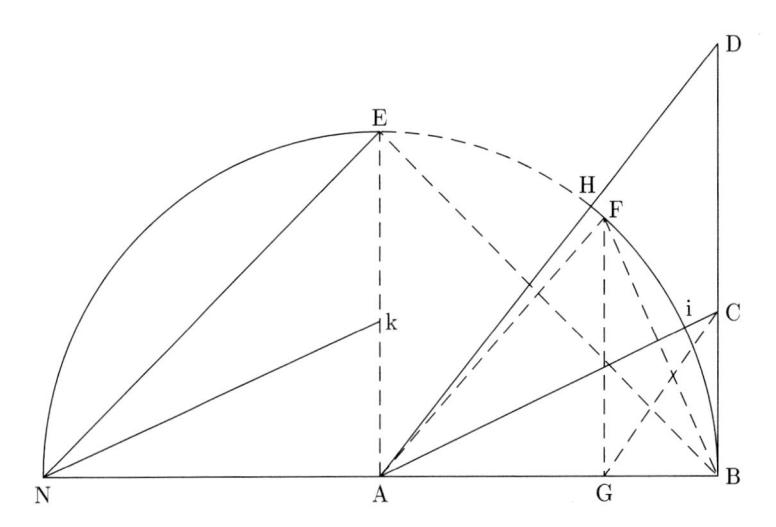

compleate the quadrant & drawe BE; from C drawe a paralel to AD, & where it cuts AB at G, erect a p[er]pend: to AB & produce it till it cut the circle at F & drawe BF. In short the prickt lines are the construction. Bi = iH by supposition.

BD.BC :: BA.BG.
BA.BG :: BEq.BFq. p[er] Lemma.
BEq = 2BAq.
BFq = ABq − BCq. for
AFq − FGq = AGq. ⎫
                  ⎬ thes added
CGq − GBq = BCq. ⎭

---

[5] In this proof and elsewhere Cavendish uses the notation BFq (introduced by Oughtred) for BF^2. He also uses a symbol which looks like a division sign but is in fact a variation on a different symbol (also devised by Oughtred) with two dots above and two below the line, to indicate proportional quantities. Cavendish's proof depends on the property we would now write as $1 - \cos\theta = 2\sin^2(\theta/2)$. In his Lemma he proves a slightly weaker version, namely that $1 - \cos\theta$ (the versed sine) is proportional to $\sin^2(\theta/2)$ (the square of the subtense). The proof of the Lemma uses elementary trigonometry and properties of ratios. In particular, if NB, BF, BG, are in geometric progression (what Cavendish would have called continued proportion, indicated by his two-dot symbol), then $NB \times BG = BF^2$. Cavendish writes NBG (*sc.* rectangle NBG) for the product $NB \times BG$. The main part of the proof is not easy to follow because Cavendish mixes up statements that are obvious with those he has yet to prove. In particular, the statement $BF^2 = AB^2 - BC^2$ is crucial but is not proved until several lines after it is given. Cavendish (or someone else, perhaps Hobbes) must have seen the difficulty, for he then re-wrote the argument in a better order. The proof is considerably more sophisticated than Cavendish's earlier attempt (see Letter 37, n. 9), or than his mathematics generally, and he eventually admitted that it was not entirely his own work (see Letter 72). A version of it was printed in Pell, *Controversiae pars prima*, pp. 52–3. For further discussion see van Maanen, 'The Refutation', pp. 333–5.

AFq − FGq + CGq − GBq = AGq + BCq. / transp:
AFq = ABq. / & AGq = CGq. for ang: GCA = DAC = GAC
[er]go: GC = AG. / [er]go: ABq − BCq = FGq + GBq = BFq. [er]go:
BD.BC :: 2BAq.ABq − BCq. which was to be proued.

Lemma. / versed sines are as the q:^{s} of theyr subtenses. BA.BG :: BEq.BFq.
NB.BF.BG÷ / [er]go: NBG = BFq. // NB.BE.BA÷ / [er]go: NBA = BEq. //
[er]go: NBA.NBG :: BEq.BFq but NBA.NBG :: BA.BG. / [er]go: BA.BG ::
BEq.BFq. which was etc: // & so in all versed sines & theyr subt[enses]
The quadr: ENA with the lines NE & Nk are described onelie for the demonstr:
of this Lemma: & the line Nk not produced to F (as it should be) for obscuring
the principall figure.

Peraduenture the demonstr: in this order would be better.
ang: GCA = DAC = CAG. / [er]go: CG = GA
BD.BC :: BA.BG. // BA.BG :: BEq.BFq p[er] Lemma
BEq. = 2BAq/ BFq. = ABq − BCq. for
(AFq)ABq − FGq = AGq ⎫
                     ⎬ these two aeq: added
(CGq)AGq − GBq = BCq ⎭
ABq − FGq + AGq − GBq = AGq + BCq. / by transp:
ABq − BCq = FGq + GBq. = BFq. [er]go:
BD.BC :: 2ABq.ABq − BCq. which was etc.

My Lemma heere is a propos: in Mersen: Balistica pag: 49[6] but not demonstrated by him; but Mr: Hobbes shewed me a demonstr: of it by himself. differing from myne; but neuer thought of appliing it to your Theorem:

I shewed Mr: Hobbes this my demonstr: of your theorem; who likes it verie well; but till I knowe howe you aproue of it I shall not rest fullie satisfied.

---

[6] Mersenne, *Cogitata physico-mathematica*, p. 49.

Letter 48
1 [/11] January [1646]
Pell to Cavendish, from Amsterdam

BL, MS Add. 4280, fo. 116^r (draft)
van Maanen P15

Right Honourable

Yours of Nov 1/11 I answered [>at large] 14 dayes after[1] that is, almost 7 weekes agoe. And therefore I hope [>it is now come to your hands.] I have since received 2 [>of yours:] the former dated Nov. 16.26 with the addresses by which I hope to convey this to you.[2] The latter dated Dec 6/16 with Mr Palliers demonstration,[3] which I had allmost determined to have sent [>backe,] transformed into ye manner of those demonstrations which I sent in my last but I resolved to let it alone till I heard how you liked those & y^t so much ye rather, because this being ye first letter sent by y^t new addresse I thought not good to adventure so much in this, till [>by experience I had] learned y^t there was no feare of miscarriage. You have heere inclosed an Extract out of ye first letter from yt freind of whom I made mention in my last letter.[4] I see ye Jesuite is of Flanders not of Brabant as I guessed,[5] The title of y^t 8[^{vo} *blotted*]

---

[1] Letters 42 and 44.

[2] This apparently refers to Letter 43 (dated 8/18 Nov.), which included Hobbes's directions for posting letters; Pell was probably confused by his own annotation on it, 'Received Novemb. 16.26'.

[3] Letter 45; on the demonstration see Letter 45, n. 1.

[4] This friend has not been identified (see Letter 44, n. 18), and the extract from the letter has not survived.

[5] Pell had previously guessed that the Jesuit was della Faille (see Letter 44, n. 15). Subsequent comments (see Letters 66, 67) suggest that the circle-squarer was Grégoire de Saint-Vincent SJ (1584–1667), who was born in Bruges, studied at the Collegium Romanum in Rome and taught mathematics at Louvain from 1621. After spending some time in Prague he returned to the Spanish Netherlands, and was living in Antwerp in 1646. On de Saint-Vincent see H. Bosmans, 'Documents inédits sur Grégoire de Saint-Vincent', *Annales de la société scientifique de Bruxelles*, 27, part 2 (1902–3), pp. 21–63; K. Bopp, 'Die Kegelschnitte des Gregorius a St. Vincentio in vergleichender Bearbeitung', *Abhandlungen zur Geschichte der mathematischen Wissenschaften mit Einschluss ihrer Anwendungen*, 20 (1907), 87–314; H. van Looy, 'Chronologie et analyse des manuscrits mathématiques de Grégoire de Saint-Vincent (1584–1667', *Archivum historicum societatis iesu* 49 (1980), pp. 279–303.

booke[6] [>almost persuades me] y^t he intends to goe y^t new way for squaring of ye Circle which I fell upon divers yeeres agoe, which produced a long [*blotted*] spec]ulation de Toreuticis ./. de figuris per mo[tum circa *blotted*] axem quiescent[em] quasi in torno factis.[7] 4 Alphabets at 2.23[*blotted*] sheetes in an Alphabet will be 184 sheetes but if each le[tter *blotted*] have 3 sheetes it will be 276 [>sheetes[8] a greate price & bulke, after ye Jesuites fashion] [It is such another [bable?] as Bettinus his apery[9] *deleted*] Another Be[ttinus *blotted*] perhaps

As for our Hamburgers, of whom you inquire, I wrote to y^m both, eleven weekes agoe. Tassius hath not answered me yet. Dr Jungius sent me [>such] an answere,[10] y^t I am resolved to write [>to] him no more till my booke is printed, for I see we doe but hinder one another. I have therefore determined to leave him to his better studies, for, me thinkes, he understands not well my paper. At first he dreamed that [>he that will use] that kind of argument, must be beholding to some Canon of Tangents and therefore told me y^t my demonstration was not αὐτάρκης y^t is selfe-sufficient, I replyed y^t my booke shewes y^t I owe not my

---

[6] Although the context suggests that Pell is referring to the forthcoming book by de Saint-Vincent (see Letter 49, n. 2), that work was a large folio volume, not an octavo, and its title did not contain any hint about such a 'new way' of squaring the circle. Pell's comment about the possible implications of the title suggests that although the information he received was primarily about de Saint-Vincent, it had become muddled with some news about the work of André Tacquet SJ, who taught mathematics at the Jesuit 'École de mathématiques' in Louvain (in 1644) and in Antwerp (1646–9): Tacquet's treatise *Cylindricorum et annularium libri IV; item de circulorum volutione per planum, dissertatio physicomathematica* would eventually be published (as a quarto, however, not an octavo) in Antwerp in 1651. On Tacquet see H. Bosmans, 'Le Jésuite mathématicien anversois André Tacquet (1612–60)', *De gulden passer – Le Compas d'or*, 3 (1925), pp. 63–87; de Vyver, 'L'École de mathématiques', p. 273; G. H. W. Vampaemel, 'Jesuit Science in the Spanish Netherlands', in M. Feingold, ed., *Jesuit Science and the Republic of Letters* (Cambridge, Mass., 2003), pp. 389–432, esp. pp. 405–12.

[7] 'Of torneutics, i.e., of shapes made by mo[tion around a?] still axis, as if on a lathe.' The ending of 'quiescent[em]' is an expanded contraction. Pell was referring here to solids of revolution, that is, solids generated by the revolution of a curve about an axis. 'De toreuticis' means, properly, 'of things pertaining to carving in relief' (from τορευτικός, an adjective derived from τόρευμα, which means 'work in relief'). However, if Pell really intended to write 'toreuticis', he was probably thinking of one occurrence of the word τόρευμα, in Euripides, *The Madness of Hercules*, line 978, where it means 'wheeling around' and should more properly be τόρνευμα, meaning 'that which is turned on a lathe'. Pell's term should therefore be 'torneuticis' (from τορνευτικός, meaning 'pertaining to that which is turned on a lathe'). Or it may be that the omission of the 'n' was simply a slip of his pen.

[8] Pell refers to the signatures of a folio volume – either a folio in fours (with each gathering consisting of two sheets) or a folio in sixes (with each consisting of three sheets). The standard continental alphabet for signatures consisted of 23 letters (with 'i' and 'j' treated as one letter, 'u' and 'v' treated as one letter, and 'w' not included). Thus '4 Alphabets' would be a book with the sequence of signatures 'A – Z', 'AA – ZZ', 'AAA – ZZZ', and 'AAAA – ZZZZ'. A folio book of 276 sheets would have 1104 pages. For the actual length of de Saint-Vincent's book, see Letter 49, n. 2.

[9] Pell refers to *Apiaria universae philosophiae mathematicae*, 2 vols. (Bologna, 1642), a huge compilation by a well-known Jesuit mathematician, Mario Bettini. (The 'Apiaria' of the title is an apiary or bee-hive; Pell's 'apery' is perhaps, uncharacteristically, a pun.)

[10] See Letter 41, n. 6, and Letter 44, n. 16.

numbers to ye Canonics he answered, 'tis true, there [>are] rules long since given, for ye making of ye Canon of Tangents & if I will take so much needlesse paines, I may by those rules calculate such tangents as I have neede of. But he sayes he [>denied] my demonstration to be selfe-sufficient for another reason which is so pittyfull y^t it is not worth ye writing. [>and yet he will have nothing altered in his censure which will [outwardly?] enough agree with mine.] I see he understands not y^t my Theoreme alone is sufficient to shew those numbers without any other mens rules adjoined. [And I feare y^t some of your brave wits of France doe not apprehend ye hundredth part of ye use of that excellent Theoreme. Nor perhaps I neither, though it may be I shall make it appeare usefull, to such purposes as men now dreame not of. In a word, it is worthy that you should attempt to find out at least one of ye innumerable wayes, by which it may be demonstrated *deleted*] I thinke not fit to interrupt him any more privately. When my booke is printed, perhaps he will wish he had contented himselfe to answer as Monsieur Robervall,[11] who had more discretion y^n to trouble himselfe with ye impertinent inquiring how I found my media:[12] [>it sufficed him y^t they are [apt?] & well put together to make a firm demonstration] and so he pronounces of it. Nor have I reason to looke for any other censure from those y^t perfectly understand it. [>Whose] small number will perhaps in a little be increased by ye reading of my booke, which notwithstanding all ye rubs it meetes withall [may perhaps *deleted* >will I hope] be worthy to see ye light one day. In ye meane time, I rest

Your Honours most humble servant
I.P.

Jan. 1. Stylo Vet.

---

[11] On Roberval's demonstration see Letter 35, n. 2.
[12] 'Means' (of demonstration).

Letter 49

15/25 January 1646

Pell to Cavendish, from Amsterdam

BL, MS Add. 4280, fo. 116 (draft)
van Maanen P16

Right Honourable

This day fortnight I directed a letter for you to Mr Hobbes,[1] in which was an extract concerning ye Jesuit [>who] is printing a booke de quadratura circuli; which, as I heare since, will be about 180 sheetes: so y^t it seemes each letter of his Alphabet hath but 2 sheetes.[2] Two dayes after I had written to you, I received a letter from Sr William Boswell containing two of yours & an extract of a letter from DesCartes to him which I have heere transcribed.[3] To yours I must answere 'Tis a good rule y^t you write of, to see y^t your propositions be convertible, where you intend regresse.[4] But 'tis not good enough, for in your [demonstration *deleted* >deduction] they were Indefinite aequations which are allways convertible, but the fault was, that having drawen 2 (the 3^d & 4th) out of one (the second) You did so joyne y^m together, that you wholly lost the praedicate of the second: or, if you will, y^e praedicate & y^e denominator of y^e subject of y^e first So y^t you made no use at all of those data namely y^t y^e divisor must be $AA - BB$ or y^t if y^e divisor be $AA - BB$ y^e quotient would be D id est tangens dupli arcûs;[5] which being so rejected you were utterly disabled to make any use of those beginnings: So [>barren &] unlucky was y^t [>incestuous] match of y^e 3^d with y^e 4th

But I have examined your demonstration which you sent me Dec 17/27 [>&

---

[1] Pell's letter, sent via Hobbes, was Letter 48.

[2] If, as seems likely (see Letters 52, 66) Pell's information was about Grégoire de Saint-Vincent's *Opus geometricum* (also known by the title on the engraved titlepage: *Problema austriacum plus ultra quadratura circuli*), then these details were misleading. The book contains 1226 pages of text; as a folio bound in fours, this would be described as roughly 306 sheets, or 153 gatherings. (Cf. Letter 48, n. 8.)

[3] On Boswell see Letter 19, n. 13. The two letters from Cavendish were Letters 46 and 47 (both annotated by Pell: 'Received Jan 3/13'). The extract from a letter from Descartes was (as Cavendish's comment in Letter 52 makes clear) the commendation of Pell's theorem which was printed in Pell, *Controversiae pars prima*, p. 59, under the heading: 'An extract from a letter written by M. Descartes in reply to a great man who, writing to him in French, enquired, among other things, what his opinion was of my little refutation' ('Excerptum ex D. Cartesii epistolâ responsoriâ ad virum magnum, qui Gallicè ad eum scribens, inter alia, quaesiverat ejus sententiam de meâ refutatiunculâ'). It is reprinted in A&T iv, pp. 342–3, where the editors speculate that the person written to by Descartes may have been Constantijn Huygens or his brother-in-law; the evidence of this letter shows that it was Boswell.

[4] See Letter 46, at n. 2.

[5] 'That is, the tangent of the double angle'.

finde it] firme & sufficient to demonstrate what you intended that is BD.BC ::
2BAq.ABq − BCq Though perhaps it would come neerer to y^t which I affirme if
you subjoine ye converse Ergo ABq−BCq.2BAq :: BC.BD So y^t your demonstration will run thus. Because AC bisects DAB and CG is parallel to AD therefore
as ABq − BCq. [>to] 2ABq so BC to BD But [>I conceive it would be] worth
[>your] while to bestow a quarter of an hour to make a second demonstration
wth ye same diagram & ye same media demonstratoria to prove y^t if BD be
to BC as 2BAq to ABq − BCq then GC will be [>equall to AG &] parallel to
AD and by consequence BAD will be equall to 2BAC, which seemes to me to
be ye right truth upon which my precept is built.[6] 'Tis true, all those, which
you have seene, doe the like; they all prove y^t the tangent of ye double arch will
be equall to such a quotient When my precept seemes rather to require y^t we
prove that such a quotient will be a tangent of a double arch. I desire y^t we may
have yours both wayes, seeing you have done ye former so well & may adde y^e
latter so easily. I suppose your diagram will fall better with ye semicircle under
ye triangle as in this loose paper inclosed. I have put your demonstration into a
fit posture for ye presse[7] & find y^t ye Lemma may as well or better be woven
into ye body of ye demonstration.

[*Sixteen lines heavily deleted; the substance of them is apparently given in what follows:*]

By honouring my theoreme with your demonstration you have obliged not
onely me but also those learned men who have done the [>like]: [>And by your]

---

[6]In modern notation the foregoing passage is as follows:

I have examined your demonstration ... and find it firm and sufficient to demonstrate what you intended, that is

$$\frac{BD}{BC} = \frac{2BA^2}{AB^2 - BC^2}.$$

Though perhaps it would come nearer to that which I affirm if you subjoin the
converse, therefore

$$\frac{AB^2 - BC^2}{2BA^2} = \frac{BC}{BD}.$$

So that your [complete] demonstration will run thus. Because AC bisects $D\hat{A}B$,
and CG is parallel to AD therefore as $AB^2 - BC^2$ to $2(AB)^2$ so is BC to BD. But
I conceive it would be worth your while to bestow a quarter of an hour to make a
second demonstration with the same diagram and the same *media demonstrato-
ria* [means of demonstration] to prove that if $\frac{BD}{BC} = \frac{2BA^2}{AB^2 - BC^2}$ then CG will
be equal to AG and parallel to AD and by consequence $B\hat{A}D = 2B\hat{A}C$, which
seems to me to be the right truth upon which my precept is built.

In other words, Pell approved Cavendish's proof but requested that he should also provide the
converse: Cavendish had proved that if $B\hat{A}D = 2B\hat{A}C$ then $\tan B\hat{A}D = \dfrac{2\tan B\hat{A}C}{1 - \tan^2 B\hat{A}C}$. Pell
now wanted him to prove that if $\tan B\hat{A}D = \dfrac{2\tan B\hat{A}C}{1 - \tan^2 B\hat{A}C}$ then $B\hat{A}D = 2B\hat{A}C$. (See also
van Maanen, 'Refutation', pp. 339–340.)

[7]Cavendish's demonstration was printed in Pell, *Controversiae pars prima*, pp. 52–3.

appearing in print [>amongst them] (as I conceive, without any dishonour to your selfe) you shall give publike & everlasting testimony [>for] y^e Mathematickes, that you [>have] found [>enough] worth & content in those unstained pleasures to persuade you to bestow so much time on y^m as to become able to demonstrate fully & solidly new truths, where professors of great fame have beene at a losse.

Whereas you desire leave to shew my demonstrations of [my *deleted* >mine own] theoreme, perhaps some will thinke, y^t you have none to shew. For those two excellent concise proofes as you call y^m are [>not demonstrations of knowen truth, but inquisitions of the unknowen habitude] which this Tangent beares to ye Radius & ye tangent of halfe ye arch: and therefore in ye conclusion I thinke I wrote (not quod erat demonstrandum, but) quod quaerebatur.[8] [And perhaps you will shew it to some who woulde *deleted*] 'tis true y^t y^e inquisitions [>are] of such a nature y^t y^e close being altered y^{ei} may both passe for demonstrations with those who consider y^t properly the [>proposed] theoreme is of solids namely y^t RRD − TTD = 2RRT and therefore amongst Geometers [>the] media demonstratoria[9] [>may be] of 3 dimensions. But they will more unquestionably passe for good demonstrations amongst those y^t consider y^e theoreme as Logisticall (where we are not like Geometers shut up in 3 dimensions, but left at liberty to speculate products composed of as many factors as we list [)].[10] And in ye inquisition you see I looke upon ye diagram onely to finde two Logisticall [habitudes *deleted* >aequations] R.T + a :: T.b. And R.R + b :: T.a which being given, a Zeteticall Logista[11] cares not to know what R.T. a or b signify. Out of those two aequations he will undertake to make you two other aequations shewing the Logisticall habitudes of a & b to R & T and this is all y^t I there doe: but if I would have [>so] demonstrated y^t if T + a be 2RRT/RR − TT y^n [>it] is ye tangent of y^t double onely [I should perhaps have wrought thus *deleted*] I must have wrought otherwise

But having heretofore shown you many examples of another forme of demonstration & inquisition peculiar to my selfe for ought I know. I thought good to give you an example or two of another [>peculiar] forme which I also use when I finde it convenient which because you seemed to li[ke *blotted*] so well in my demonstration de medio Arithmetico & Geometrico[12] I thought it not amisse to let you see y^t it was also apt for inquisition And therefore I sent you those two examples. which though they be such as y^t their workeman neede [>not care who see them as if he were] ashamed of his worke, yet if any Caprichious head [>should see them who will not understand] why I sent those rather than any other, out of that innumerable company of [>inquisitions &] demonstrations

---

[8]'That which was to be demonstrated' ... 'that which was sought'.

[9]'The means of demonstration'.

[10]The parenthesis after 'list' is missing in the MS.

[11]Viète described four stages of analysis: poristic, zetetic, rhetic, and exegetic. 'Zetetic' (from ζήτησις, 'searching', 'investigation') was the process of setting up the equations pertaining to a given problem; 'logista' means 'calculator' or 'reasoner'.

[12]'Of the arithmetical and geometrical mean' (see Letter 38).

which I pretend to have of y^t theoreme He would perhaps attempt to persuade you y^t I am not able to pen a demonstration, phrasi Geometricâ.[13] But I have beene so tired with impertinent estimators of my skill, that I am resolved to take [>no] notice of any more of them. For if I should goe on to harken & reply, I might [>weary] my selfe in private & never finde leisure to publish my booke. For which cause I desire y^t no copy of those inquisitions may be given out: Yet if you shall thinke fit to doe otherwise, it is in your power to dispose of whatsoever belongs to

    your Honours humble servant
    [John Pell *deleted*]

Amsterdam. JAN. 15.25. 1645 46.

---

[13] 'In the geometrical style'.

456   JOHN PELL (1611–1685)

Letter 50

[22 January/] 1 February 1646

Pell to Cavendish, from Amsterdam

BL, MS Add. 4280, fo. 117^r (draft)
van Maanen P17

Right Honourable
    Mr Godolphin[1] writes to me from ye Hagh that he is going to Paris, [>seeing] therefore this letter comming perhaps too late [thither *deleted* >to his lodging,] may happen to be lost, I have resolved to write nothing [>in it] whose [>missecarriage] may much trouble me, but by that meanes there will also be nothing worth your reading. So that you may now throw it away. [>For] I hope [>it is no newes to tell you] that I have written twice to you (a weeke & 3 weekes ago) both directed to Mr Hobbes.[2] I thinke not good to write a third, that way, till I heare y^t at least one of y^m hath found him out, for then I count y^m safe. In those two I answered (I thinke, fully) 4 of yours dated Novemb 8/18 Decemb 6/16. 12/22. 17/27.[3]
    I will now therefore onely adde that I hope shortly to receive a second demonstration from your selfe. I should also tell [>P] Mersenne if I had leisure, that by his endeavour I expect the judgements of 4 more, Mydorge Fermat, Toricell & Cavalier.[4] which being done, I hope I shall [>not] trouble your selfe or him more [>in that kinde.]
    It may be Mr Godolphin will make use of this opportunity to become Knowen to your Honour. He seemed also already to know Mr Hobbes.[5] Least therefore in discourse hearing y^t he hath beene heere my disciple in ye Mathematickes you [>or he] should expect something extraordinary in him, I thought good to [>give notice] y^t it is not there to be found. He was heere but 5 weekes and because we

---

[1] This was probably a son either of Francis Godolphin (1605–67), a member of a prominent Cornish gentry family who was Royalist governor of the Scilly Isles in the civil war until their surrender in 1646, or of his brother, the poet and MP Sidney Godolphin (1610–43). The letter to Pell has not survived.

[2] Letters 48, 49.

[3] Letters 43, 45, 46, 47.

[4] Pell had previously been led to expect a demonstration from Mydorge (see Letter 37, n. 8, and Letter 39, at n. 4); one was eventually supplied, and added at the last moment to Pell, *Controversiae pars prima* (pp. 90–4). Nothing was obtained from Fermat. Pell had already sent a request to Mersenne to obtain demonstrations from Torricelli and Cavalieri (see Letter 42, n. 3, and Letter 46, n. 4); Cavalieri did contribute one (printed in Pell, *Controversiae pars prima*, p. 60).

[5] Hobbes had known Sidney Godolphin (see above, n. 1), and received a bequest of £200 from him; in 1651 he would dedicate *Leviathan* to Francis Godolphin.

were allwayes uncertaine when he should goe away, I could not begin with him [>or] proceed so orderly as otherwise I [>would] have done. Some tastes he hath had and no more. Though otherwise [>he have] a capacity I thinke fit for those studies & worthy ye helpe of those y^t can better doe it than

Your [>honours most] humble servant
John Pell.

Letter 51

24 January/3 February 1646

Cavendish to Pell, from Paris

BL, MS Add. 4278, fos. 238–9 (original); fo. 240^r (enclosure, original)
van Maanen C23

Worthie S^r:

I giue you manie thankes for your Letter wherin I receiued a note concerning the booke of the quadrature of the circle[1] which I inquired after; though I doute he hath not geometricallie done it yet if he haue well handled the rest of the heads he treats of, it will be well worth the hauing./ I receiued your letter from M^r: Hobbes hand, so that I thinke you may safelie direct your letters to him, who I doute not but will safelie conuei them to me, if wee should remoue from hence which I see no greate aparence of as yet./ I doe wonder at Doctor Jungius his iudgment concerning your refutation of the Dane;[2] if I knewe him not to be an excellent man, this would make me haue a slender opinion of him; but something has diuerted him & set his iudgment awrie in this, whatsoeuer it be./ I doute you receiued not my last letter to you with a demonstration of myne of your Theoreme;[3] & least that miscaried I heere send it to you againe[4] though perhaps not verbatim yet the same in effect. I haue M^r: Hobbes & M^r: Roberualls[5] aprobation for the truth of it, but yet till I heare howe you like it I shall not rest so fullie satisfied in it.// There is a proposition in Mersennus preface to his Mechanicks pag: 3 as I take it thus, *vt summa exponentium tam in applicatis quam in diametro, ad exponentem applicatarum ita parallelogrammum ambiens ad figuram.*[6] but not demonstrated; Mersennus tells me it is Monsr: De Cartes his proposition but that he sent him not the demonstration of it; M^r:

---

[1] Letter 48.

[2] See Letter 48.

[3] That letter does appear not to have reached Pell.

[4] See the enclosure to this letter, another version (also using the same symbol for proportional quantities) of the proof he had sent a few weeks earlier (see Letter 47, n. 5).

[5] See Letter 19, n. 7.

[6] 'As the sum, of the exponent of the ordinates and that of the diameters, is to the exponent of the ordinates, so is the surrounding parallelogram to the figure.' Cavendish paraphrases a passage in the preface to the 'Tractatus mechanicus', the fifth part of Mersenne's *Cogitata physico-mathematica*, sig. a2^r. The figure in question is one of the higher parabolas, with equation (in modern notation) $y^4 = x^3$. Mersenne was claiming, correctly, that the ratio of its area to that of a circumscribed or surrounding parallelogram is 4 to 7 (as is easily proved using integral calculus). The result may have come from Descartes, as Mersenne suggested, but Fermat may also have proved results of this kind, as did Roberval.

Roberuall tells me it is Mons[r]: Fermats[7] proposition but that he neuer sawe the demonstration of it; but saies he thinkes he could doe it but that [>it] would be a verie longe demonstration./ I thinke Mersennus sent me the demonstration of it into england [>but has forgotten it];/ if for a diuersion you could spare so much time as to demonstrate it & send it me, you should [>doe] me a greate fauoure.// I understand not Herigons second quest: of his sec: rootes in his Isagoge to his Algeb:[8] I meane, not howe he limits A./ he deliuers it so intricatlie that I can not easilie fit it to that example you gaue me of an vndetermined problem[9] when I was with you at Amsterdam. if you please in your next letter a little to explaine it you should doe me a fauoure; I meane not the whole doctrine of determination; for that I suppose woulde be to large for a letter & I expect that when you shall publish your Analyticks, which worcke I confess I extreamelie desire to see; neither would I haue you trouble your self to demonstrate that proposition in Mersennus (I nowe desire) except you finde the demonstration maye be indifferent short; for I woulde not trouble you too much to demonstrate an other mans proposition./ I haue troubled you sufficientlie with this letter allreadie for the present; therefore wishing you all happiness I remaine

Your assured friend to serue you
Charles Cauendysshe

Paris Jan: 24/Feb: 3. 1646.

---

[7] See Letter 4a, n. 7.

[8] The sixth volume of Pierre Hérigone's *Cursus mathematicus* (see Letter 19, n. 9) contained an 'Isagoge de l'Algebre' (2nd pagination, pp. 74–98); Cavendish refers to the section of it entitled 'Questions des secondes racines', question 2 (p. 92). Questions with 'second roots' were those in which there were two unknown quantities, as, for example, in a pair of simultaneous equations. Hérigone's question (the second of three examples given by him) is as follows. 'Thirty people – men, women and children – have spent 30 shillings, or 360 pence, but in such a way that each man has paid 5 shillings, or 60 pence; each woman has paid 10 pence, and each child 3 pence. The question is: how many men, how many women and how many children were there?' ('Trente personnes, hommes, femmes, enfans, ont depensé 30 sols, ou 360 deniers, en sorte neantmoins que chaque homme paye 5 sols, ou 60 deniers, chaque femme 10 deniers, & chaque enfant trois deniers: la demande est, combien il y auoit d'hommes, de femmes et d'enfans?').

[9] An 'undetermined' or 'indeterminate' problem is one in which there are more unknown quantities than given conditions, and therefore potentially an infinite number of solutions. Hérigone's question (see n. 8) is of this kind: there are three unknown quantities (the numbers of men, women, and children), but only two equations (from the total number of people and the total amount spent). If we let the letters h, f, and e stand for the number of *hommes*, *femmes*, and *enfants* (as Pell does in Letter 53), the problem reduces to:

$h + f + e = 30$
$60h + 10f + 3e = 360$

The paper that Pell gave Cavendish in Amsterdam is now in BL, MS Harl. 6083, fo. 129. Pell's copy is in BL, MS Add. 4415, fo. 200, with a note that he gave it to Cavendish on 1 March 1645 [see Illustration 3]. Pell's interest in indeterminate equations almost certainly arose from his study of Diophantus. The subject was to recur many years later in *An Introduction to Algebra*, and in John Wallis's account of Pell's mathematics in *A Treatise of Algebra*, pp. 214–18.

[*addressed:*] To my worthie friend M^r: John Pell Publick Professor of the Mathematicks at Amsterdam. In T'oude Conuoy op de zee dijk in Amsterdam.

[*annotated by Pell:*] Received Feb 9/19

[*cover annotated by Birch:*] 24 Janua. 1645/6 OS

[*Enclosure to Letter 51:*]

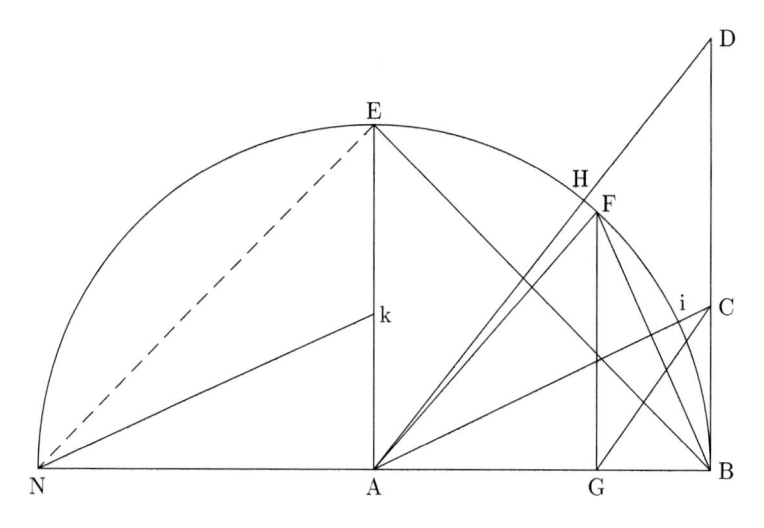

produce the arch BH to a quadr: at E. & drawe BE & AE. / from C drawe a paralel [>CG] to AD, & where it cuts AB at G, erect a p[er]pend: to AB & produce it till it cut the periphery: at F, & drawe BF & AF. the arch Bi = iH by supposition. / construction & supposition

AE = AB = AF. [er]go: AEq = ABq = AFq. / CG paralel to AD [er]go: ang: ACG = ang: CAD; which is = ang: CAG by supposi: / [er]go: ang: ACG = ang: CAG. [er]go: CG = AG. / & CGq = AGq. // BEq = AEq + BAq. that is BEq = 2BAq. / BFq = ABq − BCq. / for AFq − FGq = AGq. / & CGq − GBq = BCq. these two last aequations being added make AFq − FGq + CGq − GBq = AGq + BCq. & by transposition AFq − BCq = AGq + FGq − CGq + GBq. / but ABq = AFq : / & AG[>q:] = CG[>q:] [er]go: ABq − BCq = FGq + GBq. / but BFq = FGq + GBq. / [er]go BFq. = ABq − BCq.

  CG being paralel to AD. / BD.BC :: BA.BG. / but BA.BG :: BEq.BFq. that is BA.BG :: 2BAq.ABq − BCq. / but BD.BC :: BA.BG. / [er]go. BD.BC :: 2BAq.ABq − BCq. which was to be proued.

NB.BF.BG÷ [er]go: NBG = BFq. / NB.BE.BA÷ [er]go: NBA = BEq. [er]go: NBA.NBG :: BEq.BFq / but NBA.NBG :: BA.BG. [er]go: BA.BG :: BEq.BFq. which was etc:/ Lemma

Letter 52

6/16 February 1646

Cavendish to Pell, from Paris

BL, MS Add. 4278, fos. 241–2 (original); fo. 244^r (enclosure, original)
van Maanen C24

Worthie S^r:

manie thankes for yours of 15/25 of Jan:/[1] I vnderstand not clearlie where you write; my demonstration would run thus.[2] Because AC bisects DAB & CG is paralel to AD, therefore as ABq − BCq.2ABq :: BC.BD. but I shall not trouble you more concerning that demonstration, since you aproue of it. & though I esteeme it an honor & greate fauour from you that you would print it; which if you doe [>yet] I could wish you would not put my name to it,[3] but onelie that a friend & Scholler of yours did it at your desire; but being done for your sake I leaue it whollie to doe as you best like; for there is no cause why I should either be proude or ashamed of it; by Gods grace I shall be neither; yet I holde it the best waie to giue no occasion for an iniust censure./ I haue heere sent you the demonstration of the conuers of your Theorem as you desired; with your owne figure,[4] because I describe figures so ill./ I found it not difficult, hauing done the other before. Des Cartes Testimonie of your refutation of the Dane[5] is a verie full & true one, but I had rather he had demonstrated your theorem; which if he doe I desire you will send it me for I should desire to see what media he would vse./ M^r: [>Hobbes] Confesses Des Cartes to be a goode geometrician, & saies if he had imploied his time whollie in it, he thinkes he would haue bin inferioure to none, but nowe he prefers Roberual, Caualiero, Fermat, & Tauricel before him./ I neuer sawe yet goode demonstrations of the rules of Alligation, & Fallsoode,[6] if you knowe anye authors that haue done it, I praye you let me knowe them; or doe me the fauoure to sende me your owne, at your conuenient leasure./ I wonder I heare not from Doctor Jungius; I desire to knowe if he hath printed

---

[1] Letter 49.

[2] See Letter 49, at n. 6.

[3] The demonstration was in fact printed under Cavendish's name (Pell, *Controversiae pars prima*, pp. 52–3).

[4] See the enclosure (lacking a diagram) to this letter. In it Cavendish attempted to prove the converse of Pell's theorem, namely, that if Pell's formula holds then angle DAB is twice angle DAC etc.

[5] See Letter 49, n. 3.

[6] By 'Alligation' Cavendish means indeterminate problems of the type he had asked Pell about in Letter 51 (at n. 9). By 'Fallsoode' he means the rule of false position (see Letter 32, n. 12).

his newe Logick, & Appolonius;[7] as allso if you proceed to print Appolonius, & Diophantus./[8] I like the heads of Vincentius booke[9] so well that I intend to send for it; they saie he is an excellent man./ And so wishing you all happiness I remayne.

   Your assured friend to serue you
   Charles Cauendysshe

Paris 6/16 1645[/]1646

[*addressed:*] To my verie worthie friend M^r: John Pell, Publick Professor of the Mathematickes at Amsterdam. In T'oude Conuoy op de Zee dijke in Amsterdam.

[*annotated by Pell:*] Received Febr. 15/25

[*cover annotated by Birch:*] 6/16 Febr. 1645/6

[*Enclosure to Letter 52:*]

BD.BC :: BA.BG / AE & GF p[er]pend: to AB / AF = AB
CG, BE, BF, AF. construction

BD.BC :: 2ABq.ABq − BCq. supposition.

$\left.\begin{array}{l}(AFq - FGq = AGq) \\ (CGq - BGq = BCq)\end{array}\right\}$ added

AFq − FGq + CGq − BGq = AGq + BCq. = [*sic*]ABq − BCq = AGq + FGq − CGq + BGq.
but BFq = ABq − BCq. for BD.BC :: BA.BG.
BD.BC :: 2ABq.ABq − BCq. / BA.BG :: BEq.BFq.
[er]go: 2ABq.ABq − BCq :: BEq.BFq. / but BEq = 2ABq.
[er]go: BFq = ABq − BCq. / [er]go: BFq = AGq + FGq − CGq + BGq.
but BFq = FG[q] + BGq / [er]go: AGq + CGq. / & AG = CG. & [er]go: ang: ACG + ang: CAG / & because BD.BC :: BA.BG. [er]go: CG paralel to AD. [er]go: ang: ACG = ang: DAC. [er]go: deniq[ue] ang: DAC = ang: CAG. which was etc:[10]

---

   [7] See Letter 42, nn. 10, 11.
   [8] See Letter 14, n. 13, and Letter 23, n. 2.
   [9] See Letter 49, n. 2.
   [10] Here Cavendish attempted to prove the converse of Pell's theorem, that is, that if Pell's formula holds then angle DAB is twice angle DAC. The proof is faulty, however, since Cavendish's initial assumption that GC is parallel to AD (or, as he put it, BD : BC = BA : BG) is equivalent to saying that angle ACG = angle DAC = angle CAG, the result he was trying to prove. Pell pointed out the error in his reply (Letter 55). The = in the 6th line should be 'ergo'. In his antepenultimate line, Cavendish carelessly wrote 'FG', emended here to 'FGq'.

Letter 53

9/19 February 1646

Pell to Cavendish, from Amsterdam

BL MS Add. 4280, fo. 117^r (draft)
van Maanen P18

Right Honourable

This morning I received yours of Feb 3 St Novo.[1] Wherein I was glad to reade y^t my letters have found the way to Mr Hobbes. So that I will now make no question but y^t my last also of Jan 15.25[2] is come to his hands In it I gave you account of yours of Decemb 17.27[3] which brought your demonstration. Of which I [>may] therefore now be silent.

I have looked a little for Mersennus his proposition[4] [>but cannot] finde it; be pleased in your next to point out more distinctly y^e place where it is to be sought. [In ye meane time let me see what is to be said to y^e other which I can finde Herigon. Isagog. pag. 92.

$$\begin{array}{l|l} h = ? & 1 \mid h + f + e = 30 \\ f = ? & 2 \mid 60h + 10f + 3e = 360 \\ e = ? & 3 \mid (*) \\ & 4 \mid \textit{deleted}]^5 \end{array}$$

In ye meane time I send you heere as much of Herigon explained as I thinke you desired,[6] and rest

Your Honours most humble Servant.

Feb. 9.19.

[*postscript:*] This city hath lately multiplied me by 3/2:[7] had they done it by 3,

---

[1] Letter 51.
[2] Letter 49.
[3] Letter 47.
[4] See Letter 51, n. 6, and Letter 54, n. 4.
[5] See Letter 51, n. 8. In this deleted section of the draft letter, Hérigone's problem is set out in Pell's three-column style. The unknown quantities h, f, and e are listed on the left, and the known equations (see Letter 51, n. 9) on the right. The '(*)' in line 3 indicates that there is no third equation or condition, and that the problem is therefore indeterminate. In Pell's method, the '(*)' could be replaced at a later stage by some arbitrary condition that would allow a solution.
[6] Pell's fuller treatment of Hérigone's problem was written separately, and is not contained in this draft. Some of his work on the problem is in BL, MS Add. 4415, fo. 195^r.
[7] In other words, increased Pell's salary by half.

I should have [called over *deleted*] presently sent for all y^t I have. But as it is now, – nimium est quod intelligitur.[8]

---

[8]Pell cites a phrase from Quintilian, *The Orator's Education* [or *The Institutes of Oratory*], I.iii.17 (ed. and tr. D. A. Russell, 5 vols. (Cambridge, Mass., 2001), i, pp. 102–3), where Quintilian alludes briefly to the abuse of children by their teachers, and then breaks off with the remark: 'I will not dwell on this subject: what I am hinting at is already too much' ('Non morabor in parte hac: nimium est quod intellegitur'). Pell's purpose in using this phrase is far from clear.

Letter 54

28 February/10 March 1646

Cavendish to Pell, from Paris

BL, MS Add. 4278, fo. 245 (original)
van Maanen, C25

Worthie S^r:

I send you heere inclosed a letter to Monsr: des Cartes, which I desire you will be pleased to conuei to him./[1] I send you also a note of the same experiments which I nowe send to him, & desire he will giue a reason of the seuerall proportions of these uibrations in harde bodies of seuerall figures, I should desire the like from you but that I trouble you too much otherwayes. I giue you manie thankes for your letter[2] wherein I receiued so full an explanation of what I desired in Herigon that I nowe wonder I founde it not my self./ I neuer sawe a demonstration of Monsr: Fermats method de maximis et minimis, there are examples of it in Herigons Supplement Algeb: from page 59 to page 69./[3] if the demonstration be not longe you shall doe me a greate fauoure to demonstrate it & send it me./ The proposition in Mersennus is in his Preface to his Mechanickes pag: 2. 4th: section Generalem etiam regulam etc: to the 5th: section.[4] but I desire at this time onelie the planes & not the solides./ I hope you haue receiued my demonstration of the conuerse of your Theorem, if you haue not I shall send it you by the next after notice giuen. I am glad your citie hath increased your pension & hope they will more increase it. I longe to heare of your Diophantus[5] or your owne Analiticks. And so wishing you all happiness I remaine

    Your assured friend to serue you
    Charles Cauendysshe

---

[1] Cavendish's letter to Descartes has not survived; it contained a query about the 'centre of agitation' of suspended bodies, which Descartes answered at length in his letter to Cavendish of 30 Mar. 1646 (A&T iv, pp. 379–88).

[2] Letter 53.

[3] Fermat had first developed his method of maxima and minima in the late 1620s. His treatise 'Methodus ad disquirendam maximam et minimam' circulated from 1638 onwards and was known to Pell, and an Appendix to it was known to Cavendish (see Letter 39, n. 8), but Cavendish may not have seen or understood the original treatise. For Hérigone's work see Letter 19, n. 9; as Cavendish notes here, Hérigone gave four examples of Fermat's method of maxima and minima in his 'Supplementum algebrae', and applied it to finding tangents of conic sections (*Cursus*, vi, pp. 59–69).

[4] Mersenne, preface to 'Tractatus mechanicus', the fifth part of his *Cogitata physico-mathematica*, sigs. a1^v–a2^r: Cavendish had referred to this in Letter 51, but Pell had not been able to find it (see Letter 53, n. 4). The quoted phrase is 'A general rule, also...'

[5] See Letter 14, n. 13.

Paris Feb 28./March 10. 1645[/]1646

[*addressed:*] To my verie worthie friend M^r: John Pell publicke Professor of the Mathematickes at Amsterdam. In T'oude Conuoy op de zee dijke in Amsterdam.

[*annotated by Pell:*] Received March 8/18
    Tractatus Mechanicus Theoricus & Practicus[6]

[*cover annotated by Birch:*] 28 Febr. 1645/6

---

[6] Pell notes here the title of Mersenne's treatise: see above, n. 4.

Letter 55

2/12 March 1646

Pell to Cavendish, from Amsterdam

BL, MS Add. 4280, fos. 117–118^r (draft), 123^r (authorial fair copy of comments on Cavendish's demonstrations, referred to here as *fair copy*)
van Maanen P19

Right Honourable:

By yours of Feb. 6/16 I [>learned] y^t you had received mine of Jan 15.25.[1] I hope you have since also received ([>not onely] an extraordinary by Mr Godolphin,[2] wherein I expressed a desire of seeing ye censures of Mydorge Fermat Toricell & Cavalier)[3] but also a later of Feb 9.19 in answer to yours of Feb 3. Styl. Novo[4] So that now there remaines [>unanswered] this last of Feb 6.16. Wherein you sent me a second demonstration with a halfe-permission to print your first Which leave I shall be carefull to make use of, as becomes him y^t is [>as] jealous of your honour as any of those whom you have obliged. And therefore [>not to] give occasion to any man to say y^t I have crouded you into ye midst of a throng of pedants or made you captaine of a rout of mercenaries [whatever their birth education or genius be *deleted*]: I have resolved to make no mention of your name unlesse I can finde [>at least one] so neere your ranke y^t you may not be ashamed to be seene in publike in his company. [>Though of] our English Nobility I dare not hope it, seeing I know [>no other Englishman above] ye degree of a Baronet able to make a demonstration Except ye Earle of Leicester, who, [>as] Mr Warner told me, is a demonstrative man.[5] In this therefore [>I pray you rely upon] my care and be pleased to take y^s second demonstration once more into consideration, because I conceive it will be worth your Labour so to doe. And as yet, it falles short, of what I wished. The force

---

[1] Respectively, Letters 52 and 49.

[2] On Godolphin see Letter 50, n. 1. This 'extraordinary' letter (*sc.* one not sent by the ordinary mail) was Letter 50; it appears not to have been delivered to Cavendish (see Letter 56, at n. 2).

[3] See Letter 50, n. 4.

[4] Respectively, Letters 53 and 51.

[5] Robert Sydney, Earl of Leicester (1595–1677), Ambassador to Paris (1636–41), Chief Governor of Ireland (1641–3): Clarendon called him 'a man of great parts, very conversant in books and much addicted to the mathematics'. He was the son-in-law of Henry Percy, the ninth Earl of Northumberland and patron of the mathematician Thomas Harriot. As a young man Sydney had studied mathematics with Harriot, and in 1621, shortly before Harriot's death, he was appointed one of his executors. The other executors were Thomas Aylesbury, John Protheroe, and Walter Warner (on whom see Letter 1, n. 7), so Sydney and his 'demonstrative' (mathematical) abilities would certainly have been known to Warner.

of your former demonstration runs thus.

(1) Any BAC = CAD
(2) CG is parallel to DA
therefore GCA is = to CAD
therefore Any BAC = GCA
therefore AG = GC

which helpes to prove y^t BFq = ABq − BCq wherein lyes y^e grounds of y^e difference betweene [>y^e rest of] your demonstration & Monsieur [Palliers *deleted*] Carcavys[6] &c

If now [you *deleted* >we] would have this inverted, we must goe backward and In y^e construction (taking [>the former conclusion] for graunted y^t ABq − BCq.2ABq :: BC.BD) Seeke $\sqrt{ABq - BCq}$ and call y^t BF which inscribe &c By y^s aequation & your Lemma, demonstrate y^t BC.BD :: BG.[B]A and therefore AD, GC parallels therefore any GCA = CAD

The rest of your demonstration inverted will prove AG = CG & therefore any GCA = GAC

Out of these 2, followes: any CAD = CAG id est CAB

Quod erat demonstrandum

The fault in your [*fair copy adds:* 2d] demonstration is y^t when I desired y^t you should prove GC to be parallel to AD, you [assume *deleted* >suppose] in y^e construction y^t BD.BC :: BA.BG which is the thing to be proved, onely Expressed in other words.[7] [*fair copy adds:* For that parallelisme and this Analogy depend mutually upon one another immediately]

I hope I have [>in this loose paper inclosed][8] sufficiently shewed how I conceive y^t demonstration ought to be; [>though] not w^{th}all ye particulars, because I would have it meerely yours; It will as well deserve to be printed as any y^t I have seene, as serving for an Example of a profitable Exercise in turning of demonstrations, keeping ye diagrams unaltered.

---

[6]For Pierre de Carcavi's demonstration see Pell, *Controversiae pars prima*, p. 51.

[7]In Letter 49 Pell had already spelled out how Cavendish should prove the converse of his theorem (see Letter 49, n. 6). In this passage he goes into further detail, and points out the error in Cavendish's argument in Letter 52 (see Letter 52, n. 10). He argues as follows:

If now we would have this inverted, we must go backward and in the construction, taking the former conclusion for granted, that $\dfrac{AB^2 - BC^2}{2BA^2} = \dfrac{BC}{BD}$, seek $\sqrt{AB^2 - BC^2}$ and call that BF which inscribe etc. By this equation and your lemma, demonstrate that $\dfrac{BC}{BD} = \dfrac{BG}{BA}$ and therefore AD, GC [are] parallels [and] therefore any $G\hat{C}A = C\hat{A}D$. The rest of your demonstration inverted will prove $AG = CG$ and therefore any $G\hat{C}A = G\hat{A}C$. Out of these two follows: any $C\hat{A}D = C\hat{A}G$, that is, $C\hat{A}B$.
Quod erat demonstrandum.

The fault in your [second] demonstration is that ... you suppose in the construction that $\dfrac{BD}{BC} = \dfrac{BA}{BG}$ which is the thing to be proved ... [etc.]

[8]The text of the enclosed paper does not survive.

PART III: THE CORRESPONDENCE, LETTER 55   469

[>Last Thirsday] DesCartes came into [>our] Auditory[9] & heard me reade though when I had done he excused it, saying y^t if his [>guide] had knowen my chamber so well as my publike houre & place, he would rather have come thither to me: he went with me to my lodging: Where we had long discourse of Mathematicall matters, though I [>sought not so much] to speake, [>my selfe as to give him occasion to speake.] He thinkes it needlesse to say any thing more of my refutation [>against Longomontanus *deleted*] than he hath done & is contented y^t any body turne it into Latine & give it to ye printer.[10] I perceive he demonstrates not willingly. He sayes he hath penned very few demonstrations in his life (understand after ye style of ye Old Grecians which he affects not) THAT he never had an Euclide of his owne but in 4 dayes, 30 yeares agoe, (He is now 50 yeares old, wanting 3 months) he posted over it & so restored [>it] to him y^t had lent it him. He then thought all y^e [demonstrations very easy *deleted*] theorems [>very] manifest, but cares not to remember any more of y^m y^n 47.I de triangulo rectangulo & 4.VI de triangulis similibus.[11] He hath a slight opinion of Euclid & Apollonius for writing so largely y^t which he conceives may be put into so little roome He suspects Diophantus might be excellent in the [parts *deleted* >bookes] wch are lost,[12] but in most of the questions which he hath wrought, he finds they might have beene solved with lesse adoe [>than] Diophantus makes[13] Of all ye Ancients he magnifies none but Archimedes, who he sayes, in his bookes de Sphaera & Cylindro[14] & a piece or two more, shewes himselfe fuisse bonum Algebraicum & habuisse vere-magnum ingenium.[15] I will not trouble you with what he said of Vieta, Fermat Robervall [>& Golius][16] Of Mr Hobbes I durst make no mention to him;

[*the following paragraph in the draft is marked for insertion at this point:*] I shewed him ye titles of [>P] Vincentius his bookes.[17] He told me that there was another Jesuit (as I remember in Antwerpe) y^t threatned to write against him as soone as he came forth affirming [>y^t he had seene some theoremes false in the beginning of his second booke. Whereupon a third man sent y^m to Descartes desiring his judgement. he replyed that they were all true, but [>that] vulgar

---

[9]The lecture-hall of the Amsterdam Athenaeum.

[10]See Letter 49, n. 3.

[11]'Of the right-angled triangle' ... 'Of similar triangles'. These references are to Euclid I, proposition 47, the proof of 'Pythagoras' theorem', and VI, proposition 4, on the properties of similar triangles.

[12]Seven out of the original 13 books of Diophantus' *Arithmetica* were lost.

[13]For similar criticism of Greek mathematical writers by Descartes, see A&T iii, p. 156, and x, p. 376.

[14]Archimedes' treatise *De sphaera et cylindro*, one of his best known works, had survived in the original Greek: see Archimedes, *Opera omnia*, ed. J. L. Heiberg *et al.*, 4 vols. (Leipzig, Stuttgart, 1910–75), i, pp. 1–229.

[15]'To have been a good algebraist and to have had a truly great mind'.

[16]See, respectively, Letter 7, n. 5, Letter 4a, n. 7, Letter 19, n. 7, and Letter 23, n. 9.

[17]On Grégoire de Saint-Vincent see Letter 48, n. 5, and Letter 49, n. 2.

wits could not [>so fully] apprehend y^m as to acknowledge their truth.[18]

He says he had no other instructor for Algebra than [>ye reading of] Clavij Algebram above 30 yeares agoe[19] That, not long after comming into Denmarke, he visited Longomontanus & proffered to demonstrate to him ye ground of his error. They spent one whole day together, shut up in a chamber alone, In ye evening when they should part, he perceived y^t Longomontanus understood none of his reasons. So he thought it not worth the while to goe to him any more. He praises my way of dealing with him in rationall numbers, utterly excluding all mention or thought of surds, and [>thinkes that if] Longomontanus cannot understand y^t [>paper] he can understand nothing. And therefore wondered to heare y^t he had written twice against me.

Of Mathematicall things He is resolved to write [nothing *altered to* no more]

He says the ingenious will find them out alone, & for ye rest it is but lost labour to write. [>To] this purpose he said so much, that when he was gone I found my will to write, [>something] abated; but me thought I had a greater opinion of my skill than I had before Of which alteration you can perhaps hardly guesse the cause.

Friday comming out of the Auditory, I met him in the streete he went with me againe to my lodging & stayed with me as long as [>he had done] the day before. But [He says the Sciences are but now in incunabilis[20] *deleted*] Most of our talke tended to other matters & such as might better agree with ye capacity of a third man y^t was by

Let us therefore leave him & looke upon ye rest of your letter where (besides what you inquire of Jungius,[21] to which I can say nothing) you desire good demonstrations of the rules of Alligation & Falshood[22] But, as I remember, I have heeretofore told you, that neither of them have ever beene taught [>or confirmed] as they ought, And amongst ye taskes which I have set my selfe, that is one, To shew y^e [>grounds & supply the defects] of those rules.[23] [>Of which

---

[18] The identities of the critic and the third man are not known; but a surviving letter from Descartes (written probably in May or June 1645) to the third man gives his opinion as follows: 'Even though the propositions of the Reverend Jesuit Father, which you had taken the trouble to send to me, are very true, I do not therefore hope that he will be able to deduce from them a quadrature of the circle' ('Encore que les propositions du Reuerend Pere Iesuite que vous auiez pris la peine de m'enuoyer soient tres-vrayes, ie n'espere pas pour cela qu'il en puisse déduire la quadrature du cercle': A&T iv, p. 227).

[19] Descartes refers to C. Clavius, *Algebra* (Rome, 1608). On previous occasions (in 1637 and 1638) Descartes had similarly denied knowledge of the work of his predecessors, claiming that he had not read Viète and had never heard of Harriot. He may indeed have read little of the work of other algebraists, but given his contacts with Mersenne and the mathematical community in Paris, it seems implausible, not to say impossible, that he was unaware of mathematical ideas that had by then been in circulation for 20 to 40 years.

[20] 'In the cradle' (literally, 'in swaddling-clothes').

[21] See Letter 52, n. 7.

[22] See Letter 52, n. 6.

[23] None of Pell's work on indeterminate equations appeared in print until the publication of *An Introduction to Algebra* in 1668.

you saw a faire beginning a yeare agoe. But if the world in stead of affording me leisure, goe on with impertinent diversions to beate me off from all my intentions, My friends] must excuse

Your Honours most humble servant
J.P.

March 2/12

Letter 56

21/31 March 1646

Cavendish to Pell, from Paris

BL, MS Add. 4278, fos. 246–7 (original); fo. 254^r (enclosure, original)
van Maanen, C26

Worthie S^r:

I receiued yours of March 2/12.[1] yesterdaye; for the which I giue you manie thankes; I neuer heard (as I remember) of M^r: Godolphins comming hether, so I dout I neuer receiued that letter;[2] all the others I beleiue I haue receiued & returned my thankes for them. The censure which [>I] doute may by some be cast on me if you name me in your booke, is that I should be so ambitious as to desire to apeare in the ranke of so manie worthie Learned men; & not anie contempt for being ranked with gowne men; for I thinke few will censure so. but I refer it wholie to your pleasure; for be their censure what it will it shall not, by Gods grace make me thinke speake or doe anie thinge more than in honor & honestie I ought to doe./ I conceiue not why I may not make that analogie BD.BC :: BA.BG in the construction.[3] if BD, BC, BA, be giuen, which I suppose they are, & from that analogie to proue CG paralel to AD & the more mediatelie it followes the better;/ I further conceiue the proposition which I was to proue is that this analogie being granted 2ABq.ABq − BCq :: BD.BC. that then the ang: GAC = ang: DAC, which I suppose I did in my former; but neuer the less I haue heer sent you an other that waye you would haue.[4] I desire your censure both of this & why I may not take that [>analogie for] construction in the former./ If my former demonstration agreed with Monsr: Carcauius in anie parte of it,[5] it is more than I knowe for I assure you I plaied [>not] the Truant at all that waye, for I had it not by me when I made myne neither did I thinke of it at all, nor tooke anie copie of it, nor of anie of the rest but onelie one of Monsr: Roberualls[6] & this of Caualiers[7] which P: Mersenne gaue me to send to you, who remembers him to you. I heare no newes of the rest but onelie that

---

[1] Letter 55.

[2] See Letter 50, n. 1, and Letter 55, n. 2.

[3] Cavendish failed to understand Pell's objection in Letter 55 (at n. 7).

[4] See the enclosure to this letter. In it Cavendish gave a construction for BF and went on to prove that angle GAC = angle DAC. All this was as Pell requested in Letter 55 (see Letter 55, n. 7).

[5] See Pell's comment in Letter 55, at n. 6.

[6] On Roberval's demonstration see Letter 35, n. 2.

[7] On Cavalieri's demonstration of Pell's theorem see Letter 50, n. 4. A fair copy of Cavalieri's demonstration, in Pell's hand, is in BL, MS Add. 4278, fos. 251^r, 252^r, 253^v.

Toricel is about a Treatise of the center of grauitie.[8] I shall againe put Mersen in mynde of your desire of the censures of Fermat & the rest./[9] I am most glad that you are entred in to an aquaintance with M^r: de Cartes, who douteless is an excellent man, yet I praye you let not his discourse perswade you from writing & printing your intended worckes for I conceiue not but that the greatest witts & most learned doe much benefit on an other by publishing & communicating theyr choice thoughts one to an other, & that euen meaner witts & Clerkes gleane something from them too. I something meruaile at his censure of those olde greekes, & especiallie of Diophantus; I desire by your next your opinion of Diophantus & why your will to write was abated, & the opinion of your skill increased by M^r: de Cartes discourse;[10] I thinke you will not be the prouder for my commendations & therefore I will freelie write my thoughtes; that as far as I can guess either by Monsr: des Cartes writings or by Monsr: Roberualls discourse they are neither of them your aequalls in the Analitickes; though I finde greate cause to esteem them both excellent in it. I desire allso if you please, to let me knowe Monsr: des Cartes opinion of Monsr: Roberuall & the rest./ If it be not too much trouble I desire you will explane in Monsr: de Cartes his geometrie pag: 386 howe $\frac{q}{2y}$ is $\frac{1}{2}a\sqrt{aa+cc}$. and howe [>the value of] z pag: 387 is $\frac{1}{2}\sqrt{aa+cc}+\sqrt{\frac{1}{2}aa+\frac{1}{4}cc+\frac{1}{2}a\sqrt{aa+cc}}$ or etc: & so why x etc: I beleeu it is

---

[8]Torricelli had been working on a method to find the centre of gravity of various different figures since 1643; he would announce his solution to the problem in a letter to Cavalieri on [28 Mar./] 7 April 1646 (see P. Galluzzi and M. Torrini, eds., *Le opere dei discepoli di Galileo Galilei: carteggio (1642–1648)* (Florence, 1975), pp. 47, 54–7, 284–5).

[9]See Letter 50, n. 4, and Letter 55, n. 3.

[10]For Descartes's comment about the Greeks, and Pell's comment about himself, see Letter 55.

fals printed./[11] I will not trouble you anie more concerning the rules of Aligation & falsoode, I remember you shewed me a faire beginning concerning those rules, which I hope you will shortelie print in some of your intended worckes. which I much desire to see. S^r: I haue troubled enough for once. And so wishing you all happiness I remaine.

Your assured friend to serue you
Charles Cauendysshe

Paris, March 21/31. 1646.

[addressed:] To my verie worthie friend M^r: John Pell Publicke Professor of the

---

[11] For the passage in Descartes's 'Géométrie' see A&T vi, p. 461, lines 14–17. Pell later wrote a separate paper (sent as the enclosure to Letter 59) discussing this passage, and concluding: 'So that none of those 4 aequations were false printed, as was suspected': BL, MS Harl. 6083, fo. 128^r. (Drafts or copies of parts of this paper are in BL, MS Add. 4280, fos. 119^r, 120^r, 121^v, 122^r.) Descartes's argument was that the equation:

$$+x^4 \,[\pm]\, pxx \,[\pm]\, qx \,[\pm]\, r = 0$$

could be reduced to a product of two quadratics:

and
$$xx - yx + \tfrac{1}{2}yy \pm \tfrac{1}{2}p \mp q/2y = 0$$

$$xx + yx + \tfrac{1}{2}yy \pm \tfrac{1}{2}p \mp q/2y = 0$$

where y must satisfy the cubic equation (in y^2):

$$+y^6 \,[\pm]\, 2py^4 + (pp\,[\mp]\,4r)yy - qq = 0.$$

Applying Descartes's rule to the equation:

$$z^4 + (\tfrac{1}{2}a^2 - c^2)z^2 - (a^3 + ac^2)z + (\tfrac{5}{16}a^4 - \tfrac{1}{4}a^2c^2) = 0$$

gives the cubic:

$$y^6 + (a^2 - 2c^2)y^4 + (c^4 - a^4)y^2 - a^6 - 2a^4c^2 - a^2c^4 = 0.$$

A solution of this equation is $y^2 = a^2 + c^2$, as may easily be checked, but Descartes did not give his working, merely remarking that the equation could be solved by the method already explained ('par la methode desia expliquée'). With this value of y^2:

$$\frac{q}{2y} = \frac{a^3 + ac^2}{2\sqrt{a^2 + c^2}} = \frac{a}{2}\sqrt{a^2 + c^2}$$

(the first result questioned by Cavendish). Now the biquadratic in z factorizes into the two quadratics:

$$z^2 - \sqrt{a^2 + c^2}\,z + \tfrac{3}{4}a^2 - \tfrac{1}{2}a\sqrt{a^2 + c^2} = 0$$
$$z^2 + \sqrt{a^2 + c^2}\,z + \tfrac{3}{4}a^2 - \tfrac{1}{2}a\sqrt{a^2 + c^2} = 0$$

yielding, by the usual formula for quadratic equations:

$$z = \tfrac{1}{2}\sqrt{a^2 + c^2} \pm \sqrt{-\tfrac{1}{2}a^2 + \tfrac{1}{4}c^2 + \tfrac{1}{2}a\sqrt{a^2 + c^2}}$$

(the second result questioned by Cavendish). The manipulations are a little lengthy but Descartes's working is correct, and not 'fals printed' as Cavendish thought.

PART III: THE CORRESPONDENCE, LETTER 56   475

Mathematickes at Amsterdam. In T'oude Conuoy op de Zee diike in Amsterdam.

[*annotated by Pell:*] Received March 30 [/] April 9

[*cover annotated by Birch:*] 21 March 1645/6

[*Enclosure to Letter 56:*]

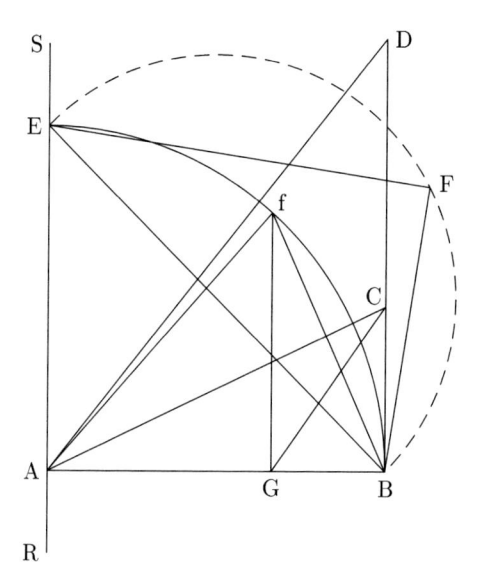

vpon BE as a diameter describe the semicircle EFB. from the point [>A] as center & rad: AC describe a circle til it cut AE produced in S. / produce EA, so that AR be made aequal to ES. & from the point E, & distance ER = AS = AC, inscribe EF in the pointed circle & ioine BF; / BFq = ABq − BCq. / for ABq + BCq added to ABq − BCq makes 2ABq. / but BEq = 2ABq & EFq + FBq = BEq. / & EFq = ACq = ABq + BCq. [er]go: BFq = ABq − BCq.
from the point B, distance BF inscribe Bf in the primarie circle BfE, & from f, let fall a p[er]pend: fG & ioine CG. by the lemma as BEq.BFq :: BA.BG. that is 2ABq.ABq − BCq :: BA.BG but by supposit: as 2ABq.ABq − BCq :: BD.BC. / [er]go. $BD.BC :: BA.BG$ // Afq = AGq + fG[q] = ABq / GCq = GBq + BCq / added Afq + GCq = AGq + Gfq + GBq + BCq / AB[q] − BCq = AGq + Gfq + GBq − GCq / but ABq − BCq = Bfq / [er]go: Bfq = AGq + Gfq + GBq − GCq. but Bfq = Gfq+GBq; [er]go: AGq = GCq. / & AG = GC. & ang: GCA = ang: GAC; but ang: GCA = ang: DAC. (for CG is paralel to AD. because $BD.BC :: BA.BG$ as hath bin formerlie proued) [er]go: ang: GAC = ang: DAC which was to be proued.[12]

---

[12] Here Cavendish gave a construction for BF and went on to prove that angle GAC = angle DAC. All this was as Pell requested in Letter 55 (see Letter 55, n. 7). In the equation immediately after the double virgule, Cavendish carelessly wrote 'fG', emended here to 'fGq'; in the following line, similarly, he wrote 'AB', emended here to 'ABq'.

Letter 57
23 March/2 April 1646
Pell to Cavendish, from Amsterdam

BL, MS Add. 4280, fo. 118^r (draft)
van Maanen P20

Right Honourable

My last was written y^s day 3 weekes.[1] Since I have received one of yours as [you may see *deleted* >appeares] by this of Des Cartes which I received yesterday[2] [>some of the] contents of it I shall hope to heare from your selfe; for as for mine owne part I have no leisure to make those experiments & it is an inviolable rule with me never to seeke cur sit ere I be sure of an sit,[3] nor much to trust other mens relations for an sit especially Mersenn's.[4] I had thought to have sent you [my explanation of *deleted*] Herigons examples De maximis & minimis wrought after my fashion[5] But of this be assured y^t Mr Fermat hath given no perfect rule [>for y^t inquiry] Des Cartes when he was heere, said y^t he [>had] never thought of any such probleme till they sent him that rule of Mons Fermat, which said he I examined & sent him instances [>of cases] wherein it would faile This perhaps you may finde in Paris. But neither the one nor the other can perfectly solve y^t Probleme.[6] I make my selfe beleeve y^t I can [>there also] say something new. But of y^s at more leisure. For if I write any more now, I shall neglect y^s poste & y^{se} will come a weeke later to you than needes. I therefore take my leave & rest

Your [>Honours most humble & most obedient] servant

Aprill. 2. Easter Munday. [*at head of letter:* [Aprill 2 Stylo Novo *deleted*]]

[*postscript:*] [>I want] ye Christen names of [>Mess] Carcavy & Pallieur.[7]

---

[1] Letter 55.

[2] Pell thus forwarded the long letter from Descartes to Cavendish, dated [20/] 30 Mar., concerning the 'centre of agitation' of suspended bodies, of which the autograph survives: A&T iv, pp. 379–88.

[3] 'Why it is' ... 'whether it is'.

[4] These discussions of the 'centre of agitation' had apparently been started by Mersenne; see Descartes's earlier letter to Mersenne on this topic, A&T iv, pp. 366–70, and the undated fragment of a letter from Mersenne to Cavendish (datable to some time after Oct. 1645) in BL, MS Add. 4278, fo. 248^r.

[5] See Letter 54, n. 3.

[6] Fermat demonstrated his method of 'maxima and minima' only for some simple cases. Hérigone's method was directly based on Fermat's. A general method of finding maxima and minima became possible only after the development of the differential calculus some 20 to 30 years later.

[7] Pell had already been correctly informed of de Carcavi's Christian name (see Letter 38, n. 19); Le Pailleur's name was Jacques.

Letter 58

18/28 April 1646

Cavendish to Pell, from Paris

BL, MS Add. 4278, fos. 249–50 (original)
van Maanen C27

Worthie S^r

    manie thankes for yours with M^r. des Cartes;[1] if you would haue a coppie of it, I shall send you one: M^r: Hobbes praised it verie much at first, but after Monsr: Roberuall douted of it, M^r: Hobbes seemeth to doute too./ Mr Roberuall is confident that neither the proposition is true nor his argumentation good for the proofe of it.[2] I desired him to write his exceptions which he saies he will. I assure you I esteeme highlie of Monsr: de Cartes let them saye what they will; & as I thinke Monsr: de Cartes hath shewn a huge wit & abilitie in this letter to me: but I dare not giue an absolute censure between so excellent men, especiallie till I fullie conceiue all M^r: Roberualls exceptions concerning this letter. M^r: Roberuall tells me that M^r: Fermats method de maximis et minimis[3] & the application of it to tangents is most assuredlie true in all aequations which are quadratique & he thinkes in higher powers too; & that though M^r: de Cartes tooke some exceptions of it at first, yet afterwardes he confessed it to be true & a better waye to finde the tangents to curue lines than his owne./[4] I am nowe reading a Tract of M^r: Fermats concerning the places of Indefinite aequations as for example if the aequation be AA = BE. where A & E are the vnknowne indetermined quantities, & B; a knowne quantitie. the place for this aequation is a parabola, whose latus rectum is B; applicata is A; & the distans in the axe from the vertex to the applicata is E./ but if the aequation be BD = AE the

---

    [1]Letter 57, with the enclosed letter from Descartes to Cavendish.

    [2]For the objections by Roberval (on whom see Letter 19, n. 7) to Descartes's theory see Letter 61, n. 2.

    [3]See Letter 54, n. 3.

    [4]The problem of finding tangents to curves exercised many mathematicians in this period, but, as with the (related) problem of maxima and minima, a full solution became possible only with the development of the calculus. Fermat's method was based on finding maxima and minima and worked well for quadratic curves. Descartes's method, outlined in the 'Géométrie', required the construction of a normal to the curve (a line at right angles to the tangent), from which the tangent itself could then be found; his idea was simple and elegant in theory, but was not easy to carry out in practice.

478   JOHN PELL (1611–1685)

place is an hyperbole AC:[5] I knowe not whether this tract be amongst my papers in england;/ but I am transcribing of it; not knowing when I shall see the other if I haue it./ I longe to heare that you print something: I rest:

   Your assured friend to serue you
   Charles Cauendysshe

P[aris *page torn*] Aprill. 18/28: 1646.

[*addressed:*] To my verie worthie friend Mr John Pell Publicke Professor of the Mathematicks at Amsterdame In T'oude Conuoy op de zee diike in Amsterdame.

[*annotated by Pell:*] Received April 25 [/] May 5

[*cover annotated by Birch:*] 18/28 Apr. 1646

---

[5]Cavendish was referring to Fermat's 'Ad locos planos et solidos isagoge', written c.1636, of which a copy survives in a collection of Cavendish's papers: BL, MS Harl. 6083, fos. 113–14. (The text was eventually published in Fermat, *Varia opera*, pp. 2–7.) In classical mathematics a 'locus' was a path traced out by a point moving under certain restraints; such 'loci' were Cavendish's 'places'. Fermat, following Apollonius, was particularly interested in cases where the path was a conic section (circle, ellipse, parabola, or hyperbola), and in 'Ad locos planos' he derived the equations of these curves algebraically. Bearing in mind that Fermat, like Viète, used A and E for his unknown or variable quantities, the equations quoted by Cavendish may be rewritten in modern notation with y instead of A for the ordinates ('applicata') and x instead of E for the distances from the vertex, or origin. They then become $y^2 = bx$, the equation of a parabola with 'latus rectum' (width at the focus) b; and $yx = bd$, the equation of a rectangular hyperbola.

Letter 59

4/14 May 1646

Pell to Cavendish, from Amsterdam

BL, MS Add. 4280, fo. 118^r (draft)
van Maanen P21

Right Honourable

If I use many words I shall lose this poste. The transcribing of this inclosed paper[1] hath cost me more time than I made account of, & the printer robbed me of the rest, for my booke is in the presse.[2]

I have sent your second to Des Cartes.[3] I desire Mons. Pallieurs name[4] & Cavalieris judgement of my refutation, though it be but 2 or 3 lines excerpted out of his letter to P. Mersenne; for it is likely that he said something more than barely *There is a demonstration of your Theoreme*.[5] I am removed & therefore I pray you now to direct your letters to

Your Honours most humble servant
In 't vergulde Hart in de nieuwe hoogstrate t' Amsterdam.[6]

May 4.14. 1646

---

[1] See Letter 56, n. 11.

[2] Pell, *Controversiae pars prima*.

[3] This letter, containing further queries from Mersenne about Descartes's theory of the 'centre of agitation' of suspended bodies, has not survived, but its contents can be partly deduced from Descartes's reply of [5/] 15 May (A&T iv, pp. 415–19).

[4] See Letter 57, n. 7.

[5] On Cavalieri's demonstration of Pell's theorem see Letter 50, n. 4. Although Cavalieri had previously sent this to Mersenne, it appears that it had not yet been forwarded to Pell. The letter from Cavalieri to Mersenne ([7/] 17 Mar. 1646: *MC* xiv, pp. 143–6) began: 'I believe you have already received my demonstration of Pell's rule...' ('Credo quod jam acceperis meam demonstrationem regulae Pellii...').

[6] 'At the gilded hart, in the Nieuwe Hoogstraat, Amsterdam'. The Nieuwe Hoogstraat (called 'new' because it was constructed as part of the expansion of the city in 1593) is a continuation of the Oude Hoogstraat, to the east of the Kloveniersburgswal (see M. van Gelder, *Amsterdamsche straatsnamen geschiedkundig verklaard* (Amsterdam, 1913), p. 54).

Letter 60

11/21 May 1646

Pell to Cavendish, from Amsterdam

BL, MS Add. 4280, fo. 118 (draft)
van Maanen P22

Right Honourable

The last weeke I [>sent] explications of those 4 [>Cartesian] aequations [in DesCartes his Geometry pag. 386 &c. as you desired in *deleted*] proposed in your letter of March 21/31[1] wherein are many other passages to be answered if I had time. But my letters emptinesse will be this once supplyed by this inclosed from Des Cartes.[2] [If Mr Robervall[3] could be brought to [>seeke] ye demonstration of Mr Fermats way to determine maxima & minima,[4] he would finde the imperfection of it & y^t for that it can doe would see how to expresse it better. Shall I ever have leisure to shew the world what holes *deleted*]

[To finde Tangents for curve lines[5] you have in [>English] an Exercitation of mine shewing 4 or 5 severall wayes to finde Tangents for curve lines *deleted*]

I give you thankes for offering me a coppy of his former:[6] but being taken up with thoughts so different from all y^t can be said of that argument, I thinke it will be better y^t I doe not so much as looke after it till I be more at leisure: In the meane time perhaps Mr Robervalls exceptions & his reply will open the whole matter more cleerely.

To determine Maxima & Minima in all speculations Mathematicall is a Probleme infinitely above all that ever I yet saw. Yet it would be worthy Mr Robervalls labour to demonstrate ye truth & the bounds of Mr Fermats method. For it is not enough for him to say, *I know it is true in aequations quadraticall & thinke y^t it is so in higher powers too.*[7]

I doe not yet wholly despaire once againe to see your Honour & those collections which you make among those nimble & able wits of France: such as that you speake of, De Locis.[8] Of which argument I have also some notions, which I beleeve will enable me to finde flawes in all that hath beene yet said of it.

---

[1] See Letter 56, n. 11, and Letter 59, at n. 1.

[2] For this letter from Descartes to Cavendish, dated [5/] 15 May 1646, see A&T iv, pp. 415–19.

[3] See Letter 19, n. 7.

[4] See Letter 54, n. 3, and Letter 58, n. 3.

[5] Pell's treatment of tangents has not been identified.

[6] The previous letter from Descartes to Cavendish: see Letter 58, n. 1.

[7] Pell puts into direct speech the opinion attributed to Roberval in Letter 58.

[8] This apparently refers to the treatise by Fermat, 'Ad locos planos': see Letter 58, n. 5.

My freind from Brussells writes that he hath beene in Lovaine with D[r] Gutischovius[9] and shewed [>him] my paper against Longomontanus[10] which he transcribed, saying, that it was ingenious. It may be heereafter we shall have some further approbation under his hand. He writes y[t] he shewed it also to P. Vincentius,[11] who read it twice or thrice over but understood it not, and confessed himselfe not versed in Algebra at all. I am much afraid he [>will give us] a monstrous Cyclometry. My old man hath printed a kind of [bable *deleted* >resverie] of 3 sheetes of paper & dedicated it to Golius of Leiden.[12] [Meane time, my booke is in the presse *deleted*] Who lately wrote to me to let Apollonius alone for him, as having determined to goe in hand with it, as soone as his Arabicke lexicon is done.[13] Yesterday his Arabian comming to see me[14] I asked him how farre they were come, he answered they were in Jîm, that is, words beginning with the fift letter. I asked when he thought it would be finished; he said, About 6 or 7 yeeres hence. I asked how old he thought Golius was; he replyed At least fifty [>yeeres old] I would I [>could persuade him to lend me] a hundred pound upon condition to receive two hundred for it, when he publishes Apollonius.[15]

For mine owne part I [>have not promised to waite his leisure. But he needes not feare that] I shall begin the publishing of any other thing, till this be finished which I have now in y[e] presse against Longomontanus:[16] to which P. Mersenne [>may perhaps have something more] to contribute, [>but he must be] admonished to send it hither in due time. [>Which would lay a] further obligation upon

Your Honours most humble & most obedient Servant.

May 11.21
In 't vergulde Hart In de nieuwe hooghstraet t' Amsterdam.

---

[9] On van Gutschoven see Letter 42, n. 6; Pell's 'friend' has not been identified.

[10] The 'refutatiuncula'. [11] Grégoire de Saint-Vincent: see Letter 48, n. 5.

[12] The 'old man' was Longomontanus; this work was his *Caput tertium*. The work consists of three quarto gatherings (A^4–C^4), paginated 73–96 (as it was intended to be bound after *Rotundi in plano*, which has 72 pp.); the 'appendicula' or short appendix attacking Pell is on pp. 92–6. On Golius see Letter 23, n. 9.

[13] On Golius's plans to translate books 5–7 of Apollonius, and his Arabic lexicon, see Letter 35.

[14] Nicolaus Petri, an Arabic-speaking weaver from Aleppo (Greek Orthodox in religion), who was brought to Europe by Christian Ravius in 1641 and later worked for Golius (see Toomer, *Eastern Wisedome*, pp. 144–5, 185). In a letter to Johannes Coccejus of 23 May 1646, Ravius referred to this meeting between Petri and Pell: see Coccejus, *Opera ἀνέκδοτα*, ii, p. 666.

[15] Pell possibly alludes here to the common practice of British travellers to the Ottoman Empire: in the words of Samuel Chew, 'A traveller before starting upon his voyage "put out" a sum of money ; if he never returned home the person or group who had assumed the risk retained the premium; but if he returned he received anywhere from two to five times the amount according to the estimate of the dangers of the voyage' (*The Crescent and the Rose: Islam and England during the Renaissance* (New York, 1937), p. 32).

[16] Pell, *Controversiae pars prima*.

Letter 61

22 May/1 June 1646

Cavendish to Pell, from Paris

BL, MS Add. 4278, fos. 255–6 (original)
van Maanen C28

Worthie S^r

I giue you manie thankes for yours of May 11/21 with which I receiued one from M^r: desCartes;[1] & doe heerin send one to him with M^r: Roberualls exceptions to his former letter to me concerning vibrations./[2] I giue you manie thankes for your explaining those aequations of M^r: desCartes his booke;[3] I see I must take heed howe I desire anie fauoure from you for I see you will not spare your self to giue me satisfaction, well I againe giue you manie thankes for it & will be carefull heereafter not to diuert you ouermuch with my quaeres. I doute I haue made the same faulte to Monsr: desCartes who douteless is a most excellent man; but I doute not both of his & your pardon & doe intend still to trouble you both nowe & then, but sparinglie./ I doute I shall not nowe send you Monsr: Palliers christian name[4] for I yet knowe it not, & the post goeing a daye sooner than vsuall, [>&] those I imploied to inquire it not knoweing that makes them thinke it time enough to tell me it to morrowe (I suppose.) but I shall not [>faile] God willing to send you [>it] by the next./ I hope you haue receiued both those demonstrations of myne of the conuerse of your theorem, which I assure my self are both true. yet I longe to heare your iudgment of them./ Mersennus is not in the Towne;[5] but I hope you haue receiued Caualieros demonstration of your Theorem which I sent you from him./[6] The post is nowe goeing so that I can no more but wishing you all happiness I remaine

    Your assured friend & seruant
    Charles Cauendysshe

Paris [Maie *deleted*] June 1. newe stile 1646.

---

[1] Letter 60 (and see Letter 60, n. 2).

[2] Roberval's letter (to Cavendish, for Descartes) is printed in A&T iv, pp. 420–8. (For Descartes's 'former letter ... concerning vibrations' see Letter 57, n. 2) From Cavendish's wording here, and from Descartes's reply (A&T iv, pp. 429–35), it appears that he also enclosed a letter of his own to Descartes; but that letter has not survived.

[3] See Letter 56, n. 11, and Letter 59, at n. 1.

[4] See Letter 57, n. 7.

[5] Mersenne was on an extended tour of south-western France; he had left Paris on [10/] 20 Apr., and would return in the last week of Aug. (*MC* xiv, pp. 212–16).

[6] See Letter 50, n. 4, and Letter 59, n. 5.

[*addressed:*] To my verie worthie friend Mr: John Pell Publick Professor of the Mathematicks at Amsterdam. In't vergulde Hart in de niewe hooghstrate. T' Amsterdam.

[*annotated by Pell:*] Received June 9 new style
Cost 24 stuyvers.[7]

[*cover annotated by Birch:*] 1 June N.S. 1646

---

[7] There were 20 stuivers in one guilder; for the value of a guilder, see Letter 63, n. 3.

Letter 62

28 May/7 June 1646

Cavendish to Pell, from Paris

BL, MS Add. 4278, fos. 257–8 (original)

Worthie S^r:

 I am ashamed to trouble you so often, & not doe you the seruice which I woulde; for I can not nowe send you Monsr: Palliers' Christian name as I promised to doe[1] but my agents failed me; & M^r: Petit[2] whom I most relied on, is gone for England, so that I dare not nowe promise anie more; I spoke to daie to M^r: Hobbes & he to M^r: Roberuall[3] concerning it; but what they will doe I knowe not. In the meane time I send you heere a demonstration of your theorem by M^r: Mydorge & of his owne hand;[4] It is Analiticall & verie compendious as he hath set it downe, but if the proofes of what he takes as graunted were to be set downe I thinke it would be as longe as some others are; but it is a goode one & so I doute not but you will esteem of it. I asked M^r: Mydorge if he would haue his demonstration printed as it is, Analiticall, & he saied I./[5] Some of the ancients indeed I thinke vse it sometimes. Pere Mersennus is not yet returned,[6] but I thinke Caualiero writ no more to him concerning your refutation & theorem, than I sent you.[7] M^r: Mydorge highlie commends your refutation, & saies your theorem is a most vsefull one, which all ingenious theoremes are not, more than to exercise ratiocination./ I wonder I neuer hearde from Doctor Jungius./[8] I longe to heare that you print your Diophantus[9] or your owne Analitickes. And so wishing you all happiness I remayne

 Your assured friend to serue you
 Charles Cauendysshe

Paris June 7. [olde *deleted* >newe] stile 1646.

[*addressed:*] To my verie worthie friend M^r: John Pell, Publicke Professor of the Mathematickes in Amsterdam. In't vergulde Hart in de nieuwe hooghstrate t' Amsterdam.

[*cover annotated by Birch:*] 7 June 1646 N.S.

---

 [1] See Letter 57, n. 7, and Letter 61, n. 4.
 [2] William Petty (see Letter 42, n. 8).   [3] See Letter 19, n. 7.
 [4] Mydorge's 'compendious' demonstration (supplemented later by a more lengthy comment) reached the printer in time to be added at the end of the book (see Letter 50, n. 4).
 [5] = Aye.   [6] See Letter 61, n. 5.   [7] See Letter 61, n. 6.
 [8] Cavendish had apparently received no reply to his letter to Jungius of [1/] 11 May 1645: see Letter 34, n. 7.
 [9] See Letter 14, at n. 13.

Letter 63

[29 June/] 9 July 1646

Pell to Cavendish, from Amsterdam

BL, MS Add. 4280, fo. 118^v (draft)

Right Honourable

I had sent this inclosed[1] sooner, if I had beene at home, but I came home but 5 dayes agoe from Breda, where ye prince of Orange [>had?] detained me & some others whom he made professors in his new Colledge & Gymnasium y^t he is erecting in that City.[2] [>So that now I am deliberating whether] I shall leave y^s which I have heere and accept of y^t new condition (1000 guilders[3] a yeare, with promises of greater matters I know not when) to be professor of Philosophy & Mathematickes in Breda. Thus the world [>goes on] to give me diversions enough, nor can I see when I shall enjoy [y^t which I do so extreamely desire, *deleted*] leisure to draw the picture of my owne soule for the use of posterity [as I desire, conceiving it to be ye onely meanes for me to doe good when I am dead *deleted*] For without doubt something I could say, not meerely unworthy the reading. But this of Breda will robbe me of all my time & torment me with most tedious pedantery under ye name of philosophicall disputations.[4] [Which if I once meddle withall, I shall fall out of mine owne favour & count my selfe ye most unprofitable burden that ye earth beares. so extreamely doe I abhorre the senselesse wranglings of vulgar philosophy, with which many a hopefull young wit is utterly marred & made incorrigible. *deleted*]

Since my last I have received 2 of yours dated June 1 & June 7.[5] I thanke Mr Midorge for his demonstration,[6] and since as you say he highly commends my refutation I could wish that he could find time to write though it were but a line or two to that purpose. Otherwise my grumbly adversaries will say that most of my adstipulators[7] acknowledge [>ye truth of] my Theoreme but like [>not] my manner of using it. And some no doubt will conceive [>y^t] the silence of so many skilfull men [>is] onely a courteous manner of [>signifying] their dislike

But let men thinke & say of it as they please, for mine owne part I intended it

---

[1] Unidentified.

[2] The Illustre School at Breda, also known as the Collegium Auriacum or College of Orange.

[3] The rate of exchange was 11.11 guilders to £1 sterling (see McCusker, *Money and Exchange*, p. 44); thus 1000 guilders was worth £90.

[4] Shortly before he took up his post at Breda, Pell's appointment was changed to a professorship of mathematics only.

[5] Letters 61, 62.

[6] See Letter 50, n. 4.

[7] An adstipulator is one who joins as an additional party in a bargain.

for a Specimen not onely of my selfe & the printer, but also of the readers, and had it found no better entertainement with others than with Longomontanus & Mr Hardy[8] I had irrevocably decreed never to have printed any thing Mathematicall. For why should I take so much paines if I were so unhappy in my expressions y^t I could make no man understand me?

But it hath found better successe and obtained ye approbations of the most skillfull and for the rest, if by printing we will needes cast our pearles before swine, We must be content to beare ye punishment y^t followes it, to see [>ourselves torne &c,] our pearles troden under foote.

You will be pleased to pardon the extravagancies of

Your Honours most humble servant

July 9

---

[8] See the comments on Hardy in Letter 24.

Letter 64

9/19 July 1646

Cavendish to Pell, from Paris

BL, MS Add. 4278, fos. 259–60 (original)
van Maanen C30

Worthie S^r:

manie thankes for yours of Juli 9th: st: No:[1] which I receiued the other daye. I am glad it is in your choise to be Professor at Breda,[2] I confess I should be sorrie so much of your time shoulde be taken from you as to hinder you from publishing your intended worckes except it were much to your aduantage other waies, & to that I should submit, but you knowe best what you haue to doe. M^r: Hobbes is goeing out of towne to a more retired place[3] for his [suties *altered to* sudies [*sic*]] so that you must not nowe direct your letters to him but to Monsr: de Bose[4] thus A Monsr: Monsr: de Bose Secretaire du Roy et valet de chambre de la Reine; Rue de Four: vis a vis d'un Apoticaire.[5] if you direct it allso to me in english as you did your last it will be a sure waie./ M^r: Mydorg is out of towne, but I haue writ to him as you desire./[6] And so wishing you all happiness I remaine

    Your assured friend to serue you
    Charles Cauendysshe

Paris Julie 9/19 1646

[*addressed:*] To my verie worthie friend M^r: John Pell Publick Professor of the mathematickes in Amsterdam. In't vergulde Hart in de nieuwe hooghstrate t' Amsterdam.

[*cover annotated by Birch:*] 9/19 July 1646

---

[1] Letter 63.

[2] See Letter 63, n. 2.

[3] Hobbes had been invited by his friend Thomas de Martel to stay with him in Montauban, in the south of France; in the event he did not go, being required instead to teach mathematics to the Prince of Wales in Paris. (See Hobbes, *Correspondence*, i, pp. 131–3, 136–7; ii, p. 851.)

[4] Charles du Bosc (d. 1659) had stayed as a young man with the second Earl of Devonshire in England, and was thus an old acquaintance of Hobbes. A courtier with strong philosophical interests, he was also a friend of Mersenne. (See Hobbes, *Correspondence*, ii, pp. 795–7.)

[5] 'M. [du Bosc], secretary to the King and *valet de chambre* to the Queen, rue [du] Four, opposite an apothecary'. The rue du Four runs off the Boulevard Saint-Germain, in what is now the 6th arrondissement.

[6] See Letter 63, at n. 6.

Letter 65

7/17 August 1646

Cavendish to Pell, from Paris

BL, MS Add. 4278, fos. 261–2 (original)
van Maanen C31

Worthie S^r:

I giue you manie thankes for your letter.[1] I thought to haue writ to M^r: des Cartes, but I thought best to defer it awhyle, till I might knowe somewhat of the opinion of some of oure learned men heer, concerning the controuersie between him & M^r: de Roberualle, touching the center of agitation;[2] I finde them hitherto lothe to giue theire iudgment, therefore it becomes me not to presume to giue anie; but I shall presume God willing shortelie to returne M^r: de Cartes manie thankes for his fauoures to me, both concerning that quaere & otherwaies, & I assure you I esteeme as highlie of him as I did before this controuersie, & more, for if there be furder considerations, yet what he hath saied is most ingenious./ I longe to heare if Vietas worckes be yet printed,[3] & vincentio of the quadrat: of the circle etc:.[4] but most of all of somewhat of yours, especiallie your analitickes. Both you & I are beholding to Monsr: Mydorge, who sent me this inclosed last weeke,[5] I expecting but a line or two of his approbation of your refutation of Longomontanus. I haue troubled you enough at this time. And so wishing you all happiness I rest

 Your assured friend to serue you
 Charles Cauendysshe

Paris Aug: 7/17 1646.

[*addressed:*] To my verie worthie friend M^r: John Pell Publicke Professor of the

---

[1] In the light of Pell's comment at the start of Letter 67, it appears that this refers to Letter 63 – to which Cavendish had already replied in Letter 64.

[2] This had become a delicate matter. Descartes's letter to Cavendish of [5/] 15 June 1646 (A&T iv, pp. 429–35), dismissing Roberval's theories on this subject, had evinced an angry reply from Roberval (ibid., pp. 502–8). Cavendish had found this reply so ill-tempered that he refused to pass it on to Descartes, and it had eventually been transmitted via Mersenne instead: see Descartes's comment in his letter to Mersenne of [23 Oct./] 2 Nov. 1646, (ibid., p. 555), and Cavendish's own account in Letter 68.

[3] See Letter 42, n. 2. Van Schooten's edition of Viète was not apparently completed until late Oct. or early Nov. 1646: see *MC* xiv, p. 585, n. 4.

[4] De Saint-Vincent's work was published in 1647 (see Letter 49, n. 2).

[5] This was a further and much lengthier comment by Mydorge, following the 'compendious' demonstration previously sent (see Letter 62, n. 4), and also incorporated in the text by him printed in Pell, *Controversiae pars prima*.

Mathematickes at Amsterdam. In't vergulde Hart, in de nieuwe hooghstrate t'Amsterdam.

[*annotated by Pell:*] Received Sept [3 *deleted*] [/] Aug 29 at Breda

[*cover annotated by Birch:*] 7/17 Aug. 1646

Letter 66

[2/] 12 October 1646

Cavendish to Pell, from Paris

BL, MS Add. 4278, fo. 263 (original)
van Maanen C32

Worthie S^r:

    it is so longe since I hearde from you, that I doute you haue not receiued my two last to you:[1] in the later of which there was a little treatise of M^r: Mydorge concerning his opinion of your refutation of Longomontanus;[2] I desire to knowe by your next if you haue receiued it. I desire you will allso doe me the fauoure to let me knowe whether vietas worckes be all yet reprinted together as was intended;[3] & the Jesuits booke at Antwerpe of the quadrature of the circle etc:.[4] but I desire most to knowe where you are & howe you doe & next what you haue readie for the press; for if this newe colledge at Breda haue not diuerted you, I assure my self you would [>haue] had somewhat in the press by this time. M^r: Hobbes reades mathematickes sometimes to oure Prince, but I beleeue he hath spare time enough besides to goe on with his philosophie;[5] I sawe latelie a booke of the Jesuit Kircher of light & shaddowe;[6] it hath so manie fine figures in it that I suspect it hath no greate matter in it, & Monsr: Gassendes doeth not much commend it, so that I haue no incouragement to buie it or reade it. I should be glad to heare of Doctor Jungius;[7] it is so longe since I writ to him, & hauing neuer receiued anie answer from him, that I am allmost discouraged from writing to him againe; yet I may not forget the fauoures I receiued from him when I was with him & shall shortelie againe returne him my thankes for them, hoping he will at last bestowe some fauoures on me at a distance as well as neere hande. I desire allso to knowe howe & what M^r: Tassius[8] doeth. I haue troubled you enough for once. And so wishing you all happiness I rest

    Your assured friend to serue you
    Charles Cauendysshe

The date of this letter is presumed to be New Style; Cavendish's previous letter took 22 days to reach Pell in Breda (via Amsterdam), so this is more likely to have taken 14 days than four.

[1] Letters 64, 65.
[2] See Letter 65, n. 5.
[3] See Letter 65, n. 3.
[4] See Letter 65, n. 4.
[5] See Letter 64, n. 3; his 'philosophie' here was *De corpore*.
[6] Athanasius Kircher, *Ars magna lucis et umbrae* (Rome, 1646).
[7] See Letter 62, n. 8.
[8] See Letter 16, n. 4.

Paris Octob: 12 1646

[*addressed:*] To my verie worthie friend M^r: John Pell Publicke Professor of the Mathematickes at Amsterdam. In't vergulde Hart in de nieuwe hooghstrate. t'Amsterdam.

[*annotated by Pell:*] Received Octob 26.16

[*cover annotated by Birch:*] Oct. 1646

Letter 67

7/17 November 1646

Pell to Cavendish, from Breda

BL, Ms Add. 4280, fo. 131^r (draft)
van Maanen P24

Right Honourable

I doe not remember that I have written to your honour since July 9.[1] Since which I have received 3 of yours dated July 9. Aug 7. and Octob,[2] by which last I see y^t none of yours have missecarried though they went first to Amsterdam & thence backe againe hither. Which to prevent heereafter, I have [>heere] sent a new addresse for your letters to me as I now first follow your new direction, though I thinke Mr Hobbes [>be] likely to abide at Paris all this winter. I left Amsterdam above [>3 months] agoe and have heere had a great deale of impertinent labour in [layeing *deleted* >helping to laye] ye foundation of this petty Vniversity & [>Orange] Colledge[3] [for Nobility ye Noblesse & gentry of these *deleted*] But His Highnesse hath beene pleasd to ease me of the philosophicall profession, sending another [>hither] to supply y^t place but a day before we began, so y^t I was then [>suddenly] put to make a new Inaugurall oration[4] [>better agreeing with] ye Mathematicall profession, which is now onely left me. I was also forced to leave my booke[5] in ye presse imperfect nor doe I well know how much of it is since printed, A day or two after I came thence I heare Mr Golius came and left his censure of our controversy[6] with ye Printer, desiring that it might be adjoined at least as an appendix. Your demonstrations [>fill] ye 52 & 53 page of my booke but I have not ye sheete to send you. I hope shortly I shall send ye whole booke Or at least so much of it as I thinke fit to put to ye presse so inconveniently at y^t distance. I desire as soone as may be to employ our new printer heere[7] upon that [>greater] & perhaps farre better part which remaines.

Vietas workes are now to be sold, I have heere a coppy lying by me, but [>my]

---

[1] Letter 63.

[2] Letters 64, 65, 66.

[3] See Letter 63, n. 2.

[4] Pell, 'Oratio inauguralis'.

[5] Pell, *Controversiae pars prima*.

[6] This demonstration by Golius (on whom see Letter 23, n. 9) was added at the end of the book: Pell, *Controversiae pars prima*, pp. 95–6.

[7] Johann van Waesberghe, who was based in Utrecht, was appointed official printer to the Illustre School at Breda (see van Alphen, 'De illustre school', p. 302).

booke seller could not tell me ye price it is above 140 sheetes[8] and as one writes to me from Amsterdam they there aske 5 Reicks-dalers[9] for it. [>Kircher of light & shadow[10] I have not yet seene. Nor doe I heare] that Vincents quadrature is yet finished.[11] I [>should have] thanked Mons Midorge [for his good affection though *deleted*] as much for a line or two as for y^t long discourse,[12] but we must [>give] our freinds leave to express their affections as themselves like best.

Your letter to Des Cartes I sent to Sir W. Boswell,[13] who without question can convey it safely to him.

Be pleased [>heereafter] to direct your letters to

your honours most humble servant
John Pell

Ten huyse van Mr Buck in den gulden Eenhoren op de markt in BREDA.[14]

Novemb. 7.17. 1646.

---

[8] Viète, *Opera mathematica*, is a quarto containing 560 pages (hence Pell's '140 sheetes': Pell has apparently mistaken it for a folio).

[9] In sterling, approximately £1 1s 5d – a very high price for a book of this size. (For the rate of exchange see Letter 21, n. 5)

[10] See Letter 66, n. 6.

[11] See Letter 65, n. 4.

[12] See Letter 65, n. 5.

[13] This letter, which has not survived, was presumably enclosed with Letter 66. Its contents can be deduced from Descartes's reply, written on [23 Oct./] 2 Nov. (A&T iv, pp. 558–62): Cavendish had made some observation that prompted Descartes to make further comments on his disagreement with Roberval, and had also raised a query about Descartes's *Principia*, part 4, art. 153. On Boswell see Letter 19, n. 13.

[14] 'At the house of Mr Buck, [at the sign of] the golden unicorn, in the market-place at Breda'. This house has been identified as no. 11, Grote Markt (see van Maanen, 'The Refutation', p. 343, n. 56).

Letter 68

27 November/7 December 1646

Cavendish to Pell, from Paris

BL, MS Add. 4278, fos. 265–6 (original)
van Maanen C33

Worthie S^r:
  manie thankes for yours of the 7/17 of Nouemb:[1] by which I perceiue you haue receiued all myne & in one of them Monsr: Mydorge his discourse of your Theorem,[2] which I confess was more than I expected from him, for I beleeue he studies not Mathematiques much nowe & therefore we are the [>more] obliged to him, though as you write a fewe lines of his approbation would haue satisfied you [>as] well, yet wee must take our friends expressions as they are pleased to giue them, especiallie proceeding from an affectionate care to oure full satisfaction as I dare saie his did. & I conceiue no fault in it but that it would make your booke bigger than you meane it, if you should print it with it./ I esteeme it a greate honor you doe me to admit my demonstrations in your booke[3] though I desire as I formerlie did, not to be named, otherwise than a friend scholler & countrieman of yours & this I desire, to auoyde the impertinent expressions which foolish toungs maye produce; but I remit it to your better iudgment./ I am glad to heare your [*sic*] are so well at Breda, for I hope it is much to your aduantage as well in profit, as in honor. I longe much for your booke, but extreamlie for the second part of it for there I suppose wee shall haue some what of your Analitiques. I longe allso to heare whether you proceed with your Appollonius, & Diophantus.[4] I am in despaire of Goleas his Apollonius,[5] at least for manie yeares./ Vietas worckes are newlie come hither;/[6] you are obliged to P: Mersen for Monsr: Pallieurs christian name, whose note to Mersen I heere send

---

[1] Letter 67.
[2] See Letter 65, n. 5.
[3] Pell, *Controversiae pars prima*, pp. 52–3.
[4] See Letter 14, n. 13, and Letter 23, n. 2.
[5] See Letter 35, and Pell's comment in Letter 60.
[6] See Letter 65, n. 3.

PART III: THE CORRESPONDENCE, LETTER 68   495

you;/[7] I hope M[r]: de Cartes hath receiued my letter of thankes[8] for his fauoures; I haue desisted from trobling him more concerning the businesse of uibration in question between him & Mons[r]: Roberuall, for the expression of M[r]: Roberualls last replie was (I conceiue) too sharpe, so that I remitted it to P: Mersenne to send if he so pleased; which he did, & M[r]: de Cartes hath replied to it accordinglie;[9] the business is too difficult for me to iudge of, for it puts oure learned men heere to the gaze./ Regius followes M[r]: de Cartes in his philosophie allmost verbatim so far as he hath writ, & I suppose he hath got the rest (of trees & animalls) from him by waie of discourse.[10] Pere Vincentioes booke[11] is not yet ariued heere that I knowe of, but dayelie expected./ M[r]: Hobbes his iournie to Montauban was staied, being imploied to reade Mathematickes to oure Prince;[12] My Lord Jerman[13] did (I beleeue) doe him that fauoure & honor; for his friends heer I am confident had no hand in it. M[r]: Hobbes his lodging being vncertaine I repent not you direct your letters to Mons[r]: du Bose[14] whose abode is constant. I am latelie aquainted with your friend M[r]: Gilbert,[15] who seemes to be an able & honest man; I am verie glad of his acquaintance./ I heare M[r]: Oughtreds Clauis is nowe reprinted with some additions, as dialling, & Euclides tenth booke

---

[7] In a belated response to Pell's request (see Letter 57, at n. 7), Cavendish had finally obtained, via Mersenne, the information that Le Pailleur's name was Jacques; this came in the form of a letter from Le Pailleur to Mersenne, which Cavendish enclosed with this letter (BL MS Add. 4278, fo. 264, printed in *MC* xiv, pp. 658–9).

[8] See Letter 67, n. 13.

[9] See Letter 65, n. 2.

[10] Henricus Regius (Hendrick de Roy) (1598–1679), who taught medicine at Utrecht, had been an enthusiastic Cartesian in the early 1640s, but broke publicly with Descartes in 1647, asserting that the mind might be nothing more than a mode of a corporeal substance. Cavendish refers here to his recently published work *Fundamenta physices* (Amsterdam, 1646). Descartes himself had complained in a letter to Mersenne ([25 Sept./] 5 Oct. 1646: A&T iv, pp. 508–13) that Regius had taken material from his published works, adding to it ideas which Regius had obtained from him personally in both direct and indirect ways (p. 510).

[11] See Letter 49, n. 2.

[12] See Letter 64, n. 3.

[13] Henry Jermyn, first Baron Jermyn of St Edmundsbury, later first Earl of St Albans (d. 1684), had been in the service of Queen Henrietta Maria since 1628; having been appointed Governor of Jersey in 1644, he was employed by the Queen to bring the Prince of Wales to Paris in 1646. He became the dominant figure among the advisors at her court-in-exile.

[14] See Letter 64, nn. 4, 5.

[15] The mathematician William Gilbert (d. 1654), who graduated from Lincoln College, Oxford, in 1616, corresponded with Henry Gellibrand, and was described by Hartlib in 1634 as 'wholly spending himself in a Verulamian Philosophi for the Natural part' (see Webster, *Great Instauration*, p. 128). He worked as a surveyor of plantations in Ireland, and served as MP for Dublin University.

after his manner;[16] & I hope something more of analitickes though I heard not that mentioned; but I expect the great addition to Analitickes by yourself. I doute M^r: Hobbes will not finish & publish his phisickes this tweluemonth./[17] I longe to heare what Doctor Jungius doeth. & allso M^r: Tassius./[18] M^r: Roberuall hath halfe promised to polish the geometrie by Indiuisibles which Caualiero hath begun, for he saies he inuented & vsed that waie before Caualieros booke was published;[19] & that he can deliuer that doctrine much easier & shorter; & shew the vse of it in diuers propositions which he hath inuented by the help of it; but I doute it will be longe before he publish it; though I assure my self he is verie skillfull in it./ S^r: I haue troubled you enough at this time, therefore wishing you all happiness, I commit you & vs all to Gods holie protection, & remaine

    Your assured friend to serue you
    Charles Cauendysshe

Paris Decemb: 7 newe stile 1646.

[*addressed:*] To my verie worthie friend M^r: John Pell Publicke Professor of the Mathematickes at Breda, Ten huyse van M^r: Buck in den gulden Eenhoren, op de markt, in Breda.

[*annotated by Pell:*] Received. decemb. 6/16.

[*cover annotated by Birch:*] 7. Decemb. 1646 N.S.

---

[16] Oughtred's *Clavis* (1631) was issued in English as *The Key of the Mathematicks new Forged and Filed* in 1647 (see Letter 32, n. 4). The four appendices included in this edition were (i) on the numerical solution of equations, (ii) on the calculation of interest, (iii) on the rule of false position (see Letter 32, n. 12), and (iv) a treatise on sundials, written by Oughtred when he was 22 and translated into English by Christopher Wren. In the Latin edition published in 1648, the treatise on sundials was omitted, being replaced by two further appendices: one on the tenth book of Euclid, the other on Archimedean solids. So far as the text itself was concerned, the 1647 and 1648 editions were identical, both carrying a number of revisions to the original 1631 text (hence the phrase 'new Forged and Filed' in the 1647 title). The evidence of Cavendish's letter might be taken to show that the '1647' edition was published before the end of 1646; but his reference to the inclusion of the appendix on the tenth book of Euclid suggests that his information came from someone involved in the preparation of both editions (possibly Oughtred himself), not someone who had just obtained a printed copy of the English edition. There were political motives behind the publication of the 1647 edition, since Oughtred, a staunch Royalist, needed to placate those who had instigated sequestration proceedings against him. The edition carried an obsequious dedication to Richard Onslow, sequestrator for Surrey, the frontispiece depicted Oughtred in sober Puritan dress, and the acknowledgements to Cavendish that had appeared in the first edition were discreetly omitted.

[17] Hobbes, *De corpore*.

[18] See Letter 16, nn. 4, 5, and Cavendish's comment in Letter 62, at n. 8.

[19] Cavalieri's method was published in his *Geometria indivisibilibus promota* in 1635. Roberval's methods and results on quadrature emerged during his correspondence with Fermat on the subject during 1636 (see Mahoney, *Mathematical Career of Fermat*, pp. 218–21). In his dispute with Torricelli in 1646–7, Roberval claimed to have used the same method five years before Cavalieri's publication and there is no reason to doubt his claim, but he was accused of dishonesty by Torricelli. (For a discussion of the relationship between the theories of Cavalieri and Roberval see Walker, *A Study of the Traité des Indivisibles*, pp. 15–16, 45–8, 142–3.)

Letter 69

11/21 June 1647

Cavendish to Pell, from Paris

BL, MS Add. 4278, fos. 267–8 (original)
van Maanen C34

Worthie S^r:

    it is so longe since I hearde from you that I can not chuse but inquire howe you doe & where you are; It is longe since I sent you M^r: Palliers christian name, I hope you haue receiued my letter that conueied it.[1] Oure Learned men heere publish nothing, onelie M^r: Gassendes hath sent his booke of Epicurus his life; & philosophie,[2] (most of it I beleeue his owne.) I was latelie at his chamber where he shewed me his lectures of the Sphaere last yeare, nowe in printing,[3] though theere be nothing newe in it yet he explicates things so well that it will be worth the reading./ I haue not reade much of vincentioes booke[4] but M^r: Roberuall[5] tells me that he hath not done the deed; for he supposes that maye be done, which is as harde to doe as the quadrature it self./ I haue latelie reade againe some of vieta & Hariot of the constitution of aequations especiallie cubick, & methinkes all the cases are not fullie prosecuted, to knowe infalliblie in euerie aequation howe manie rootes & of what nature;[6] but I hope one daie your self will perfit that, & the rest of the analitickes according to those heades which you were pleased to write [>in a letter] to me to Hamburg;[7] I desire to knowe howe & what Doctor Jungius does; I dare not present my seruice to him, fearing I am not in his good opinion, though I knowe not in what I haue disobliged him; but hauing writ to him two yeares since[8] & receaued no answeare, makes me suspect it. yet if you holde correspondencie with him, if you please to remember

---

[1] See Letter 68, n. 7; Cavendish refers here to Letter 68.

[2] P. Gassendi, *De vita et moribus Epicuri libri octo* (Lyon, 1647).

[3] P. Gassendi, *Institutio astronomica iuxta hypotheseis tam veterum quam Copernici et Tychonis Brahei* (Paris, 1647).

[4] See Letter 49, n. 2.

[5] See Letter 19, n. 7.

[6] In general, a cubic equation has three roots, but in some cases two of those roots are a pair of complex conjugates (numbers of the form $a \pm bi$); such cases in the 16th and 17th centuries were considered 'irreducible' or 'impossible'. Thus when Cavendish asked about 'howe many rootes and of what nature', he meant how many *real* roots, and whether positive or negative, single or repeated. Viète's *De numerosa potestatum resolutione* and *De aequationum recognitione*, and Harriot's *Artis analyticae praxis*, like all previous texts, treated only positive roots, though Harriot in manuscript handled both negative and complex roots.

[7] Letter 32.

[8] See Letter 34, n. 7, and Cavendish's comment in Letter 62, at n. 8.

my respects to him & my desire to heare from him you shall doe me a greate fauoure; my seruice to M^r: Tassius[9] allso; I desire you will let me knowe if they haue printed anie thinge. I desire your opinion of Schootens booke, especiallie of cubick aequations;[10] & of M^r: Oughtreds addition of affected aequations.[11] S^r: I haue troubled you enough for one time; And therefore wishing you all happiness I remayne

 Your assured friend to serue you
 Charles Cauendysshe

Paris June. 11/21 1647.

[*addressed:*] To my worthie friend M^r: John Pell, Publicke Professor of the Mathematickes at Breda, Ten huise van M^r: Buck in den gulden Eenhoren op de markt, in Breda.

[*annotated by Pell:*] Received June 20 [/]30

[*cover annotated by Birch:*] 11/21 June 1647

---

[9] See Letter 16, n. 4.

[10] Frans van Schooten (1615–60), son of a famous mathematics teacher at Leiden (to whose post he succeeded in 1646), worked for many years as editor of Viète and translator of Descartes. The work referred to here is his *De organica conicarum sectionum in plano descriptione tractatus ... cui annexa est appendix de cubicarum aequationum resolutione* (Leiden, 1646). On van Schooten see J. E. Hofmann, *Frans van Schooten der jüngere* (Wiesbaden, 1962).

[11] An affected equation is one in which there is more than one term (unlike a simple equation of the form $x^n = c$, which requires only straightforward root extraction). One of the appendices to the 1647 English version (and subsequent editions) of Oughtred's *Clavis* (see Letter 68, n. 16) was a treatise 'Of the Resolution of Adfected Aequations in Numbers' (pp. 121–69), in which Oughtred dealt with cubic, quartic, and quintic equations by the methods first expounded by Viète in *De numerosa potestatum resolutione*.

Letter 70

1/11 July 1647

Pell to Cavendish, from Breda

BL, MS Add. 4280, fo. 131 (draft)
van Maanen P25

Right Honourable

When I received Monsieur Pallieurs christen name in your last but one,[1] I was made beleeve y^t I should, soone after, have some of ye coppies of my booke[2] to send into France. But it is not long since I had [>ye first compleate coppy] for my selfe; and am now [>about] to give order to my printer to send over ten coppies, [>to – I know not whom, in Paris to be] distributed according to ye names in y^s list.[3] For it [>were not fit] to put your Honour to all y^t trouble & I have nobody else there to [>whose care] I [>may with] confidence recommend y^m. It may be, before I shall finde a convenient conveyance of y^m they will be to be sold at some of ye shops in Paris; [>Especially of those y^t have commerce with my printer.] If y^{ei} be, y^{ei} are to be inquired for, by y^s title, Controversia de Verâ Circuli Mensura inter Ch. Longomontanum &c & Jo. Pellium &c Pars prima. Amstelodami apud Jo Blaeu, A booke of halfe a quire of paper [>in quarto.]

I have sent a coppy [>of it] to Padre Vincentio, whose booke [>nor Mr Oughtreds] I have not yet seene,[4] [>as being not yet come to this towne.]

Our printer hath printed my inaugurall oration amongst ye rest,[5] with ye statutes of our Colledge & petty University heere. I suppose it will be to be sold within ye fortnight, but it will not be worth your reading. Dr Regius (whose [>Fundamenta physices] you have seene) hath lately published his Fundamenta Medica, a booke of 38 sheetes in quarto. After which, shall follow his Medicationes or practise of physicke,[6] part of which I have already seene.

From our Hamburgers I heere nothing: Dr Young[7] must be let alone in his

---

[1] Letter 68 (n. 7).

[2] Pell, *Controversiae pars prima*.

[3] The list accompanying this letter does not survive. The only relevant names in Pell's general distribution list for this book (BL, MS Add. 4416, fo. 102^r) are of those contributors to the volume who resided in France: Cavendish, Hobbes, Roberval, Le Pailleur, Mydorge, de Carcavi, and Mersenne. Possibly the ten copies were intended to be distributed to these seven people, with the surplus copies going to Mersenne for forwarding elsewhere.

[4] See Letter 49, n. 2, and Letter 68, n. 16.

[5] See Letter 67, n. 4.

[6] H. Regius, *Fundamenta medica* (Utrecht, 1647); that work was reissued, with the first edition of the other work mentioned here, as *Medicina et praxis medica, medicationum exemplis demonstrata* (Utrecht, 1648). On Regius see Letter 68, n. 10.

[7] Jungius (Jung); the other 'Hamburger' was Tassius (see Letter 16, nn. 4, 5).

pedanticall morosity, till he [>be in a] better moode. I easily beleeve y^t Vieta Harriot &c doe not satisfy you in cubicall aequations.[8] I shall be able to make good y^t period which escaped me in my inaugurall oration heere; Qui hanc artem, veri investigatricem, nostro etiam seculo perfectissimè callere, ejusque vi *nullum non problema solvere posse* dicebantur, nec *Analyticam* suam plenè tradidêre, nec illius matrem divinam illam *syntheticam* attigisse videntur. *Utriusque* autem restitutioni *meos* etiam conatûs plurimum conferre posse puto.[9]

This with some other passages [>of] y^t oration which I was forced to print in spite of my teeth, may perhaps raise me a [>hideous storme] of envy & contradiction. But what remedy? It seemes ye fate of

your Honours most humble servant
John Pell.

July 1. 11. 1647
Breda.

[*address noted as follows:*] A Monsieur
　Monsieur de Cavendish, Chevalier Anglois a Paris
Recommandee a Monsieur
　Monsieur de Bose Secretaire du Roy et Valet de Chambre de la Reine
　Rue de Four vis a vis d'un Apoticaire.

---

[8] See Cavendish's comment in Letter 69, at n. 6.

[9] 'Those people who, in our age, were said to be extremely well versed in that art of investigating the truth, and by its means to be able to "leave no problem unsolved", have not fully taught its *analytic* art, nor do they seem to have attained its divine mother, the *synthetic* art. I believe that I can contribute very much to the restitution of both arts' ('Oratio inauguralis', in *Inauguratio illustris scholae*, p. 180). The phrase 'nullum non problema solvere', 'to leave no problem unsolved', was the final dictum of Viète's *Isagoge*, encapsulating his hope that algebra would open up previously intractable problems. The 'analytic art' was Viète's description of algebra (see Letter 1, n. 6). For the relationship between analysis and synthesis, see Letter 44, n. 6.

Letter 71

6/16 August 1647

Cavendish to Pell, from Paris

BL, MS Add. 4278, fos. 269–70 (original)
van Maanen C35

Worthie S^r:
    manie thankes for yours of the 1/11 of Julie[1] which I receiued at S^t: Germaines;[2] I defered writing till my returne hither; [>I] made inquirie for your booke & found [>them] (as you supposed) to be solde; I bought one; wherein I finde you haue done me the honor to put in my demonstration of your Theorem;/[3] I hope Longomontanus will rest satisfied; if not, I suppose all others will./ I am verie glad that you haue finished this; for I hope nowe er longe wee shall haue Appollonius or Diophantus,[4] or rather your owne Analitickes (at least some parte of it) which I much desire to see, for I expect not much more than wee haue till you publish yours./ Monsr: Gassendes hath latelie printed the summ of his astronomicall lectures last yeare, but I hope wee shall shortelie haue the life of Epicurus & I beleeue with it some of his owne philosophie,[5] which douteless will be verie goode; excepting Doctor Jungius[6] & M^r: Hobbes, I esteem him the best I knowe./ If Doctor Jungius hath printed anie thing I praie you let me knowe. I shall take your aduise & forbeare writing to him for a whyle. I heare no news of those ten copies of your booke, when I doe (if you please) M^r: Hobbes & I shall endeauoure to haue them disposed [>of] according to your direction./ And so committing you & vs all to Gods holie protection I remayne
    Your assured friend to serue you
    Charles

Paris Aug: 6/16. 1647

[*addressed:*] To my verie worthie friend M^r: John Pell publick Professor of the Mathematicks at Breda, Ten huyse van M^r: Buck in den gulden Eenhoren, op de markt, in Breda.

[*added to address in another hand:*] poort de paris[7]

[*annotated by Pell:*] Received Aug 14 [/]24

[*cover annotated by Birch:*] 6/16 Aug. 1647.

---

[1] Letter 70.
[2] Saint Germain-en-Laye, where Queen Henrietta Maria and the Prince of Wales resided.
[3] See Letter 68, n. 3, and Pell's comment about the sale of the book in Paris in Letter 70.
[4] See Letter 14, n. 13, and Letter 23, n. 2.   [5] See Letter 69, nn. 2, 3.
[6] See Letter 16, n. 5.   [7] 'Paris gate'.

Letter 72

4/14 February 1648

Cavendish to Pell, from Paris

BL, MS Add. 4278, fos. 271–2 (original); MS Add. 4280, fo. 92^r (copy of a section of the letter, in Pell's hand)

van Maanen C36

Worthie S^r:

it is so longe since I writ to you & [>longer] since I receiued anie from you that I doute my last letter (about 3 moneths since) miscaried;[1] I haue not yet heard anie newes of those 10 [>bookes] of yours which you intended to send hither;[2] but I must nowe acknowledg both my sin to God & my fault to you in that I writ you concerning my owne demonstration, that I was not helped by Monsieur Palliers demonstration,[3] for that was a flat lie, for in the unfolding of my aequation, I imitated him & followed his steps; & for the demonstration of the proposition in Mersennus his booke:[4] M^r: Hobbs (as I writ formerlie) gaue me the hint of it. so that onelie the application of that proposition, & the applieing the proportion of the tangents, vpon the radius, & some little more was my owne; it seems I was in no goode moode when I writ you that letter, I hope I am nowe in a better, which is to aske humblie pardon of God for this & all other sinns of whose pardon in my Sauiour I am verie confident; the next is yours which I dout not of, as allso of your fauourable construction of this confession; for I beleeue if some knew it they would account it an excellent maddnesse, or at the verie least verie ridiculous; but I had rather be so esteemed, than reallie haue so foule a blotch within me, & not indeauoure by a free confession through Gods grace to rid my self of it; but I haue troubled you enough with this./ There is a booke by a Polander latelie come hither of the Phases of the moone; & of a motion of libration in the moone;[5] I haue not read it but I asked M^r: Gassendes how he liked it; who commended it, but methought not extreamelie; it seems

---

[1] Cavendish's 'doubt' (*sc.* suspicion) appears to have been correct; no letter from that date survives among Pell's papers. Cf. Pell's comment on this in Letter 73.

[2] See Letter 70, n. 3.

[3] Cavendish apparently refers to his comment in Letter 47: 'it depends not of anie thinge I haue seene in anie demonstration of the same Theorem.' For Le Pailleur's demonstration see Letter 45, n. 1 (and note also Cavendish's comment in that letter that he had found himself following Le Pailleur's 'way' in one of his own attempts to demonstrate Pell's theorem).

[4] This refers to the proposition from Mersenne's 'Ballistica' used by Cavendish in his 'lemma' to the demonstration he enclosed with Letter 47: see Letter 47, n. 6.

[5] The book by the astronomer Johannes Hevelius (1611–87), *Selenographia: sive, lunae descriptio* (Danzig, 1647), discusses the 'motion of libration' of the moon on pp. 236–50, 341, 421–2.

that Polander hath a verie good perspectiue glass but I hope not so goode as Fontanus glass; for wee haue that heer; presented to my brother from S^r: Kenelm Digbie,[6] but wee haue not yet looked in it, the Tube for it being not yet come; but wee dailie expect it; as I remember S^r: Kenelme writ it was 28 palmes in length, a palme is taken for an indiferent span; M^r: Gassendes tolde me the exact measure but I haue forgot it; but doutless it is much longer than anie I haue seen & I beleeue much better; for it is they say Fontanus best glass by which I beleeue he made those obseruations which he published, which peraduenture you haue seene. I thinke I writ to you of 3 glasses my brother hath of Eustacio Diuinos[7] whom some esteem a better worckman than Fontanus; the best of these 3 is (as I thinke) better than myne, & longer as I take it by a yarde; or there abouts;/ my brother hath allso 2 of the famous Tauricelli but wee haue not yet tried them./ he hath allso another of 37 palmes at least as I take it, made by Eustacio Diuino, which if it be well wrought, will douteless excell all the rest;[8] being of so great a length; but not yet tried./ he hath yet an other to come from P: Reieta,[9] which I heare is ariued in the Lowecountries & hath bin there as I remember [>they writ] tried & saied to be verie rare./ So that wee are & shall be prettie well furnished with perspectiues./ I desire your opinion of P: vincentio his booke of the quadrature of the circle;/[10] though M^r: Roberuall[11] saie he hath not done it, & that he supposeth that to be granted which is as difficult as the

---

[6] Francesco Fontana (d. 1656) was a well-known instrument-maker in Naples; he published his own astronomical observations as *Novae coelestium terrestriumque rerum observationes* (Naples, 1646). The scientist and philosopher Sir Kenelm Digby (1603–65) had probably acquired this telescope in Italy, on one of the two missions he undertook to Rome (to negotiate with the Pope on behalf of Queen Henrietta Maria) between 1645 and 1648. It is described in a list of the Marquess of Newcastle's telescopes as 'Fontanus his Glass, marked 22 Palmes which is 16 foote, and a halfe but it must draw eighteene foote' (Nottingham University Library, MS Pw 1 668).

[7] Eustachio Divini or de Divinis (1620–95), instrument-maker and astronomer in Rome; he and Giuseppe Campani were the most famous telescope-makers in Italy. He also published his own astronomical observations, and enjoyed close relations with the Jesuit scientist Honoré Fabri (see Daumas, *Scientific Instruments*, pp. 30, 63–6).

[8] This telescope was described as 'Eustatio Divino, his Greate Glass, for the greate Tube, being marked with 37 Palmes and a halfe, is to draw 28 Foote, and 8 Inches, besides foure Inches for the convex Glass, at the Eye' (Nottingham University Library, MS Pw 1 668). This may also have been supplied by Sir Kenelm Digby: Divini would later write that he made many telescopes of 15, 24, and 36 palms' length, and that Digby had taken six of them when he left Rome (E. Divini ['de Divinis'], *Brevis annotatio in Systema Saturnium Christiani Huygenii* (The Hague, 1660), pp. 2–3). The entire collection of seven telescopes referred to here by Cavendish (one by Fontana, four by Divini, two by Torricelli – on whom see Letter 35, n. 17) was sold (together with six other telescopes) to Hobbes later in 1648 (Nottingham University Library, MSS Pw 1 668 and Pw 1 406); Hobbes in turn sold the collection to the third Earl of Devonshire in 1659 (Chatsworth, MSS Hobbes E 3 and Hardwick 33, entries for Apr. 1659).

[9] From the fact that this telescope by Rheita (see Letter 13, n. 4, and Letter 21, n. 15) was not included in the collection sold to Hobbes (see above, n. 8), it may be deduced either that it never arrived, or that it was judged superior to all the others and retained by Newcastle.

[10] See Letter 49, n. 2.

[11] See Letter 19, n. 7.

504   JOHN PELL (1611–1685)

quadrature it self; I confess I vnderstand not M^r: Roberualls meaning cleerlie in saying so. for though I haue not read the booke with that diligence that I ought, neither peraduenture without an instructor am capable to vnderstand it throughlie & perfectlie;/ yet [>so far as I conceiue] graunting his waie of finding the rates of magnitudes, which I perceiue not anie defect in, I must graunt he hath done it; though not so as I wish it, that is, as the parabola is, in numbers;/ or if it be incommensurab[le, *page torn*] in irrationalle numbers;[12] it makes me suspect it is ue[ry *page torn*] difficult if possible to giue the proportion in anie num[bers *page torn*] his waye; because he hath not at all atempted i[t *page torn*] that I can finde, & beleeu so much of his abiliti[es *page torn*] it would be a hard taske to doe more on his gr[ound *page torn*] than he hath done, for douteless he is an excellent geometrician: I longe to haue your opinion of hi[s *page torn*] booke:/ this that I write is of his first waye of [the *page torn*] quadrature of the circle; the other waies I vnderst[and *page torn*] not [>yet] so well./ I desire allso your iudgment of M^r: oughtred english clauis, especiallie of his resolution of affected aequations; & his short demonstration of the rule of falshoode[13] which I like verie well./ I haue forgot if I sent you an adresse for your letters, but to be sure I send you this. A Monsr: Du Bose Secretaire du Roy et valet de chambre de la Reyne, Rue S^t: Honorè prez la Palais Royal deuant le mouton rouge chez Monsr: Bocan. s.[14] I longe verie much stil to see something of yours in print, especiallie analiticks; I thinke I haue sufficientlie tired you; therefore wishing you all happiness I commit you & vs all to Gods holie protection, & remayne.

   Your assured friend to serue you,
   Charles Cauendysshe

Paris Feb: 4/14 1647.

[*addressed:*] To my verie Worthie Friend M^r: John Pell publick Professor of the Mathematicks at Breda; Ten huyse van M^r: Buck, in den gulden Eenhoren, op de markt, in Breda.

[*cover annotated by Birch:*] 4/14 Febr. 1647/8

---

[12] For 'rates of magnitudes' read 'ratios of magnitudes'. The quadrature (squaring) of the circle is impossible by traditional ruler and compass methods, or, as Cavendish hoped for it, in rational or surd (irrational) numbers.

[13] See Letter 68, n. 16, and Letter 69, n. 11; the appendix on 'The rule of false Position' was on pp. 173–4 of the English translation of Oughtred's *Clavis*. For Cavendish's earlier interest in the rule of false position, see Letter 52, at n. 5.

[14] 'To M. du [Bosc], secretary to the King and *valet de chambre* of the Queen, rue Saint-Honoré, near the Palais Royal, in front of the [sign of the] red sheep, at the house of M. Bocan s[eigneur?]'. On du Bosc see Letter 64, n. 4; M. Bocan has not been identified.

Letter 73

1/11 March 1648

Pell to Cavendish, from Breda

BL, MS Add. 4280, fos. 131^v–132^r (draft)
van Maanen P26

Right Honourable.

Since my last of July 1.11[1] I have received [>two letters] from your Honour bearing date Aug 6.16 & Febr 4.14.[2] So y^t ye letter written 3 months before this last,[3] seemes to have beene misse-carried. I much wonder y^t those 10 bookes are not yet come to your hands.[4] The Amsterdammer, to whom I last recommended ye care of sending them,[5] sent me word long since that he had sent them. But by whom he sent them, he hath not yet told me. I shall enquire who & where he is y^t undertooke to deliver y^m to Monsieur du Bose[6] for your Honour.

The Polander mentioned in your Letters I suppose to be Hevelius[7] a Senator of Dantisc whose booke is to be sold at Amsterdam but ye greatnesse of ye price deterres me from sending for it So y^t I have not seene it. Fontana's booke I have, sent me from Sr William Boswell.[8] I should be glad to heare that with some of [>your] 8 perspicills[9] you discover more than Hevelius & Fontana have done.

About 7 months agoe I received a letter from P. Vincentio thanking me for my coppy, [>and praising it] obliquely desiring my judgement of his great booke,[10] promising me a coppy of an appendix shortly to be added & concluding thus, Mathesim unicè tibi, meque commendo & rogo ut amicorum tuorum albo adscribere velis.[11] But [>I had not then] seene his booke; About 3 months after,

---

[1] Letter 70.
[2] Letters 71, 72.
[3] See Letter 72, n. 1.
[4] See Letter 72, n. 2.
[5] Probably Blaeu (cf. the reference in Letter 70 to 'my printer').
[6] See Letter 64, n. 4.
[7] See Letter 72, n. 5.
[8] On Fontana see Letter 72, n. 6; on Boswell see Letter 19, n. 13.
[9] The eight telescopes mentioned in Letter 72.
[10] On de Saint-Vincent and his book see Letter 48, n. 5, and Letter 49, n. 2; 'my coppy' refers to a copy of Pell, *Controversiae pars prima*. Pell's copy of de Saint-Vincent's letter ([9/] 19 July 1647, from Ghent) is BL, MS Add. 4431, fo. 408; Pell's annotation on it records that he received it at Breda on 8/18 Aug. Three drafts survive of Pell's reply to it (dated 13/23 Aug. 1647): BL, MSS Add. 4280, fo. 181^r; Add. 4423, fo. 347^r; Add. 4426, fo. 220^r.
[11] 'I commit myself, and my mathematics, only to you, and I ask that you might be willing to count them among your friends.'

My Lord Culpepers eldest sonne[12] sent it me from Antwerpe. I sent it presently to binding where it lay so long, y^t (all my leisure, which I had reserved for it, being past [>before it was bound]) I was fallen upon other businesse so y^t I have not yet found time to examine it. But [>I intend very] shortly to take it in hand. Though I have no great reason to [>hope that I shall finde] more in it than all those Mathematicians have done who have had it in their hands almost a yeare. I perceive by your [>former] letters, that Mr Robervall hath had it these 9 or 10 months,[13] and yet neither he nor any other that I can heare of, hath found [>in it] what [>all men] expected, that is, the delineatory or supputatory [>habitude of] a circles diameter [>to] his periphery [>or to] ye side of an equalling square.[14]

Yet Dr Gutischovius,[15] Mathematick Professor at Lovan, told Mr Culpeper that he finds no fault in his arguments but thinkes he hath done the businesse. And I see your last speakes to ye same purpose. [>I shall not faile to] give your Honour account of what I finde [>in him] after diligent examination. I heare y^t Oughtredi clavis is come forth [>at London] anew in Latine,[16] but have not yet seene it. [My Conicall endeavours are not likely to see ye light as comments upon Apollonius *deleted*] As for Apollonius Neither Mr Golius his manuscript nor mine hath more than 7 bookes,[17] but [>I heare that] Pater Richardo, whose Euclide I have not yet seene, came this winter [>from Madrid] to Antwerpe with a compleate coppy of 8 bookes by him translated [>(I thinke out of Greeke)] and illustrated wth a very large comment.[18] He left it there in the hands of a freind

---

[12] John Colepeper ('Lord Culpeper') (1600–60), created Baron Colepeper of Thoresway in 1644, was an influential advisor to Charles I. He accompanied the Prince of Wales to Paris in 1646. His eldest son, Alexander, was born in 1629 or 1630; he married Catherine Ford (aged 12) at Calais in Sept. 1648, and died in Mar. 1649. It appears that Alexander Colepeper had been a pupil of Pell's at Breda; a letter from him to Pell, dated [14/] 24 Oct. [1647] and written perhaps in Antwerp, states: 'I have changd my resolution of returning to Breda, & have chose rather to go directly from hence to France where I shall be in a few dayes ... I am very sensible of your many favours to me' (BL, MS Add. 4278, fo. 356^v).

[13] See Letter 69, at n. 5.

[14] Finding the ratio of the diameter of a circle to either the circumference or the area was the classical problem of the quadrature of the circle; see Letter 72, n. 12. 'Delineatory or supputatory' means 'geometric or arithmetic' (see Letter 32, n. 6).

[15] See Letter 42, n. 6.

[16] This 2nd edition of the Latin text was published by Thomas Whitaker (London, 1648); see Letter 68, n. 16.

[17] See Letter 23, nn. 2, 9, and Letter 35, n. 10.

[18] The French Jesuit Claude Richard (1588–1664) had been Professor of Mathematics at the University of Madrid since 1624. His edition of Euclid, *Euclidis elementorum geometricorum libros tredecim*, was published at Antwerp in 1645. His translation of Apollonius (but only of the first four books, which survived in Greek) was eventually published there in 1655: *Apollonii Pergaei conicorum libri IV, cum commentariis*. The false rumour that he was about to publish all eight books sprang perhaps from Mersenne's comment (in his *Novarum observationum physico-mathematicarum ... tomus III* (Paris, 1647), p. 71), that Richard would soon publish 'the four latter books' ('4. posteriores libros') of Apollonius.

[the brother of John de la failly[19] ye Jesuite till *deleted* >Because] the printer (Hieronymus Verdussen)[20] told him that till ye peace be published,[21] he would not begin to print it.

Since my wife & family came hither, I am removed from ye house where I first lodged, so y^t that addresse may be left out, nor needes [>there] any new one in its place. [>Since] ye poste so well knowes

   your Honours most humble servant
   John Pell

March. 1/11 1647[/]1648. Breda

[*additional note:*] on the side of the letter

For the right Honourable Sr Charles Cavendysshe Knight Recommandee a Monsieur Monsieur du Bose, Secretaire du Roy et Valet de Chambre de la Reyne. Rue St Honorè, prez la Palais Royal devant le mouton rouge chez Monsr Bocan a PARIS

[*postscript:*] In a loose paper

You [>did not well to give] your selfe y^e [>LIE] unlesse you were better able to prove it. For whereas you [>now] tell me that in ye unfolding of your aequation you imitated Monsieur Pallieur & followed his steps and had from Mr Hobbes ye hint of your demonstration for a proposition in Mersennus booke.[22] This doeth not at all contradict any thing that ever you wrote to me. For you never told me y^t you demonstrated my Theoreme without all manner of helpe from men more exercised. Onely in a loose paper inclosed in my letter just 2 yeares agoe I wrote that [>in] $BFq = ABq - BCq$ lyes the ground of y^e difference betweene y^e rest of your demonstration & Mons. *Carcavy's*[23] to which you answered 19 dayes after, in this manner 'If my former demonstration agreed with Monsieur *Carcauins* in any part of it it is more than I know, for I assure you I played not the Truant at all y^t way: for I had it not by me when I made mine, neither did I thinke of

---

[19] The source of Pell's information was a letter from Richard Wake in Madrid to Sir William Boswell, dated [29 Dec. 1647/] 8 Jan. 1648: a copy in Pell's hand (made, presumably, from one forwarded to him by Boswell) is in BL, MS Lansdowne 751, fo. 57^r. This also furnishes the name of della Faille's brother: 'In the meane time the coppy of the booke is in the hands of a friend of mine named Vincentio de la falie, who has a Brother a Jesuite that waits upon Don Jan de Austria one of the rarest mathematicians in the world.' On Jean Charles della Faille, see Letter 44, n. 15.

[20] An Antwerp printer and bookseller, later also active in Vienna: see J. Benzing, *Die Buchdrucker des 16. und 17. Jahrhunderts im deutschen Sprachgebiet*, 2nd edn. (Wiesbaden, 1982), pp. 382, 491.

[21] The Treaty of Münster, ending the Eighty Years' War between Spain and her former provinces in the northern Netherlands, the United Provinces, was agreed in January 1648; however, ratification by the States-General was delayed for several months, and the peace was finally proclaimed at The Hague on 26 May/5 June 1648.

[22] See Letter 72; whether through charity or through forgetfulness, Pell disregards the comment made in Letter 47 (see Letter 72, n. 3).

[23] See Letter 55.

it at all, nor tooke any copie of it nor of any of ye rest, but onely one of Monsr Robervalls & this of Cavaliers which P. Mersenne gave me to send [>to] you.'²⁴

Now because I see not how y^s passage [>of Carcavy] contradicts ye former, [>of Pallieur & Hobbes] I will take ye boldnesse to tell you once more y^t you wronged your selfe in telling me y^t you had written a flat lie to me.

This paper is saucy & fearelesse knowing that ye worst you can doe to it is to throw it into the fire, to which it was destined and for y^t end separated from ye rest of my letter. that if you thought good to keepe that, yet this might not remaine as a record against you or an Example [>shewing] how much weake memoryes doe sometimes disturbe tender consciences. GOD KEEPE YOU!

[*additional note, perhaps not intended for inclusion in the letter or paper:*] My 38th Birthday.²⁵

---

[24] See Letter 56.

[25] Pell was born on 1/11 Mar. 1611. When he described 1/11 Mar. 1648 as his '38th Birthday' he presumably meant that he was entering his 38th year. (Three years earlier he had similarly described 1 Mar. 1645 as his 35th birthday; see BL, MS Add. 4415, fo. 200 (Illustration 3) and Letter 51, n. 9).

Letter 74

[23 July/] 2 August 1648

Cavendish to Pell, from Rotterdam

BL, MS Add. 4278, fos. 273–4 (original)
van Maanen C37

Worthie S^r:

Comming so neere you I can not but trouble you with this letter, since I doute I shall not haue the happiness to see you at this time, our staie heere beeing as I suppose but verie short; hastening all wee may, after our noble Prince./[1] I left all my learned & worthie acquaintance at Paris in theyr accustomed health, M^r: Gassendes though infirme, yet proceeds with his Epicurean Phylosophie; the half of which, I dout, is not yet printed./[2] M^r: Hobbes hath nowe leasure to studie & I hope wee shall haue his within a twelue moneth./[3] M^r: de Cartes wee left at Paris, to whom I am much obliged for his fauoures, I hope he will ere verie longe publish what remaines of his phisickes for he tolde me he had (as I remember) satisfied himself concerning the nature & properties of Animalls./[4] M^r: de Cartes & M^r: Hobbes haue met & had some discourse, & as they agree in some opinions so they extremelie differ in others, as in the nature of hardness; M^r: Hobbes conceiuing the cause of it to be an extream quicke motion of the atomes or minute pa[r]tes of a bodie which hinders an other bodie from entring. & M^r: de Cartes conceiues it a close ioining of the partes at rest, which apeares to me more reasonable./[5] M^r: Roberuall imploies himself so much in teaching that I doute it will [>be] longe ere he publish anie thing,[6] though he haue manie excellent things, as of mechanicks, especiallie of the center of grauitie, much of which he hath founde by Indiuisibles; in which he is verie expert, & I hope he

---

[1] The Prince of Wales had travelled from Paris to The Hague in June 1648, to take command of a group of warships that had deserted the Parliamentarian navy; Newcastle and Cavendish followed soon afterwards (their departure having been delayed by the need to settle their most pressing debts in Paris), and took lodgings in Rotterdam.

[2] See Letter 18, n. 4. As usual in Cavendish's writing, to 'doubt' means to suspect.

[3] The cause of Hobbes's leisure was the departure of his pupil, the Prince of Wales; the work referred to here was *De corpore* (not in fact published until 1655).

[4] Descartes's work on biology, *De homine*, was published posthumously in Leiden in 1662; previously unpublished material on physics was issued under the title *Le Monde* in Paris in 1664.

[5] In 1641 Hobbes and Descartes had conducted, via Mersenne, an increasingly ill-tempered correspondence on the subject of the refraction of light; disagreement about the cause of hardness arose there in relation to their accounts of the transmission of light through different media (see Hobbes, *Correspondence*, i, pp. 54–79, 86–113).

[6] Roberval (see Letter 19, n. 7) was Ramus Professor at the Collège Royal; he also retained a teaching post at the Collège de Maître Gervais.

will one daie write a treatise of Indiuisibles & theyre vse, with diuers examples of curious propositions which he hath found by them./[7] I haue heere by me a treatise of Analiticks in Manuscript by M[r]: Chauau[8] a Mathematician at Paris; there are somethings in it which I haue not mett with or at least obserued formerlie. but what I meet with of that subiect, I esteem it but as a diuertisment to pass some time in, till you publish yours./ I sawe a booke at Paris of the excellent Caualieros[9] latelie printed, concerning Indiuisibles;[10] whom you knowe was the Inuentor or Restorer of that kinde of Geometrie; I had not time to reade it before I came awaye, & they are not to be bought; M[r]: Carcaui[11] comming latelie from Italie brought this with him./ I hope by the grace of God wee shall haue some resting place; & then I shall doe as I nowe doe, trouble you with a letter, & desire your iudgment of Pere Vincentios quadrature of the circle[12] & peraduenture other quaeres. And so wishing you all happiness committing you & vs all to Gods holie protection I remaine

Your assured friend & seruant
Charles Cauendysshe

Roterdam: Aug: 2: 1648.

[*addressed:*] To my verie worthie friend M[r]: John Pell Publick Professor of the Mathematickes in Breda. giue this

[*annotated by Pell:*] Received Aug. 10 or July 31.

[*cover annotated by Birch:*] 2 Aug. 1648

---

[7] Roberval's 'Traité des indivisibles' was eventually published posthumously, in *Divers ouvrages de mathématique et de physique, par Messieurs de l'Académie Royale des Sciences* (Paris, 1693), pp. 190–245; see the translation and analysis of this work in Walker, *A Study of Roberval's Traité*.

[8] On Chauveau and his treatise see Letter 16, n. 21.

[9] The words 'excellent Caualieros' have been underlined in red ink by Pell.

[10] This new book was B. Cavalieri, *Exercitationes geometricae sex* (Bologna, 1647). Cavalieri's classic work on indivisibles was his *Geometria indivisibilibus promota*, to which Pell had previously referred (see Letter 29, at n. 5).

[11] On de Carcavi see Letter 38, n. 19.

[12] See Letter 49, n. 2.

Letter 75

7/17 August 1648

Pell to Cavendish, from Breda

BL MS Add. 4280, fo. 132^v (draft)
van Maanen P27

Right Honourable

A weeke agoe I received a letter from your Honour dated Aug 2.[1] expressing so short a stay in Roterdam, that I could not but feare ye losing of my labour in answering y^t letter or [>in] comming my selfe to waite upon your Honour.

And therefore I should not write [>now], had I not thought upon Sr William Boswell,[2] to whom I intend to recommend this, y^t so I may have it againe if it get not to Rotedam [sic] before my Lord of Newcastle be gone to sea.[3]

My last to your Honour was 5 months agoe,[4] [>the misse-carriage of] which (especially because of ye inclosed paper) I much [>feared.] I finde no mention of it in [>your] last: and yet I hope you had it because [>your] superscription [>to mee] was without particular addresse according to y^t which I had written in y^t last of mine.

I am glad to heare of those [>new] pieces newly come out & shortly to be published.[5] I live heere in a corner y^t not onely disenables me to print any thing my selfe but also keepes me ignorant of all things printed by others.

To morrow it will be a [>full leape] yeere since I received P. Vincentio's letter: [>wherein he] promised to send me a coppy of another [>booke] quem brevi, saith he, priori subjungere intendo.[6] But I heare nothing of it. I suppose ye argument of it should be ductus planorum in plana sed in orbem[7] Or else novae speculationes quae aliâ viâ quam quae proportionalitates implicat, circuli

---

[1] Letter 74.

[2] See Letter 19, n. 13.

[3] When Newcastle reached Rotterdam he found that the Prince of Wales had already put to sea with the English warships (see Letter 74, n. 1); his first impulse was to take a ship and try to catch up with the Prince, but in the event he was dissuaded by his wife. (See Trease, *Portrait of a Cavalier*, p. 159.)

[4] Letter 73.

[5] The works by Gassendi, Hobbes, and Cavalieri referred to in Letter 74.

[6] 'Which I intend to add soon to the earlier one': see Letter 73, nn. 11, 12.

[7] 'The drawing of planes into planes, but [this time] into a sphere'. In book 6 of his *Opus geometricum*, entitled 'De ductu plani in planum' (pp. 703–864), de Saint-Vincent explained a method of producing solids by 'drawing' one plane figure into another – for example, drawing one triangle into another to give a square-based pyramid (p. 707). At the end of his treatise, on p. 1226, he promised a further book covering those things he had neglected, amongst them 'ductuum planorum in plana sed in orbem' (in other words, applying the same technique to produce a sphere).

& hyperbolae tetragonismos exhibere possunt.[8] Of both which he speakes in ye Epilog of y^t great volume.

So that it seemes an infinite businesse to [>scan all] his conatûs Cyclometricas.[9] [>For By that time that] we have done [>wth] this, We are in danger to have as much more to examine.

[>By his hugenesse he deterres most from buying or reading him. & by these] reserves he makes it impossible [>finally & totally to refute him & to shew] y^t he cannot square ye Circle, unlesse [>we] demonstrate y^t it cannot be squared Which, you know, very [>famous Geometers have] affirmed but [>(as I remember) they] never proved it

Longomontanus hath beene dead about a yeare.[10] They say he lay a dying, when my booke arrived in Copenhaven, so that he never saw ye coppy which I sent him.

Had it not hung so long in ye presse it might perhaps have persuaded him to publish [>a recantation or at least to pen one] to be published after his death Which now, for ought I can heare, he hath not done

The conclusion of your letter mentions hopes of a resting place Of wch hope, at y^s distance, I see [>not the grounds] But when soever it shall happen, no man shall more gladly see [>it] or heare of it than

    Your Honours most humble servant
    John Pell

Breda. Aug. 7.17. (1648)

[*addressed:*] For the right Honourable Sr Charles Cavendysshe Knight &c
    At my Lord Marquis of Newcastles lodgings in Roterdam.

---

[8] 'New speculations which may exhibit quadratures of the circle and the hyperbola, by a method different from the one that involves proportionalities'. Like the previous quotation, this is from p. 1226 of de Saint-Vincent, *Opus geometricum*; this particular passage is marked in Pell's own copy of the book (Busby Library, Westminster School, pressmark K F 6).

[9] 'Attempts to measure the circle'.

[10] Longomontanus died on [28 Sept./] 8 Oct. 1647.

Letter 76

9/19 August 1648

Cavendish to Pell, from Rotterdam

BL, MS Add. 4278, fos. 275–6 (original)
van Maanen C38

Worthie S^r:

manie thankes for your last letter;[1] which worthie S^r: Williame Boswell[2] was pleased carefullie to send in one of his owne to me; but I must excuse my negligence in not giuing you thankes for your last letter & note,[3] writ to me at Paris; all I can saie is, I am not verie sure whether I writ you my thankes for it from Paris or not; but you receiuing none it seems I did not; for letters seldom faile. howsoeuer, I now giue you manie thankes both for that letter & the note in it which fauoure & all your others I shall [>by] Gods grace, euer acknowledge, & indeuoure to requite to my power. I knowe not whether you haue yet read P: Vincentios greate booke,[4] but I guess by your letter you finde some faulte in his quadrature & woulde be loath to ingage to take the like paines to discouer his error in his next voluminous booke, but yet methinkes he hath so manie good propositions & new to me, that I wish he woulde publish an other volume. I desire you will doe me the fauoure to let me know if he haue failed; wherin, I doute would be too longe for a letter; therfore onelie the first if you please. you touch in your letter (as I conceiue) as if the Impossibilitie of the quadrature of the circle might be demonstrated, which next to the demonstration of the quadrature, & the finding of two meane proportionalls, & some few other difficult propositions, I should esteem of; for it would saue a greate deale of pretious time./ M^r: du Cartes shews in his geometrie probablie that 2 meane proportionalls can not be found by the circle & the straight line[5] but I conceiue it not so full a demonstration, but that men may still hope to finde it; therefore if that allso might be demonstrated impossible, it would saue allso the like pretious time./ I knowe none likelier than your self to demonstrate either the truth or the impossibilities of these propositions, as allso the trisection of an angle.[6] I desire you will in your letter

---

[1] Letter 75.

[2] See Letter 19, n. 13.

[3] Letter 73 (and its postscript, which was written on a separate piece of paper).

[4] See Letter 49, n. 2, and Pell's comments in Letter 75.

[5] See A&T vi, pp. 443–4. The problem of finding two mean proportionals is equivalent to the classical problem of doubling the cube.

[6] In this passage Cavendish has touched on all three classical problems of mathematics: squaring the circle, doubling the cube, and trisecting an angle. All three are impossible using Euclidean 'ruler and compass' methods, but there was as yet no proof of this.

be pleased to giue a touch of these things./ I send you heer inclosed a proposition of M^r: Roberualls,[7] though to send propositions to you be I thinke (according to the prouerb,) to send owles to Athens, but for aduertisment I suppose it will not be vnwellcome./ I hope you haue a Printer at Breda,[8] if not, there is a store not far of. so I hope that will be no hinderance to your publishing of anie thinge; which I doe as I euer haue done much longe for./ I thought wee should not haue staied so longe heer, neither doe we yet knowe howe longe it will be, but I suppose longe enough to heare from you, if your occasions will permitt you to doe me that fauoure. And so wishing you all happiness I commit you & vs all to Gods holie protection & remaine

Your assured freind & seruante
Charles Cauendysshe

Roterdam Aug: 9/19. 1648

[*addressed:*] To my verie worthie friend M^r: John Pell publicke Professor of the Mathematicks at Breda. giue this
[*added to address in another hand:*] Breda. 6s[9]

[*annotated by Pell:*] Received Aug 19. 29.

[*cover annotated by Birch:*] 9/19 Aug. 1648.

---

[7] This copy of a proposition by Roberval (on whom see Letter 19, n. 7), in Cavendish's hand, concerning the area of a spherical triangle, is BL, MS Add. 4278, fo. 277^r. The proposition also appears, attributed to both Roberval and Harriot, in a collection of Cavendish's notes: BL, MS Harl. 6002, fo. 32^r.

[8] See Letter 67, n. 7.

[9] 6 stuivers (see Letter 61, n. 7).

Letter 77

14/24 October 1648

Cavendish to Pell, from Antwerp

BL, MS Add. 4278, fos. 278–9 (original)
van Maanen C39

Worthie S^r:

douting you receiued not my Last letter to you from Roterdam,[1] wherein I gaue you thankes for your manie fauoures & particularlie for your letter & note I receiued longe since at Paris; which I heere repeate againe, least I should seeme vnthankfull. I allso thought it not amiss to let you knowe wee are nowe at Antwerp; & likelie to remaine heere till it shall please God to reduce the affaires of England to such a condition of peace or warr as may become honest men to return home./[2] I haue heere Toricells worckes[3] & M^r: Oughtreds Clauis both english & latin;[4] I suppose you haue them; otherwise if I knew how to communicate to send them to you I woulde; & desire your iudgment of them, especiallie of Toricells curue indiuisibles & of Oughtreds resolution of affected aequations, & his demonstration of the rule of falsoode.[5] They tell me P: Vincentio hath nothing yet more in the press heere;[6] but that P: Richards Apollonius will shortlie be in the press./[7] I wish M^r: Oughtred would publish his exact table of Rumbes[8] which he tolde me of longe since./ I desire to knowe what bookes of Nauigation you esteem best;/ I like Snellius his Typhis[9] for the guiding of a

---

[1] Letter 76.

[2] Cavendish's brother did remain in Antwerp until the Restoration.

[3] Torricelli, *Opera geometrica* (see Letter 46, n. 9). This contained a simplified version of Cavalieri's work on indivisibles, and was the book through which Cavalieri's ideas became more generally known.

[4] See Letter 68, n. 16.

[5] See Letter 68, n. 16, and Letter 69, n. 11.

[6] On de Saint-Vincent see Letter 48, n. 5, and Letter 49, n. 2.

[7] Claude Richard's edition was not in fact published until 1655: see Letter 73, n. 18.

[8] William Oughtred published a brief treatise on navigational mathematics, *An Addition unto the Use of the Instrument Called the Circles of Proportion, for the Working of Nauticall Questions* (London, 1633) (a supplement to his *The Circles of Proportion and the Horizontall Instrument* (London, 1632)). Ch. 3 of this treatise, 'Of the Mariners Compasse, and Rumbes or points thereof', discussed the points of the compass, known as 'rhumbs' by mariners, and presented a 'Table of Rumbes' (p. 19); Oughtred noted that the traditional division of the compass was into 32 rhumbs, but suggested that other, more numerous divisions would be useful.

[9] Willebrord Snel, *Tiphys batavus, sive histiodromice, de navium cursibus et re navali* (Leiden, 1624).

ship/ P: Fournier[10] is more general for the building etc: of a ship:/ but I desire your opinion whom you esteem best./ as allso for Fortification; Doguens Moderne Militarie Architecture[11] I sawe at Roterdam but had not time to reade it then & would not buy [>it] till I heare it aproued of by some of iudgment. it is published this yeare;/ Fridach[12] hath bin esteemed the best hitherto.// M^r: Oughtred (as I remember) tolde me he had a waye to finde the proportionall part in anie Table[13] verie exactlie; as for example in the Table of sines,[14] you knowe the differences doe not increase or decrease proportionablie & yet to finde [>the] proportionall [>part] without anie sensible error, in anie part of the Table. I desire to knowe if you haue found anie such waye, if you haue not, I dout not but you may when you please.// S^r: I haue troubled you too much at one time, therefore wishing you all happiness & praieing for a happie peace in England; which God Allmightie of his goodnesse send I remayne

>Your affectionate friend & humble seruant
>Charles Cauendysshe

Antwerp. Octob: 14[/]24. 1648.

[*addressed:*] To my verie worthie friend M^r: John Pell publick Professor of the Mathematicks at Breda. giue this at Breda

[*annotated by Pell:*] Received Octob. 22 [/] Novemb. 01.

[*cover annotated in another hand:*] Algebra nova P Luneschos Patavii 1644 in folio Canicularia Bainbridgij[15]

[*cover annotated by Birch:*] 14/24 Oct. 1648.

---

[10] Georges Fournier SJ, *Hydrographie, contenant la théorie et la practique de toutes les parties de la navigation* (Paris, 1643).

[11] Matthias Dögen (1605/6–62), a German engineer, lived in Amsterdam and was employed by the Dutch Admiralty. His *Architectura militaris moderna* was published there in 1647; French and German translations were published (also in Amsterdam) in 1648.

[12] Adam Freitag, *Architectura militaris nova et aucta, oder newe vermehrte Fortification* (Leiden, 1631); in the French translation (*L'Architecture militaire* (Leiden, 1635)) the author's name is given as 'Freitach'.

[13] The words 'in anie Table' are underlined in red by Pell. Cavendish was discussing here the interpolation of trigonometrical tables. Henry Briggs had written extensively on interpolation, or subtabulation, in *Arithmetica logarithmica*, and also in his *Trigonometria britannica* (Book 1, pp. 35–60). Pell himself had been interested in tables and the interpolation of tables since his student days, and had corresponded with Briggs on this subject in 1628. However, it was not a simple subject to explain to Cavendish, and Pell did not attempt to do so (see his brief comments in Letter 78).

[14] The words 'Table of sines' are underlined in red by Pell.

[15] This annotation was apparently added mistakenly to this letter; it refers to two works mentioned in the next letter from Pell to Cavendish (see Letter 78, nn. 13, 17).

Letter 78

[25 October/] 4 November 1648

Pell to Cavendish, from Breda

BL, MS Add. 4280, fo. 133^r (draft); MS Add. 4278, fo. 282^r (enclosure: Pell, authorial copy)

van Maanen P28

Right Honourable

Since my last,[1] I have received 2 from your Honour the one from Roterdam, the other 3 dayes agoe from Antwerp.[2] To both which I now answere thus.

1 I never saw any thing of Toricello,[3] Fournier[4] or Döguius.[5] Of which last, when I lived at Amsterdam, I heard some say that he was likely to be but an exscriber[6] for the most part; yet having had all Hortensius[7] his papers, perhaps he might in them finde something singular & worth the reading. Fr. Schooten tells us that he is writing *circa optimam muniendi rationem*.[8] In the meane time Freitag[9] hath gotten the name, but neither he nor any other satisfy me.

2 I give your Honour many thankes for Mr Robervalls demonstration of the Area of a sphaericall triangle.[10] Albert Girard wrote of it Anno 1629.[11] But our owne countryman Mr Harriot had done it long before. Of which we have Mr Brigges his testimony printed in Dr Hakewills Apology for Gods providence, where Mr Brigges, reckoning up Mathematicall inventions makes y^t the 7th 7° *Aream*, saith he, *trianguli sphaerici* vel quantitatem anguli solidi *invenisse* primus docuit peritissimus Geometra Thomas Harriotus, quum ante eum nemo hoc sit assequutus

---

[1] Letter 75.

[2] Letters 76, 77.

[3] See Letter 77, n. 3.

[4] See Letter 77, n. 10.

[5] See Letter 77, n. 11.

[6] One who copies things out (from the works of others).

[7] See Letter 11, n. 2.

[8] 'About the best method of fortification'. Since 1646 Frans van Schooten (on whom see Letter 69, n. 10) had lectured at the school for engineers attached to Leiden University; the science of fortification was one of the main subjects taught there. However, his work in this field was never published. (See also Pell's comment in Letter 89.)

[9] See Letter 77, n. 12.

[10] See Letter 76, n. 7.

[11] Albert Girard (1595–1632), a mathematician and engineer from Lorraine who studied and worked in the Netherlands, had discussed the area of spherical triangles in his *Invention nouvelle en algèbre* (Amsterdam, 1629).

&c.[12] This of Mr Brigges puts me in minde of a little treatise of his Colleague Dr Bainbridge De Canicularibus printed in octavo at Oxford this yeare;[13] which it may be your Honour hath not yet seene.

3 I have Oughtredi Clavis Latin & English, but both imperfect. The Latine ends with ye sheete O & begins at the sheete B and therefore perhaps wants onely the title sheete A.[14] The English booke wanted some dialling diagrams, which I have supplyed by the helpe of a freinds perfecter coppy.

4 In the rule of fals position[15] no man satisfyes me, much lesse hee, who touches onely that poore parcell of it that reaches not so high as a quadratic aequation. The exaltation of it, to higher aequations of many dimensions, hath beene attempted by some that were not their crafts-masters & therefore were not able fully to determine how farre it would serve us, where & when it would faile us. Much lesse have they taught the whole doctrine of the false positions; of which perhaps they never thought.

5 I am also certaine that in those 3 sheetes (from pag 121 to pag 169) he hath not given us the compleate doctrine concerning the resolution of adfected aequations in numbers;[16] though he tell us that 28 (the number of his precepts) is a perfect number. But it may be Padua hath supplyed all, that I would or he should have said of that argument. For 3 or 4 dayes agoe, one told me that he had seene at the Haghe, a great booke in folio, printed at Padua within these 2 yeares, called Algebra nova, written in Latine by a Spanish Jesuite, whose name, he sayes, is

---

[12]'Thomas Harriot, that most skilled geometer, was the first person who taught how to find the area of a spherical triangle or the quantity of a solid angle; no one had managed to do this before him.' George Hakewill's *An Apologie of the Power and Providence of God in the Government of the World* was first published in Oxford in 1627; it argued against the idea that human nature and the world were in decline, collecting examples of achievements by the moderns that surpassed those of the ancients. Henry Briggs, Professor of Mathematics at Oxford, sent Hakewill a letter listing eight such achievements of modern mathematicians, including Copernican astronomy, logarithms, and this discovery by Harriot: Hakewill inserted it in the 2nd edition of his book (*An Apologie or Declaration of the Power* ...[etc.] (Oxford, 1630), here p. 264: the printed text has 'invenire', not 'invenisse', and 'cum', not 'quum').

[13] John Bainbridge, *Canicularia*, ed. John Greaves (Oxford, 1648). John Bainbridge (1582–1643) had been Professor of Astronomy at Oxford. This publication consisted mainly of an unfinished treatise by Bainbridge on the theories of classical astronomers about Sirius, the 'dog star', and the 'dog days', the period of the year when Sirius is in the ascendant. To this John Greaves (1602–52), Bainbridge's successor as Professor of Astronomy, added a demonstration of the heliacal rising of Sirius, and notes on other stars drawn from a Persian MS (Bodl. MS Gravius 5) of the star catalogue of Ulug Beg, the grandson of Timur Leng (Tamerlane).

[14] See Letter 68, n. 16; the gatherings of the 1648 Latin edition do begin with 'A', and end with 'O'.

[15] See Letter 32, n. 12.

[16] See Letter 69, n. 11.

.... Luneschlos.[17] It is likely that some of the best furnished shops of Antwerp have it.

6 Of Navigation I have seene no writer that comes neere that which I thinke might be done in that kinde. Mr Oughtred in his treatise of that subject, tells us of tables of Rumbs, but I did not thinke he had made any such.[18] Nor can I guesse at his way of correcting the proportionall part so exactly as he speakes of. For the Rumbs, he needes only Secants, whose differences are all additive, not alternè additive & subtractive as of sines, and therefore if ever he speake publikely of these things, perhaps he will say nothing of Newes, Sines or Tangents. Howsoever sure I am of that, if he said he could insert intermedialls *in any table*, he undertooke more than he or any man else can performe. Of which it is easy to give instances innumerable. I have not Mr Briggs his Trigonometria Britannica; but, as I remember, he hath there taught undoubted artifices for inserting intermediall adscripts in his Trigonometricall Canons.[19] But, I feare, they are not many that understand his meaning or see the necessary certainety of his precepts. But as for Rumbs, although the Seamen had tables of them calculated to a haires breadth they *would* not (I was about to say they *could* not) use them. Of which, if neede be, more may be said another time.

This inclosed paper was lately sent me from one that had beene my auditor at Amsterdam.[20] It may serve for an exercise of your Zetetica.[21] What you finde may be compared with my answer to it; which I shall send, whensoever your Honour shall call for it. In the meane time I continue

---

[17] Joannes de Luneschlos, *Thesaurus mathematum reseratus per algebram novam tam speciebus quam numeris declaratam et demonstratam* (Padua, 1646). The author was neither Spanish nor a Jesuit. He was German, born in Solingen in 1620; he matriculated at Padua University in 1646 and received his doctorate there in 1648. Thereafter he spent some time in Scandinavia before becoming Professor of Philosophy at Heidelberg in 1651. (See the text accompanying the portrait of Luneschlos in Anon., *Parnassus Heidelbergensis omnium illustrissimae huius academiae professorum icones exhibens* (Heidelberg, 1660); T. Weigle, 'Die deutschen Doktorpromotionen in Philosophie und Medizin an der Universität Padua von 1616–1663', *Quellen und Forschungen aus italienischen Archiven und Bibliotheken*, 45 (1965), pp. 325–84; here p. 357; D. Drüll, *Heidelberger Gelehrtenlexikon, 1652–1802* (Berlin, 1991), pp. 93–4.) Pell's eventual judgement of Luneschlos's work was recorded in his letter to Hartlib of 9/19 July 1657, from Zurich: 'Surely he understood but a little, when he printed a booke, at Padua, in folio, concerning Algebra, dedicated to Louys de Geer. A friend bought it at the Hague & gave it me Anno 1649. It then seemed, to me, almost alltogether transscribed out of others. I remember some few propositions, which had faults enough in them, to make me suspect they were his owne' (BL, MS Add. 4364, fo. 150^v). Pell's copy of Luneschlos's *Thesaurus mathematum reseratus* is in the Busby Library, Westminster School, pressmark I.F.27; it contains a loose leaf in Pell's hand (between pp. 302 and 303), pointing out errors in the text.

[18] See Letter 77, n. 8.

[19] 'Adscripts' to the circle were circumscribed or inscribed lines, that is, sines or tangents. Pell is referring to the interpolation of trigonometrical tables: see Letter 77, n. 13.

[20] See the enclosure to this letter; the identity of the 'auditor' is not known.

[21] See Letter 49, n. 11.

your Honours most humble & faithfull servant
John Pell.

Novemb. 4.

[*postscript:*]

1 Mr Shaw,[22] the English merchant knowes Simon Caters[23] a carman, that comes hither 3 times every weeke, by whom my booke &c may be safely sent from thence hither

2 That P. Mersenne is dead,[24] is no newes, I suppose.

[*addressed:*] For the Right Honourable Sr Charles Cavendysshe Knight &c.
At my Lord Marquis of Newcastle's lodgings in Antwerp.

[*Enclosure to Letter 78:*]

Habet aliquis pecuniam, quam [>numerum, quem] si per 280 multiplices, facto addas numerum quadratum, ex aggregato $\sqrt{}$ quad: extrahas. prodibit numerus 807

$\sqrt{280a + bb} = 807$
$\therefore 280a + bb = 651249$
$\therefore 280a = 651249 - bb$
$\therefore a = \dfrac{651249 - bb}{280}$

Hic aqua mihi haeret. Datum tantùm unum est, quaeruntur autem duo. nec quid assumere possim video.[25]

---

[22] Unidentified.

[23] Unidentified.

[24] Mersenne had died on [22 Aug./] 1 Sept. 1648, in Paris.

[25] 'Someone has a sum of money [>a number] which, if you multiply it by 280, add a square number to the result, and extract the square root of the total, will produce the number 807 [...] Here I dry up. There is only one given quantity, but two are sought; nor do I see what I might assume.'

Letter 79

[3/] 13 November 1648

Cavendish to Pell, from Antwerp

BL, MS Add. 4278, fos. 280–1 (original)
van Maanen C40

Worthie S^r:

   manie thankes for your letter;[1] I am glad our Countrieman had the honor to demonstrate [>first] that proposition I sent you; but it shews M^r: Roberuall is a goode mathematician that hath demonstrated the same proposition.[2] I expressed not my self so fullie as I might in mentioning M^r: Oughtreds exact Table of Rumbs, as assuming my self he had done them, for that is more than I knowe;[3] but he tolde me manie yeares since he had an infallible way to calculate them; which if he haue not yet done I wish he would shew the waye, & I dout not but some woulde take the paines to calculate them; & though as you wrote I doute oure mariners would hardlie be brought to vse them, yet exactness if it may be is best./ I am glad you conceiue some what may be done in nauigation more than is yet, I hope you will one daie publish it, & an exact Algebra, if anie thing in this worlde can be exact; for I expect it from none but your self; yet I haue made some inquirie of that newe Algebra you mention[4] but can not yet meete with it./ I hope M^r: Schooten will aduance Fortification,[5] for you confirme me in that I suppose it may be yet much needed./ I receiued a letter from my Lo: widdrington[6] who is at the Hague that he had some discourse with a sea Captayne that tolde him there was a Polander[7] who had latelie discouered (with a goode perspectiue I suppose) some spotts in the moone perpendicular (as my Lo: writes) one to an other, by which [the meanes *deleted*] the longitude may easilie be found out. which I confess I much dout of; though I thinke if the [>motion of the] starrs about Jupiter were well knowne & Tables Calculated some goode might be done. M^r: Oughtred first & M^r: Warner after tolde me in

---

[1] Letter 78.

[2] See Letter 78, n. 12 (the 'Countrieman' is Harriot) and Letter 76, n. 7.

[3] See Letter 77, n. 8, and Pell's comment in Letter 78, at n. 18.

[4] Luneschlos, *Thesaurus mathematum reseratus* (see Letter 78, n. 17).

[5] See Letter 78, n. 8.

[6] Widdrington (see Letter 25, n. 6) had travelled with the Prince of Wales to the Netherlands in June 1648 (Cokayne, *Complete Peerage*, xii, part 2, p. 626).

[7] Hevelius (see Letter 72, n. 5).

effect this manie yeares since & Galileo had the same conceit./[8] M^r: Bond an olde Mathematician at London [>was] in hope to finde it by the Loadestone;[9] he is an humble man & speakes verie meanelie of himself, & yet he found an easie & shorte demonstration of that proposition concerning sphaericall triangles which M^r: Oughtred demonstrated first, who tolde me M^r: Bonds demonstration was shorter, & would haue had me send it to them whom I had sent his formerlie too; but I tolde him he taking the paines first shoulde haue the honor of it.// manie thanks for your inclosed paper;[10] at first I thought I might haue taken 66 at pleasure but then a, would be of seuerall values & haue a fraction which I suppose was not the meaning, & therefore I blotted out what I had tried as you may see on the topp of the paper./[11] after considering that $g > b$ & that $gg - bb$ was produced by $g + b$ in $g - b$ & etc: as within the scrawled lines on the right side I made me supose $g - b = c$. & etc: as in the paper.// S^r: I must aske your pardon for sending you such a blurd expression, & no less homelie & perplext explanation; but I assure my self your friendship will pardon all./ I desire you will let me knowe if I haue done it, & not mistake the quare; as allso that you will be pleased to send me your resolution of this quaere, & to lend me D^r: Bainbridg de canicularibus./[12] I heere send you the excellent Torricell[13] & should be ashamed to desire you to returne it when you haue at your leasure read it, but that it is verie difficult to be had, & it was the gift of my worthie kinde friend P: Mersen who to my sorrow I heare is dead./[14] yet if you please keep it till either you or I can get an other, & if I get one I shall send it you, & then desire you will returne this I nowe send you./ S^r: I haue troubled you enough for once therefore committing you & vs all to Gods holie protection I remaine

    Your assured friend & humble seruant
    Charles Cauendysshe

Antwerp Nov: 13. 1648.

---

[8] On Galileo's suggestion (in 1612) that the motion of the moons of Jupiter (or 'Medicean stars') could be used as an astronomical clock, thus making possible the calculation of longitude on earth, see Débarbat and Wilson, 'The Galilean Satellites of Jupiter', p. 145.

[9] Henry Bond (d. 1678) had taught navigation to London mariners in the 1630s. Prompted by Henry Gellibrand's work on the variation of the compass, he conducted research on this and the angle of inclination of the needle, and proposed using these phenomena to determine longitude. His claims would eventually be investigated by a Royal Commission (including Pell) in 1675. (See Bryden, 'Magnetic Inclinatory Needles'.)

[10] See Letter 78, n. 20.

[11] Cavendish enclosed with this letter his attempt to solve the mathematical problem: BL, MS Add. 4278, fo. 282^v. His method involved little more than trial and error, assigning specimen values to a.

[12] See Letter 78, n. 13.

[13] See Letter 77, n. 3.

[14] See Letter 78, n. 24.

[*addressed:*] To my verie worthie friend M^r: John Pell publick professor of the Mathematickes at Breda giue this

[*annotated by Pell:*] Received Nov. 7[/]17

[*cover annotated by Birch:*] 13 Nov. 1649.

Letter 80

4/14 November 1648

Pell to Cavendish, from Breda

BL, MS Add. 4280. fo. 133^v (draft)
van Maanen P29

I give your Honour many thankes for sending Torricellius,[1] with leave to reade him over at leisure & with as much attention as he deserves.

I have considered your answere to our probleme[2] & finde that your numbers were artificially found out, & are altogether such as were required. But that is not sufficient in such questions as this. For we cannot be sure that a solution will satisfy the proposer, till it be so perfect that a Demonstration can be adjoined, proving y^t ye proposed conditions can not be found in any other numbers, than those which are exhibited by the answerer. Such a solution was that wch I sent backe to Amsterdam. Of which I should now, according to promise, send a coppy to your Honour. But I hoped that You had rather first take it in hand once againe, endeavouring to worke out such a full solution. Which, in my opinion, will be worth the attempting. And perhaps, when all is done, there will be something left in my solution worthy to be considered, though at the first sight the question itself appear but a slender one.

It is likely that the proposer would have said thus.
Si $\sqrt{280a + bb} = 807$, quantus est a, quantus b?
Innumerabiles erunt quaesitorum valores, si surdis, fractis, negatis locus detur. At quot erunt, si ijs exclusis, uterque (id est, tam a, quàm b) sit effabilis, integer & affirmatus?[3]

Your answere was, a = 1334
b = 527

Novemb. 14. 24

[*annotated by Pell:*] (Sent wth Dr Bainbridges book)[4]

---

[1] See Letter 77, n. 3, and Letter 79, at n. 13.

[2] See Letter 79, nn. 10, 11.

[3] 'If $\sqrt{280a + bb} = 807$, how much is a, and how much is b? The values of the things sought will be innumerable, if we admit surds, fractions, and negative numbers. But how many will there be, if those are excluded, and each of the things sought (i.e. both a and b) is a natural number, an integer and positive?'

[4] See Letter 78, n. 13.

Letter 81

[27 November/] 7 December 1648

Cavendish to Pell, from Antwerp

BL MS Add. 4278, fos. 283–4 (original)
van Maanen C41

Worthie S^r:

I returne you your book with manie thankes, I am glad to see there is yet so able a mathematician in oxford, for methinks M^r: greaues hath done his task well, & Doctor Bainbridges is a handsom discours[1] & shews finelie howe the heliacall rising of that star euerie 5th: yeare goes on a daie in the Aegyptican calendar./ the obseruations of so grate [sic] a Prince[2] is to be esteemed a raritie though they were not exact; but I suppose these are goode./ I haue heere sent you what I haue found towards the solution of your problem;[3] & as I giue you thankes for the exercice this problem hath giuen me; so nowe I shall giue you more thankes if you please to send me your solution of it./ M^r: Hariot pag: 101. in his 6th: section giues some examples; where the addition & subtraction of binomiall cubick rootes are expressed by numbers; I suppose there is a generall method extant of extracting such binomiall rootes;[4] it there be I haue [>either] forgot it or neuer reade it; I desire your information heerin./ And so wishing you all happiness I remaine

    Your assured friend to serue you
    Charles Cauendysshe

Antwerp. Decemb. 7. 1648

[addressed:] To my verie worthie friend Mr John Pell publick professor of the mathematicks at Breda giue this

[cover annotated by Birch:] 7 Decemb. 1648

---

[1] Bainbridge, *Canicularia* (which contained also the work by Greaves): see Letter 78, n. 13.
[2] Ulug Beg: see Letter 78, n. 13.
[3] For the problem, see Letters 78, 80; Cavendish's enclosure is BL MS Add. 4278, fo. 285^r.
[4] The formula for solving cubic equations requires the calculation of cube roots of binomials, that is, cube roots of numbers of the form $a \pm \sqrt{b}$, a non-trivial problem since, except in special cases, the attempt to find such a root leads straight back to the original cubic equation. The examples given on p. 101 of Harriot, *Artis analyticae praxis*, are carefully chosen (it is not difficult, for example, to check by direct calculation that a cube root of $\sqrt{108} + 10$ is $\sqrt{3} + 1$), but there is no straightforward general method of finding such cube roots.

Letter 82

19/29 December [1648]

Pell to Cavendish, from Breda

BL, MS Add. 4280, fo. 133^v (draft)
van Maanen 30

Right Honourable

The solution, which I sent back to Amsterdam, was but of a few lines; not giving any account how I sought it, as this inclosed doeth.[1] Which I desire may be sent hither againe, because I have no coppy of it. If your Honour shall thinke it worthy the transcribing & want a scribe, I shall be heereafter more at leisure to doe it, than I am at this present. Yet I thought it better to send it now after this rude manner, than to torture you a weeke longer with an expectation of a thing of so small moment.

Your Honours most humble servant

Dec. 19.29.

[*annotated by Pell at head of letter:*] (sent with halfe a sheete of paper written on one side)

---

[1] For the problem, see Letters 78, 80. Cavendish's copy of this enclosure is BL, MS Harl. 6083, fo. 139. It is headed 'The question What Integer Affirmd numbers A & B, if $\sqrt{280a + bb} = 807$'; it includes five 'tabellae' (tables) in which different possible solutions are tested.

Letter 83

[26 December 1648/] 5 January 1649

Cavendish to Pell, from Antwerp

BL, MS Add. 4278, fo. 286 (original)
van Maanen C42

Worthie S^r:

    manie thankes for your letter with the solution of the problem there inclosed;[1] which I heere send back according to your desire; I haue transcribed it myself; & nowe see those numbers artificiallie found which I could not finde but by trieing; some few onelie excepted; yet I perceiued that all the numbers of B must be odde but not particularlie [>to end in] 3 [& *deleted* >or] 7. yet I hope you haue receiued that imperfect attempt of soluing that problem not for it owne sake but because it was infolded in the packet with Doctor Bambrigs booke[2] which I returned with manie thankes for it./ but to returne to your problem I confess I conceiue not yet why the adding of 70 to those 4 values of B should be all & the onelie values of B./ much less doe I yet conceiue the demonstration of those concinnities you afterward mention./ but the finding of the value of BB to be 249. or those by adding of 280 to it; is an excellent example, & I doute not but it may be made vse of in questions of the like nature. I doute not but I shall like all the rest of the process of the worck as well when I vnderstand it./ I should be glad to heare of the publishing of your Diophantus,[3] somewhat to pacifie owre hungrie appetites till you publish your [>owne] analitical worcke./ I haue not yet seene that newe Algebra print[ed *page torn*] at Padua;[4] if you haue, I desire your opinion of it when [you *page torn*] haue reade it./ S^r: I haue troubled you enough, therefore commit[ting *page torn*] you & vs all to Gods protection I rest

    Your assured friend to serue you
    Charles Cauendysshe

Antwerp: Jan: 5. 1648.

[*addressed:*] To my verie worthie friend M^r: John Pell publick Professor of the Mathematicks at Breda, giue this at Breda

[*annotated by Pell:*] Received Jan 2[/]12 1648[/]9

[*cover annotated by Birch:*] 5 Janua. 1648/9

---

[1] See Letter 82.
[2] See Letter 78, n. 13, and Letter 81, nn. 1, 3.
[3] See Letter 14, n. 13.
[4] See Letter 78, n. 17.

528   JOHN PELL (1611–1685)

Letter 84

24 January/3 February [1649]

Pell to Cavendish, from Breda

BL, MS Add. 4280, fo. 133^v (draft)
van Maanen P31

Right Honourable
 I confesse that I ought neither to have fogotten to give notice that I had received my Bainbridge with your letter & inclosed paper;[1] Nor to have put you to the trouble to coppy out mine. But want of leisure having beene the cause of both I will not doubt of your Honours pardon.
 I see by your last that you fully conceive my way of finding that b must end in 3 or 7 & that bb can be no other than those which stand in tabella prima[2] continued &c. I doubt not also but you see to ye bottom of ye device of Tabella secunda,[3] by a few additions avoiding 17 extractions of ye square roote of ye numbers in tabella prima. [>If] you have [>not yet] found [>some way to] demonstrate that all those & onely those which are in tabellâ tertiâ[4] are such values of b as were required; perhaps the narrow hide-bound demonstrations of this adjoined paper[5] will satisfy you. It had beene as easy to have given demonstrations of a larger reach which should have proved more generall conclusions, whereof these are but consectaries.[6] [>But those you will find by the help of these.]
 [Des Cartes tells us (pag 389)[7] that it does us not so much good to reade demonstrations penned by another, as to finde them out our selves by suspecting & examining the truth of undemonstrated assertions. *deleted*]
 If we lived farre asunder, I should feare that this long harvest of a little corne would not be thought worthy the postage of so many letters as have passed to & fro about it. All that I could say of & upon occasion of this probleme would make a greater bulke than I have leisure to write. And there are in your [>former] letters [>other] passages of [>more consequence; to] which I have as yet given no answere. Of some of which I intend to write by ye next opportunity [>&] in y^e meane time remaine

---

[1] See Letter 83, at n. 1; for the work by Bainbridge, see Letter 78, n. 17.
[2] 'The first table'. On this and the other tables mentioned here, see Letter 82, n. 1.
[3] 'The second table'.
[4] 'The third table'.
[5] This enclosed paper has not survived.
[6] = consequences, logical deductions.
[7] A reference to his 'Géométrie', book 8 (A&T vi, p. 464).

Your Honours most humble Servant
John Pell.

Feb 3. stylo novo

[*postscript:*] The paper needes not be sent backe, I have a coppy of it.

[*annotated by Pell at head of letter:*] (sent wth another halfe sheete of paper written on one side)[8]

---

[8] This 'halfe sheete', in which Pell answered Cavendish's query in Letter 83 about the values of B, is BL, MS Harl. 6083, fo. 144^r.

530   JOHN PELL (1611–1685)

Letter 85

12/22 March 1649

Cavendish to Pell, from Antwerp

BL, MS Add. 4278, fos. 287–8 (original)
van Maanen C43

Worthie S^r:

it is so longe since I receiued that letter from you wherein you included the demonstration of some quaeres I made of your Problem formerlie sent,[1] that I am ashamed I returned not thankes sooner; all I can saie is that besides the late sadd accident,[2] I am loath to trouble you too oft, especiallie since you spare not yourselfe to satisfie my quaeres; yet I shall be bolde to make some nowe; first if there be an uniuersall waye of extracting cubick rootes etc: of binomials,[3] as there is of the q: rootes of the sixe binomialls & residualls./[4] I suppose van cullen hath done something of this subiect,[5] but I haue him not nowe by me;/ next whether by the table of sines the rootes of all cubick aequations & higher powers may not be found, as Herigon pag. 42. Supplem: Algebra finds of some cubicke aequations;/[6] as I remember my olde worthie learned friend M^r: Warner tolde me, that by the Logorithmes or his analogick numbers (I haue forgot which) he thought a waie might be found to finde the roote of anie affected aequation./[7] Thirdlie whether P: vincentioes quadrature of a circle[8] (anie of his waies) require the solution of this problem. Datis tribus quibuscumque magnitudinibus, rationalibus vel irrationalibus, datisque duarum ex illis, logarithmis tertiae logorithmum

---

[1] Letter 84.

[2] This probably refers to the execution of Charles I on 30 Jan. [/9 Feb.] 1649.

[3] See Letter 81, n. 4.

[4] 'q: rootes' are square roots; the six binomials and residuals are number of the form: $a \pm \sqrt{b}$, $\sqrt{a} \pm b$, $\sqrt{a} \pm \sqrt{b}$, the 'binomials' being connected by a plus sign and the 'residualls' by a minus.

[5] The Dutch mathematician Ludolf van Ceulen had discussed binomials (but not in relation to cube roots) in his *Fundamenta arithmetica et geometrica*, tr. W. Snel (Leiden, 1615), pp. 13–23.

[6] The method of solving cubic equations by tables of sines depends on the triple-angle identity: $\sin 3\theta = 3\sin\theta - 4\sin^3\theta$. A cubic equation of the form $3p^2 x - x^3 = 2q^3$ is easily solved using the substitution $x = 2p\sin\theta$, yielding $\sin 3\theta = q^3/p^3$ and hence θ (provided $q < p$). This was the method used by Hérigone in his 'Supplementum algebrae' (*Cursus mathematicus*, vi, 2nd pagination, p. 42). Pell hoped that similar method might be found for equations of higher degree.

[7] On Warner and his 'analogick numbers' (antilogarithms) see Letter 1, n. 7. The method for solving an affected (*sc.* polynomial) equation by logarithms consists of approximating the graph of the equation locally by $y = x^r$, then 'straightening' it using the relationship $\log y = r \log x$ and interpolating between known values by simple use of ratios.

[8] See Letter 49, n. 2.

geometricè inuenire./[9] I finde this in my worthie kinde friend P: Mersennus his booke printed 1647./[10] & this problem accounted by him & M[r]: Roberuall[11] as harde to solue as the quadrature of the circle it self./ I desire your opinion of this & of the whole booke it self at your conuenient leasure when you haue examined it./ I must nowe desire a fauoure from you in the behalfe of my noble friend my Lorde Widdrington, who hath two sonns[12] heere that haue made some progress in the mathematicks, & some entrance I beleeue into Algebra; my Lordes greate desire is that they should be your schollers, especiallie for Algebra, & desires that you woulde for the present be pleased to direct theyr studies therin, either by prescribing what bookes they should reade, or such instructions of your owne as you shall thinke fitt for younge beginners; my Lords intention being that they shall remaine hear; if you can conuenientlie doe it at such a distance you will much oblige a noble Lorde who I dout not will requite your fauoures & I shall esteem it an addition of your fauoures to me; these youthes are aboute 15. & 16 yeares of age./ I haue troubled you enough if not too much at one time. Therefore wishing you all happiness I commit you & vs all to Gods holie protection & remaine,

Your assured friend to serue you
Charles Cauendysshe

Antwerp: Mar: 12/22 1648.

[addressed:] To my verie worthie friend M[r]: John Pell publick Professor of the Mathematickes at Breda. giue this at Breda

[cover annotated by Birch:] 12/22 March 1648/9

---

[9]'Given any three magnitudes whatsoever, rational or irrational, and given the logarithms of two of them, find the logarithm of the third by geometrical means.'

[10]The passage just quoted is from Mersenne, *Novarum observationum* ... *tomus III*, p. 72, where Mersenne introduces it with the statement that it is 'a problem that perhaps requires a much more difficult solution than the squaring of the circle itself' ('problema, quodque forsan longè difficiliorem, quàm ipsa quadratura, solutionem requirit'), and argues that de Saint-Vincent's method of squaring the circle depends on it. In the modern understanding of logarithms, log 1 is taken to be 0, and once a base a has been chosen, the logarithm of any number y is defined as x, where $y = a^x$. Conversely, given any number and its logarithm it is possible to determine a and hence to find the logarithm of any other number to the same base. When Mersenne was writing in the 1640s it was not yet conventional to let log 1 = 0, so two numbers and their logarithms were needed to fix a logarithmic scale, but Mersenne was not sure that this could always be done, hence his question (see R. P. Burn, 'Alphonse Antonio de Sarasa and Logarithms', *Historia mathematica*, 28 (2001), pp. 1–17, here p. 3).

[11]See Letter 19, n. 7.

[12]Widdrington (see Letter 25, n. 6, and Letter 79, n. 6) had eight sons (and two daughters); one of the two mentioned here was probably the eldest, William (c.1631–75), who succeeded to the barony in 1651 and would accompany the Marquess of Newcastle to England at the Restoration in 1660.

Letter 86

4/14 April [1649]

Pell to Cavendish, from Breda

BL, MS Add. 4280, fos. 133^v–134^r (draft)
van Maanen P32

Right Honourable

To the demands of your last letter,[1] give me leave at this time to reply in generall, diferring the fuller answeres till more leisure.

1 The extraction of the Cubicall rootes of binomialls, requires artifices much unlike those which sufficed for their Quadr. rootes.[2] When your Honour first desired it, I thought it would be no long worke to transcribe what I had beene doing concerning that probleme 9 yeares agoe, when the famousest Mathematicians in Holland tooke it in hand upon occasion of a quarrell betweene two lesse skilfull teachers,[3] who had fallen out about it so eagerly that we heard ye noise of it into England. But now comming to compare those my papers with their bookes, I finde considerable defects in their rules wherein they seeme to acquiesce; Though none of them have demonstrated their truth, nor indeed can they of all, without correction. This collation & endeavour to discover & supply their defects hath held me in a longer processe than at the first I expected: so that I must crave a little more respit before I let it goe out of my hand.

2 Your next is a greater & farre more usefull desideratum; To make use of the tables of sines for solution of affected aequations.[4] I have long had a designe to shew vaste uses (hitherto unknowen, at least untaught) of Mathematicall tables;

---

[1] Letter 85.

[2] See Letter 85, n. 3.

[3] Ten years earlier a young mathematician in The Hague, Johan Stampioen, had published a treatise on algebra, *Algebra ofte nieuwe stel-regel* (The Hague, 1639). Soon afterwards another teacher of mathematics, Jacob van Wassenaer (a surveyor from Utrecht), published a scathing attack on it ('I. a Wassenaer', *Anmerckingen op den nieuwen stel-regel van Iohan Stampioen* (Leiden, 1639), describing Stampioen as a charlatan and singling out his treatment of the extraction of cube roots of binomials (*Algebra*, pp. 25–7) as especially incompetent. Stampioen published six pamphlet-replies to van Wassenaer, and made a formal complaint to the Rector of Leiden University, calling on the University to adjudicate. Judgment was eventually given on [14/] 24 May 1640 by the two Professors of Mathematics at Leiden, Golius and van Schooten, ruling in van Wassenaer's favour. (See van Wassenaer ['I. a Waessenaer'], *Den on-wissen wis-konstenaer I. I. Stampioenius ontdeckt door sijne ongegronde weddinge ende mis-lucte solutien van sijne eygene questien* (Leiden, 1640), esp. pp. 9–18 (Stampioen's complaints), 46–56 (further criticisms of Stampioen on cube roots and binomials), 81–7 (the judgment).)

[4] See Letter 85, n. 4.

as well of those which the bungling Astronomers use onely for their Canonicall trigonometrie, as of the later Canons of Nepero-Briggian & Warnero-Pellian numbers.[5] Amongst these uses, this desideratum of yours, is none of the least. And perhaps one day I shall have opportunity to tell the world this my dreame at large. In the meane time, having resolved to lay hold upon any occasion to say something of it in publike, I intended to make a digression concerning it, in the remaining part of my reply to Longomontanus.[6] Who had never beene misseled into those Labyrinths wherein he died; if some skilfull professor of pure Mathematickes had shewen him the grounds, construction & use of those Canons before his Astronomicall maisters taught him that grosse handling which serves their turne. But he being now dead, I know not when I shall publish the continuation of that controversy.

3. You mention a Logarithmicall Probleme spoken of by Mersennus (in a booke printed 1647 which I have not seene) and by him accounted *as hard* to solve as the quadrature of the circle. He might have said it is *harder*, for I thinke I could prove it *impossible* as you propound it:[7] Whereas I never yet saw the quadratures impossibility demonstrated. But I have not hitherto seene, that Vincentio any where requires that probleme as necessary to any of his quadratures.[8]

4. To your last I know not what to say. I should be glad to serve that noble Lord[9] wherein I may. Yet unlesse I knew what his sonnes can doe already, it will not be easy for me to give apt advice for their studies. But I will thinke on it, and in the meane time remaine

Your Honours most humble Servant

Breda April 4.14.

---

[5] Henry Briggs adapted and expanded the tables of logarithms first published by John Napier: see Briggs, *Arithmetica logarithmica*. On the tables of antilogarithms compiled by Warner and Pell, and their fate, see Letter 1, n. 7, and Letter 14, at nn. 8, 9.

[6] This part was never published.

[7] See Letter 85, nn. 9, 10. Pell seems to have misunderstood and underestimated what was being asked. Given $\log y_1$ and $\log y_2$, it is easy to find the logarithm of a third number y_3 in terms of the other two if it is related to y_1 and y_2 in some obvious way, for instance if $y_3 = y_1 y_2$ or $y_3 = y_1/y_2$. No such information about y_3 was given in Mersenne's formulation, leading Pell to suppose that the problem was impossible.

[8] See Letter 49, n. 2, and Cavendish's query in Letter 85, at nn. 8–10.

[9] Widdrington; see Letter 85, at n. 12.

Letter 87

6/16 August 1649

Pell to Cavendish, from Breda

BL, MS Add. 4280, fo. 134^r (minute)
van Maanen P32(bis)

Aug. 6. 16. Munday
I sent him rolled up upon a stick my expanse demonstration of reduced biquadratics[1] & in y^e covert I wrote onely thus.
 Si xxxx(∗) − 35xx + 90x = 56; tum x = cui?[2]

Van Maanen mistakenly dates this minute to 16/26 Aug.

[1] The 'expanse demonstration' is almost certainly the sheet on biquadratic equations in Pell's hand that is to be found amongst Cavendish's papers in BL, MS Harl. 6083, fos. 100^v–101.

[2] 'If $xxxx(*) - 35xx + 90x = 56$, then x is equal to what?' The symbol (∗) here indicates that there is no term in x^3. This convention for denoting missing terms was introduced by Descartes, and probably suggested to Pell the idea of using the same symbol in his three-column method to denote missing conditions.

Letter 88

[13/] 23 August 1649

Cavendish to Pell, from Antwerp

BL, MS Add. 4278, fos. 289–90 (original)
van Maanen C44

Worthie S^r
   manie thankes for your late fauoures when I was with you,[1] & for your paper of resoluing those qq: aequations into two q: aequations; which I receiued last fridaie;[2] I see by the process of the worck, (those two qq: aequations being aequall) there is produced the cubick aequation y^6: etc: but howe $xx + yx$ & $xx - yx$ are assumed with the addition [& *deleted*] or subtraction of 11 & 12 aequation; I knowe not; for I suppose it is those q: aequations are sought for; for by the multiplication of them by one an other the qq: aequation giuen is produced as I conceiue; I haue not De Cartes geometrie by me, when I haue I shall better studie your paper & hope to finde the reason of your process.[3] The more I looke on this paper of yours, the more I like your waye of omitting no case; I remember I tolde you I conceiued not howe some cubick aequations were produced by

---

[1] Pell's summary diary for 1648–9 includes the following entries: '[July] 25. [/Aug.] 4 S^r Ch Cavendysshe L Widdrington came'; '[July] 28 [/Aug] 7 they went backe to Antwerp' (Busby Library, Westminster School, loose sheet inserted at the end of Busby, 'Accounts and Memoranda').

[2] The words 'your paper' and 'I receiued' are underlined in red by Pell; 'q: aequations' are quadratic equatiions and 'qq: aequations' are biquadratic equations. See Letter 87, n. 1.

[3] For Descartes's rule for writing a biquadratic equation as a product of two quadratics see Letter 56, n. 11.

multiplication, of supposing 3 rootes as for example this[4] $a^3 + S^{pl}$: $a = z^s$:/ thus I can produce it. $b - a = c$./ [er]go: $b^3 - 3ab^2 + 3a^2b - a^3 = c^3$. & by transp: $a^3 - 3ba^2 + 3b^2a = b^3 - c^3$./ $-a + b = c$ [times] $3ba$ / $-3ba^2 + 3b^2a = 3bca$ [er]go: $a^3 + 3bca = b^3 - c^3$./ which is like to $a^3 + S^{pl}$:$a = z^s$: & so this aequation $a^3 - Ra^2 = z^s$:/ thus I can produce; suppose $g > f$ / $gf > q$ of: $\frac{1}{2}f$. [er]go: $a^2 + fa + gf = 0$. is an impossible aequation & hath no rootes at all, not of/ $a^2 + fa + gf$ [times] $a - g = a^3 + fa^2 + gfa - ga^2 - gfa - g^2f$. suppose $g - f = d$/ [er]go: $a^3 - da^2 = g^2f$./ like $a^3 - Ra^2 = z^s$:/ I desire you will doe me the fauoure to send me the waye howe to finde those resolutions of those cubick aequations which Cardan found; but shewes not howe he found them;/[5] I desire allso at your conuenient leasure some examples in Analyticks, that shews the vse of substitution, with some generall directions howe to substitute; for the particular & complete doctrine of it, I conceiue would be so much as to make a booke; which I shall expect when you publish your Analitickes; in the meane time I wish the excuse of wanting a well furnished printer taken awaye that wee might haue

---

[4] In the passage that follows, Cavendish twice uses Harriot's rectangular bracket to represent the multiplication of two terms; in the transcript and in the modern notation given below this is indicated by adding the word 'times'.

> I remember I told you I conceived not how some cubic equations were produced by multiplication, of supposing 3 roots as for example this:
>
> $a^3 + S_{plano}a = z_{solido}$

(Cavendish is here writing the equation in Viète's style in which dimensional homogeneity is preserved by making the coefficient of a a 'plane' term and the constant or given term a 'solid'.)

> Thus I can produce it.
> $b - a = c$ so $b^3 - 3ab^2 + 3a^2b - a^3 = c^3$ and by transposition $a^3 - 3ba^2 + 3b^2a = b^3 - c^3$.
> $-a + b = c$, times $3ba$, gives $-3ba^2 + 3b^2a = 3bca$
> so $a^3 + 3bca = b^3 - c^3$ which is like to $a^3 + S_{plano}a = z_{solido}$.
>
> And so this equation $a^3 - Ra^2 = z_{solido}$ thus I can produce.
>
> Suppose $g > f$
> $gf >$ square of $(f/2)$ so $a^2 + fa + gf = 0$ is an impossible equation and has no roots at all, nor has $(a^2 + fa + gf)$ times $(a - g)$, which gives $a^3 + fa^2 + fga - ga^2 - gfa - g^2f$.
> Suppose $g - f = d$ so $a^3 - da^2 = g^2f$, like $a^3 - Ra^2 = z_{solido}$.

[5] Letter 90 (at n. 2) shows that Cavendish was thinking of a reference by Descartes (see A&T vi, pp. 471–3) to the rules for solving equations of the form $z^3 \pm pz \pm q = 0$, given by the physician, mathematician, and philosopher Girolamo Cardano (1501–76). The rules are in Cardano, *Ars magna*, fos. 29^v–33; Cardano gave a demonstration of each one, but Descartes quoted them without proof.

your Diophantes./⁶ I send you heere Dr Fabry de motu locali,⁷ to reade at your leasure./ And so committing you & vs all to Gods holie protection I remayne.....

   Your assured friend to serue you
   Charles Cauendysshe

Antwerp: Aug: 23. 1649

[*addressed:*] To my verie worthie friend M^r: John Pell publick Professor of the Mathematicks at Breda
   Breda.

[*annotated by Pell:*] Received Aug 26

[*cover annotated by Birch:*] 23 Aug. 1649.

---

⁶See Letter 14, n. 13.

⁷Fabri (see Letter 47, n. 2) had been reluctant to publish the lecture-courses he gave at Lyon, despite the fame they had brought him; the initiative was eventually taken by his pupil Pierre Mousnier (at Gassendi's insistence). The work referred to here was thus published as *Tractatus physicus de motu locali*, 'auctore P. Mousnerio, cuncta excerpta ex praelectionibus R. P. H. Fabri' ('by P. Mousnier: a collection of extracts from the lectures of the Reverend Father H. Fabri') (Lyon, 1646). (See Gassendi, *Opera omnia*, vi, p. 239; E. Caruso, 'Honoré Fabri Gesuita e scienzato', *Quaderni di Acme* 8 (1987), pp. 85–126, esp. pp. 90–1.)

Letter 89

21/31 August 1649

Pell to Cavendish, from Breda

BL, MS Add. 4280, fo. 134 (draft)
van Maanen P33

Right Honourable

I give your Honour thankes for sending Fabry which I received with your letter[1] 5 dayes agoe: but have [>not] beene since at sufficient leisure to answere your demands. Nor perhaps is it necessary so to doe. Till we see ye contrary let us hope that [>all our desiderata will be supplied by] Cartesiana Geometria cum commentarijs Francisci à Schooten & D^i de Beaune newly printed in Latine at Leyden; I thinke, by ye Elzevirs[2] Monsieur van Schooten himselfe told me, 4 dayes agoe, that it is, as I remember, 46 sheetes of paper. I asked him concerning his promised treatises, He saith y^t he intends shortly to give ye printer his treatise de locis planis:[3] and, till y^t be finished, not to meddle wth y^t other de optima muniendi ratione.[4] I doubt not but you will easily finde it amongst those [>Antwerp] booksellers that correspond wth those of Leyden. When we have read it we shall see what remaines to be supplyed by

Your Honours most humble servant

Breda. Aug 21[/]31. 1649

[*postscript:*] My humble service to my Lord Widdrington

[*second postscript:*] In that expansa[5] of mine, You seeme to sticke onely at the 2 last aequations of every collation, They are made of the 11th & 12th aequations according to ye direction of y^t forme tabellae secundae[6] which stands quoted in

---

[1] Letter 88 (n. 7: Fabri).

[2] The translation of Descartes's 'Géométrie' by Frans van Schooten (see Letter 69, n. 10), with notes by himself and Florimond de Beaune (see Letter 35, n. 16), was in fact printed by Descartes's publisher, Jean Maire (Leiden, 1649).

[3] This restitution of Apollonius' *Loca plana*, entitled 'Apollonii Pergaei loca plana restituta', was eventually published as book 3 of van Schooten's *Exercitationum mathematicarum libri quinque* (Leiden, 1657).

[4] This work was never published (see Letter 78, n. 8).

[5] The 'expanse demonstration' referred to in the minute of Letter 87 (see also Letter 88, n. 2).

[6] 'Of the second table'. The second table, in the top margin of Pell's 'expansa', shows which cases of $xx - yx \pm M$ and $xx + yx \pm N$ (the 11th and 12th equations) must be multiplied to form a given biquadratic equation. There is a reference to the relevant part of this table in a marginal note next to the second equation in each of Pell's worked examples.

y^e margent of the second aequation. For that marginall note allwayes shewes to which of those 4 formes I had respect, in each inquisition.

Si xxxx($*$) − 35xx + 90x = 56; tum x = cui?[7] It will be worth y^e while to apply y^e precepts of my expansa to y^s question

---

[7]See Letter 87, n. 2.

Letter 90

[25 September/] 5 October 1649

Cavendish to Pell, from Antwerp

BL, MS Add. 4278, fos. 291–2 (original)
van Maanen C45

Worthie S^r:

I thought to haue forborne writing to you till I had seen the comment on M^r: de Cartes geometrie;[1] which is likelie enough might haue giuen me occasion of more quaeres than nowe I shall trouble you with; I shall nowe onelie desire the demonstration of those rules of Cardans of resoluing [>those] cubicall aequations which de Cartes recites in his geometrie[2] & to returne to M^r: Chauaux treatise of Algebra,[3] & the Jesuits booke which vindicates P: vincentio,[4] if you haue no further vse of them; & allso anie treatises which came from england to you which I had left with you;[5] but all this at your conuenient leasure;/ this more; if the coefficient & homogeneall in an aequation be integers why the rootes [>must] be integers./[6] in Schootens Appendix[7] pag: 105. $x^2 - ax - 2bb = 0$ quae vlterius diuidi nequit;[8] yet the rootes are $x = \sqrt{2b^2 + \frac{1}{4}a^2} + \frac{1}{2}a$ // $x = \frac{1}{2}a$ mi: $\sqrt{2b^2 + \frac{1}{4}a^2}$. / & pag: 114 $x^2 + ax - 2bb = 0$. // $x = \sqrt{2b^2 + \frac{1}{4}a^2} - \frac{1}{2}a$ / $x =$ mi: $\sqrt{2b^2 + \frac{1}{4}a^2} - \frac{1}{2}a$. / are the rootes, though he saye quae vlterius reducibilis non est[9] it may be he meanes the rootes are no integers; which is true./ I vnderstand not allso verie well howe by his fig: pag: 103. he findes the 2 other rootes in the

---

[1] The commentary by van Schooten and de Beaune: see Letter 89, n. 2.

[2] See Letter 88, n. 5.

[3] See Letter 74, n. 8.

[4] This probably refers to a work by the Flemish Jesuit Alphonse Antoine de Sarasa, *Solutio problematis a R. P. Mersenno Minimo propositi* ... [with] *Quòd quadratura circuli à R. P. Gregorio a S^{to} Vincentio exhibita, abeat illud necdum solutum problema* (Antwerp, 1649), the first part addresses the problem posed by Mersenne concerning the logarithmic scale (see Letter 85, n. 10); the second part defends de Saint-Vincent's quadrature of the circle. (On de Sarasa see Burn, 'Alphonse Antonio de Sarasa'.) Pell's notes on this book are in BL, MS Add. 4421, fos. 82–3.

[5] This apparently refers to MSS which Cavendish had given to Pell for safe-keeping in the period 1642–3, and which had remained in the custody of Pell's wife in London until she joined Pell in Breda in 1647 or 1648.

[6] It is not clear where Cavendish got this idea, since it is clearly not the case that every equation with integer coefficients has integer roots.

[7] See Letter 69, n. 10.

[8] 'Which cannot be divided any further', that is to say, no further simplification is possible.

[9] 'Which is not reducible any further'.

PART III: THE CORRESPONDENCE, LETTER 90 541

aequation $x^3 = \dfrac{a^2}{b^2} \text{ in } x. / +2abc$ // nor by his figures pag. 112 & 113. the roote/
thoug [sic] of this latter aequation $x^3 = \dfrac{a^2}{b^2} \text{ in } x / -2abc$. he saies somewhat,
which giues me a guess to finde the roote.// I confess these quaeres are fitter for discourse in a winter night [>by a fire] than for a letter; but I vse my accustomed freedom with you;/ I receiued a letter latelie from M^r: Hobbes; which puts me in hope wee shall haue his philosophie[10] printed the next springe; he writes to me of hopes to finde a right line aequall to a parabolick line;[11] I haue no more to trouble you with at this time, but commit you & vs all to Gods holie protection.

   Your assured friend to serue you
   Charles Cauendysshe

Antwerp. Oct: 5th. 1649.

[*postscript:*] my Lo: Widdrington[12] presents his seruice to you; I praye you myne to M^r: Bruerton.[13]

[*addressed:*] To my verie worthie friend M^r. John Pell; publick Professor of the Mathematicks at Breda. giue this at Breda

[*cover annotated by Birch:*] 5 Oct. 1649.

---

[10] *De corpore.*

[11] A fragment of this letter from Hobbes survives, in which Hobbes writes that 'I shall in my chapter of Motion giue you a straight line equall to any semiparabolicall line, or the lines of any of the [other *deleted*] cubique, quadratoquadratique or other bastard parabolas' (Hobbes, *Correspondence*, ii, p. 776). These were examples of the problem of rectification, or finding a straight line equal to a given curved line. Descartes (following Archimedes) claimed that the rectification of geometric curves (which he defined as those expressed by an equation in two variables) was impossible (A&T vi, p. 412). The problem was not solved until 1657, for the semicubical parabola, by William Neile in Oxford and (independently) by Hendrick van Heuraet in the Netherlands.

[12] See Letter 25, n. 6.

[13] William Brereton, later third Baron Brereton (1631–80), was studying mathematics under Pell at Breda. He would later become an original Fellow of the Royal Society, and would be one of Pell's most loyal patrons.

Letter 91

3/13 October 1649

Pell to Cavendish, from Breda

BL, MSS Add. 4280, fos. 134^v (rough draft) 149 (authorial copy); Add. 4418, fo. 110^r (second draft)

van Maanen P34

[*rough draft and authorial copy:*]

Right Honourable,

By your last of Octob. 5[1] I perceive y^t you have not yet gotten Cartesii geometriam in Latine.[2] I have it, but not yet bound up. I have been sent the title of it, y^t so you may ye more easily [>find] it or cause it to be sent for. [>In it You will finde resolutions for some of your quaeres, and [>perhaps] occasions of many new quaeres. Which according to my leisure, I shall be ready to satisfy.

[*rough draft:* In your last you enquire, why the rootes must be integers, if the coefficient & homogeneall in an aequation be integers. I answere [>If any man have written so, he hath taught a manifest falshood,[3] and perhaps meant to have written this]

In all aequations if any one of ye rootes be a fraction, all the coefficients and ye Homogeneum comparationis[4] cannot be intire numbers and therefore

If [>all] the coefficients and ye Homogeneum be integers, none of the rootes are fractions

---

MS Add. 4418, fo. 110^r is a second draft, half-way between the rough draft and the authorial fair copy. In it, the final passage of the authorial copy, 'This other paper I send...', is inserted after 'I shall be ready to satisfy.' Most of the interlineations in the rough draft are incorporated in the second draft. There are two significant differences. In the long deleted section of the rough draft (which is still included in the half-way draft) the '6' is altered to '5'; the entry in that list for 'Incommensurable possible = 0' is deleted, and the last three items are renumbered accordingly. Earlier, in the paragraph beginning 'In all aequations if any one...', after 'be intire numbers' the following is added:

'as $a = \frac{1}{2}$ $a = \frac{3}{2}$ $\therefore aa - 2a + \frac{3}{4} = 0$ the homogeneum is not intire.

$a = \frac{1}{2}$, $a = 4$ $\therefore aa - 4\frac{1}{2}a + 2 = 0$ the coefficient is not intire

and therefore,'

[1] Letter 90.

[2] See Letter 89, n. 2.

[3] Pell was correct (see Letter 90, n. 6) and tried to explain why, in a deleted section of the letter and an enclosure (see n. 10 below).

[4] Literally, the homogene of comparison, or what we would now call the numerical or constant term of a polynomial equation. (The description was Viète's, for whom algebraic homogeneity was essential.)

Therefore

An aequation being proposed wth [>one or more] fractions among the data cleere it of fractions by y^e old knowen rules[5]

[*long section deleted, as follows:*
No fraction affirmative or negative can be a value of y^e roote of an aequation so cleared.

But y^e roote will be one of these 6

Commensurable to ye Homogeneum $\begin{cases} \text{integer} > 0 & /1 \\ \text{integer} < 0 & /2 \end{cases}$

Incommensurable $\begin{cases} \text{possible} \begin{cases} = 0 & /3 \\ > 0 & /4 \\ < 0 & /5 \end{cases} \\ \text{impossible} \quad\quad 0 \quad\quad\quad /6 \end{cases}$

Of [>some] of these 6 kindes you shall have as many rootes as there are dimensions in the highest power of any aequation whatsoever provided that the coefficients & homogeneum be unbroken.

But The poste makes me breake off heere. The next weeke I intend to send backe Chauveau[6] [>wth] the Catalog of y^r bookes & papers.[7] In ye meane time I remaine

Your Honours most humble servant
J. P.

Octob. 3.13. *deleted*]

[*postscript:*] Mr Brereton[8] presents his service to your Honour & to my Lord Widdrington.[9]]

[*authorial copy: after first paragraph above ('...ready to satisfy'), continues:*]

---

[5] Several 16th-century texts taught the rules for clearing equations of fractions. Perhaps the clearest exposition, and one certainly known to Pell, was Stevin's, in *L'arithmetique ... aussi l'algebre* (Leiden, 1585), pp. 271–9.

[6] See Letter 74, n. 8.
[7] See Letter 90, n. 5.
[8] See Letter 90, n. 13.
[9] See Letter 25, n. 6.

This other paper[10] I send, in stead of an answere, to one of ye quaeres in your last: Of Chauveau &c the next weeke, God willing. In ye meane time I remaine

Your Honours most humble servant
John Pell

Octob. 3.13.

[*postscript:*] Mr Brereton prefers his service to your Honour & to my Lord widdrington.

[*annotated by Pell:*] The paper heere mentioned is coppied out on the other side of this.

[*Enclosure to Letter 91:*]

In all aequations
either the homogeneum comparationis or at least one of the coefficients will be a fraction if any of the rootes be a fraction
    and therefore
If all the coefficients and the homogeneum be integers, none of the rootes is a fraction
    therefore
An aequation being proposed with one or more fractions among y^e data; it will be good to cleare it of fractions by ye old knowen rules
    For
No fraction (affirmative or negative) can be a value of the roote of an aequation so cleared
    But
The roote will be one of these 5 kindes;

Commensurable to the Homogeneum $\begin{cases} \text{Integer} & > 0 \quad /1 \\ \text{integer} & < 0 \quad /2 \end{cases}$

Incommensurable to y^e Homogeneum $\begin{cases} \text{possible} \begin{cases} > 0 \quad /3 \\ < 0 \quad /4 \end{cases} \\ \text{impossible} \quad 0 \quad \;\;/5 \end{cases}$

---

[10]Pell claims that a polynomial equation with integer coefficients cannot have a rational (commensurable) root. Although he does not say so, he is assuming that the equation is monic, that is, the leading term has coefficient 1 (otherwise the equation $4x^2 - 9 = 0$ would be an obvious counter-example). Such (monic) polynomials, he says, have roots of five kinds: commensurable (i) positive integers (ii) negative integers, or incommensurable (iii) positive irrationals (iv) negative irrationals, and (v) complex (impossible) roots. (In the deleted draft he had included a sixth case where some roots are zero.) By 'unbroken' in the last line he means 'not fractions'.

And

Of some of these 5 kindes you shall have as many rootes as there are dimensions in the highest power of any aequation whatsoever, Provided that the coefficients & homogeneum be unbroken.

Letter 92

[22 January/] 1 February 1650

Cavendish to Pell, from Antwerp

BL, MS Add. 4278, fos. 293–4 (original)
van Maanen C46

Worthie S^r:

after a longe silence I am nowe bolde to trouble you againe; wee haue latelie heer a booke of the passions of the soule in french, writ by M^r: de Cartes, printed at Amsterdam;[1] so that I doute not but you haue seen it, I like it verie well, but I beleeue M^r: Hobbes will not aproue of what is grounded vpon his Metaphysiques, which he vtterlie mislikes./ wee haue nothing heere in the press of philosophie or mathematiques that I heare of; but I heare M^r: Gotsco at Louaine is in hand with a comment on M^r: de Cartes his philosophie & I hope on his geometrie,[2] though M^r: Schooten & De Baune haue done verie well;[3] in the meane time I shall desire you will doe me the fauoure to explicate that which I writ concerning the aequations to finde a tangent to a curue line,[4] which I thinke they haue omitted./ I desire allso those bookes & papers,[5] if you make no furder vse of them, but if you doe I praye you keepe them still; for I expect them at last with vsurie, or rather what is yours, the principall, & then, but the vse;/ I desire to knowe where Mr: de Cartes is./[6] I commit you & vs all to Gods holie protection; & remaine

    Your assured friend to serue you
    Charles Cauendysshe

Antwerp Feb: 1 1649

[*addressed:*] To my verie worthie friend M^r: John Pell Publick Professor of the Mathematiques at Breda. giue this at Breda

---

[1] R. Descartes, *Les Passions de l'âme* (Amsterdam, 1649). This work was printed by Louis Elzevier in Amsterdam; some copies, however, were printed by him under an arrangement with a Parisian bookseller, and bear the imprint 'A Paris, chez Henri Le Gras'. The first copies had been distributed in Nov. 1649 (see A&T xi, p. 293).

[2] The words 'Mr: Gotsco' are underlined in red by Pell. On van Gutschoven ('Gotsco') see Letter 42, n. 6. No such work was published by him, however.

[3] See Letter 89, n. 2.

[4] This writing by Cavendish has not survived. For previous discussions of the topic of finding tangents to curves, see Letter 58, n. 4, and Letter 60, n. 5.

[5] See Letter 90, n. 5.

[6] Descartes was in Stockholm, at the court of Queen Christina, having left Holland in Sept. 1649.

[*annotated by Pell:*] Received. Febr. 6. [/] Jan 27. 1650 [/] 1649
[*cover annotated by Birch:*] 1 Febr. 1649/50

Letter 93

[19 February/] 1 March 1650

Cavendish to Pell, from Antwerp

BL, MS Add. 4278, fos. 295–6 (original)
van Maanen C46

Worthie S^r:

manie thankes for your letter, & bookes;[1] you haue set such a marcke in your letter on the manuscript, that makes me not esteem Monsr: de Baune so excellent an Analyst as I did, for I doute diuers of those places you haue corrected are not the fault of the Transcriber but mistakes of the Author.[2] though he be held one of the best Analysts in France & one who vnderstandes Monsr: de Cartes geometrie exceeding well./ I expected Monsr: Fermats manuscripts when you returned the rest,[3] & doe still when you shall returne those yet in your handes, for I hope you haue some of his. but whatsoeuer you haue [>of] myne I pray you peruse them as longe as you please & then returne them./ A Freind of myne who likes M^r: de Cartes booke of the passions of the soule[4] exceedinglie well; desired your opinion of it, which if you will be pleased to doe you shall doe me a fauoure. I haue not yet M^r: Gassendes his Epicurean philosophie;[5] nor haue heard from M^r: Hobbes a longe time but S^r: william Dauenant latelie sent my Brother a Preface, to an intended Poem of his not yet printed; but the preface printed & directed to

---

[1] This letter has not survived, being one of three non-extant letters (according to Pell's own numbering) written by him between Letters 91 and 95 (see the Textual Introduction); the identity of the books is therefore unknown.

[2] This refers to a MS copy of the short commentary on Descartes's 'Géométrie' by Florimond de Beaune, the 'Notes briefves' (written by 1639 at the latest: see A&T ii, pp. 510–19). The MS copy owned by Cavendish (BL, MS Harl. 6796, fos. 267–90) was probably sent to him from Paris in 1641–2; he had passed it then to Pell, who had just now returned it to Cavendish. By the time he returned it, Pell had had the opportunity to compare it with the translation of this commentary printed in the Latin edition of the 'Géométrie' (see Letter 89, n. 2), as his own annotation on the first page of the MS shows. Pell annotated the MS, not only comparing it with the printed Latin version, but also correcting several errors in the calculations.

[3] Pell had evidently been responding to the request made by Cavendish in Letter 90 (at n. 5). Two MS copies of works by Fermat survive among Cavendish's papers: 'Ad locos planos et solidos isagoge' (BL MS Harl. 6083, fos. 113–14) and 'Ad methodum de maxima et minima appendix' (ibid., fos. 116^v, 330^v).

[4] See Letter 92, n. 1.

[5] Gassendi, *Animadversiones*. The printing of this book was not finished until Sept. 1649: Hobbes informed Gassendi of the arrival of the first copies in Paris in a letter of the 22nd of that month (Hobbes, *Correspondence*, i, p. 178).

M^r: Hobbes with M^r: Hobbs his answear to it likewise printed & bound together.[6] I haue not yet read them; I had rather read his philosophie[7] which I hope he will ere longe publish./ And so committing you & vs all to Gods holie protection I remaine

    Your assured friend to serue you
    Charles Cauendysshe

Antwerp: March 1. 1649.

[*addressed:*] To my verie worthie friend M^r: John Pell, Publicke Professor of the Mathematickes at Breda. giue this
    At Breda

[*cover annotated by Birch:*] 1 March 1649/50

---

[6] Sir William Davenant, *A Discourse upon Gondibert ... with an Answer to it by Mr. Hobbs* (Paris, 1650). Hobbes's 'Answer', written in the form of a letter to Davenant, is dated [31 Dec. 1649/] 10 Jan. 1650. On Davenant see Letter 97, n. 9.

[7] Hobbes's *De corpore* (and, perhaps, *De homine*).

Letter 94

[3/] 13 May 1650

Cavendish to Pell, from Antwerp

BL MS Add. 4278, fos. 298–9 (original)
van Maanen C48

Worthie S^r:

thinking to haue seen you at Breda, made me to forbeare writing; but being vncertaine nowe when I shall see you giues you this trouble. My man tolde me he deliuered the instrument & booke;[1] I doute the instrument is such, that it will not be vsefull; I heere send an other broken parcell of it./ I latelie looked on Schootens Resolution of cub: aequations, & pag: 105,[2] in the latter end he findes the values of the rootes 2a; & −a; −a. which I thinke I conceiue: but pag: 98 in an aequation like graduated & affected: & where the semicircle is diuided in to 3 aequall partes he findes the roote to be the side of an aequilaliter triangle inscribed, & mentions no other rootes; which difference from pag: 105. I conceiue not, but desire your explication at your conuenient leasure; or rather by our doctrine to knowe in a cubicall aequation when it is possible & when not; & if possible, howe manie rootes & of what kinde.[3] S^r: if you please to doe me this fauoure I shall esteeme it an addition to the manie fauoures receiued from you, & shall indeauoure to requite them to my power./ I desire allso at your leasure to answeare those quaeres of S^r: Th: Alesburies./[4] I desire to knowe what you heare of M^r: de Cartes his papers;[5] I latelie receiued [a *deleted* >his] booke of the passions,[6] directed to me by his owne hande which I esteem as a legacie, & shall euer honoure the memorie of so worthie a person & so friendlie a man./ I heare Keircher the Jesuit hath latelie printed [>a] booke at Roome wherein he teaches howe in a little time to make anie man of reasonable Capacitie to make

---

[1] On the nature of this 'instrument' see Letter 95, n. 2; the book was an anonymous work written against Longomontanus (see Letter 95, n. 8). The 'man' may perhaps have been Robert Nash (see Letter 96, n. 5).

[2] Cavendish refers to a passage in the 'Appendix' to van Schooten's *De organica conicarum sectionum in plano descriptione tractatus* (see Letter 69, n. 10).

[3] See Letter 69, n. 6.

[4] These queries have not survived. Sir Thomas Aylesbury (1576–1657) had held various public offices under James I and Charles I, including Surveyor of Ships, Master of Requests, and Master of the Mint. He was created a baronet in 1627. He had been an executor of Harriot's will, and was a generous patron to Walter Warner, taking an active interest in his scientific work. He left England in 1649, settling first in Antwerp (where, evidently, he made contact with Cavendish), then (1652) in Breda.

[5] Descartes had died in Stockholm on [1/] 11 Feb. 1650.

[6] See Letter 92, n. 1.

songes of all sortes, & aires as well as expert Musitians./[7] But none of his bookes are yet ariued heere that I knowe of. S[r]: I haue troubled you enough for once; So committing you & vs all to Gods holie protection I rest

Your assured friend to serue you
Charles Cauendysshe

Antwerp May: 13. 1650

[*addressed:*] To my verie worthie friend M[r]: John Pell Publicke Professor of the Mathematickes at Breda. giue this at Breda.

[*annotated by Pell:*] Received May 5[/]15.

[*cover annotated by Birch:*] 13 May 1650

---

[7] Athanasius Kircher, *Musurgia universalis, sive ars magna consoni et dissoni in X. libros digesta*, 2 vols. (Rome, 1650).

Letter 95

7/17 May 1650

Pell to Cavendish, from Breda

BL, MS Add. 4280, fo. 135^r (authorial copy)
van Maanen P38

Right Honourable

    I have received your letter[1] with the ivory style in it. What besides is wanting to make up your instrument[2] as it was at first, I shall supply as soone as I can finde a workeman. For I long to see what monstrous figures it will make. It seemes Mr Gilbert[3] had forgotten the shape of that, which himselfe had used so often in Ireland and understood not the reason of it so well as to finde that in his understanding, which was slipt out of his memory. I beleeve his intention was to make an instrument by which we might make a figure *like* to one given in ratione datâ.[4] But this, at the best, was such that it necessarily allwayes made its draughts *unlike* the coppy. Perhaps it will be as good cheape to make a new one, as to amend this. He had ill hap to make a toole that will not once performe that for which he made it. But some are not much happier in their rules & precepts. A few dayes agoe I found a precept in Metius his Geometria practica; Datis duabus rectis, duas medias proportionales invenire Geometricè.[5] I examined it, & found that amongst the innumerable rationes which the datae[6] may have, there are two & but two, wherein his precept will performe what it promiseth. So it is not allwayes false.

    It may be Van Schooten was halfe a sleepe, when he wrote that passage which you mention pag. 98. Sure I am that it is false. And therefore in my booke I have cancelled 9 lines, beginning lin [*sic*] 13. And in their stead I have written farre other wise. Of which correction I have heere sent you a coppy.[7]

---

[1] Letter 94.

[2] Precise details of this instrument with an ivory 'style' (stylus) are lacking, but Pell's other comments in this letter suggest that it was a version (constructed, somewhat defectively, by William Gilbert) of Scheiner's pantograph (see Letter 6, n. 4).

[3] See Letter 68, n. 15.

[4] 'In a given proportion'.

[5] A. Metius, *Geometriae practicae* (reissued as part of his *Arithmeticae libri duo, et geometriae lib. VI*). Pell refers to part 1, ch. 4, art. 11 (pp. 29–30), which is entitled 'Datis duabus rectis, duas medias proportionales invenire' ('Given two straight lines, to find their two mean proportionals'); Pell adds 'Geometricè' ('by geometrical means'). On Metius see Letter 16, n. 22. On the problem of finding two mean proportionals see Letter 76, n. 6.

[6] 'Proportions' ... 'given [quantities]'.

[7] For the passage in question, see Letter 94, at n. 2. Pell's copy of his correction does not survive.

With your instrument came that Anonymus contra Longomontanum printed at Paris, Anno 1644.[8] I would endeavour to make him see the defects of his reasoning, if I knew which way to convey a paper to him. I doe not intend to write so, as if I suspected Monsieur Hardy to be the Author, because we are not sure of it. Yet I beleeve he very well knowes who wrote it.

Of Kirchers musicke[9] I have heard, severall wayes; but know not how long it will be, ere I shall come to see it.

Of Des Cartes his papers,[10] I doe not yet heare any thing.

Mr Ravensberg, professor Physicae & Matheseos at Utrecht, died a fortnight agoe.[11] Nor can we yet heare what course is taken with his papers & great Library.

What I have written De Cubicarum aequationum determinatione,[12] is not yet fit to be seene by any body but my selfe. If I be not forced from it by other businesse, I shall soone polish it: and then I hope I shall not neede to be afraide of a skillfull reader.

your Honours most humble servant.

---

[8] Cavendish seems to have acquired this pamphlet before his departure from Paris in 1648: a list of books drawn up by him, apparently recording items he owned at the end of his stay in Paris, includes 'Ciclom: Seuer: elenchus' (BL MS Harl. 6083, fo. 141^r). An exemplum of this anonymous printed work (a single quarto gathering: A^4) survives among Pell's papers: BL, MS Add. 4421, fos. 67–70. As it is not listed in the *National Union Catalog*, nor in the catalogues of the British Library, the Bibliothèque Nationale, or the Royal Library, Copenhagen, its title and publication details deserve to be recorded here in full: Anon., *Cyclometriae a Christiano Severini Longomontano mathematicarum superiorum in Academia Hafniensi regio professore, repetitis vicibus, atque adeo tribus in libris, anno 1612 Hafniae, anno 1627 Hamburgi, hoc anno 1644 Amstelodami editis, publicatae elenchus. Quo ostenditur, posita diametro circuli 10.000.000, & circumferentia 31.418.596, circumferentiam circuli maiorem esse ambitu figurae centum nonaginta trium laterum circulo circumscriptae* (Paris: apud Robertum Sara, 1644). Elsewhere Pell noted this title and commented: 'The author's name is omitted. Petrus Scavenius, in the catalogue of books he had bought, says that the author of that refutation was Claude Hardy' ('Autoris nomen non est appositum. Claudium Hardy fuisse illius Elenchi Autorem affirmat Petrus Scavenius in Catalogo librorum à se conquisitorum'): BL, MS Add. 4414, fo. 218^r). (This refers to P. Scavenius, *Designatio librorum in qualibet facultate, materia & lingua rariorum ... conquisitorum* (Copenhagen, 1665), p. 376, where Scavenius does not list it as a work he has bought, but merely includes it in a listing of anonymous and pseudonymous publications.) On Hardy see Letter 24, n. 24.

[9] See Letter 94, n. 7.

[10] See Letter 94, at n. 5.

[11] Jacob Ravensberg or Ravensperger (1615–50) had become Professor of Philosophy at the University of Utrecht in 1644, and Professor of Mathematics and Physics in 1648 (see C. Burman, *Traiectum eruditum, virorum doctrina inlustrium, qui in urbe Trajecto, et regione traiectensi nati sunt, sive ibi habitarunt, vitas, facta et scripta exhibens* (Utrecht, 1738), pp. 281–2, and Dibon, *La Philosophie néerlandaise*, pp. 211–12). He died either on [12/] 22 Apr. (according to Burman), or on [11/] 21 Apr. (according to G. J. Loncq, *Historische schets der Utrechtsche Hoogeschool tot hare verheffing in 1815* (Utrecht, 1886), p. 81).

[12] 'On the determination of cubic equations'. Despite the disclaimer here, Pell seems to have sent at least a sample of his work to Cavendish: see Letter 99, n. 4. But his treatment of cubics was never published.

May 7.17. (1650)

[*postscript:*] I send heerewith the remainder of that Manuscript.[13] I have now in my hands no manuscripts of yours but in folio & 5 printed bookes in quarto.[14]

---

[13] Cavendish's comment in Letter 96 shows that this was a manuscript by Roberval – probably the treatise 'De locis' referred to in Letter 101, at n. 4.
[14] Unidentified.

Letter 96

[15/] 25 May 1650

Cavendish to Pell, from Antwerp

BL, MS Add. 4278, fos. 300–1 (original)
van Maanen C49

Worthie S^r:

manye thankes for your letter[1] & note vpon Schooten;[2] & the rest of M^r: Roberualls manuscript.[3] I longe till you haue polished your notes of determining cubicall aequations,[4] where I doute not but to finde what I haue not yet mett with. I haue sent you by this bearer Robert Nash[5] a smalle token as an acknowledgment of your fauoures; & haue no more at this time to trouble you with, but remayne

    Your assured friend to serue you
    Charles Cauendysshe

Antwerp. May: 25. 1650.

[*addressed:*] To my verie worthie Friend M^r: John Pell Publick Professor of the Mathematickes at Breda.

[*annotated by Pell:*] Received May 26

[*cover annotated by Birch:*] 25 May 1650

---

[1] Letter 95.
[2] See Letter 95, n. 7.
[3] See Letter 95, n. 13.
[4] See Letter 95, at n. 12.
[5] Probably a household servant of the Cavendishes: he was named as a witness, together with other household servants, in a legal document relating to the Cavendish family estates in 1652: Nottinghamshire Archives, Nottingham, MS DDP.1.1.6.51. (Information kindly supplied by Dr Lucy Worsley.)

Letter 97

16/26 May 1650

Pell to Cavendish, from Breda

BL, MS Add. 4280, fo. 136ʳ (draft)
van Maanen P39

Right Honourable

Since my last, I have seene heere 3 coppies of Mʳ Gassendus his philosophia Epicurea.[1] It costs 24 guilders[2] bound in 2 volumes. I have received a letter from one Isaac Gruter of Middleburg. It is he that translated Sʳ Henry Savils english notes upon Tacitus &c.[3] He tells me that Sʳ William Boswell[4] did put into his hands Dʳ Gilberts Physiologia nova; that he is fitting it for the presse;[5] and that a printer hath beene allready at some cost for the cutting of diagrams for it. Yesterday samuel Sorbier was at my house. It is he that published Gassendi Metaphysicam Exercitationem contra Cartesium and since translated Mʳ Hobbes de Cive into French.[6] He told me that the most of the figures and diagrams,

---

[1] See Letter 93, n. 5.

[2] See Letter 63, n. 3.

[3] Isaac Gruter (1610–80) had taught at the Latin school in his native town of Middelburg from 1633 to 1640, but in the 1640s lived mainly in The Hague. He later became rector of Latin schools in Nijmegen and Rotterdam. His edition of H. Savile, *In Taciti histor.* [&] *Agricolae vitam, et commentarius de militaria romana*, was published in Amsterdam in 1649.

[4] See Letter 19, n. 13.

[5] Gruter's edition of this MS by William Gilbert (the writer on magnetism, not the person referred to in Letters 68 and 95) was printed (with a generous acknowledgement to Sir William Boswell on the titlepage) by Elzevier: *De mundo nostro sublunari philosophia nova* (Amsterdam, 1651). Pell has mistakenly substituted the word 'physiologia', influenced perhaps by the full title of Gilbert's most famous work, *De magnete, magneticisque corporibus, et de magno magnete tellure, physiologia nova, plurimis & argumentis, & experimentis demonstrata* (London, 1600). Boswell had previously consulted Pell about the publication of this MS. In January 1649 he wrote to Pell from The Hague (making the same mistake in the title of the MS): 'I pray, doe you think Gilbertj Physiologia MS (wᶜʰ I believe you haue seen in yᵉ Kings library at Sᵗ. James) would make anything to his, or yᵉ renowne of oʳ nation, if printed in these wild tymes. I find he had the start of many of oʳ. moderne mad-cappes, & so may challenge precedence!' (BL, MS Lansdowne 751, fo. 58: annotated by Pell, 'Received Jan 27/ Feb. 6').

[6] Samuel Sorbière (1615–70), a Huguenot with scientific and philosophical interests, had been a member of Mersenne's circle in Paris before moving to the Netherlands in 1642. He organized the printing there of Gassendi's *Disquisitio metaphysica*, and the second edition of Hobbes's *De cive* (Amsterdam, 1647). His own translation of *De cive*, entitled *Elemens philosophiques du citoyen*, was published in Amsterdam in 1649. At the time of this encounter with Pell he was practising as a physician in The Hague; later in 1650 he became Rector of the Protestant Academy at Orange. He converted to Roman Catholicism in 1653, and spent most of the rest of his life in Paris.

belonging to Mr Hobbes his Philosophy, are already graven in Copper at Paris.[7] He also said, that almost all the coppies of Gassendi philosophia are sold: so that they talke already of a new impression in France.

To day Mr Nash brought me a letter with 16 ducatoons.

Your Honours most humble & most thankfull servant.[8]
J. Pell.

May 16[/]26

[*postscript:*] They say that Sr W. Davenant is taken at sea, with all that he had; Gondibert and all.[9]

[*additional note by Pell:*] Munday in Whitsun weeke I restored his Fabri de Motu[10] to himselfe.

---

[7] This refers to Hobbes, *De corpore*, and (probably) also to diagrams for Hobbes, *De homine*.

[8] On Nash see Letter 96, n. 5; the ducatoon was a Dutch silver coin worth 63 stuivers (approx. 5s 7d in sterling) (see McCusker, *Money and Exchange*, pp. 4, 9).

[9] The poet William Davenant (1606–68), who had served under Newcastle as Lieutenant-General of the Ordnance during the first year of the civil war, had later raised supplies of arms in France for the Royalist cause, and had entered the service of Queen Henrietta Maria. In early 1650 he had been sent as a Royalist official to Virginia (and/or Maryland); his ship was seized in the Channel by a Parliamentarian frigate, and he was imprisoned on the Isle of Wight. (See M. Edmond, *Rare Sir William Davenant* (Manchester, 1987), esp. pp. 103–4.) His heroic-philosophical poem *Gondibert*, to which he had already published the preface (see Letter 93, n. 6), was published in London in 1651.

[10] See Letter 88, n. 7.

Letter 98

[29 May/] 8 June 1650

Pell to Cavendish, from Breda

BL, MS Add. 4280, fo. 136^v (draft)
van Maanen P40

Right Honourable

    These bookes cost one & twenty gilders.[1] I have looked y^m over & am sure they are complete & in order, so y^t they may presently be sent to a Bookebinder.

    Your Honours most humble servant
    John Pell.

June 8.

---

[1] Comments in Letters 97 and 99 show that the 'bookes' sent with this letter were the unbound volumes of Gassendi, *Animadversiones*. For the value of a guilder see Letter 63, n. 3.

Letter 99

[31 May/] 10 June 1650

Cavendish to Pell, from Antwerp

BL, MS Add. 4278, fos. 302–3 (original)
van Maanen C50

Worthie S^r:

manie thankes for M^r. Gassendes booke;[1] which I euen nowe receiued; I haue sent you by the same man that brought it 21 guilders.[2] I giue you allso manie thankes for M^r: Hobbes booke[3] which my brother nowe hath, & reades, & seemes to like it as well, as formerlie he desired it [>much;] though at Breda he seemed not so earnest of it;/ I can not but againe thanke you for your paper of cubicall aequations,[4] which I vnderstande so much of, that I longe much for the compleat doctrine of them, which I hope you will finish as your occasions will giue you leaue. I am ashamed I did not visit M^r: Bruerton when I was at Breda,[5] I praye you my seruice to him. And so committing you & vs all to Gods holie protection I remayne

> Your assured friend to serue you
> Charles Cauendysshe

Antwerp: June 10. 1650

[*addressed:*] To my verie worthie friend M^r: John Pell Publicke Professor of the Mathematicks at Breda giue this at Breda

[*annotated by Pell:*] Received June 4. 14

[*cover annotated by Birch:*] 10 June 1650

---

[1] See Letter 93, n. 5, Letter 97, at n. 1, and Letter 98.

[2] See Letter 63, n. 3.

[3] The identity of this book is not clear. It was presumably not the 'Answer' to Davenant, which Newcastle already possessed (see Letter 93, n. 6); nor *Of Humane Nature*, of which Pell seems to have been unaware until he received the information which he passed on in his next letter (Letter 100). The most likely candidate is the newly published *De corpore politico* (London, 1650: Thomason's copy is dated 4 [/14] May), the unauthorized printing of the second half of *The Elements of Law*. Alternatively, it may have been Sorbière's translation of *De cive* (see Letter 97, n. 6).

[4] Among Cavendish's papers there is one small leaf in Pell's hand entitled 'CUBICARUM Æquationum Radices determinatae' ('The roots of cubic equations determined'): BL, MS Harl. 6083, fo. 152^r. This was possibly sent as a sample of the work mentioned in Letter 95 (at n. 12).

[5] On Brereton see Letter 90, n. 13; on Cavendish's visit to Breda see Letter 88, n. 1.

Letter 100

4/14 June 1650

Pell to Cavendish, from Breda

BL, MS Add. 4280, fo. 136^v (draft)
van Maanen P41

Right Honourable

Just a weeke agoe I sent Gassendus by the carman.[1] I have since received a letter out of England with this postscript. 'I suppose you have seene a piece of Mr Hobbes Of humane nature. printed at Oxford, this yeare.[2] The next we looke for is ye first part of his philosophie De corpore.'[3]

I have also since heard y^t John Jansson of Amsterdam[4] (who is Typographus Regius[5] in Sweden) is commanded by ye Queene to print all Vossius his [>yet-unpublished treatises] for she bought his library.[6] It is thought [>she will doe the like for Grotius and Cartesius] if there be any thing fit for ye presse found amongst y^{eir} [>papers]

Last Munday night a servant of ye Princess Royall[7] brought her word y^t the day before, y^t is Sunday, betweene 10 & 11 o'clocke he had seene ye King goe o' shipboard[8] & that [>he] tarried on ye shore till y^t ship was quite out of ken. [Before he went he was assured y^t for his welcome in Edenburg the [*two words*

---

[1] See Letter 98, n. 1.

[2] T. Hobbes, *Of Humane Nature: Or, the Fundamental Elements of Policie* (Oxford, 1650): this was an unauthorized printing of the first half of Hobbes's *The Elements of Law*. The titlepage described it as: 'London, Printed by T. Newcomb, for Fra: Bowman of Oxon.'

[3] The quotation, marked here by quotation marks, is indented, with a vertical line, in the MS. The original of this letter has not survived; its author is unknown.

[4] Johannes Janssonius (Jan Jansz.) (1588–1664), one of the leading booksellers in Amsterdam: see M. M. Kleerkooper and W. P. van Stockum, *De boekhandel te Amsterdam voornamelijk in de 17e eeuw: biographische en geschiedkundige aantekeningen*, 2 vols. (The Hague, 1912–14), i, pp. 295–301; ii, pp. 1315–24.

[5] 'Royal printer'.

[6] G. J. Vossius (Voskens) (see Letter 11, n. 4) died on [7/] 17 Mar. 1649; his son Isaac, Queen Christina's librarian, arranged the sale of his entire library, including his MSS, to the Queen, for 20 000 guilders (see Rademaker, *Leven en werk van Vossius*, pp. 258–9). Nothing came, however, of the plan mentioned here.

[7] Princess Mary, daughter of Charles I, who had married William II of Orange.

[8] This news was false; there had been some discussion of a projected voyage by Charles II to Edinburgh to treat with the Scottish Covenanters, but the plan was abandoned.

deleted] in stead of tapistry was hanged wth Montrosian quarters[9] *deleted*]

y. H. m. h. s.[10]

June 4.14

[*postscript:*] After y^s letter was seald up, [I brake it open againe *deleted*] to give notice y^t ye carman had brought me your letter wth 21 guilders.[11]

---

[9] James Graham, fifth Earl and first Marquis of Montrose (1612–50), who had been the leading Royalist commander in Scotland, had visited Charles II in The Hague, and had been appointed by him Lieutenant-Governor of Scotland in March 1649. He had travelled first to Denmark and then to Sweden, to gather money and supplies for a landing in Scotland. This took place in the following year: his force was defeated (on 27 Apr. [/7 May] 1650), and he was executed soon thereafter.

[10] 'Your Honour's most humble servant'.

[11] See Letter 99, n. 2.

Letter 101

[27 August/] 6 September 1650

Cavendish to Pell, from Antwerp

BL, MS Add. 4278, fos. 304–5 (original)
van Maanen C51

Worthie S^r:
    after a longe silence I am nowe bolde to trouble you. I haue rather turned ouer the leaues & onelie read the sum of Epicurus his philosophie[1] than anie other wise; but as far as I can guess my worthie friend M^r: Gassendes hath both maintained & opposed Epicurus when he ought, most excellentlie. God send the goode man his health & I hope he will yet publish more of his excellent pieces, especiallie of Astronomie./[2] I haue heard nothing from M^r: Hobbes a great whyle but meane God willing shortlie to write to him, & when I heare anie thing of his Philosophie[3] I shall let you knowe. I should be verie glad to heare that you were publishing anie thing of those manie which I assure my self you haue, though peraduenture not all ordered to your minde; but I hope a smalle time woulde doe that, if your leasure would permit it./ M^r: Roberualls Treatise de locis[4] in my coppie ends with 10^a: aequatio per Parabolam et circulum; the last words et omnia positione dantur.[5] I desire to know if yours ends so./ If your leasure haue permitted you to correct M^r; Chauaus Algebra[6] I desire you would be pleased to send it me by the next oportunitie & vieta de Recognitione etc: aequationum./[7] If you haue the uariation of the compass & Latitude at Breda; I desire you will do me the fauoure to sende [it *deleted*] me them./ I desire to knowe if there be anie Sphaers made in Hollande according to Copernicus his hypothesis./ but I haue troubled you too much for once, therefore shall defer more desires & quaeres to an other time./ M^r: Bruerton,[8] did me the honor to come hither & deliuer your

---

[1] See Letter 93, n. 5.

[2] Cf. Letter 69, n. 3.

[3] *De corpore* (and, perhaps, *De homine*).

[4] This was apparently the MS treatise by Roberval which Pell had returned to Cavendish: see Letter 95, n. 13, and Letter 96, n. 3. It may perhaps have been a version of the treatise which was eventually published as 'De geometrica planarum et cubicarum aequationum resolutione' ('On the geometric resolution of plane and cubic equations') (*Divers ouvrages*, pp. 136–89).

[5] 'The tenth equation, for a parabola and a circle' ... 'all things are given by position.'

[6] See Letter 74, n. 8, and Letter 90, at n. 3.

[7] F. Viète, *De aequationum recognitione*. William Oughtred made notes in a copy of this book owned by Cavendish. The notes were copied by Pell and are preserved in BL, MS Add. 4423, fos. 146–9.

[8] See Letter 90, n. 13.

PART III: THE CORRESPONDENCE, LETTER 101   563

remembrance to me; I praye you my seruice to him./ And so God keepe you & vs all

your assured friend to serue you
Charles Cauendysshe

Antwerp: Sep: 6: 1650

[*addressed:*] To my verie worthie Friend M^r: John Pell publick Professor of the Mathematicks at Breda. giue this
At Breda

[*annotated by Pell:*] Received Sept 10

[*cover annotated by Birch:*] 6 Sept. 1650

Letter 102

[22 October/] 1 November 1650

Pell to Cavendish, from Breda

BL, MS Add. 4280. fo. 138^r (authorial copy)
van Maanen P42

Right Honourable

By this young gentleman[1] I thought fit to send nothing but a heape of excuses for my silence all this while. And indeed it would make a great heape, if I should at large signify what diversions I have since had. Your letter[2] found me standing by the death-bed of one of my Colleagues,[3] whom I loved so well as few men doe their brethren. Which I expressed so immoderately, that my Physician sent me from his grave directly to my bed, from whence some thought I should not have risen. But the malignity of the fever, that accompanied that excessive griefe, was so abated by a timely sweate and some other Physicke, y^t I kept not the house long for it. As soone as I went abroad, besides other unusuall diversions, I found so much untoward businesse to be done for the widow[4] (every one casting the burthen upon mee) that it made me very unfit to thinke of such things as your letter makes mention of. Yet I must not deferre my answere any longer, though I reply not at this time to your whole letter.

Your Treatise de locis[5] reached no farther than that passage which you quote out of it. The numbers, which I set to ye pages of your coppy, will shew whether any intermediall leaves be lost.

I know no man heere that hath made any observation of the Latitude or Variation, or that cares to know either of them. But the common opinion is, that the Latitude is 51.37.

As for the Copernican Sphaer's; I doubt not but your Honour hath seene Blaeu his Institutio Astronomica, de usu Globorum, & sphaerarum, in octavo, translated into Latine by my predecessor Hortensius and printed at Amsterdam,

---

[1] Unidentified.

[2] See Letter 101.

[3] Johannes Brosterhuysen (1596–1650), Professor of Greek and Botany at the College of Orange at Breda, died on 10 [/20] Sept. 1650 (the day on which Pell received Letter 101), after an illness lasting eight days (see van Seter, 'Prof. Johannes Brosterhuysen', p. 142).

[4] Brosterhuysen had married his housekeeper, Amalia Rhode, the 35-year-old widow of Johan de Wit, in 1648 (van Seter, 'Prof. Johannes Brosterhuysen', pp. 137, 140).

[5] See Letter 95, n. 13, Letter 96, n. 3, Letter 101, n. 4.

Anno 1640.[6] Wherein he tells us of two Copernican Spheres; the one Generall for the fixeds & all the planets; the other particular, excluding the five planet-starres. Both which are described in that booke & are both to be sold apud Joannem Blaeu, Amstelodami; where I have divers times seene them.

I hope I shall be able, the next weeke to send something else by our old Carman. In the meane time, I pray your Honour to call to minde, whether ever you saw any mans solution of the Arithmeticall question heerein inclosed.

your Honours most humble servant.

Novemb. 1. (1650)

[*annotated by Pell:*] The paper inclosed was onely thus[7]

a = ?	aa + bc = P, ut 16
b = ?	bb + ac = S, ut 17
c = ?	cc + ab = T, ut 18

---

[6]Willem Jansz. Blaeu (1571–1638), founder of the printing firm and father of Johan Blaeu, had studied astronomy in his youth under Tycho Brahe, and was famous as a maker of astronomical and terrestrial globes. His treatise on that subject, *Tweevoudigh onderwys van de hemelsche en aerdische globen* (Amsterdam, 1620), was translated by Hortensius (on whom see Letter 11, n. 2) as *Institutio astronomica de usu globorum & sphaerarum caelestium ac terrestrium* (Amsterdam, 1634); a French version was also published (Amsterdam, 1642). (See P. J. H. Baudet, *Leven en werken van W. J. Blaeu* (Utrecht, 1871), esp. pp. 35–49.)

[7]Pell later wrote that this problem was brought to him by 'Mr. William Brereton of Breda anno 1649' (BL, MS Add. 4413, fo. 52). Pell persisted with the problem over many years, and in 1662 he and John Wallis produced a solution, eventually printed in Wallis's *A Treatise of Algebra*, pp. 225–56.

Letter 103

[30 October/] 9 November 1650

Cavendish to Pell, from Antwerp

BL, MS Add. 4278, fos. 306–7 (original)
van Maanen C52

Worthie S^r:

manie thankes for your letter;[1] & I am verie glad you are so well recovered from your dangerous sickness; & [>yet] can not but condole the loss of your colleague,[2] your choice of friendship answering me he was a worthie man; but Gods will be done in all./ manie thankes for the problem you sent;[3] I finde it too hard for me, therfore desire your solution of it at your leasure; I remember not anie solution of it./ I haue Kircher of musicke & haue looked a little of [*sic*] it; he hath something of combinations[4] but intends a compleate doctrine of it, which (to my aprehension) I neuer yet sawe but expect one heerafter from you, when it shall please God to giue leasure & oportunitie to you to inable you to publish it, & [>giuing] life & abilitie to me to reade it. Kircher repeates much what others haue writ; but I beleeue he hath somewhat new; & for some of his tables I confesse I doe not yet vnderstand them; I desire your opinion of him; for I suppose you haue the booke./ no more at this time but that I remayne,

   Your assured friend to serue you
   Charles Cauendysshe

Antwerp: Nouemb: 9. 1650

[*addressed:*] To my verie worthie friend M^r: John Pell publicke Professor of the Mathematickes at Breda. giue this
   At Breda.

[*cover annotated by Birch:*] 9 Nov. 1650

---

[1] Letter 102.

[2] See Letter 102, n. 3.

[3] See Letter 102, at n. 7.

[4] The brief discussion of musical combinatorics is in Kircher, *Musurgia*, book 8, part 1 (vol. ii, pp. 3–27); Cavendish's notes on this book include notes on that particular section (BL, MS Harl. 6083, fo. 58^r).

Letter 104

[30 October/] 9 November 1650

Pell to Cavendish, from Breda

BL MS Add. 4280, fo. 138^v (authorial copy)
van Maanen P43

Right Honourable

In my last,[1] sent 8 dayes agoe by Sr William Vavasour's sonne,[2] I did make halfe a promise to send *something else* the next weeke by our old Carman. That *something then intended* was such as you expected from me, not this unexpected newes of the death of my M^r, the Prince of Orange. Who fell sicke of the small pockes at the Haghe on Munday the last of October and the next Sunday night about 9 o'clocke died there.[3] The newes came but yesterday morning to this towne: Where all mens eyes are upon the behaviour of those that depended more immediately upon him. Some of my Colleagues tell mee, that I am well inough: because they know that, about ten weekes agoe, two of the Magistrates of Utrecht were heere with me, sent from the whole senate of that city to offer me 200 guilders a yeare more than they had given to Mr Ravensberg,[4] yet requiring no more service of mee than hee had done. I then answered, that the Prince gave me more than they offered me; and that his Highnesse had never given me notice that he had rather dismisse mee than keepe me any longer heere. About foure weekes after, they were sent to pray the Prince to let them have me. I have beene told that his Highnesse, in stead of graunting their request, sent them home with a very harsh answere. How true this is, I know not. Sure I am that they neither came nor sent to mee, since they spake with the Prince. It may be they will now thinke that businesse more feasible, because none is in their way, whose

---

[1] Letter 102.

[2] William Vavasour, of Copmanthorpe, Yorkshire, was created a baronet in 1643 (G. E. Cokayne, *The Complete Baronetage*, 5 vols. (Exeter, 1900–9), ii, p. 212). In the latter part of that year he was the commander of Royalist forces in Gloucestershire (see Anon., *A True Relation of a Wicked Plot Intended and still on Foot against the City of Glocester, to Betray the same into the Hands of the Cavaliers* (London, 1644), pp. 10, 15, 18–19, 22). He was closely associated with Prince Rupert, who appointed him Field Marshal General in 1644 (P. R. Newman, *The Old Service: Royalist Regimental Colonels and the Civil War, 1642–46* (Manchester, 1993), p. 105). He was banished from England in December 1645, and entered the service of the King of Sweden. The identity of this son (a child of Sir William's first wife, who was Dutch) is not known; he must have pre-deceased Sir William, who died without any male heir in 1659 (Cokayne, *op. cit.*, ii, p. 212).

[3] William II died on [27 Oct./] 6 Nov. 1650. The Prince was Pell's 'M^r' (Master), as Pell was employed by the Illustre School at Breda.

[4] On Ravensberg see Letter 95, n. 11; on the attempts by the authorities at Utrecht to recruit Pell, see Kernkamp, *Acta et decreta senatus*, i, pp. 254, 256–8, 268–9.

displeasure they regard. But in the meane time I am in the same praedicament with my Collegues; uncertaine what will become of this Schola Auriaca[5] and how long it will be ere all the Professors shall be warned to seeke their fortunes elsewhere.

I am not out of hope, that, because of this extraordinary accident, I may yet a weeke longer diferr the obeying of your commaunds, without losing the title of

Your Honours most humble & obedient servant.

Novemb. 9. stylo novo.

---

[5]'School of Orange' (the Illustre School, Breda).

Letter 105

4/14 December 1650

Pell to Cavendish, from Breda

BL, MS Add. 4280, fo. 138$^\text{v}$ (authorial copy)
van Maanen P44

Right Honourable.

My last was Novemb 9.,[1] written perhaps at y$^\text{e}$ same time that your Honour was writing to me: For 3 dayes after, I received one of yours dated also Novemb. 9.[2] In which because you make mention of Kircher,[3] I thought fit to send you Des Cartes on the same argument.[4] I transcribed it hastily out of an imperfect & ill-written coppy. If you have not seene a perfecter, you will perhaps thinke it fit to take a transcript of this, before it be sent backe againe. But if you have all-ready a better coppy, I pray you let it accompany mine in its returne hither.

What Kircher hath done more than others I know not, having never seene his booke; though Mr Brereton[5] did his best to get it. But could never yet come to see it, or heare where it was to be sold.

Your Vieta[6] hopes to find the way home, this day 7 night.

your Honours most humble servant.

Decemb. 4.14.

---

[1] Letter 104.

[2] Letter 103.

[3] See Letter 103, n. 4.

[4] Descartes's short treatise, *Compendium musicae* (A&T x, pp. 89–141), which attempts a mathematical analysis of musical harmony, was an early work, written in 1618 and presented by him as a New Year's gift to Isaac Beeckman in 1619. When Descartes later demanded its return, Beeckman retained a copy. Recent research, however, suggests that Pell's copy (which is BL MS Add. 4388, fos. 70–83) and another MS copy made by Frans van Schooten (now in Groningen University Library) were made not from Beeckman's copy but from another source – probably a MS owned by Alphonse Pollot, a friend of Descartes and of Constantijn Huygens who lived in Utrecht between 1639 and 1642. (See M. van Otegem, 'Towards a Sound Text of the *Compendium musicae*, 1618–1683, by René Descartes (1596–1650)', *Lias: Sources and Documents relating to the Early Modern History of Ideas*, 26 (1999), pp. 187–203. Van Otegem's statement that Pollot visited Breda in 1648 is, however, incorrect: that visit was made by Huygens, at Pollot's request (see C. Huygens, *Briefwisseling*, iv, p. 479).) By the time Pell wrote this letter, the work had just been published (under the title *Musicae compendium* (Utrecht, 1650)). Its editor was probably Godefroid van Haestrecht, who was a friend of Pollot and probably also derived his text from Pollot's copy.

[5] See Letter 90, n. 13.

[6] See Letter 101, n. 7.

Letter 106

[6/] 16 December 1650

Cavendish to Pell, from Antwerp

BL, MS Add. 4278, fos. 308–9 (original); MS Add. 4422, fo. 213^r (copy of the final section of the letter, beginning 'I shall adde...', in Pell's hand)
van Maanen C53

Worthie S^r:

I receiued [>this day] your letter with Monsr: des Cartes Compendium of Musick;[1] for both which I giue you manie thankes; I must aske your pardon for not answearing your last letter before this,[2] sooner; though the sad newes wee had at that time might pleade some excuse. God be praysed oure news is nowe somewhat better; onelie the loss of the most noble Prince of orange remaynes, which yet is most hopefullie supplied by the birth of his younge son./[3] I hope there is no cause of your thinking of remouall from Breda, for I suppose that which was founded & continued by those worthie Princes, will not soone be demolished; but if business goe well with our King, which I trust in God it will, I hope there maye be occasion fitting to call you home./ The manuscript you sent me is printed, Traiecti ad Rhenum, Typis Gisberti à Zijll, et Theodori ab Ackersijck, 1650.[4] one of which I haue, by my worthie friend M^r: Le Stranges[5] guift who tolde me theer was few in this towne; if I can finde none in the towne, I will send you myne to peruse at your leasure; I shall keep it a fewe dayes till I haue read it, & then send both it & your manuscript; Kircher is copious[6] & as I thinke hath done verie well; I wish Monsr: de Cartes had bin as copious or at least more than he is in this subiect./ I shall adde onelie one quaere to my former; which is the number & kindes of Sphaericall Triangles; rectangle acute, obtuse, quadrantall etc:/ I haue no booke of the doctrine of triangles by me;

---

[1] See Letter 105, n. 4.

[2] Letter 104.

[3] William III was born on [4/] 14 Nov. 1650, eight days after his father's death.

[4] This was the edition of Descartes, *Compendium musicae*, printed at Utrecht ('Traiecti ad Rhenum'): see Letter 105, n. 4.

[5] This probably refers to Roger L'Estrange (1616–1704), the author (and future Licenser of the Press, knighted in 1685). Following his involvement in the abortive Kentish rising of 1648, he left England, publishing an account of it 'in Flanders' (see his statement in *To the Right Honorable, Edward Earl of Clarendon ... The Humble Apology of Roger L'Estrange* (London, 1661), p. 5; the work published in Flanders was *L'Estrange his Vindication to Kent, and the Justification of Kent to the World* (n.p., 1649)). He was a keen viol-player with a life-long interest in music: see G. Kitchin, *Sir Roger L'Estrange* (London, 1913), pp. 39–40). (Help with this identification was kindly supplied by Dr Lynn Hulse.)

[6] See Letter 94, n. 7.

neither doe I remember that I haue seene it in anie, If in anie, in Steuin;[7] but I desire it from you who I assure my self hath it exactlie./ our Noble Duke of Yorcke passed by vs this daye,[8] I should haue troubled some of his followers with this to you; but receiuing yours after they were gone; I must nowe send it the ordinarie waye.

God protect you & vs all

Your assured friend to serue you
Charles Cauendysshe

Antwerp: De: 16 1650

[addressed:] To my verie worthie friend M[r]: John Pell Publick Professor of the Mathematickes at Breda.
giue this At Breda

[annotated by Pell:] Received Dec. 20.

[cover annotated by Birch:] 16 Decemb. 1650

---

[7] See Stevin, *Les Oeuvres mathematiques*, ii, part 1, book 3 (pp. 28–90), 'Des triangles spheriques'.

[8] James, Duke of York (later James II) had left his mother's court-in-exile in Paris, against her wishes, in Oct. 1650, and moved first to Brussels, then (in Jan. 1651) to The Hague.

Letter 107

[10/] 20 December 1650

Cavendish to Pell, from Antwerp

BL, MS Add. 4278, fo. 310 (original)
van Maanen C54

Worthie S^r:

    I heere send you a printed coppie of the manuscript you sent me;[1] which I allso returne; you may dispose of the booke printed, as you please, for I keepe M^r: Le Stranges guift[2] I will trouble you no more at this time but committing vs all to Gods holie protection, remaine

    Your assured friend to serve you
    Charles Cauendysshe

Antwerp: [>De.] 20. 1650

[*addressed:*] [*page torn*]y verie worthie friend M^r: John Pell [*page torn*] Professor of the Mathematicks at Breda, [*page torn*] this at Breda

---

[1] Descartes, *Compendium musicae*; see Letter 105, n. 4.
[2] Another copy of the book by Descartes; see Letter 106, at n. 5.

Letter 108

[10/] 20 January 1650 [/1651]

Cavendish to Pell, from Antwerp

BL, MS Add. 4278, fos. 311–12 (original)
van Maanen C55

Worthie S^r:

I doute my last letter[1] came not to you; the loss is not much; more than an acknoweledgment of the receite of yours with manie thankes; & the loss of the manuscript of De Cartes Musick, & a printed booke of the same;[2] which two bookes makes it the more considerable;/ I will trouble you with no more quaeres [>nowe at present] than formerlie I haue done, but wishing you all happiness, remaine

Your assured friend to serue you
Charles Cauendysshe

Antwerp: Jan: 20 1650

[*addressed:*] To my verie worthie friend M^r: John Pell Publicke Professor of the Mathematicks at Breda giue this
At Breda.

[*cover annotated by Birch:*] 20 Janu. 1650/1

---

[1] Letter 107.
[2] See Letter 107; as Letter 109 shows, these fears were unfounded.

Letter 109

14/24 January 1651

Pell to Cavendish, from Breda

BL, MS Add. 4280, fo. 257^r (draft)
van Maanen P45

Right Honourable

    My answere to your demaund concerning sphaericall triangles might have accompanyed this your Vieta,[1] if I had kept it a little longer. But it was high time to give you account of my receiving Des Cartes both manuscript & printed.[2] for which I give you many thankes, & remaine

    Your Honours most humble servant

Jan. 14.[/]24. 1650[/]51

---

[1] For the demand see Letter 106, at n. 7; for the work by Viète see Letter 101, n. 7.
[2] See Letter 107.

Letter 110
[24 February/] 6 March 1651
Cavendish to Pell, from Antwerp

BL, MS Add. 4278, fos. 313–14 (original)
van Maanen C56

Worthie S[r]:

I should haue returned you thankes for your letter[1] which I receiued with my vieta[2] longe since; I was in hope to haue heard from M[r]: Hobbes; but haue not yet: I hope he deferres writing, being busie in putting his philosophicall worckes to the press.[3] I desire you will doe me the fauoure to satisfie my quaere of Sphaericall Triangles,[4] & to solue that problem you sent me,[5] which I finde too hard meate for me. I haue no more to trouble you with at this time. And so committing you & vs all to Gods holie protection I remayne

Your assured friend to serue you
Charles Cauendysshe

Antwerp. March: 6: 1650

[addressed:] To my verie worthie friend M[r]: John Pell Publicke Professor of the Mathematicks at Breda. giue this
At Breda

[cover annotated by Birch:] 6 March 1650/1

---

[1] Letter 109.

[2] For this work, sent with Letter 109, see Letter 101, n. 7.

[3] Cavendish's concern, as previously, seems to have been with Hobbes's 'philosophy', i.e. his *De corpore* (and, perhaps, *De homine*). He appears to have been unaware of Hobbes's recent work on *Leviathan*, which, at this time, Hobbes was just completing (its dedicatory epistle is dated 15/25 Apr. 1651).

[4] See Letter 106, at n. 7.

[5] See Letter 102, at n. 7.

Letter 111

[5/] 15 May 1651

Cavendish to Pell, from Antwerp

BL, MS Add. 4278, fos. 315–16 (original)
van Maanen C57

Worthie S^r:

manie thankes for the booke[1] which I receiued by the hand of your worthie scholler.[2] who with his worthie compaignon were pleased to visit me./ I haue not yet read [your *deleted* >this] booke you nowe send me; I suppose he hath facilitated or added somewhat to his former comment; but I expect not much in Analiticks till you publish something in that matter; in the meane time I desire you will doe me the fauoure to send me the solution of the problem[3] you sent me a whyle since

$a = ?$ $\quad\|\quad$ $aa + bc = P$, vt 16
$b = ?$ $\quad\|\quad$ $bb + ac = S$, vt 17
$c = ?$ $\quad\|\quad$ $cc + ab = T$, vt 18

& the enumeration of all the kindes of Sphaericall triangles[4] if it be not too great a trouble to you./ I haue not heard from M^r: Hobbes a longe time nor of M^r: Gassendes or the rest of the learned men theere./ I haue no more at this time but committing you & vs all to Gods holie protection; remayne

   Your assured friend to serue you
   Charles Cauendysshe

[*page torn*]rp [*added by Pell:* >Antwerp] May: 15 1651

[*addressed:*] To my verie worthie friend M^r: John Pell Publicke Professor the

---

[1] Cavendish's remark later in this letter, 'I suppose he hath facilitated or added somewhat to his former comment', makes it possible to identify this work on 'Analiticks' (algebra) as F. van Schooten, *Principia matheseos universalis, seu introductio ad geometriae methodum Renati Des Cartes*, ed. E. Bartholin (Leiden, 1651). Cf. Cavendish's phrase in Letter 89 (at n. 1): 'the comment on M^r: de Cartes geometrie'. This work by van Schooten was not, however, a continuation or improvement of his previously published commentary on Descartes's 'Géométrie'; it was an elementary introduction to Cartesian algebra, written up by van Schooten's pupil Erasmus Bartholin from lectures given by van Schooten many years before (see Hofmann, *Frans van Schooten*, p. 5).

[2] Neither the scholar nor the companion has been identified.

[3] See Letter 102, at n. 7.

[4] Cavendish repeats a request first made in Letter 106, at n. 7.

PART III: THE CORRESPONDENCE, LETTER 111   577

Mathematickes at Breda. giue this
   At Breda

[*annotated by Pell:*] Received May 6[/]16

[*cover annotated by Birch:*] 15 May 1651

Letter 112

20/30 August 1651

Cavendish to Pell, from Antwerp

BL MS Add. 4278, fos. 317–18 (original)
van Maanen C58

Worthie S^r:

it is so longe since I hearde from you, & hauing latelie heard of your indisposition of health, makes me nowe trouble you, to inquire of your health;/ hauing forgot whether I sent you this proposition of a geometricall spirall;[1] to be sure I doe nowe sende it you, desiring, as your health & occasions will permit to finde a demonstration of what is heere onelie propounded for worthie Mersenne of whom I had it gaue me no more than this, & I beleeue he had no more sent him./ I desire allso an answeare allso of my former quaeres[2] as your leasure will permit you.

God Allmightie protect you & vs all

Your assured friend to serue you:
Charles Cauendysshe

Antwerp: Aug: 20/30 1651

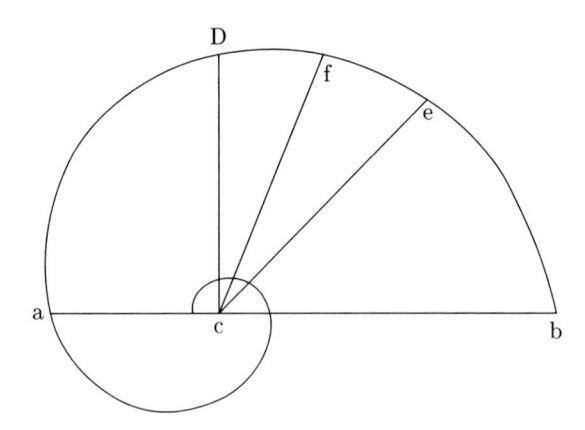

Esto quaelibet recta linea ab, secta vtcumque in c; deinde erigatur perpendicular:

---

[1] Cavendish reproduces here (with minor alterations) part of a text by Torricelli sent as an enclosure in a letter from him to Roberval of [27 June/] 7 July 1646 (*MC* xiv, pp. 356–7). A copy of the original text (in a scribal hand) survives among Cavendish's papers: BL MS Harl. 6083, fo. 338, together with Pell's comment (fo. 339^r: see Letter 113, enclosure).

[2] See Letter 111.

cD; quae sit media inter bc, ca; iterum recta ce, media sit inter Dc, cb; secetque Dc, ce, secet bifariam angulum Dcb; Amplius recta cf, media inter Dc, ce, secet bifariam angulum Dce; et sic fiat semper, ducendo medias proportionales quae bifariam secent angulos, habebimusque puncta b, e, f, D, a; et quotcunque alia voluerimus per quae transibit quaedam linea curua, quam spiralem geometricam appellamus./ sed aliter eam possumus definire, si recta linea aB, manente eius extremo a, circumagatur in plano, aequali semper velocitate; et eodem tempore, aliquod punctum c, eiusdem lineae super eâdem moueatur, ea lege vt spatia aequalibus temporibus peracta inter se proportionalia sint eande[*page torn*] proportionem seruant quam habebunt distantiae [*page torn*]iusdem puncti à centro a; Iterum eandem spirale[*page torn*] geometricam habebimus.

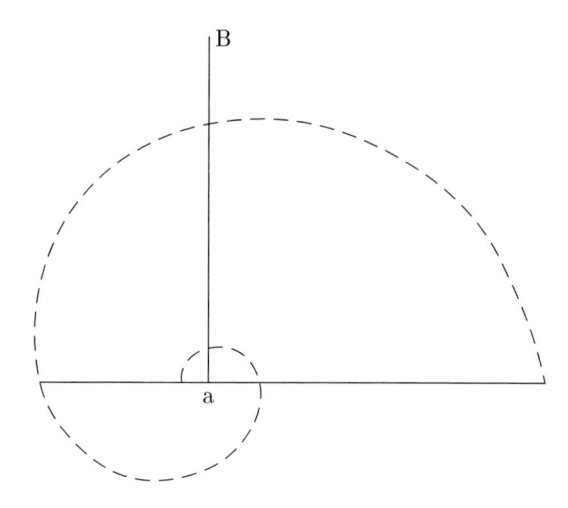

Peculiare hoc habet haec linea quod antequam perueniat ad centrum, infinitas reuolutiones circa ipsum absoluere debebit; attamen longitudo eius, non solum quo ad partes, sed et vniuersa nota est: nam si fuerit centrum a, vnusque ex radiis aB, ipsa vero tangens BC, et angulus BaC rectus; erit tangens BC, aequalis vniuersae spirali BDa; Dato verò quocunque arcu BJ, facillimè ipsi recta linea aequalis abscinditur, demonstraturque.

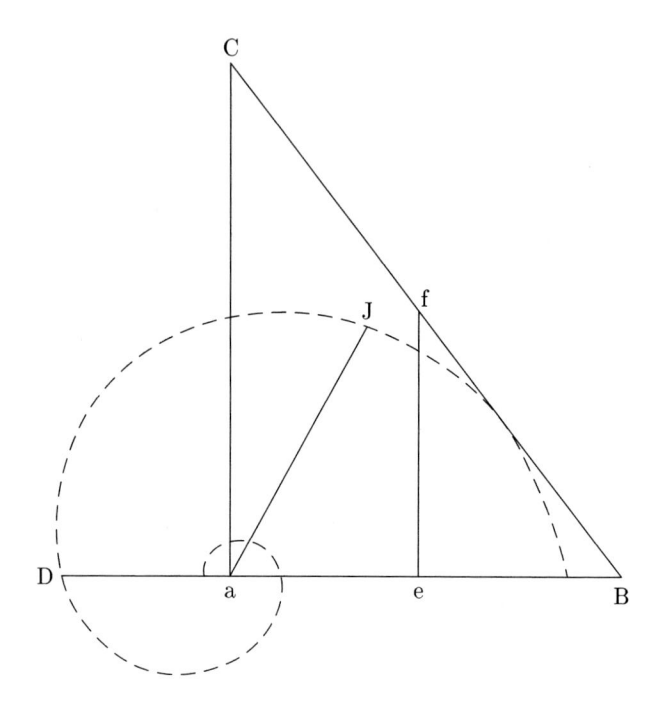

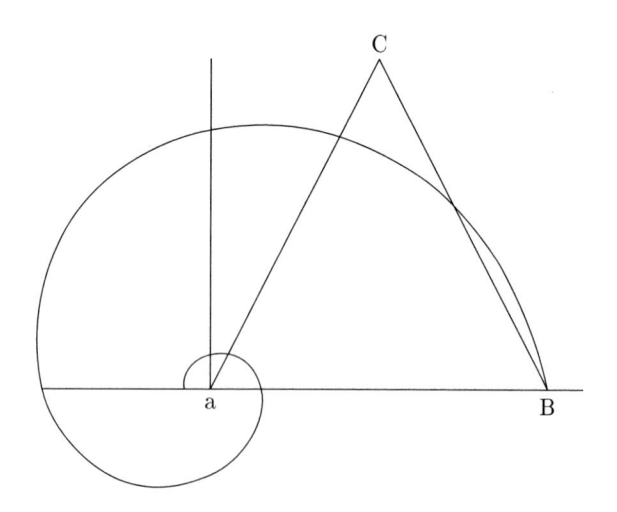

tangens BC. fiat super basi Ba. Triangu^{lum} aequicrurum aBC. erit aequale vniuerso spatio sub spirali et recta aB contento.

Haec omnia more ueterum demonstrantur per duplicem positionem; sine doctrina indiuisibilium.³

[*addressed:*] To my verie worthie friend M^r: John Pell publicke Professor of the Mathematiques at Breda.

giue this at Breda

[*annotated by Pell:*] Received Aug. 21. 31.

[*cover annotated by Birch:*] 20/30 Aug. 1651.

---

³'Let there be any straight line *ab*, intersected anywhere at *c*; then let there be erected the perpendicular *cD*, being the mean between *bc* and *ca*. Again let the straight line *ce* be the mean between *Dc* and *cb*; *ce* also intersects *Dc*, let it bisect the angle *Dcb*; further, let the straight line *cf*, the mean between *Dc* and *ce*, bisect the angle *Dce*; and so let it continue, always drawing the mean proportionals that bisect the angles; so that we shall have points *b*, *e*, *f*, *D*, and *a*; and as many others as we wish; through which will pass a certain curved line, which we may call a geometric spiral.

'But we can define it in a different way. If the straight line *aB*, keeping the end *a* fixed, is moved round in a plane, always at the same speed; and, at the same time, if any point *c* on that line is moved along it in such a way that the spaces traversed in equal periods of time are proportional to one another, keeping the same proportion that the distances of that point have from the centre *a*; once again we shall have the same geometric spiral.

'This spiral line has this peculiar property: before it reaches the centre, it must complete an infinite number of revolutions around it; nevertheless its length is known – not only that of its parts, but in total. For if there were the centre *a*, one of the radii *aB*, the tangent *BC* and the right angle *BaC*; the tangent *BC* would be equal to the entire spiral *BDa*. And, indeed, given any arc *BJ*, a straight line equal to it can be very easily cut off and demonstrated.

'Let the tangent *BC* be on the base *Ba*; the isosceles triangle *aBC* will be equal to the entire space contained between the spiral and the straight line *aB*.

'All these things are demonstrated in the manner of the Ancients, by double position, without the doctrine of indivisibles.'

Up to '... recta linea aequalis abscinditur, demonstraturque' ('... a straight line equal to it can be very easily cut off and demonstrated'), this is a quotation (with very minor verbal changes) from the Torricelli text. The next sentence is not in Torricelli. The final sentence is a paraphrase of part of the next paragraph in Torricelli's text; 'the manner of the Ancients, by double position' was the method of exhaustion with *reductio ad absurdum* applied twice.

This spiral is now known as the equiangular or logarithmic spiral.

Letter 113

[22 August/] 1 September 1651

Pell to Cavendish, from Breda

BL, MS Add. 4280, fo. 137^r (authorial copy); BL, MS Harl. 6083, fo. 339^r (enclosure: original, sent to Cavendish); BL, MS Add. 4280. fo. 319^r (enclosure: authorial copy)

van Maanen P46

Right Honourable

One of these inclosed papers is a slight beginning to consider the Diagrams proposed in your last,[1] which I received yesterday night. The other[2] had beene sent you long since, if various diversions had not crouded it out of my memory. Nor am I at this time at leisure to make long excuses for my long silence, but am constrained, in some haste rudely to beg a pardon for

> Your Honours most humble servant
> John Pell.

Breda Sept. 1. stylo novo, 1651

[*Enclosure to Letter 113:*]

Harum spiralium species sunt innumerabiles. Innumerae enim possunt sumi rationes ipsius ac ad cb.

Si verò recta ab, quàm ille jubet secari *utcunque*, fuerit secta *aequaliter*, ratio rectae ac ad cb erit ut 1 ad 1; ideoque et dc ad cb erit ut 1 ad 1 (Nam si extremae sunt aequales, aequales erunt et omnes mediae proportionales) Et proinde *prima* harum spiralium species erit exactè *circularis*, cujus centrum est c. Ad quod nunquam illa perveniet, etiamsi innumeras revolutiones absolverit. Ideoque hîc circularis peripheria infinitiès in se revoluta et proinde infinita censebitur. Neque aliter circulari conveniet illud praeceptum de Longitudine universae spiralis in-

---

[1] Cavendish's 'last' was Letter 112; Pell's first 'paper' is printed here as the enclosure to this letter.

[2] As Cavendish's reference to this in Letter 114 shows, the second enclosure was a text by Pell on spherical triangles (a long-delayed response to Cavendish's request in Letter 106). Among Cavendish's papers there is a small leaf in Pell's hand, beginning: 'Of Sphaericall Triangles, I find 16 sorts...' (BL, MS Harl. 6083, fo. 29^r). Pell's own copy of this brief text is BL, MS Add. 4422, fo. 209^r (annotated by him: 'I sent a coppy of this to Sr Charles Cavendyshe Septemb. 1. stylo novo 1651'). The same MS contains a page of diagrams to accompany that text (fo. 210^r); a longer text, on folio paper, demonstrating how many sorts of spherical triangles there are (fos. 211–212^r, 214); and a draft of a reply to Cavendish's query, perhaps intended as part of this letter but not in fact used (fo. 215^r: 'You tell me in your letter, that you desire an exact enumeration...').

veniendâ; Nam peripheriae tangens erit infinita. Sed alterum praeceptum circulo minimè convenit, nisi illius aream infinitam dixeris. Nam, si fecero angulum CAB aequalem angulo CBA, rectae CA et CB nunquam concurrent.

Sed de hâc et reliquis harum spiralium speciebus, plura posthac.[3]

[*authorial copy annotated by Pell:*] A coppy of this was sent to Sr Charles Cavendysshe Sept. 1. 1651 stylo nov.

---

[3]'The types of these spirals are innumerable. For it is possible to take innumerable ratios between *ac* itself and *cb*.

'If, indeed, the straight line *ab*, which he orders to be intersected *anywhere*, had been cut *into equal parts*, the ratio between the straight lines *ac* and *cb* would be as 1 to 1; and so the ratio of *dc* to *cb* would also be as 1 to 1. (For if the extremes are equal, all the mean proportionals will also be equal.) And therefore the *first* type of these spirals will be exactly *circular*, with centre at *c*. Which it will never reach, even if it has completed innumerable revolutions. So in this case the circular circumference will be considered infinitely many times turned upon itself, and therefore to be infinite. Otherwise that precept about finding the length of a universal spiral will not apply to a circular one; for the tangent to the circumference will be infinite. But the other precept can hardly be applied at all to a circle, unless you say that its area is infinite. For, if I make the angle CAB equal to the CBA, the straight lines CA and CB will never meet.

'But of this and other types of these spirals, more later.'

Letter 114

[26 September/] 6 October 1651

Cavendish to Pell, from Antwerp

BL, MS Add. 4278, fos. 321–2 (original)
van Maanen C59

Worthie Sr:

manie thankes for the fauoure of your last letter,[1] & the consideration you haue so well begun of the geometricall spirall;[2] with the varietie of sphaericall triangles./[3] S^r: Th: Alesburie[4] remembers him to you & desires to knowe if you would be pleased to shew the vse of M^r: Hariots doctrine of triangulare numbers; which if you will doe, he will send you the originall; I confess I was so farr in loue with it that I coppied it out;[5] though I doute I vnderstand it not all; much less the many vses which I assure myself you will finde of it./ till this be printed I shall esteem of my owne coppie./ If you haue had the leasure to note & explicate M^r: Chauaus Treatise of Algebra[6] I desire you will be pleased to send it./ The worthye bearer heerof M^r: Doctor Morleie[7] though not so speedie yet a sure messenger. I haue no more at this time;/ God protect you & vs all.

    Your assured friend to serue you
    Charles Cauendysshe

Antwerp: Octo: 6. 1651.

[*addressed:*] To my verie worthie friend M^r: John Pell Publick Professor of the Mathematicks at Breda, giue this

[*annotated by Pell:*] Received Octob. 8. stylo novo. 1651

[*cover annotated by Birch:*] 6 Octob. 1651.

---

[1] Letter 113.   [2] See Letter 113, n. 1.   [3] See Letter 113, n. 2.   [4] See Letter 94, n. 4.

[5] Aylesbury was an executor of Harriot's will; this letter is of some importance in showing that he still held some of Harriot's original papers 30 years after Harriot had died. Harriot's unpublished treatise on triangular numbers entitled 'De numeris triangularibus et inde de progressionibus arithmeticis. Magisteria magna T. H.' is now BL, MS Add 6782, fos. 107–46; the copy made by Cavendish is in BL, MS Harl. 6083, fos. 403–55.

[6] See Letter 74, n. 8, and Letter 90, n. 3.

[7] George Morley DD (1597–1684), Oxford theologian, close friend of Edward Hyde and Edmund Waller, and a man of wide intellectual interests, had been ejected from Christ Church, Oxford, in 1648; he joined the court-in-exile in Paris in 1649, moved to The Hague in 1650 and then settled in Antwerp. After the Restoration he became Bishop of Worcester (1660), then Bishop of Winchester (1662).

Letter 115

14/24 October 1651

Pell to Cavendish, from Breda

BL, MS Add. 4280, fo. 137^r (authorial copy)
van Maanen P47

Right Honourable

Having done with this manuscript of yours[1] I made no haste to send it backe, because I knew you had another copy; But had I considered that this is written in a larger hand than that other, and is therefore lesse troublesome to read, I had not kept it so long from you. The Diagrams in it, which are without letters, ought not to have any added, because there are no directory letters in the text adjoined. But the like Diagrams, made on other paper, may have full explications set by them. Which I cannot send heerewith, as not having done them. Indeed I had rather finish my owne thoughts of that argument, than explain the obscurities and supply the innumerable defects of another.

I have begun to review my owne papers, & to write them out faire; having to that purpose, lately procured from Amsterdam 2 reames (40 quires) of paper. If God graunt me health and leysure, I hope I shall not give over till I have made most of them legible enough for a Printer.[2] Yet if Sr Thomas Ailesbury be pleased to send me that tract Of triangular numbers,[3] I shall not thinke it a stop in my way. Nay rather, peradventure it may come very seasonably; My papers which I have now under hand, tending (I thinke) to the same ends that those speculations of Mr Harriots doe.[4] But I speake but by conjecture, I shall be able to say more when I see them. In the meane time I remaine

Your Honours most humble servant
John Pell

Breda Octob. 14. 24. 1651

---

[1] This probably refers to the MS treatise by Chauveau (see Letter 90, n. 3, and Cavendish's renewed request in Letter 114).

[2] How far Pell had proceeded with this task before he left Breda in June 1652 is not known. None of these fair copies survives, and none appeared in print.

[3] See Letter 114, nn. 4, 5.

[4] Harriot's treatise on triangular numbers contains several tables constructed by the method of constant differences. This was a topic that Pell and Warner had also explored in some detail (see Letter 77, n. 13, on Pell's interest in the interpolation of tables).

[*postscript:*] I doubt not but that you have seene M[r] Hobbes his Leviathan[5] & D[r] Harvey's new booke de generatione Animalium.[6]

---

[5] Hobbes's *Leviathan* was published in May 1651.

[6] William Harvey, *Exercitationes de generatione animalium* (London, 1651).

List of manuscripts

This list is confined to those MS items that have been cited in the text or notes of this book. The descriptions given here are summary descriptions; in the case of composite volumes, they describe not the entire contents of the MS but only the specific items that have been cited. An entry of the form 'A to B' means a letter or letters from A to B.

Bakewell, Derbyshire

Chatsworth House

Hardwick 33: Third Earl of Devonshire, accounts
Hobbes B 3: Gunter, treatise on 'Gunter's Rule'
Hobbes E 3: record of purchase of telescopes

Cambridge

Cambridge University Library [CUL]

Add. 9597/5/1: 'Diophanti arithmetica'
Add. 9597/13/1/41: Brancker to Collins
Add. 9597/13/1/42: Collins to Brancker
Add. 9597/13/1/80–81: Collins to Pell
Add. 9597/13/1/86: Collins to Pell
Add. 9597/13/1/89: Collins to Pell
Add. 9597/13/1/99: Derand to Cavendish
Add. 9597/13/1/219–219a: Pell to Collins
Add. 9597/13/1/220: text by French author on biquadratic equations
Baker Oo. VI. 113: Jungius, 'Protonoeticae philosophiae sciagraphia'
RGO 1/50: Flamsteed memorandum

Pepys Library, Magdalene College

2612: Pepys, Christ's Hospital papers

St John's College Archives

C 7.16: College Letter Book

Chelmsford, Essex

Essex Record Office [ERO]

microfiche D/P 278/1/1: Laindon parish register
microfiche D/P 414/1/2: Fobbing parish register, 1654–80
Q/SR 422/5: indictment of John Pell Jr
Q/SR 460/37: indictment of Thresher
microfiche SOG/12: Hearth Tax returns, 1662
microfilm T/A 772, item 1: Fobbing 'parish book', 1630–1700

Chichester, Sussex

West Sussex Record Office [WSRO]

microfilm 608: Bishop's Transcripts, Southwick, 1606–1812
microfilm 874: Bishop's Transcripts, Steyning, 1591–1812

The Hague

Algemeen Rijksarchief

ref. 1.08.11 (Nassause Domeinsraad, 1581–1811), 7989 [formerly 1061]: Illustre School, Breda, invitations; draft statutes; *Reglement ende ordonnantie*; *Waerschouwinghe uyt Breda*; resolutions of curators

Hamburg

Staats- und Universitätsbibliothek [SUBH]

Pe 5: Pell, *Idea mathesews*
Pe. 65: Pell to Jungius; Jungius to Pell; Tassius to Jungius; Pell to Tassius
Sup. ep. 97 (= Pe. 1a): Cavendish to Jungius
Sup. ep. 100: Hartlib to Tassius
Suppl. 1 and Suppl. 2: Jungius, restitution of Apollonius
Wo. 22: Jungius, 'Statica natantium'
Wo. 30: Jungius, 'Logica didactica'
Wo. 31: Jungius, 'Logicae hamburgensis praestantia'

Leeuwarden

Rijksarchief in de Provincie Friesland

van Eysinga – Vegelin van Claerbergen papers, 67, second folder, 'Brieven van geleerde Lieden': Mersenne to Vegelin

Leiden

Leiden University Library

BPL 293.1: Rivet to de Willem

Cod. 51 . Gol.: Heron of Alexandria, $M\eta\chi\alpha\nu\iota\kappa\acute{\alpha}$

Huygens 37: Dauber to Constantijn Huygens

Lewes, Sussex

East Sussex Record Office [ESRO]

Archdeaconry wills, W/A 15 (microfilm)

'Eastbourne baptisms, 1558–1898' (typescript transcript)

'Eastbourne burials, 1558–1843' (typescript transcript)

London

British Library [BL]

Add. 4255: Pell, note on Webster MS; Pell, extract from Lower, *Tractatus*

Add. 4278: Cavendish to Pell; Pell to Collins; Mersenne to Pell; Pell to Mersenne (with enclosure: refutation of quadrature); Brereton to Pell; Pell to Brancker; Brewster to Pell; Pell, note on payments to Brewster; Brancker to D. Hartlib; Brancker to Pell; Collins to Pell; Makin receipt; Sheldon, license; anon. to Cavendish; Hobbes, theorem on tangents; Hobbes, diagram; Hobbes, note; Mersenne to Cavendish; Le Pailleur to Mersenne; Colepeper to Pell; Roberval, proposition

Add. 4279: Wolzogen to Pell; Jungius to Pell; Tassius to Pell; Pell to Tassius; Mersenne to Pell; van Diest to Brosterhuysen; Petty to Pell; Makin to Pell; Thorndike to Pell; Pell, memorandum and note on Thorndike and Warner's papers; *Gründtliche Beschreibung*, proofs; Stucki to Pell; Rahn to Pell; Hartlib to Pell; Thompson to Pell; J. Nye to Pell; Pell to Brouncker?; Tonge to Pell; Pitt to Pell; Pell to Pitt; Morland to Pell; Winthrop to Brereton; Pell to Hastings; Thresher to Pell; W. Raven to Pell; Hastings to Pell

Add. 4280: Pell to Cavendish; Pell to Wingate; Pell to Hartlib; Pell to Leake; Pell to Petty; Pell, 'refutatiuncula'; J. Ravius to Pell; Pell to Jungius; Pell, *Controversiae* (draft material); Hortensius to Brosterhuysen; Brosterhuysen, notes on botany; Pell to I. Pell; I. Pell to Pell; Pell to Makin; Pell, valedictory speech to Zurich; articles of agreement between Pell and Adams; I. Pell funeral expenses; Pell to John Pell Jr; John Pell Jr to Pell; John Pell Jr to E. Pell; Sheldon document; Pell, petition; S. Nye to Pell; Pell to Gildredge; Brereton to Pell; Wallis, 'Animadversions' on Hobbes; Oldenburg to Brereton; Pell to Mersenne

Add. 4299: Haak to Pell; Huygens to Oldenburg

Add. 4363: Pell to Thurloe

Add. 4364: Pell to Hartlib; Cromwell, instructions to Pell; Pell and Dury to Thurloe; Pell to Thurloe; Thorndike to Pell; Pell to Thorndike; Buxtorf to Worthington; Worthington to Hartlib

Add. 4365: Pell, astronomical text; Pell, notes on compilation of antilogarithms; Pell, notes on observations of comet; Dury to Thurloe; Pell, translation of Waldensians' request; Rahn to Pell; Pell to Haak; Pell, observation of solar eclipse; Pell to R. H.; Dury to Pell; Ravius to Pell; Ravius, *Synopsis chronologiae biblicae infallibilis*; R. H. to Pell; Pell, notes on 'F. C.', prescription

Add. 4377: Pell, notes on Arabic language and translations from Koran; Morstyn to Hartlib; Dalgarno to Hartlib; Dalgarno, *Character Universalis* and *A New Discovery of the Universal Character*; Dalgarno, specimen of 'character'; Hartlib to Pell; Dalgarno to Pell; Pell, notes on Dalgarno; Pell to Dalgarno; Dalgarno to Brereton; Dalgarno, printed broadside

Add. 4381: Pell, Breda oration, draft

Add. 4384: H. Reynolds, 'Architectiones'; H. Reynolds, key to 'Architectiones'

Add. 4387: Pell to Tapp; Pell, treatise on tables for sundials; Pell. 'Horologiographia'

Add. 4388: Pell, text on method, logic, and teaching; Pell, notes on Birchensha; Descartes, 'Compendium musicae'

Add. 4393: Pell, minutes of commission on longitude; Flamsteed to Pell

Add. 4394: Pell, notes on Webbe; Collins, inventory of Warner MSS; Warner, fragment on hydraulics; Pell, list of books owned; Rahn to Pell; Pell, notes on Chinese book; Pitt, broadside

Add. 4395: Warner and Aylesbury, observations of refraction; Warner to Cavendish

Add. 4396: Aylesbury to Percy

Add. 4397: Pollard, astronomical text; Pell, 'Tabulae directoriae'; Pell, notes on apology for Vieta; Pell, 'Eclipticus prognosta'

Add. 4398: Briggs to Pell; Pell, 'Progymnasma' and notes on it; Pell, notes on Cardinael problem; Johnson to Pell; Schönauer to Ulrich; Potter, *Explicatio numeri bestiae*, proofs; Pell to Stucki; Pitt to Brancker; Rahn to Pell; Royal Society, summons to Pell to meeting; Fermat, 'Methodus'

LIST OF MANUSCRIPTS

Add. 4399: Pell, note on *Idea*

Add. 4400: Pell, notes on Pitiscus; Pell, *Controversiae*, draft material; Huygens to Oldenburg; de Sluse to Oldenburg

Add. 4401: Pell, treatise on quadrant; Pell, fair copy of material for *Introduction to Algebra*

Add. 4403: Pell, 'Almanacke'; Pell, 'Prognostication'; H. Reynolds, 'Nuncius volucris'

Add. 4404: Pell, note of dues for Laindon and Fobbing

Add. 4407: Pell, criticisms of Alsted; Pell, note on improvement of sight; Pell, note of conversation with Warner; Cavendish to Warner; Pell, note on problem from Leake; Browne, 'serpentine scale' (design and text); Pell, Breda oration, draft; testimonials for Föge; Pell, testimonial for Föge; Pell, list of mathematical books in Zurich library; Oldenburg, note; Newton to Collins; Pell, memorandum on 'Alhazen's problem' correspondence; Huygens to Oldenburg; Pell, 'Imitatio Nepeira'; Pell, notes on Apollonius; de Beaugrand, treatise on tangents; Chauveau, 'Traicté d'algèbre'; Fermat, 'Methodus'

Add. 4408: Pell, notes on Gellibrand; Pell, notes for 'Comes mathematicus'; Pell, page-proof of *Idea* (1650); Wolzogen to Hartlib; Pell, note on Stampioen; Lloyd to Pell; Pell to Lloyd; certificate of return of books to Illustre School, Breda; Hartlib to Pell; Wolzogen to Hartlib; Pell to Leybourn

Add. 4409: Pell, notes for treatise on 'Blagrave's Jewel'; H. Reynolds, *Magnae Britanniae chronographia*; Pell, notes on philosophical language; Pell, notes on analysis and synthesis; Pell, comments on Oughtred and Harriot; Pell, notes on Viète and Harriot; Pell to Hartlib(?); Pell, 'Trigonometria logistica' (fragment); Pell, note on 12-tome scheme; Pell, note on reason and revelation; Pell, note on Thomas and Judith Kirk; Pell to Brereton; Pell, plans for treatises on trigonometry

Add. 4410: Pell, notes on calculation of Easter

Add. 4411: Wallis or Pell, notes on solution of Brereton's problem; Wallis, solution of Titus's problem; Pell, translation of Archimedes, *Psammites* (fragment)

Add. 4412: Pell, notes on Brereton's problem

Add. 4413: Pell, note of visit by White; Pell, notes on Brereton's problem; Pell, notes on logarithmic tables; Pell, mathematical problem; Pell, notes on Cardano, Stevin, and Viète; Pell, 'Of Aequations'; Pell, notes on equations; Pell, notes on Apollonius

Add. 4414: Pollard to Pell; Pell to Pollard; Pell to Collins; Collins, draft preface; *Introduction to Algebra*, proofs corrected by Pell, and draft title page; Pell, notes on logarithmic tables; Pell, notes on Briggs; Pell, notes on Diophantus; Brancker, draft preface; Pell, version of Brouncker's solution to Fermat's problem; Pell, note on Hardy pamphlet

Add. 4415: Pell, 'The Section of angles'; Pell, note on 'Turris Babel'; Pell, 'Logistica generalis' and 'Mechanica generalis'; Pell, self-examination; Pell to Goad; Pell, notes on Harington; Pell, notes on refraction and Cartesian lenses project; Pell, notes on telescopes; Pell, algebraic difference tables; Pell, 'Posoteticall and Logisticall habitudes'; Pell, 'Progymnasma' and notes on it; Pell, 'Strife of Analytica and Synthetica'; Pell, mathematical problems; Pell, notes on Apollonius; Pell, note on Mercator; Pell, notes on logarithms; Pell, mathematical problem (for Cavendish); Pell, indeterminate equations

Add. 4416: Reynolds, *Magnae Britanniae chronographia*; M. Rogers to I. Pell; ? to Pell, cover-sheet; von Franckenberg to Hartlib; W. Beecher to Sir William Beecher; Pell, distribution list for *Controversiae*; Pell, memorandum; Pell, title page for Lansberg, 'Everlasting Tables'; Pell, notes on value of π; Pell, notes on habitudes; Pell, note of books borrowed from Titus

Add. 4417: Pell, notes on Viète; Warner, treatise on tables of refraction (fragment), and note on it by Pell; Viète, 'Harmonicon coeleste' fragment; Pell, notes on Mercator; Pell, 'Quadraticall Aequations solved by Delineation'; Pell, notes for 'refutatiuncula'; Pell, notes on Diophantus; Pell, note on Dary's problem; Pell, table for biquadratic equation

Add. 4418: Pell to ?; Pell, notes on Byrd; ? to Pell; Pell, note on antilogarithmic tables; Pell, notes for 'refutatiuncula'; Pell, notes on squares

Add. 4419: Pell to Hartlib; Pell, memorandum on Aylesbury's problem; Stucki to Pell; Pell, note on Harriot; Pell, notes on Diophantus

Add. 4420: Pell, 'Logistica'; Pell, algebra of knowledge; Pell, notes on Jungius

Add. 4421: Warner, notes on mechanics; Pell, lecture (fragment); Pell to C. Ravius; Pell, list of names; Olearius to Pell; Pell, *Controversiae*, draft material; Stucki to Pell; Pitt, broadside; Pell, notes on de Sarasa

Add. 4422: Pell, note on Warner, mathematical problems; Pell, note on Collins, mathematical problem; Pell, note on Hooke's multiplying engine

Add. 4423: Pell, note on philosophical language for place-names; Pell, notes on Vieta; Pell, note on Kalthoff's device; Pell, note on *Idea*; Pell, notes on copy of Viète annotated by Oughtred; Pell, notes on Descartes and Warner on refraction; Pell, memorandum on use of paper; Pell, notes on Campion; Pell to de Saint-Vincent; Rahn to Pell; Pell, memorandum on Baker

Add. 4424: Pell to Croply; Pell, note on Rahn; Pell, note on observing lunar eclipse; Pell, notes on *Tabula numerorum quadratorum*; Oldenburg to Pell; de Sluse to Oldenburg; Pell, table of sines; Royal Society, summons to Pell to meetings; Pell, notes on Warner–Pell antilogarithms; Pell, table of differences

Add. 4425: Pell to Hartlib; Pell, notes for letter to Hartlib; Hobbes, demonstration and fragment addressed to Brouncker; Pell, memorandum on Hobbes; Pell, refutation of Hobbes; Dary to Collins; Wallis, solution to Brereton's problem; Pell, notes on Brereton's problem; Pell, notes on logarithms

Add. 4426: Pell, prefatory material for treatise on 'Proportioned Line'; H. Reynolds, *Magnae Britanniae chronographia*; Pell, note on wife's debts and funeral arrangements; Pell, household accounts; Royal Society, summons to Pell to Council meeting; Pell, note on preparation of antilogarithmic tables

Add. 4427: Pell to Collins; Pell, notes for *Introduction to Algebra*

Add. 4428: Pell to James Pell; Pell, 'Methodus docendi'; Pell, notes on Helvicus; Pell to ?; Pell to I. Pell; Pell, notes on refraction and Cartesian lenses project; Pell to Gildredge; Wren to Oldenburg; Pell, table for equation

Add. 4429: Pell, notes on Dury; Pell, notes on concepts of measuring; anon., list of inventions; Alsted, 'Fundamenta disciplinarum mathematicarum', with criticisms by Pell; Pell, 'Geometricall problems'; Pell, notes on construction of 14-sided polygon; Mountagu to Pell; Pell to Mountagu; Pell, notes on squares

Add. 4430: Pell, mathematical problems; Pell, notes on logarithms, roots, and Trithemius; Pell, *Controversiae*, draft material; Pell, notes on Diophantus

Add. 4431: Pell, prefatory material for treatise on 'Proportioned Line'; Pell to cousin; H. Reynolds, key to 'Architectiones'; Pell, notes on Hérigone; Pell, notes on Gemma Frisius; Pell, notes on Girard; Pell, notes on Viète; Pell, notes for 'Manual of Logarithmicall Trigonometry'; I. Pell to Pell; Pell, note on Gosselin problems for Cavendish; Pell, notes on refraction; Pell, notes on lunar eclipse, 'Phaenomena lunaria'; Pell, Breda oration, draft; de Saint-Vincent to Pell; Pell, notes for sermons; Pell, 'Labyrinthus ingenii'; Pell, draft astronomical tables; Pell, 'Logarithmes for Asscripts of a Circle'

Add. 4441: Pell, scheme for 'General College'; Pell, list of Royal Society committees

Add. 4443: Haak to Pell

Add. 4444: Cavendish to Warner

Add. 4458: Pell, notes on Lescarbot

Add. 4474: Pell, note on Long's telescope; Pell, note on Beecher; Pell, note on Warner; Lodewijk Hendrik to Illustre School, Breda; Pell, 'Responsionis Capita'; Pell, notes for sermons; Pell, note on Viète proposal; Collins, 'Of some improvements of Algebra'

Add. 6269: Hartlib, petition to Parliament; Hartlib to Worthington

Add. 6782: Harriot, treatise on triangular numbers

Add. 6783: Harriot, equations

Add. 24850: Haak to Pell

Add. 41846: Harrison, 'Arca studiorum'

Add. 70499: Andrews to Newcastle; Cavendish to Bates

Egerton 2711: Wyatt and Surrey poems; Harington notes; Harington to Selden

Harl. 379: d'Ewes to his mother

Harl. 3360: Hobbes, optical treatise

Harl. 6002: Cavendish, note on Harriot and Roberval, proposition

Harl. 6083: Cavendish to Pell; Pell, summary of roots in cubic equations; 'Elemens des indivisibles'; Chauveau, 'Traicté d'algèbre'; Jungius, 'Phoranomica'; Fermat, 'Ad methodum appendix'; Fermat, 'Ad locos planos'; Pell, mathematical problem (for Cavendish); Pell, demonstration; Cavendish, notes on Kircher; Torricelli text on spirals, with comment by Pell; Pell, comments on spherical triangles; Harriot, treatise on figurate numbers (copy by Cavendish)

Harl. 6754: Warner, treatise on analysis of bullion; Warner, treatise on money and exchange

Harl. 6755: Cavendish, notes on Warner's treatise on alloys

Harl. 6756: Warner, 'De loco imaginis'

Harl. 6796: Pell to Cavendish; de Beaugrand, treatise on tangents; de Beaune, 'Notes briefves'

Harl. 7012: Pell to Vossius

Lansdowne 684: H. Reynolds, 'Macrolexis, sive nuncius volucris'; H. Reynolds, 'Magnae Britanniae chronographia'

Lansdowne 745: Pell to Thurloe

Lansdowne 746: Pell to Thurloe

Lansdowne 747: Pell to Morland

Lansdowne 748: Pell to Morland

Lansdowne 751: Pell, petitions; Boswell to Pell; Brereton to Pell; Council of State, decision; Cromwell, letter of accreditation for Pell; Pell to Moriaen; Wake to Boswell

Lansdowne 752: Pell, translations of German documents

Lansdowne 754: Pell, autobiographical note and chronology; Pell, draft petition; Pell, petition; Pell, memorandum; Thurloe to Pell; Cromwell to Pell; invitation to Cromwell's funeral

Lansdowne 1238: Norgate to Williams

Sloane 417: Hübner to Hartlib; Hübner to Bisterfeld

Sloane 639: Hübner to Comenius; Hübner to Herbert; Hübner to Gronovius

Sloane 648: von Franckenberg to Hartlib

Sloane 649: Brookes document

Sloane 652: Wolzogen, 'Magnum opus mathematicum'

Sloane 653: Hartlib, list of desiderata; Hartlib, list of works under preparation

Sloane 1466: Webbe to Hartlib; Brookes to Hartlib

Sloane 1536: Moleyns prescriptions

City of Westminster Archives Centre

St Margaret's, Westminster, parish registers, vol. 6: marriages, 1653–8

St Margaret's, Westminster, parish registers, vol. 7: marriages, 1664–84

Guildhall Library

12,873/1: Christ's Hospital, Royal Mathematical School, minute and memoranda book

International Genealogical Index (microfiche)

London Metropolitan Archives

microfilm X019/009: Consistory Court of London, calendar of wills and administrations, 1670–1720

microfilm X24/66: St Dunstan's, Stepney, baptisms 1568–1656

microfilm X105/022: St Giles-in-the-Fields, parish register, burials, 1668–92

London University Library

Carlton Shorthand Collection, item 1615: B. Reynolds, 'The invention of Radiography' (engraving)

Mercers' Company Archives, Mercers' Hall [MCA]

Acts of Court, vol. 4 (1595–1629)

Renterwarden's Accounts, 1618–1629

Renterwarden's Accounts, 1629–1639

Willson notes, file 1, 'Guide to Sources of the History of Collyer's School, Horsham, Part One' (typescript)

Public Record Office, The National Archives [PRO]

C 24/654 (Town depositions, Chancery): Ithamar Pell deposition; J. Guillim deposition; T. Meade deposition

IND 1/6066: Court of King's Bench, docket books, 1680

KB 27/2007: Court of King's Bench, Plea Rolls, Trinity 1680

KB 27/2008: Court of King's Bench, Plea Rolls, Trinity 1680

Royal Society

81 (Commercium epistolicum): Leibniz to Oldenburg
Classified Papers, VIII (1): Pell, observation of solar eclipse
Classified Papers, XXIV (1): Collins, draft review of Pell–Rahn
Classified Papers, XXIV (3): Collins, extract from Pell letter
Classified Papers, XXIV (13): Collins, 'discourse' on roots of equations
Domestic V: minutes of agricultural committee

Westminster School, Busby Library

Busby, 'Accounts and Memoranda', loose sheet inserted at end: Pell, diary of events
pressmark I D 2: Cardano, *Ars magna*, with notes and inserts by Pell
pressmark I D 9: MS copy of Viète, *De potestatum resolutione*
pressmark I F 10: Stevin, *Oeuvres*, with notes and inserts by Pell
pressmark I F 21: Diophantus, *Arithmeticorum*, with notes and inserts by Pell
pressmark I F 27: Luneschlos, *Thesaurus*, with inserts by Pell
uncatalogued: 1687 catalogue of Pell's library

New Haven, Conn.

Beinecke Library, Yale University

Osborn Collection, 16792: Isham to Hartlib

Northampton

Northamptonshire Records Office

IL 3422, bundle VI: Warner papers

Nottingham

Nottinghamshire Archives

DDP.1.1.6.51: Cavendish estates document

Nottingham University Library

Pl E 12/10/1/9/1: Earl of Newcastle, rentals book
Pw 1 406: record of loan for telescopes
Pw 1 668: list of telescopes

Oxford

Bodleian Library [Bodl.]

Add. C. 308: Sheldon, letter-book

Arch. o. c. 3: Apollonius, *Conics*

Ashmole 423: Streete testimonials

Aubrey 6: Aubrey, notes on Pell

Aubrey 10: Aubrey, notes on education; Aubrey, copy of Pell's version of problem by Cardinael; Aubrey, note on Lewis

Aubrey 13: Pell to Haak

Ballard 14: Aubrey to Wood

e Mus. 203: correspondence deciphered by Wallis

Gravius 5: Ulug Beg, star catalogue

Rawl. A 261: Cromwell, letter of accreditation for Pell

Rawl. A 328: Cromwell, letter of recommendation for Dury; payment records for Dury; Cromwell, order for payment for Pell

Rawl. Letters 83: Boswell to Vossius

Selden supra 109: Langbaine to Selden

Thurston 3: Apollonius, *Conics*

Wood F 39: Aubrey to Wood

Worcester College

63: Pell, problems from L. and T. Digges, *Stratioticos*

64: Collins, notes on Pell's mathematics; Aubrey, copy of Pell's version of problem by Cardinael; Aubrey, note on Pell; Pell, summary of algebraic operations; Davenant, treatise on algebra; Collins, letter to Baker? on cubic equations; Pell, table for equation

Rome

Ponteficia Università Gregoriana

557: Vegelin to Kircher

St Andrews

St Andrews University Library

31009: Collins to Gregory

San Marino, Calif.

Huntington Library

Hastings Collection, Literature, box 1, folder 1: Makin, elegy
MS 8799: Makin to Countess of Huntingdon
MS 8800: Makin to Countess of Huntingdon
MS 8801: Makin to Countess of Huntingdon

Sheffield

Sheffield University Library

Hartlib Papers (CD-ROM edition: Ann Arbor, Mich., 1995; 2nd edn. 2002) [HP]
HP 2/9/15: Dury to Hartlib
HP 7/11/1A: Hartlib to Dury
HP 7/12/2: Hartlib to Dury
HP 7/16/1B: Hartlib to Dury
HP 8/4A–B: Pell to Hartlib
HP 8/60: Hartlib, list of inventions
HP 13/345: Hartlib to Culpeper
HP 14/1/6A: Hartlib, mock titlepage of Pell, *Idea*
HP 22/3/1: Dury and Pell, comments on Brookes
HP 22/6/1–6: Anon., 'An Imperfect Enumeration'
HP 22/8/1A: Hartlib, list of desiderata
HP 22/10/1A: Hartlib, list of desiderata
HP 22/15/1A: Pell letter or memorandum
HP 23/2/20A: Hartlib, accounts
HP 28/2/1–26: Ephemerides, 1651
HP 28/2/27–44: Ephemerides, 1652
HP 28/2/45–82: Ephemerides, 1653
HP 29/2/1–65: Ephemerides, 1634
HP 29/3/1–65: Ephemerides, 1635
HP 29/7/1–16: Ephemerides, 1658
HP 29/8/1–9: Ephemerides, 1659
HP 30/4/1–36: Ephemerides, 1639
HP 30/4/37–68: Ephemerides, 1640

HP 30/4/69–80: Ephemerides, 1641
HP 31/12/14A: Pell, queries for Mercator
HP 35/5/1–180: Comenius, 'Methodus linguarum novissima'
HP 36/1/3: Pell to Hartlib
HP 37/5A: Moriaen to Hartlib
HP 37/7A–8B: Moriaen to Hartlib
HP 37/13B: Moriaen to Hartlib
HP 37/21B: Moriaen to Hartlib
HP 37/40B: Moriaen to Hartlib
HP 37/41A: Moriaen to Hartlib
HP 37/50A: Moriaen to Hartlib
HP 37/112B–113A: Moriaen to Hartlib
HP 37/114A: Moriaen to Hartlib
HP 37/116A: Moriaen to Hartlib
HP 37/117A: Moriaen to Hartlib
HP 37/123A: Moriaen to Hartlib
HP 37/146B: Moriaen to Hartlib
HP 37/166A: Moriaen to Hartlib
HP 43/74A: Pell to Hartlib
HP 44/1/2: Jonston to Hartlib
HP 44/1/21: Jonston to Hartlib
HP 46/6/8–9: Speed to Hartlib
HP 46/6/15A: Speed to Hartlib
HP 46/6/20A: Speed to Hartlib
HP 47/2: Dury, 'Proposalls'
HP 55/20/1–5: Andreae, 'Leges Societatis Christianae'
HP 71/8A: Pell, typographical invention

Washington, DC

Folger Shakespeare Library

V. a. 296: John Ward, notebook

Zurich

Staatsarchiv [SAZ]

E II 457a: Letter of recommendation for Dury; Cromwell, letters of recommendation for Dury; Dury to Ulrich; Pell, speech; Dury, speech; 'Judicium Theologorum Basileensium'

E II 457b: Cromwell, letter of accreditation for Pell; Cromwell, letter of recommendation for Dury; 'Acta der Conferenz zu Araw'

E II 457c: Dury to Ulrich

E II 457d: Dury to Ulrich

E II 457f: Stouppe to Ulrich; Stouppe, baptismal certificate

Zentralbibliothek

B 102: Cromwell, letter of accreditation for Pell

C 114a: Rahn, 'Algebra Speciosa'

C 114b: Rahn, 'Solutio problematum Diophanti'

F 71: Rahn to Hottinger

F 72: Stucki to Hottinger

Bibliography

This bibliography is confined to those works that have been cited in the text or notes of this book. The alphabetical order is of the first element of the author's name that bears a capital letter.

Abbott, W. C., *The Writings and Speeches of Oliver Cromwell*, 4 vols. (Cambridge, Mass., 1937–47.

Adam, C., 'Descartes et ses correspondants anglais', *Revue de littérature comparée*, 17 (1937), pp. 437–60.

Adamson, J. W., *Pioneers of Modern Education, 1600–1700* (Cambridge, 1905).

Adelung ['Adelungk'], W. H., *Die annoch verhandene hamburgische Antiquitäten oder Alterthums-Bedächtnisse* (Hamburg, 1696).

van Alphen, G., 'De Illustre School te Breda en haar boekerij', *Tijdschrift voor geschiedenis*, 64 (1951), pp. 277–314.

Alston, R. C., *Treatises on Short-hand* (Leeds, 1966).

Anon. [C. Hardy?], *Cyclometriae a Christiano Severini Longomontano mathematicarum superiorum in Academia Hafniensi regio professore, repetitis vicibus, atque adeo tribus in libris, anno 1612 Hafniae, anno 1627 Hamburgi, hoc anno 1644 Amstelodami editis, publicatae elenchus. Quo ostenditur, posita diametro circuli 10.000.000, & circumferentia 31.418.596, circumferentiam circuli maiorem esse ambitu figurae centum nonaginta trium laterum circulo circumscriptae* (Paris, 1644).

— *A True Relation of a Wicked Plot Intended and still on Foot against the City of Glocester, to Betray the same into the Hands of the Cavaliers* (London, 1644).

— *Look to it London, Threatned to be fired by Wilde-fire-zeal, Schismatical-faction, & Militant-mammon. Discovered July 15 1648 in a Discourse with one Croply and Hide* (n.d. [London], 1648).

— *The Resolutions of the Army, against the King, Kingdome and City. July 15 1648* (n.d. [London], 1648).

— *A True State of the Case of the Commonwealth of England, Scotland, and Ireland, and in Reference to the Late Establish'd Government by a Lord Protector and a Parliament* (London, 1654).

— *Gründtliche Beschreibung der neuen Regiments-Verfassung in dem gemeinen Wesen Engelland, Schott- und Irrland samt den zugehörigen Eyländern und andern Landschafften, unter dem Herren Protector und dem Parlament* (Schaffhausen, 1657).

— *Parnassus Heidelbergensis omnium illustrissimae huius academiae professorum icones exhibens* (Heidelberg, 1660).

— [J. Collins], Review of Rahn and Pell, *Introduction to Algebra*, in *Philosophical Transactions of the Royal Society*, no. 35 (18 May 1668), pp. 688–90.

— [J. Collins?], Review of Pell, *Tabula numerorum quadratorum*, in *Philosophical Transactions of the Royal Society*, no. 82 (22 Apr. 1672), pp. 4050–2.

— [M. Lewis?], *An Essay to revive the Antient Education of Gentlewomen* (London, 1673).

— *Catalogus librorum tam impressorum quam manuscriptorum bibliothecae publicae Universitatis Lugduno-Batavae* (Leiden, 1716).

— *The Tryal of John Hampden, Esq. ... in the Great Case of Ship-Money* (London, 1719).

— 'Select Documents XXXIX: A List of the Department of the Lord Chamberlain of the Household, Autumn, 1663', *Bulletin of the Institute for Historical Research*, 19 (1942–3), pp. 13–24.

Apollonius, *Opera*, tr. J.-B. Memus (Venice, 1537).

— *Conicorum libri quattuor*, ed. and tr. F. Commandino (Bologna in 1566; 2nd edn. Paris, 1626).

— *Conicorum libri IV, cum commentariis*, ed. and tr. C. Richard (Antwerp, 1655).

— *Conicarum sectionum libri V, VI & VII*, ed. and tr. C. Ravius (Kiel, 1669).

— *Conicorum libri octo*, tr. E. Halley (Oxford, 1710).

— *Les Coniques*, ed. and tr. P. ver Eecke (Bruges, 1923).

— *Conics, Books V to VII: The Arabic Translation of the Lost Greek Original in the Version of the Banū Mūsā*, ed. and tr. G. J. Toomer, 2 vols. (New York, 1990).

Arber, E., ed., *A Transcript of the Registers of the Company of Stationers of London, 1554–1640 AD*, 5 vols. (London, 1875–94).

Archimedes, *Opera omnia*, ed. J. L. Heiberg *et al.*, 4 vols. (Leipzig, Stuttgart, 1910–75).

Armstrong, E. V., and H. S. Lukens, 'Lazarus Ercker and his "Probierbuch": Sir John Pettus and his "Fleta Minor"', *Journal of Chemical Education*, 16 (1939), pp. 553–62.

Attree, F. W. T., and J. H. L. Booker, 'The Sussex Colepepers', *Sussex Archaeological Collections*, 47 (1904), pp. 47–81, and 48 (1905), pp. 65–98.

Atwell, G., *The Faithfull Surveyour* (Cambridge, 1658).

Aubrey, J., *Miscellanies* (London, 1696).

— *'Brief Lives', chiefly of Contemporaries, set down by John Aubrey, between the years 1669 & 1696*, ed. A. Clark, 2 vols. (Oxford, 1898).

Aylmer, G. E., *The King's Servants: The Civil Service of Charles I, 1625–1642* (London, 1961).

— *The State's Servants: The Civil Service of the English Republic, 1649–1660* (London, 1973).

Baillet, A., *La Vie de Monsieur Descartes*, 2 vols. (Paris, 1691).

Baily, F., *An Account of the Revd. John Flamsteed* (London, 1835).

Bainbridge, J., *Canicularia*, ed. J. Greaves (Oxford, 1648).

Baker, T., *Clavis geometrica catholica* (London, 1684).

Balthasar, J. A., ed., 'Eidgenössiche Gesandtschaft an Cromwell im Jahr 1653', in his *Helvetia: Denkwürdigkeiten für die XII Freistaaten der schweizerischen Eidgenossenschaft*, i, Heft 4 (Zurich, 1823), pp. 561–98.

Barbour, P. L., 'Introduction' to 'B. Makin' (attrib.), *An Essay to Revive the Antient Education of Gentlewomen* (Los Angeles, 1980).

Barlaeus [van Baerle], C., *Mercator sapiens sive oratio de conjungendis mercaturae et philosophiae studiis* (Amsterdam, 1632).

— *Epistolarum liber* (Amsterdam, 1667).

Barnard, T. C., *Cromwellian Ireland: English Government and Reform in Ireland, 1649–1660*, 2nd edn. (Oxford, 2000).

Barnett, P. R., *Theodore Haak (1605–1690): The First German Translator of Paradise Lost* (The Hague, 1962).

Batten, J. M., *John Dury: Advocate of Christian Reunion* (Chicago, 1944).

Baudouin de Montarcis, P., *Traité des fondemens de la science générale et universelle* (Paris, 1651).

Beierlein, P. R., *Lazarus Ercker: Bergmann, Hüttenmann und Münzmeister im 16. Jahrhundert* (Berlin, 1955).

Bell, E. T., *The Last Problem*, 2nd edn., ed. U. Dudley (n.p., 1990).

Bell, G. M., *A Handlist of British Diplomatic Representatives, 1509–1688* (London, 1990).

Bennett, J. A., *The Mathematical Science of Christopher Wren* (Cambridge, 1982).

Benzing, J., *Die Buchdrucker des 16. und 17. Jahrhunderts im deutschen Sprachgebiet*, 2nd edn. (Wiesbaden, 1982).

Beresford, J., *The Godfather of Downing Street: Sir George Downing, 1623–1684* (London, 1925).

Berghaus, G., *Die Aufnahme der englischen Revolution in Deutschland, 1640–1669* (Wiesbaden, 1989).

Bernhardt, J., 'Une Lettre-programme pour "l'avancement des mathématiques" au XVIIe siècle: l' "idée générale des mathématiques" de John Pell', *Revue d'histoire des sciences et de leurs applications*, 24 (1971), pp. 309–16.

Bettini, M., *Apiaria universae philosophiae mathematicae*, 2 vols. (Bologna, 1642).

Biblia sacra polyglotta, ed. B. Walton, 6 vols. (London, 1653-7).

de Billy, J., *Nova geometriae clavis algebra* (Paris, 1643).

de Bils, L., *Kopye van zekere ampele acte van Jr. Louijs de Bils, Heere van Koppensdamme, Bonem, &c. rakende de wetenschap van de oprechte anatomie des menselijken lichaams* (Rotterdam, 1659).

— , tr. J. Pell, *The Coppy of a Certain Large Act (Obligatory) of Yonker Louis de Bils, Lord of Koppensdamme, Bonen, &c. Touching the Skill of a Better Way of Anatomy of Man's Body. Printed (in Low Dutch) at Rotterdam* (London, 1659).

Bion, N., *The Construction and Principal Uses of Mathematical Instruments*, tr. E. Stone, 2nd edn. (London, 1758).

Birch, T., ed., *A Collection of the State Papers of John Thurloe*, 7 vols. (London, 1742).

— *The History of the Royal Society of London*, 4 vols. (London, 1756-7).

Blaeu, W. J., *Tweevoudigh onderwys van de hemelsche en aerdische globen* (Amsterdam, 1620).

— *Institutio astronomica de usu globorum & sphaerarum caelestium ac terrestrium*, tr. M. Hortensius (Amsterdam, 1634)

Blagden, C., *The Stationers' Company: A History, 1403-1959* (London, 1960).

Blagrave, J., *The Mathematical Jewel, Shewing the Making and most Excellent Use of a Singular Instrument so called* (London, 1584).

Blaydes, F. A., ed., *The Visitations of Bedfordshire, annis Domini 1566, 1582, and 1634*, Harleian Society Publications, xix (London, 1884).

Blekastad, M., *Comenius: Versuch eines Umrisses von Leben, Werk und Schicksal des Jan Amos Komenský* (Oslo, 1969).

— *Menneskenes sak: den tsjekkiske tenkeren Comenius i kamp om en universal reform av samfunnslivet* (Oslo, 1977).

Blencowe, R. W., 'Extracts from the Journal and Account Book of the Rev. Giles Moore, Rector of Horstead Keynes, Sussex, from the year 1655 to 1679', *Sussex Archaeological Collections*, 1 (1848), pp. 65-127.

Blok, F. F., *Caspar Barlaeus: From the Correspondence of a Melancholic* (Assen, 1976).

— *Isaac Vossius and his Circle: His Life until his Farewell to Queen Christina of Sweden, 1618-1655* (Groningen, 2000).

Bom, G. D., *Het hooger onderwijs te Amsterdam van 1632 tot onze dagen: bibliographische bijdragen* (Amsterdam, 1882).

Bohatec, J., *Die cartesianische Scholastik* (Leipzig, 1912).

Bolton, R., *The History of the Several Towns, Manors, and Patents of the County of Westchester, from its First Settlement to the Present Time*, 2nd edn., 2 vols. (New York, 1881).

Bombelli, R., *L'algebra* (Bologna, 1572).

Bonfante, G., 'Una descrizione linguistica d'Europa del 1614', *Paideia*, 10 (1955), pp. 22–7.

Bopp, K., 'Die Kegelschnitte des Gregorius a St. Vincentio in vergleichender Bearbeitung', *Abhandlungen zur Geschichte der mathematischen Wissenschaften mit Einschluss ihrer Anwendungen*, 20 (1907), 87–314.

Borch, O., *Itinerarium, 1660–1665*, ed. H. D. Schepelern, 4 vols. (Copenhagen, 1983).

Borrell, J., *Logistica, quae et arithmetica vulgo dicitur in libros quinque digesta* (Lyon, 1559).

Bosmans, H., 'Documents inédits sur Grégoire de Saint-Vincent', *Annales de la société scientifique de Bruxelles*, 27, part 2 (1902–3), pp. 21–63.

— 'Le Jésuite mathématicien anversois André Tacquet (1612–1660)', *De gulden passer – Le Compas d'or*, 3 (1925), pp. 63–87.

Bosse, A., *La Maniere universelle de Mr Desargues ... pour poser l'essieu & placer les heures & autres choses aux cadrans au soleil* (Paris, 1643).

— *Sentimens sur la distinction des diverses manières de peinture, dessein & graveure, & des originaux d'avec leurs copies* (Paris, 1649).

Boulliau ['Bullialdus'], I., *Philolai, sive dissertationis de vero systemate mundo, libri IV* (Amsterdam, 1639)

— ed., *Theonis ... eorum, quae in mathematicis ad Platonis lectionem utilia sunt, expositio* (Paris, 1644).

— *Astronomia philolaica, opus novum, in quo motus planetarum per novam ac veram hypothesim demonstrantur* (Paris, 1645).

Boyle, R., *The Correspondence*, ed. M. Hunter, A. Clericuzio, and L. M. Principe, 6 vols. (London, 2001).

Braddick, M. J., and M. Greengrass, eds., 'The Letters of Sir Cheney Culpeper (1641–1657)', *Camden Miscellany*, xxiii (Camden Society, ser. 5, vol. 7) (London, 1996), pp. 105–402.

Brambora, J., *Znižní dílo Jana Amose Komenského: studie bibliografická* (Prague, 1954).

Brancker, T., *Doctrinae sphaericae adumbratio, unà cum usu globorum artificialium* (Oxford, 1662).

Brandt, F., *Thomas Hobbes' Mechanical Conception of Nature*, tr. V. Maxwell and A. I. Fausbøll (Copenhagen, 1928).

Brann, N. L., *Trithemius and Magical Theology: A Chapter in the Controversy over Occult Studies in Early Modern Europe* (New York, 1999).

Brauer, K., *Unionstätigkeit John Duries unter dem Protektorat Cromwells: ein Beitrag zur Kirchengeschichte des siebzehnten Jahrhunderts* (Marburg, 1907).

von Braunmühl, A., *Christoph Scheiner als Mathematiker, Physiker und Astronom* (Bamberg, 1891).

Brerewood, E., *Enquiries touching the Diversity of Languages and Religions through the Cheife Parts of the World* (London, 1614).

Brett, A., *A Model for a School for the Better Education of Youth* (n.p., n.d. [London?, c.1672]).

Briggs, H., *Arithmetica logarithmica* (London, 1624).

— *Trigonometria britannica, sive de doctrina triangulorum libri duo*, ed. H. Gellibrand (Gouda, 1633).

Brink, J. R., 'Bathsua Makin: Educator and Linguist (1608?–1675?)', in J. R. Brink, ed., *Female Scholars: A Tradition of Learned Women before 1800* (Montreal, 1980), pp. 86–100.

— 'Bathsua Reginald Makin: "Most Learned Matron"', *Huntington Library Quarterly*, 54 (1991), pp. 313–26.

von Brockdorff, C., *Des Charles Cavendish Bericht für Joachim Jungius über die Grundzüge der Hobbes'schen Naturphilosophie*, Veröffentlichungen der Hobbes-Gesellschaft, iii (Kiel, 1934).

Brugmans, H., J. H. Scholte, and P. Kleintjes, eds., *Gedenkboek van het Athenaeum en de Universiteit van Amsterdam, 1632–1932* (Amsterdam, 1932).

Bryden, D. J., 'Evidence from Advertising for Mathematical Instrument Making in London, 1556–1714', *Annals of Science* 49 (1992), pp. 301–36.

— 'Magnetic Inclinatory Needles: Approved by the Royal Society?', *Notes and Records of the Royal Society of London*, 47 (1993), pp. 17–31.

Budgen, W., *Old Eastbourne: Its Church, its Clergy, its People* (London, n.d. [c.1913]).

Burke, A. M., *Memorials of St. Margaret's Church Westminster, Comprising the Parish Registers, 1539–1660 and the Churchwardens' Accounts, 1460–1603* (London, 1914).

Burman, C., *Traiectum eruditum, virorum doctrina inlustrium, qui in urbe Traiecto, et regione traiectensi nati sunt, sive ibi habitarunt, vitas, facta et scripta exhibens* (Utrecht, 1738).

Burn, R. P., 'Alphone Antonio de Sarasa and Logarithms', *Historia mathematica*, 28 (2001), pp. 1–17.

Burner, S. A., *James Shirley: A Study of Literary Coteries and Patronage in Seventeenth-Century England* (New York, 1988).

Butler, A. M., *Steyning, Sussex: The History of Steyning and its Church from 700–1913* (Croydon, n.d.).

Butler, E. H., *The Story of British Shorthand* (London, 1951).

Cagnolati, A., *Il circolo di Hartlib: riforme educative e diffusione del sapere (Inghilterra, 1630–1660)* (Bologna, 2001).

Cajori, F., *A History of the Logarithmic Slide Rule and Allied Instruments* (London, 1909).

Calendar of State Papers, Domestic, 1611–1618, ed. M. A. E. Green (London, 1858).

Calendar of State Papers, Domestic, 1619–23, ed. M. A. E. Green (London, 1858).

Calendar of State Papers, Domestic, 1636–7, ed. J. Bruce (London, 1868).

Calendar of State Papers, Domestic, 1640, ed. W. D. Hamilton (London, 1880).

Calendar of State Papers, Domestic, 1640–1, ed. W. D. Hamilton (London, 1882).

Calendar of State Papers, Domestic, 1641–3, ed. W. D. Hamilton (London, 1887).

Calendar of State Papers, Domestic, 1644, ed. W. D. Hamilton (London, 1888).

Calendar of State Papers, Domestic, 1652–3, ed. M. A. E. Green (London, 1878).

Calendar of State Papers, Domestic, 1653–4, ed. M. A. E. Green (London, 1879).

Calendar of State Papers, Domestic, 1670, ed. M. A. E. Green (London, 1895).

Calendar of State Papers ... in the Archives and Collections of Venice, 1653–4, ed. A. B. Hinds (London, 1929).

Calvisius [Kalwitz], S., *Melopoeia sive melodiae condendae ratio, quam vulgo musicam poeticam vocant* (Erfurt, 1592).

— *Compendium musicae pro incipientibus* (Leipzig, 1594).

— *Musicae artis praecepta nova et facillima* (Leipzig, 1612).

Cameron, E., *The Reformation of the Heretics: The Waldensians of the Alps, 1480–1580* (Oxford, 1984).

Caramuel y Lobkowitz, J., *Steganographia necnon claviculae ... Ioannis Trithemii genuina, facilis, dilucidaque declaratio* (Cologne, 1635).

— *Mathesis audax, rationalem, naturalem, supernaturalem, divinamque sapientiam ... substruens exponensque* (Louvain 1644).

— see also Rheita.

Cardinael, S. H., *Practijck des landmetens* (Amsterdam, 1614).

Cardano, G., *Artis magnae, sive de regulis algebraicis liber* (Nuremberg, 1545).

Caruso, E., 'Honoré Fabri Gesuita e scienzato', *Quaderni di Acme* 8 (1987), pp. 85–126.

Cavalieri, B., *Directorium generale uranometricum* (Bologna, 1632).

— *Specchio ustorio overo trattato delle settioni coniche* (Bologna, 1632).

— *Geometria indivisibilibus continuorum nova quadam ratione promota* (Bologna, 1635).
— *Compendio delle regole dei triangoli colle loro dimostrazioni* (Bologna, 1638).
— *Centuria di varii problemi per dimostrare l'uso e la facilità dei logaritmi* (Bologna, 1639).
— *Nuova pratica astrologica di fare le direttioni secondo la via rationale* (Bologna, 1639).
— *Annotazioni all'opera* (Bologna, 1639).
— *Appendice della nuova pratica astrologica* (Bologna, 1640).
— *Exercitationes geometricae sex* (Bologna, 1647).

Cavendish, M., *The Life of the thrice Noble, High and Puissant Prince William Cavendishe, Duke, Marquess, and Earl of Newcastle* (London, 1915).

Cerutti, F. F. X., et al., *Geschiedenis van Breda*, revd edn., 3 vols. (Schiedam and Breda, 1977–90).

van Ceulen, L., *Fundamenta arithmetica et geometrica*, tr. W. Snel (Leiden, 1615).

Chamberlayne, E., *Angliae notitia; Or, the Present State of England* (London, 1669).

Chew, S., *The Crescent and the Rose: Islam and England during the Renaissance* (New York, 1937).

Chmaj, L., *Bracia polscy: ludzie, idee, wpływy* (Warsaw, 1957).

Christianson, J. R., *On Tycho's Island: Tycho Brahe and his Assistants, 1570–1601* (Cambridge, 2000).

Clavius, C., *Algebra* (Rome, 1608).

Clerke, A., 'John Pell', in *The Dictionary of National Biography*, 2nd edn., 22 vols. (London, 1909), xv, pp. 706–8.

Clifton, G., 'The Spectaclemakers' Company and the Origins of the Optical Instrument-making Trade in London', in R. G. W. Anderson, J. A. Bennett, and W. F. Ryan, eds., *Making Instruments Count: Essays on Historical Scientific Instruments presented to Graham L'Estrange Turner* (Aldershot, 1993), pp. 341–64.

— *Directory of British Scientific Instrument Makers, 1550–1851* (London, 1995).

Clucas, S., 'In Search of "The True Logick": Methodological Eclecticism among the "Baconian Reformers"', in M. Greengrass, M. Leslie, and T. Raylor, eds., *Samuel Hartlib and Universal Reformation: Studies in Intellectual Communication* (Cambridge, 1994), pp. 51–74.

— 'Walter Warner', in A. Pyle, ed., *The Dictionary of Seventeenth-Century British Philosophers*, 2 vols. (London, 2000), ii, pp. 858–62.

— 'Corpuscular Matter Theory in the Northumberland Circle', in C. Lüthy, J. E. Murdoch and W. R. Newman, eds., *Late Medieval and Early Modern Corpuscular Matter Theories* (Leiden, 2001), pp. 181–207.

Coccejus, J., *Opera ἀνέκδοτα theologica et philologica*, 2 vols. (Amsterdam, 1706).

Cokayne, G. E., *The Complete Baronetage*, 5 vols. (Exeter, 1900–9).

— ['G. E. C.'], *The Complete Peerage*, ed. V. Gibbs, G. H. White, and R. S. Lea, 12 vols. (London, 1912–59).

Colie, R., *Light and Enlightenment: A Study of the Cambridge Platonists and the Dutch Arminians* (Cambridge, 1957).

Collins, J., 'An Account concerning the Resolution of Equations in Numbers', *Philosophical Transactions*, 4 (1669), pp. 929–34.

— 'To Describe the Locus of a Cubick Aequation', *Philosophical Transactions*, 14 (1684), pp. 575–82.

— *Commercium epistolicum D. Johannis Collins, et aliorum de analysi promota: jussu Societatis Regiae in lucem editum* (London, 1712).

— see also Anon.

Colomesius, P., ed., *Gerardi Joan. Vossii et clarorum virorum ad eum epistolae*, 2 vols. (London, 1690).

Comenius, J. A., *Janua linguarum reserata* (Leszno, 1631).

— *Porta linguarum trilinguis reserata et aperta*, ed. J. Anchoran (London, 1631).

— *Januae linguarum reseratae vestibulum* (Leszno, 1633).

— *Conatuum comenianorum praeludia ex bibliotheca S.H.* (Oxford, 1637).

— *Pansophiae prodromus* (London, 1639).

— *Linguarum methodus novissima* (Leszno, 1648).

— *Lux in tenebris* (Amsterdam, 1657).

— *Via lucis* (Amsterdam, 1668).

— *Mutterschule*, ed. A. Richter (Leipzig, 1891).

— *School of Infancy*, ed. W. S. Monroe (n.p., n.d. [1895?]).

— *Veškeré spisy*, vols. i, iv, vi, ix, x, xv, xvii, xviii (Brno, 1911–38).

— *Dva spisy vševědné: Two Pansophical Works*, ed. G. H. Turnbull (Prague, 1951).

Commelin, C., *Beschryvinge van Amsterdam* (Amsterdam, 1693).

Cook, H. J., 'Time's Bodies: Crafting the Preparation and Preservation of Naturalia', in P. H. Smith and P. Findlen, eds., *Merchants and Marvels: Commerce, Science, and Art in Early Modern Europe* (New York, 2002), pp. 223–47.

Cooper, M. A. R., 'Robert Hooke, City Surveyor: An Assessment of his Work as Surveyor for the City of London in the Aftermath (1667–74) of the Great Fire' (City University, London, PhD dissertation, 1999).

Coote, E., *The English Schoole-Maister* (London, 1596).

Curtis, M. H., *Oxford and Cambridge in Transition, 1558–1642* (Oxford, 1959).

Dalgarno, G., *Character Universalis* (n.p., n.d. [1657]).

— *A New Discovery of the Universal Character* (n.p., n.d. [1657]).

— *Ars signorum, vulgo character universalis et lingua philosophica* (London, 1661).

Dansk biografisk leksikon, 3rd edn., 16 vols. (Copenhagen, 1979–84).

Daumas, M., *Scientific Instruments of the Seventeenth and Eighteenth Centuries and their Makers*, ed. and tr. M. Holbrook (London, 1972).

Davenant, Sir William, *A Discourse upon Gondibert ... with an Answer to it by Mr. Hobbs* (Paris, 1650).

— *Gondibert* (London, 1651).

Débarbat, S., and C. Wilson, 'The Galilean Satellites of Jupiter from Galileo to Cassini, Römer and Bradley', in R. Taton and C. Wilson, eds., *Planetary Astronomy from the Renaissance to the Rise of Astrophysics, Part A: Tycho Brahe to Newton* (Cambridge, 1989), pp. 144–57.

Debus, A. G., *Science and Education in the Seventeenth Century: The Webster-Ward Debate* (London, 1970).

Dechales, C.-F. M., *Cursus seu mundus mathematicus* (Lyon, 1674).

DeMott, B., 'Comenius and the Real Character in England', *Publications of the Modern Language Association of America*, 70 (1955), pp. 1068–81.

Descartes, R., *Discours de la méthode et les essais* (Leiden, 1637).

— *Meditationes* (Paris, 1641).

— *Principia philosophiae* (Amsterdam, 1644).

— *Specimina philosophiae* (Amsterdam, 1644).

— *Geometria*, ed. and tr. F. van Schooten (Leiden, 1649).

— *Les Passions de l'âme* (Amsterdam, 1649).

— *Musicae compendium* (Utrecht, 1650).

— *De homine* (Leiden, 1662).

— *Le Monde* (Paris, 1664).

— *Oeuvres*, ed. C. Adam and P. Tannery, rev. edn., 11 vols. (Paris, 1974).

Dibon, P., *La Philosophie néerlandaise au siècle d'or*, i (Paris, 1954).

— *Regards sur la Hollande du siècle d'or* (Naples, 1990).

— and F. Waquet, *Johannes Fredericus Gronovius: pèlerin de la République des Lettres* (Geneva, 1984).

Dickinson, H. W., *Sir Samuel Morland, Diplomat and Inventor, 1625–1695* (Cambridge, 1970).

Dickson, D. R., *The Tessera of Antilia: Utopian Brotherhoods & Secret Societies in the Early Seventeenth Century* (Leiden, 1998).

The Dictionary of National Biography, 2nd edn., 22 vols. (London, 1909).

The Dictionary of Scientific Biography, 18 vols. (New York, 1970–90).

Digby, Sir Kenelm, *Two Treatises: In the one of which, the Nature of Bodies, in the other, the Nature of Mans Soule, is looked into* (Paris, 1644).

Digges, L., and T. Digges, *An Arithmeticall Militare Treatise named Stratioticos* (London, 1579).

Dijksterhuis, E. J., 'John Pell in zijn strijd over de rectificatie van den cirkel: bijdrage tot het jubileum van de Universiteit van Amsterdam', *Euclides: tijdschrift voor de didactiek der exacte vakken*, 8 (1931–2), pp. 286–96.

Diophantus, *Arithmetica*, ed. and tr. W. Holzmann ['G. Xylander'] (Basel, 1575).

— *Arithmeticorum libri sex*, ed. and tr. C. G. Bachet de Méziriac (Paris, 1621).

— *Arithmeticorum libri sex*, ed. and tr. C. G. Bachet de Méziriac, with notes by P. Fermat, ed. S. Fermat (Toulouse, 1670).

Divers ouvrages de mathématique et de physique, par Messieurs de l'Académie Royale des Sciences (Paris, 1693).

Divini ['de Divinis'], E., *Brevis annotatio in Systema Saturnium Christiani Huygenii* (The Hague, 1660).

Dodson, J., *The Anti-logarithmic Canon, Being a Table of Numbers, Consisting of Eleven Places of Figures, Corresponding to all Logarithms under 100,000* (London, 1742).

Dögen, M., *Architectura militaris moderna* (Amsterdam, 1647).

Drüll, D., *Heidelberger Gelehrtenlexikon, 1652–1802* (Berlin, 1991).

Dury, J., *The Reformed Librarie-Keeper with a Supplement to the Reformed-School* (London, 1650).

— *The Reformed School*, 2nd edn. (London, 1651).

Edmond, M., *Rare Sir William Davenant* (Manchester, 1987).

van Eerde, K. S., *John Ogilby and the Taste of His Times* (London, 1976).

Elias, J. E., *De vroedschap van Amsterdam, 1578–1795*, 2 vols. (Haarlem, 1903–5).

Elichmann, J., *Tabula Cebetis graece, arabice, latine* (Leiden, 1640).

Elmer, P., *The Library of Dr. John Webster: The Making of a Seventeenth-Century Radical* (London, 1986).

Elsner, B., ed., *"Apollonius Saxonicus": die Restitution eines verlorenen Werkes des Apollonius von Perga durch Joachim Jungius, Woldeck Weland und Johannes Müller* (Göttingen, 1988).

Encyclopaedia mathematica ad agones panegyricos anni 1640 ... in Claromontano Parisiensi Societatis Jesu Collegis (Paris, 1640).

Ercker, L., *Beschreibung allerfürnemsten mineralischen Ertzt und Berckwercks Arten* (Prague, 1574).

'Espinasse, M., *Robert Hooke* (London, 1956).

Euclid, *Elementorum geometricorum libros tredecim*, ed. C. Richard (Antwerp, 1645).

— *The Thirteen Books of the Elements*, ed. and tr. Sir Thomas Heath (Cambridge, 1908; reprinted New York, 1956).

Evans, R. C., *Jonson and the Contexts of his Time* (London, 1994).

Evelyn, J., *Numismata: A Discourse of Medals, Ancient and Modern* (London, 1697).

— *Sculptura, with the Unpublished Second Part*, ed. C. F. Bell (Oxford, 1906).

Evenius, S., *Methodi linguarum artiumque compendiosioris scholasticae demonstrata veritas* (Wittenberg, 1621).

d'Ewes, Sir Simonds, *The Autobiography and Correspondence of Sir Simonds d'Ewes, Bart., during the Reigns of James I and Charles I*, ed. J. O. Halliwell, 2 vols. (London, 1845).

Fabian, B., *Der Gelehrte als Leser: über Bücher und Bibliotheken* (Hildesheim, 1998).

Fabri, H., *Theses de universa mathematica* (Lyon, '1646' [1645]).

— *Tractatus physicus de motu locali*, ed. P. Mousnier (Lyon, 1646).

della Faille, J. C., *Theoremata de centro gravitatis partium circuli et ellipsis* (Antwerp, 1632).

Fale, T., *Horologiographia: The Art of Dialling* (London, 1593).

Feingold, M., *The Mathematicians' Apprenticeship: Science, Universities and Society in England, 1560–1640* (Cambridge, 1980).

— 'Isaac Barrow: Divine, Scholar, Mathematician', in M. Feingold, ed., *Before Newton: The Life and Times of Isaac Barrow* (Cambridge, 1990), pp. 1–104.

— 'The Humanities', in N. Tyacke, ed., *Seventeenth-Century Oxford (The History of the University of Oxford*, iv) (Oxford, 1997), pp. 211–357.

— 'The Mathematical Sciences and New Philosophies', in N. Tyacke, ed., *Seventeenth-Century Oxford (The History of the University of Oxford*, iv) (Oxford, 1997), pp. 359–448.

— 'Science as a Calling? The Early Modern Dilemma', *Science in Context*, 15 (2002), pp. 79–119.

Fermat, P., *Varia opera*, ed. S. Fermat (Toulouse, 1679).

— see also Diophantus.

Firth, C. H., *The Last Years of the Protectorate, 1656-1658*, 2 vols. (London, 1909).

Flamsteed, J., *The Correspondence of John Flamsteed, The First Astronomer Royal*, ed. E. G. Forbes, L. Murdin, and F. Willmoth (Bristol, 1995–).

Fletcher, H., *The Perfect Politician: Or, a Full View of the Life and Actions (Military and Civil) of O. Cromwel*, 2nd edn. (London, 1680).

Fontana, F., *Novae coelestium terrestriumque rerum observationes* (Naples, 1646).

Forbes, E. G., 'The Origins of the Greenwich Observatory', *Vistas in Astronomy*, 20 (1976), pp. 39–50.

Forster, L. W., *Georg Rudolf Weckherlin: zur Kenntnis seines Lebens in England*, Basler Studien zur deutsche Sprache und Literatur, 2 (Basel, 1944).

— 'Aus der Korrespondenz G. R. Weckherlins', *Jahrbuch der deutschen Schillergesellschaft*, 4 (1960), pp. 182–97.

Foster, J., *Alumni oxonienses: The Members of the University of Oxford, 1500–1714*, 4 vols. (Oxford, 1891–2).

Foster, S., and J. Twysden, *Miscellanies or Mathematical Lucubrations* (London, 1659).

Fournier, G., *Hydrographie, contenant la théorie et la practique de toutes les parties de la navigation* (Paris, 1643).

Fowler, D., 'An Approximation Technique and its Use by Wallis and Taylor', *Archive for the History of Exact Sciences*, 41 (1990), pp. 189–233.

von Franckenberg, A., *Briefwechsel*, ed. J. Telle (Stuttgart, 1995).

Freitag, A., *Architectura militaris nova et aucta, oder newe vermehrte Fortification* (Leiden, 1631).

— ['Freitach'], *L'Architecture militaire* (Leiden, 1635).

Fück, J., *Die arabischen Studien in Europa bis in den Anfang des 20. Jahrhunderts* (Leipzig, 1955).

Galluzzi, P., 'Gassendi and *l'affaire Galilée* of the Laws of Motion', in J. Renn, ed., *Galileo in Context* (Cambridge, 2001), pp. 239–75.

— and M. Torrini, eds., *Le opere dei discepoli di Galileo Galilei: carteggio (1642–1648)* (Florence, 1975).

Gassendi, P., *Disquisitio metaphysica seu dubitationes et instantiae adversus Renati Cartesii metaphysicam et responsa* (Amsterdam, 1644).

— *De proportione qua gravia accidentia accelerantur epistolae tres, quibus ad totidem epistolas R. P. Cazrei ... respondetur* (Paris, 1646).

— *De vita et moribus Epicuri libri octo* (Lyon, 1647).

— *Institutio astronomica iuxta hypotheseis tam veterum quam Copernici et Tychonis Brahei* (Paris, 1647).

— *Animadversiones in decimum librum Diogenis Laertii, qui est de vita, moribus, placitisque Epicuri*, 3 parts in 2 vols. (Lyon, 1649).

— *Opera omnia*, 6 vols. (Lyon, 1658).

— see also Rheita.

van Gelder, M., *Amsterdamsche straatsnamen geschiedkundig verklaard* (Amsterdam, 1913).

Gellibrand, H., *A Discourse Mathematicall on the Variation of the Magneticall Needle* (London, 1635).

Gemma, R., *Arithmeticae practicae methodus facilis* (Antwerp, 1540).

Gentles, I., *The New Model Army in England, Ireland and Scotland, 1645–1653* (Oxford, 1992).

Getaldus [Getaldić, Ghetaldi], M., *De resolutione et compositione mathematica libri quinque*, ed. A. Brogiotti (Rome, 1631).

Giggeius, A., *Thesaurus linguae arabicae*, 4 vols. (Milan, 1632).

Gilbert, W., *De magnete, magneticisque corporibus, et de magno magnete tellure, physiologia nova, plurimis & argumentis, & experimentis demonstrata* (London, 1600).

— *De mundo nostro sublunari philosophia nova* (Amsterdam, 1651).

Girard, A., *Invention nouvelle en algèbre* (Amsterdam, 1629).

Glorioso [or Gloriosi], G. C., *Exercitationum mathematicarum decas prima in qua continentur varia et theoremata et problemata*, 2 vols. (Naples, 1627–35).

Godet, M., et al., *Dictionnaire historique et biographique de la Suisse*, 8 vols. (Neuchatel, 1921–34).

Godwin, F., *The Man in the Moone; Or, a Discourse of a Voyage thither by D. Gonsales* (London, 1638).

Golius, J., *Lexicon arabico-latinum* (Leiden, 1653).

Goodchild, P., '"No phantasticall Utopia, but a reall place": John Evelyn, John Beale and Backbury Hill, Herefordshire', *Garden History* 19 (1991), pp. 105–27.

Gosselin, G., *De arte magna, seu de occulta parte numerorum, quae algebra & almucabala vulgo dicitur* (Paris, 1577).

Gouge, W., *A Learned and Very Useful Commentary on the Whole Epistle to the Hebrews* (London, 1655).

Gouk, P., *Music, Science and Natural Magic in Seventeenth-Century England* (New Haven, Conn., 1999).

Graunt, J., *Natural and Political Observations upon the Bills of Mortality* (London, 1666).

Greengrass, M., 'Samuel Hartlib and Scribal Communication', *Acta comeniana*, 12 (1997), pp. 47–62.

— 'George Dalgarno', in A. Pyle, ed., *The Dictionary of Seventeenth-Century British Philosophers*, 2 vols. (London, 2000), i, pp. 233-4.

Grell, O. P., *Dutch Calvinists in Early Stuart London: The Dutch Church in Austin Friars, 1603-1642* (Leiden, 1989).

Grimble, I., *The Harington Family* (London, 1957).

Grundy, W. G. C., ed., *Manes verulamiani* (London, 1950).

Guldin, P., *De centro gravitatis*, 4 vols. (Vienna, 1635-41).

Gunter, E., *A Canon of Triangles: Or, A Table of Artificial Sines, Tangents, and Secants* (London, 1620).

— *The Description and Use of the Sector Crosse-Staffe and other Instruments. With a Canon of Artificial Signes and Tangents* (London, 1624; 2nd edn., London, 1636).

van Gutschoven, G., *Arithmeticae practicae regulae brevissimae* (Antwerp, 1654).

Hakewill, G., *An Apologie of the Power and Providence of God in the Government of the World* (Oxford, 1627; 2nd edn. Oxford, 1630).

Hall, M. B., *Henry Oldenburg: Shaping the Royal Society* (Oxford, 2002).

Halliwell, J., ed., *A Collection of Letters Illustrative of the Progress of Science in England from the Reign of Queen Elizabeth to that of Charles the Second* (London, 1841).

Hardy, C., *see* Anon.

Harriot, T., *Artis analyticae praxis, ad aequationes algebraicas nova, expedita, & generali methodo, resolvendas: tractatus*, ed. W. Warner (London, 1631).

Hartlib, S., *Samuel Hartlib his Legacie: Or, an Enlargement of the Discourse of Husbandry used in Brabant and Flaunders, wherein are Bequeathed to the Common-Wealth of England more Outlandish and Domestick Experiments and Secrets in Reference to Universall Husbandry* (London, 1651; 2nd edn. 1652).

— *The Reformed Common-wealth of Bees, Presented in Several Letters and Observations to Samuel Hartlib* (London, 1655).

Harvey, W., *Exercitationes de generatione animalium* (London, 1651).

Hayne, T., *Linguarum cognatio, seu de linguis in genere, & de variarum linguarum harmonia dissertatio* (London, 1639).

Hayward, J. C., 'New Directions in Studies of the Falkland Circle', *The Seventeenth Century*, 2 (1987), pp. 19-48.

Heesakkers, C. L., 'Foundation and Early Development of the Athenaeum Illustre at Amsterdam', *Lias: Sources and Documents relating to the Early Modern History of Ideas*, 9 (1982), pp. 3-18.

Helvicus, C., *Libri didactici, grammaticae universalis ... in usum scholarum editi* (Giessen, 1619).

Henisch ['Henischius'], G., *Arithmetica perfecta & demonstrata, doctrinam de numero triplici, vulgari, cossico & astronomico nova methodo per propositiones explicatam continens, libris septem* (Augsburg, 1609).

Hennessy, G. L., *Chichester Diocese Clergy Lists* (London, 1900).

Herbert, E., *De veritate* (Paris, 1624).

Hérigone, P., *Cursus mathematicus*, 6 vols. (Paris, 1632–42).

Heron of Alexandria, *Opera quae supersunt omnia*, ed. and tr. L. Nix and W. Schmidt, 5 vols. (Leipzig, 1899–1914).

Hervey, H., 'Hobbes and Descartes in the Light of some Unpublished Letters of the Correspondence between Sir Charles Cavendish and Dr John Pell', *Osiris*, 10 (1952), pp. 67–90.

Hess, H.-J., 'Bücher aus dem Besitz von Christiaan Huygens (1629–1695) in der Niedersächsischen Landesbibliothek Hannover', *Studia leibnitiana*, 12 (1980), pp. 1–51.

Hevelius, J., *Selenographia: sive, lunae descriptio* (Danzig, 1647).

Hine, W., 'Athanasius Kircher and Magnetism', in J. Fletcher, ed., *Athanasius Kircher und seine Beziehungen zum gelehrten Europa seiner Zeit* (Wiesbaden, 1988), pp. 79–97.

Hitchens, W. J., A. Matuszewski, and J. Young, eds., *The Letters of Jan Jonston to Samuel Hartlib* (Warsaw, 2000).

Hobbes, T., *De cive*, 2nd edn. (Amsterdam, 1647).

— *Elemens philosophiques du citoyen*, tr. S. Sorbière (Amsterdam, 1649).

— *De corpore politico* (London, 1650).

— *Of Humane Nature: Or, the Fundamental Elements of Policie* (Oxford, 1650).

— *Leviathan* (London, 1651).

— *Elementorum philosophiae sectio prima de corpore* (London, 1655).

— *ΣΤΙΓΜΑΙ... Or, Markes of the Absurd Geometry, Rural Language, Scottish Church-Politicks and Barbarismes of John Wallis* (London, 1657).

— *Elementorum philosophiae sectio secunda de homine* (London, 1658).

— *Dialogus physicus* (London, 1661).

— *De principiis et ratiocinatione geometrarum* (London, 1666).

— *Critique du* De mundo *de Thomas White*, ed. J. Jacquot and H. W. Jones (Paris, 1973).

— *The Correspondence*, ed. N. Malcolm, 2 vols. (Oxford, 1994).

Hobbs, M., 'Drayton's "Most Dearely-Loved Friend Henery Reynolds Esq"', *Review of English Studies*, n.s., 24 (1973), pp. 414–28.

Hofmann, J. E., *Frans van Schooten der jüngere* (Wiesbaden, 1962).

Holzach, F., 'Über die politischen Beziehungen der Schweiz zu Oliver Cromwell', *Basler Zeitschrift für Geschichte und Altertumskunde*, 4 (1905), pp. 182–245, and 5 (1906), pp. 1–58.

Hooke, R., *Philosophical Collections*, no. 5, Feb. 1682 (London, 1682).

— *The Diary of Robert Hooke, 1672–1680*, ed. H. W. Robinson and W. Adams (London, 1935).

van der Horst, K., 'A "Vita Casparis Barlaei" written by himself', *Lias: Sources and Documents relating to the Early Modern History of Ideas*, 9 (1982), pp. 57–83.

Hotson, H., *Johann Heinrich Alsted, 1588–1638: Between Renaissance, Reformation, and Universal Reform* (Oxford, 2000).

Hottinger, J. H., *Schola tigurinorum Carolina* (Zurich, 1664).

Hudde, J., 'De maximis et minimis', in R. Descartes, *Geometria*, ed. and tr. F. van Schooten, 2nd edn., 2 vols. (Amsterdam, 1659–61), i, pp. 507–16.

Huguet, E., *Dictionnaire de la langue française du seizième siècle*, 7 vols. (Paris, 1925–67).

Hunter, M., *John Aubrey and the Realm of Learning* (London, 1975).

— *The Royal Society and its Fellows, 1660–1700: The Morphology of an Early Scientific Institution* (Chalfont St Giles, 1982).

— *Establishing the New Science: The Experience of the Early Royal Society* (Woodbridge, 1989).

Hutton, S., 'Sir Charles Cavendish', in A. Pyle, ed., *The Dictionary of Seventeenth-Century British Philosophers*, 2 vols. (London, 2000), i, pp. 165–6.

Huygens, Christiaan, *Oeuvres complètes*, 22 vols. (The Hague, 1888–1950).

Huygens, Constantijn, *De briefwisseling*, ed. J. A. Worp, 6 vols. (The Hague, 1911–17).

Huygens, L., *The English Journal, 1651–1652*, ed. A. G. H. Bachrach and R. G. Collmer (Leiden, 1982).

Hyde, E., *The Life of Edward Earl of Clarendon*, 3 vols. (Oxford, 1827).

Inwood, S., *The Man who Knew Too Much: The Strange and Inventive Life of Robert Hooke, 1635–1703* (London, 2002).

Israel, A., 'Das Verhältnis der Didactica magna des Comenius zu der Didaktik Ratkes', *Monatshefte der Comenius-Gesellschaft*, 1 (1892–3), pp. 173–95, 242–74.

Jackson, W. A., ed., *Records of the Court of the Stationers' Company, 1602 to 1640* (London, 1957).

Jacquot, J., 'Sir Charles Cavendish and his Learned Friends', *Annals of Science*, 3 (1952), pp. 13–27, 175–91.

— 'Harriot, Hill, Warner and the New Philosophy', in J. W. Shirley, ed., *Thomas Harriot, Renaissance Scientist* (Oxford, 1974), pp. 107–28.

Jardine, L., *The Curious Life of Robert Hooke: The Man who Measured London* (London, 2003).

Jesseph, D., 'Of Analytics and Indivisibles: Hobbes on the Methods of Modern Mathematics', *Revue d'histoire des sciences*, 46 (1993), pp. 167–74.

Johnson, J., and S. Gibson, *Print and Privilege at Oxford to the Year 1700* (Oxford, 1946).

Jones, H. W., 'Der Kreis von Welbeck', in J.-P. Schobinger, ed., *Grundriss der Geschichte der Philosophie, begründet von Friedrich Ueberweg: Die Philosophie des 17. Jahrhunderts*, Bd iii, *Die Philosophie des 17. Jahrhunderts: England*, 2 vols. (Basel, 1988), i, pp. 186–91.

Jones, K., *A Glorious Fame: The Life of Margaret Cavendish, Duchess of Newcastle, 1623–1673* (London, 1988).

Jonson, B., *Works*, ed. C. H. Herford, P. Simpson, and E. Simpson, 11 vols. (Oxford, 1925–52).

Jungius, J., *Geometria empirica* (Rostock, 1627; 2nd edn. Hamburg, 1642).

— *Logica hamburgensis* (Hamburg, 1635; 2nd edn. Hamburg, 1638).

— *Phoranomica, id est de motu locali* (Hamburg, 1650).

— *Doxoscopiae physicae minores, sive isagoge physica doxoscopa*, ed. M. Fogel (Hamburg, 1662).

— *Historia vermium e mss. schedis ... a Johanne Vagetio communicata* (Hamburg, 1691).

— *Logicae hamburgensis additamenta*, ed. W. Risse (Göttingen, 1977).

— see also Elsner.

Jusserand, J. J., *A French Ambassador at the Court of Charles the Second: Le Comte de Cominges, from the Unpublished Correspondence* (London, 1892).

Juynboll, W. M. C., *Zeventiende-eeuwsche beoefenaars van het arabisch in Nederland* (Utrecht, 1931).

Kalwitz: *see* Calvisius.

Kangro, H., *Joachim Jungius' Experimente und Gedanken zur Begrundung der Chemie als Wissenschaft: ein Beitrag zur Geistesgeschichte des 17. Jahrhunderts* (Wiesbaden, 1968).

Kargon, R., *Atomism in England from Hariot to Newton* (Oxford, 1966).

Kayser, W., *Hamburger Bücher, 1491–1850, aus der Hamburgensien-Sammlung der Staats- und Universitätsbibliothek Hamburg*, Mitteilungen der Staats- und Universitätsbibliothek Hamburg, vii (Hamburg, 1973).

Kempenaars, C. M. P. M., 'Some New Data on Gerard Kinckhuysen (*c.*1625–1666)', *Nieuw archief voor wiskunde*, 4 (1990), pp. 243–50.

van Kempen-Stijgers, T., and P. Rietbergen, 'Constantijn Huygens en Engeland', in H. Bots, ed., *Constantijn Huygens: zijn plaats in geleerd Europa* (Amsterdam, 1973), pp. 77–141.

Kepler, J., *Gesammelte Werke*, ed. M. Caspar and W. von Dyck (Munich, 1938–).

Kernkamp, G. W., ed., *Acta et decreta senatus: vroedschapsresolutiën en andere bescheiden betreffende de Utrechtsche Academie*, 3 vols. (Utrecht, 1936–40).

Kersey, J., *The Elements of that Mathematical Art commonly called Algebra* (London, 1673).

Kiel, I., 'Technology Transfer and Scientific Specialization: Johann Wiesel, Optician of Augsburg, and the Hartlib Circle', in M. Greengrass, M. Leslie, and T. Raylor, eds., *Samuel Hartlib and Universal Reformation: Studies in Intellectual Communication* (Cambridge, 1994), pp. 268–78.

— *Augustanus opticus: Johann Wiesel (1583–1662) und 200 Jahre optisches Handwerk in Augsburg* (Berlin, 2000).

Kieper ['Kyperus'], A., *Institutiones physicae*, 2 vols. (Leiden, 1645–6).

Kinckhuysen, G., *De grondt der meet-konst* (Haarlem, 1660).

— *Algebra ofte stel-konst* (Haarlem, 1661).

Kircher, A., *Ars magnesia* (Würzburg, 1631).

— *Magnes, sive de arte magnetica* (Rome, 1641; 2nd edn. Cologne, 1643).

— *Ars magna lucis et umbrae* (Rome, 1646).

— *Musurgia universalis, sive ars magna consoni et dissoni in X. libros digesta*, 2 vols. (Rome, 1650).

Kitchin, G., *Sir Roger L'Estrange* (London, 1913).

Kleerkooper, M. M., and W. P. van Stockum, *De boekhandel te Amsterdam voornamelijk in de 17e eeuw: biographische en geschiedkundige aantekeningen*, 2 vols. (The Hague, 1912–14).

Knowlson, J., *Universal Language Schemes in England and France, 1600–1800* (Toronto, 1975).

Kordes, U., *Wolfgang Ratke (Ratichius, 1571–1635): Gesellschaft, Religiosität und Gelehrsamkeit im frühen 17. Jahrhundert* (Heidelberg, 1999).

Korr, C. P., *Cromwell and the New Model Foreign Policy: England's Policy towards France, 1649–1658* (Berkeley, Calif., 1975).

Kronk, G. W., *Cometography: A Catalogue of Comets* (Cambridge, 1999–).

Kvačala, J., ed., *Die pädagogische Reform des Comenius in Deutschland bis zum Ausgange des XVII Jahrhunderts*, 2 vols., Monumenta Germaniae paedagogica, xxvi, xxxii (Berlin, 1903–4).

Labrousse, E., *Conscience et conviction: études sur le XVIIe siècle* (Paris, 1996).

Lacey, T. A., *Herbert Thorndike, 1598–1672* (London, 1929).

Langedijk, D., ' "De Illustre School ende Collegium Auriacum" te Breda', *Taxandria: tijdschrift voor noordbrabantsche geschiedenis en volkskunde*, ser. 5, year 1 (1934), pp. 257–70, 290–300, 328–36; ser. 5, year 2 (1935), pp. 28–39, 72–98, 128–35.

Larcom, T. A., ed., *History of the Cromwellian Survey of Ireland A.D. 1655–6, commonly called 'the Down Survey'* (Dublin, 1851).

Le Neve, J., *Fasti ecclesiae anglicanae 1541–1857*, ii, 'Chichester Diocese', compiled by J. M. Horn (London, 1971).

Lennard, R., 'English Agriculture under Charles II: The Evidence of the Royal Society's "Enquiries"', *Economic History Review*, 1st ser., 4 (1932–4), pp. 23–45.

van Lennep, D. J., *Illustris Amstelodamensium Athenaei memorabilia* (Amsterdam, 1832).

Lenoble, R., *Mersenne ou la naissance du mécanisme* (Paris, 1943).

Leslie, M., 'The Spiritual Husbandry of John Beale', in M. Leslie and T. Raylor, eds., *Culture and Cultivation in Early Modern England: Writing and the Land* (Leicester, 1992), pp. 151–72.

L'Estrange, R., *L'Estrange his Vindication to Kent, and the Justification of Kent to the World* (n.p., 1649).

— *To the Right Honorable, Edward Earl of Clarendon ... The Humble Apology of Roger L'Estrange* (London, 1661).

Lewis, R., 'John Wilkins's *Essay* (1668) and the Context of Seventeenth-Century Artificial Languages in England' (Oxford University D.Phil. dissertation, 2003).

Lindeboom, J., 'Johannes Duraeus en zijne werkzaamheid in dienst van Cromwell's politiek', *Nederlandsch archief voor kerkgeschiedenis*, n.s., 16 (1921), pp. 241–68.

Linemannus, A., *Disputatio ordinaria continens controversias physico-mathematicas quam publicae ventilationis subjicit M. Albertus Linemannus, respondente Alberto Kieper Regiom: Prusso* (n.p. [Königsberg], 1636).

Littledale, W. A., ed., *The Registers of St. Vedast, Foster Lane, and of St. Michael le Quern, London*, 2 vols., Publications of the Harleian Society, Registers, vols. 29–30 (London, 1902–3).

Loncq, G. J., *Historische schets der Utrechtsche Hoogeschool tot hare verheffing in 1815* (Utrecht, 1886).

Long, P. O., 'The Openness of Knowledge: An Ideal and its Context in 16th-Century Writings on Mining and Metallurgy', *Technology and Culture*, 32 (1991), pp. 318–55.

Longomontanus [Lomberg, Langberg], C. S., *Cyclometria ex lunulis reciproce demonstrata* (Copenhagen, 1612).

— *Astronomia danica* (Amsterdam, 1622).

— *Cyclometria vere ac absolute in ipsa natura circuli* (Hamburg, 1627).

— *Fundamentum cyclometriae rationalis* (Copenhagen, 1628).

— *Geometrica quaesita tredecim* (Copenhagen, 1631).

- *Inventio quadraturae circuli* (Copenhagen, 1634).
- *Coronis problematica* (Copenhagen, 1637).
- *Problemata duo geometrica* (Copenhagen, 1638).
- *Literulae supplices ad S. R. Daniae* (Copenhagen, 1639).
- *Inventio theorematis nobilissimi* (Copenhagen, 1643).
- *Problema quod, tam aequationibus in numeris, quam comparatione ad alia, diversimodo quidem inventa ... absolutam circuli mensura praestat* (Copenhagen, 1643).
- *Admiranda operatio trium numerorum* (Copenhagen, 1644).
- *Rotundi in plano, seu circuli, absoluta mensura, duobus libellis comprehensa* (Amsterdam, 1644).
- *Ἐλέγξεως Joannis Pellii contra Christianum S. Longomontanum de mensura circuli Ἀνασκευή. In Pellium & Guldinum verae Cyclometriae adversantes* (Copenhagen, 1644).
- *Ἐνέργεια proportionis sesquitertiae pro certis figuris circulo adscriptis, & per numeros determinatis, data circuli diametro* (Copenhagen, 1644).
- *Controversia inter Christianum Longomontanum & Johannem Pellium, de vera circuli mensura, ubi defectus canonis trigonometrici, sub initium ejusdem, ostenditur* (Copenhagen, 1645).
- *Caput tertium libri primi de absolute mensura rotundi plani ... una cum ... appendice de defectu canonis trigonometrici sub hujus initium, ubi circulo rite mensurando, duntaxat inserviret* (Copenhagen, 1646).

van Looy, H., 'Chronologie et analyse des manuscrits mathématiques de Grégoire de Saint-Vincent (1584–1667)', *Archivum historicum societatis iesu* 49 (1980), pp. 279–303.

Lower, M. A., *The Worthies of Sussex* (Lewes, 1865).

de Luneschlos, J., *Thesaurus mathematum reseratus per algebram novam tam speciebus quam numeris declaratam et demonstratam* (Padua, 1646).

van Maanen, J. A., 'The Refutation of Longomontanus' Quadrature by John Pell', *Annals of Science*, 43 (1986), pp. 315–52.

- *Facets of Seventeenth Century Mathematics in the Netherlands* (Utrecht, 1987).

McClure, N. E., ed., *The Letters of John Chamberlain*, 2 vols. (Memoirs of the American Philosophical Society, xii, parts 1, 2) (Philadelphia, 1939).

McCusker, J. J., *Money and Exchange in Europe and America, 1600–1775* (Chapel Hill, NC, 1978).

Macfarlane, A., *The Family Life of Ralph Josselin, a Seventeenth-Century Clergyman: An Essay in Historical Anthropology* (New York, 1977).

Mahl, M. R., and H. Koon, eds., *The Female Spectator: English Women Writers before 1800* (Bloomington, Indiana, 1977).

Maierù, L., *La teoria e l'uso delle coniche nel Cinquecento: intersezione fra la conoscenza dei testi di Apollonio e dei 'Veteres' e il senso delle loro traduzioni* (Caltanisetta, 1996).

Makin, B., *see* B. Reynolds.

Malcolm, N., 'The Publications of John Pell, F.R.S. (1611–1685): Some New Light, and some Old Confusions', *Notes and Records of the Royal Society of London*, 54 (2000), pp. 275–92.

— 'Six Unknown Letters from Vegelin to Mersenne', *The Seventeenth Century*, 16 (2001), pp. 95–122.

— *Aspects of Hobbes* (Oxford, 2002).

— 'Hobbes, the Latin Optical Manuscript, and the Parisian Scribe', *English Manuscript Studies*, ed. P. Beal and J. Griffith, 13 (2003), forthcoming.

de Marandé, L., *Le Théologien français*, 3 vols. (Paris, 1641).

Matthews, A. G., *Calamy Revised* (Oxford, 1934).

— *Walker Revised* (Oxford, 1948).

Maurolico, F., *Emendatio et restitutio Conicorum Apollonii Pergaei* (Messina, 1654).

Mazerolle, F., 'Nicolas Briot: tailleur général des monnaies (1606–1625)', *Revue belge de numismatique*, 60 (1904), pp. 191–203, 295–314.

Meinel, C., ed., *Der handschriftliche Nachlass von Joachim Jungius in der Staats- und Universitätsbibliothek Hamburg* (Stuttgart, 1984).

— *Die Bibliothek des Joachim Jungius: ein Beitrag zur Historia litteraria der frühen Neuzeit* (Göttingen, 1992).

Meinsma, K. O., *Spinoza en zijn kring: over Hollandse vrijgeesten* (The Hague, 1896).

Menk, G., *Die Hohe Schule Herborn in ihrer Frühzeit (1584–1660)* (Wiesbaden, 1981).

Mercator, N., *Logarithmotechnia* (London, 1668).

Mersenne, M., *Harmonicorum libri* (Paris, 1636).

— *Harmonie universelle* (Paris, 1636).

— *Cogitata physico-mathematica* (Paris, 1644).

— *Universae geometriae mixtaeque mathematicae synopsis* (Paris, 1644).

— *Novarum observationum physico-mathematicarum ... tomus III* (Paris, 1647).

— *La Correspondance*, ed. C. de Waard *et al.*, 17 vols. (Paris, 1933–88).

— *see also* Niceron.

Metius, A., *Arithmetica et geometriae practica* (Franeker, 1611).

— *Geometriae practicae pars quarta, continens munitionum delineandarum muniendarumque genuinam & propriam institutionem* (Franeker, 1625).

— *Arithmeticae libri duo, et geometriae libri VI* (Leiden, 1626).

Michel, G., *Die Welt als Schule: Ratke, Comenius und die didaktische Bewegung* (Berlin, 1978).

Miller, P., 'The "Antiquarianization" of Biblical Scholarship and the London Polyglot Bible (1653–57)', *Journal of the History of Ideas*, 62 (2001), pp. 463–82.

Milton, J., *Complete Prose Works*, ed. D. W. Wolfe *et al.*, 8 vols. (New Haven, Conn., 1953–86).

Molhuysen, P. C., ed., *Bronnen tot de geschiedenis der Leidsche Universiteit*, 7 vols. (The Hague, 1913–24).

de Molina Cano, A., *Descubrimientos geometricos* (Antwerp, 1568).

— tr. N. Jansonius [Janszoon], *Nova reperta geometrica* (Arnhem, 1620).

Molnar, A., A. Armand-Hugon, and V. Vinay, *Storia dei Valdesi*, 3 vols. (Turin, 1974–80).

Morland, S., *The History of the Evangelical Churches of the Valleys of Piemont* (London, 1658).

— *Elevation des eaux par toute sorte de machines reduite à la mesure, au poids, à la balance, par le moyen d'un nouveau piston, & corps de pompe, & d'un nouveau mouvement cyclo-elliptique* (Paris, 1685).

— *Hydrostaticks: Or, Instructions Concerning Water-Works. Collected out of the Papers of Sir Samuel Morland*, ed. J. Morland (London, 1697).

Mouton, G., *Observationes diametrorum solis et lunae apparentium* (Lyon, 1670).

Müller, J., 'Zur Bücherkunde des Comenius: chronologisches Verzeichnis der gedruckten und ungedruckten Werke des Johann Amos Comenius', *Monatshefte der Comenius-Gesellschaft*, 1 (1892–3), pp. 19–53.

Mullick, I., *The Poetry of Thomas Randolph* (Bombay, 1974).

Mundy [Munday], A., *Chruso-thriambos: The Triumphs of Gold*, ed. J. H. P. Pafford (London, 1962).

Munk, W., *et al.*, eds., *The Roll of the Royal College of Physicians of London* (London, 1861–).

Murdin, L., *Under Newton's Shadow: Astronomical Practices in the Seventeenth Century* (Bristol, 1985).

Murphy, D., *Comenius: A Critical Reassessment of his Life and Work* (Dublin, 1995).

Mydorge, C., *Prodromi catoptricorum et dioptricorum, sive conicorum operis ... libri primus et secundus* (Paris, 1631).

— *Prodromi catoptricorum et dioptricorum: sive conicorum operis ... libri quatuor priores* (Paris, 1639).

Napier, J., *Mirifici logarithmorum canonis descriptio* (Edinburgh, 1614).

Napoletani, P. D., 'Galilei e due matematici napoletani: Luca Valerio e Giovanni Camillo Gloriosi', in F. Lomonaco and M. Torrini, eds., *Atti del convegno 'Galilei e Napoli', 12–14 aprile 1984* (Naples, 1987), pp. 159–90.

Nedham, M., *The Case of the Commonwealth of England, Stated* (London, 1650).

Nellen, H. J. M., *Ismaël Boulliau (1605–1694): astronome, épistolier, nouvelliste et intermédiaire scientifique* (Amsterdam, 1994).

Newman, P. R., *The Old Service: Royalist Regimental Colonels and the Civil War, 1642–46* (Manchester, 1993).

Newton, I., *Correspondence*, ed. H. W. Turnbull et al., 7 vols. (Cambridge, 1959–77).

— *The Mathematical Papers of Isaac Newton*, ed. D. T. Whiteside, 8 vols. (Cambridge, 1967–81).

Niceron, J. F., *La Perspective curieuse* (Paris, 1638).

— and M. Mersenne, *La Perspective curieuse du R. P. Niceron ... avec l'Optique et la catoptrique du R. P. Mersenne*, ed. G. P. de Roberval (Paris, 1652).

Nieuw nederlandsch biografisch woordenboek, 10 vols. (Leiden, 1911–37).

Noble, M., *The Lives of the English Regicides*, 2 vols. (London, 1798).

O'Brien, J. J., 'Commonwealth Schemes for the Advancement of Learning', *British Journal of Educational Studies*, 16 (1968), pp. 30–42.

Odložilík, O., 'Z pansofických studií J. A. Komenského', *Časopis Matice Moravské*, 52 (1928), pp. 125–98.

Oldenburg, H., *The Correspondence*, ed. A. R. Hall and M. Boas Hall, 13 vols. (Madison, Wis., and London, 1965–86).

Olearius, A., *Vermehrte neue Beschreibung der Muscowitischen und Persischen Reyse* (Schleswig, 1656).

Ollard, R., *Cromwell's Earl: A Life of Edward Mountagu, 1st Earl of Sandwich* (London, 1994).

Ormerod, G., *The History of the County Palatine and City of Chester*, 2nd edn., 3 vols. (London, 1882).

d'Orville, G. [or W.], *De constantia in adversis* (Amsterdam, 1666).

van Otegem, M., 'Towards a Sound Text of the *Compendium musicae*, 1618–1683, by René Descartes (1596–1650)', *Lias: Sources and Documents relating to the Early Modern History of Ideas*, 26 (1999), pp. 187–203.

Otto, F., 'Zacharias Rosenbach', in *Allgemeine deutsche Biographie*, 56 vols. (Leipzig, 1875–1912), xxix, pp. 199–200.

Oughtred, W., *Arithmeticae in numeris et speciebus institutio: quae tum logisticae, tum analyticae, atque adeo totius mathematicae, quasi clavis est* (London, 1631; 2nd edn. London, 1648).

— *To the English Gentrie ... The Just Apologie of Wil: Oughtred, against the Slanderous Insimulations of Richard Delamain, in a Pamphlet called Grammeologia* (n.p., n.d. [1631]).

— *The Circles of Proportion and the Horizontall Instrument* (London, 1632).

— *An Addition unto the Use of the Instrument Called the Circles of Proportion, for the Working of Nauticall Questions* (London, 1633).

— *The Key of the Mathematicks new Forged and Filed* (London, 1647).

— *Trigonometria,* with *Canones sinuum, tangentium, secantium: et logarithmorum pro sinubus et tangentibus* (London, 1657).

Palissy, B., *Discours admirables, de la nature des eaux ... des pierres, des terres, du feu et des emaux* (Paris, 1580).

Palmerino, C. R., 'Two Jesuit Responses to Galileo's Science of Motion: Honoré Fabri and Pierre Le Cazre', in M. Feingold, ed., *The New Science and Jesuit Science: Seventeenth-Century Perspectives* (Dordrecht, 2003), pp. 187–227.

Parry, M., ed., *Chambers Biographical Dictionary of Women* (London, 1996).

Pascal, B., *Essai sur les coniques* (Paris, 1640).

Pastine, D., *Juan Caramuel: probabilismo ed enciclopedia* (Florence, 1975).

Patočka, J., 'Komenského názory a pansofické literární plány od spisů útěšných ke "Všeobecné poradě"', in D. Gerhardt *et al.*, eds., *Orbis scriptus: Dmitrij Tschižewskij zum 70. Geburtstag* (Munich, 1966), pp. 594–620.

Pearce, E. H., *Annals of Christ's Hospital*, 2nd edn. (London, 1908).

Pell, J., *The English Schoole* (London, 1635).

— *[An Idea of Mathematics]* (n.p. [London], n.d. [1638]).

— *[Idea mathesews]* (n.p. [London], n.d. [1638]).

— *Controversiae de vera circuli mensura ... pars prima* (Amsterdam, 1647).

— 'Oratio Inauguralis', in *Inauguratio illustris scholae ac illustris collegii auriaci, à celsissimo potentissimoque Arausionensium principe, Frederico Henrico in urbe Breda erectorum, cum orationibus solemnibus ipsâ inaugurationis die & seqq. aliquot habitis* (Breda, 1647), pp. 168–83.

— 'An Idea of Mathematics', in J. Dury, *The Reformed Librarie-Keeper with a Supplement to the Reformed-School* (London, 1650), pp. 33–46.

— Preface to Anon., *Gründtliche Beschreibung der neuen Regiments-Verfassung in dem gemeinen Wesen Engelland, Schott- und Irrland samt den zugehörigen Eyländern und andern Landschafften, unter dem Herren Protector und dem Parlament* (Schaffhausen, 1657).

— *Easter not Mis-timed: A Letter written out of the Country to a Friend in London, concerning Easter-Day* (London, 1664).

— *Tabula numerorum quadratorum decies millium ... A Table of Ten Thousand Square Numbers* (London, 1672).

— 'Idea mathesews' in R. Hooke, *Philosophical Collections*, no. 5, Feb. 1682 (London, 1682), pp. 127–34.

— see also Rahn.

Pell, R., 'Thomas Pell: First Lord of the Manor of Pelham, Westchester Co., New York', *Pelliana*, n.s., 1, no. 1 (1962).

— 'The Story of Dr John Pell', *Pelliana*, n.s., 1, no. 2 (Oct. 1963), pp. 1–48.

— 'Sir John Pell, Second Lord of the Manor of Pelham', *Pelliana*, n.s., 1, no. 2 (Oct. 1963), pp. 49–67.

Pepys, S., *Private Correspondence and Miscellaneous Papers, 1679–1703*, ed. J. R. Tanner, 2 vols. (London, 1926).

— *Diary*, ed. R. Latham and W. Matthews, 11 vols. (London, 1970–83).

Petau ['Petavius'], D., ed., *Uranologion* (Paris, 1630).

— *Dogmata theologica*, 4 vols. (Paris, 1644–50)

Peters, H., *Mr Peters Report from the Army, to the Parliament, made Saturday the 26 of July 1645* (London, 1645).

Petri [Pieterszoon], N., *Practique om te leeren reckenen* (Amsterdam, 1583).

Pettus, Sir John, *Fleta Minor: The Laws of Art and Nature, in Knowing, Judging, Assaying, Fining, Refining and Inlarging the Bodies of Confin'd Metals* (London, 1683).

Petty, W., *The Advice of W. P. to Mr. Samuel Hartlib for the Advancement of some Particular Parts of Learning* (London, 1648).

Pfister, R., *Kirchengeschichte der Schweiz*, 3 vols. (Zurich, 1964–84).

Pingré, A.-G., *Annales célestes du dix-septième siècle*, ed. G. Bigourdan (Paris, 1901).

Pitiscus, B., *Trigonometria*, 3rd edn. (Frankfurt, 1612).

Pitt, M., *The Cry of the Oppressed, Being a True and Tragical Account of the Unparallel'd Sufferings of Multitude of Poor Imprisoned Debtors, in Most of the Gaols of England ... Together with the Case of the Publisher* (London, 1691).

Placcius, V., *De arte excerpendi, vom gelehrten Buchhalten, liber singularis* (Stockholm and Hamburg, 1689).

Pliny, *Natural History*, ed. and tr. H. Rackham, W. H. S. Jones, and D. E. Eichholz, 10 vols. (London, 1967).

Plomer, H. R., *A Dictionary of the Booksellers and Printers who were at work in England, Scotland and Ireland from 1641 to 1667* (London, 1907).

Plot, R., *The Natural History of Oxford-shire* (Oxford, 1677).

Potter, F., *An Interpretation of the Number 666* (Oxford, 1642).

— *Interpretatio numeri 666* (Amsterdam, 1677).

Pratt, W., *The Arithmeticall Jewell* (London, 1617).

Prins, J. L. M., *Walter Warner (ca. 1557–1643) and his Notes on Animal Organisms* (Utrecht, 1992).

Probst, S., *Die mathematische Kontroverse zwischen Thomas Hobbes und John Wallis* (Hanover, 1997).

Pumfrey, S., '"These 2 hundred years not the like published as Gellibrand has done de Magnete": the Hartlib Circle and Magnetic Philosophy', in M. Greengrass, M. Leslie and T. Raylor, eds., *Samuel Hartlib and Universal Reformation: Studies in Intellectual Communication* (Cambridge, 1994).

Purver, M., *The Royal Society: Concept and Creation* (Cambridge, Mass., 1967).

Quintilian, *The Orator's Education*, ed. and tr. D. A. Russell, 5 vols. (Cambridge, Mass., 2001).

Rademaker, C. S. M., *Leven en werk van Gerardus Joannes Vossius (1577–1649)* (Hilversum, 1999).

Rahn, J. H., *Teutsche Algebra, oder algebraische Rechenkunst, zusamt ihrem Gebrauch* (Zurich, 1659).

— ['Rhonius'], and J. Pell, *An Introduction to Algebra*, tr. T. Brancker (London, 1668).

Ramus [de La Ramée], P., *Arithmeticae libri duo et algebrae totidem* (Frankfurt, 1559).

— *Geometriae libri septem et viginti* (Basel, 1569).

Ravius, C., *Dissertatio mathematica proponens novum summumque totius matheseos inventum problemata omnia certa methodo inveniendi & inventa solvendi* (Paris, 1639)

— *Panegyrica prima orientalibus linguis dedicata* (Utrecht, 1643).

— *Synopsis chronologiae biblicae infallibilis* (Berlin, 1670)

Raylor, T., 'Newcastle's Ghosts: Robert Payne, Ben Jonson, and the "Cavendish Circle"', in C. J. Summers and E.-L. Pebworth, eds., *Literary Circles and Cultural Communities in Renaissance England* (Columbia, Mo., 2000), pp. 92–114.

Regius [de Roy], H., *Fundamenta physices* (Amsterdam, 1646).

— *Fundamenta medica* (Utrecht, 1647).

— *Medicina et praxis medica, medicationum exemplis demonstrata* (Utrecht, 1648).

Reglement ende ordonnantie by Syne Hoocheyt gestatueert voor Rector Magnificus, Professoren, Regens, Sub-Regens, collegisten, ende buyten-studenten van de Illustre Schole ende Collegium Auriacum tot Breda (The Hague, 1646).

Reynolds, B., *Musa virginea graeco-latino-gallica, Bathsuae R. (filiae Henrici Reginaldi gymnasiarchae et philoglotti apud londinenses) anno aetatis suae decimo sexto edita* (London, 1616).

Reynolds, H., *Magnae Britanniae chronographia* (n.p. [London], 1625).

Reynolds, H., *Mythomystes* (London, 1632).

Reynolds, M., *The Learned Lady in England, 1650–1760* (New York, 1920).

Rheita ['Reita'], A. M. Schyrle von, *Novem stellae circa Iovem, circa Saturnum sex, circa Martem non-nullae, A P. Antonio Reita detectae & satellitibus adiudicatae* (Louvain, 1643).

— *Oculus Enoch et Eliae, sive radius sidereomysticus: pars prima* (Antwerp, 1645).

— *Oculi Enoch et Eliae pars altera sive theo-astronomia* (Antwerp, 1645).

— and P. Gassendi, *Novem stellae circa Jovem visae, et de eisdem P. Gassendi judicium* (Paris, 1643).

—, P. Gassendi, and J. Caramuel y Lobkowitz, *Novem stellae circa Jovem, circa Saturnum sex, circa Martem non-nullae, a P. Antonio Reita detectae & satellitibus adiudicatae, de primis ... D. Petri Gassendi iudicium D. Ioannis Caramuel Lobkowitz eiusdem iudicii censura* (Louvain, 1643).

Ricci, M., *Exercitatio geometrica de maximis et minimis* (Rome, 1668).

Riccioli, G. B., *Almagestum novum* (Bologna, 1651).

Rigaud, S. J., ed., *Correspondence of Scientific Men of the Seventeenth Century*, 2 vols. (Oxford, 1841).

Rivington, S., *The History of Tonbridge School* (London, 1925).

Robinson, C. J., *A Register of the Scholars Admitted into Merchant Taylors' School, from A.D. 1562 to 1874*, 2 vols. (Lewes, 1882).

Rørdam, H. F., *Kjøbenhavns Universitets historie fra 1537 til 1621*, 4 vols. (Copenhagen, 1868–74).

— 'Aktstykker til Universitets historie i tidsrummet 1621–60', *Dansk magazin, indeholdende bidrag til den danske histories og det dansk sprogs oplysning*, 5th ser., vol. 1 (1887–9), pp. 36–72, 133–57, 198–222, 332–55, and vol. 2 (1889–92), pp. 1–28, 126–52, 217–42, 320–49.

Rosenbach, Z., *Moses omniscius, sive omniscientia mosaica, sectionibus VI* (Frankfurt am Main, 1633).

— *Methodus omniscientiae Christi, cum specimine omniscientiae Gentilis, & indice corollarium physicorum Novi Testamenti* (Herborn, 1634).

Rossi, G., *Francesco Maurolico e il risorgimento filosofico e scientifico in Italia nel secolo XVI* (Messina, 1888).

Rossi, P., *Logic and the Art of Memory: The Quest for a Universal Language*, tr. S. Clucas (London, 2000).

Rostenberg, L., *The Library of Robert Hooke: The Scientific Book Trade of Restoration England* (Santa Monica, Calif., 1989).

Rouse Ball, W. W., and J. A. Venn, *Admissions to Trinity College, Cambridge*, 5 vols. (London, 1911–16).

Rozbicki, M., *Samuel Hartlib z dziejów polsko-angielskich związków kulturalnych w XVII wieku* (Warsaw, 1980).

Rusterholz, S., 'Abraham von Franckenberg', in H. Holzhey et al., eds., *Grundriss der Geschichte der Philosophie, begründet von Friedrich Ueberweg: Die Philosophie des 17. Jahrhunderts*, Bd iv, *Das Heilige Römische Reich, deutscher Nation, Nord- und Ostmitteleuropa*, 2 vols. (Basel, 2001), i, pp. 85–95.

Rymer, T., *Foedera*, 3rd edn., ed. G. Holmes, 10 vols. (The Hague, 1739–45).

Sadler, J. E., *J. A. Comenius and the Concept of Universal Education* (London, 1966).

de Saint-Charles ['à Sancto Carolo'], L. J., *Bibliographia parisina, hoc est, catalogus omnium librorum Parisiis, annis 1643 & 1644 inclusivè excusorum* (Paris, 1645).

de Saint-Vincent ['à Sancto Vincentio'], G., *Opus geometricum quadraturae circuli et sectionum coni decem libris comprehensum* (Antwerp, 1647).

Salmon, E., 'Southwick', *Sussex Archaeological Collections*, 63 (1922), pp. 87–111.

Salmon, V., 'Joseph Webbe: Some Seventeenth-Century Views on Language-Teaching and the Nature of Meaning', *Bibliothèque d'humanisme et renaissance: travaux et documents*, 23 (1961), pp. 324–40.

— 'An Ambitious Printing Project of the Early Seventeenth Century', *The Library*, 3rd ser., 16 (1961), pp. 190–6.

— 'Problems of Language Teaching: A Discussion among Hartlib's Friends', *Modern Language Review*, 59 (1964), pp. 13–24.

— 'The Family of Ithamaria (Reginolles) Reynolds Pell', *Pelliana*, n.s., 1, no. 3 (1965), pp. 1–24, 105–6.

— *The Study of Language in 17th-Century England* (Amsterdam, 1979).

de Sarasa, A. A., *Solutio problematis a R. P. Mersenno Minimo propositi . . .* [with] *Quòd quadratura circuli à R. P. Gregorio a S^{to} Vincentio exhibita, abeat illud necdum solutum problema* (Antwerp, 1649).

Sassen, F. L. R., *Het wijsgerig onderwijs aan de Illustre School te Breda (1646–1669)*, Mededelingen der Koninklijke Nederlandse Akademie van Wetenschappen, afd. letterkunde, n.s., vol. 25, no. 7 (Amsterdam, 1962).

— 'Levensberichten van de hoogleraren der Illustre School te Breda', *Jaarboek van de geschied- en oudheidkundige kring van stad en land van Breda "De Oranjeboom"'*, 19 (1966), pp. 123–57.

Savile, H., *In Taciti histor. [&] Agricolae vitam, et commentarius de militaria romana*, ed. and tr. I. Gruter (Amsterdam, 1649).

Scavenius, P., *Designatio librorum in qualibet facultate, materia & lingua rariorum ... conquisitorum* (Copenhagen, 1665).

Schaller, K., *Die Pädagogik des Johann Amos Comenius und die Anfänge des pädagogischen Realismus im 17. Jahrhundert*, 2nd edn. (Heidelberg, 1967).

Scheiner, C., *Pantographice seu ars delineandi res quaslibet per parallelogrammum lineare seu cavum mechanicum mobile* (Rome, 1631).

de Schickler, F., *Les Églises du refuge en Angleterre*, 3 vols. (Paris, 1892).

Schlosser, H., *Die Piscatorbibel: ein Beitrag zur Geschichte der deutschen Bibelübersetzung* (Heidelberg, 1908).

Schlueter, P., and J. Schlueter, *An Encyclopedia of British Women Writers* (New York, 1988).

Schnyder-Spross, W., *Die Familie Rahn von Zürich* (Zurich, 1951).

van Schooten, F., *De organica conicarum sectionum in plano descriptione tractatus ... cui annexa est appendix de cubicarum aequationum resolutione* (Leiden, 1646).

— *Principia matheseos universalis, seu introductio ad geometriae methodum Renati Des Cartes*, ed. E. Bartholin (Leiden, 1651).

— *Exercitationum mathematicarum libri quinque* (Leiden, 1657).

van Schurman, A. M., *Opuscula hebraea, graeca, latina, gallica prosaica & metrica* (Leiden, 1648).

— *Whether a Christian Woman should be Educated and Other Writings from her Intellectual Circle*, ed. and tr. J. L. Irwin (Chicago, 1998).

Scott, J. F., *The Scientific Work of René Descartes (1596–1650)* (London, n.d. [1952]).

Scriba, C. J., 'Mercator's Kinckhuysen-Translation in the Bodleian Library at Oxford', *British Journal for the History of Science*, 2 (1964), pp. 45–58.

— 'John Pell's English Edition of J. H. Rahn's *Teutsche Algebra*', in R. S. Cohen, J. J. Stachel, and M. M. Wartofsky, eds., *For Dirk Struik: Scientific, Historical and Political Essays in Honor of Dirk J. Struik* (Dordrecht, 1974), pp. 261–74.

Series lectionum publicarum, in Illustri Schola Auriaco-Bredana, inchoandarum secundo die Octobris 1646 (n.p., n.d. [Breda, 1646]).

Serjeantson, R., 'Herbert of Cherbury before Deism: The Early Reception of the *De veritate*', *The Seventeenth Century*, 16 (2001), pp. 217–38.

van Seters, W. H., 'Prof. Johannes Brosterhuysen (1596–1650), stichter en opziener van de Medicinale Hof te Breda', *Jaarboek van de geschied- en oudheidkundige kring van stad en land van Breda "De Oranjeboom"*, 6 (1953), pp. 106–51.

Shapin, S., and S. Schaffer, *Leviathan and the Air-Pump: Hobbes, Boyle and the Experimental Life* (Princeton, NJ, 1985).

Shapiro, B. J., *John Wilkins, 1614–1672: An Intellectual Biography* (Berkeley, Calif., 1969).

Sharpe, K., *The Personal Rule of Charles I* (New Haven, Conn., 1992).

Sharpe, R., 'The Naming of Bishop Ithamar', *English Historical Review*, 117 (2002), pp. 889–94.

Shaw, W. A., *The Knights of England*, 2 vols. (London, 1906).

Sherwood, R., *The Court of Oliver Cromwell* (London, 1977).

Shirley, J. W., 'The Scientific Experiments of Sir Walter Ralegh, the Wizard Earl, and the Three Magi in the Tower, 1603–1617', *Ambix*, 4 (1949), pp. 52–66.

Shumaker, W., *Renaissance curiosa* (Binghampton, NY, 1982).

Simpson, A. D. C., 'Richard Reeve – the "English Campani" – and the Origins of the London Telescope-Making Tradition', *Vistas in Astronomy*, 28 (1985), pp. 357–65.

Skinner, Q., *Visions of Politics*, 3 vols. (Cambridge, 2002).

van der Sluijs, P., 'Constantijn Huygens en de Muiderkring', in H. Bots, ed., *Constantijn Huygens: zijn plaats in geleerd Europa* (Amsterdam, 1973), pp. 188–309.

Smith, D. E., *Rara arithmetica: A Catalogue of the Arithmetics Written before the Year MDCI*, 2 vols. (Boston, 1908).

Smith, H., *The Ecclesiastical History of Essex under the Long Parliament and the Commonwealth* (Colchester, n.d. [1932]).

Smith, L. P., *The Life and Letters of Sir Henry Wotton*, 2 vols. (Oxford, 1907).

Smitskamp, R., *Philologia orientalis: A Description of Books Illustrating the Study and Printing of Oriental Languages in 16th- and 17th-Century Europe* (Leiden, 1992).

Snel, W., *Apollonius Batavus seu exsuscitata Apollonii Pergaei περὶ διωρισμένης τομῆς geometria* (Leiden, 1608).

— *Tiphys batavus, sive histiodromice, de navium cursibus et re navali* (Leiden, 1624).

— *Doctrina triangulorum* (Leiden, 1627).

Somerset, E., *A Century of the Names and Scantlings of such Inventions, as at Present I can Call to Mind to have Tried and Perfected* (London, 1663).

Sorbière, S., *Relations, lettres, et discours* (Paris, 1660).

Spalding, R., ed., *The Diary of Bulstrode Whitelocke, 1605–1675*, Records of Social and Economic History, n.s., xiii (Oxford, 1990).

Spalding, R., *Contemporaries of Bulstrode Whitelocke, 1605–1675*, Records of Social and Economic History, n.s., xiv (Oxford, 1990).

Speidell, J., *New logarithmes* (London, 1619).

Sprat, T., *The History of the Royal Society of London* (London, 1667).

Stampioen, J., *Algebra ofte nieuwe stel-regel* (The Hague, 1639).

— *Solutie op sijn eygen problema* (The Hague, 1640).

Stedall, J. A., 'Catching Proteus: The Collaborations of Wallis and Brouncker. II: Number Problems', *Notes and Records of the Royal Society of London*, 54 (2000), pp. 317-31.

— 'Rob'd of Glories: The Posthumous Misfortunes of Thomas Harriot and His Algebra', *Archive for the History of Exact Sciences*, 54 (2000), pp. 455-97.

— *A Discourse Concerning Algebra: English Algebra to 1685* (Oxford, 2002).

— *The Greate Invention of Algebra: Thomas Harriot's Treatise on Equations* (Oxford, 2003).

Steiner, H., *Der Zürcher Professor Johann Heinrich Hottinger in Heidelberg, 1655-1661* (Zurich, 1886).

Stephens, J. E., ed., *Aubrey on Education: A Hitherto Unpublished Manuscript by the Author of Brief Lives* (London, 1972).

Stevin, S., *L'Arithmétique contenant les computations ... aussi l'algebre* (Leiden, 1585).

— *Les Oeuvres mathematiques*, ed. A. Girard (Leiden, 1634).

Stieg, M. F., ed., *The Diary of John Harington, M.P., 1646-53*, Somerset Record Society, lxxiv (Old Woking, 1977).

Stouppe, J.-B., *A Collection of the Several Papers sent to his Highness the Lord Protector ... Concerning the Bloody and Barbarous Massacres ... Committed on many Thousands of Reformed, or Protestants Dwelling in the Vallies of Piedmont* (London, 1655).

Strauss, E., *Sir William Petty: Portrait of a Genius* (London, 1954).

Streete, T., *Astronomia carolina: A New Theory of the Coelestial Motion* (London, 1661).

Stubbs, M., 'John Beale, Philosophical Gardener of Herefordshire. Part I: Prelude to the Royal Society (1608-63)', *Annals of Science*, 39 (1982), pp. 463-89.

— 'John Beale, Philosophical Gardener of Herefordshire. Part II: The Improvement of Agriculture and Trade in the Royal Society (1663-83)', *Annals of Science*, 46 (1989), pp. 323-63.

Sturm, E., 'Pansophie und Pädagogik bei Jan Amos Komenský', in K. Gossmann and C. T. Schielke, eds., *Jan Amos Comenius, 1592-1992: theologische und pädagogische Deutungen* (Gütersloh, 1992), pp. 101-25.

Suter, H., *Die Mathematiker und Astronomen der Araber und ihre Werke* (Leipzig, 1900).

Tacquet, A., *Cylindricorum et annularium libri IV; item de circulorum volutione per planum, dissertatio physicomathematica* (Antwerp, 1651).

Tannenbaum, R., 'Thomas Pell', in *American National Biography*, 24 vols. (New York, 1999), xvii, pp. 263–4.

Tannery, P., 'Sur le Mathématicien français Chauveau', *Bulletin des sciences mathématiques*, 19 (1895), pp. 34–7.

Tapp, J., *The Path-Way to Knowledge; Containing the Whole Art of Arithmeticke* (London, 1613).

Targioni-Tozzetti, G., *Notizie degli aggrandamenti delle scienze fisiche accaduti in Toscana nel corso di anni LX. del secolo XVII*, 3 vols. (Florence, 1780).

Tassius, J. A., *Disputatio de rebus astronomicis et geographicis* (Hamburg, 1635).

Taton, R., *L'Oeuvre mathématique de G. Desargues* (Paris, 1951).

— 'Desargues et le monde scientifique de son époque', in J. Dhombres and J. Sakarovitch, eds., *Desargues et son temps* (Paris, 1991), pp. 25–53.

Tautz, K., *Die Bibliothekare der churfürstlichen Bibliothek zu Cölln an der Spree: ein Beitrag zur Geschichte der Preussischen Staatsbibliothek im siebzehnten Jahrhundert* (Leipzig, 1925).

Taylor, E. G. R., 'Robert Hooke and the Cartographic Projects of the Late Seventeenth Century', *The Geographical Journal*, 90 (1937), pp. 529–40.

— '"The English Atlas" of Moses Pitt, 1680–83', *The Geographical Journal*, 95 (1940), pp. 292–9.

— *The Mathematical Practitioners of Tudor and Stuart England* (Cambridge, 1954).

Tazbir, J., *Stanisław Lubieniecki: przywódca ariańskiej emigracji* (Warsaw, 1961).

Teague, F., *Bathsua Makin, Woman of Learning* (Lewisburg, PA, 1998).

Thewes, A., *Oculus Enoch ... Ein Beitrag zur Entdeckungsgeschichte des Fernrohrs* (Oldenburg, 1983).

— 'Ein Ordensbruder und Astronom aus dem Kapuzinerkloster Passau', *Ostbairische Grenzmarken: Passauer Jahrbuch für Geschichte, Kunst und Volkskunde*, 30 (1988), pp. 95–104.

Thijssen-Schoute, C. L., *Nederlands cartesianisme*, 2nd edn. (Utrecht, 1989).

Thomas, Sir Keith, 'The Life of Learning', *Proceedings of the British Academy*, 117 (2002), pp. 201–35.

Todd, H. J., ed., *Memoirs of the Life and Writings of the Right Rev. Brian Walton*, 2 vols. (London, 1821).

Todd, J., ed., *Dictionary of British Women Writers* (London, 1989).

Toomer, G. J., *Eastern Wisedome and Learning: The Study of Arabic in Seventeenth-Century England* (Oxford, 1996).

— see also Apollonius.

Torricelli, E., *Opera geometrica* (Florence, 1644).

Trease, G., *Portrait of a Cavalier: William Cavendish, First Duke of Newcastle* (London, 1979).

van Tricht, H. W., *Het leven van P. C. Hooft* (The Hague, 1980).

Trithemius, J., *Polygraphiae libri VI* (Basel, 1518).

— *Steganographia, hoc est, ars per occultam scripturam animi sui voluntatem absentibus aperiendi certa* (Frankfurt am Main, 1606).

von Tschirnhaus, E. W., 'Nova methodus auferendi omnes terminos intermedios ex data aequatione', *Acta eruditorum* (1683), pp. 204–7.

Turberville, A. S., *A History of Welbeck Abbey and its Owners*, 2 vols. (London, 1938).

Turnbull, G. H., *Samuel Hartlib: A Sketch of his Life and his Relations to J. A. Comenius* (Oxford, 1920).

— *Hartlib, Dury and Comenius: Gleanings from Hartlib's Papers* (London, 1947).

— 'Samuel Hartlib's Influence on the Early History of the Royal Society', *Notes and Records of the Royal Society of London*, 10 (1952–3), pp. 101–30.

— 'Johann Valentin Andreae's *Societas Christiana*', *Zeitschrift für deutsche Philologie*, 73 (1954), pp. 407–14.

Turnbull, H. W., ed., *James Gregory Tercentenary Memorial Volume* (London, 1939).

Turner, A. J., 'Mathematical Instruments and the Education of Gentlemen', *Annals of Science*, 30 (1973), pp. 51–88.

Turner, G. L'E., *Elizabethan Instrument Makers: The Origins of the London Trade in Precision Instrument Making* (Oxford, 2000).

Valerio, L., *De centro gravitatis solidorum libri tres* (Rome, 1604).

— *Quadratura parabolae* (Rome, 1606).

Vampaemel, G. H. W., 'Jesuit Science in the Spanish Netherlands', in M. Feingold, ed., *Jesuit Science and the Republic of Letters* (Cambridge, Mass., 2003), pp. 389–432.

Vaughan, R., *The Protectorate of Oliver Cromwell, and the State of Europe during the Early Part of the Reign of Louis XIV, Illustrated in a Series of Letters*, 2 vols. (London, 1839).

Venn., J., and J. A. Venn, eds., *Alumni cantabrigienses*, part 1, 2 vols. (Cambridge, 1922).

Venning, T., *Cromwellian Foreign Policy* (London, 1995).

Victoria County History, County of Chester, 3 vols. (London, 1979–87).

Victoria County History of Sussex (London, 1905–).

Viète [Vieta], F., *In artem analyticam isagoge* (Tours, 1591).

— *Zeteticorum libri quinque* (Paris, 1593).

— *Apollonius Gallicus, seu exsuscitata Apollonis Pergaei περὶ ἐπαφῶν geometria* (Paris, 1600).

— *De numerosa potestatum ad exegesin resolutione*, ed. M. Getaldus [Getaldić, Ghetaldi] (Paris, 1600).

— *Ad angularium sectionum analyticen theoremata*, ed. A. Anderson (Paris, 1615).

— *De aequationum recognitione et emendatione tractatus duo*, ed. A. Anderson (Paris, 1615).

— *In artem analyticam isagoge, ad logisticen speciosam notae priores* (Paris, 1631).

— *Opera mathematica* (Leiden, 1646).

Viviani, V., *De maximis et minimis* (Florence, 1659).

Vlacq, A., *Het tweede deel van de nieuwe stelkonst* (Gouda, 1627).

Vossius, G. J., *De quatuor artibus popularibus, de philologia, et scientiis mathematicis* (Amsterdam, 1650).

van de Vyver, O., 'L'École de mathématiques des Jésuites de la province flandro-belge au XVIIe siècle', *Archivum historicum societatis iesu*, 49 (1980), pp. 265–78.

de Waard, C., 'Martinus Hortensius', in *Nieuw nederlandsch biografisch woordenboek*, 10 vols. (Leiden, 1911–37), i, cols. 1160–4.

— 'John Pell', in *Nieuw nederlandsch biografisch woordenboek*, 10 vols. (Leiden, 1911–37), iii, cols. 961–6.

— 'Un Écrit de Beaugrand sur la méthode des tangentes de Fermat à propos de celle de Descartes, *Bulletin des sciences mathématiques*, ser. 2, 17 (1918), pp. 157–77.

— 'Wiskundige bijdragen tot de pansophie van Comenius', *Euclides: tijdschrift voor de didactiek der exacte vakken*, 25 (1949–50), pp. 278–89.

Waer-schouwinghe uyt Breda (n.p., n.d. [Breda, 1646]).

Walker, E., *A Study of the Traité des indivisibles of Gilles Persone de Roberval* (New York, 1932).

van der Wall, E. G. E., *De mystieke chiliast Petrus Serrarius (1600–1669) en zijn wereld* (Leiden, 1987).

Wallace, J., 'The Engagement Controversy, 1649–1652: An Annotated List of Pamphlets', *Bulletin of the New York Public Library*, 68, no. 6 (1964).

— *Destiny his Choice: The Loyalism of Andrew Marvell* (Cambridge, 1968).

Wallis, J., *Elenchus geometriae hobbianae* (Oxford, 1655).

— *Arithmetica infinitorum* (Oxford, 1656).

— *Hobbius heauton-timoroumenos, Or, A Consideration of Mr Hobbes his Dialogues* (Oxford, 1662).

— 'Logarithmotechnia Nicolai Mercatoris', *Philosophical Transactions*, 3 (1668), pp. 753–64.

— 'De rationum et fractionum reductione', appended to J. Horrocks, *Opera posthuma*, 2nd edn. (London, 1678).

— *A Treatise of Algebra* (London, 1685).

— *Opera mathematica*, 3 vols. (Oxford, 1693–9).

Wallis, P. J., 'An Early Mathematical Manifesto – John Pell's *Idea of Mathematics*', *The Durham Research Review*, 5, no. 18 (Apr. 1967), pp. 139–48.

— 'John Pell', in *The Dictionary of Scientific Biography*, 18 vols. (New York, 1970–90), x, pp. 495–6.

Ward, J., *The Lives of the Professors of Gresham College* (London, 1740).

Wasmuth, M., *Idea astronomiae chronologiae restitutae* (Kiel, 1678).

van Wassenaer, J. ['I. a Wassenaer'], *Anmerckingen op den nieuwen stel-regel van Iohan Stampioen* (Leiden, 1639).

— ['I. a Waessenaer'], *Den on-wissen wis-konstenaer I. I. Stampioenius ontdeckt door sijne ongegronde weddinge ende mis-lucte solutien van sijne eygene questien* (Leiden, 1640).

Watson, F., 'Dr Joseph Webbe and Language Teaching (1622)', *Modern Language Notes*, 26 (1911), pp. 40–6.

Weber, J., *Gerechnet Rechenbüchlein auf Erfurtischen Wein- und Tranks-Kauff* (Erfurt, 1570; 2nd edn. Erfurt, 1583).

Webster, C., *Samuel Hartlib and the Advancement of Learning* (Cambridge, 1970).

— *The Great Instauration: Science, Medicine and Reform, 1626–1660* (London, 1975).

Webster, J., *The Displaying of Supposed Witchcraft* (London, 1677).

Weigle, T., 'Die deutschen Doktorpromotionen in Philosophie und Medizin an der Universität Padua von 1616–1663', *Quellen und Forschungen aus italienischen Archiven und Bibliotheken*, 45 (1965), pp. 325–84.

Wertheim, G., 'Die Algebra des Johann Heinrich Rahn (1659) und die englische Übersetzung derselben', *Bibliotheca mathematica: Zeitschrift für Geschichte der mathematischen Wissenschaften*, ser. 3, no. 3 (1902), pp. 113–26.

Westfall, R., *Never at Rest: A Biography of Isaac Newton* (Cambridge, 1980).

Westin, P. G., 'Brev från John Durie åren 1636–1638', *Kyrkhistorisk årsskrift*, 33 (1933), pp. 193–349.

Westlake, H. F., and L. E. Turner, eds., *The Register of St. Margaret's, Westminster, London, 1660–1675*, Publications of the Harleian Society, lxiv (London, 1934).

Whitaker, E. A., 'Selenography in the Seventeenth Century', in R. Taton and C. Wilson, eds., *Planetary Astronomy from the Renaissance to the Rise of Astrophysics: Part A: Tycho Brahe to Newton* (Cambridge, 1989), pp. 119–43.

White, R., *Hemisphaerium dissectum: opus geometricum in quo obiter tractatur de maximis inscriptibilibus, & minimis circumscribentibus* (Rome, 1648).

White, T., *De mundo dialogi* (Paris, 1642).

Wichmann, E. H., *Heimatskunde: topographische, historische und statistische Beschreibung von Hamburg und der Vorstadt St. Georg* (Hamburg, 1863).

Wijnman, H. F., 'De amsterdamsche rekenmeester Sybrandt Hansz Cardinael', *Het boek*, n.s., 22 (1933–4), pp. 73–94.

Wilkins, J., *An Essay towards a Real Character* (London, 1668).

Willems, A., *Les Elzevier: histoire et annales typographiques* (Brussels, 1880).

Willis, T., *De anima brutorum* (London, 1672).

Willmoth, F., *Sir Jonas Moore: Practical Mathematics and Restoration Science* (Woodbridge, 1993).

Willson, A. N., *A History of Collyer's School* (London, 1965).

Wingate, E., *The Use of the Rule of Proportion* (London, 1626).

— *The Construction and Use of the Line of Proportion* (London, 1628).

— *Arithmetique made Easie* (London, 1630).

— *Of Natural and Artificial Arithmetic* (London, 1630).

Winthrop, R. C., ed., *Correspondence of Hartlib, Haak, Oldenburg, and Others of the Founders of the Royal Society, with Governor Winthrop of Connecticut, 1661–1672* (Boston, 1878).

Witkam, J. J., *Jacobus Golius (1596–1667) en zijn handschriften*, Oosters genootschap in Nederland; x (The Hague, 1980).

Wolf, R., *Biographien zur Kulturgeschichte der Schweiz*, 4 vols. (Zurich, 1838–62).

Wood, A., *Fasti oxonienses; or, Annals of the University of Oxford*, 2 vols., appended to vols. ii and iv of A. Wood, *Athenae oxonienses*, ed. P. Bliss, 4 vols. (London, 1813–20).

— *The Life and Times of Anthony Wood, Antiquary, of Oxford, 1632–1695, Described by Himself*, ed. A. Clark, 5 vols. (Oxford, 1891–1900).

Woodward, H., *A Light to Grammar and All Other Arts and Sciences* (London, 1641).

Woolrych, A., *Britain in Revolution, 1625–1660* (Oxford, 2002).

Worden, B., *The Rump Parliament, 1648–1653* (Cambridge, 1974).

Worsley, L., 'The Architectural Patronage of William Cavendish, Duke of Newcastle, 1593–1676' (University of Sussex D.Phil. dissertation, 2001).

Worthington, J., *The Diary and Corrrespondence*, ed. J. Crossley, 3 vols., Chetham Society, vols. 13, 36, 94 (1847–86).

Wotton, Sir Henry, *The Elements of Architecture* (London, 1624).

Wren, C., *Parentalia: Or, Memoirs of the Family of the Wrens* (London, 1750).

Young, J. T., '"To Leave No Problem Unsolved": The New Mathematics as a Model for Pansophy', *Acta comeniana*, 12 (1997), pp. 85–95.

— *Faith, Medical Alchemy and Natural Philosophy: Johann Moriaen, Reformed Intelligencer, and the Hartlib Circle* (Aldershot, 1998).

Young, P., *Marston Moor, 1644: The Campaign and the Battle* (Moreton-in-Marsh, 1997).

Young, R. F., *Comenius in England* (London, 1932).

Ziggelaar, A., 'Die Erklärung des Regenbogens durch Marcantonio de Dominis, 1611: zum Optikunterricht am Ende des 16. Jahrhunderts', *Centaurus*, 23 (1979), pp. 21–50.

Index

Note: all names are ordered by the first element of the name that bears a capital letter (with the sole exception of The Hague, listed under 'H'). Works referred to in the text are listed under the author's name, if known; anonymous works are listed by title. Page-references here may include references to material in the notes, which has been indexed on the following basis: names and topics appearing in the notes are included, but the titles of works cited, and the names of their authors, are not. Throughout, 'JP' stands for John Pell.

Aarau 155
Abbott, W. C. 6
van Ackersijck, Theodore 570
Acta eruditorum 321
Adams, William 184
agriculture 51, 168, 171, 224
Aleppo 415
algebra 56–9, 73, 76, 92, 99–100, 132–3, 204–9, 217–23, 239–40, 249, 268–80, 289, 299, 319, 336, 345–6
  Harriot's 272–8, 317, 320, 326
  of knowledge 265
algebraic geometry 279, 326
Alhazen's problem 220, 243
Alsted, Johann Heinrich 41, 44, 46, 51
  works: 'Fundamenta disciplinarum mathematicarum' 57
Alting, Hendrik 131
Amalia, Princess of Orange 125–7, 137
Amsterdam 71, 73, 76, 98–100, 102–7, 110–11, 113, 118–20, 124, 152, 352–3, 357–61, 372–6, 379–81, 384–7, 388–92, 394, 398–400, 405–9, 414–18, 423–6, 431–2, 437–42, 449–57, 459, 463–4, 467–71, 476, 479–81, 485–6, 492–3, 505, 524, 526, 546
Amsterdam Athenaeum 77, 93, 100, 102–6, 108, 111, 121, 128, 286, 292, 352, 359–60, 362–8, 384, 463–4, 469, 519
analysis 56, 241, 454
  and synthesis 262, 267, 422, 439, 444, 500
  see also: algebra
analytic art 250, 267, 336, 500
anatomy 180
Anchoran, John 41
Anderson, Alexander 248
Andreae, Johann Valentin 27, 41
Andrewes, Lancelot 15

angle, trisection of 39, 295, 321, 349, 513, 530
Anglo-Dutch War 136, 149–50
Anslo, Gerbrandt 76
Anstey, Peter 9
'Antilia', 'Antilians' 27, 41
antilogarithms 84–5, 92–3, 142–3, 166, 238, 280–6, 336, 338, 350, 354, 358, 369, 530, 533
Antwerp 132, 397, 399–401, 410–11, 506, 515–17, 519, 521–3, 525, 527, 530–1, 535–7, 540–1, 546–51, 555, 559, 562–3, 566, 570–3, 575–81, 584
Apollonius 107, 110, 113, 119, 131, 175, 243, 248–9, 292–4, 341–2, 358, 365–6, 469, 478, 538
  works: *Conics* 110, 119, 229, 292–4, 384–8, 393, 397, 414–16, 419, 421, 428, 446, 462, 481, 494, 501, 506, 515
Archimedes 248, 257–8, 365–6, 385, 496, 541
  works: *De sphaera et cylindro* 469; *Psammites* 319
Aristotle 117, 271
arithmetic
  and geometry 406–7, 506
  operations of 258–60
Arnhem 172, 175
Arundel House 88–9
astronomy 19, 21, 97, 105, 110, 140, 190–1, 193, 251–5, 355, 369, 373–6, 380, 399, 410, 502–3, 521
Atwell, George 22
Aubrey, John 13, 23, 39, 84, 86, 164, 182, 192, 210, 230–2, 235, 239–42, 252, 270, 280–1, 290, 301, 315–16, 320, 322–3, 326
  works: 'Idea of Education' 240; *Miscellanies* 231

Augsburg 344, 376, 380, 393, 395, 397, 401
Autolycus of Pitane 365–6
Aylesbury, Sir Thomas 64, 79, 81–5, 273–6, 280, 284, 317, 327, 550, 584–5

Bachet de Méziriac, Claude Gaspar 205, 208, 248, 312–13
  works: *Diophanti libri sex* 289–90, 312–13
Bacon, Francis, Baron Verulam 17, 40, 45, 77, 124
  works: *Novum organum* 45
Baconianism 28, 41, 52, 67, 69, 97
Baden 175
Baillet, Adrien 128
  works: *La Vie de M. Descartes* 128
Bainbridge, John 516, 518
  works: *Canicularia* 516, 518, 522, 524–5, 527–8
Baker, Thomas 241, 248, 311–12, 316, 323
  works: *Clavis geometrica catholica* 241, 311–12, 319
Barlaeus, Caspar 102, 104–7, 124, 130, 131, 352
  works: *Mercator sapiens* 105
Barlow, Thomas 115, 222–3
Barrow, Isaac 17, 196, 248, 322
Bartholin (Bartholinus), Erasmus 248, 576
  see also: van Schooten, *Principia matheseos*
Basel 87, 155, 175
Basildon 186–7
Baudry, Guillaume 91
Beale, John 170–1, 180, 189, 191, 213
de Beaugrand, Jean 248, 341, 372–3, 377
de Beaune, Florimond 115, 288, 296, 416–17, 538, 540, 546, 548
  works: 'Notes briefves' 548
  see also: Descartes, *Geometria*
Beckers, Danny 9
Bedlam 38
Beecher, William 81, 133, 365
Beecher, Sir William 80–2, 85, 91, 364–5
Beeckman, Isaac 569
Bellamy, John 42
Bern 155, 156
Bettini, Mario 450
Bible 42, 44–5, 47, 140
Biblia sacra polyglotta 140
Billericay 234

Billingshurst 33, 39
de Billy, Jacques 81, 115, 248, 288, 364
  works: *Nova geometriae clavis* 364–5, 409
de Bils, Lodewijk 180–1
  works: *The Copy of a Certain Large Act* 180–1; *Kopye van zekere ampele acte* 180–1
binomial theorem 306
binomials and residuals 530, 532
  cube roots of 525, 530, 532
Birch, Thomas 5–6, 23, 242, 331
Birchensha (Birkenshaw), John 192
Bisterfeld, Johann Heinrich 51, 63, 71, 109
Blaeu, Johan 105, 110–11, 120, 229, 286, 360, 385, 414–15, 499, 505, 565
Blaeu, Willem Janszoon 564–5
  works: *Institutio astronomica* (tr. Hortensius) 564–5
Blagrave, John 21–2
Bocan, M. 504
Bodinus, Elias 40
Böhme, Jakob 71
Bombelli, Raffaele 249–50, 272, 289, 360
  works: *L'algebra* 289
Bond, Henry 93, 225–6, 522
Borch, Ole 194
Born (Bornius), Hendrik 124, 128, 135
Borrell, Jean 279
du Bosc, Charles 88, 487, 495, 500, 504–5
Bosse, Abraham 142, 365
  works: *Sentimens sur la distinction* 142
Boston 39
Boswell, Sir William 77–8, 97–8, 100, 108, 123, 131–2, 374, 410, 414, 420, 427, 444, 452, 493, 505, 507, 511, 513, 556
Boughton (or Broughton), Stephen 234–5
Boulliau (Bullialdus), Ismaël 72, 115, 365–6, 394, 414, 416
  works: *Astronomia philolaica* 366, 414, 419; (ed.) *Theonis ... expositio* 365
Bowie, Angus 9
Boyle, Robert 137, 171, 180–1, 189, 191, 198, 215, 217, 227, 232–3
Brahe, Tycho 110, 565
Brancker, Thomas 199–207, 212, 232, 243, 248, 306–8
Brandt, Frithiof 2
Bray, William 42

INDEX 641

Breda 106, 119, 120–38, 139, 146, 182–3, 294–6, 301, 485, 492–3, 499–500, 505–8, 511–12, 514, 517–20, 524, 526, 528–9, 532–4, 538–9, 542–5, 550, 552–4, 556–62, 564–5, 567–9, 574, 582–3, 585–6
Brereton Hall 198–9, 207, 213, 242–3
Brereton, William, third Baron Brereton 133, 137, 139, 171–3, 177–9, 185, 187, 189, 191, 198–9, 203, 209–10, 212–13, 215–16, 231, 266, 295–7, 301, 314, 316, 327, 541, 543–4, 559, 562–3, 565, 569
Brereton's problem 295–6, 300–1, 313–4, 316, 320, 565, 576
Brerewood, Edward 49
Brett, Arthur 232
Brewster, Nathaniel 188
Bridgewater 400
Brieg 26
Briggs, Henry 19, 58, 248, 250, 254–7, 280, 297, 303–4, 321, 326–7, 349, 351, 390, 398, 516–18, 533
　works: *Arithmetica logarithmica* 58, 250, 256, 303; *Trigonometria britannica* 256, 519
Briot, Nicolas 81, 91
Brooke, Lord: *see* Greville
Brookes, William 28, 45–7
Brosterhuysen, Amalia 564
Brosterhuysen, Johan 124, 126, 130–1, 135, 564, 566
Brouncker, William, Viscount Brouncker 179, 189, 191, 193, 195, 220–1, 223–7, 248, 300, 320
Browne, Thomas 92–3, 282
Bruce, Alexander, Earl of Kincardine 190
Brussels 481
de Bruyn, Johannes 136
Buck, Mr 493, 496, 498, 501, 504
Buckden 33
Burgh, Albert Coenratszoon 78, 100, 102, 111–12, 360
Burton family 15, 39
Burton, Pelham 15
Burton, William 15, 62, 131
Busby, Richard 131, 185, 197, 205, 224, 230, 238, 241–3, 247, 286, 312
Buxtorf, Johannes 171
Byrd, William 130

Cabeljau, Joannes 104–5, 352
Calamy, Edmund 149

calculus 221, 306, 327
calendars, Gregorian and Julian 8, 186, 251
calendrical calculations 251–2, 254, 305
Calvisius (Kalwitz), Seth 395, 397–8, 401, 403
Cambridge University 17, 18, 26, 78, 149
Campion, Thomas 130
Canterbury, Archbishop of: *see* Sheldon
Caramuel y Lobkowitz 355, 375–6, 380
　works: *Mathesis audax* 375–6; (ed.) *Steganographia Trithemii* 375
de Carcavi, Pierre 115–16, 367, 425, 432, 434, 476, 510
　works: demonstration of Pell's theorem 427–8, 468, 472, 507–8
Cardano, Girolamo 248–50, 272
　works: *Ars magna* 249–50, 272, 536, 540
Cardinael, Sibrant Hanszoon 132–3
　works: *Practijck des landmetens* 239–40
Carnaby, Sir William 394–5, 398
Cartesianism 124, 133–5
Cassegrain, M., 216–17
Caters, Simon 520
Cavalieri, Bonaventura 115–16, 118, 288, 296, 326, 339, 368, 395, 397–8, 401, 417, 441, 444, 461, 496
　works: demonstration of Pell's theorem 456, 467, 472, 479, 482, 484, 508; *Directorium generale* 368, 398; *Exercitationes* 510–11; *Geometria indivisibilibus promota* 296, 368, 398, 496
Cavendish, Charles (son of second Earl of Devonshire) 87
Cavendish, Sir Charles 57, 85–93, 96–9, 102, 108–9, 111, 113–22, 127, 129, 131–2, 136, 139, 168, 181, 190, 211, 268, 278, 288–91, 327, 331–3, 335–586 *passim*
　works: first demonstration of Pell's theorem 288, 422, 431, 433, 437–40; second demonstration of Pell's theorem 288, 446–8, 452–4, 458, 460–1, 463, 492, 494, 502
Cavendish, Sir Charles (Sr) 86, 87
Cavendish, Elizabeth ('Bess of Hardwick') 86
Cavendish, Margaret (née Lucas; Marchioness, later Duchess, of Newcastle) 87, 139

Cavendish, William, Earl (then Marquess, later Duke) of Newcastle 84, 86–7, 89, 97–8, 181, 355, 361, 393–4, 397, 503, 509, 511–12, 531, 548, 557, 559
Cavendish, William, first Earl of Devonshire 86, 89
Cavendish, William, second Earl of Devonshire 87–9
Cavendish, William, third Earl of Devonshire 503
centre of agitation 465, 476, 479, 488
centre of gravity 473, 505
van Ceulen, Ludolf 248, 257, 287, 311, 530
Chamberlain, John 87
Chapman, William 34–5
Charles I 400, 506, 530
Charles II 86, 130, 211, 225–6, 301, 393, 487, 490, 495, 506, 509, 511, 560–1, 570
  household of 213
  restoration of 181
Chauveau, Jean-Baptiste 248, 367, 371, 379, 405, 510
  works: 'Traicté d'algèbre' 367, 510, 540, 543–4, 562, 584–5
Chelsea College 97
chemistry 51, 102, 198–9
Chichester 29–32, 33–4
Child, Robert 141
child truths 263–4, 439,
Christ Church, Oxford 86
Christina, Queen, of Sweden 135, 546, 560
Christ's Hospital 224–5
Church of England 24, 62, 79, 182–3
circle, squaring the 110–11, 173, 195, 211, 216, 286–8, 296–8, 311, 327, 504, 506, 512–13, 531, 533, 540
Civil War 93, 97–8
Clarke, Samuel 140
Clavius, Christopher 470
  works: *Algebra* 470
Clerke, Agnes 5
Clerkenwell 87, 98
clocks: *see* horology
Clodius, Frederick 185, 198–9, 241
Clotworthy, John, Viscount Massereene 233
Colepeper family 39
Colepeper, Alexander 506
Colepeper (or 'Culpeper'), John, Baron Colepeper 506
Colepeper, Sir Cheney: *see* Culpeper
Colepeper, Sir Thomas 39, 62
Collège de Maître Gervais 509

College of Orange 122
  see also: Illustre School, Breda
College of Physicians 235
Collège Royal 509
Collegiants 107
Collegium Auriacum 122
  see also: Illustre School, Breda
Collins, John 58, 93–4, 108, 187, 194, 199–200, 203–13, 216–23, 229, 238, 240, 248, 268, 286, 303, 305, 307–9, 312–14, 316, 318, 320–7
  works: 'An Account concerning the Resolution of Equations' 313; 'Improvements of Algebra' 325; 'Narrative about Aequations' 324; 'To describe the Locus' 323
Collyer, Richard 25
Collyer's School 25
Cologne 87, 153, 376, 380
Colvius, Andreas 98
Colwall, Daniel 224–5
Comenianism 67, 69, 71, 240
Comenius (Komenský), Jan Amos 28, 32, 40–2, 46–51, 55–6, 67, 70, 95–7, 107, 109, 162, 171
  works: *Conatuum praeludia* 49–51, 67, 95; 'Didaktika Česká' 48; *Informatorium der Mutterschule* 47–8, 96–7; *Janua linguarum* 41, 162; *Januae ... vestibulum* 47; 'Libellus Bohem[icus]' 48; *Linguarum methodus novissima* 48; *Lux in tenebris* 171; *Pansophiae prodromus* 51, 74; 'Paradoxa Astronomica' 50; 'Praecognita' 49–50
Commandino, Federico 248, 292–3, 384–5, 414–15
Commercium epistolicum 320
complex (impossible) numbers 299, 317–20, 497, 543–4
'conatus' 67
conic sections 292–4, 478
Connecticut 39, 215
Constantinople 76
Cooper, Sir Anthony Ashley 81
Coote, Edmund 43
  works: *The English Schoole-Maister* 43
Copenhagen 110, 114, 512
Copenhagen University 110, 424
Copernicus 410, 562, 564–5
Coutereels, Jan 248

INDEX

Cradock, Walter 147
Crick 94
Cromwell, Oliver 146–52, 155, 158–60, 175, 177
Cromwell, Richard 177
Croone, William 224
Croply (Thomas?) 62
cryptography 38, 54
cube, doubling the 195–6, 513, 552
Cudworth, Ralph 149
Culpeper, Sir Cheney 39, 71, 81, 96, 346
cycloid 75

Dacres, Arthur 197
Dalgarno, George 169–70
Danzig (Gdansk) 505
Dary, Michael 248, 278, 305, 327
Dauber, Johann Heinrich 123, 126, 128, 135
Davenant, Edward 315–16
Davenant, John 315
Davenant, Sir William 548–9, 557
  works: *A Discourse* 548–9; *Gondibert* 557
Dechales, Claude François Milliet 248
  works: *Cursus* 301
Deptford 52
Derand, François 88
Desargues, Girard 76, 365, 367, 414
Descartes, René 54–5, 59, 63–4, 72, 80–1, 85, 91, 96, 99, 109, 115–16, 130–1, 133–5, 137, 164, 168–9, 171, 205, 248, 265, 271, 288–90, 294–6, 313, 326, 335, 337, 340–2, 344–6, 357–8, 364, 367, 369, 373, 377–8, 382, 395, 397, 413, 427, 444, 458, 461, 465, 469–70, 473, 476–7, 479–80, 482, 488, 493, 495, 509, 513, 534, 536, 541, 546, 548, 550, 553, 560, 569
  works: *Compendium musicae* 569–74; *De homine* 509; *Discours de la méthode* [and] *Essais* 75, 131, 134, 265, 353, 445; 'Dioptrique' 81, 85, 134, 335, 340; *Geometria* (ed. and tr. van Schooten and de Beaune) 296, 310, 324, 538, 540, 542, 546, 576; 'Géométrie' 64, 134, 265, 279, 290, 473–4, 480, 513, 528, 535, 546; judgement on Pell's theorem 444, 452, 461, 469; *Le Monde* 509; *Lettres* 210; *Les Passions de l'âme* 546, 548;

Principia [and] *Specimina* 353–5, 357, 362–3, 368, 371–2, 377–8, 382, 395, 410, 493, 546
  see also: Cartesianism
van Diest, Hendrik 131
difference methods 282–4, 301–5, 313, 324–6, 516, 585
Digby, Sir Kenelm 95, 503
  works: *Two Treatises* 410
Digges, Leonard and Thomas 240
  works: *Stratioticos* 240
Digne 366
Dinet, Jacques 367
Diodati, Theodore 224
Diophantine problems 289, 299, 311–14, 326, 360
Diophantus 103, 105, 107, 109–10, 113, 128, 131, 205, 207–9, 216, 242–3, 248–9, 289–90, 299, 311, 313–14, 326, 367, 459
  works: *Arithmetic* 207, 289–90, 292, 359–60, 369, 382, 384–5, 393, 397, 413, 419, 421, 428, 446, 462, 465, 469, 473, 484, 494, 501, 527, 537
Divini (or de Divinis), Eustachio 503
division sign: *see* notation
Docemius, Johann 46
Dodson, James 93
Dögen, Matthias 516–17
  works: *Architectura militaris* 516
de Dominis, Marc'Antonio 80
Dorislaus, Isaac 153
Dort, Synod of 104
Downing, George 159, 181
Downs, battle of the 136
Drayton, Michael 34–5
Dulaurens, François 321
Duport, James 17
Durham College 146
Dury, John 28, 30–2, 40, 46, 71, 96–7, 107, 145, 149, 151–6, 158, 168, 183, 189, 216, 228–9, 233, 271, 315
  works: 'De cura paedagogica' 32
Dymock, Cressy 141

Eastbourne 14, 26, 39
Eastdean 18–19
Easter, calculation of 186
Edinburgh 560
educational theory 28, 39–48, 56, 240
Elbing 26–9, 107
'Elemens des indivisibles' 339
Elichmann, Johann 72, 94
  works: *Tabula Cebetis* 94

644 INDEX

Elizabeth, Princess (daughter of Charles I) 36
Elizabeth, Princess (daughter of Elector Palatine) 98, 107, 116, 164, 374
Elizabeth, Princess (Electress Palatine, 'Winter Queen' of Bohemia) 27, 130
Elzevier family 98–100, 279, 538
Elzevier, Louis 353, 546, 556
Encyclopaedia mathematica 409
encyclopaedism 41–6
Engagement 145
Epicurus 497, 562
equations 57–9, 320–6, 542–5
  for angle trisection 295, 321
  biquadratic: *see* quartic
  cubic 295, 311, 318, 323, 497–8, 500, 525, 530, 535–6, 540–1, 550, 553, 555, 559, 562
  indeterminate 290–2, 463, 520, 524 *see also*: indeterminate problems
  linear 278–9
  'Pell's' 320
  polynomial ('affected' or 'adfected') 57, 209, 249, 272, 276–7, 310, 313, 320–6, 530, 542–5
  quadratic 275
  quartic 208, 294–5, 316, 320, 474, 534–5, 538
  solution, methods of 217–18, 221–3, 279, 299, 311, 320–326, 496, 498, 504, 515, 518, 530, 532–3, 542–5
Ercker, Lazarus 236
  works: *Beschreibung* 236–7
Erickson, Amy 9
Erpenius, Thomas 94
An Essay to Revive ... Education of Gentlewomen 232
Euclid 242–3, 248, 271, 365–6, 385, 469
  works: *Elements* 242–3, 326, 422, 439, 444, 469, 496, 506
Euclidean algorithm 257
Euler, Leonhard 320
Evelyn, John 142, 189, 191, 224
  works: *Kalendarium hortense* 210
Evenius, Sigismund 46
Evertszoon, Evert 368, 372
d'Ewes, Sir Symonds 34, 37
Exeter College, Oxford 200

Fabri, Honoré 446
  works: *Theses* 446; *Tractatus physicus* (ed. Mousnier) 537–8, 557

della Faille, Jean Charles 434, 440, 449, 507
della Faille, Vincent ('Vincentio') 507
Fale, Thomas 22, 252
  works: *Horologiographia* 22, 252
false position, rule of 407, 461, 470, 473, 496, 504, 515, 518
'Falus Redivivus' (pseudonym of JP) 22
'Feege, Marcus' (pseudonym of JP) 153
Feingold, Mordechai 9
Fell, John 230
de Fermat, Pierre 99, 115, 118, 208, 216, 248, 288, 312–13, 326, 341–2, 345, 358, 377, 397, 416, 419, 428, 456, 459, 461, 465, 467, 469, 473, 476–8, 480
  works: 'Ad locos planos' 477–8, 480, 548; 'Ad methodum de maxima et minima' 428, 548; 'De contactibus sphaericis' 248; 'Methodus ad disquirendam' 428
de Fermat, Samuel 208, 216
Fire of London 204
Fisher, Samuel 149
Flamsteed, John 226
Flushing (Vlissingen) 176
Fobbing 166, 183–8, 191–3, 231
Föge, Marcus 152, 161
Folkington (or Fowington) 39, 62
Fontana, Francesco 503, 505
fortification 516, 521
Foster, Samuel 248, 297, 300
Fournier, Georges 516–17
  works: *Hydrographie* 516
Fox, George 177
France, policy towards 150–1, 159–60
von Franckenberg, Abraham 39, 71
Frank, Mark 187
Frankfurt 153, 189
Frederick V, Prince, Elector Palatine 27
Frederik Hendrik, Prince of Orange 121–2, 134, 485, 492
Freher, Mr 185
Freitag, Adam 516
  works: *Architectura militaris* 516
Frenicle de Bessy, Bernard 248, 367
Frisius, Gemma: *see* Gemma
Friston 18–19
Frost, Gualter 54, 63–4, 82
Frost, Gualter (son of the above) 148

Gale, Thomas 229–30
Galilei, Galileo 89, 124, 400, 427, 441, 522
  works: *Della scienza mecanica* 89

Galois, Évariste 322
Gascoigne, William 369
Gassendi, Pierre 97–8, 109, 124, 355,
 365–6, 373–6, 380, 395, 421,
 445, 490, 497, 501–3, 509, 537,
 562, 576
 works: *Animadversiones* 371, 378–9,
 382, 548, 556 60, 562; *De
 proportione* 427; *De vita et
 moribus* 497, 501, 509, 511;
 Disquisitio metaphysica 373,
 382, 387–8, 394, 556; *Institutio
 astronomica* 497, 501;
 Judicium 355, 373–6, 380
Gataker, Thomas 149
de Geer, Louis 107, 519
Gellibrand, Henry 52, 61, 83, 225, 248,
 495, 522
 works: *Discourse Mathematicall* 52, 256
Gemma 'Frisius', Regnier 56, 407
 works: *Arithmeticae practicae
 methodus* 409
'General College' (proposed by JP) 143–5
Geneva 159
geometry 52, 89, 206, 270–1, 373
Getaldus, Marinus 409
 works: *De resolutione* 409
Gibson, Thomas 248
Giessen 40
Giggeius, Antonius 94
Gilbert, William (mathematician) 495,
 552
Gilbert, William (writer on magnetism)
 87
 works: *De mundo* 556
Gildredge, Nicholas 191–2
Girard, Albert 56, 248, 289–90, 360, 517
 works: edn. of Stevin, *L'Arithmétique*
 289
Gissop (or Jessup), Dr 87
Glauber, Johann Rudolf 102
Glorioso (or Gloriosi), Giovanni Camillo
 288, 417, 441
Goad, Thomas (Professor of Law or
 theologian) 65–6, 70
goat's blood 178
Goddard, Jonathan 173–4, 179, 189–90,
 224
Godolphin, Francis 456
Godolphin, Sidney 456
Godolphin, Mr (son of one of the above)
 106, 456–7, 467, 472
Godwin, Francis 124
 works: *The Man in the Moone* 124

Golius (Gool), Jacob 99–100, 109, 114,
 116, 118–20, 131, 266, 292–4,
 384–6, 414–16, 419, 421, 469,
 481, 494, 506, 532
 works: demonstration of Pell's theorem
 492; *Lexicon* 416, 481
Gosselin, Guillaume 90, 248, 278
 works: *De arte magna* 278, 290
Gouge, William 172
 works: *A Learned Commentary* 172
Gouldingham, William 213–14
Gowers, Christopher 184
Graham, James, fifth Earl and first
 Marquis of Montrose 561
Graunt, James 210
 works: *Natural and Political
 Observations* 210
Gravett, Richard 14, 16
Great Burstead 213
Greaves, John 518, 525
Gregory, James 210, 218–22, 248, 312,
 314, 321–2, 326–7
Gresham College 20, 52, 116, 179–80,
 191–3, 196–7
Greville, Robert, Lord Brooke 32–3, 96
Grinken, Jeremiah 20
Grinken, Robert 20
Groningen 116, 133
Gronovius, Johann Friedrich 72, 94–5
Grotius, Hugo 560
*Gründtliche Beschreibung der neuen
 Regiments-Verfassung* 161–3
Gruter, Isaac 556
Guildford 88
Guillim, John 101
Guisony, Pierre 181
Guldin, Paul 112–13, 219, 388–9, 418
 works: *De centri gravitatis inventione*
 219, 388
Gunter, Edmund 20–1, 89, 248, 250, 254,
 257
 works: *A Canon of Triangles* 250, 253;
 *The Description and Use of
 the Sector Crosse-Staffe* 252
van Gutschoven, Gerard 118, 132, 433,
 440, 481, 506, 546

Haak, Theodore 28, 61–3, 67, 71–2, 74,
 77, 79, 94–6, 100, 103, 131,
 148, 172–4, 177, 185, 188,
 191–3, 196, 199, 210–11, 213,
 215–16, 224, 228–9, 241, 297,
 312, 341–2
Haarlem 292
habitudes 259–60, 270

Hackney 32
van Haestrecht, Godefroid 569
Hague, The 104, 116, 137, 152–3, 361, 364, 456, 518, 521, 567
Hakewill, George 517–18
 works: *An Apologie* 517–18
Halle 46
Halley, Edmond 193, 292, 384–5
Halliwell, James Orchard 2
Hamburg 46, 64, 108, 216, 352, 354–7, 361–2, 369–72, 377–8, 382–3, 392–3, 395–7, 401–4
Hamilton, Gen. Sir Alexander 399–400, 403
Hammond, Lt-Gen. Thomas 400, 403
Handson, Ralph 253
Harderwijk 133–4
hardness, cause of 509
Hardy, Claude 113, 176, 358, 390–1, 486, 553
 works: *Cyclometriae* 550, 553
Harington, John 80, 82, 278
Harriot, Thomas 56–7, 59, 64, 73, 82–3, 85, 88, 132–3, 248, 272–8, 291, 314, 316–17, 320, 326–7, 345, 467, 470, 497, 500, 514, 517, 521, 536, 550, 584–5
 works: *Artis analyticae praxis* 56, 82, 88, 133, 243, 260, 272–6, 409, 525; 'De numeris triangularibus' 584–5
Harrison, Thomas 94–6
Hartlib, Daniel 198
Hartlib, Georg 27, 41, 198
Hartlib, Samuel, 26–34, 39–55, 57–8, 60–86, 94–9, 106–7, 109, 121, 131, 135, 137, 140–6, 166–72, 180–3, 185, 189–91, 240, 255–6, 261–2, 265, 275–6, 280, 294, 297, 315, 346, 380, 519
 works: *The Reformed Common-wealth of Bees* 141; *Samuel Hartlib his Legacie* 141, 191
Harvey, William 586
 works: *Exercitationes* 586
Hastings family (Earls of Huntingdon) 36
Hastings, George 233, 236
Hatton, Sir Christopher 81
Hausmann, Johann Jakob 125, 127
Heidanus, Abraham 131
Heidelberg University 27, 41, 163, 519
Helwig, Christoph 40
Henisch (Henischius), Georg 344, 346
 works: *Arithmetica perfecta* 344, 346, 348, 350

Henrietta Maria, Queen 495, 557
Henrion, Denis 248
Heraclitus 345
Herbert, Edward, Baron Herbert of Cherbury 51, 63, 65, 77, 91
Herborn 41, 44, 133
Hérigone, Pierre 56, 248, 291, 321, 327, 336, 373, 377, 476
 works: *Cursus* 248, 268, 295, 373, 377, 459, 463, 465, 530
Heron (or Hero) of Alexandria 416
 works: $Βαρουλκός$ 416
Hervey, Helen 2
van Heuraet, Hendrick 541
Hevelius, Johannes 140, 191, 193, 502–3, 505, 521
 works: *Selenographia* 191, 502, 505
Heywood, William 186
Hill, Abraham 210, 224, 236
Hobbes, Thomas 84, 89–90, 109, 115–16, 131, 137, 139, 172–3, 195–6, 217, 271, 288, 297, 341, 345, 369, 373, 378, 382, 395, 397, 413–14, 416, 419, 421, 425–6, 431–2, 434, 436, 442, 448, 452, 456, 458, 461, 463, 469, 477, 484, 487, 490, 492, 495–6, 501–3, 507–9, 541, 546, 548–9, 556–7, 559, 562, 575–6
 works: 'Answer' 549, 559; 'Conicall demonstration' 412, 419, 425; *De corpore* 172–3, 397, 412–13, 496, 509, 511, 541, 549, 556–7, 560, 562, 575; *De corpore politico* 559; *De homine* 549, 556–7, 562, 575; demonstration of Pell's theorem 419, 425, 427–8; *De principiis* 210; *Dialogus physicus* 195; *Elemens philosophiques* 556, 559; *The Elements of Law* 560; *Leviathan* 575, 586; 'A Minute ... of the Optiques' 434; $ΣΤΙΓΜΑΙ$ 172; *Of Humane Nature* 560; treatise on optics 366
Hodges, Thomas 46
van Hogelande, Cornelis 72, 115
Holden, Henry 410
Holder, William 236
Holland, Mary 14
Holland, William 14–15
Holybushe: see de Sacrobosco
Hooft, Pieter 106

INDEX

Hooke, Robert 191, 196–8, 223–30, 235, 236, 242, 312, 314
  works: *Lampas* 224
Horne, Thomas 46
horology 178, 190
Horsham 25–6
Hortensius, Martinus (Maarten van den Hove) 77, 100, 102, 105, 130–1, 352, 517, 564
Hoskins, Sir John 228, 236
Hottinger, Johann Heinrich 163–4, 178
Hotton, Godefroid 107, 131, 152
Howard, Thomas, second Earl of Arundel 88–9
Howard, William, Viscount Stafford 89
Hübner, Joachim 50, 63, 71, 92, 94–7, 166
  works: 'Idea politicae' 67
Hudde, Johann 106, 323–4
Hudde's rule 323–4
Hulse, Lynn 9
Hunt, Henry 228
Hutton, Sir Richard 95
Huygens, Christiaan 107, 128, 211, 217, 219, 220, 227, 248, 296, 380
Huygens, Constantijn 106, 108, 124, 128, 130, 132, 135, 569
Huygens, Constantijn (son of the above) 107
Hyde, Edward (Earl of Clarendon) 82, 86, 584
hydraulics 84

'idea' 50, 67
Illustre School, Breda 121–38, 294–6, 485, 487, 490, 492, 499, 568
'incomposites': *see* prime numbers
indeterminate problems 290–2, 310–11, 313–14, 320, 459, 461, 463, 470, 534
indivisibles 296, 326, 339, 368, 496, 509–10, 515
infinite series 305, 327
instruments, mathematical 20–1, 23, 93
interpolation: *see* tables
Ireland 146, 495, 552
irenicism, Protestant 28, 97, 107, 145, 149, 183, 216
Isham, Sir Justinian 81, 141, 192
ben Israel, Menasseh 107, 152–3

Jamaica 181
James I of England and VI of Scotland 27, 35
James II of England and VII of Scotland 82, 571

Janszoon (Janssonius), Jan 229, 560
Jeffreys, John 16
Jennings, John 194
Jermyn, Henry, Baron Jermyn 495
Jesus College, Cambridge 14, 17
jewels, French Queen Mother's 156–7
Johanniskirche 370
Johnson, Samson 116, 123, 130, 132, 137, 140, 183–5, 188
Joly, M. 226
Jones, Richard 189
Jonson, Benjamin 34
  works: *An Entertainment at the Blackfriars* 87; *Masque of Queens* 34
Jonston, Jan 42, 67
  works: 'Idea historiae' 67
Josselin, Ralph 184
Jungius, Joachim 64, 108–9, 116–17, 363–4, 371, 378, 388–9, 395, 401, 403, 405–6, 413, 419, 423, 427–8, 432, 434–5, 440–1, 443, 446, 450–1, 458, 461–2, 470, 490, 496–501
  works: 'Apodictica' 364; 'Apollonius saxonicus' 401, 434, 446, 462; *Geometrica empirica* 363; 'Logica didactica' 434, 446, 462; 'Logicae hamburgensis praestantia' 434, 446, 462; 'Meditationes de insectibus' 401; *Phoranomica* 401, 435; 'Protonoeticae philosophiae sciagraphia' 364; 'Statica natantium' 401

Kalthoff, Caspar 65, 194
Kargon, Robert 6
Keller, Balthasar 175, 200, 311
Kempenaars, C. M. P. M. 294
Kepler, Johann 83, 110, 365–6
van Kerckhove, Johan, heer van Heenvliet 122, 125, 128, 132
de Kerouaille, Louise 226
Kersey, John 207, 248, 305
Keylway, Robert 339
Kieper (Kyperus), Albert 124, 126, 131, 135
  works: *Institutiones physicae* 124
Kinckhuysen, Gerard 106, 206, 243, 248, 291–2, 294–5, 316, 327
  works: *Algebra, ofte stel-konst* 206, 291–2, 295, 308; *De grondt der meet-konst* 206, 308
King's Bench prison 233, 235

648   INDEX

King's College, Cambridge 88
Kircher, Athanasius 74, 377, 566
    works: *Ars magna lucis* 490; *Magnes* 377; *Musurgia* 550–1, 553, 566, 569
Kirk, John 188
Kirk, Judith (née Pell, daughter of JP) 39, 78, 187–8, 236
Kirk, Thomas 188, 236
Kirk, Thomas (son of above) 188
Kirk, William 188
Komenský: *see* Comenius
Königsberg 26–7, 124
Koran 94
Kyburg 164

de Lacombe, Jean 367
Laindon 186–7, 212, 214–15, 231
Lamplugh, William 46
Langbaine, Gerard 140, 149
van Langren, Michael Florent 426, 440
language
    natural 49, 55
    philosophical 54–5
    universal 38, 55, 265
Lansberg, Philip 248, 255, 257
League, Franco-Swiss 150–1, 156
Leake, John 93, 111, 116, 194, 225, 268
Le Cazre, Pierre 427
Leibniz, Gottfried Wilhelm 221–3, 312, 322, 325
Leiden 538
Leiden University 28, 50, 63, 72, 75, 102, 124, 133–4, 135, 415, 532
lenses: *see* optics; telescopes
Le Pailleur, Jacques 116, 414, 428, 443, 476, 479, 482, 484, 494–5, 497, 499, 502, 507
    works: demonstration of Pell's theorem 443, 449, 468
Lescarbot, Marc 152
    works: *Le Tableau de la Suisse* 152
Leslie, Alexander, first Earl of Leven 403
Leslie, David, Baron Newark 403
L'Estrange, Roger 570–1
Leszno 40, 47–8, 70, 96, 171
Lewis, Mark 232
Leybourn, William 194
Lincoln, Bishop of: *see* Williams, John
Lincoln's Inn 80
Lister, Martin 223
Lloyd, Sir Charles 132–3
Lloyd, William 229–30
loci 477–8, 480

Lodewijk Hendrik, Count, of Nassau 133–4
logarithms 20–1, 23, 52, 58, 93, 200, 250, 253–6, 302–5, 323, 326, 398, 530–1, 533
    see also: antilogarithms
logic 50
London 19–20, 32–3, 44, 61–6, 71, 74, 78, 83–4, 88–9, 91, 93, 97–8, 101, 116, 129, 139, 141, 179–80, 184–5, 194, 197, 213, 215, 221, 337–8, 344–5, 349–50
London, Bishop of: *see* Sheldon
Long, Robert 79–80, 82, 86
longitude 64, 193, 225–6, 426, 521–2
Longomontanus, Christian Severinus (Christian Sørensen Lomberg or Langberg) 93, 110–21, 248, 286–8, 299, 360, 369, 371, 378, 388–91, 398, 401, 403, 405, 409, 416–18, 423–5, 431–2, 440–1, 458, 469–70, 481, 486, 501, 512, 533, 553
    works: *Astronomia danica* 110–11, 360; *Caput tertium* 119, 481; *Controversia* 115, 423–5; Ἐλέγξεως 112–13, 117, 388–90, 423; *Problema* 388; *Rotundi in plano* 110–11, 117, 286, 360, 423
Louis XIV, King of France 159, 218–19, 237
Louvain 480, 546
Louvain University 440
Lowden, Margaret 228
Lower, M. A. 13
Lower, Richard 241
    works: *Tractatus de corde* 241
Lubieniecki, Stanisaw 107, 132
Lull, Ramon 56
Luneschlos, Johann 168, 248, 518–19
    works: *Thesaurus* 516, 518–19, 527

van Maanen, Jan 2, 3, 9, 331
Macclesfield, Earls of, library of 303
McEwen, Dorothea 9
Madrid 506
Magdalen College, Oxford 17
Magdalene College, Cambridge 14, 17, 159
magic squares 75
magnetism 52, 75, 225, 522
Makin, Bathsua (née Reynolds or Reginalds) 34, 36–8, 100–1, 131, 140, 174–5, 212, 232

INDEX

Makin, Richard 100–1, 174
Mandeville, Lord: *see* Montague
Manes verulamiani 17
Mansfield (or Mansell), Sir Robert 335
de Marandé, Léonard 367
Marburg 123
Marr, John 93
Marr, William 194
Marshal, Stephen 149
Marston Moor, battle of 87, 108, 355, 369, 400
de Martel, Thomas 487
Martin, Dr 140
Mary, Princess (daughter of Charles I) 127, 130, 560
Maurolico, Francesco 365–6, 428
  works: *Emendatio et restitutio* 428
maxima and minima 428, 465, 476–7, 480
Mayfield 53
Meade, Thomas 101
mean proportional 273
  see also: cube, doubling the
means, arithmetic and geometric 412, 426, 429, 454
mechanics 51, 84, 223, 270, 509
  see also: motion, perpetual
Mede, Joseph 30
Meinel, Christoph 67
Melanchthon, Phillip 359
Memus, Johannes Baptista 292, 384
Menelaus of Alexandria 365–6
Mengoli, Pietro 248
Mercator, Nicolaus 106, 168, 192, 206, 210, 217, 221, 239, 248, 305–6
  works: *Logarithmotechnia* 248, 305–6
Mercers' Company 25
Mersenne, Marin 52, 54–5, 61, 67, 69, 72–6, 88, 90, 91, 95, 108, 115, 117–18, 128, 268, 292–3, 296, 332, 341–2, 348–9, 351, 358, 365–8, 371–5, 379–80, 382, 405, 417, 419, 421, 427–9, 431–3, 436, 441, 444, 446, 456, 458–9, 470, 472–3, 476, 479, 481–2, 484, 494–5, 506, 508, 520, 522, 533, 540, 556, 578
  works: *Cogitata physico-mathematica* 365, 371, 374–5, 448, 458, 463, 465–6, 502, 507; *Harmonie universelle* 55; *Novarum observationum* 530–1, 533; *Universae geometriae synopsis* 108, 365–8, 371, 393
Merton College, Oxford 19, 173–4
metallurgy 83, 89, 102

Metius, Adrianus (Adriaen) 248, 257, 315, 367, 552
  works: *Geometriae practicae* 552
Michaelson, John 186
Middelburg 556
Milan 398
Mizzi, Eddie 67, 69
Moleyns (or Moulins), James (or Jacobus) 142
de Molina Cano, Juan Alfonso 390
Montague, Edward, Viscount Mandeville 96
de Montarcis, Pierre Baudouin 367
Montauban 487, 495
Moore, Sir Jonas 178, 193–4, 225–6
Moorfields 179
Moray, Sir Robert 193, 214–15
More, Henry 178
More, William 30, 255
Moriaen, Johann 72–4, 76–8, 92, 98–100, 102–3, 107, 113, 116, 121, 131, 135, 153, 172, 175–6, 178, 211, 276, 279–80
Moriaen, Odilia (née von Zeuel) 153
Morin, Jean-Baptiste 226
Morland, Joseph 237
Morland, Sir Samuel 147, 158–9, 181, 211, 219, 225, 237
Morley, George 182–3, 584
Morrell, William 36
Morstyn, Faustus 169
Moseley, Humphrey 346
Moslye (or Mosley), Mr 337, 346
motion, perpetual 65
Mountagu, Edward, Earl of Sandwich 147, 176, 178, 189, 193–4
Mousnier, Pierre 537
  see also: Fabri
Mouton, Gabriel 221, 325–6
  works: *Observationes diametrorum* 221, 325
Moxon, Joseph 302
Muiderkring 106, 124
Münster, Treaty of 507
music 124–5, 130, 192, 550–1, 566, 569–70
Mydorge, Claude 88, 96, 115, 118, 169, 248, 288, 365–6, 412, 414, 416, 419, 421, 427, 484–5, 487–8, 493–4
  works: demonstration of Pell's theorem 421, 456, 467, 484–5, 488, 490, 493, 494; *Prodromi catoptricorum* 88, 412

Napier, John 250, 254, 390, 398, 533

works: *Mirifici logarithmorum canonis descriptio* 250
Nash, Robert 550, 555, 557
Naudé, Gabriel 355
naval architecture 89, 516
navigation 52, 515–16, 519, 521, 526–8
Neale (or Neile), Sir Paul 179, 189, 214, 300
Neile, William 541
de Neufville, Gerhard 248
Neumann, Peter 9
New College, Oxford 306–7
New England 181
Newchurch 204
Newton, Sir Isaac 17, 206, 217–18, 221–2, 306, 327
Niceron, Jean-François 366, 373, 380, 382
  works: *La Perspective curieuse* 366, 377, 380, 382
Norgate, Edward 125–6
notation 265, 269, 274, 299
  division sign 201, 269
  Harriot's 260, 274
  Oughtred's 261
Nottingham 87
number theory 326
Nuñez, Pedro 248
Nye, John 187
Nye, Philip 26, 149, 183, 187
Nye, Richard 25
Nye, Stephen (or Steven) 187, 234

Offerman, Barthold 357
Office of Address 145
Ogle, John 129, 130
Oldenburg, Dora Katherina (née Dury) 216, 227
Oldenburg, Henry 153, 180, 189, 191, 205, 210–11, 216–17, 219–24, 227–8, 238, 308, 322
  children of 216, 227–8
  see also: Philosophical Transactions of the Royal Society
Olearius, Adam 116
Onslow, Richard 496
optics 75, 80, 83–5, 89–92, 137, 190, 339–40, 343, 346, 365, 369
Orange, Prince of: *see* Frederik Hendrik; William II; William III
Orange, Princess of: *see* Amalia
Orsett 186
d'Orville, Guillaume (Wilhelmus) 361
d'Orville, Jacques Philippe 361
d'Orville, Jean 361–2, 371, 376, 378, 379, 392

Oughtred, William 56–7, 59–60, 86, 88–9, 99, 115, 140, 182, 240, 248, 250, 262, 273–5, 314, 326, 339, 349, 351, 368, 496, 515–16, 519, 521–2
  works: *An Addition* 515, 519; *Clavis* 56, 59, 88, 199, 248, 257, 261, 268, 270, 275, 297–8, 405–6, 409, 495–6, 506, 515, 518; *The Key* 495–6, 498–9, 504, 515, 518; *Trigonometria* 302
Owen, John 149
Oxford 'club' 179–80
Oxford University 18, 23, 27, 50, 62, 149, 169, 297

π, rational approximations to 257, 314–15, 320, 506, 513
Padua 518
Padua University 519
le Pailleur: *see* Le Pailleur
Palatinate, Palatine cause 27, 77
Palatine 'collections' 62, 79
Palissy, Bernard 141
pansophy 41–2, 48, 50, 56, 66, 78, 94–6
Panton, Henry 16
Pappus 218, 243, 248, 292, 366, 385, 416
  works: *Collections* 292
Paris 72, 74–5, 88, 109, 369, 412–13, 419–22, 424–5, 433–6, 443–8, 456, 458–62, 465–6, 472–5, 477–8, 482–4, 487–91, 494–9, 501–4, 509–10, 513, 515
Parliament 96–8, 107
Pascal, Blaise 210, 300, 414
  works: *Lettres de A. Dettonville* 300
Pasor, Georg 44
Pasor, Mathias 27, 116, 131
Payne, Robert 86, 88–90
pedagogy: *see* educational theory
Pelham Manor 15
Pell, Bathshua (daughter of JP) 100–1, 177, 188
Pell, Elizabeth (aunt of JP) 15
Pell, Elizabeth (daughter of JP) 78, 177, 186, 188
Pell, Ithamar (née Reynolds or Reginalds) 33–5, 62, 100–1, 129, 137, 139, 174–6, 184–5, 352, 442
Pell, James (son of JP) 100
Pell, James (uncle of JP) 15, 17, 21, 26, 39, 62, 139
Pell, Joan (née Gravett) 14

INDEX

Pell, John
  Ambassador to Protestant cantons, appointed 147–8
  Arabic studies 94, 140, 384–5
  birth 13
  death 241
  dispute with Longomontanus 111–21, 286–8
  doctorate in Divinity 186
  education 16–23
  Fellow of Royal Society, elected 188
  financial problems 63–6, 173–5, 177–8, 181–2, 184–5, 187–8, 231–5
  'General College' plan 143–5
  German, knowledge of 161–2
  imprisonment 233–5
  marriage 33
  medical problems 241
  ordination as priest 182
  Professor of Mathematics, Amsterdam, appointed 102
  Professor of Mathematics, Breda, appointed 121–3
  Rector of Fobbing, instituted 182
  religious views 24, 79, 183
  schoolmaster 25–6, 29–33, 39, 61–2
  tangent theorem 114, 194, 287–8, 296, 299, 300, 311, 378, 389–91, 425, 431
  three-column method 92, 165, 207, 240, 268–70, 289, 293, 299–300, 311, 314, 320, 327, 336, 463, 534
  translations: Archimedes, *Psammites* 319; Comenius, *Informatorium der Muttershule* (revised) 47–8, 243; Descartes, *Discours de la méthode* 75, 243; Ercker, *Beschreibung* 236–7, 243
  Vicar of Laindon, instituted 186
  Vice-President of Royal Society, appointed 227
  works (projected and/or drafted): 'analyticall worke' 92, 109, 131, 243, 276–8, 289, 336, 339–42, 346, 351, 354–5, 358, 369, 377, 393, 397, 403, 405, 410, 413, 419, 421, 428, 442, 446, 459, 465, 470, 474, 484, 487–8, 494, 496, 501, 521, 527, 535, 576; 'Apologia pro Francisco Vieta' 121; 'The best Way & manner' 21, 251–2; 'De canone mathematico' 221–2, 243; 'Comes mathematicus' 58, 68, 69, 94, 243, 266; 'Consiliarius Mathematicus' 68, 266; 'The Construction and Use of the Proportioned Line' 251, 253–4; 'de demonstratione' 425; demonstration(s) of his double-angle theorem 444–5, 454–5; 'De numero et ordine disciplinarum' 50; 'The Description and Use of the Quadrant' 21, 251–2; 'Discourse' on Gellibrand on magnetic variation 52, 243, 256; 'Eclipticus prognosta' 255; edition of Apollonius 110, 131, 292–4, 384–8, 393, 397, 414–16, 419, 421, 446, 462, 470, 474, 481, 487, 494, 501, 506; edition of Diophantus 109–10, 131, 289–90, 359–60, 369, 382, 384, 393, 397, 413, 421, 428, 446, 462, 465, 470, 473, 474, 484, 487, 494, 501, 527, 535; edition of Huygens–de Sluse controversy 220; 'The Everlasting Tables ... by Philip Lansberg' 255; 'General College' scheme 143–5; 'Geometricall problems' 271; 'Grammar for ye N.T.' 45; 'Horologiographia' 22, 251–3; 'Imitatio Nepeira', 251, 254; 'Index' of New Testament 44; 'Labyrinthus ingenii', 251, 254; 'The logarithmes for the Asscripts' 255 'Logistica' 57, 258–60, 265; 'The Manuall of Logarithmicall Trigonometry' 58, 243, 255–6; 'Mathematicus αὐτάρκης' 68, 94, 166, 266; 'Mechanica generalis' 270–1 on method, logic, and 'teaching of disciplines' 50; 'Methodus docendi' 31; 'Methodus metricae nostrae' 32; 'A New Almanacke' 21, 251; 'Notatu digniora' 251, 253; 'Of Aequations' 277–8; 'Pandectae Mathematicae' 68, 266; 'A Prognostication' 21, 251; 'Progymnasma Analytico-logisticum' 58, 81, 260–2, 299; 'Quadraticall Aequations' 279, 299; scheme for 12-tome work 66; second part of reply to

Longomontanus 121, 533; 'The signe Leo' 19; 'The Strife of Analytica and Synthetica' 262–5; table of antilogarithms 92–3, 109, 142–3, 166, 238, 243, 280–6, 323, 336, 338, 350, 358, 369, 533; 'The Tables' 239, 243; 'Tabulae directoriae' 21, 251, 254; 'Trigonometria logistica' 58, 256

see also: Apollonius; Diophantus; Euclid; Pappus

works (published): *Controversiae pars prima* 119–21, 164–5, 173, 247, 288, 479, 481, 492, 494, 499, 501–2, 505, 512; 'Day-Fatality of Rome' 230–1; *Easter not Mis-timed* 186; *The English Schoole* 42–4; *Gründtliche Beschreibung* (Preface) 161–3; *Idea of Mathematics* 31, 67–76, 99, 127, 145, 240, 243, 265–7; *Idea mathesews* 67, 69, 71–6, 127, 145, 166, 409; *Introduction to Algebra* (by Rahn and Pell) 199–208, 218, 229, 247, 290, 306–12, 326; 'Oratio inauguralis' 126–7, 499–500; 'refutatiuncula' 111–12, 114–16, 286–8, 299, 360, 362–3, 369, 371–2, 378, 388, 401, 405, 423, 432, 444; tables of areas and volumes 237–8; *Tabula numerorum quadratorum* 219, 247

Pell, John, of King's Lynn 14
Pell, John (Jr, son of JP) 100, 177, 185–6, 213–15, 232–4
Pell, John (Sr, father of JP) 13–15
Pell, Judith: see Kirk, Judith
Pell, Katherine 129
Pell, Mary: see Raven, Mary
Pell, Mary (née Holland) 14
Pell, Richard 39, 78
Pell, Susan (née Burton) 15
Pell, Thomas 14–15, 39, 182, 214–15
Pell, William 100
Pepys, Samuel 142, 159, 178, 198, 225
Percy, Henry, ninth Earl of Northumberland 82, 273, 467
Perkins, Adam 9
Petau, Alexandre 72
Petau, Denis 72, 367

'Peters, Adrian' (pseudonym of Thurloe) 153
Petri, Nicolaus (Arabic scribe) 481
Petri (Pieterszoon), Nicolaus (mathematician) 320
  works: *Practique* 320
Pettus, Sir John 228, 236
Petty, William 114, 131, 146, 189, 198, 223, 233, 434, 441–2, 484
Philemon, Johannes 124, 135
Philosophical Transactions of the Royal Society 205, 210, 219, 305, 313
Piedmont 157
Pitiscus, Bartholomaeus 56, 85, 240, 248, 253
  works: *Trigonometria* 253
Pitt, Moses 200, 202–5, 211, 229–30, 235, 307, 312
  works: *Cry of the Oppressed* 230; English Atlas 229–30
Placcius, Vincent 95
plague 198
Plato 394
Pollard, Arthur 18–19
  works: 'Pro Latitudine' 251–2
Pollot, Alphonse 569
della Porta, Giambattista 224
Potter, Francis 163, 192
  works: *An Interpretation of 666* 163
Pratt, William 20
precept of synthetica ('the great precept') 263–4, 266–8, 270
prime numbers ('incomposites') 175, 200, 205, 256–7, 299, 311
prime truths 264
Principe, Larry 9
problema Austriacum 305
progressions (arithmetic, geometric, and others) 261, 270, 305, 318
Progymnasma (or the 'great' or 'grand' problem) 58, 81, 259, 260–2, 264, 270–1, 299, 310, 421, 428
Protheroe, John 467
Ptolemy 248, 385
Puteanus, Erycius 355, 376
Pym, John 82, 96

quadrant 251–2
Quakers 183
Quendon 187
ibn Qurra, Thābit 292, 386

Rahn, Johann Heinrich 163–5, 199–202, 229, 248, 298–300, 327

works: *Introduction to Algebra* (by
    Rahn and Pell) 199–208, 218,
    229, 247, 290, 306–12; 'Solutio
    problematum Diophanti' 207;
    Teutsche Algebra 165,
    199–200, 298–300, 306, 309–11,
    318–19, 324, 326
Ramus, Petrus (de La Ramée, Pierre)
    248, 272, 365–6
  works: *Arithmeticae* 272
Randolph, Thomas 17
Ranelagh, Lady 233
ratio 440, 458, 504
Ratke, Wolfgang 40
Raven, Charles 242
Raven, Henry 174, 176
Raven, Mary (née Pell) 39, 78, 174,
    176–7, 187–8, 234
Raven, Miles 187, 215, 233–5, 242
Raven, Roger 174, 187–8, 215, 231, 233–4,
    242
Raven, William 235, 242
Ravensberger (or Ravensberg), Jacob 135,
    553, 567
Ravius, Christian 53, 72, 75–6, 107, 113,
    121, 175–6, 229, 292, 358, 384,
    390–2, 415, 481
  works: *Dissertatio mathematica* 76
Ravius, Johann 76, 114
Rawlinson, Richard 298
Ray, John 17
Raylor, Timothy 9
'real characters' 38
Recorde, Robert 248
rectification of curves 541
Reeve (or Reeves), Richard 91 335–40,
    343–4, 346, 348–9, 351
Reeve (or Reeves), Richard (son of the
    above) 337
Reeve, Thomas 184
refraction: *see* optics
Regius, Henricus (Hendrik de Roy) 132,
    495, 499
  works: *Fundamenta medica* 499;
    Medicina et praxis 499
Reinalds, Mr 66
Remonstrants 104, 107
van Renesse, Lodewijk Gerard 122–3, 126,
    135
Reynardson, Lady 200, 232
Reynardson, Sir Abraham 200
Reynolds, Henry (poet and courtier) 34
  works: *Mythomystes* 34
Reynolds (or Reginalds), Henry 33–5,
    37–9, 53, 174

works: 'Architectiones' 35, 38; *Magnae
    Britanniae chronographia* 35;
    'Nuncius volucris' (or
    'Macrolexis' and 'Scenolexis')
    35, 38, 39
von Rheita, Anton Maria Schyrle 97–8,
    355, 374–6, 383, 393, 395,
    397–9, 401, 403, 405, 410, 421,
    503
  works: *Novem stellae* 97–8, 355, 374–6,
    383; *Oculus Enoch et Eliae*
    410, 429, 433, 436, 441, 445
Ricci, Michelangelo 208
Riccioli, Giovanni Battista 168
  works: *Almagestum novum* 168
Richard, Claude 506, 515
Rivet, André 72, 74, 122, 125, 128, 130,
    135
'Robertson, John' (pseudonym of Dury)
    153
de Roberval, Gilles Personne 99, 109,
    115–18, 131, 168–9, 358, 367,
    372–3, 377, 380, 397, 412,
    425–6, 432, 434, 451, 458–9,
    461, 469, 472–3, 477, 480, 482,
    484, 488, 497, 503–4, 514, 517,
    521, 578
  works: 'De locis' 554–5, 562, 564;
    demonstration of Pell's
    theorem 414, 429, 451, 508;
    'Traité des indivisibles' 510
Rogers, Francis 36, 100–1
Rogers, Mespira (formerly Morrell, née
    Reynolds or Reginalds) 35–6,
    100
Rome 382, 550
Rooke, Lawrence 179, 189, 298, 300
Rosenbach, Zacharias 44
  works: *Methodus omniscientiae Christi*
    44; *Moses omniscius* 44
Roth, Sir John 185
Rotterdam 132, 410, 509–11, 513–16
Rouen 421
'roulette': *see* cycloid
Rous, Francis 145
Royal Society 105, 139, 143, 179–80,
    188–93, 195–6, 198, 209–11,
    220–1, 223–30, 236, 242–3, 312,
    316
  committees of 190–1
Rulice, Johann 107, 131
Rupert, Prince 399, 567
Rye 66

INDEX

de Sacrobosco, Johannes (John
        Holybushe or Holywood) 320
Sadler, John 106, 131, 145
St Giles-in-the-Fields 241–2
St James's Library 147
St John, Oliver 146
St John's College, Cambridge 87
St Margaret's, Westminster 184
de Saint-Charles, Louis Jacob 364
de Saint-Pierre, sieur 226
de Saint-Vincent, Grégoire 132, 249,
        296–7, 434, 440, 449–50, 481,
        499, 511, 515, 533, 540
  works: *Opus geometricum* 296–7, 452,
        458, 462, 469–70, 488, 490, 493,
        497, 499, 503, 510–13, 530, 533
Salisbury House 195–6
Salmasius, Claudius (Claude de
        Saumaise) 131
Salmon, Ernest 13
Salmon, Vivian 37
Salusbury, Oliver 224
sand, vitrifying 190–2
Sandbach (or Sambach), Mrs 212
Sanderson, Robert 182
de Sarasa, Alphonse Antoine 540
  works: *Solutio* 540
Sassen, Ferdinand 6
Savile Library 242
Savile, Sir Henry 19, 556
  works: *In Taciti hist.* (tr. Gruter) 556
Savoy 87, 157–60
Savoy, dowager Duchess of 157–8
Savoy Hospital 97
Scarborough, Charles 179, 189, 300, 316
Scavenius, Petrus 553
Schaffhausen 162
Scheiner, Christoph 344, 552
  works: *Pantographice* 344, 346, 348–50
Schönauer, J., 163
van Schooten, Frans (the elder) 249
van Schooten, Frans (the younger) 99,
        114, 128, 132, 165, 205, 249,
        280, 296, 310–11, 324, 517, 521,
        532, 538, 540, 546, 552, 569
  works: *De organica conicarum
        sectionum* 498, 540–1, 550,
        552, 555; *Exercitationum* 311,
        538; *Principia matheseos* (ed.
        Bartholin) 165, 310, 576
  see also: Descartes, *Geometria*; Viète,
        Opera
Schurman, Anna Maria 36
Schwarts, George 30, 255
Seaman, Lazarus 149

Sedan 123
Seekers 183
Selden, John 80, 96, 106, 140
Serenus of Antinoupolis 365–6
Serrarius, Peter 107
Shaw, Mr 520
Sheldon, Gilbert 182, 186–8, 199, 216,
        232–3, 305, 307
Ship Money 95, 394
al-Shirazi, 'Abd al-Malik 292, 385–6, 388,
        415
Shortgrave, Richard 194
shorthand 37–8, 53
Slingsby 87
de Sluse, René-François 217–18, 220, 222
Smith, Eddie 9
Smithson, Hunter 90
Snel, Willebrord 515
  works: *Tiphys batavus* 515–16
Sorbière, Samuel 556–7
Sorö 114
Southwick 13
Speed, William 29–30, 32, 183
Speidell, John 250
  works: *New Logarithmes* 250
spherical triangles 522, 524, 517–8, 570,
        574–5, 576, 582, 584
de Spinoza, Baruch 107
spiral, logarithmic 578–84
Sprat, Thomas 179
  works: *History of the Royal Society*
        179, 210
square numbers: see tables
Stampioen, Jan 107–8, 222, 249, 532
Stanhope, Lord 125
Stanier, Mr 79
Statencollege, Leiden 104
Stationers' Company 42–3
steganography: see cryptography
stenography: see shorthand
Stevin, Simon 248–9, 272, 278, 360, 407
  works: *L'Arithmétique* 272; *Les
        Oeuvres* 571
Steyning grammar school 14–16
Stockar, Johann Jakob 149, 162
Stockholm 546
Stouppe (or Stuppa, or Stoppa),
        Jean-Baptiste 154–5, 158, 160
Streete, Thomas 193–4
  works: *Astronomia carolina* 193
Strode, Thomas 248
Stucki, Johann Rudolf 163, 177
Suter, Johann Kaspar 162
Swann, Utricia (née Ogle) 129–30
Swann, William 129

INDEX

Swart, Steven 229
Sweden 560
Switzerland, Protestant cantons of 148–52, 154–7, 159–61
Sydenham, Col. William 145
Sydney, Robert, Earl of Leicester 467
Symner, Miles 146
synthesis 56, 241
   see also: analysis

tables, mathematical 21–2, 57–8, 84, 219, 239, 243, 249, 318
  of antilogarithms 280–6
    see also under: Pell, works (projected and/or drafted); Warner, works
  astronomical 251
  interpolation of 282–4, 301–5, 313, 324, 516, 519, 585
  of roots of equations 299, 325
  of sines 516, 530, 532
  of square numbers 301, 306
  of tangents 112, 115, 117, 119, 287, 389, 401, 450
  trigonometrical 253, 516, 519
  see also: antilogarithms
Tacquet, André 249, 450
Talbot, Gilbert 89
tangent theorem: *see* Pell, tangent theorem
  demonstrations of 416–7, 419, 421–2, 424–5, 427–9, 431–3, 436–41, 443–56, 460–2, 467–8, 472, 475, 479, 482, 484, 486, 488, 494, 502, 507
tangents to curves 220, 341, 477, 480
tangents, trigonometrical: *see* tables
Tapp, John 19, 21, 249, 252–3, 320
  works: *The Pathway to Knowledge* 320
Tassius, Johann Adolf 64, 70, 95, 99, 109, 116–17, 363–4, 371, 378, 392, 395, 401, 405, 418, 428, 440, 450, 490, 496, 498–9
  works: judgement on Pell's theorem 432
Taylor, Eva 22
telescopes, telescopy 80, 91–2, 216–17, 335–6, 338, 369, 374–6, 380, 383, 393, 395, 397, 399–401, 403, 410, 433–4, 436, 503
Tesselschade, Maria 106
Thābit ibn Qurra: *see* ibn Qurra
Theodosius of Bithynia 365–6
Theon of Smyrna 365, 394
Thompson, Anthony 179–80
Thorndike, Herbert 17–18, 24, 109, 142–3, 166, 238, 286

Thresher, Abraham 234
Thurloe, John 148, 153–7, 160, 175, 177, 183
Thynne, Charles 141
Tilston 212
Titus, Silius (or Silas) 209, 214, 222, 226, 301, 307, 314–16, 327
Tonge (or Tongue), Ezerel 194, 232
Toomer, Gerald 2, 9
Torricelli, Evangelista 115, 118, 288, 417, 441, 444, 461, 467, 473, 496, 503, 517, 578, 581
  works: *Opera geometrica* 296, 427, 433, 441, 444–5, 515, 522, 524
Torricellian experiment 190
Tottenham 200, 203, 232
Tovey, Nathaniel 142, 166
trigonometry 253–4, 256, 266, 302
Trinity College, Cambridge 16–18, 20, 24
Trinity College, Dublin 146
Trithemius, Johannes 53–4, 375
  works: *Steganographia* 54
A True State of the Case of the Commonwealth 161–2
von Tschirnhaus, Ehrenfried Walter 238, 312, 321–2
  works: 'Nova methodus' 321
Tschirnhaus transformations 321
Turnbull, G. H. 31
Turnbull, Herbert 1
Turner, Francis 306–7
Turner, Mr (mathematical assistant) 285, 350
Turner, Thomas 307
Twysden, John 300
typography 53, 224

Ulrich, Johann 154
Ulrich, Johann Jakob 154–5, 177, 189, 211
  sons of 156, 177
Ulug Beg 518, 525
utopianism 27
Utrecht 124, 153, 567
Utrecht University 135–6, 358, 553, 567

Valerio, Luca 365–6
Vaughan, Robert 1–2
Vavasour, Sir William, Bt 567
  son of 567
Vegelin, Philip Ernst 74, 76, 118
Verdussen, Hieronymus 507
Vernon, Francis 207
Vernon, Richard 15, 39

Viète (Vieta), François 57, 59, 73, 89, 98–100, 121, 127, 164, 249–50, 261, 265–6, 272, 274, 276, 279–80, 320–1, 325, 336, 345–6, 360, 366–7, 372, 378, 406, 408, 415, 454, 469, 478, 497, 500, 542
  works: *De aequationum* 89, 272, 562, 569, 574–5; *De numerosa resolutione* 498 'Harmonicon coeleste' 99; *Isagoge* 250, 500; *Opera mathematica* (ed. van Schooten) 279–80, 296, 415–16, 433, 488, 490, 492–3; *Zeteticorum libri quinque* 289
viticulture: *see* agriculture
Vlacq, Adriaen 58, 250, 255
Vossius, Gerardus Joannes 77–8, 102, 104, 106–7, 119, 131, 229, 319, 352, 560
  works: *De scientiis mathematicis* 319
Vossius, Isaac 72, 229, 560

de Waard, Cornelis 5–6
Wadham College, Oxford 179
van Waesberghe, Johann 492
Wake, Richard 507
Waldensians 157–60
Waller, Edmund 584
Wallis, John 172–3, 192, 195–6, 205, 210–11, 218, 221, 226, 238, 249, 261, 268, 275, 280, 286, 297–8, 300–1, 305, 312–20, 323–4, 327, 339, 368, 565
  works: *Arithmetica infinitorum* 297; 'Circuli Quadraturam' 173, 297–8; *A Treatise of Algebra* 313–20, 326; 'Logarithmotechnia Nicolai Mercatoris' 305
Wallis, P. J. 5–6
Walton, Brian 140
Ward, Alderman 224
Ward, John 198–9
Ward, Seth 178, 226, 297–8, 405
Warner, Walter 18, 56, 64, 80, 82–5, 89–92, 99, 109, 141–2, 273–6, 280–1, 284–6, 295, 317, 323, 325, 327, 336, 338, 358–9, 366, 467, 521, 530, 550, 585
  works: table of antilogarithms 84, 92, 109, 142–3, 166, 238, 280–6, 323, 336, 338, 354–5, 358, 369, 533; 'Table of refractions' 84; 'Tract for ye table of refraction' 90; treatise on optics 366
Wasmuth, Matthias 236
van Wassenaer, Jacob 249, 532
Watts, Edward 285, 350
Webbe, Joseph 28–9, 40, 45, 54
  works: 'Tables of Variation' 54
Weber, Johann 407
Webster, Charles 9
Webster, John 223
  works: *The Displaying of Supposed Witchcraft* 223
Weckherlin, Georg Rudolf 77, 82
Welbeck Abbey 87, 90
Welden, William 199
Wellingor 87, 335–6, 339, 346–8, 351
Wendelin, Gottfried 376
Westchester County 214
Westminster 188, 340–1
Westminster School 41, 43, 185, 197, 205, 230, 242, 286, 312, 352
Westminster, Treaty of 152
Whalley, Thomas 18
Whistler, Daniel 235
Whitaker, Thomas 506
White, Richard 132, 140
White, Thomas 132
Whitechapel 235
Whitelaw, David 189
Whitell, John 25
Whitelocke, Bulstrode 146, 159
Widdrington, William, first Baron Widdrington 393, 521, 531, 533, 538, 541, 543
Widdrington, William, second Baron Widdrington 531, 533
Wiesel, Johann 380
de Wilhem, David 123, 135
Wilkins, John 149, 169, 179, 189, 191, 193, 196–8, 210–11
  works: *Essay towards a Real Character* 169, 211
William II, Prince of Orange 127, 567, 570
William III, Prince of Orange 570
Williams, John 33, 78, 79, 96–7, 183
Williams, Lady Rebecca 188
Williamson, Sir Joseph 222–3
Willingdon 21, 39, 62
Willis, Thomas 216
  works: *De anima brutorum* 216
Wils, Peter 294
Wimbourne 234
Wingate, Edmund 20, 23, 58, 249, 250, 255

works: *Arithmetique* 20, 58; *The Construction and Use of the Line of Proportion* 23, 250, 253–4; *The Use of the Line of Proportion* 23
Wingate, Edward 20, 250, 254
Winthrop, John 214–15
Wise, Mr 224
Wolzogen, Baron Johann Ludwig 70–1, 107, 116, 119, 132, 166–8
works: 'Magnum opus mathematicum' 70, 166
Wolzogen, Baroness 107
Wood, Anthony 13, 23, 239, 319
Wood, Robert 101, 298
Woodward, Hezekiah 45–6
Worsley, Benjamin 146
Worsley, Lucy 9

Worthington, John 149, 171, 178, 182, 216
Wotton, Sir Henry 87
works: *The Elements of Architecture* 124
Wren, Sir Christopher 69, 141, 179, 189, 209, 211, 225–7, 229, 298, 496
Wyche, Sir Cyril 239

Xylander (Holzmann), Wilhelm 249, 360

von Zeuel, Adam 153
von Zeuel, Peter 172
van Zijll, Gijsbert 570
Zollicoffer, Tobias 178, 211
Zurich 154–8, 160–5, 168, 170–1, 175, 189–90, 211, 297–8
city library 163